Multi Talent

ATIS cable robot

- inspects visually and documents 100% of the cable surface using moving-, still- and panorama images

- inspects thermographically

- measures coat thickness and creates 3D surface laser scans

- applies durable corrosion protection ATIS cableskin®

- examines magneto-inductively for wire breaks and corrosion

- welds PE ducts and helical fillets subsequently

For more information **www.cablerobot.de**

Alpin Technik und Ingenieurservice GmbH • Plautstraße 80 • D - 04179 Leipzig • www.alpintechnik.de • info@alpintechnik.de

Literature for bridge building by Ernst & Sohn

KLAUS IDELBERGER
Fuß- und Radwegbrücken
Beispielsammlung
2011.
182 S., 351 Abb., Br.
€ 49,90*
ISBN 978-3-433-02937-4

KLAUS IDELBERGER
The World of Footbridges
From the Utilitarian to
the Spectacular
2011.
183 pages, 351 figures.
Softcover.
€ 69,–*
ISBN 978-3-433-02943-5

■ Structural engineers are often more interested in road and rail bridges of spectacular construction and enormous spans than in relatively narrow footbridges built for modest loads. City planners and landscape architects, on the other hand, see inner-city pedestrian and cycle bridges as important architectural elements and generally invite bridge builders to compete to find the winning design. As there are no official guidelines for the design of footbridges, the building techniques and performance of existing bridges are an important source of information for planners.

This book contains over 100 examples of bridges built worldwide over the last ten years: open pedestrian and cycle bridges, cattle and utility bridges and some enclosed skyways. The collection is arranged according to load-bearing structure. For each bridge there is a brief description of the location and the structural system, including special construction details, and illustrated by photographs, plans and elevations. "Footbridges" is a treasure trove for structural engineers.

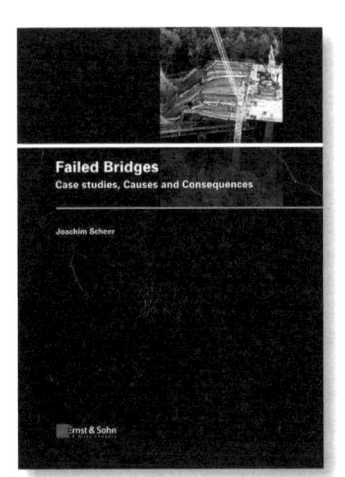

JOACHIM SCHEER
Failed Bridges
Case Studies, Causes
and Consequences
2010.
321 pages. 120 fig. 15 tab.
Hardcover.
€ 79,–*
ISBN 978-3-433-02951-0

■ When bridges fail, often with loss of human life, those involved may be unwilling to speak openly about the cause. Yet it is possible to learn from mistakes. The lessons gained lead to greater safety and are a source of innovation.

This book contains a systematic, unprecedented overview of more than 500 bridge failures assigned to the time of their occurrence in the bridges' life cycle and to the releasing events. Primary causes are identified. Many of the cases investigated are published here for the first time and previous interpretations are shown to be incomplete or incorrect. A catalogue of rules that can help to avoid future mistakes in design analysis, planning and erection is included.

A lifetime's work brilliantly compiled and courageously presented – a wealth of knowledge and experience for every structural engineer.

Ernst & Sohn
Verlag für Architektur und technische
Wissenschaften GmbH & Co. KG

Kundenservice: Wiley-VCH
Boschstraße 12
D-69469 Weinheim

Tel. +49 (0)6201 606-400
Fax +49 (0)6201 606-184
service@wiley-vch.de

Ernst & Sohn
A Wiley Company

Online-Bestellung: www.ernst-und-sohn.de

* Der €-Preis gilt ausschließlich für Deutschland. Inkl. MwSt. zzgl. Versandkosten. Irrtum und Änderungen vorbehalten. 0196500006_dp

mageba
Switzerland www.mageba.ch

Safety **Reliability** **Durability**

engineering connections®

Incheon Grand Bridge, South Korea
Equipped with 73 TENSA®MODULAR expansion joints. Largest expansion joint type LR25 (movement capacity: 2,000mm) with ROBO®GRIP anti-skid surface, ROBO®SLIDE high grade sliding material and a ROBO®CONTROL remote monitoring system.

- structural bearings
- expansion joints
- seismic devices
- monitoring & services

Modular expansion joint | Monitoring scheme | ROBO®CONTROL Box

---

BRIDON

Bridge Cable Specialists

Global technology leader in the manufacture of cable solutions for the world's most demanding applications.

- Customised solutions supporting unique architecture
- High strength, lightweight, visually elegant solutions
- Minimised life-time maintenance and cost
- Seamless service from concept to completion

Bridon International Ltd
Balby Carr Bank, Doncaster,
DN4 5JQ
Tel: +44 1302 565100
structures@bridon.com

Bridon International GmbH
Magdeburger Straße 14 a,
D-45881 Gelsenkirchen
Tel: +49-209-8001 0
info@bridon.de

STYLITE®

ECCS – European Convention for Constructional Steelwork / Associacao Portuguesa de Construcao Metalica e Mista (eds.)

Design of Steel Structures

Eurocode 3: Design of steel structures.
Part 1-1: General rules and rules for buildings.

April 2010
XVI, 446 pages, 295 figures, 105 tables, Softcover.
€ 70,–*
ISBN 978-3-433-02973-2

■ This book introduces the basis design concept of Eurocode 3 for current steel structures, and their practical application. Numerous worked examples will facilitate the acceptance of the code and provide for a smooth transition from earlier national codes to the Eurocode.

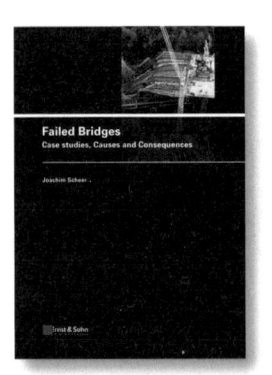

JOACHIM SCHEER

Failed Bridges

Case Studies, Causes and Consequences.
Foreword by Christian Menn

May 2010
XIV, 307 pages, 120 figures, 15 tables, Hardcover.
€ 79,–*
ISBN 978-3-433-02951-0

■ A systematic overview of more than 400 failures evaluated according to the time of their occurence in the life cycle of the bridge and the primary cause of the collapse. Including a catalogue of rules to help prevent errors in design, planning and erection.

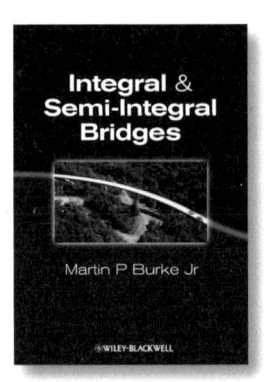

MARTIN P. BURKE JR.

Integral and Semi-Integral Bridges

July 2009. ca. 272 pages, Hardcover.
€ 89,90*
ISBN 978-1-4051-9418-1

■ Worldwide, integral type bridges are being used in greater numbers in lieu of jointed bridges because of their structural simplicity, economy, and durability. Written by a practicing bridge design engineer from the USA who has spent his career involved in the origination, evaluation and design of such bridges, this book shows how the analytical complexity due to the elimination of movable joints can be minimized to negligible levels so that most moderate length bridges can be easily and quickly modified or replaced with either integral or semi-integral bridges.

Steel Construction

Design and Research

Volume 5, 2012, 4 issues per year.
Editor in chief: Dr.-Ing. Karl-Eugen Kurrer

Annual subscription print
ISSN 1867-0520
for companies € 154,–*
for libraries € 520,–*

Annual subscription print + online
ISSN 1867-0539
for companies € 178,–*
for libraries € 598,–*

Subscription prices are net-prices exclusive of VAT but inclusive postage and handling charges.

ECCS – European Convention for Constructional Steelwork / Associacao Portuguesa de Construcao Metalica e Mista (eds.)

Fire Design of Steel Structures

EC1: Actions on structures.
Part 1-2: Actions exposed to fire.
EC3: Design of steel structures.
Part 1-2: Structural fire design.

May 2010
XXIV, 428 pages, 134 figures, 21 tables, Softcover.
€ 70,–*
ISBN 978-3-433-02974-9

■ This publication sets out the design process in a logical manner giving practical and helpful advice and easy to follow worked examples that will allow designers to exploit the benefits of the new approach given in the Eurocodes to fire design.

STEFAN NIXDORF

StadiumATLAS

Technical Recommendations for Grandstands in Modern Stadia

February 2008
368 pages, 695 figures, Hardcover.
€ 79,–*
ISBN 978-3-433-01851-4

■ This StadiumATLAS is a building-type planning guide for the construction of spectator stands in modern sports and event complexes. The principles of building regulations and the guidelines of important sports associations are analyzed and interrelated.

TONY BRYAN

Construction Technology

Analysis and Choice

2. Edition – March 2010
464 pages, Softcover.
€ 89,90*
ISBN 978-1-4051-5874-9

■ The second edition of Construction Technology: analysis & choice is to be expanded to include commercial buildings, giving a single textbook covering all the basic forms of construction studies on professional courses.

■ Steel Construction unites in one journal the holistic approach to steel construction. In the interests of "construction without depletion", it skilfully combines steel with other forms of construction employing concrete, glass, cables and membranes to form integrated steelwork systems.

The scientific and technical papers in Steel Construction are primary publications. This journal is aimed at all structural engineers, whether active in research or practice.

The ECCS – European Convention for Constructional Steelwork and Ernst & Sohn have agreed that the journal Steel Construction founded by Ernst & Sohn in 2008 is to be the official journal for ECCS members from 2010 onwards. You will find more information about membership on the ECCS homepage, www.steelconstruct.com.

Ernst & Sohn
Verlag für Architektur und technische Wissenschaften GmbH & Co. KG

Customer Service: Wiley-VCH
Boschstraße 12
D-69469 Weinheim

Tel. +49 (0)6201 606-400
Fax +49 (0)6201 606-184
service@wiley-vch.de

Ernst & Sohn
A Wiley Company

Structural Protection Systems

Good Vibrations

MAURER Cable Dampers

- **Adaptive dampers for stay cables** – active adaptation to the vibration behaviour, if required with integrated measurement of the cable tension forces and monitoring
- **Integrated elastomeric or friction dampers**
- **External elastomeric or friction dampers**
- **External viscous dampers**

Maurer Söhne cable dampers are based on longstanding research and highly reliable technology. They are individually adapted to structural requirements.

Maurer Söhne GmbH & Co. KG
Frankfurter Ring 193, 80807 Munich/Germany
Phone +49/89/32394-0
Fax +49/89/32394-306
ba@maurer-soehne.de
www.maurer-soehne.com

MAURER SÖHNE
forces in motion

Since 1876

Holger Svensson
CABLE-STAYED BRIDGES
40 Years of Experience Worldwide

Ernst & Sohn
A Wiley Company

Holger Svensson

CABLE-STAYED BRIDGES

40 Years of Experience Worldwide

Ernst & Sohn
A Wiley Company

Dedicated to my revered Mentor Prof. Fritz Leonhardt
and my dear wife Meg in appreciation of their support

Introduction

Cable-stayed bridges are currently in a fast development, worldwide. While in 1986 about 150 major cable-stayed bridges were known, their number has increased to more than 1000 today. Their spans have also increased by leaps. From 1975, when the span record was 404 m, it jumped to 856 m in 1995 and today has reached 1104 m. A span limit is not yet in sight, and there are already designs for cable-stayed bridges with main spans up to 1800 m. As the economic range for suspension bridges is limited for very long spans, cable-stayed bridges are a focus of interest for bridge engineers worldwide.

This book addresses experienced engineers as well as students. It has been developed from the scripts of the 'Lectures on cable-stayed bridges', which have been given by the author at the University of Dresden for the 7th and 8th semester for bridge engineering students since 2009.

It aims to cover all aspects of the design, construction planning and execution on site, dealing with principles that appear important to the author, based on his 40-year experience as a bridge engineer. Additional details are given in the nearly 350 references.

Key aspects of the book are:
- the historical development from precursors up to the present day,
- the structural details of beams, towers and especially the stay cables which are a crucial component of cable-stayed bridges,
- the preliminary design of cable-stayed bridges which provides the best understanding of the flow of forces and permits initial sizing and independent checking of a design, and
- the erection of cable-stayed bridges, which is equally as important as their final stage.

With the exception of the cable sizing, codes are consciously not referred to. The investigated bridges are from all over the world and have been designed in accordance with various codes, including DIN, Eurocode, AASHTO, British Standard and others. The governing factor for the designs are ultimately the laws of nature which are identical worldwide.

The selection of the cable-stayed bridges highlighted here is quite subjective and consists of:
- bridges in which the author participated or in which at least Leonhardt, Andrä and Partners (LAP) were involved,
- bridges with unusual structural details, and
- bridges with record spans.

In general only completed bridges are treated. The discussion of proposals, as interesting as they may be, would go beyond the scope of this book.

The book contains nearly 1300 figures, mainly in color, in accordance with the conviction of the author that a good picture is more meaningful than any description. This is especially true for the appearance and for the construction.

The author has named the engineers involved in as many bridges as possible. This should help to improve the public standing of engineers which is otherwise often neglected. The responsible engineers are also recognized as authors of the given references. A bridge is, of course, never the work of an individual – a whole design team is required, which works in mutual trust with the engineers of the client and with the engineers on site in order to successfully complete a bridge project.

At the end of the book two DVDs are included, which record the 30 lectures that the author gives at Dresden University.

Holger Svensson, Zeuthen, 2012

THE ART OF ENGINEERING

2nd Orinoco River Bridge at Ciudad Guayana | Venezuela

Rhine Bridge Wesel | Germany

Elbe Bridge Schönebeck | Germany

Sava Bridge Belgrade | Serbia

Fehmarn Belt Crossing connecting Denmark and Germany

Leonhardt, Andrä und Partner

www.lap-consult.com

Acknowledgement

The author is grateful to the University of Dresden for the opportunity to hold lectures on cable-stayed bridges subsequently to his active work, initially arranged by Professor Dr.-Ing. Jürgen Stritzke. The continuing support by Professor Dr.-Ing. Manfred Curbach, Head of the Institute for Concrete Structures, and by Professor Dr.-Ing. Richard Stroetmann, Head of the Institute for Steel and Timber Structures, as well as Dipl.-Ing. Peter Deepe and Dr.-Ing. Uwe Reuter for the preparation of the video records of the lectures is highly appreciated.

The author is indebted to the following colleagues who generously provided him with information: Dr. Charles Birnstiel, USA, on the Nienburg Bridge, M. Jacques Combault on the Rion-Antirion Bridge, M. Jean-Marie Crémer on the Ben Ahin and Millau Bridges, Professor Dr. Guido Morgenthal on the Stonecutters and Sutong Bridges, Dr.-Ing. Herbert Schambeck on the Metten and Flößer Bridges, Dr.-Ing. Klaus Stiglat on early cable-stayed bridges in France and M. Michel Virlogeux on the Normandy Bridge. Information on stay cables was provided by Dipl.-Ing. Werner Brand (DYWIDAG-Systems International, Germany), Mr W.S. Cheung (Cabletek, South-Korea), Dr. Hans Rudolf Ganz (VSL, Switzerland), M. Erik Mellier (Freyssinet, France), Dr. Christian Braun (Maurer, Germany), Dr. Marcel Poser (BBR, Switzerland), Mr Friedhelm Rentmeister (Bridon, UK) and Dr. Yoshito Tanaka (Shinko, Japan). The other material used, especially drawings and photographies, originates from the archives of the author and LAP.

Special thanks go to the author's long-term colleagues and friends Dr.-Ing. Imre Kovacs for the preparation of the sections on dynamics and Dr.-Ing. E.h. Dipl.-Ing. Reiner Saul for the section on cable sizing.

The publishers Wilhelm Ernst & Sohn and Wiley-Blackwell provided helpful support through the editor, Dipl.-Ing. Claudia Ozimek, supported by Ms. Sophie Bleifuß for the layout and Ms. Uta Beate Mutz for preparing the print.

The author translated the book into English himself with the exception of the sections on dynamics, which were done by Professor Dr. Guido Morgenthal. The translation was reviewed and improved by Mr Paul Beverley, UK. The complete German and English text was typed by Ms. Eva Gassmann and all graphics and their conversion into English was done by Ms. Mirela Beutel-Anistoroaiei; to all these the author is grateful for their indispensable skill and patience.

Cable-Stayed Bridges. 40 Years of Experience Worldwide. First Edition. Holger Svensson.
© 2012 Ernst & Sohn GmbH & Co. KG. Published 2012 by Ernst & Sohn GmbH & Co. KG.

The Author

Holger Svensson has extensive experience in the design, construction engineering and on-site supervision of cable-stayed and other long-span bridges all over the world.

In Germany he was involved in the checking of the Kocher Valley Bridge and the detailed design of the cable-stayed Flehe Bridge.

In the USA he designed several cable-stayed bridges: Pasco-Kennewick (concrete), East Huntington (concrete), Sunshine Skyway (composite alternate), Burlington (composite) and the Houston Ship Channel Crossing at Baytown (composite).

In Norway he was in charge of the design for the Helgeland Bridge and in Scotland he advised on the design of the Leven River Bridge, both concrete cable-stayed bridges.

In Sweden he was checking engineer for the Höga Kusten suspension bridge (main span 1210 m) and the Sunningesund and Ume Älv composite cable-stayed bridges.

In Australia he advised on the design of the cable-stayed Glebe Island concrete bridge in Sidney and on the My Tuan Bridge in Vietnam for AusAID.

For the Asian Development Bank he reviewed the design and construction of the record-breaking cable-stayed composite Yang Pu Bridge (main span 602 m) in Shanghai, China.

In Hong Kong he advised on the design and construction of the Kap Shui Mun Bridge to Lantau Airport.

He was also responsible for the design of several major girder and arch bridges in Germany and elsewhere.

Registrations
Current: Member of the Chamber of Engineers in Berlin, Germany; PE in the USA; CEng, FIStructE in Great Britain.
Formerly: MSAICE in South Africa; PEng in Canada; MHKIE in Hong Kong, China; RPEQ in Australia; TPEng. in Malaysia.

Memberships
2003 to 2011 Vice-President of IABSE.
Member of the Bridge Advisory Board of the German Railways.
Member of the Jury for the German Bridge Prize.
Member of the Jury for the German Structural Engineering Prize.
Member of the German Convent for Building Culture.

Publications
More than 100 publications and 180 verbal presentations.
Author of the book 'Cable-Stayed Bridges – 40 Years of Experience Worldwide', German Edition by Ernst & Sohn 2011, English Edition by Wiley-Blackwell 2012.

Honours
1999 James Watt Medal, Institution of Civil Engineers, London.
2000 Henry Husband Prize, Institution of Structural Engineers, London.
2011 Emil Mörsch Commemoration Medal, German Concrete Society.

Prof. Dipl.-Ing. Holger Svensson
PE, CEng, FIStructE,

born 1945 near Hamburg, Germany
1969 Dipl.-Ing., Stuttgart University
2009–2011 Lecturer and
since 2012 Professor for cable-stayed bridges at the University of Dresden, Germany

1970–1971 Contractor Grinaker
in South Africa and Botswana
1972–2009 Leonhardt, Andrä and Partners, Consulting Engineers
Design and checking of major, mainly cable-stayed bridges
1992–2009 Executive Director
1998–2008 Speaker of the Executive Board
2009 Chairman of the Board
Since 2010 Independent Consulting Engineer

VSL: cable-stayed bridges around the world

www.vsl.com

VSL's services

VSL offers specialist bridge construction solutions for all phases of a project, from the initial concepts right through until construction is complete:
- Design, detailing, manufacturing and installation of stay cables and post-tensioning
- Design, detailing, manufacturing, commissioning and operation of specialised bridge erection equipment and formwork for piers and pylons. Systems include solutions for free-cantilever construction, span-by-span erection, precast segmental erection, precast beam methods, full-span precast methods, incremental launching, whether cast-in situ or precast.
- Design, detailing and implementation of specialised heavy-lifting applications.
- Construction engineering, including stage by stage analysis, geometry control, method statements and detailing of construction sequences.
- Construction and project management, for packages of work or as the main contractor.

VSL's Stay Cable systems

The VSL SSI Saddle
This is the first stay cable saddle to combine the advantages of an injected saddle with full strand-by-strand installation and replacement, proven fatigue performance and exceptional durability. Its compact size with direct load transfer allows the design of more slender and aesthetic bridge pylons.

VSL Friction dampers and VSL Gensui dampers for stay cables
Both damping systems combine the highest efficiency with outstanding durability and minimum maintenance requirements. Minimising the number of moving parts reduces wear and tear. VSL's damping systems are versatile and can be used for new construction as well as in retrofit applications.

The VSL SSI 2000 Stay cable system
High fatigue resistance, durability and redundant protection through multiple protective barriers are the main features of the system. Three different versions are available to meet project-specific aerodynamic requirements.
- The standard SSI 2000 system has an optimised stay pipe to control rain-wind induced vibration and minimise wind drag.
- The SSI 2000-C system is more compact, with a smaller stay pipe diameter to reduce the wind load effects on the structure.
- The SSI 2000-D has active corrosion protection using permanent dehumidification combined with even smaller stay pipe diameters.

VSL

CREATING SOLUTIONS TOGETHER

Table of contents

1	**Introduction** 16
1.1	**Design fundamentals** 17
1.1.1	General 17
1.1.2	Overall system 19
1.1.2.1	Cable arrangement 19
1.1.2.2	Cable stiffness 20
1.1.2.3	Geometry 21
1.1.2.4	Support conditions 21
1.1.3	Tower shapes 23
1.1.3.1	Two outer cable planes 23
1.1.3.2	One central cable plane 23
1.1.3.3	Spread central cable planes 24
1.1.4	Beam cross-sections 24
1.1.4.1	Steel cross-sections 24
1.1.4.2	Concrete cross-sections 25
1.1.4.3	Composite cross-sections 25
1.1.4.4	Hybrid beams (steel/concrete) 26
1.1.4.5	Double deck cross-section 26
1.1.5	Stay cables 26
1.1.5.1	Systems 26
1.1.5.2	Cable anchorages 26
1.2	**Aesthetic guidelines for bridge design** 30
1.2.1	Introduction 30
1.2.2	Aesthetic guidelines 30
1.2.2.1	Guideline 1: Clear structural system 30
1.2.2.2	Guideline 2: Good proportions 31
1.2.2.3	Guideline 3: Good order 33
1.2.2.4	Guideline 4: Integration into the environment 34
1.2.2.5	Guideline 5: Choice of material 35
1.2.2.6	Guideline 6: Coloring 36
1.2.2.7	Guideline 7: Space above the bridge 38
1.2.2.8	Guideline 8: Recognizable flow of forces 38
1.2.2.9	Guideline 9: Lighting 41
1.2.2.10	Guideline 10: Simplicity 41
1.2.3	Collaboration 42

2	**The development of cable-stayed bridges** 46
2.1	**The precursors of cable-stayed bridges** 47
2.1.1	Introduction 47
2.1.2	Historical development 47
2.1.2.1	Historical designs 47
2.1.2.2	First examples and failures 48
2.1.2.3	John Roebling and stiffened suspension bridges 51
2.1.2.4	Transporter bridges 52
2.1.2.5	Approaching the modern form 55
2.2	**Steel cable-stayed bridges** 58
2.2.1	Introduction 58
2.2.2	Beginnings 58
2.2.3	The Düsseldorf Bridge Family 59
2.2.4	Further Rhine river bridges 62
2.2.5	Special steel cable-stayed bridges 70
2.2.6	Cable-stayed bridges with record spans 76
2.3	**Concrete cable-stayed bridges** 80
2.3.1	General 80
2.3.2	Development of concrete cable-stayed bridges 81
2.3.3	Bridges with concrete stays 92
2.3.3.1	Riccardo Morandi's bridges 92
2.3.3.2	Later examples 92
2.3.3.3	Bridges with concrete walls 94
2.3.4	Cable-stayed bridges with thin concrete beams 94
2.3.5	Record spans 98
2.4	**Composite cable-stayed bridges** 101
2.4.1	General 101
2.4.2	Cross-sections 101
2.4.3	Special details 104
2.4.4	Economic span lengths 104
2.4.5	Beginnings 105
2.4.6	Record spans 105
2.4.7	Latest examples 111
2.5	**Special systems of cable-stayed bridges** 118
2.5.1	Series of cable-stayed bridges 118
2.5.1.1	Load transfer 118
2.5.1.2	Intermediate piers 118
2.5.1.3	Stiff towers 118
2.5.1.4	Stayed towers 118
2.5.1.5	Frames 121
2.5.1.6	Accommodation of longitudinal deformations 121
2.5.1.7	Examples 123
2.5.2	Stayed beams 130
2.5.2.1	Stayed from underneath 130
2.5.2.2	Stayed from above (extradosed) 130
2.5.3	Cable-stayed pedestrian bridges 133

Cable-Stayed Bridges. 40 Years of Experience Worldwide. First Edition. Holger Svensson.
© 2012 Ernst & Sohn GmbH & Co. KG. Published 2012 by Ernst & Sohn GmbH & Co. KG.

3	**Stay cables** 140		**3.7**	**Cable sizing** 160
3.1	**General** 141		3.7.1	General 160
			3.7.2	Sizing by permissible stresses 160
3.2	**Locked coil ropes** 141		3.7.2.1	Permissible stresses for static loads 160
3.2.1	System 141		3.7.2.2	Permissible fatigue range 160
3.2.2	Fabrication 142		3.7.2.3	Permissible stresses during cable exchange 161
3.2.3	Modern corrosion protection systems 142		3.7.3	Sizing in ultimate limit state 161
3.2.3.1	General 142		3.7.3.1	Ultimate limit state 161
3.2.3.2	Galvanizing of the wires 142		3.7.3.2	Fatigue 161
3.2.3.3	Filling 142		3.7.3.3	Cable exchange 162
3.2.3.4	Paint 143		3.7.3.4	Service limit state 162
3.2.4	Inspection and maintenance 143		3.7.4	Summary 162
3.2.5	Damage 143			
3.2.5.1	Köhlbrand Bridge 143		**3.8**	**Cable dynamics** 163
3.2.5.2	Maracaibo Bridge, Venezuela 145		3.8.1	General 163
3.2.5.3	Flehe Rhine River Bridge 146		3.8.2	Fundamental parameters 164
3.2.5.4	Lessons from the damage 146		3.8.2.1	Static wind load 164
			3.8.2.2	Natural frequencies 165
3.3	**Parallel bar cables** 146		3.8.3	Dynamic excitation 165
			3.8.3.1	Galloping oscillations 165
3.4	**Parallel wire cables** 147		3.8.3.2	Anchorage excitation 166
3.4.1	System 147		3.8.3.3	Parametric resonance 168
3.4.2	Corrosion protection 148		3.8.3.4	Buffeting 168
3.4.2.1	Polyethylene (PE) pipes 148		3.8.3.5	Vortex-induced vibrations 169
3.4.2.2	Wrappings 150		3.8.4	Countermeasures 169
3.4.2.3	Grouting 150		3.8.4.1	Dampers 169
3.4.2.4	Damage 151		3.8.4.2	Surface profiling 174
3.4.2.5	Petroleum wax 151		3.8.4.3	Cross ties 174
3.4.3	Fabrication 152			
			3.9	**Cable installation** 175
3.5	**Parallel strand cables** 153		3.9.1	General 175
3.5.1	General 153		3.9.2	Locked coil ropes 175
3.5.2	System 153		3.9.2.1	General 175
3.5.3	Corrosion protection 153		3.9.2.2	Example 175
3.5.3.1	Traditional 153		3.9.3	Parallel wire cables 178
3.5.3.2	With dry air 153		3.9.3.1	General 178
3.5.4	Fabrication 154		3.9.3.2	Example 178
3.5.5	Durability tests 154		3.9.4	Parallel strand cables 179
3.5.5.1	Tensile strength and fatigue strength 154		3.9.4.1	General 179
3.5.5.2	Water tightness 154		3.9.4.2	Example 179
3.5.5.3	Sustainability 154		3.9.5	Cable calculations 184
3.5.6	Monitoring 154		3.9.5.1	Cable deformations 184
			3.9.5.2	Measuring of cable forces 184
3.6	**Cable anchorages** 156			
3.6.1	General 156			
3.6.2	Support of anchor heads 156			
3.6.3	Anchorage at the tower 158			
3.6.3.1	Continuous 158			
3.6.3.2	Composite cable anchorages at tower head 158			
3.6.3.3	Cable anchorage in concrete 158			

4	**Preliminary design of cable-stayed bridges** 186	**4.4**	**Protection of bridges against ship collision** 248	
4.1	**Action forces for equivalent systems** 187	4.4.1	Introduction 248	
4.1.1	General 187	4.4.2	Collision forces 248	
4.1.2	System geometry 187	4.4.3	Protective structures 252	
4.1.3	Normal forces of articulated system 188	4.4.3.1	General 252	
4.1.4	Live loads on elastic foundation 189	4.4.3.2	Out of reach 252	
4.1.4.1	Beam on elastic foundation 189	4.4.3.3	Artificial islands 253	
4.1.4.2	Buckling – non-linear theory 190	4.4.3.4	Guide structures 253	
4.1.5	Permanent loads on rigid supports 192	4.4.3.5	Independent protective structures 258	
4.1.5.1	Dead load 192	4.4.3.6	Strong piers 261	
4.1.5.2	Post-tensioning 193	**4.5**	**Preliminary design calculations** 266	
4.1.5.3	Shrinkage and creep 193	4.5.1	General 266	
4.1.6	Towers 195	4.5.2	Typical cable-stayed concrete bridge 266	
4.1.6.1	In the longitudinal direction 195	4.5.2.1	System and loads 266	
4.1.6.2	In the transverse direction 195	4.5.2.2	Normal forces for articulated system 267	
4.1.7	Stay cables 197	4.5.2.3	Bending moments 269	
4.2	**Action forces of actual systems** 197	4.5.3	Typical cable-stayed steel bridge 270	
4.2.1	Permanent loads 197	4.5.3.1	General 270	
4.2.1.1	General 197	4.5.3.2	System 270	
4.2.1.2	Concrete bridges 198	4.5.3.3	Section properties and loads 270	
4.2.1.3	Steel bridges 199	4.5.3.4	Beam moments from live load 270	
4.2.1.4	Towers 202	4.5.3.5	Permissible beam moments 271	
4.2.2	Live loads 202	4.5.3.6	Moments from dead load for articulated system 271	
4.2.3	Kern point moments 204	4.5.4	Cable-stayed bridge with side spans on piers 272	
4.2.4	Non-linear theory (second order theory) 206	4.5.4.1	System and loads 272	
4.2.5	Superposition 208	4.5.4.2	Cable forces of articulated system 273	
4.2.6	Temperature 208	4.5.4.3	Bending moments for beam 274	
4.2.7	Eigenfrequencies 210	4.5.5	Cable-stayed bridge with harp arrangement 275	
4.3	**Bridge dynamics** 211	4.5.5.1	With regular side spans 275	
4.3.1	General 211	4.5.5.2	With side spans on piers 276	
4.3.2	Overview of wind effects 213	4.5.6	Cable-stayed bridge with longitudinal A-tower 276	
4.3.3	Wind profile, turbulence and turbulence-induced oscillations 214	4.5.6.1	System and loads 277	
4.3.3.1	Wind parameters 214	4.5.6.2	Normal forces for articulated system 277	
4.3.3.2	Natural modes of vibration of structures 216	4.5.6.3	Cable sizing 278	
4.3.3.3	Section forces under turbulent excitation 218	4.5.6.4	Bending moments for beam 278	
4.3.4	Vortex-induced vibrations 222	4.5.6.5	Post-tensioning 278	
4.3.5	Self-excitation and other motion-induced effects 224	4.5.7	Slender cable-stayed concrete bridge 279	
4.3.5.1	General description, background 224	4.5.7.1	System and loads 279	
4.3.5.2	Practical examples of bending-type galloping 226	4.5.7.2	Stay cables 280	
4.3.5.3	Practical examples of torsional galloping 228	4.5.7.3	Beam moments 283	
4.3.5.4	Flutter 232	4.5.7.4	Aerodynamic stability 286	
4.3.6	Damping measures 236	4.5.7.5	Towers 287	
4.3.7	Wind tunnel testing 240			
4.3.7.1	General 240			
4.3.7.2	Overview of important types of wind tunnel testing 240			
4.3.8	Earthquake 244			

5	**Construction of cable-stayed bridges** 290		**6**	**Examples for typical cable-stayed bridges** 326
5.1	**Examples** 291		**6.1**	**Cable-stayed concrete bridges with precast beams** 327
5.1.1	General 291		6.1.1	General 327
5.1.2	Tower construction 291		6.1.2	Pasco-Kennewick Bridge 327
5.1.2.1	Steel towers 291		6.1.2.1	General layout 327
5.1.2.2	Concrete towers 291		6.1.2.2	Construction engineering 332
5.1.2.3	Composite towers 291		6.1.2.3	Completed bridge 344
5.1.3	Beam construction 292		6.1.3	East Huntington Bridge 346
5.1.3.1	General 292		6.1.3.1	General design considerations 346
5.1.3.2	Concrete beam 293		6.1.3.2	Construction 346
	Free cantilevering 293		6.1.3.3	Completed bridge 349
	Launching 299			
	Rotating 299		**6.2**	**CIP concrete cable-stayed bridge**
	Rotating on scaffolding 302			Helgeland Bridge 352
5.1.3.3	Steel beams 304		6.2.1	General layout 352
	Free cantilevering 304		6.2.1.1	Introduction 352
	Launching 304		6.2.1.2	Bridge system 353
	Transverse shifting 305		6.2.2	Construction 358
5.1.3.4	Composite beam 305		6.2.2.1	Climate 358
	Free cantilevering 305		6.2.2.2	Towers 358
	Launching 309		6.2.2.3	Beam 360
			6.2.2.4	Stay cables 362
5.2	**Construction engineering** 312		6.2.2.5	Instrumentation 366
5.2.1	General 312		6.2.2.6	Completed bridge 367
5.2.2	Construction engineering by dismantling 312		6.2.3	Summary 367
5.2.2.1	General 312			
5.2.2.2	Dismantling from $t = \infty$ to $t = 1$ 313		**6.3**	**Cable-stayed steel bridge**
5.2.2.3	Dismantling of bridge 313			Strelasund Crossing 369
	With floating crane 313		6.3.1	Design considerations 369
	Dismantling with derrick 314		6.3.1.1	Bridge alternates 369
5.2.2.4	Aerodynamic stability 315		6.3.1.2	Optimizing the cable-stayed solution 371
5.2.3	Example for construction engineering 316		6.3.1.3	Structural details 371
5.2.3.1	Forward construction 316		6.3.2	The cable-stayed bridge 371
5.2.3.2	Construction engineering 316		6.3.2.1	Span lengths 371
5.2.3.3	Construction manual 316		6.3.2.2	Beam cross-section 371
5.2.3.4	Control measurements 316		6.3.2.3	Wind barriers 372
5.2.4	Design of auxiliary stays 320		6.3.2.4	Tower 372
5.2.4.1	Symmetrical auxiliary stays for towers 320		6.3.2.5	Stay cables 373
5.2.4.2	One-sided auxiliary stays for towers 321		6.3.2.6	Aerodynamic investigation 373
5.2.4.3	Auxiliary stays for beam 321		6.3.3	Construction 375
5.2.4.4	Without auxiliary stays for beam 321		6.3.3.1	Construction engineering 375
5.2.5	Auxiliary tie-backs for travelers 323		6.3.3.2	Construction of the main bridge 376
			6.3.3.3	Completed bridge 386

6.4	**Composite cable-stayed bridge**		
	Baytown Bridge 387		
6.4.1	General layout 387		
6.4.1.1	Bridge system 387		
6.4.1.2	Composite beam 388		
6.4.1.3	Towers 390		
6.4.1.4	Stay cables 390		
6.4.1.5	Aerodynamic stability 391		
6.4.2	Construction 391		
6.4.2.1	Foundations 391		
6.4.2.2	Towers 391		
6.4.2.3	Beam 392		
6.4.2.4	Stay cables 398		
6.4.2.5	Completed bridge 398		
6.4.3	Summary 401		

Index 428
Bridge Index 428
References 431
Figure Origins 439
List of Advertisers 441

Appendix: 40 years of experience with major bridges all over the world 443
Beginnings 443
Bridges in Germany 443
Cable-stayed bridges abroad 445
New developments by competition 446
Checking of bridges 449
Participation in Code Commissions 452
Current projects 452
Summary 453
References 454

6.5	**Hybrid cable-stayed bridge**		
	Normandy Bridge 402		
6.5.1	Design considerations 402		
6.5.1.1	Structural design 402		
6.5.1.2	Cable dynamics 404		
6.5.2	Construction 404		
6.5.2.1	Tower 404		
6.5.2.2	Concrete approach bridges 406		
6.5.2.3	Steel main span 408		
6.5.2.4	Cable installation 408		
6.5.2.5	Completed bridge 410		

Lectures on cable-stayed bridges on DVD 458

6.6	**Series of cable-stayed bridges** 411		
6.6.1	Millau Bridge 411		
6.6.1.1	General 411		
6.6.1.2	Design 412		
6.6.1.3	Construction 412		
6.6.1.4	Completed bridge 416		
6.6.2	Rion-Antirion Bridge 418		
6.6.2.1	General 418		
6.6.2.2	Design 418		
6.6.2.3	Construction 421		
6.6.2.4	Completed bridge 425		

7 Future development 426

1 Introduction

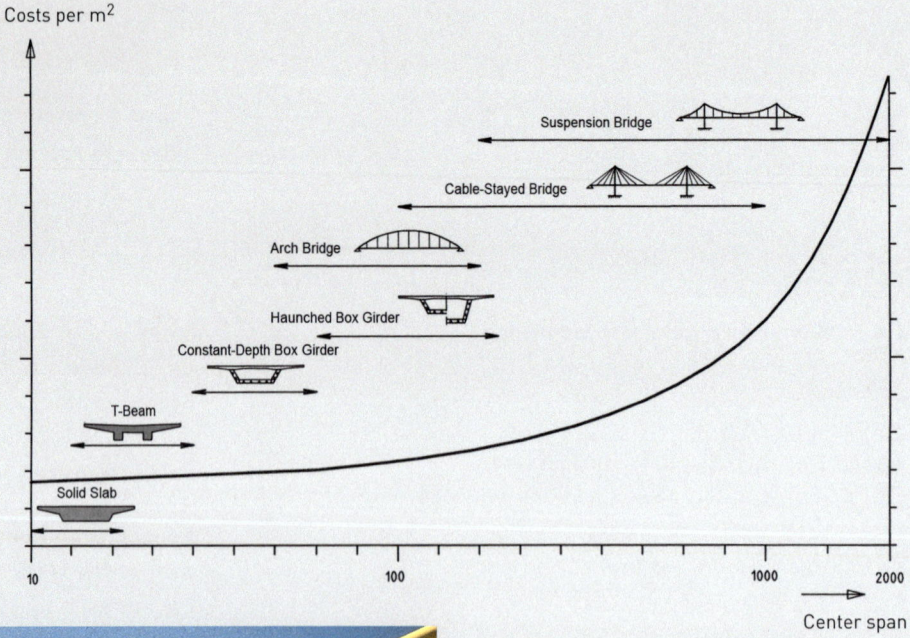

Cable-Stayed Bridges. 40 Years of Experience Worldwide. First Edition. Holger Svensson.
© 2012 Ernst & Sohn GmbH & Co. KG. Published 2012 by Ernst & Sohn GmbH & Co. KG.

1.1 Design fundamentals
1.1.1 General

The position of cable-stayed bridges within all bridge systems is given in Fig. 1.1. Their spans range between continuous girders and arch bridges with shorter spans at one end, and suspension bridges with longer spans at the other.

The economic main span range of cable-stayed bridges thus lies between 100 m with one tower and 1100 m with two towers.

What are the special advantages of cable-stayed bridges?

First of all the bending moments are greatly reduced by the load transfer of the stay cables, Fig. 1.2. By installing the stay cables with their predetermined precise lengths the support conditions for a beam rigidly supported at the cable anchor points can be achieved and thus the moments from permanent loads are minimized, Fig. 1.3.

Even for live loads the bending moments of the beam elastically supported by the stay cables remain small.

Negative live load moments may occur over the vertical bearings at the towers. They can be avoided by supporting the beam by the stay cables only, including in the tower region. The biggest positive and negative moments occur in the side spans near the hold-down piers, which may require special measures. Shear forces remain small.

Large compression forces in the beam are caused by the horizontal components of the inclined stay cables. The normal forces in the main and side span equal one another so that only uplift forces have to be anchored in the abutments which act as hold-down piers.

The largest horizontal components belong to the backstay cables, Fig. 1.3. The beam compression forces thus increase quickly at the beam ends and increase less quickly near the towers where they reach a maximum. For concrete beams this compression force is beneficial. Only in the bridge center, where the compression forces from cables are small, is additional longitudinal post-tensioning required, which also has to overcome tension from friction, braking, etc.

The cross-section of steel beams requires only little additional material to cover the overall compression forces.

A second important advantage of cable-stayed bridges is their ease of construction, Fig. 1.4:
– Arch bridges with large spans are not stable during erection until the arch is closed and the horizontal support forces are anchored. They have to be temporarily supported, e.g. by auxiliary piers or temporary tie-backs.
– Self-anchored suspension bridges, which may be required when their horizontal cable component cannot economically be anchored due to bad soil conditions, need temporary supports of their beams until the main cables are installed.
– In cable-stayed bridges, however, the same flow of forces is present during free-cantilever construction stages as after completion. This is true for free cantilevering to both sides of the tower as well as for free cantilevering the main span only.

The third main advantage lies in the fact that cable-stayed bridges are inherently stiffer than suspension bridges. This is especially true

Figure 1.1 (left side above) Relation between construction costs per m^2 and main span length

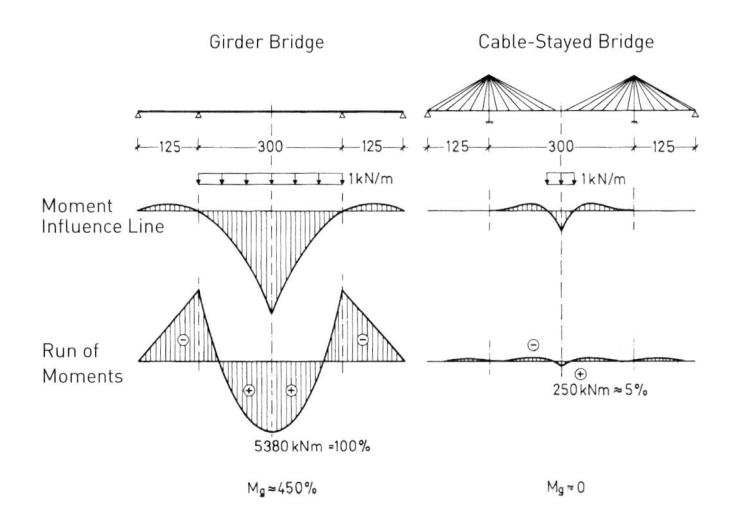

Figure 1.2 Comparison of beam moments

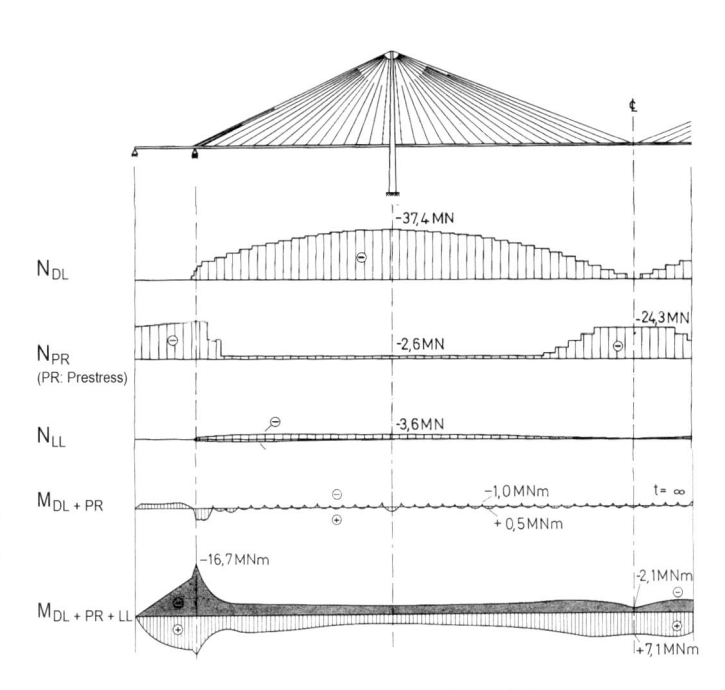

Figure 1.3 Bending moments and compression forces in beams of cable-stayed bridges

Figure 1.4 Erection systems for arches, self-anchored suspension bridges and cable-stayed bridges

a) suspension bridge

b) cable-stayed bridge with Twin-towers

c) cable-stayed bridge with A-towers

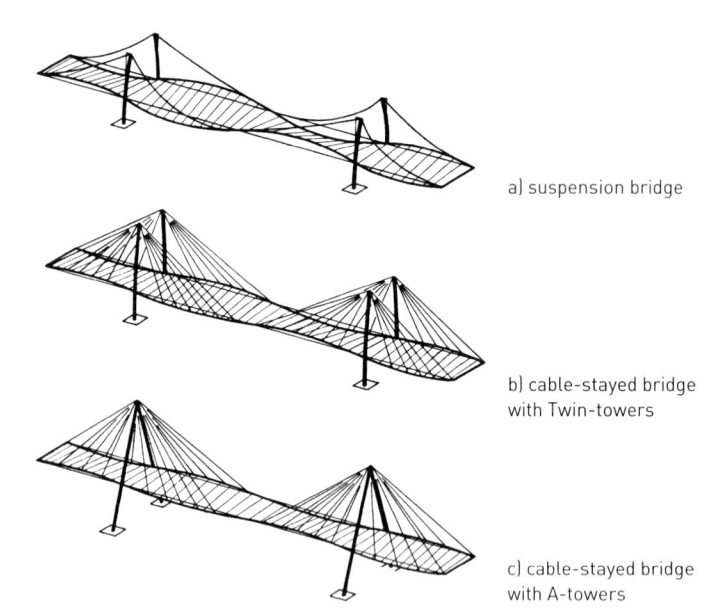

Figure 1.5 Eigenmodes of suspension bridges and cable-stayed bridges

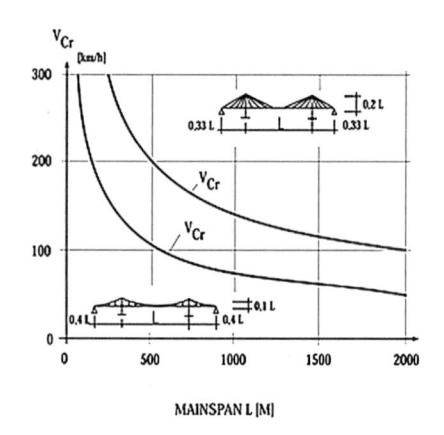

Figure 1.6 Critical wind speeds of suspension bridges and cable-stayed bridges

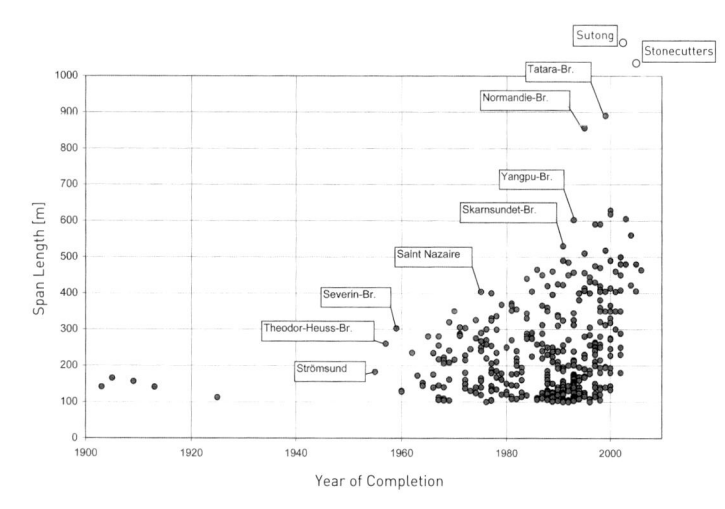

Figure 1.7 Development of main span lengths of cable-stayed bridges

for longitudinally eccentric loads, for which the main cables of a suspension bridge find a new equilibrium without increase of cable stresses (stressless deformations), whereas stay cables always have to be additionally stressed in order to carry any load.

Suspension bridges are thus generally not suitable for high railway loads: their deformations become too large.

The eigenfrequencies of cable-stayed bridges are significantly higher than those of suspension bridges. The torsional frequencies, which are especially important for the aerodynamic safety against flutter, can further be increased by using A-towers which omit the elasticity of the backstay cables, Fig. 1.5.

The critical wind speed for the onset of flutter is thus higher for cable-stayed bridges than for suspension bridges. The diagram in Fig. 1.6 shows that for a main span of 500 m the critical flutter wind speed for a typical cable-stayed bridge is about 200 km/h while for a corresponding suspension bridge it is only 100 km/h.

These advantages of cable-stayed bridges have become generally known worldwide since the 1970s, and since that time their number has increased strongly, Table 1.1. The total number of major cable-stayed bridges worldwide has currently reached about 1000.

The main span length of cable-stayed bridges has also increased dramatically, the longest currently being 1088 m, Fig. 1.7 [1.1].

Fritz Leonhardt and his collaborators have disseminated the above findings in lectures and publications all over the world since the 1970s, [1.2 – 1.9].

Table 1.1 Number of cable-stayed bridges until 2000

	1900–1950	1960	1970	1980	1990	2000	Σ
Japan				1	8	12	21
China						17	17
Germany		2	5	6	2		15
USA				2	7	4	13
Netherlands				2	1	2	5
France				2		2	4
USSR				1	1	2	4
Great Britain				1	1	2	4
other				3	9	21	33
Σ	0	2	5	18	29	62	116

1.1.2 Overall system
1.1.2.1 Cable arrangement

Cable-stayed bridges have at least one forestay and one backstay cable per tower. The forestays act like two piers omitted in the main span.

The first cable-stayed bridge across the River Rhine in the 1950s used relatively few stay cables, which comprised bundles of locked coil ropes, so that the distances between the cable anchorages at the beam were 40 to 60 m. Their anchorages and corrosion protection were difficult, and auxiliary tie-backs were required to overcome the large distances between the cables during construction.

The bridge designer, Hellmut Homberg (see Fig. 2.78), used many individual stay cables with distances of only 4.5 m for the Bonn North Bridge, which simplified decisively the structural detailing and the construction, Fig. 1.8. This design was trendsetting for the further development of cable-stayed bridges.

The number of stay cables is chosen advantageously in such way that for each stay only one cable is required so that the anchorages at the tower and at the beam become most simple.

With cable capacities now up to 20 MN, even for very wide bridges the distances between the cable anchorages at the beam can be 5 to 15 m, and even up to 20 m. Such small cable distances allow free cantilevering to be used without the need for auxiliary tie-backs during construction.

The reduction of cable distances at the beam over the years is shown in Fig. 1.9.

There are three basic cable arrangements in bridge elevation. With the fan arrangement, Fig. 1.10, all cables join at the tower head. With the harp arrangement, Fig. 1.11, all cables run parallel and are anchored over the height of the tower. The true fan arrangement can not be realized practically if the condition has been set that, in the case of damage, each individual cable must be able to be exchanged by reversing the installation procedure. This exchangeability requires individual cable anchorages to have certain minimum distances at the tower head.

In such way an intermediate cable arrangement between harp and fan is created, Fig. 1.12, which should be as close as possible to the fan arrangement due to its economical advantages. The harp arrangement, however, is aesthetically advantageous because there are no visual intersections of the stay cables, even for two cable planes in a skew view.

The arrangement of stay cables is also influenced by the relationship between side span and main span, and the location of the backstay anchorages, e.g. by connecting them directly above the piers in the side spans. The harmonic cable arrangement plays an important role in the aesthetic appearance of a cable-stayed bridge and has thus to be studied carefully from the very beginning of a design.

In addition to the basic arrangements mentioned above special configurations have been used.

The early South Elbe River Bridge used a star-shaped cable arrangement for which the few cables were separated at the towers in

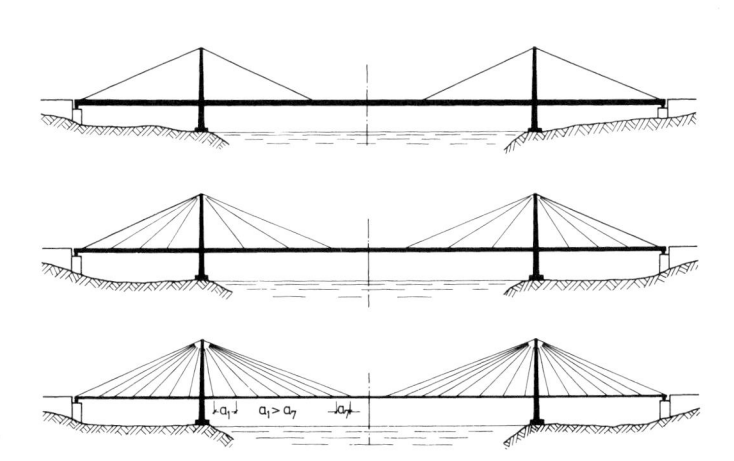

Figure 1.8 Development to multi-stay cable bridges

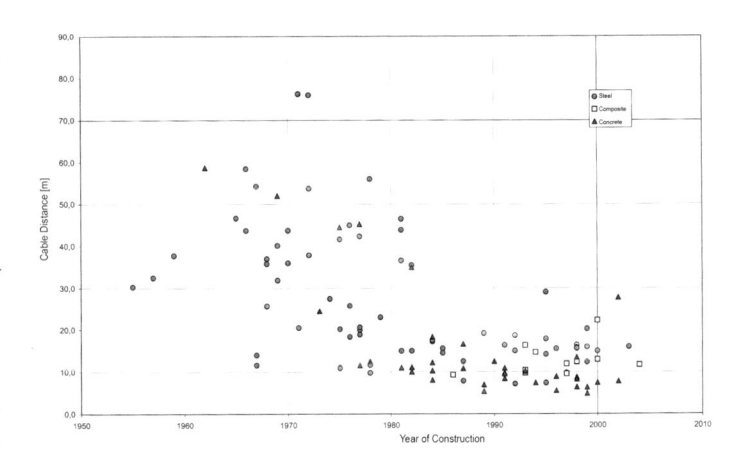

Figure 1.9 Development of cable distances at the beam

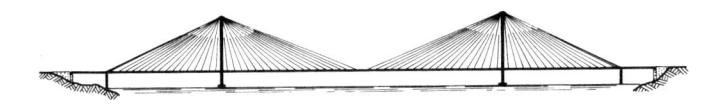

Figure 1.10 Fan system

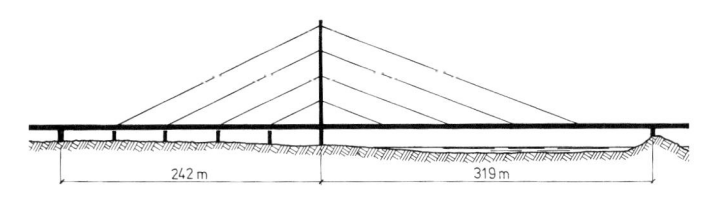

Figure 1.11 Harp system

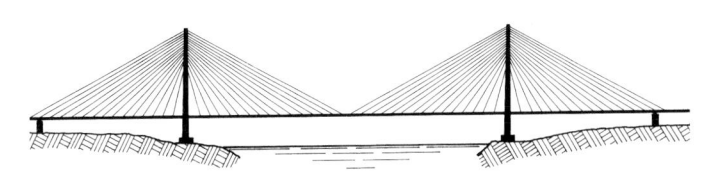

Figure 1.12 Intermediate system fan–harp

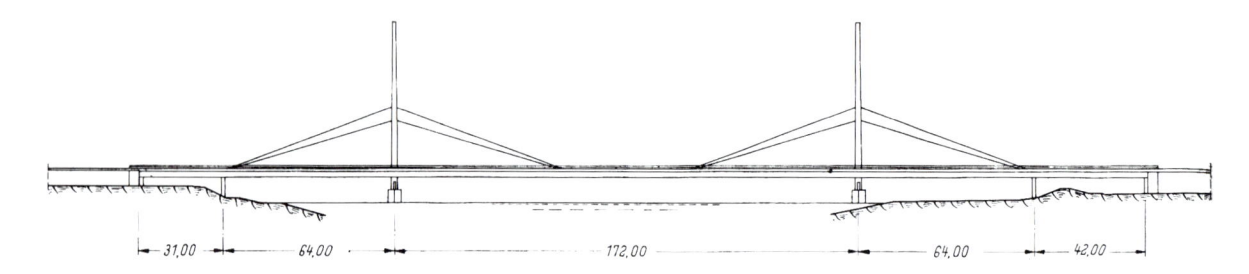

Figure 1.13 South Elbe River Bridge, star-shaped cable arrangement

order to simplify their anchorages, Fig. 1.13 [1.10]. Since the stays are still anchored together at the beam, this solution was not repeated.

For open cross-sections with low torsional stiffness two outer cable planes are required to carry eccentric live loads by a force couple. Hellmut Homberg (see Fig. 2.78) used mainly steel box girders for his designs. They have a high torsional rigidity and permit a central cable plane and also have a high bending stiffness so that a beam supported at the towers allows the removal of the first cables at both sides of the towers to some greater distance. This arrangement is called "Homberg's window", Fig. 1.14.

Figure 1.14 Bonn North Bridge, Homberg's window

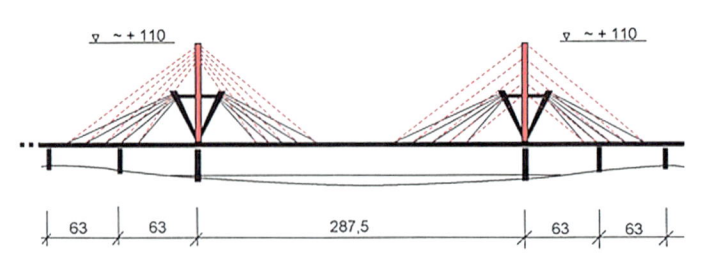

The high negative bending moments above the tower piers can be carried by the box girder. This transfer of loads near the towers via shear and bending directly into the foundations is more economic than supporting the beam only by the cables.

The Rhine River Bridge at Ilverich is located in the vicinity of Düsseldorf airport. The normally required tower height of 110 m could thus not be realized because of air traffic safety requirements.

In order to avoid cables at too uneconomically flat an angle, each tower was split into two V-shaped legs and the backstay and forestay cable forces are carried by a tie, Fig. 1.15 [1.12].

Figure 1.15 Rhine River Bridge at Ilverich, shortened towers

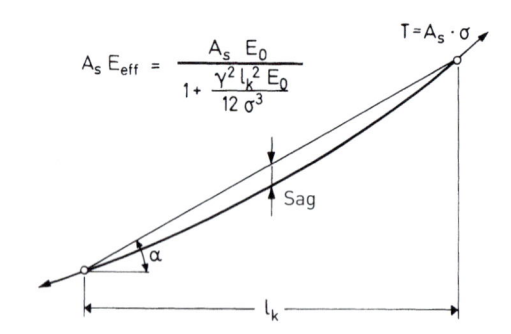

1.1.2.2 Cable stiffness

The stiffness of cable-stayed bridges is governed by the stiffness of the stay cables, which is reduced by the cable sag, as given in the formula (1-1) by Ernst [1.13] (Fig. 1.16):

$$A_s E_{eff} = \frac{A_s E_0}{1 + \frac{\gamma^2 l_k^2 E_0}{12 \sigma^3}} \quad (1\text{-}1)$$

A_s Cable Steel Area
E_0 Modulus of Elasticity for straight Cables
E_{eff} E-Modulus of Cable with Sag
γ Specific Weight of Cable including Corrosion Protection
l_k Horizontal Length of Cable
σ Tensile Stress in Cable

It gives the effective modulus of elasticity of a stay cable in relation to a straight vertical cable E_0, the specific weight of the cable γ, the horizontal projection of length l_k, and the steel stress σ. The steel stress is of major importance and occurs in the formula as the third power, Fig. 1.16.

The influence of the various parameters on the cable stiffness is shown in Fig. 1.17.

In order to achieve a high cable stiffness, a high cable stress is required. Therefore, high-strength steel as developed for prestressing tendons has to be used. The high dead weight of a concrete beam is advantageous in this respect. Assuming strands with a steel of 1660/1860 N/mm² and assuming a factor of safety for the ultimate limit state of 2.2, a service load stress of 845 N/mm² is reached. Assuming further a ratio of live load to dead load of 0.2, a cable stress of about 680 N/mm² can be used for the permanent load stage which governs the deflections of a cable-stayed bridge.

Figure 1.16 Effective stiffness of stay cables

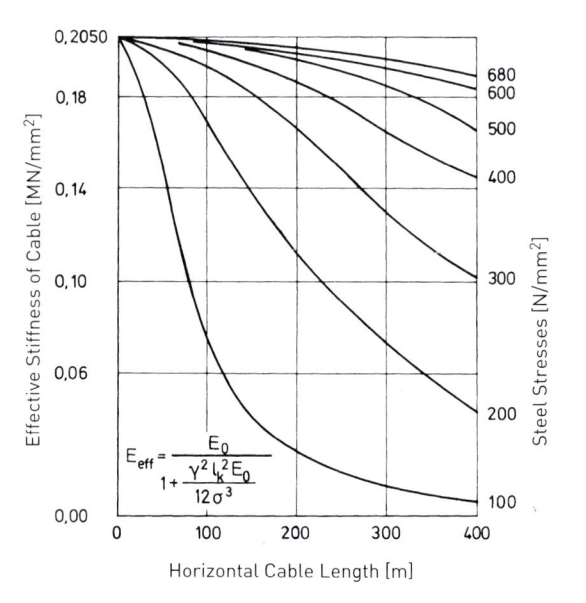

Figure 1.17 Relation between $E_{Effective}$ of cables, span lengths and steel stress

Fig. 1.17 shows that for such steel stresses the loss of cable stiffness is small.

1.1.2.3 Geometry

Main spans

In a first approximation the length of a side span comes to about 40% of the main span for a concrete beam under road traffic loads. For a steel beam with railway loading this ratio may be reduced to 30%.

The governing factor for the span ratio is the stress in the backstay cables. For the cable size required for the biggest backstay force with live load in the main span, the fatigue stress range for live load in the main and side span should not exceed the permissible range of, for example, 200 N/mm2, Fig. 1.18.

Tower heights

High towers reduce the required amount of cable steel and the compression forces in the bridge beam up to a cable inclination of 45°, but increase the costs for the towers themselves.

Optimizations indicate that tower heights of about 1/5 of the main span above deck achieve minimum costs for the whole bridge, Fig. 1.19.

Stiff towers

If the geometrical conditions require two equal spans with only one tower at the center, the tower height should be chosen so that the flattest cables have the usual inclination of about 27°. This gives a ratio of tower height to each main span of about 0.5.

Since there are no typical backstays, which restrain the tower head, the tower has to be stiff by itself. This is best achieved by towers that are A-shaped in the longitudinal direction, Fig. 1.20.

1.1.2.4 Support conditions

Longitudinal

The longitudinal support conditions for the beams of cable-stayed bridges are determined by the changes in lengths due to temperature, shrinkage and creep as well as longitudinal forces such as braking, wind and earthquake. A survey of the different support conditions with their advantages and disadvantages is given in Table 1.2 [1.14] for railway bridges which are especially sensitive in this regard.

For railway bridges high above ground the main problem is the high braking loads which are best distributed equally to both towers. A rigid connection to each tower, however, may create too high tower moments from the beam changes of lengths due to temperature and shrinkage and creep.

A compromise between these contradictory requirements are hydraulic buffers, which act as rigid connections between beam and towers for short-term loads such as braking or earthquake, but offer no resistance for slowly applied loads such as temperature, shrinkage and creep. Their disadvantage is that they require monitoring and

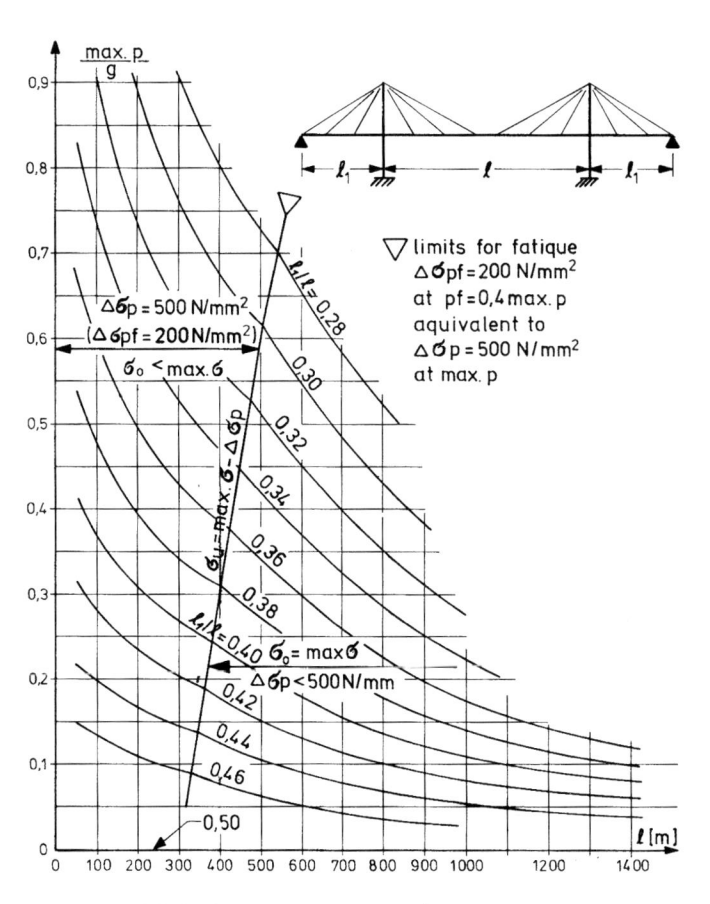

Figure 1.18 Relation between span ratios and backstay steel stresses

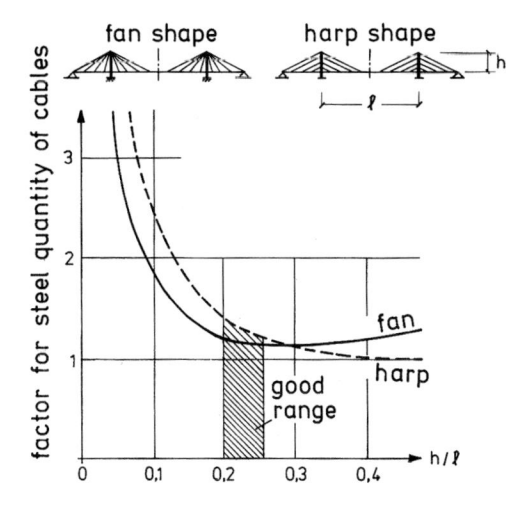

Figure 1.19 Relationship between tower height and amount of cable steel

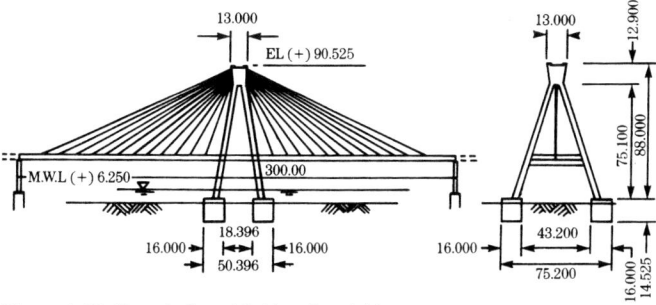

Figure 1.20 Olympic Grand Bridge, Seoul, Korea

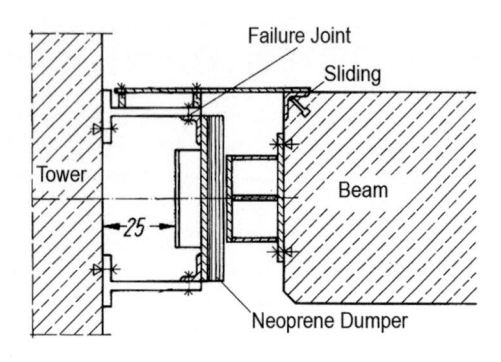

Figure 1.21 Transverse buffers at tower

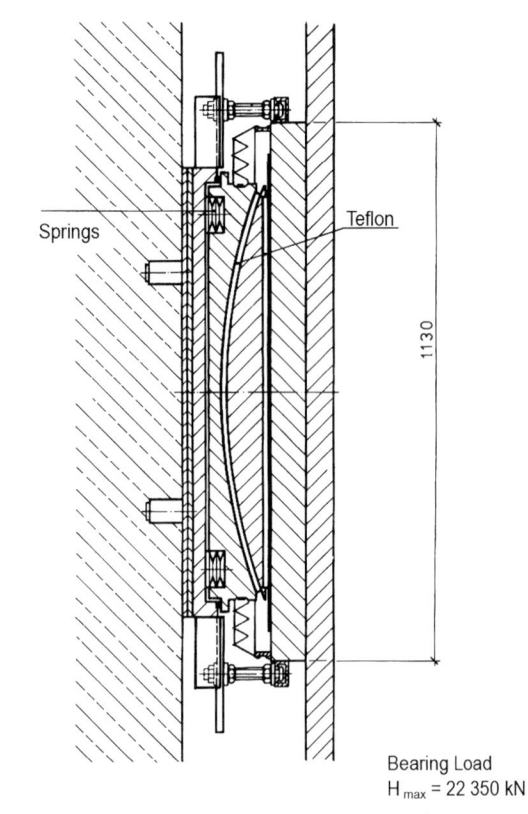

Figure 1.22 Post-tensioned transverse tower bearing

careful maintenance. A practical solution is to use high neoprene bearings which follow the beam deformations by inclination.

Another possibility, chosen in Japan, is to connect both towers with long cables to the beam so that the cables are elastically restrained by the beam deformations and thus both towers are equally loaded.

Transverse

The governing load for cable-stayed bridges in the transverse direction is wind on the cables, the towers and the beam. In many cases transverse buffers at the towers with little play can take the transverse beam loads, Fig. 1.21 [1.15].

Transversely post-tensioned bearings have been used where transverse movements are to be avoided, Fig. 1.22 [1.16].

In order to cater for the horizontal bridge rotations at the abutments due to wind, a transversely fixed rotational bearing can be located in the center of the end cross girder. Should there be piers in the side span, their bearings can normally slide in the transverse direction in order to avoid undue restraints from the main span.

Anchor piers

At the anchor piers (hold-down piers) the uplift forces in the backstay cables from unbalanced loads in the main span have to be anchored. The simplest solution is to connect the beam directly to the anchor pier with the help of a rotational connection, Fig. 1.23. The anchor piers are stressed down into the foundations and have to be flexible enough to follow the longitudinal changes of length in the beam [1.17].

For a sliding support of the beam at stiff abutments pendulums have traditionally been used which extend deep into the abutments and are articulated at both ends. Especially with steel bridges it has to be taken into account that these pendulums may obtain both ten-

Table 1.2 Longitudinal support conditions

No.	System	Characteristics	Advantages	Disadvantages
1	Floating bridge beam	Free deformations due to change of temperature Introduction of breaking forces into cables and towers	Same stresses and deformations for both halves of bridge	Large bending moments in towers and large horizontal movements of beam due to breaking and unsymmetrical live loads
2	One fixed bearing	Free deformations due to change of temperature All horizontal forces act on one hold-down pier	Less deformations than in system 1	Strong forces on hold-down pier Longitudinally unsymmetric Large temperature movements in other hold-down pier
3	Hydraulic buffers at both towers	Two fixed bearings for short-term loads Floating system for long-term loads	Small deformation of the symmetrical system Transfer of breaking forces onto both towers	System with varying support conditions depending on period of loading

1.1 Design fundamentals

sile and compression forces and thus have to be designed for fatigue. This is especially important for the end articulations.

The fatigue problem can be avoided if the beam is post-tensioned through the bearings against the anchor piers and thus a minimum compression force on the bearings is always ensured, Fig. 1.24 [1.18]. The longitudinal movements have again to be accounted for.

1.1.3 Tower shapes

The first cable-stayed bridges used steel towers. Since towers are mainly loaded by compression, concrete towers are more economical and, therefore, mainly used today. Only if extremely bad foundation conditions would require very long piles, are the lighter steel towers used today.

1.1.3.1 Two outer cable planes

The beams of cable-stayed bridges should normally be supported at the outsides in order to restrain the rotational deformations most effectively. This requires two cable planes, supported by two tower legs, Fig. 1.25. For small and medium spans vertical tower legs may be used. To stiffen them transversely they can be connected with cross beams. For longer spans the two tower legs should lean towards one another above deck in an A-shape, which provides increased torsional rigidity to the beam and stiffens the towers for transverse loads.

For bridge beams with a high clearance the two foundations for a straight A may be uneconomic. The two legs can then be pulled together underneath the beam to permit a common foundation.

1.1.3.2 One central cable plane

Only if one central cable plane is used, the cables for short and medium spans are anchored at one single tower at the bridge center

Figure 1.23 Articulated hold-down pier

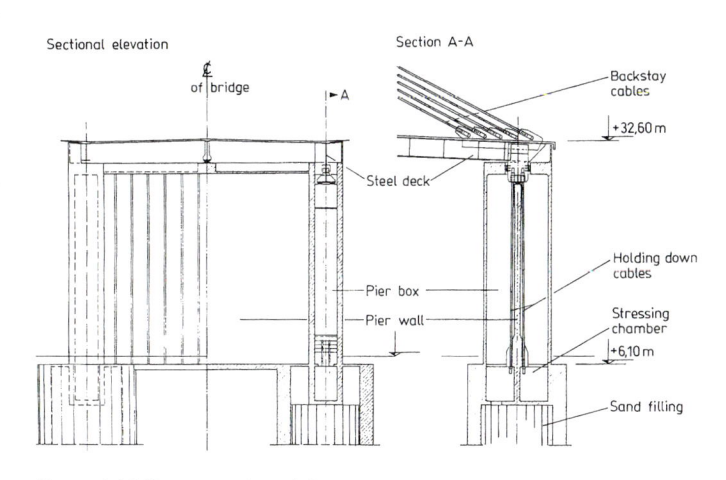

Figure 1.24 Prestressed pendulum

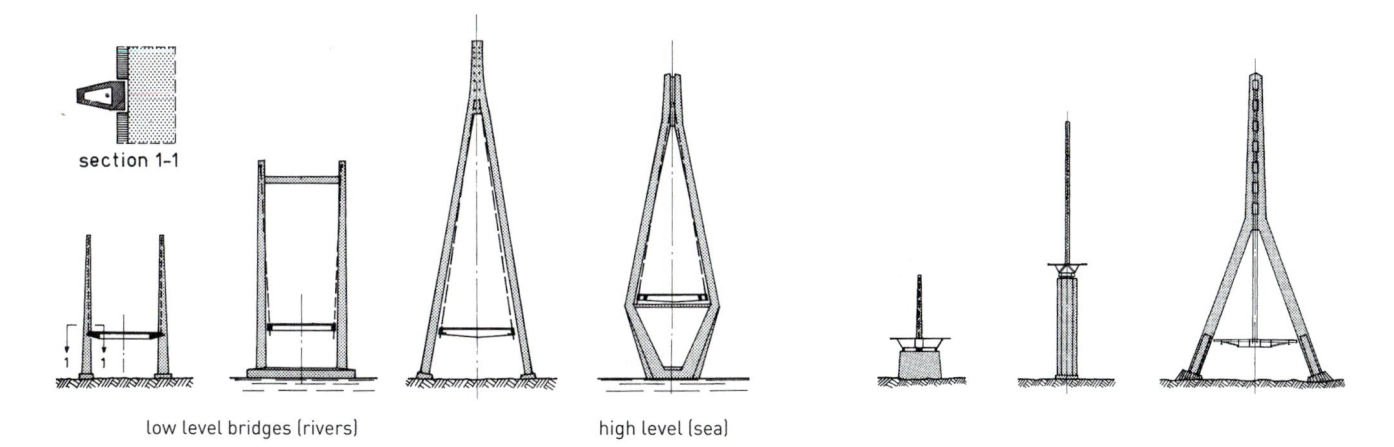

Figure 1.25 Towers for two cable planes

Figure 1.26 Towers for one cable plane

which is designed against lateral buckling, Fig. 1.26. For large spans the central plane can be anchored in the extended tip of an A-tower.

1.1.3.3 Spread central cable planes

If a cable-stayed bridge has to carry outer roadway lanes and central railway tracks it can be expedient to place the two tower legs in between the rail tracks and the roadway lanes. This results in partially spread tower legs, Fig. 1.27 [1.19]. The two cable planes are arranged parallel to the tower legs.

1.1.4 Beam cross-sections

Small cable distances result in small bending moments from the dead load in the beam. The live load moments are mostly restraint moments which decrease with the depth of the beam. Therefore, it is economic to choose a small depth for the beam, which is nearly independent of the span length. The limiting condition is the safety against buckling of the beam.

A further condition is to ascertain a minimum deflection radius under concentrated loads. This can easily be achieved for traffic loads even with slender beams, but railway bridges require larger beam depth.

1.1.4.1 Steel cross-sections

Steel cross-sections comprise an orthotropic deck supported by the main and cross girders. All beam members act together for the local and global loads.

Two outer cable planes
Open cross-sections: For short to medium spans the torsion from eccentric live loads can be carried by a couple in the outer cable planes. Open cross-sections with little torsional resistance can thus be used, Fig. 1.28 [1.20].

Box girders: Long spans may require the use of a box girder even with two outer cable planes in order to achieve the required torsional stiffness. An example is the current world record span holder, the Sutong Bridge in China, with a main span of 1088 m, Fig. 1.29 [1.21].

Separate beams: The Stonecutters Bridge in Hong Kong has a central tower with two outer cable planes. In order to provide the necessary space for the tower, two separate beams connected by cross girders are used, Fig. 1.30 [1.22].

One central cable plane
A central cable plane requires a box girder cross-section in order to carry the torsional moments. A typical example is the beam for the Rhine River Bridge at Ilverich, Fig. 1.31 [1.12].

Spread central cable planes
Open cross-sections: For the two spread central cable planes at the Rhine River Bridge Mannheim-Ludwigshafen two box girders provide the required torsional stiffness, and transverse bending is

Figure 1.27 Tower for partially spread central cable planes

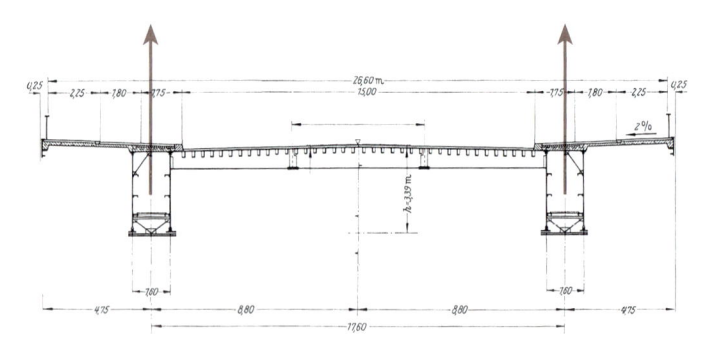

Figure 1.28 Open steel cross-section

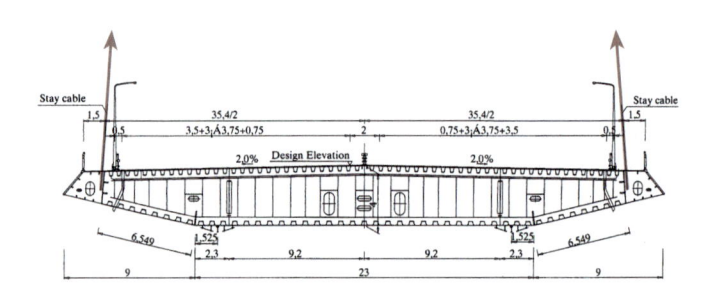

Figure 1.29 Steel box girder

Figure 1.30 Separate steel beams connected by cross girders

1.1 Design fundamentals

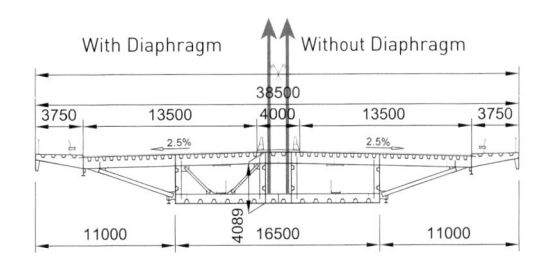

Figure 1.31 Steel box girder

carried by the cross girders, Fig. 1.32 [1.19]. The area between the two boxes is used by two tramway tracks which are lowered in one end region.

Box girders: The railway tracks in the center of the Sava River Bridge Ada Ciganlija, Fig. 1.33 [1.23], remain on top. The two spread central cable planes are anchored at the outside of the central box. For introduction of the cable forces into the 45 m wide beam additional webs are used.

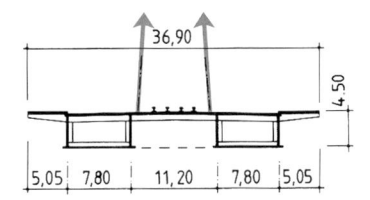

Figure 1.32 Separate steel boxes

1.1.4.2 Concrete cross-sections

The shape of concrete cross-sections is determined from similar considerations to steel cross-sections. It has, however, to be taken into account that concentrated tensile forces have to be carried by tendons.

Two outer cable planes

Open cross-sections: The open cross-section of the Pasco-Kennewick Bridge has triangular boxes at the outsides for aerodynamic reasons, Fig. 1.34 [1.15]. Transverse bending is carried by cross girders which are post-tensioned across the whole beam width. Additional short tendons transfer the cable forces via the inclined slabs into the main girders.

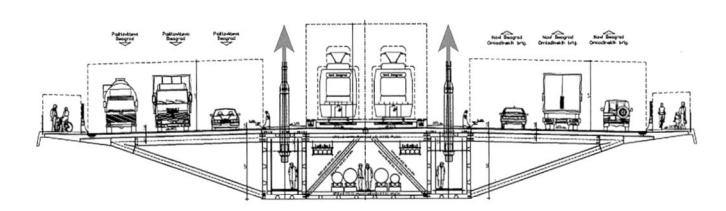

Figure 1.33 Steel box girder with wide cantilevers

Box girders: For two cable planes and medium spans, box girders are generally not required.

For the Posadas-Encarnación Bridge in Argentina a box girder was selected due to the eccentric railway track, Fig. 1.35 [1.24]. The transverse stiffening of the box was achieved by individually post-tensioned transverse stiffeners from precast elements which carry tension and compression.

Solid cross-sections: The elastic support of the beam from the stay cables provides a good distribution of heavy single traffic loads. For smaller spans a solid slab may be used, Fig. 1.36 [1.25]. In this case the transverse slenderness ratio comes to 1 : 23.5.

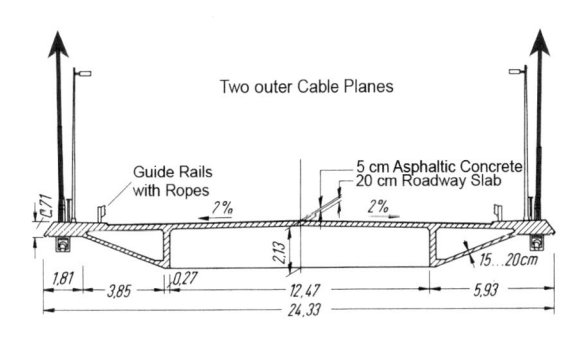

Figure 1.34 Open concrete cross-section

One central cable plane

A central cable plane also requires, for concrete beams, a box girder to carry the torsional moments from eccentric live loads.

For the Brotonne Bridge in France the cable forces are transmitted to the webs by post-tensioned diagonals, Fig. 1.37 [1.26].

The transverse distribution can also be achieved with cross girders, but larger quantities are required.

Spread central cable planes

The cables of the two spread central cable planes of the Second Main River Bridge at Hoechst are directly anchored to the webs of the central box girder, Fig. 1.38 [1.27].

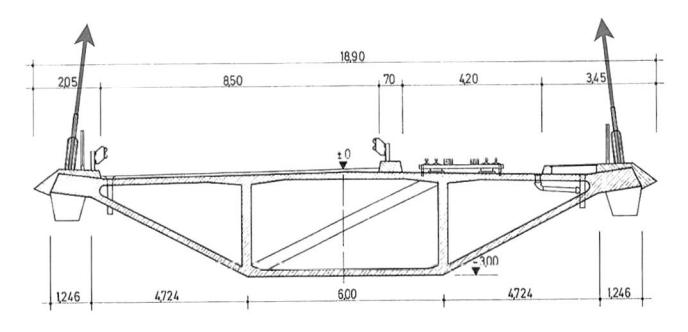

Figure 1.35 Concrete box girder

1.1.4.3 Composite cross-sections

Composite cross-sections use a concrete roadway slab as the top flange of the steel main and cross girders, connected by shear studs.

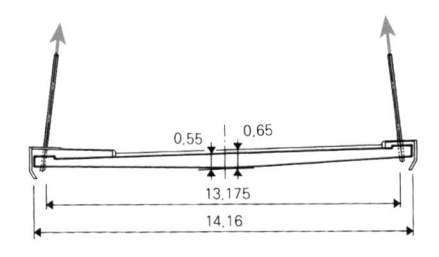

Figure 1.36 Solid cross-section

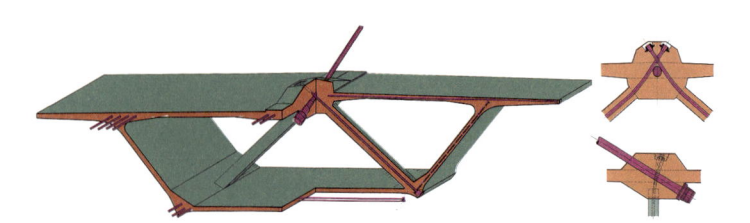

Figure 1.37 Concrete box girder

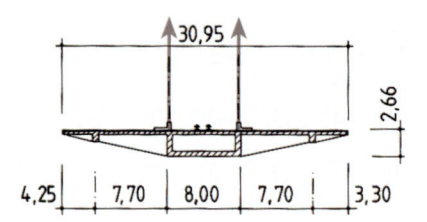

Figure 1.38 Concrete box girder

The concrete slab should always be under compression in both directions: longitudinally by the compression forces from the inclined cables, transversely as the top flange of a single-span girder between the two outer cable planes. Central cable planes should not be used for composite cross-sections.

The roadway slab can be built cast-in-place or from precast elements. The latter have the advantage that their shrinkage and creep is smaller after installation due to their advanced age. It has to be ascertained that the joints do not form weak spots during the life of the bridge. A typical composite cross-section is given in Figs 1.39 and 1.40 [1.28].

1.1.4.4 Hybrid beams (steel/concrete)
Hybrid cable-stayed bridges comprise a steel beam in the main span and a concrete beam in the side spans. The heavier concrete beam serves as a counterweight to the lighter steel main span. Both cross-sections are coupled at or near the tower with shear studs and tendons if the compression force from the cables is not sufficient to overcome local tensile forces from bending.

Two outer cable planes
The Normandy Bridge uses a flat steel box girder for aerodynamic stability in the main span of 856 m which is balanced by a similarly shaped concrete beam in the side spans, Fig. 1.41 [1.29]. The concrete beams extend some 100 m on both sides into the main span, Fig. 1.42.

One central cable plane
For the Rhine River Bridge, Flehe, steel and concrete box girders are used with a central cable plane, Figs 1.43 and 1.44 [1.16]. They join in the tower axis.

The side spans rest on outside piers. The support reactions are reduced by the central backstays in the cable anchorage region.

1.1.4.5 Double deck cross-section
If a multi-lane freeway has to be carried together with two railway tracks the railway may be placed on the lower level. This requires about an 8 m deep beam, for which a truss is often selected in order to permit a free view for the railway passengers during the long ride.

Two outer cable planes
For the Öresund Bridge in Denmark the cable forces are introduced by triangles into the lower and upper chord of the truss, Fig. 1.45 [1.30].

These are arranged as outriggers so that the beam can pass between the tower legs, Fig. 1.46.

One central cable plane
The third Orinoco Bridge carries only a single railway track on the lower chord. The truss can thus be supported by a single cable plane attached to the short upper cross girders between the directional traffic lanes, Fig. 1.47 [1.31].

The cable forces are transmitted by bending in the cross girders into the truss diagonals, the required torsional stiffness being provided naturally by the truss.

1.1.5 Stay cables
Stay cables are the characteristic structural components of cable-stayed bridges. Their performance governs the behavior of the complete bridge, not only in the final stage but also during construction. The durability of the stay cables determines the robustness of a cable-stayed bridge.

1.1.5.1 Systems
The currently used stay cable systems are locked coil ropes, parallel wire cables and parallel strand cables as outlined in Table 1.3.

For economic reasons parallel strand cables are mostly used worldwide today. Even in Germany, where until now mainly locked coil ropes have been used, the latest two cable-stayed bridges were provided with parallel strand cables. They are fabricated rather similarly by all major fabricators worldwide, Fig. 1.48 [1.32].

The strands are anchored with profiled wedges in anchor blocks. The depths of the wedge teeth increase gradually so that they anchor the strands safely, but reduce their fatigue strength only slightly.

The corrosion protection for every single strand comprises the following:
- galvanizing of every single wire in the strand
- filling the interstices between the single wires with grease
- surrounding each strand with a directly extruded PE-sheath.

Such strands are called monostrands.

A number of these monostrands are combined into a cable, for which all strands are covered additionally by a PE pipe, Fig. 1.48. At the transition from the free length to the anchorage region the strands spread out, the radial forces are contained by a clamp. A damper may be positioned at this location.

The installation of these stay cables takes place on site by assembling the individual components. The monostrands are pulled into the PE pipe. Each strand is individually stressed in such way that after complete cable assembly all strands have the same stress. It is possible to restress the complete cable with a large jack. Single strands may be exchanged later individually.

1.1.5.2 Cable anchorages
The stay cables typically run through a steel pipe to their anchor points. There split shims prevent sliding of the anchor heads. The lengths of the cables and thus their forces can be adjusted by changes in the thickness of the shims.

Support nuts around the anchor head can also be used for retaining the anchor head and adjustment of cable lengths.

DYWIDAG-SYSTEMS INTERNATIONAL

Sae Poong Bridge, Gwangyang, South Korea

DSI - Supplies Solutions to the Construction Industry

Approved Quality **On Time Delivery** **Excellent Service**

DSI is the global market leader in the development, production and supply of innovative Post-Tensioning Systems. In line with our strong service approach, we are always committed to satisfying our customers' demands.

- DYWIDAG Bonded Post-Tensioning Systems
- DYWIDAG Unbonded Post-Tensioning Systems
- External Prestressing
- DYNA Bond® Stay Cable Systems
- DYNA Grip® Stay Cable Systems
- DYNA Force® Monitoring System
- Engineering and Design
- Construction Methods
- Structural Repair and Maintenance

www.dsi-posttensioning.com

North America, USA
www.dsiamerica.com

South America, Brazil
www.dywidag.com.br

EMEA, Germany
www.dywidag-systems.com

APAC/ASEAN, Australia
www.dsimingproducts.com/au

Precast Concrete Structures

■ The book reflects the current situation in precast concrete construction. Besides general observations regarding building with precast concrete elements, the book focuses first and foremost on the boundary conditions for the design of precast concrete structures, loadbearing elements and façades. Connections and specific structural and constructional issues are covered in detail and stability of precast concrete structures is another central theme. The requirements brought about by the emergence of the European Single Market are explained and the diverse possibilities for façade design are presented. A chapter on the production processes provides the reader with an indispensable insight into the characteristics of this form of industrialised building.
The book is a practical tool for engineers, but certainly also architects and students.

HUBERT BACHMANN, ALFRED STEINLE
Precast Concrete Structures
2011.
272 pages, 263 fig. 15 tables.
Softcover
€ 79,–*
ISBN 978-3-433-02960-2

One of the authors intentions is to demonstrate to engineers and architects the possibilities that factory prefabrication can offer, and hence pave the way towards the economic application and ongoing development of precast concrete construction.

Online-Order: www.ernst-und-sohn.de

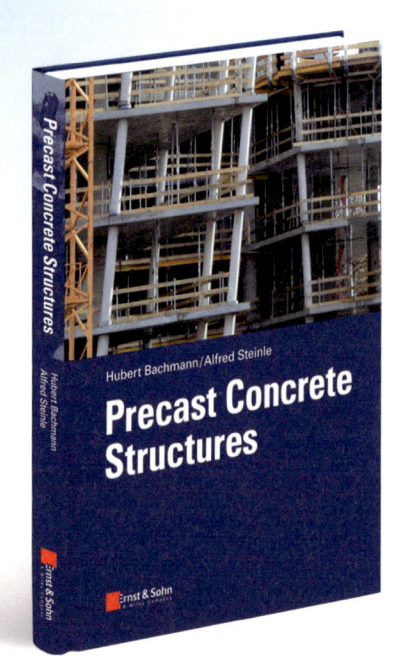

Ernst & Sohn
Verlag für Architektur und technische
Wissenschaften GmbH & Co. KG

Customer Service: Wiley-VCH
Boschstraße 12
D-69469 Weinheim

Tel. +49 (0)6201 606-400
Fax +49 (0)6201 606-184
service@wiley-vch.de

Ernst & Sohn
A Wiley Company

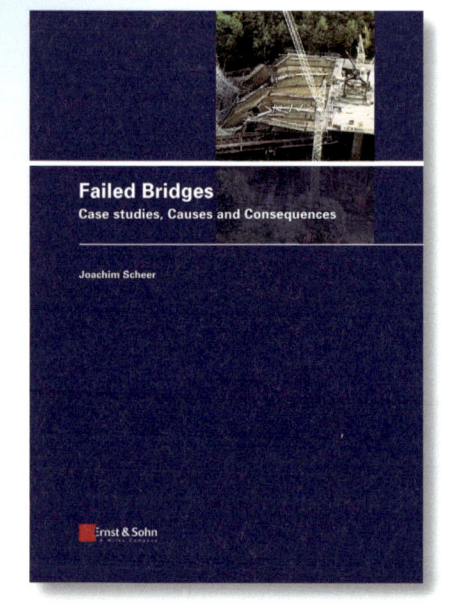

Failed Bridges

■ This is a systematic overview of more than 400 bridge failures in a new form of presentation. The incidents of damage and collapse have been assessed and assigned to the time of their occurrence in the life cycle of the bridge – during construction or in service – and to events such as collision impact or earthquake. The primary causes are identified: human error, inadequate stiffening, material flaw, overload etc. The book contains detailed analyses of bridge failures hitherto neglected in engineering literature and cases which, in the author's opinion, have been incompletely or incorrectly interpreted.

JOACHIM SCHEER
Failed Bridges
Case studies, Causes and Consequences
2010. 307 pages,
120 figures, 15 tables.
Hardcover.

€ 79,–*
ISBN 978-3-433-02951-0

A lifetime's work brilliantly compiled and courageously presented. A wealth of knowledge and experience for every civil engineer in the field or at university. The book includes a catalogue of rules that can help to avoid future mistakes in design, planning and erection and closes with some good advice for professors.

Order online! www.ernst-und-sohn.de

Ernst & Sohn
Verlag für Architektur und technische
Wissenschaften GmbH & Co. KG

Customer Service: Wiley-VCH
Boschstraße 12
D-69469 Weinheim

Tel. +49 (0)6201 606-400
Fax +49 (0)6201 606-184
service@wiley-vch.de

Ernst & Sohn
A Wiley Company

1.1 Design fundamentals

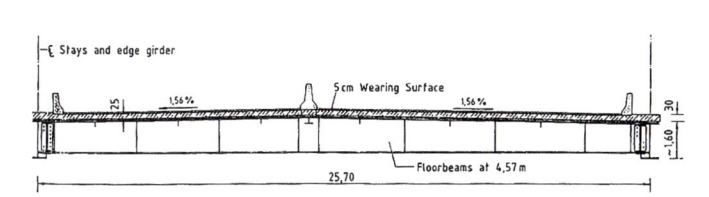

Figure 1.39 Typical open composite cross-section

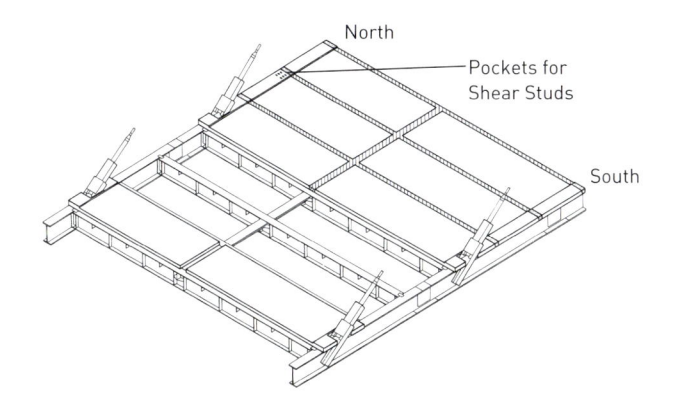

Figure 1.40 Typical composite beam

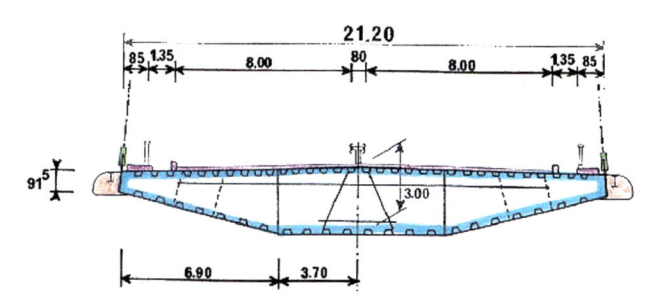

Figure 1.41 Main span steel cross-section of the Normandy Bridge, France

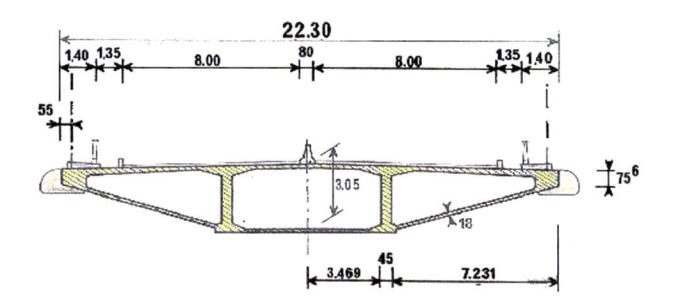

Figure 1.42 Side span concrete cross-section of the Normandy Bridge, France

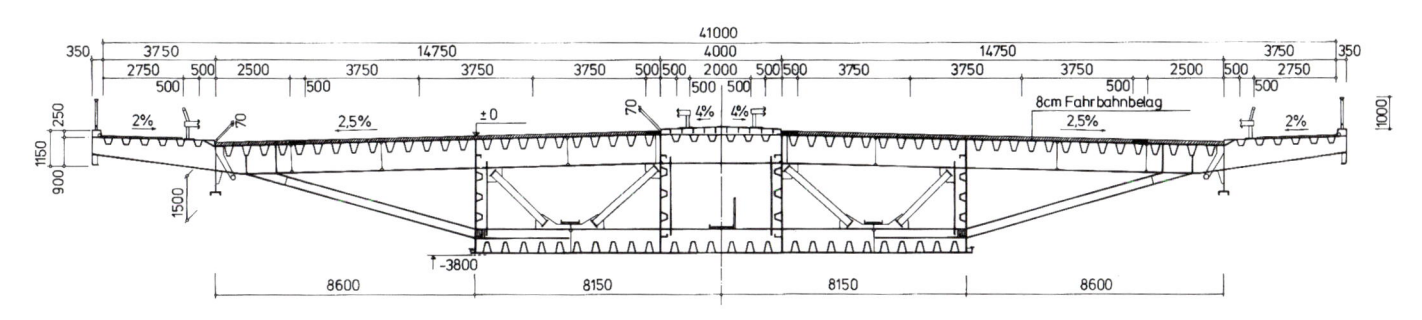

Figure 1.43 Main span steel cross-section for the Rhine River Bridge, Flehe, Germany

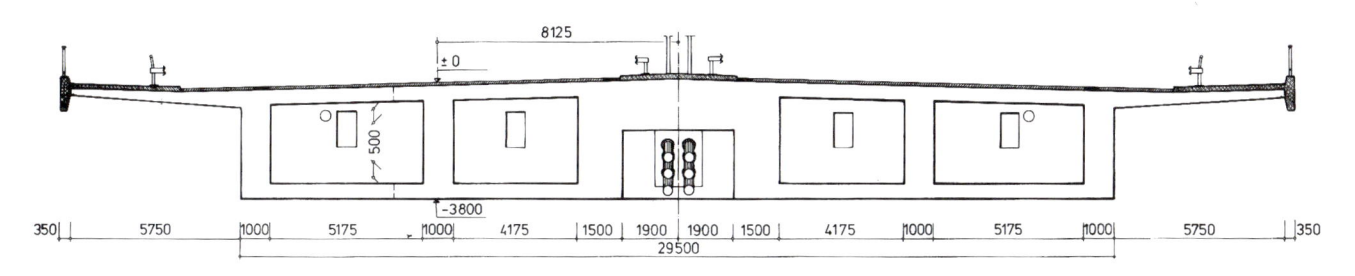

Figure 1.44 Side span concrete cross-section for the Rhine River Bridge, Flehe

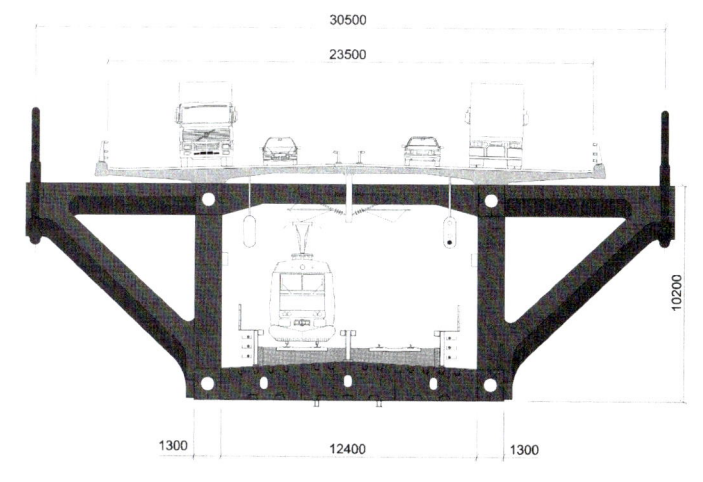

Figure 1.45 Truss beam of the Öresund Bridge, Denmark–Sweden

Figure 1.46 Cable anchorage of the Öresund Bridge

At beam

In concrete: A typical cable anchorage at a concrete beam is shown in Fig. 1.49 [1.15]. The cable forces are directly introduced into the main girder. The anchor head is supported by shims on pressure plates welded to the end of the steel pipe which transfer the cable forces either directly or via the steel pipe and shear rings into the concrete, where the tensile forces are covered by reinforcement.

At the tip of the steel pipe the cable is centered and possibly dampened by an elastomeric ring. The weather-resistant closure of the steel pipe is ensured by a neoprene boot clamped to the steel pipe and the stay cable.

In steel: A direct introduction of the cable forces into the main girder web is shown in Fig. 1.50 [1.33]. The web is extended with a butt weld through an opening in the top flange. A steel pipe is welded into this web extension to which the anchor head is attached with split shims or a support nut.

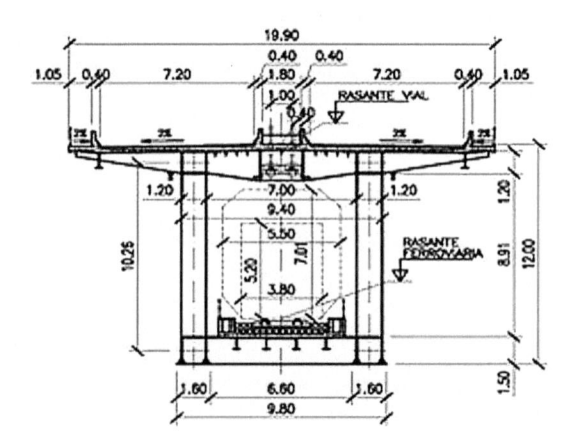

Figure 1.47 Truss for the third Orinoco Bridge, Venezuela

At the tower

In concrete: The simplest way to anchor individual cables at the tip of a concrete tower is to overlap them and thus to anchor them by compression, Fig. 1.51.

Usually the stay cables are anchored inside a tower box section, similar to that shown for a concrete beam. The tensile forces between forestays and backstays can be covered by short tendons, Fig. 1.52.

In steel: If the ideal fan shape of the cables is to be achieved they have to be split sideways into several planes. An example is the steel anchor head in Fig. 1.53, within which the cable forces are introduced into the longitudinal plates.

Usually the cables are spread vertically in their planes in order to obtain sufficient space for vertically aligned anchorages. This leads to a modified fan cable arrangement in elevation, Fig. 1.54. The tensile forces are also transferred into the longitudinal plates.

Table 1.3 Current stay cable systems

Characteristics		Modern locked coil rope	Parallel wire cable	Parallel strand cable
$E \cdot 10^{-6}$	[N/mm^2]	0.170	0.205	0.195
f_u	[N/mm^2]	1470	1670	1870
$\Delta\sigma$	[N/mm^2]	150	200	200

1.1 Design fundamentals

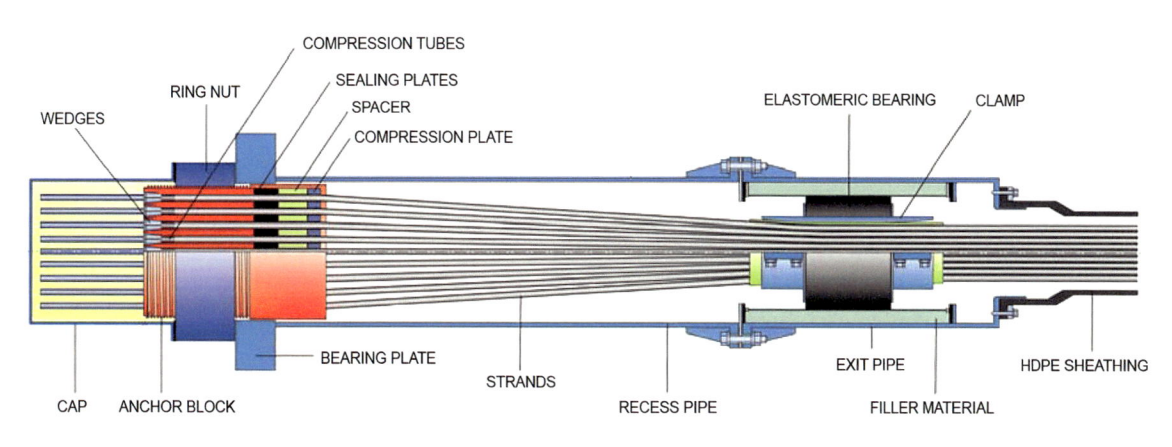

Figure 1.48 Typical parallel strand cable (DSI)

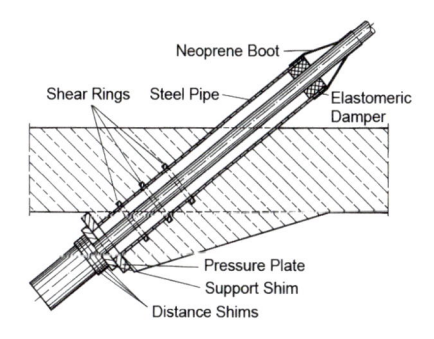

Figure 1.49 Typical cable anchorage in concrete

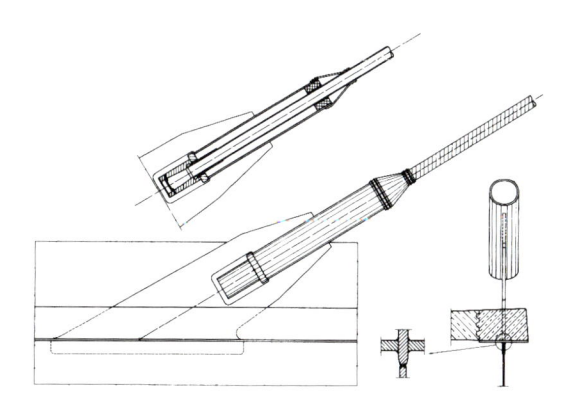

Figure 1.50 Typical cable anchorage in steel

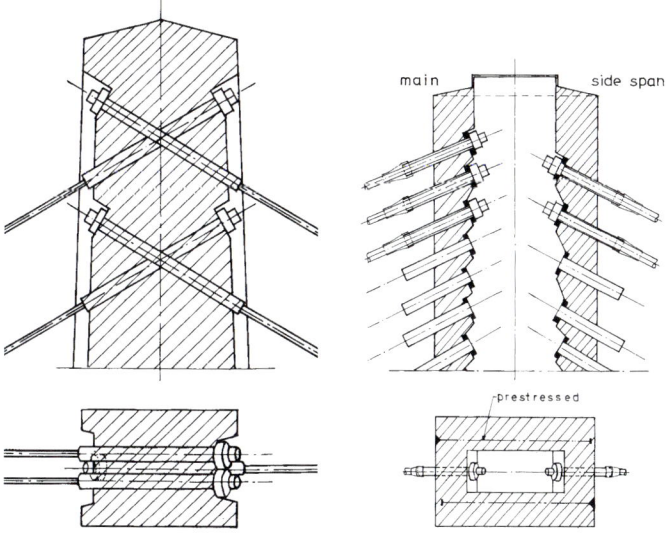

Figure 1.51 Cable anchorage by overlapping

Figure 1.52 Cable anchorage inside a box

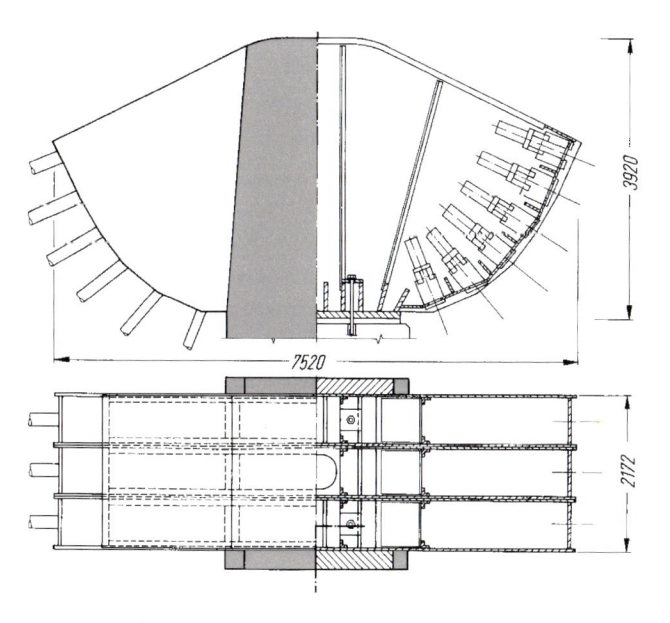

Figure 1.53 Steel tower head for fan arrangement

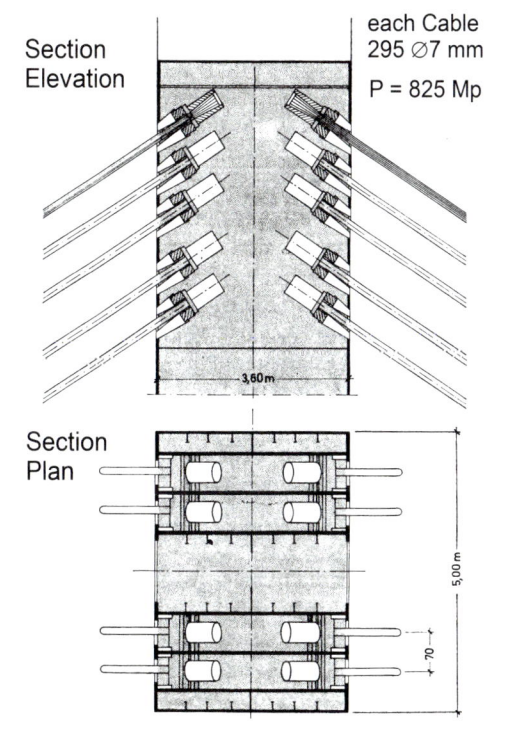

Figure 1.54 Modified fan arrangement

1.2 Aesthetic guidelines for bridge design
1.2.1 Introduction

In his ten books on architecture the Roman architect Marcus Vitruvius Pollio stated in Latin brevity and precision the three most important requirements for a structure as *firmitas, utilitas* and *venustas*. We will deal here with the *venustas*, the beauty.

There are no special aesthetic rules for cable-stayed bridges; they have to fulfill the same criteria as all other bridges. Their long spans, however, require stiff structures which can withstand, for example, heavy wind loads. Cable-stayed bridges, therefore, have to be designed taking carefully into account their aesthetics. A comparison with other types of bridges is interesting.

It is generally true that there are good-looking bridges which contribute favorably to the appearance of a city or a landscape and which are admired by most people, and there are ugly bridges which disturb the environment so that many people wish they had never been built or could be demolished. Why is one bridge considered beautiful and another one ugly?

It is true that the majority of people will judge one particular bridge in a group to be beautiful and another one to be ugly. Bridges have aesthetic characteristics which achieve the same reaction in different people. This effect depends on the individual's sensitivity and aesthetic education.

The competence of judging aesthetics develops only by repeated evaluation of consciously perceived values, and by training in visual appreciation. Aesthetic judgment thus has to be learned.

In his book *Bridges* [1.34], Fritz Leonhardt has pointed out that good-looking bridges are not designed by chance. There is no difficulty in defining 'feeling' which is particular only to a selected few, but there are aesthetic guidelines by which a design can consciously be reviewed.

Traditionally, there are two aesthetical points of view in structural engineering. The one, more technical point of view, is based on rational considerations and can be summarized with 'form follows function'. The history of bridges has, however, shown that a functionally rational design is not necessarily beautiful.

The other point of view refers to more intuitive impressions which are generally considered to be individually different. The catchwords are *'De gestibus non est disputandum'* – meaning 'Beauty lies in the eye of the beholder'. In his treatise *On Grace and Dignity* [1.35] the German poet Schiller has combined these two apparent contradictions at a higher level, so to say, dialectically. He shows that classical beauty fulfills rational, conscious requirements as well as intuitive subconscious requirements. In the following we will deal with the conscious requirements.

Ten guidelines for the aesthetic design of bridges are introduced. But these guidelines should not limit the freedom of intuition and fantasy. Good designs always depend on the talent of the designer, their sensitivity for aesthetics and their training in evaluation of the appearance.

The guidelines can, however, support the rational evaluation of designs. Good designs are normally the result of such critical analysis, which, step by step, may lead to improvements.

In the following, cable-stayed bridges are shown as examples for each aesthetic guideline, but their relevance for other bridge types is also demonstrated.

1.2.2 Aesthetic guidelines
1.2.2.1 Guideline 1: Clear structural system
'Choice of a convincing and simple load-bearing system'
A bridge must look trustworthy and stable.

A mixture of different statical systems often leads to a bad appearance. The number of different load-bearing elements should be reduced to a minimum. A statically correct system, on the other hand, is not necessarily beautiful.

The Caroni River Bridge for road and rail near Ciudad Guyana in Venezuela has the remarkable main span of 214 m, Fig. 1.55. Although the girder requires a structural depth of 14 m above the piers, the bridge does not appear clumsy due to the dynamic run of its strongly curved underside.

This is possible because road and rail are both located on the deck and the railway does not run inside the box as often done elsewhere. Therefore, a structural depth of 4.8 m in the center is sufficient. The beam depth corresponds to the run of moments for a continuous three-span girder. Only two load-bearing structural members are used, the beam and the piers, both strong and trustworthy.

The Weitingen Bridge, Fig. 1.56, crosses the Neckar Valley high above ground as a multi-span continuous girder. The constant beam depth underlines its continuity. In addition to the two main load-bearing structural members, the beam and the piers, the 216 m long side spans are stayed from underneath with an air strut. This emphasizes the high moments at these locations.

Prof. Hans Kammerer acted as architectural advisor to the bridge engineer Fritz Leonhardt.

The competitive design for an arch bridge across the Svinesund between Norway and Sweden has a convincing structural system. The arch is fixed to the rock slopes at both sides of the valley, its structural depth increasing with the run of moments from the center to the outsides, Fig. 1.57.

The roadway is hung from the arch. Its slenderness offers a contrast to the strong arch. The hangers are visually intersecting although they are vertical. This is due to the inclined arches – a definite disadvantage of inclined arches.

The competitive design for the Stonecutters Bridge in Hong Kong shows how transparently a cable-stayed bridge can be designed, even for a record span of 1007 m, Fig. 1.58.

A suspension bridge with a main span of 1210 m like the Höga Kusten Bridge in Sweden, Fig. 1.59, cannot be outclassed in clarity, simplicity and elegance. This aesthetic advantage of suspension bridges over cable-stayed bridges is undeniable.

The main cables carry the loads from the slender beam via the hangers and the concrete towers to the abutments. Four structural elements are used: Beam, hangers, main cables and towers.

1.2.2.2 Guideline 2: Good proportions
'Good proportions in all three dimensions between the structural members or between lengths and heights of the bridge spans'
A balanced ratio must exist between the structural elements, the lengths and heights of the spans, between the lit and shadowed areas as well as the masses of beam, piers and abutments.

Good proportions between span lengths and heights of the piers are important. They have to be distinctly different, resulting in lying or standing rectangles – squares appear dull.

For a bridge across a valley the ratio between span lengths and heights should be approximately kept constant.

One example is the spans for the Mosel Valley Bridge at Winningen, Fig. 1.60. Their span lengths and height above ground increase from the outside to the center. The main span rectangle with 240 m span and 120 m height has the aesthetically attractive ratio of 2 : 1.

The up to 180 m high piers of the Kocher Valley Bridge with spans of 130 m have a parabolic shape in the transverse direction. This outline provides an elegant appearance, Fig. 1.61. In spite of the large dimensions a monumental impression is avoided because the bridge fits the deep valley.

The attractive appearance of the haunched three-span girder of the Cologne-Deutz Bridge, Fig. 1.62, results from the good ratio of beam depths over the piers and in the bridge center of 2 : 1. This is also because, at the bridge ends, the beam depths are equal to those in the bridge center, and because the undersides of the beam run parallel to the bridge deck. Adhering to these rules will nearly always result in a good-looking bridge.

This is also apparent for the bridge across the Severn River in the USA, Fig. 1.63. For its 90 m main span a haunched continuous girder was also selected with a depth ratio of about 2 : 1 and a continuous transition from main span to side spans. The beam depth in the bridge center is equal to those of the side spans. This clear system underlines the main span and provides the otherwise tranquil bridge with a certain dynamic.

V-piers require a roadway gradient with just the right height above ground to achieve good-looking triangles, Fig. 1.64. The beam of the railway bridge across the Main River as part of the new high-speed railway line between Hannover and Würzburg is slightly haunched in order to emphasize the 130 m main span.

Arch bridges require distinctly different depths between the arch proper and the beam, at least in a ratio of 2 : 1 or 1 : 2. Equal structural depths between beam and arch appear dull. For the tied arch of the Elbe River Bridge at Tangermünde the bending stiffness is provided by the composite beam, Fig. 1.65, which carries the beam depths of the long approaches across the main span. The steel arches can thus be slender.

Figure 1.55 Third Caroni River Bridge for road and rail, Venezuela

Figure 1.56 Weitingen Bridge across Neckar Valley, Germany

Figure 1.57 Competitive design (LAP) for the Svinesund Bridge between Norway and Sweden

Figure 1.58 Runner-up design for the Stonecutters Bridge, Hong Kong (LAP)

Figure 1.59 Höga Kusten Bridge, Sweden

Figure 1.60 Mosel Valley Bridge at Winningen, Germany

Figure 1.61 Kocher Valley Bridge, Germany

Figure 1.62 Rhine River Bridge, Cologne-Deutz, Germany

Figure 1.63 Annapolis Bridge, USA

Figure 1.64 Railway Bridge Gemünden across the Main River at Würzburg

Figure 1.65 Elbe River Bridge at Tangermünde, Germany

Figure 1.66 Bridge across Roosevelt Dam, Arizona, USA

1.2 Aesthetic guidelines for bridge design

The fixed arch across the Roosevelt Dam in Arizona carries the bending stiffness in the strong, haunched arches proper. The roadway slab can thus be very slender, Fig. 1.66.

The beam of the Pasco-Kennewick Bridge has a depth of only 2.1 m for a main span of 300 m and thus appears slender, Fig. 1.67. This impression is heightened by the only 40 cm deep white-colored edge girder. The actual beam depth nearly disappears behind the shadowed inclined underside. The impression of a ribbon spanning the river is created.

1.2.2.3 Guideline 3: Good order
'Good order of all lines and edges of a structure which determine the appearance – the number of directions should be minimized.'
The number and directions of the structural elements should be minimized, especially for trusses. Symmetry and repetition of equal elements should lead to good order as long as monotony does not prevail.

The truss for the Hammerbrook Railway Bridge in Hamburg comprises only rising and falling diagonals which create an orderly impression even for a skew view, Fig. 1.68.

The railway bridge across the Main River near Nantenbach, Fig. 1.69, comprises only rising and falling diagonals as well, but with no cross frames in order to reduce the member directions to the minimum of two. The inclination of the diagonals remains nearly constant in spite of the changing depth of the girder so that no glancing intersections are present.

Although this is the railway bridge with the longest span in Germany, 207 m, it appears light.

In contrast typical 19th century trusses give a bewildering impression with a multitude of truss members in all directions. An example is the Howrah Bridge across the Hooghly River in Calcutta, Fig. 1.70.

The new Hooghly River Bridge in Calcutta, Fig. 1.71, by contrast, is designed as a cable-stayed bridge with clear lines. Even the regular arrangement of the longitudinal and transverse girders shows good order in a view from underneath, and this is continuously repeated without creating monotony.

The shape of the towers and the cable arrangements determine the appearance of a cable-stayed bridge. Visual intersections of the stay cables play an important role.

A parallel arrangement of the stay cables in a harp shape shows the best order as demonstrated by the Oberkassel Bridge, Fig. 1.72. One cable plane naturally does not provide intersections. During the initial design phase of the Düsseldorf Bridge Family in the early 1950s the city architect Friedrich Tamms insisted that all three bridges have stay cables in harp arrangements. Although the structural systems of the bridges are quite different, the common harp-shaped cable arrangement visually unites them into a bridge family.

Even two harp-shaped cable planes do not show intersections, Fig. 1.73, independent of whether the planes are vertical or inclined.

Figure 1.67 Pasco-Kennewick Bridge across the Columbia River, WA, USA

Figure 1.68 Hammerbrook Railway Bridge, Hamburg, Germany

Figure 1.69 Railway Bridge across Main River at Nantenbach, Germany

Figure 1.70 Howrah Bridge, Calcutta, India

Figure 1.71 Hooghly Bridge, Calcutta, India

Figure 1.72 Rhine River Bridge Oberkassel, Germany

Figure 1.73 Heinola Bridge, Finland

Figure 1.74 Stonecutters Bridge, Hong Kong, China

If the stay cables are arranged with changing inclinations in a fan shape, visual intersections can only be avoided if a single central plane of cables is used, as for example for the Stonecutters Bridge in Hong Kong, Fig. 1.74. Dissing and Weitling from Denmark were the architectural advisors to the bridge engineer, Ian Firth.

With two fan-shaped cable planes visual intersections can be mitigated by the use of an A-tower with inclined cable planes, Fig. 1.75.

This disadvantage of intersections can further be reduced if the number of cables becomes very large, Fig. 1.76, and the impression of a veil is created.

The bridge across the Houston Ship Channel in Texas, Fig. 1.77, requires eight lanes with full shoulders. The most economic system comprises two independent cable-stayed bridges with two diamond shaped towers, interconnected at beam level. This results in good order for the bridge with four cable planes. The many cables also approach the appearance of a veil and the visual cable intersections are somewhat mitigated.

1.2.2.4 Guideline 4: Integration into the environment
'Harmonic integration of the bridge into its environment'
The material and the scale of the bridge in comparison to its immediate surroundings are the subject of this guideline.

An impressive example is the Upper Havel River South Bridge, Fig. 1.78. The piers extend upwards on both sides of the bridge and use the same brownstones as the surrounding residential buildings. Prof. Walter Nöbel from Berlin was the architectural advisor for this project.

A further example for adapting a bridge design to its immediate urban surroundings is the Humboldt Harbor Bridge in Berlin, Fig. 1.79. The deviation from the normal use of materials – concrete for compression members and steel for tension and bending – was chosen here to adjust the bridge appearance to the adjacent steel-glass structure of the new Main Railway Station. Both bridge and station have been jointly designed by the architect Volkwin Marg and the engineer Jörg Schlaich.

The second freeway bridge across the Oder River near Frankfurt was built for aesthetic reasons as a series of arches similar to the existing arch bridge.

Although this bridge system is not economic today, it fits the beautiful undisturbed landscape along the Oder River, Fig. 1.80.

The bridge across the Severn River in front of the famous Annapolis Naval Academy won a US nationwide design competition, Fig. 1.81. The bridge was designed attractively curved but rather subdued in order not to compete visually with the historic buildings of the academy in the background. Due to its simple shape the bridge won the competition against more involved bridge designs with structures above the deck such as arches and stay cables.

The Dittenbrunn Bridge of the German Railways follows the natural contours and seems to grow out of the slope and thus fits in with the surroundings, Fig. 1.82.

The Schillersteg pedestrian bridge in Stuttgart connects the upper and the middle public Schloßpark. The very slender steel beam is bifurcated at one end in order to follow the natural footpaths, Fig. 1.83.

Finally, the purely technical design of the Glebe Island Bridge seems to blend in with the industrial surroundings of Sydney Harbour, Fig. 1.84.

1.2.2.5 Guideline 5: Choice of material

'Correct selection of materials in regard to their load-bearing capacity and appearance'

In general, concrete should be selected for struts, and steel for ties. A lining of large areas of piers and abutments with natural stones can be advantageous; large concrete surfaces can receive a texture from the forms or by chiseling. Large surfaces should generally be roughened, while for small areas smooth surfaces are appropriate.

In accordance with the only building material available at that time the Ilm River Bridge near Oberroßla, Fig. 1.85, was built from natural stones in 1848. Since no tensile forces can be carried by these stones, arches were selected which carry their loads in compression. After an in-depth rehabilitation and replacement of the top slab the bridge serves today for the modern high-speed railway.

Timber appears to be the natural material for the pedestrian bridge, Fig. 1.86, with the remarkable span of 60 m. The design aims at lightness and transparency. The end frames, which have to carry the transverse wind loads, would have required heavy timber sections. In order to avoid the impression of heaviness there, slender steel sections were used. The difference in material is made apparent by using different colors for steel and timber.

The 800 m long freeway bridge across the Warnow valley lies only 10 m above ground, Fig. 1.87. Concrete is the appropriate material for the relatively short spans.

The Annapolis Bridge in the USA was designed with slender octagonal concrete piers without cross girders in order to emphasize the lightness of the bridge, Fig. 1.88. These slender haunched twin piers rest on common granite clad foundations. The floating impression of the superstructure is supported by the upwards striving piers and the strong foundations.

The modern tied three-span arch bridge across the Elbe River near Pirna uses concrete for the arches in compression and steel for the girder as tie and bending member, Fig. 1.89. The concrete roadway slab distributes the local wheel loads.

The governing characteristic of the beam for the Helgeland Bridge in Norway is its compression resistance, Fig. 1.90. It was, therefore, built in concrete. A depth of 1.2 m for the main span of 425 m is sufficient to carry the bending moments and to ensure safety against buckling. This results in a record slenderness of 1:355. It becomes evident that concrete bridge girders are not necessarily clumsy.

The Rhine River Bridge at Flehe uses a good combination of materials with steel for the main span beam and concrete for the approach beam as counterweight for the main span, Fig. 1.91.

Figure 1.75 East Huntington Bridge across the Ohio River, WV, USA

Figure 1.76 Yang Pu Bridge, Shanghai, China

Figure 1.77 Houston Ship Channel Crossing, USA

Figure 1.78 Upper Havel River South Bridge, Berlin, Germany

Figure 1.79 Railway bridge across Humboldt Harbor, Berlin, Germany

Figure 1.80 Oder River Bridge, Frankfurt, Germany

Figure 1.81 Annapolis Bridge, USA

Figure 1.82 Dittenbrunn Railway Bridge along slope, Germany

The delicate appearance of the once record span Normandy Bridge confirms the use of steel as material for the very long spans, Fig. 1.92.

1.2.2.6 Guideline 6: Coloring
'An important element for pleasing appearance is the color.'
Not only steel bridges but also concrete bridges should occasionally be provided with color so that they fit their surroundings. For large areas loud colors should be avoided.

For the concrete Schatten Bridge in Stuttgart a light green was selected for the beam and a dark green for the piers in accordance with the surrounding forest, Fig. 1.93.

The assembly for the open tunnel in Ravensburg is an example how to use a strong red for slender members such as the truss diagonals. The same color was chosen for the truss of the open tunnel, for the towers of the pedestrian cable-stayed bridge, Fig. 1.94, and the curved pedestrian truss bridge, Fig. 1.95. The uniform color combines the different structures into a unit.

For the Elbe River Bridge at Tangermünde, Fig. 1.96, the coloring emphasizes the load-bearing function of the different bridge members. The bending stiffness is provided by the relatively deep beam which is provided with a strong, light blue. The arch slenderness is underlined by its light gray color.

The two nearly identical bridges across the Paraná River between the cities of Zárate and Brazo Largo near Buenos Aires in Argentina with main spans of 330 m were the first major cable-stayed bridges for road and rail, Fig. 1.97. Their unique coloring comprises the red steel diagonals at the tower tops and the white cables. Both emphasize the main span of the long bridge which is otherwise provided with uniform grey concrete color.

For the Burlington Bridge across the Mississippi off-white cables were selected, Fig. 1.98, similar to many other cable-stayed bridges. The light color underlines the floating impression of a cable-supported structure.

A special white concrete from precast elements was selected for the Dee River Bridge in Wales which gives the whole bridge an ethereal appearance, Fig. 1.99.

The yellow stay cables of the Houston Ship Channel Bridge seem to fit the strong sun and the dark blue sky in Texas, Figs 1.100 and 1.101. These yellow cables give a strong impression to the bridge users.

For the stay cables of the Rhine River Bridge at Flehe a red lead coloring was selected, Fig. 1.102.

Air traffic control requested that the towers of the Loire River Bridge at St. Nazaire be provided with a striped hazard coloring, Fig. 1.103. This unsuitable solution has nowhere else been repeated, normally hazard beacons at the tower tops are considered sufficient.

1.2 Aesthetic guidelines for bridge design

Figure 1.83 Schillersteg pedestrian bridge, Stuttgart, Germany

Figure 1.84 Glebe Island Bridge, Sydney, Australia

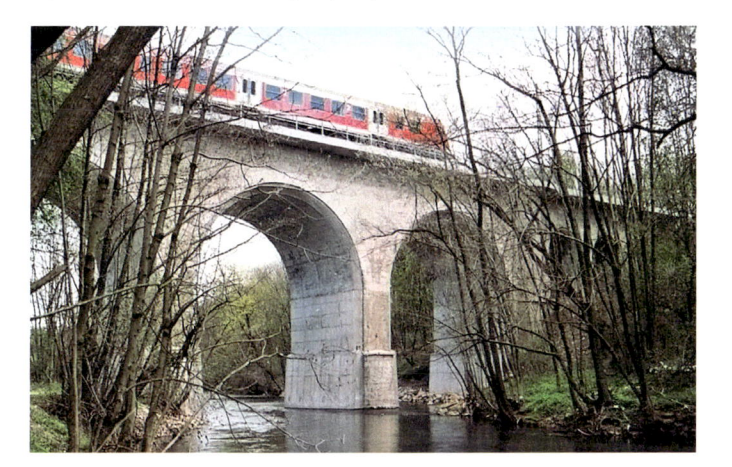

Figure 1.85 Ilm River railway bridge, Oberroßla, Germany

Figure 1.86 Pedestrian bridge across Dahme River near Berlin, Germany

Figure 1.87 Freeway bridge across the Warnow Valley, Rostock, Germany

Figure 1.88 Annapolis Bridge, USA

Figure 1.89 Elbe River Bridge, Pirna, Germany

Figure 1.90 Helgeland Bridge, Norway

Figure 1.91 Rhine River Bridge, Flehe, Germany

Figure 1.92 Normandy Bridge, France

Figure 1.93 Schatten Bridge, Stuttgart, Germany

Figure 1.94 Open tunnel and cable-stayed bridge, Ravensburg, Germany

Figure 1.95 Pedestrian truss bridge, Ravensburg, Germany

1.2.2.7 Guideline 7: Space above the bridge

'The space above the bridge should be shaped in such way that the driver experiences the bridge and gets a comfortable feeling.'

It is a good idea to show the beginning and the end of a bridge, if no bridge members extend above the bridge deck.

Old bridges have traditionally been decorated with columns or statues of saints as with the Karls Bridge in Prague which shows St Nepomuk, the patron saint of bridges, Fig. 1.104.

In this tradition the piers of the Upper Havel River South Bridge, protrude on both sides of the bridge above deck, Fig. 1.105. They support a special lighting system for this bridge, which is mainly used by pedestrians.

The Broad Street Bridge in Columbus, Ohio, was consciously designed as a slow inner city bridge, for example, by marking the ends of the bridge with columns, Fig. 1.106. This solution was found in workshops that received strong public participation.

The two cable planes of cable-stayed bridges with A-towers form a tent-like roof which provides users with a feeling of safety. A typical example is the slender Helgeland Bridge, Fig. 1.107.

The Paraná River Bridge between the cities of Posadas and Encarnación in Argentina, with one eccentric railway track, also has two cable planes with A-towers, Fig. 1.108.

From this point of view cable-stayed bridges with just one central cable plane are aesthetically questionable, as shown by the not-realized design for the Kocher Valley Bridge. Whereas the bird's eye view appears quite attractive, Fig. 1.109, a driver's view on the bridge ahead is inadequate, Fig. 1.110. The driver only sees a mast in front, whose meaning is not immediately apparent. This creates a feeling of uncertainty.

1.2.2.8 Guideline 8: Recognizable flow of forces

'A bridge must be designed in such way that the flow of forces is evident even for a casual observer.'

Even an uninitiated observer has an instinctive feeling whether the transfer of forces in a bridge appears plausible. This rule has recently been frequently violated by spectacular designs under the catchwords 'signature bridge' or 'icon bridge' which have been developed for appearance only, without regard to the efficient flow of forces. Very often in these cases the design was created by architects against whom the engineers could not or did not want to succeed.

The haunches of a bridge should rest optically secure on strong piers, and the pier heads should be wider than the underside of the beam.

The Elbe River Bridge at Torgau uses a single haunch only, the concentration of forces at this location being emphasized by a single strong pier with granite cladding which has the opposite side inclination to the other piers ('A' instead of 'V'), Fig. 1.111.

The tower of the Leven River Bridge in Scotland is transversely slightly unsymmetrical and thus reflects the curvature of the bridge beam, Fig. 1.112.

1.2 Aesthetic guidelines for bridge design

Figure 1.96 Elbe River Bridge, Tangermünde, Germany

Figure 1.97 One Zárate-Brazo Largo Bridge, Argentina; for road and rail

Figure 1.98 Mississippi Bridge, Burlington, USA

Figure 1.99 Dee River Bridge, Wales, GB

Figures 1.100 and 1.101 Houston Ship Channel Crossing, USA; stay cables

Figure 1.102 Rhine River Bridge, Flehe, Germany

Figure 1.103 St. Nazaire Bridge, France

Figure 1.104 Karls Bridge, Prague, Czech Republic

Figure 1.105 Upper Havel River South Bridge, Berlin, Germany

Figure 1.106 Broad Street Bridge, Columbus, Ohio, USA

Figure 1.107 Helgeland Bridge, Norway

Figure 1.108 Posadas-Encarnación Bridge, Argentina; for road and rail

Figure 1.109 Design Kocher Valley Bridge, Germany – bird's eye view

The towers of a cable-stayed bridge have to transfer loads from the cables into the foundations. The simplest and most effective shape is the already mentioned straight A-towers, e.g. Fig. 1.112.

If, due to the overall geometry of the bridge, the distance between the tower legs becomes too big for ship traffic or if the required two independent foundations are too expensive, the tower legs can be pulled together again underneath the deck. This leads to a diamond shape.

An early example is the alternative steel design for the Sunshine Skyway Bridge near Tampa, Florida, Fig. 1.113. The flow of forces is apparent from the widths of the tower legs in the transverse direction. The governing load is transverse wind on the stay cables which is carried in truss action by tension and compression above the beam, the tower legs above deck thus have constant widths. The deviation forces at the kink below deck create tension in the cross beam and are carried by post-tensioning. The pulled-in V-shaped tower legs further down have to carry the transverse forces in bending. Their widths thus increases downwards. The hold-down piers have similar V-shape.

The Houston Ship Channel Crossing at Baytown, Texas, USA, requires two separate beams. The most effective flow of forces is given by two diamond-shaped towers connected at beam level, Fig. 1.114. These now form a truss over their full height so that the transverse wind forces are carried by tension and compression only, also underneath the beam. The slender tower legs can thus have a constant width of only 2.1 m.

The nearby Düsseldorf airport required a height limitation for the towers of the Rhine River Bridge Ilverich to less than the usual 1/5 of the main span. In order to achieve the economic inclination of the stay cables, the towers were formed V-shaped in the longitudinal direction, Fig. 1.115. The tensile forces between the forestay and backstay cables are directly connected with a strong tie, so the flow of forces is clearly apparent.

The transfer of loads for the Rama 8 Bridge, Fig. 1.116, is quite apparent: the loads from the main span are concentrated in a large concrete counterweight at the end of the bridge.

In order to show the force transfer from the stay cables into the beam, the cables in Fig. 1.117 are led through a concrete beam and are supported on corbels underneath. The clearly visible cable anchor heads show the force transition.

The concentrated backstay cables carry the biggest loads in a cable-stayed bridge. Their load transfer should, therefore, be highlighted. In Fig. 1.118 the stiffeners parallel to the stay cables indicate the force introduction from the stay cables into the main girder web. The uplift forces are then transferred from the beam into the hold-down piers with the help of a strong 30 cm diameter pin.

The vertical anchorage of the backstays is shown very clearly as an architectural feature in Fig. 1.119.

1.2 Aesthetic guidelines for bridge design

For series of cable-stayed bridges with more than one main opening the main design problem is how to stabilize the tips of the towers, because the usual concentrated backstay cables are missing.

The center tower of the Ting Kau Bridge in Hong Kong is stayed to both outside towers at beam level, Fig. 1.120. The strong tieback cables are distinguished by their larger sag against the regular cables and thus show the flow of forces. Visual intersections between the different types of cables have been accepted.

Another solution with a more direct flow of forces was possible for the Second Orinoco River Bridge. Two main spans of 300 m each with a distance in between were required for navigational clearance. Two regular cable-stayed bridges back to back were the solution, Fig. 1.121. The hold-down pier in the center is designed as an A-frame in order to also carry the important breaking forces of the railway over the full length of the bridge.

The Alamillo Bridge in Spain by Calatrava shows an unusual flow of forces, Fig. 1.122. The permanent loads of the main span are balanced by the inclined tower. Transient loads in the main span have to be carried by bending in the tower because the usual backstay cables are missing.

1.2.2.9 Guideline 9: Lighting
'Aesthetic lighting can enhance the appearance of a bridge at night.'
The modern way of lighting a bridge uses floodlights which cover whole areas instead of using individual lights which only follow the contours.

The Kap Shui Mun Bridge in Hong Kong is lit in a pale blue, Fig. 1.123. Interesting coloring effects are achieved by the transition from strong lighting at beam level to darker colors higher up.

The Nan Pu Bridge in Shanghai uses a warm yellow light above the beam and a pale blue underneath, Fig. 1.124.

The Houston Ship Channel Bridge, USA, uses a warm yellow floodlight for the towers and the cables, located at the beam, Fig. 1.125.

The competitive design for the Stonecutters Bridge in Hong Kong uses white floodlighting for cables and towers as shown in the visualization, Fig. 1.126.

An interesting combination of colors during sunset is shown for the Heinola Bridge, Fig. 1.127. In front of the reddish sky the tower, floodlit in a pale yellow, can be clearly seen in front of a series of street lights in the main span. The individual lights at the tower are required for navigational safety.

1.2.2.10 Guideline 10: Simplicity
'Above all: simplicity'
Simplicity and refinement to the pure structural shape is most important. All additions, such as ornaments, decorations or architectural extras, should only be used as an exception. The shape of a bridge is mature if nothing can be left out.

Figure 1.110 Design Kocher Valley Bridge, Germany – driver's view

Figure 1.111 Elbe River Bridge, Torgau, Germany

Figure 1.112 Leven River Bridge, Scotland, GB

Figure 1.113 Alternative design for the Sunshine Skyway Bridge, Tampa, USA

Figure 1.114 Baytown Bridge, Houston, Texas, USA

Figure 1.115 Rhine River Bridge, Ilverich, Germany

Figure 1.116 Rama 8 Bridge, Bangkok, Thailand

Figure 1.117 Cable anchorage at the beam of the Pasco Bridge, USA

The Stöbnitz Valley Bridge is an integral post-tensioned railway bridge, Fig. 1.128. The beam has a solid cross section and rests on cylindrical concrete piers. This bridge convinces by its simplicity.

The single central piers of the Weitingen Bridge have octagonal cross-sections and a taper of 1:70, Fig. 1.129. These 120 m high single piers carry a 30 m wide freeway beam, Fig. 1.130. Piers cannot be designed more simply.

The bracing between the arches of the Tangermünde Bridge is reduced to a minimum, Fig. 1.131. Instead of the usual truss with members in different directions, only steel pipes rectangular to the arches were used.

An even simpler solution is shown for the Harbor Bridge Riesa, Fig. 1.132. The two arches are spaced widely and protrude only a little above the roadway. A bracing at their top would have led to a horizontal rectangle which would look unfavorable. Therefore, no bracing at all was used but the individual arches were stiffened against lateral buckling.

From the pedestrian bridge at Villingen-Schwenningen also nothing can be taken away, Fig. 1.133. Four forestays in a central plane support the main span and are anchored in parallel to the abutment. The slender tower is fixed to the beam, which indicates its high bending forces at this location by a haunch.

The Enz River pedestrian bridge at Mühlacker, designed by Leonhardt himself, comprises only one load-bearing element, a concrete arch, embellished with a light red railing, Fig. 1.134. The bridge shows an absolute minimum of structural elements and is of striking beauty.

1.2.3 Collaboration

Finally some remarks on the collaboration between architects and engineers.

For the planning of buildings the architect is in charge of the design and the engineer advises on technical issues. The collaboration between the architect Volkwin Marg and the engineer Jörg Schlaich in Germany shows how a congenial work can be created, [1.37].

For the design of bridges the engineer is in charge and the architect advises him on questions of shape. This collaboration for bridges can also be fruitful. The lead should always remain with the engineer because a bridge is primarily a technical structure. To design a bridge requires special knowledge and experience. Therefore, not every architect is suited to being an advisor for bridges, but experienced architects with an understanding for the flow of forces are required. The key for a successful collaboration is that each side listens carefully to the arguments of the other and tries to understand them.

The famous architect Prof. Tamms has commented [1.38] that the architectural advisor for engineering structures is well advised to hold back in all questions of the basic layout, the construction and the material, in order to attain to his intrinsical task of influencing the shape of the structure.

1.2 Aesthetic guidelines for bridge design

Figure 1.118 Anchorage of the backstays at the Baytown Bridge

Figure 1.122 Alamillo Bridge, Spain

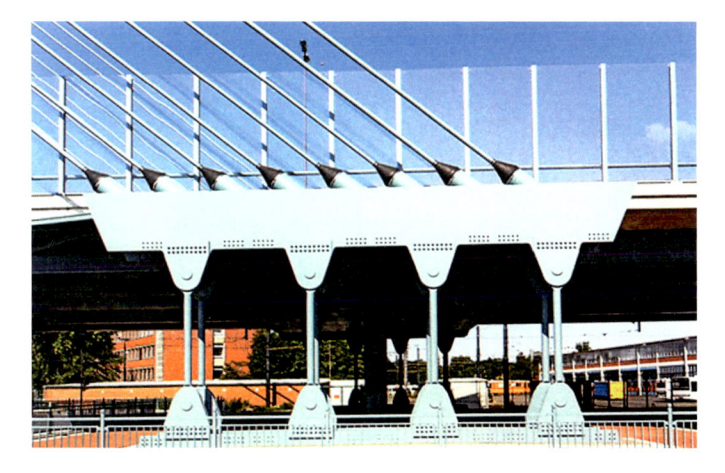

Figure 1.119 Tie-downs of the backstays

Figure 1.123 Kap Shui Mun Bridge, Hong Kong, China

Figure 1.120 Ting Kau Bridge, Hong Kong, China

Figure 1.124 Nan Pu Bridge, Shanghai, China

Figure 1.121 Second Orinoco River Bridge, Venezuela; for road and rail

Figure 1.125 Houston Ship Channel Crossing, USA

Figure 1.126 Visual rendering of the Stonecutters Bridge at night, China

Figure 1.127 Heinola Bridge, Finland

Figure 1.128 Stöbnitz Valley railway bridge, Germany

Figures 1.129 and 1.130 Weitingen Bridge, Germany; overview and center tower

Figure 1.131 Elbe River Bridge, Tangermünde, Germany

Figure 1.132 Harbor Bridge, Riesa, Germany

Figure 1.133 Pedestrian bridge at Villingen-Schwenningen, Germany

Figure 1.134 Enz River pedestrian bridge at Mühlacker, Germany

Prof. Leonhardt always appreciated the collaboration with specially qualified architects, including Prof. Tamms, Prof. Bonatz and his successor Gerd Lohmer, and Prof. Kammerer, to name only a few.

The author has also successfully collaborated with architects for bridges – Tom Piotrowski from H2L2 in the USA, Ronald Yee in the UK, Prof. Burghardt from Hamburg and Prof. Nöbel from Berlin. The Danish consulting architects Dissing & Weitling with Poul Ove Jensen have advised on outstanding cable-stayed bridge designs worldwide.

The jury for design competitions should always contain a majority of engineers in order to avoid the winning design not being realized for technical or economic problems.

As a result of the above considerations it can be stated that a well-designed bridge has a clear structural system, has harmonic proportions in good order, fits the surroundings and is as simple as possible.

For such a well-designed bridge technical and economical requirements are not contradictory to beauty. Contrary to widespread belief, a beautiful bridge is not necessarily more expensive than an ugly one. A bridge becomes, however, more expensive if the flow of forces is wrong. A beautiful bridge requires only an additional effort of technical and architectural creativity.

2 The development of cable-stayed bridges

Cable-Stayed Bridges. 40 Years of Experience Worldwide. First Edition. Holger Svensson.
© 2012 Ernst & Sohn GmbH & Co. KG. Published 2012 by Ernst & Sohn GmbH & Co. KG.

2.1 The precursors of cable-stayed bridges

2.1 The precursors of cable-stayed bridges
2.1.1 Introduction

The principle of supporting a beam or a mast by taught cables goes far back in time. Early examples are bridges from natural materials such as bamboo for the beam and lianas for ties, Fig. 2.1.

The masts of sailing vessels were always transversely stayed by shrouds, Fig. 2.2. Interestingly, this analogy still exists today in the French name 'pont a haubans' and the Russian name 'wantowyj most' for cable-stayed bridges.

A basic outline on the precursors of cable-stayed bridges can be found in the book of Troitsky [2.1], which has also been used by later books. New outlines can be found by Pelke [2.2, 2.3]; Birnstiel has recently investigated the Saale River Bridge at Nienburg [2.8]; and Stiglat has investigated early French stayed bridges [2.19], especially transporter bridges.

What is the difference between the precursors and modern cable-stayed bridges? Modern cable-stayed bridges have stay cables with well-defined and tuned cable forces which transfer their loads directly, Fig. 2.3. Their horizontal components are introduced as compression forces into the beam. The forcstay cables of the main span are tied back by the backstay cables to the ends of the bridge where they are anchored in hold-down (or anchor) piers. There each cable force is split into two components, the vertical force which is anchored to the ground and the horizontal force which is transmitted into the beam in compression. The compression forces from the forestays and the backstays equal one another out and reach their maximum at the towers.

The flow of forces of the precursors of modern cable-stayed bridges was different, for example by also anchoring the horizontal backstay component to abutments as in suspension bridges or by connecting forestay cables from both sides in the center of the main span. Furthermore, the stay cables are not control-tuned. By the time of World War II the era of modern cable-stayed bridge design was just about to start.

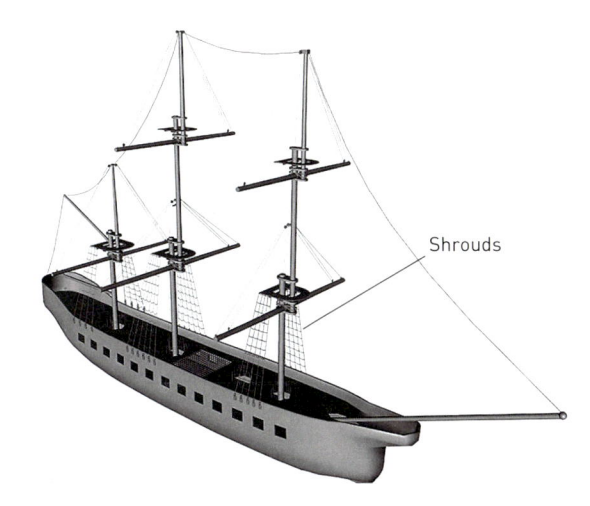

2.1.2 Historical development
2.1.2.1 Historical designs

An early example for the design of cable-stayed bridges is the drawing by Faustus Verantius, Fig. 2.4 [2.4], in which a timber beam is directly supported by eye bars in combination with a suspension cable, Fig. 2.5. The parallel arrangement of the ties is in some ways similar to a harp shape. The eye bars are anchored to hinges in the beam, similar to the modern approximation of a hinged beam.

In 1784 the German carpenter, Immanuel Löscher from Freiburg, designed a 'tie-back bridge' in which the 44.3 m main span was tied back with three timber ties to two timber towers which were restrained by timber backstays, Fig. 2.6 [2.5].

The French engineer Poyet [2.6] presented designs for cable-supported bridges with spans of 50 m in 1787. He proposed hanging a timber beam with straight iron rods from high towers, Fig. 2.7. The

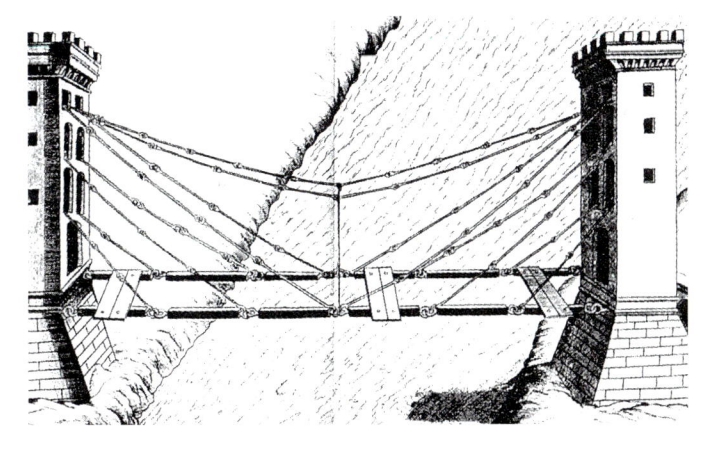

Figure 2.1 Bamboo bridge in Indonesia
Figure 2.2 'Shroud bridge'

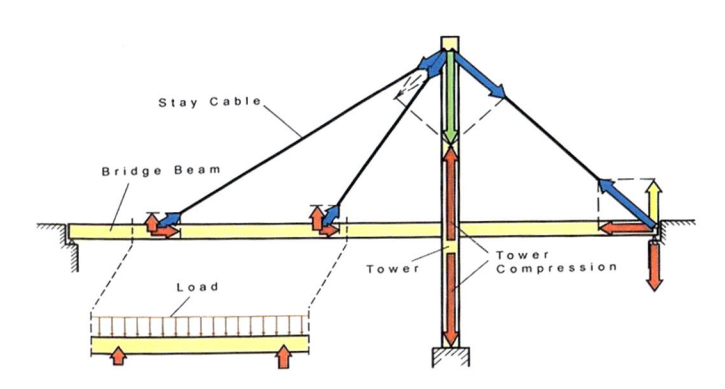

Figure 2.3 Flow of forces in modern cable-stayed bridges

Figure 2.4 Faustus Verantius, Croatian engineer (1551–1620)

Figure 2.5 Design of an eye bar supported bridge by Verantius, 1617

hanger arrangement is similar to a fan shape. The French authorities, however, did not permit their construction. All these designs have not been realized.

2.1.2.2 First examples and failures

The first proven permanent bridge with inclined ties was the Kings Meadow Bridge [2.7], in England in 1817, Fig. 2.8. Its designers were two Scottish ironmongers: James Redpath (1772–1846) and John Brown (1792–1852).

The fan-shaped stays consist of 8 mm diameter iron wires which are tied back into the foundations with 19 mm diameter bars, and the ties were adjustable with bolts. Unfortunately, during the winter of 1922/23, the bridge partially collapsed and was restored with additional 16 stays, Fig. 2.9. In 1954 the bridge finally fell prey to an extraordinary flood.

The development of stayed bridges was interrupted by the collapse of two early examples, the first Dryburgh Abbey Bridge in 1818, and the Nienburg Bridge in 1824.

In 1817 John and William Smith designed the first bridge across the River Tweed in Scotland near Dryburgh Abbey [2.7], with a span of 79.3 m, and a similar design to the Kings Meadow Bridge. The bridge deck with a width of only 1.22 m displayed strong vibrations under pedestrian loads as well as under wind. After only 6 months the bridge collapsed in 1818 during a storm after a tie-back chain broke. A possible cause is the failure of the couplings between the round bars, which were only bent back and clamped by ferrules. In the same year the Second Dryburgh Abbey Bridge was built, Fig. 2.10, this time as a suspension bridge with stiffening diagonal chains in the outer quarters of the main span.

By that time the forces in the main cables of the suspension bridge could be calculated; the stays were only added for structural reasons, but were not specifically tensioned. This was an important change in the direction of suspension bridges with stiffening stays.

The first German cable-stayed bridge, across the Saale River at Nienburg, was designed by Gottfried Bandhauer in 1824, Figs 2.11 and 2.12. The American engineer Dr. Charles Birnstiel has recently

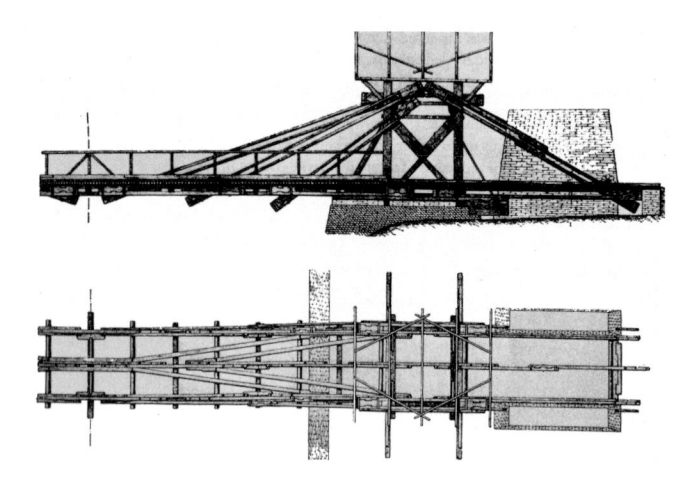

Figure 2.6 Tie-back timber bridge by Löscher, 1784, main span: 44.3 m

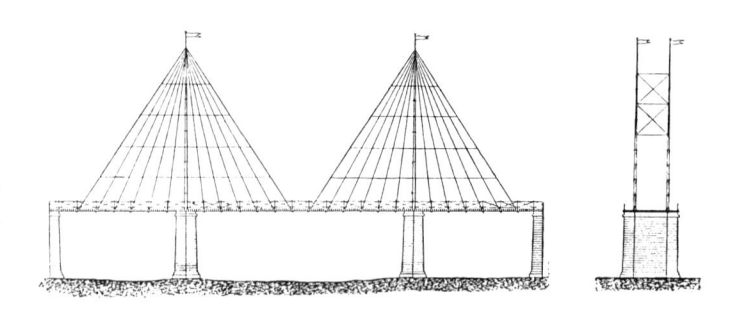

Figure 2.7 Cable-supported bridge by Poyet, 1787

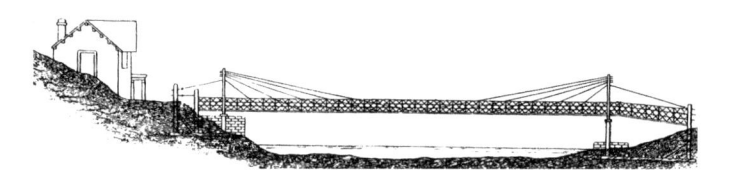

Figure 2.8 Kings Meadow Bridge, 1817

Figure 2.9 Kings Meadow Bridge, 1922

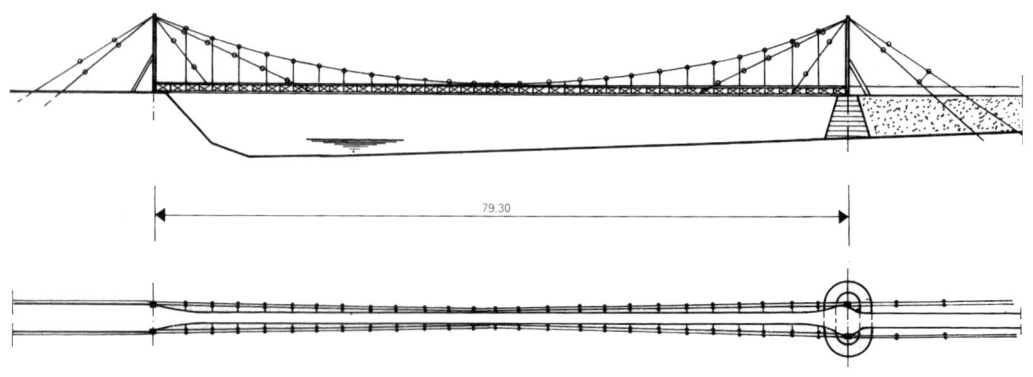

Figure 2.10 Second Dryburgh Abbey Bridge, 1818

2.1 The precursors of cable-stayed bridges

investigated the collapse of this bridge and has published his findings [2.8a and 2.8b], which we follow here. The Nienburg Bridge actually comprised two separate cable-stayed bridges with one tower each, connected by a bascule bridge in the center with a 3.5 m span which was opened for the passage of ships with high masts. This combination of a cable-stayed bridge and a bascule bridge is unique.

The main span had a length of 82 m with a 7.6 m wide bridge beam. Each tower had five pairs of forestays and three pairs of soil-anchored backstays. The stay cables were coupled with ratchet tooth joints retained by ferrules, Fig. 2.13. The anchorage of the stays to the cross girders used iron straps connected to the stays also with ratchet teeth and ferrules, Fig. 2.14.

The Nienburg Bridge is remarkable in several aspects:
- longest and widest precursor to modern cable-stayed bridges
- first use of backstays and forestays in a fan arrangement
- first cable-stayed bridge for full road traffic, i.e. horse-drawn carts
- unique bascule bridge in the center

The bridge had, however, several flaws:
- the stays were underdesigned for their actual strength
- the bridge was too flexible and thus vulnerable to oscillations
- some details were not technically mature, especially the cable connections
- the whole design was too advanced for the local artisans.

The similar sizing of all stays, although differently loaded, indicates that the actual flow of forces was not completely understood.

The stays, made from malleable iron, showed a lack of quality from the very beginning. They were tested individually with tensile machines developed by Bandhauer for the project, but 40 % of the stays did not reach the required strength and had to be reworked. The complete bridge was also test-loaded several times, the last on 27 August 1825, when a 6 t horse-drawn cart was taken across the bridge. The bridge was then opened and served the regular traffic without complaints. On 5 December 1825, the Duke of Anhalt-Köthen visited Nienburg. In his honor a torchlight procession on the bridge was ordered, which comprised more than 300 persons. When some high-spirited visitors tried to oscillate the bridge, three southern backstays failed, causing the collapse of the southern half of the bridge and 55 people drowned in the River Saale.

Investigation of the failed joints in the stays found that the iron contained a disturbed microstructure, air bubbles and inclusions of furnace slag. The high dynamic loads together with the low quality of the iron for the stays were probably jointly responsible for the failure. Brandhauer was accused of having caused the accident because his design concept was too new and because he had not prevented the procession [2.9].

Bandhauer retaliated with the publication of all his design papers. An investigation of the remaining northern half of the bridge did not show any weak spots. Bandhauer was thus completely absolved in May 1829.

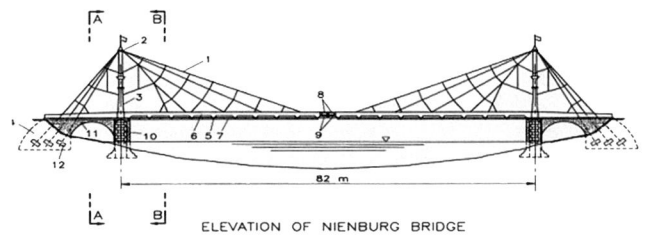

Figure 2.11 Nienburg Bridge, 1824

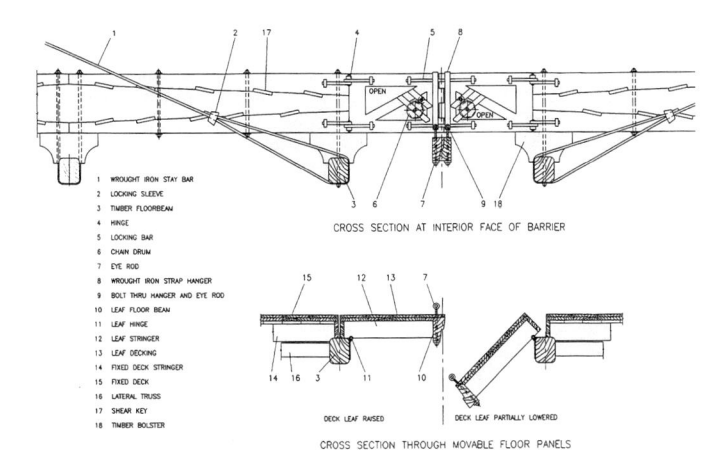

Figure 2.12 Nienburg Bridge, central bascule bridge

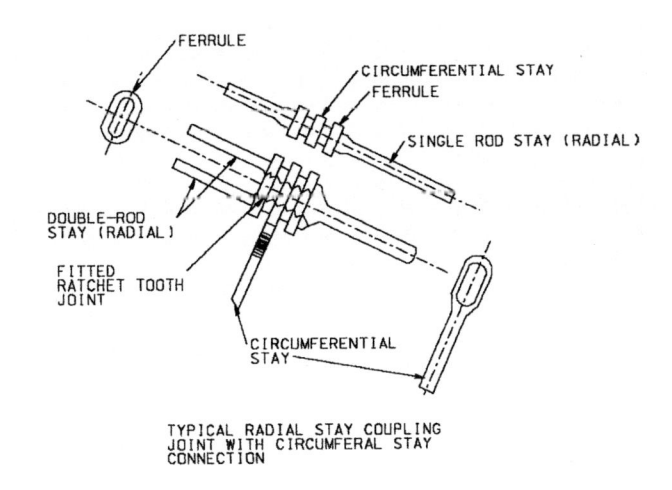

Figure 2.13 Nienburg Bridge, cable couplers

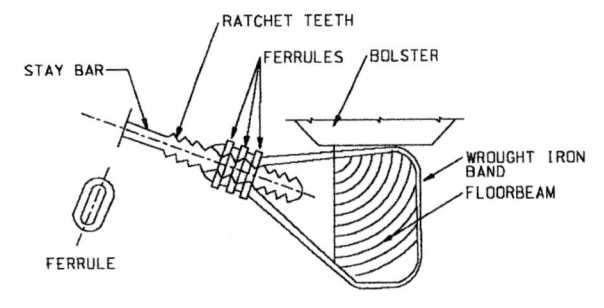

Figure 2.14 Nienburg Bridge, cable anchorage at beam

This terrible accident was the reason that the development of stayed bridges was held up in Germany for about 125 years until the Düsseldorf Bridge Family in the early 1950s.

In his publication *Descriptions of Bridges of Suspension* [2.7], in 1821, Robert Stephenson dealt with the state of the art of cable-supported bridges, Fig. 2.15, and evaluated the collapse of the First Dryburgh Abbey Bridge.

The famous French mathematician and bridge engineer Henri Navier, Fig. 2.17, based his comprehensive treatise *Memoires sur les ponts suspendus*, Fig. 2.16 [2.6], of 1832 mainly on Stephenson's publication, Fig. 2.15. Navier compared fan and harp arrangements of the stays with the result that for a given span the required steel quantities for the towers and the stays are similar, Fig. 2.18. His stay cables are still soil anchored and thus not in accordance with modern cable-stayed bridges.

Navier shows that stayed bridges cannot undergo stressless deformations like suspension bridges to adjust to changing load positions, but that stayed bridges are a stiff system which can only be deformed by elastic changes in the length of the stay cables.

In modern terminology, he realized that the highly statically indeterminate system of a cable-stayed bridge could not be calculated with the methods of his time, unlike the suspension bridge, which is basically only simply statically indeterminate. He therefore rejected cable-stayed bridges, and so the development of cable-stayed bridges in France was interrupted for nearly 50 years until Ferdinand Arnodin designed his Pont des Saint Ilpize, Fig. 2.28.

The development of stayed bridges continued in England, however. The engineer Thomas Motley developed modern ideas for a stiff bridge which should also be suitable for railways, which he realized in 1837 with the Twerton Bridge, Fig. 2.19 [2.10].

Its stiff beam comprises a deep cast-iron truss, supported by stays and hangers from round bars of malleable iron. The stiff beam carries an important part of the live loads directly to the abutments. The bridge anticipates some characteristics of modern 'extra-dosed' cable-stayed bridges.

The uplift forces at the bridge ends were anchored with threaded bars to the abutments. The construction also follows modern principles: free cantilevering from the main towers to both sides and installation of the central drop-in girder with the help of a timber auxiliary frame.

In this way the development of stayed bridges in England reached its culmination and temporarily did not develop further. The newly developed cables from higher-strength steel wires were more economic for suspension bridges, and the ideas of Motley were for some time forgotten.

Further progress took place by using stiffening stays for suspension bridges.

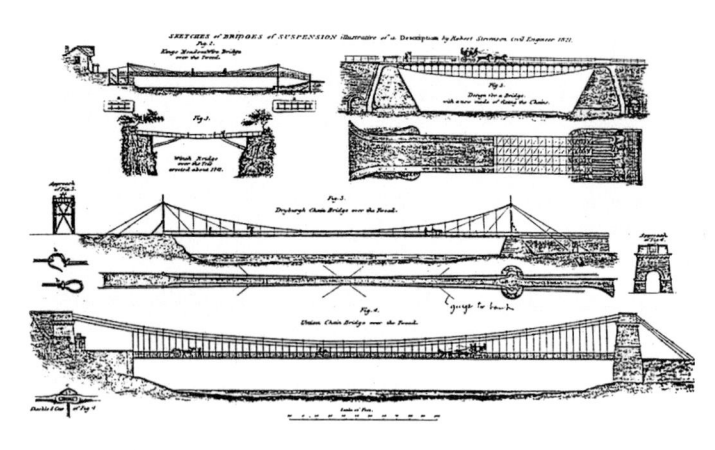

Figure 2.15 'Descriptions of bridges of suspension' by Robert Stephenson

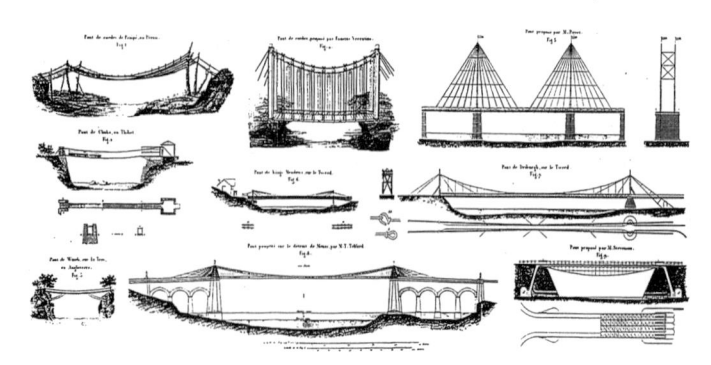

Figure 2.16 'Memoires sur les ponts suspendus' by Henri Navier

French mathematician and bridge engineer
Born 1785 in Paris, France
1818 Professor in Paris
1821 Navier-Stokes equation of fluid mechanics
1821 General theory of elasticity
1830 Professor at the École Nationale des Ponts et Chaussées
1832 Investigation of cable-supported bridges
Died 1836 in Paris

Figure 2.17 Henri Navier [2.137]

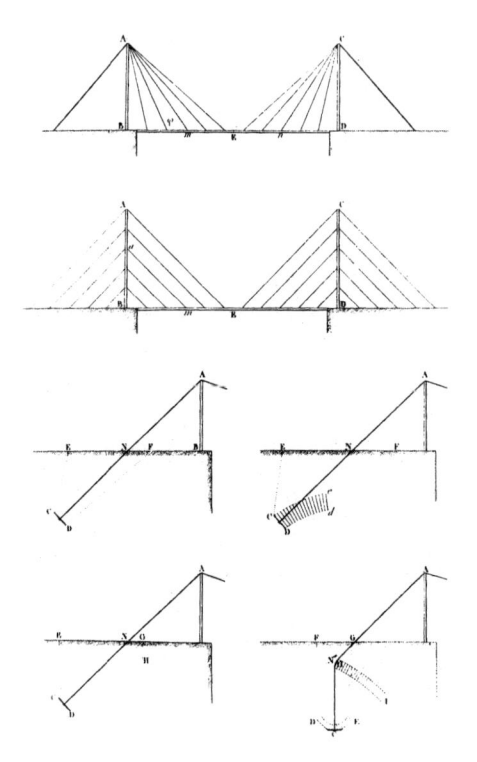

Figure 2.18 Stayed bridge systems investigated by Navier

2.1.2.3 John Roebling and stiffened suspension bridges

John Roebling was one of the most visionary bridge engineers in the 19th century, Fig. 2.20 [2.11], and he studied structural engineering at the Polytechnikum in Berlin and thereby became familiar with the treatise of Navier on cable-suspended bridges. Due to the lack of work in Germany he emigrated to the USA in 1831.

After some years as a farmer he took up structural engineering again. He first developed the fabrication of coil ropes and later of parallel wire cables, initially to pull vessels up the lift sections of canals, later for use as bridge cables. He developed the required machinery himself, Fig. 2.21 [2.12]. The factory founded by him in Trenton, Ohio, existed until recently.

Roebling intended to design major suspension bridges, an idea he had already developed back in Germany. In 1841 he published his basic findings on cable-suspended bridges [2.13]. He pointed out the superior strength and elasticity of parallel wire cables as against chains from malleable iron which had been used until then. As a consequence of some recent spectacular collapses of suspension bridges caused by wind oscillations he voiced his perception that additional inclined stays would stiffen suspension bridges and thus render them insensitive to wind oscillations. In addition, the deflections from heavy live loads would be reduced.

After several successful smaller bridges he achieved his breakthrough in 1851 with the Niagara Falls Bridge, Fig. 2.22. The main problem for this suspension bridge with a record span of 251.5 m was to achieve the necessary stiffness for the high railway loads and to stiffen the exposed beam high above water against dangerous wind oscillations. He reached these objectives in two ways. First he used a 6 m deep timber truss in order to stiffen the beam, and second used inclined stays in the outer thirds of the main span. In some ways he thus continued the perceptions of Motley.

Further similarly designed stiffened suspension bridges followed, for example the Allegheni River Bridge in Pittsburgh and the Cincinnati Bridge.

His highest achievement as a bridge engineer was the Brooklyn Bridge in New York with a new record span of 486 m, Figs 2.23 and 2.24. He started the design in 1865. Fig. 2.25 shows a design drawing for the stay cables prepared by himself. He tried to distribute at least the dead loads equally between the stay cables and the main suspension cables. The calculation of the highly statically indeterminate mixed system was not possible with the tools of his time, but his intuitive grasp of the flow of forces is admirable.

John Roebling had completed the design for the Brooklyn Bridge before he died as a result of an accident on site. The construction of the bridge was continued by his son Washington. When he became paralyzed after too fast decompression in a caisson foundation, his remarkable wife, Emily Warren Roebling, continued to supervise the site and completed the bridge successfully. Her outstanding accomplishments have been recognized in her biography, *Silent Builder* [2.13a].

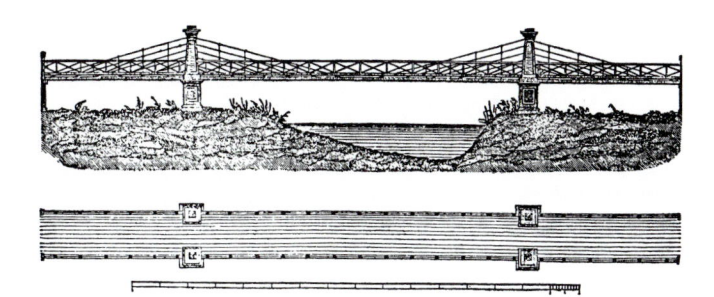

Figure 2.19 Twerton Bridge, 1837

German-American bridge engineer
Born 1806 in Mühlhausen, State of Thuringia, Germany
1826 Completion of Studies at the Polytechnikum in Berlin
1831 Emigration to the USA
1842 US patent for iron and steel wire ropes
1849 Factory for steel wire ropes in Trenton, Ohio
1855 Niagara Falls Bridge
1866 Cincinatti Bridge across Ohio River
1869 Brooklyn Bridge
Died 1869 in a bridge accident

Figure 2.20 John Roebling (Johann August Röbling)

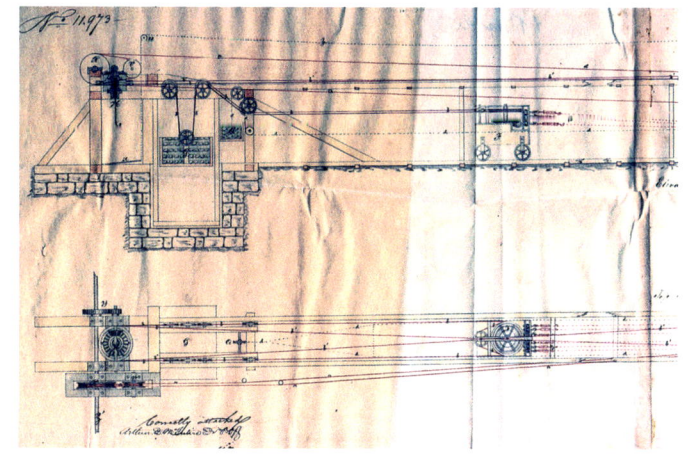

Figure 2.21 Roebling´s bridge ropes

Figure 2.22 Niagara Falls Bridge, 1855

The Franz Josef Bridge across the Moldau River in Prague by Ordish and LeFeuvre was completed in 1868, Fig. 2.26 [2.15]. A new mixed system was used in which inclined bars run from the quarter points of the main girder to the tower tips. These tension members were supported by hangers from the main suspension cables in order to reduce their sag. The two inner stays are continuous between the towers and thus do not carry their tension back to the abutments.

The Albert Bridge across the Thames River in London, with a main span of 122 m, was designed by Ordish as well, Fig. 2.27 [2.16]. Together with the Brooklyn Bridge this is the best preserved and best known suspension bridge with stiffening stays.

Ferdinand Arnodin is a famous French bridge engineer, who has forwarded the cable-stayed concept in various ways, Fig. 2.28. Similarly to Roebling he fabricated spiral bridge ropes developed by himself.

In his bridge designs he used a mixed system in which the outer quarters of the bridge beam are supported by stays and the central half is only supported by the suspension cables. The earliest example is the Pont de Saint Ilpize of 1879, Figs 2.29 and 2.30 [2.17].

Three further major bridges were designed according to this principle. The bridge across the Saone River in Lyon has a main span of 121 m, Fig. 2.31 [2.17].

The Bridge across the Rhône River in Avignon looks more monumental but follows the same principles.

The Bonhomme Bridge across the Blavet River with a main span of 163 m varies the load-carrying system in so far as the main span is divided into three sections of which the two outer ones are suspended by stay cables, Fig. 2.32 [2.18].

Also more recently, stay cables have been applied to strengthen suspension bridges. The Tamar Bridge, Fig. 2.33, was widened by adding two outer lanes. In order to carry the additional loads, stay cables were installed, of course using modern exact statical calculation and precise cable adjustment on site, Fig. 2.34.

2.1.2.4 Transporter bridges

Between 1893 and 1920 at least 18 major transporter bridges were built, which are described by Stiglat [2.19]. They served the transportation of men and freight across major waterways instead of using ferries, which would be subject to tides and would hinder ship traffic. Conventional bridges could not be used because the masts of seagoing vessels require a navigational clearance in excess of 30 m. The corresponding ramps would require too much space in crowded harbor areas. The first major transporter bridge in Potugalete, with a main span of 160 m, was designed jointly by the Spaniard Alberto de Palacio and the French Ferdinand Arnodin in 1893 [2.20].

The high-level beam which supports the transporter gondola is supported at the outsides by stay cables and in the center part by the suspension cables in accordance with Arnodin's principle, Fig. 2.35.

Similar transporter bridges were designed by Arnodin across the harbor of Bizerta, across the Seine at Rouen and across the Charente

Figure 2.23 Brooklyn Bridge, USA, 1869

Figure 2.24 Brooklyn Bridge, Stiffening stay cables

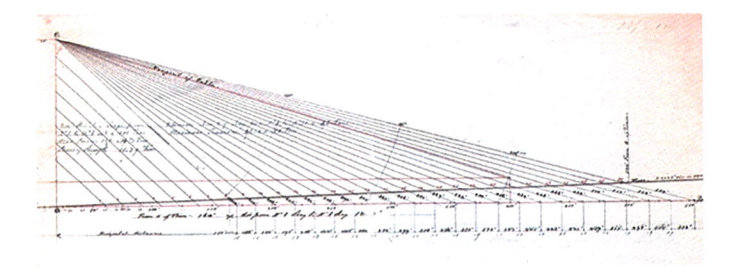

Figure 2.25 Preliminary design of the Brooklyn Bridge by Roebling

Figure 2.26 Franz Josef Bridge, Prague, 1868

River at Rochefort. For the transporter bridge across the River Loire at Nantes, Fig. 2.36, only 25 m were available for the back spans on each side [2.21]. This excluded the stiffened suspension bridge system used until then. Consequently Arnodin selected a system supported by stay cables only and used a free cantilevering construction. The 35 m long drop-in girder with a weight of 46 t was lifted from a pontoon by two cranes supported on each cantilever tip. At the end of the short cantilevers tie-down cables were anchored to enormous masonry blocks. This design and construction was a spectacular achievement by Arnodin.

With his following transporter bridge in Newport, UK, Arnodin reached the new record span of 197 m, but the structural system went back to the earlier mixed designs. The concentrated load of up to 117.5 t required a continuous girder without hinges or drop-in girder. The bridge was closed down in 1985, but reopened ten years later after in-depth rehabilitation.

The transporter bridge across the old harbor in Marseille is the final example of Arnodin's work, Fig. 2.37 [2.22]. The structural system is similar to that of the Nantes Bridge, including exclusive use of stay cables, free cantilevering construction and drop-in girder, Fig. 2.38.

Transporter bridges were built upto the 1920s, but not as cable-stayed bridges anymore. They represent the second trial to reach the modern stay cable system, this time rather more more successful.

Figure 2.27 Albert Bridge, London, 1873

French bridge engineer
Born 1845 in Sainte-Foy-lés-Lyon, France
1872 Factory for 'bridge ropes spun in opposite directions'
1887 Patent for transporter bridges
Design of 9 of the existing 18 transporter bridges
Design of several suspension bridges
Died 1924 in Chateauneuf-sur-Loire

Figure 2.28 Ferdinand Arnodin

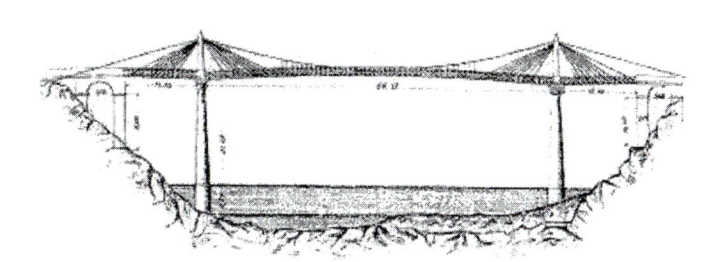

Figure 2.29 Pont de Saint Ilpize, France, 1879

Figure 2.30 Pont de Saint Ilpize, today

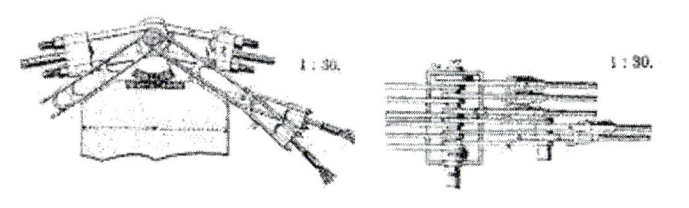

Figure 2.31 Saone River Bridge in Lyon, France 1888

Figure 2.32 Bonhomme Bridge across the Blavet River near Marbiant, 1904

Figure 2.33 Tamar Bridge, England, 2001; before widening

Figure 2.34 Tamar Bridge after widening

Figure 2.35 Transporter bridge in Portugalete/Bilbao, 1893

Figure 2.36 Transporter bridge at Nantes, France, 1903

Figure 2.37 Transporter bridge across the old harbor in Marseille, 1905

Figure 2.38 Construction of the transporter bridge in Marseille, France

2.1.2.5 Approaching the modern form

A light and simple-to-erect cable-stayed bridge was developed by the American engineers Edwin Runyon and William Flinn, Fig. 2.39 [2.16]. In order to keep the roadway beam light, the central stay cables were not anchored to the beam and tied back; rather they were continuous underneath the deck to the next tower as done in the early French examples. In this way the compression forces in the beam are reduced, a continuous horizontal rope within the beam assisting the construction of the light structure [2.16]. There are no suspension ropes anymore.

The first consistent design for cable-stayed bridges which was sufficiently stiff and economical was developed by the French engineer Albert Gisclard in 1899, Fig. 2.40. His structural system consists of inclined and horizontal cables which form a geometrical stable truss from ropes. The stay cables transfer their horizontal components not via compression in the bridge beam but via tension in the adjacent cables and the second tower into soil anchors. Strictly speaking the subsystems are three-hinged arches with tensile diagonals of cables [2.23].

The Gisclard system is suitable for railway bridges, and was often used in France and in the colonies. A typical example is the Cassagne Bridge, Fig. 2.41, with a main span of 156 m [2.24].

During a test loading a few days before bridge opening in 1909, the train derailed and killed six people, including Gisclard himself. Gaston Leinekugel Lecocq, Fig. 2.42, Gisclard's son-in-law, saved himself with a courageous leap from the test train.

Leinekugel Lecocq further developed Gisclard's system [2.24] by anchoring the forestays backwards, thereby initiating compression in the beam. Only the central forestays were anchored in tension to the other tower. A typical example is the Lézardrieux Bridge across the Trieaux River in France, completed 1925, Figs 2.43 [2.25] and 2.44 [2.26].

With this bridge Leinekugel Lecocq realized the modern principle of forming the individual load-bearing subsystems from stable triangles which use the beam and the towers as compression members and the stay cables as tie. The backstay cables were anchored to the beam. All that was missing for a modern cable-stayed system was the controlled stressing of the stays.

The final step was taken by the Spanish engineer Eduardo Torroja, Fig. 2.45 [2.27], with the Tempul Aqueduct in 1926. This bridge was originally planned as a continuous girder with two piers in the main span. When it became clear that these piers could not be built due to bad soil conditions, he replaced them by single forestays and backstays.

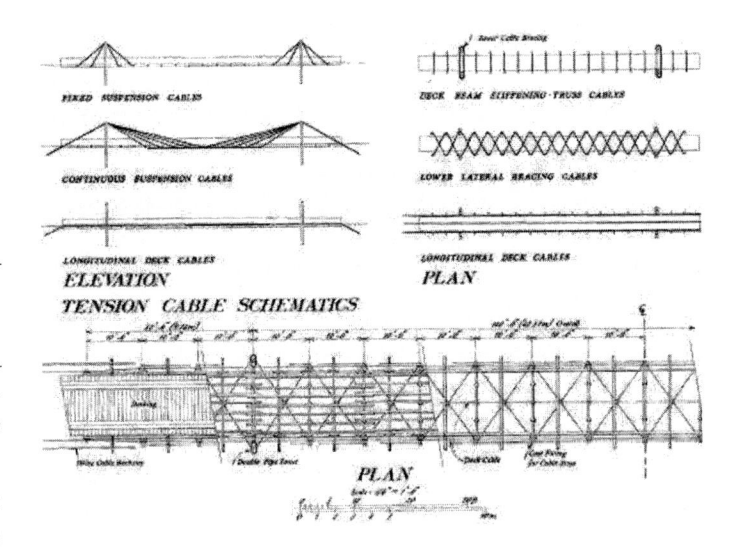

Figure 2.39 Bluff Dale Bridge, 1899

For the first time he used high-strength post-tensioning bars for the stays and stressed them to their calculated forces by raising the tower heads with hydraulic jacks against the dead weight of the water trough from reinforced concrete, Figs 2.46–2.49. Finally, the stays were encased in concrete as corrosion protection [2.27] and [2.28].

In 1938 the German engineer Franz Dischinger, Fig. 2.50, designed a high-level railway bridge across the Elbe River near Hamburg. He recognized again that a suspension bridge would be too flexible for the heavy railway loads.

His final solution corresponds with that of Roebling and others by adding stays at both sides of the main span, Fig. 2.51. He published his results in 1949 [2.29], pointing out the governing influence of the stay cables on the deformation characteristics, and developed formulas to calculate the loss of stiffness due to the change of sag. He also recognized the advantage of high-strength steel for the stays and the necessity to tune the stay cables to their precise forces. He also confirmed what Roebling had already found that these additional stays improve the aerodynamic stability. Unlike Roebling he consciously relinquished the overlapping of the stay cables with the suspension hangers in order to clarify the flow of forces, Fig. 2.51.

With Dischinger's publication on the characteristics of stay cables the development of the precursors of cable-stayed bridges before World War II was completed.

French bridge engineer
Born 1844 in Nîmes, France
Developed the Gisclard suspension bridge system
Died 1909 in a site accident

Figure 2.40 Albert Gisclard

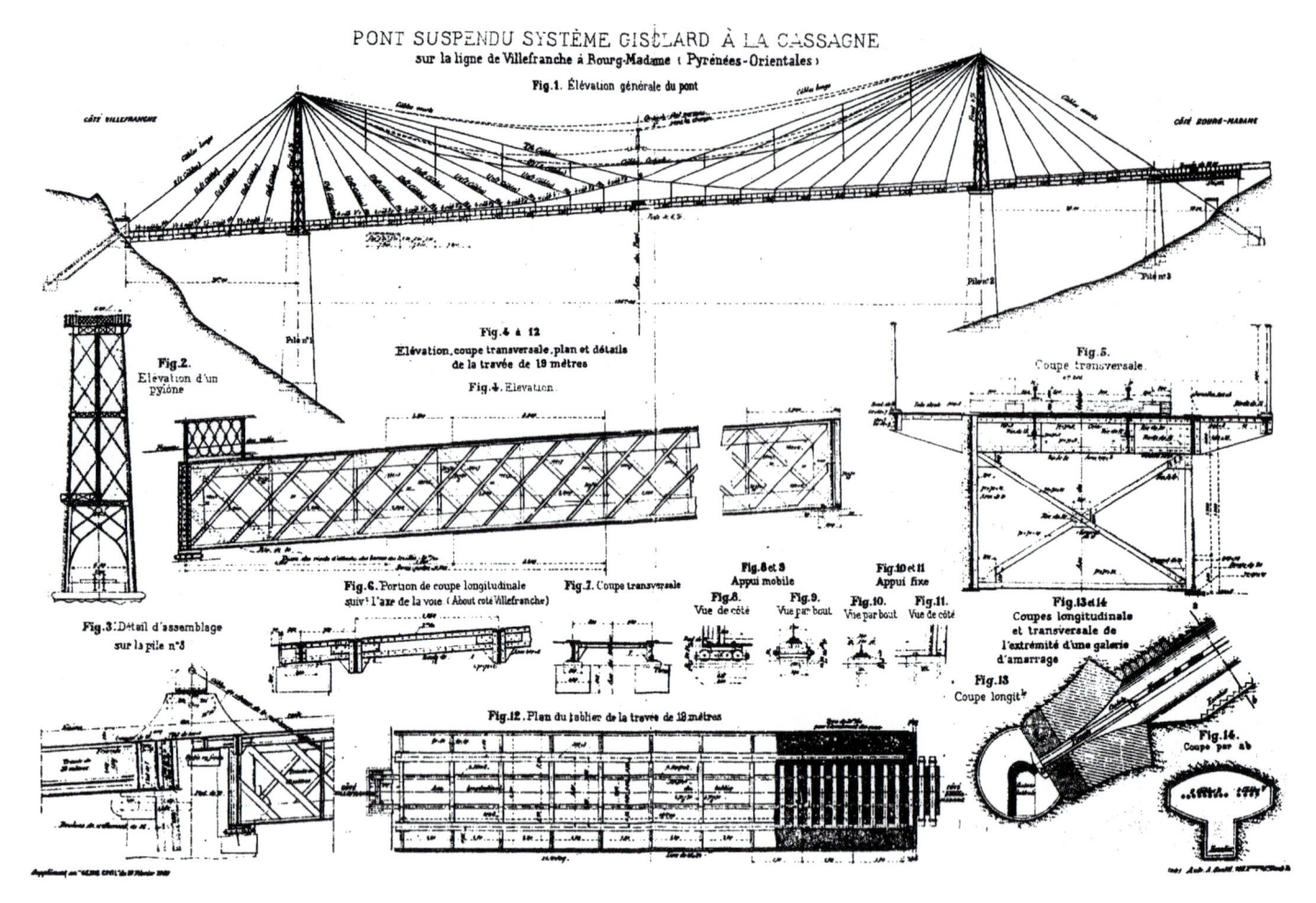

Figure 2.41 Cassagne Bridge system by Gisclard, 1909

Son-in-Law of Ferdinand Arnodin
Born 1867 in Cambrai, France
1890 PhD at École Polytechnique in mathematics and hydrodynamics
1900 Bridge designs using the Gisclard system in the company of Arnodin
1911 Book in two volumes on cable-supported bridges
1914 Bridge designs for the French army
1922 Steel contractor in Larche en Corrèze
1924 Acquisition of the steel contactor Arnodin
Died 1965

Figure 2.42 Gaston Leinekugel Lecocq

Figure 2.44 Pont de Lézardrieux, today

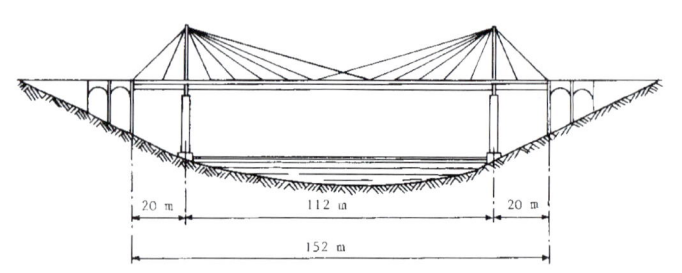

Figure 2.43 Pont de Lézardrieux, France, 1925

2.1 The precursors of cable-stayed bridges

Spanish bridge engineer
Born 1899 in Madrid, Spain
1923 Diploma as structural engineer
1929 Tempul aqueduct
1939 Professor, Escuela de Caminos in Madrid
1959 Co-founder of International Association for Shell and Spatial Structures (IASS)
Died 1961 in Madrid

Figure 2.45 Eduardo Torroja

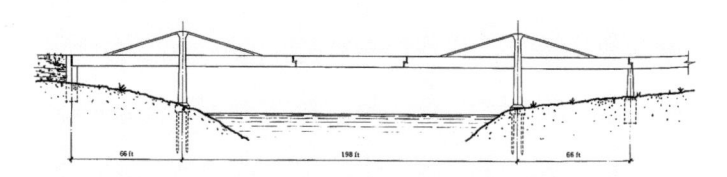

Figure 2.46 Tempul Aqueduct, 1929, 60 m

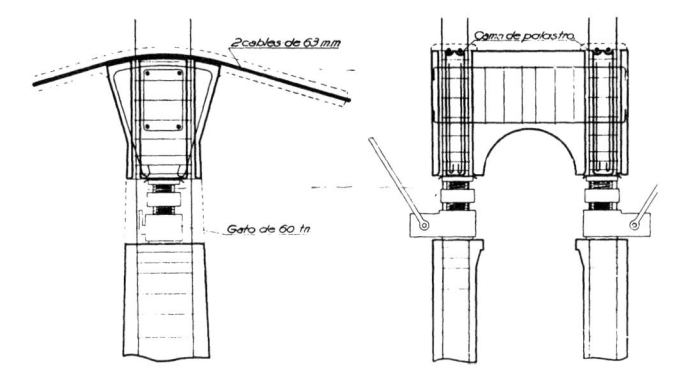

Figure 2.47 Towerhead with lifting arrangement

Figure 2.48 Tempul Aqueduct, view

Figure 2.49 Support system

Born 1887 in Heidelberg, Germany
1911 Dipl.-Ing. TH Karlsruhe
1913 Dyckerhoff & Widmann
1929 Dr.-Ing. under K. Beyer on theory of shells at Dresden University
1933 Professor for reinforced concrete, Berlin Technical University
Post-tensioning without bond: Aue Bridge
Investigation of shrinkage and creep
1938 Suspension bridges with inclined stays
Died 1953 in Berlin

Figure 2.50 Franz Dischinger [2.137]

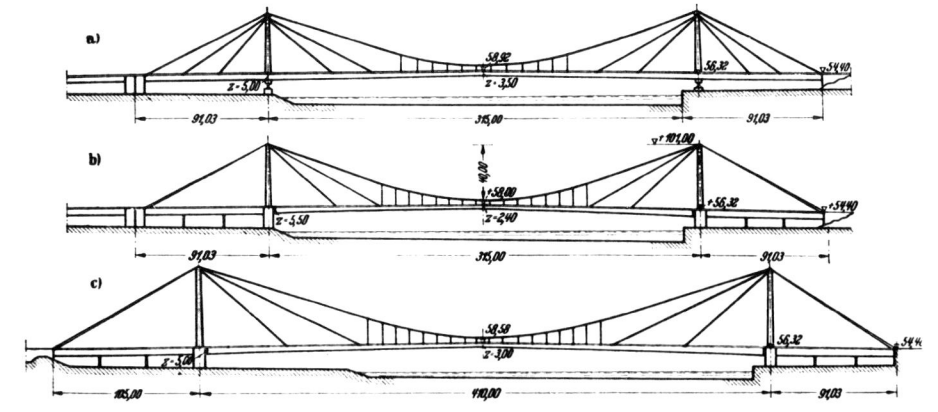

Figure 2.51 System Dischinger for the high-level Elbe River Bridge at Hamburg

2.2 Steel cable-stayed bridges
2.2.1 Introduction

The success of modern cable-stayed bridges as the governing system for long spans started with Dischinger's publication *Suspension Bridges for very heavy Loads* in 1949 [2.29]. In it Dischinger gives for the first time the design basis for stay cables.

The development of high-strength steel for the stays and methods to precisely calculate the forces and the appearance of hydraulic jacks which permit the stressing of the stay cables to their exact pre-calculated forces on site, finally overcome Navier's rejection of cable-stayed bridges due to uncertain flow of forces from 1823 [2.6].

After an initial tentative trial for the construction of suspension bridges (Duisburg-Homberg), the first modern cable-stayed bridges were built in Belgium at Donzière, in Sweden at Strömsund, and in Germany at Büchenau. The comprehensive design of the Düsseldorf Bridge Family started the development in earnest.

While initially the main spans of cable-stayed bridges appeared to be limited to less than 500 m and longer spans were supposed to be in the domain of suspension bridges, with time the cable-stayed spans increased up to 1088 m today for the Sutong Bridge in China, and even longer spans appear possible. A very helpful compilation of most of the cable-stayed bridges built to date can be found at Ewert [1.1]. While shorter spans are often built in concrete, the very large bridges have steel main spans.

2.2.2 Beginnings

Decisive for the success of steel cable-stayed bridges was the development of the orthogonal anisotropic lightweight steel deck (orthotropic deck) by Wilhelm Cornelius [2.30]. Until World War II the different members of the main girders acted independently of the others, the roadway slab now acts together with all other members, Fig. 2.52.

Important characteristics of the orthotropic deck are, Fig. 2.53: Plate thickness at least 14 mm and distance between stiffeners 300 mm. The development of the orthotropic deck is outlined in [2.31].

A comprehensive synopsis on the development of steel cable-stayed bridges is given by Weitz [2.32].

The Rhine River Bridge of Duisburg-Homberg, completed in 1954, has a main span of 260 m supported by partial suspension cables in the shape of a catenary, Fig. 2.54 [2.33, 2.34]. These partial suspension cables are anchored to the beam as in a self-anchored suspension bridge. The horizontal cable components balance one another out at the towers, Fig. 2.55.

The bridge beam comprises two slender boxes with integral orthotropic deck and shows all characteristics of a modern bridge beam, Fig. 2.56.

The disadvantage of this stayed suspension bridge becomes apparent during the difficult construction. Since the bridge is stable only in the final stage, the beam has to be supported by temporary

Figure 2.52 Old beam cross-section and orthotropic deck

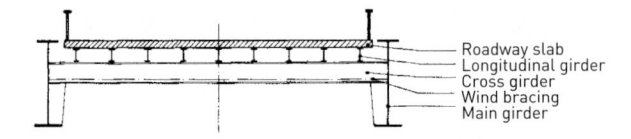

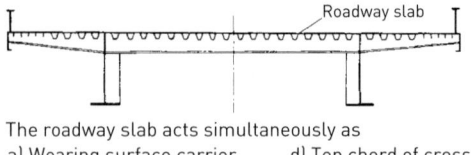

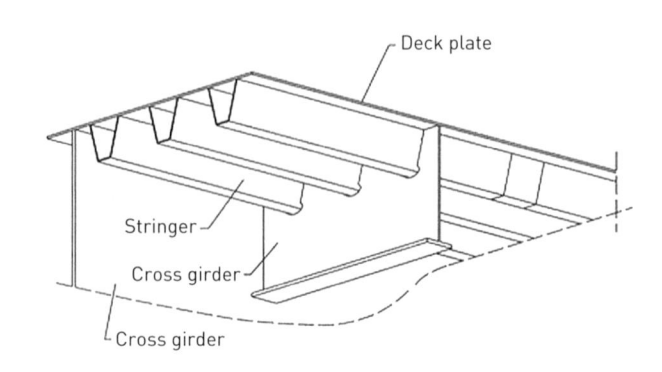

Figure 2.53 Orthotropic deck

Figure 2.54 Partial suspension bridge, Duisburg-Homberg, Germany, 1954

stays during construction across the Rhine River, with its heavy ship traffic, Fig. 2.57.

It is obvious that it would be more economic to use these temporary stays also in the final stage and thus to omit the stayed suspension cables. For this reason the Duisburg-Homberg Bridge system has not been repeated.

Dischinger advised the German steel contractor, Demag, on the design of a bridge near the city of Strömsund in Sweden in 1953. The result was the first consistent cable-stayed bridge design with a main span of 182 m, Fig. 2.58, [2.35, 2.36]. The design and construction engineering already uses most of the current modern methods.

The Strömsund Bridge is generally looked upon as the first modern cable-stayed steel bridge, because the concrete roadway distributes only local wheel loads and is not composite with the steel beam. The concrete slab thus does not participate in carrying the overall beam moments and normal forces, so the Strömsund Bridge is treated under the heading of steel bridges. The beam of the completed bridge has rather too great a depth from a modern point of view. This impression is exaggerated because the beam is unusually close to the water. The Strömsund Bridge represents a quantum leap in the development of cable-stayed bridges with the exception of its non-composite roadway slab.

2.2.3 The Düsseldorf Bridge Family

In 1952 the city of Düsseldorf started to plan three cable-stayed bridges across the Rhine with responsibility given to the architect Friedrich Tamms, Fig. 2.59, and the engineer Erwin Beyer, Fig. 2.60. Tamms requested that the bridges be 'delicate and light, slender and transparent and not overwhelmingly dominating'.

Tamms also requested the harp arrangement for stay cables, so that they are not only parallel in elevation but also in a skew view. In this way visual intersections of the stays are completely omitted. In addition, the towers were not to be connected at their tops, but the tower legs should be free-standing to distinguish them from the heavy portal frames of conventional suspension bridges [2.37].

Although the structural systems of all three bridges forming the 'Düsseldorf Bridge Family' are distinctly different, the harp arrangement and the free-standing tower legs connect them visually.

Further common characteristics of all three bridges are the use of orthotropic decks for the beam and the construction by free cantilevering without auxiliary piers and thus without interruption of the dense ship traffic on the Rhine. A general outline of the Düsseldorf Bridges can be found in [4.6]. Fig. 2.61 shows a model with the North Bridge in the background, the Oberkassel Bridge in the middle and the Knie Bridge in the foreground. The original designs were not changed although 24 years passed between their completion.

The tender design by the City of Düsseldorf for the North Bridge was completed in 1952 by Fritz Leonhardt, Fig. 2.62 [2.38], Wolfhart Andrä, Fig. 2.63, Hans Grassl, Fig. 2.64 [2.40], and Louis Wintergerst, Fig. 2.65.

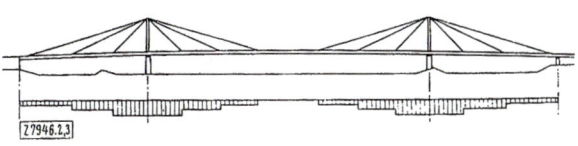

Girder bridge with pretensioned stay-cables

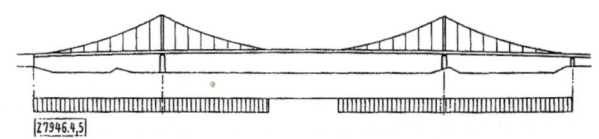

Girder bridge with pretensioned partial suspension cables

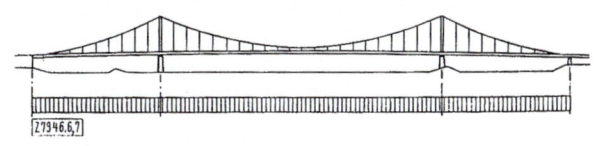

Self-anchored suspension bridge

Figure 2.55 Comparison of compression forces in the beam

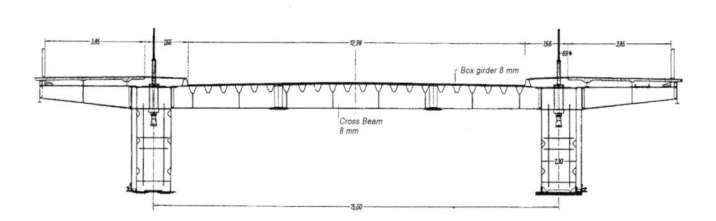

Figure 2.56 Cross-section

Figure 2.57 Free cantilevering

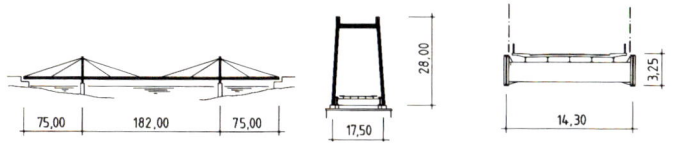

Figure 2.58 Strömsund Bridge, Sweden, 1956

The special structural analysis problems of cable-stayed bridges were basically solved with design of the North Bridge, Fig. 2.66, for example: the development of their influence lines, including stress influence lines (Kern point moments); the non-linear effects from compression with bending in the beam; the free selection of the run of moments under permanent loads; and the structural analysis of the free-cantilevering construction. Also the problems of buckling of the free-standing towers, loaded and elastically supported by the stay cables were conclusively solved. For further details see Sections 4.1 and 4.2.

The structural analysis and the construction of the North Bridge are outlined in detail in [2.41]. The principal solutions found for the North Bridge are still valid today.

All three bridges were built by the foremost German steel contractor at that time, Hein Lehmann AG, with Karl Lange, Fig. 2.67 [2.42] in charge and Fritz-Reinhard Weitz, Fig. 2.68 as chief engineer, supported by Erwin Volke, Fig. 2.69 and Carl-Heinz Rademacher, Fig. 2.70.

Immediately following the North Bridge the same engineers designed the Knie Bridge, Fig. 2.71, and the Oberkassel Bridge, Fig. 2.72. The Knie Bridge was designed with Fritz Leonhardt in charge [1.20]. In order to achieve a visual counterpoint to the high-rise buildings on the right bank of the bend in the Rhine, Tamms requested that the towers be placed on the left bank. This resulted in a main span of 320 m, structurally similar to a bridge with a 640 m main span. (This span was only exceeded by the Normandy Bridge in 1995.)

As Tamms again insisted on omitting the cross beam between the two 114 m high slender steel towers, they were designed with a T-cross-section for which the center of gravity lies as close as possible to the beam. Together with the external cable anchorage at the beam, only very small deviating cable forces acted on the towers.

The stiffness of the main span was increased by connecting each forestay to a backstay that was directly anchored to the beam above the piers in the side span. These piers serve as hold-downs with heavy ballasted concrete footings.

Although the design for the Knie Bridge was completed in 1954, it was built only in 1969 with hardly any design changes. For the construction of the main span, Hein Lehmann initially planned to use an auxiliary pier, but the costs of this pier, which would have had to be protected against ship collision, were calculated as being so high that finally a free-cantilevering construction with a span of 320 m was executed. This was unprecedented in its time and realized for the second time in the Rhine River Bridge Flehe in 1978.

The third Düsseldorf Bridge, the Oberkassel Bridge, Fig. 2.72, was then designed with Hans Grassl in charge. Because the dense inner-city traffic could not be interrupted for an extended period of time, it had to be built alongside the existing 'temporary' (since 1946!) bridge at a distance of 47.5 m, Fig. 2.73, and then transversely shifted, Fig. 2.74. For this purpose, the single tower was placed in the center

Figure 2.59 Friedrich Tamms

Born 1904 in Schwerin, Germany
1923 – 29 Study
1929 – 34 Bridge Authority Berlin
1935 – 45 Consulting Architect for Motorways
1948 – 69 Head of City Planning Office, Düsseldorf
Died 1980 in Düsseldorf

Figure 2.60 Erwin Beyer

Born 1920 in Stennweiler/Saar, Germany
1943 Dipl.-Ing. University Munich
since 1946 City of Düsseldorf
since 1968 Head of Bridge and Tunnel Authority, Düsseldorf
Planning and supervision on site of the Düsseldorf Bridge Family
1977 Dr.-Ing. E. h. University of Braunschweig
Died 1991

Figure 2.61 Düsseldorf Bridge Family, Germany

2.2 Steel cable-stayed bridges

Figure 2.62
Fritz Leonhardt

Born 1909 in Stuttgart, Germany
1931 Dipl.-Ing. Stuttgart University
1936 Promotion Stuttgart University
since 1934 Motorway Bridge Department, Berlin
1938–41 Design and Construction Supervision
Cologne-Rodenkirchen
Rhine River Suspension Bridge
1939 Foundation Office in Munich
1948–51 Cologne-Mühlheimer Suspension Bridge
1950–57 Düsseldorf Bridge Family
Many further cable-stayed bridges
1956 Television tower in Stuttgart
1957–74 Professor at Stuttgart University
Died 1999 in Stuttgart

Figure 2.64
Hans Grassl

Born 1908 in Vienna, Austria
1931 Dipl.-Ing. Vienna University
1932–36 Waagner, Vienna
1937–47 Rheinstahl-Eggers/Kehrhau, Hamburg
since 1947 Independent Consulting Engineer
Theodor-Heuss Bridge
Oberkassel Bridge
Flehe Bridge
Köhlbrand Bridge
Died 1980 in Innsbruck

Figure 2.63
Wolfhart Andrä

Born 1914 in Gera, Germany
1939 Dipl.-Ing. Stuttgart University
1939–41 Rodenkirchen Bridge
1941 Öresund Bridge
since 1945 further Collaboration with Leonhardt
1953 Partner in 'Leonhardt and Andrä'
1963 Promotion Stuttgart University
Schillersteg Stuttgart
Development of parallel wire cables
Knie Bridge
Mannheim North Bridge
Transverse shifting of Oberkassel Bridge
Weitingen Bridge
Died 1996 in Stuttgart

Figure 2.65
Louis Wintergerst

Born 1913 in Esslingen, Germany
1936 Dipl.-Ing. Stuttgart University
1938–41 Rhine River Bridge Rodenkirchen
1948 Deutz Bridge, Düsseldorf-Neuß Bridge
since 1949 Independent Consulting Engineer
Theodor-Heuss Bridge
Rhine River Bridge Maxau
Knie Bridge
Rhine River Bridge Speyer
Died 1977 in Stuttgart

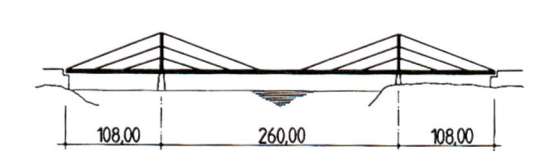

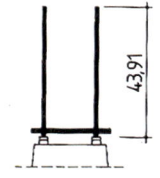

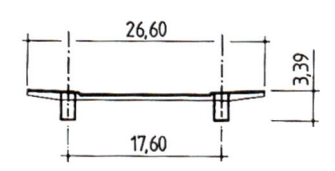

Figure 2.66 Düsseldorf North Bridge, Germany, 1957, 260 m

between the abutments and the four backstays were directly anchored with pendulums to the side span piers, similarly as with the Knie Bridge [2.45].

For shifting, the stay cables were restressed in such a way that the complete 10 000 t dead weight of the bridge was centered on the tower pier and the connections of the backstays to their tie-downs was released. A surface of polished steel was placed onto the pier on which the PTFE-underside of the tower bearing was sliding, Fig. 2.74.

Using hydraulic jacks with pull strands, the complete bridge was transversely shifted during 7 and 8 April 1976. The responsible engineer for the transverse shifting was Wolfhart Andrä. For further details see Section 5.1.3.3.

This completed all three bridges of the Düsseldorf Bridge Family, designed during 1953 and 1954 and completed in 1957, 1969 and 1976 respectively. Their design, fabrication and construction initiated the development of cable-stayed bridges worldwide.

2.2.4 Further Rhine river bridges

The Rhine is the most important European waterway between highly industrialized countries and the North Sea. It is highly frequented by cargo vessels. No piers are permitted in the navigable part of the river, requiring main spans of between 200 and 350 m built by free-cantilevering without auxiliary piers. All existing Rhine river bridges have been built since World War II and represent a unique collection of cable-stayed bridges, Table 2.1 and Fig. 2.75. Their number increases steadily, the latest addition being the cable-stayed bridge at Wesel.

In the following some further Rhine river bridges with special characteristics are introduced.

The Severins Bridge in Cologne, Fig. 2.76, [2.46], shows more variations of the cable-stayed system:
– The unsymmetrical fan arranged stay cables were placed as a counterpoint to Cologne Cathedral on the opposite river bank.
– Inclined cable planes mitigate the visual intersections of the fan stay cables in a skew view.
– Separate support of the tower and the bridge beam is used.

While the towers of the Düsseldorf North Bridge are fixed to the beam and thus form a continuous tower-beam system, the beam of the Severins Bridge runs freely between the tower legs, which are fixed to independent foundations. In this way the development of the idea of floating beams was initiated.

The Rhine River Bridge Leverkusen, Fig. 2.77, was the first freeway cable-stayed bridge and used a box girder supported by one central cable plane only. The beam thus uses a box girder to carry the eccentric live loads [2.47].

A large step forward in the development of cable-stayed bridges took place with the introduction of the multi-stay cable system for the Rhine River Bridge Bonn North, Fig. 2.79 [1.11], by Hellmut Homberg, Fig. 2.78, in 1963.

Figure 2.67 Karl Lange

Born 1911
Dipl.-Ing. TH Aachen
Executive Director of Contractor
Hein, Lehmann AG
1972 Dr.-Ing. e. h. Karlsruhe University
1957 Düsseldorf North Bridge
1965 Rhine River Bridge Leverkusen
1967 Bonn North Bridge
1969 Knie Bridge
1973 Oberkassel Bridge
1974 Köhlbrand Bridge
Died 1982 in Baden-Baden

Figure 2.68 Fritz-Reinhard Weitz

Born 1921 in Siegen, Germany
1950 Dipl.-Ing. Darmstadt University
1975 Dr.-Ing. Darmstadt University
1976 Honorary Professor University Darmstadt
1950–75 Contractor Hein, Lehmann AG
1967–83 Contractor Thyssen-Klönne
Düsseldorf Bridge Family
Bonn North Bridge
Mannheim-Ludwigshafen Bridge
Köhlbrand Bridge
Died 1983

Figure 2.69 Erwin Volke

Born 1926 in Fritzlar, Germany
1955 Dipl.-Ing. Darmstadt University
1956–88 Head of Design Department,
Hein, Lehmann AG
1988–93 Executive Director, Schüßler Plan
1999 Dr. Techn Vienna University
Bonn North Bridge
Mannheim-Ludwigshafen Bridge
Oberkassel Bridge
Died 2006 in Kaarst

Figure 2.70 Carl-Heinz Rademacher

Born 1930 in Salzwedel, Germany
1957 Dipl.-Ing. Berlin University
1960–88 Head of Construction Department,
Hein, Lehmann AG
Knie Bridge
Oberkassel Bridge
Mannheim-Ludwigshafen Bridge
Died 2011 in Erkrath

2.2 Steel cable-stayed bridges

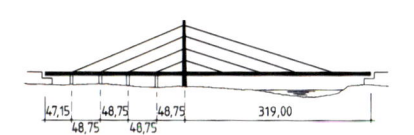

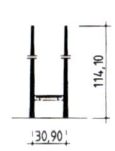

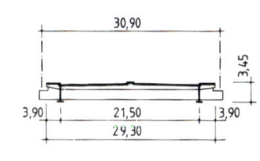

Figure 2.71 Knie Bridge, Germany, 1969, 319 m

Figure 2.72 Oberkassel Bridge, Germany, 1976, 258 m

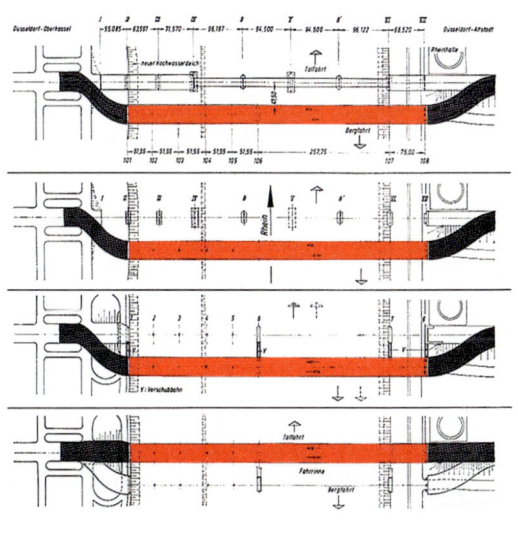

Figure 2.73 Traffic arrangement

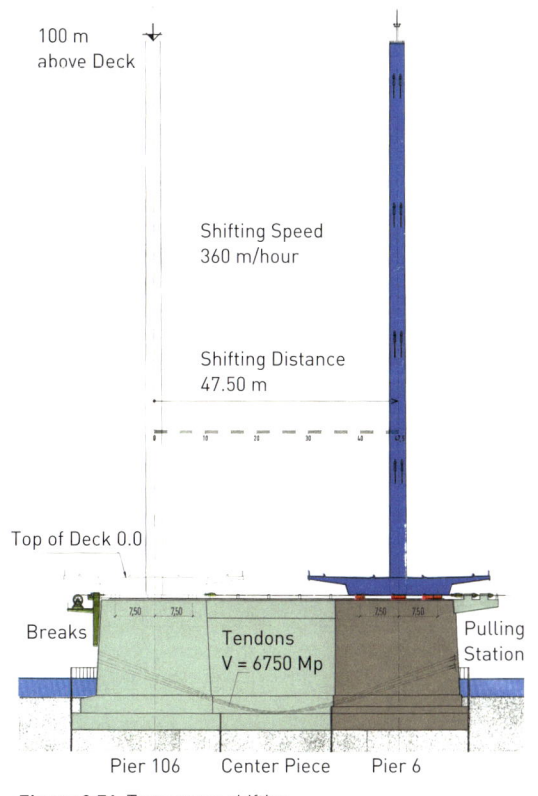

Figure 2.74 Transverse shifting

Table 2.1 Rhine River Bridges

Name	Year completed	Type
Kleve-Emmerich	1965	Suspension bridge, Federal road
Rees-Kalkar	1967	Cable-stayed bridge, Federal road
A 42	1990	Cable-stayed bridge, Freeway bridge
Duisburg-Neuenkamp	1970	Cable-stayed bridge, Freeway bridge
Ilverich	2002	Cable-stayed bridge, Freeway bridge
Düsseldorf North Bridge	1957	Cable-stayed bridge, Inner-city road
Oberkassel	1976	Cable-stayed bridge, Inner-city road
Knie Bridge	1969	Cable-stayed bridge, Inner-city road
Flehe Bridge	1979	Cable-stayed bridge, Freeway bridge
Leverkusen	1965	Cable-stayed bridge, Freeway bridge
Cologne-Mühlheim	1951	Suspension bridge, Inner-city road
Severins Bridge	1959	Cable-stayed bridge, Inner-city road
Cologne-Rodenkirchen	1954/90	Suspension bridge, Freeway bridge
Bonn North	1967	Cable-stayed bridge, Freeway bridge
Neuwied	1978	Cable-stayed bridge, Freeway bridge
Mannheim-Ludwigshafen	1971	Cable-stayed bridge, Federal road
Speyer	1974	Cable-stayed bridge, Freeway bridge
Maxau	1966	Cable-stayed bridge, Federal road

Figure 2.75 Rhine river bridges

Instead of using only a few concentrated stays consisting of a group of locked coil ropes, he designed many closely spaced individual stays for supporting the beam. Thus the beam on individual elastic supports became a beam on continuous elastic foundations.

The anchorage of the individual stays and their corrosion protection were simplified because, during free-cantilevering, auxiliary tiebacks for bridging the large distances between the concentrated cables were not required any more. Each beam section with a length of 4.5 m – equal to the cable distances at the beam – could be directly connected to its corresponding cable. The progress of computer calculations made it possible to calculate the forces in this 83-times statically indeterminate system.

For the first time, a floating system was used, in which the towers are independent of the beam and fixed directly to the foundations. The beam has no distinct longitudinally fixed bearing but is centered by the cables themselves. This system was groundbreaking for the further development of cable-stayed bridges.

A detailed appraisal of the work of Hellmut Homberg, one of the leading post-war German bridge engineers, has recently been published in [2.48].

2.2 Steel cable-stayed bridges

Figure 2.76 Severins Bridge, Germany, 1959, 301 m

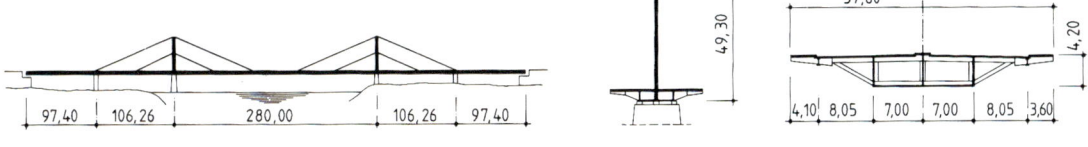

Figure 2.77 Rhine River Bridge Leverkusen, Germany, 1965, 280 m

The Rhine River Bridge at Mannheim-Ludwigshafen is the first example using a hybrid beam [1.19]. In hybrid cable-stayed bridges the loads from the steel main span are counterbalanced by a heavier concrete beam in the side span. The economic advantages of such a system are obvious. In general, this system is only feasible if piers at short distances can be placed in the side spans which support the excess weight of the concrete beam. (The only exception so far is the Sava River Bridge at Ada Ciganlija, Belgrade, Fig. 2.96).

The approach beam construction is adapted to small pier distances and takes place by spanwise self-launching formwork or by the incremental launching method. Since the construction of the side span runs ahead, auxiliary piers may be required to carry the initially missing uplift forces from the backstays. It is economic to protrude the concrete side span by some 100 m into the main span, if the costs for an additional construction method do not outweigh the savings in material. An example is the Normandy Bridge, Fig. 6.274.

The early cable-stayed bridges with hybrid beams up to 1999 are listed in Table 2.2.

The Mannheim-Ludwigshafen (Kurt-Schuhmacher) Bridge, Fig. 2.80, is a cable-stayed bridge with only one tower, carrying four traffic lanes and two tramway tracks across the Rhine. The beam width increases from 36.9 m on the Ludwigshafen side to 51.9 m for six traffic lanes on the Mannheim side. The 287 m steel main span is balanced by a post-tensioned concrete side span resting on piers at 60 and 65 m spacing. The A-tower with a height of 71.5 m above deck has partially spread legs with two corresponding cable planes to permit the placement of the two tramway tracks inside the tower legs and the 2 · 2 traffic lanes outside. Parallel wire cables were used for the first time in Germany with a unique corrosion protection from polyurethane and zinc chromate inside the PE-pipes, which had later to be replaced.

The construction took place by free cantilevering, partly with the use of an additional tie-back system to overcome the large cable distances, Fig. 5.87.

For his design of the Rhine River Bridge at Speyer, Fig. 2.81, Louis Wintergerst, Fig. 2.65, used concentrated backstays for the first time [2.49]. The side spans are directly supported by piers, and the beam is fixed to the abutment to which the backstays are anchored.

Another new type of cable-stayed bridge is the Rhine River Bridge at Neuwied, Fig. 2.82, by Hellmut Homberg, Fig. 2.78. The island in the middle of the Rhine lends itself to a central tower with two nearly equal spans, one on each side. In order to restrain the tower head an A-trestle in the longitudinal direction was selected. The bridge was built alongside an existing obsolete one and then transversely shifted by 16.25 m. The 11,400 t bearing load represents 93 % of the total bridge weight [2.50]. (Compare the Rhine River Bridge Oberkassel, Figs 2.72 – 2.74.)

The hybrid Rhine River Bridge at Flehe near Düsseldorf carries a six-lane freeway with a 364 m steel main span and a 780 m concrete approach bridge. The backstays are anchored in the first 240 m of the approach bridge which serves as counterweight for the main span, Fig. 2.83 [1.16, 2.51 – 2.54]. The 145 m high concrete tower has the shape of an inverted Y; the center plane of cables is anchored at the top.

The concrete and steel beam meet at a deep steel cross girder fixed to the tower legs so that a 1148 m continuous beam is created

Table 2.2 Hybrid Cable-Stayed Bridges up to 1999

Name Location	Year completed	Total length	Deck width	Deck depth	Steel deck/ length	% of main span	Tower type	Cable planes
Mannheim North Rhine River Bridge	1972	287 m	36.90 m	4.50 m	287 m	100 %	1 A, asymmetric	2
Flehe Bridge Rhine River Bridge	1979	368 m	41.50 m	3.80 m	368 m	100 %	1 inverted-Y	1
Tjörn Bridge Sweden	1982	366 m	15.75 m	3.0/3.0 m	386 m	105 %	2 H-vertical	2
Emscher Bridge Rhine River Bridge	1990	310 m	41.00 m	3.68 m	310 m	100 %	2 centre masts	1
Ikuchi Bridge Japan	1991	10 m	24.10 m	2.48 m	490 m	100 %	2 diamond	2
Normandy Bridge Seine River Bridge	1995	856 m	22.30 m	3.05 m	624 m	73 %	2 inverted-Y	2
Kap Shui Mun Bridge Hong Kong	1997	430 m	35.20 m	7.46 m	387 m Beton	90 %	2 H-inclined	2
Tatara Bridge Japan	1999	890 m	30.80 m	2.70 m	1.312 m	147 %	2 diamond	2

2.2 Steel cable-stayed bridges

Born 1909 in Wuppertal, Germany
1938 Promotion University Berlin
1938 Independent Consulting Engineer
Tables on grids, slabs and plates
1963 North Elbe River Bridge Hamburg
1965 Rhine River Bridge Leverkusen
1965 Rhine River Bridge Emmerich
1967 Rhine River Bridge Rees-Kalkar
1967 Rhine River Bridge Bonn North

1969 Masséna Bridge Paris
1977 Brotonne Bridge across River Seine, France (steel alternate)
1978 Rhine Bridge Neuwied
1979 Godsheide Bridge across Albert Canal, Belgium
1979 Kessock Bridge, Scotland
1987 Chao-Phraya River Bridge, Bangkok
1989 Thames River Bridge at Dartford
Died 1990 in Hagen

Figure 2.78 Hellmut Homberg

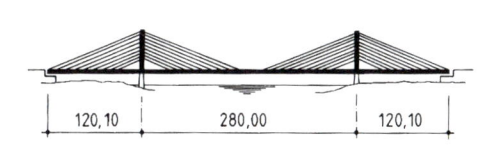

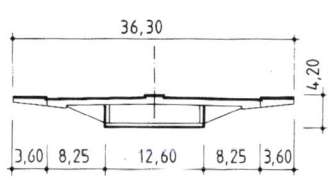

Figure 2.79 Bonn North Bridge, Germany, 1967, 280 m (Friedrich-Ebert-Bridge)

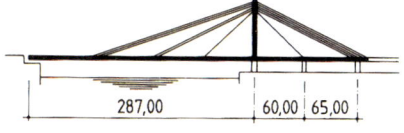

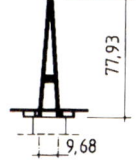

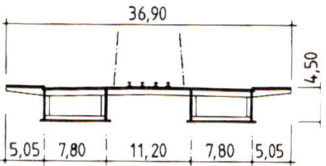

Figure 2.80 Rhine River Bridge at Mannheim-Ludwigshafen, Germany, 1972, 287 m

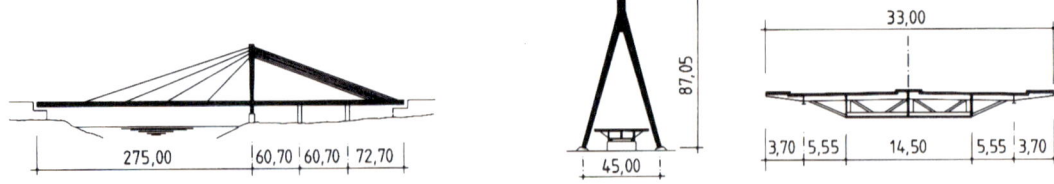

Figure 2.81 Rhine River Bridge Speyer, Germany, 1974, 275 m

Figure 2.82 Rhine River Bridge Neuwied, Germany, 1978, 212 m + 235 m

2.2 Steel cable-stayed bridges

Figure 2.83 Rhine River Bridge Flehe, Germany, 1978, 368 m

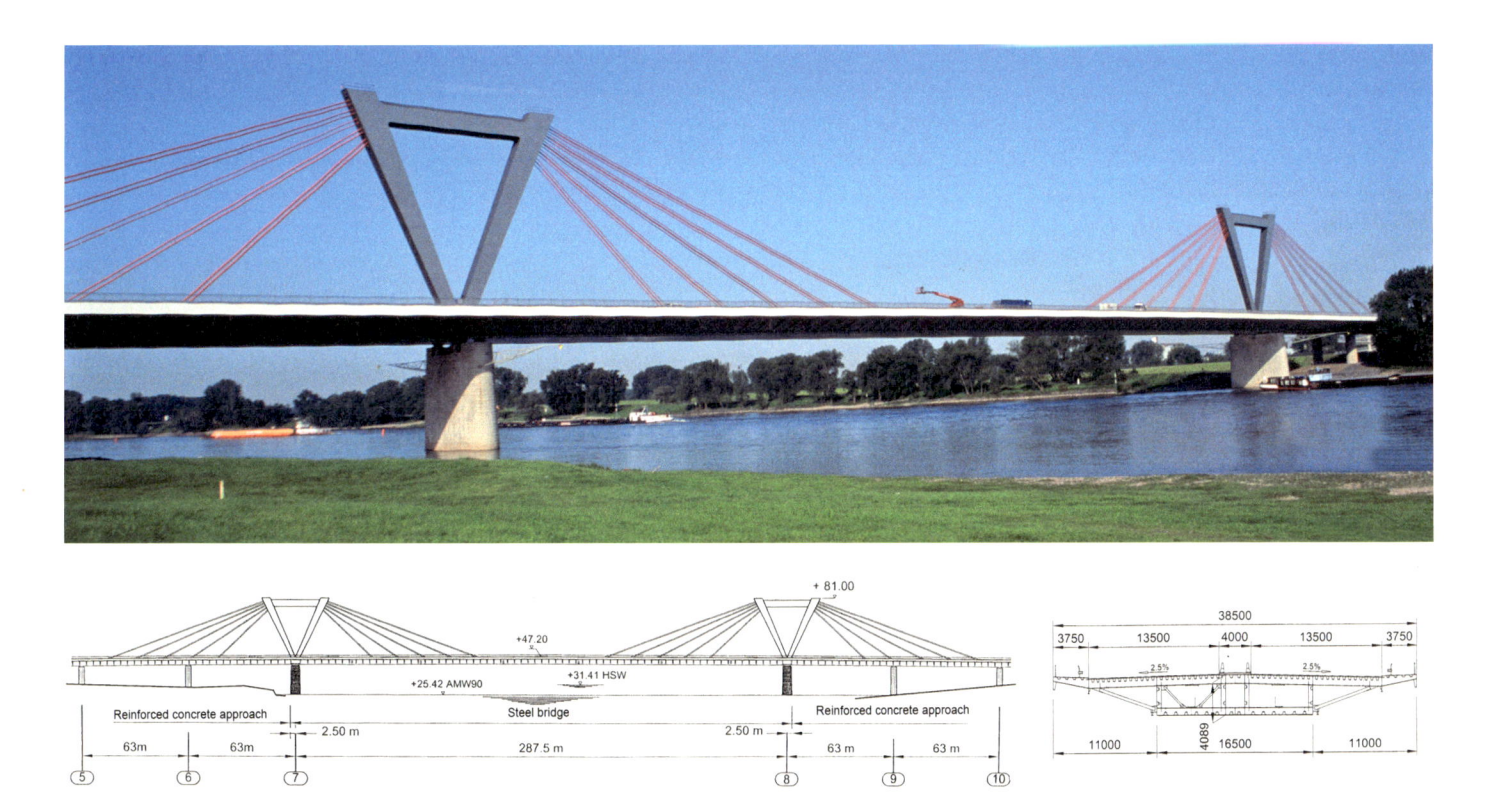

Figure 2.84 Rhine River Bridge at Ilverich, Germany, 2002, 287 m

with its fixed point at the tower. The constant beam depth of 3.80 m would mean a slenderness of 1:200 for a double bridge with two towers.

The stays each comprise 19 locked coil ropes with 60–90 mm diameter. Each rope is continuous from mainstay anchorage to backstay anchorage and is diverted in the tower on cable saddles. In the region near the tower the approach beam is provided with an additional inner central beam for the cable anchorages, Fig. 5.38.

The Rhine River Bridge at Ilverich, Fig. 2.84, is located near Düsseldorf airport and this limited the tower heights to 81 m. The river traffic required a horizontal clearance of 287.5 m which would normally require towers with a height of 110 m, Fig. 1.15. These two contradictory requirements, together with an economic minimum cable inclination, led to the unusual V-shaped tower [1.12].

The hybrid Rhine River Bridge at Wesel, with a total length of 772 m, has a steel main span of 335 m with just one tower, Fig. 2.85 [2.55, 2.56]. The 130 m high tower has an inverted Y-shape, similar to the Flehe Bridge. The stays use parallel strand cables for the second time in Germany after the Strelasund Bridge.

2.2.5 Special steel cable-stayed bridges

The first railway bridge with single forestays and backstays was opened in 1967 across the Neckar River at Untertürkheim, Fig. 2.86 [2.57]. The stays are welded I-sections.

The tower of the cable-stayed bridge at Ludwigshafen across a railway station, Fig. 2.87, is a four-legged trestle to which the 2·19 cables are anchored which carry the symmetric beam halves across a railway yard and a road intersection with spans of 140 m, a similar layout to the Neuwied Bridge. The deviation forces from the road curvature are carried by an A in the transverse direction, and the unbalanced traffic loads are carried by an A in the longitudinal direction [2.58].

The outstanding characteristic of the Bratislava Bridge, Fig. 2.88, is the backwards-inclined tower which carries a public restaurant at the top. This special impressive structure forms a contrast to the big mountain with a castle on top opposite to the tower. This unique bridge demonstrates the versatility in the design of cable-stayed bridges [2.59]. The 80 m high tower is A-shaped, and in its box legs are a high-speed elevator, safety stairs and all the service installations for the restaurant.

The two cable-stayed bridges – identical except for the foundations – across the Paraná between the cities of Zárate and Brazo Largo near Buenos Aires, Argentina, have a main span of 330 m and two side spans of 110 m. They are unique in so far as they are the first cable-stayed bridges to carry a full railway load. In addition to four traffic lanes, one railway track for the high Cooper 80 loadings is placed eccentrically at the upstream beam edge. This means that the bridge is strongly asymmetric in the transverse direction, Fig. 2.89 [2.60–2.62]. Other special characteristics of the bridges are:

- first cable-stayed bridge with steel beam and concrete tower
- first use of shop-fabricated parallel wire cables for a major bridge
- first use of large-diameter bore piles, up to 70 m deep, with possible scour depths of up to 13 m
- first use of hydraulic buffers for cable-stayed bridges for transferring the breaking forces equally to both towers
- first free cantilever construction from the tower to both sides using derricks on the deck, 50 m above water.

The engineer in charge for the design, construction engineering and supervision on site was Reiner Saul, Fig. 2.90.

The bridge beam has vertical bearings only at its ends; in between, the beam is only elastically supported by the stay cables in two outer planes at a distance of 22 m. The cables transfer the loads via steel tower heads into the concrete tower legs.

The cross-section comprises two outer triangular boxes, each 3.9 m wide, connected by cross girders and an orthotropic deck. The bottom flanges of the outside boxes are connected by a truss for improved torsional stiffness.

In the transverse direction the beam is supported by steel retainers at the anchor piers and by sliding neoprene bearings at the towers. Longitudinal forces, especially from breaking of the railway, are transmitted by hydraulic buffers equally into the cross beams of the two towers. For slowly acting longitudinal forces due to temperature the hydraulic buffers do not react, and the beam is centered by the stay cables only.

Due to the eccentric location of the railway, twin stay cables are used at the railway side and single cables on the other side. These asymmetric cable arrangements complicated the construction engineering and the erection itself.

The railway bridge across the Save River in Belgrade, Fig. 2.91, has a total length of 2000 m [2.63]. Dominating is the central cable-stayed bridge with a main span of 254 m. The beam consists of two outside box girders, connected by an orthotropic deck located 90 cm below the top flanges of the main girders, which carries the ballast and the two railway tracks.

The most startling bridge recently completed is the Alamillo Bridge across the Guadalquivir, which gives access to the Expo '92 in Seville, designed by the Spanish-Swiss architect and engineer Santiago Calatrava Valls, Fig. 2.92. The 200 m main span is supported by one central cable plane.

The bridge, Fig. 2.94, was designed in accordance with the request by the Alcalde of Seville: 'We want to build such a bold structure that future generations will wonder how we could dare to initiate it.' [2.64]. The bridge elevation was developed by Calatrava from the shape of a flying crane, Fig. 2.93. As a consequence the bridge comprises only one tower without backstays.

The main span dead load is balanced by the 30° backwards inclined tower while live loads in the main span have to be carried by bending in the tower into the foundations, a rather uneconomic solution for the pile foundations, given the bad soil conditions.

We build bridges

DSD Brückenbau GmbH

Henry-Ford-Str. 110 · D-66740 Saarlouis
Tel.: +49 (0) 6831 / 18-2661
Fax: +49 (0) 6831 / 18-2662
E-Mail: brueckenbau@dsd-steel.com
Internet: www.dsd-steel.com

Journals for the entire field of Structural and Civil Engineering

Beton- und Stahlbetonbau
Volume 107, 2012
Impact factor 2010: 0,265

Structural Concrete
Volume 13, 2012
Journal for fib members – International Federation for Structural Concrete

Stahlbau
Volume 81, 2012
Impact factor 2010: 0,234

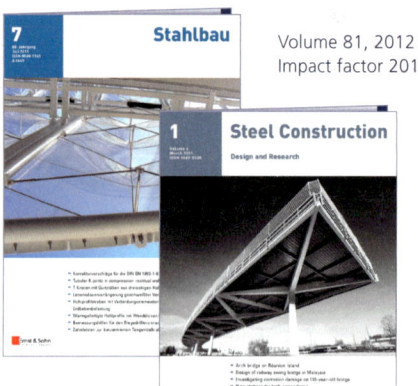

Steel Construction
Design and Research
Volume 5, 2012
Journal for ECCS members – European Convention for Constructional Steelwork

Bautechnik
Zeitschrift für den gesamten Ingenieurbau
Volume 89, 2012
Impact factor 2010: 0,141

UnternehmerBrief Bauwirtschaft
Fachzeitschrift für Führungskräfte der Bauwirtschaft
Volume 35, 2012

geotechnik
Volume 35, 2012
Official journal of the DGGT
Deutsche Gesellschaft für Geotechnik e. V.
German Geotechnical Society

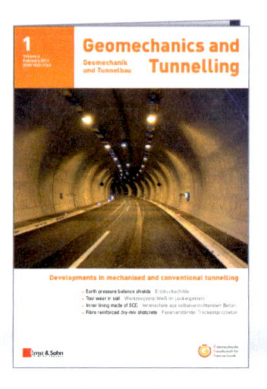

Geomechanics and Tunnelling
Geomechanik und Tunnelbau
Volume 5, 2012
Journal for ÖGG members
Österreichische Gesellschaft für Geomechanik

Mauerwerk
Zeitschrift für Technik und Architektur
Volume 16, 2012

Bauphysik
Wärme I Feuchte I Schall I Brand I Licht I Energie
Volume 34, 2012
Impact factor 2010: 0,173

Journal Online Subscription

All Ernst & Sohn specialist journals (from 2004 to the present) are available by online subscription.

WILEY ONLINE LIBRARY
www.wileyonlinelibrary.com

Order a free sample copy: www.ernst-und-sohn.de/journals

Ernst & Sohn
Verlag für Architektur und technische Wissenschaften GmbH & Co. KG

Customer Service: Wiley-VCH
Boschstraße 12
D-69469 Weinheim

Tel. +49 (0)6201 606-400
Fax +49 (0)6201 606-184
service@wiley-vch.de

Ernst & Sohn
A Wiley Company

2.2 Steel cable-stayed bridges

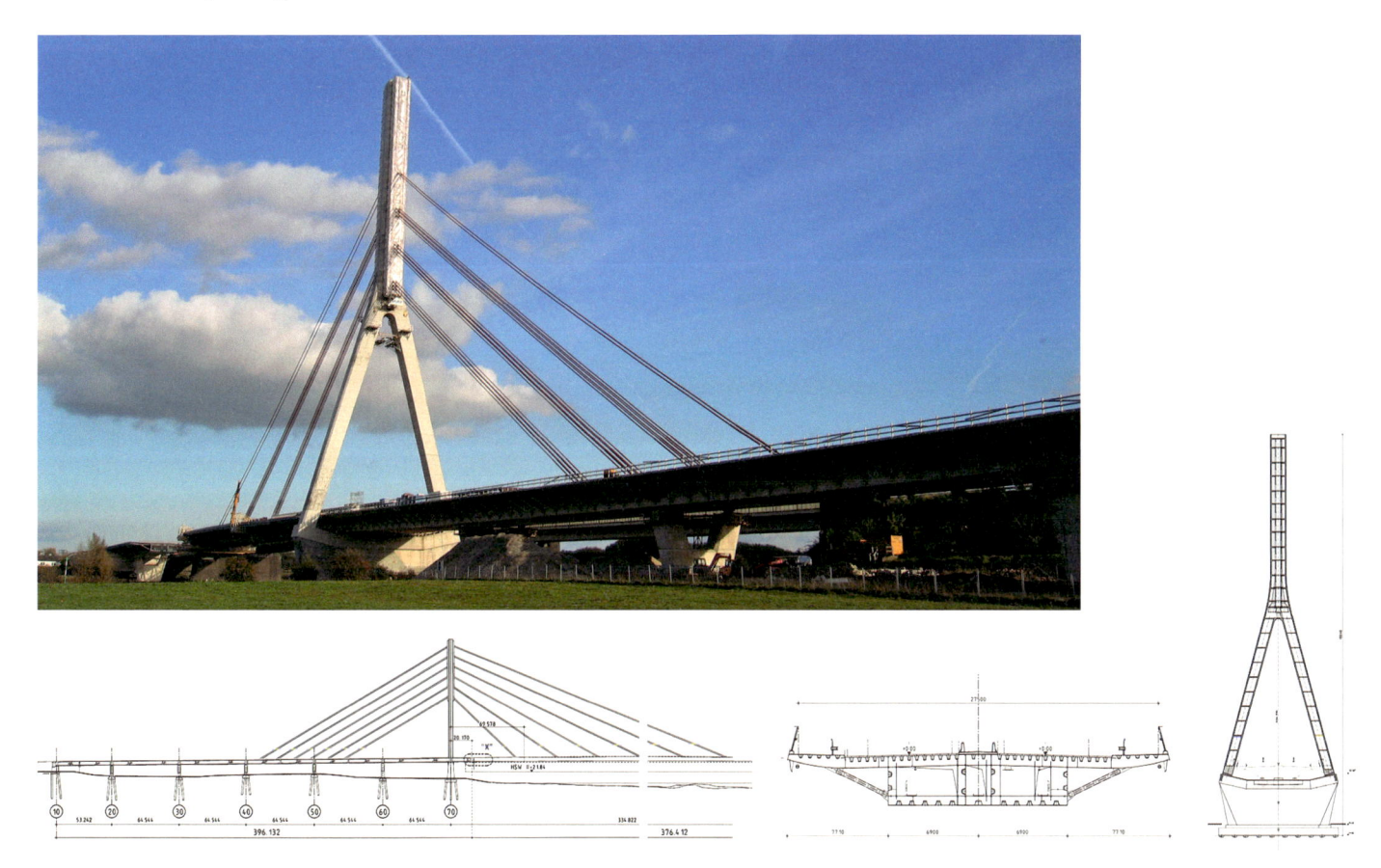

Figure 2.85 Rhine River Bridge Wesel, Germany, 2009, 335 m

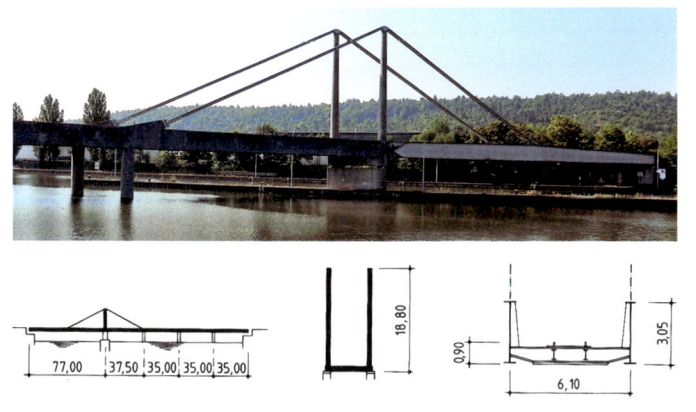

Figure 2.86 Neckar River Bridge, Untertürkheim, Germany, 1967, 77 m

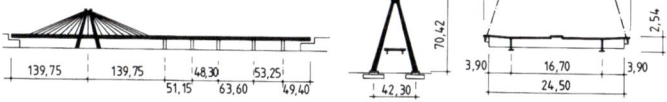

Figure 2.87 Railway station bridge, Ludwigshafen, Germany, 1968, 2 x 140 m

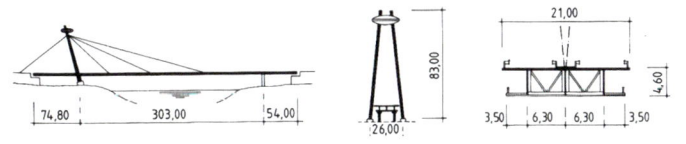

Figure 2.88 Bratislava Bridge, Slovakia, 1972, 303 m

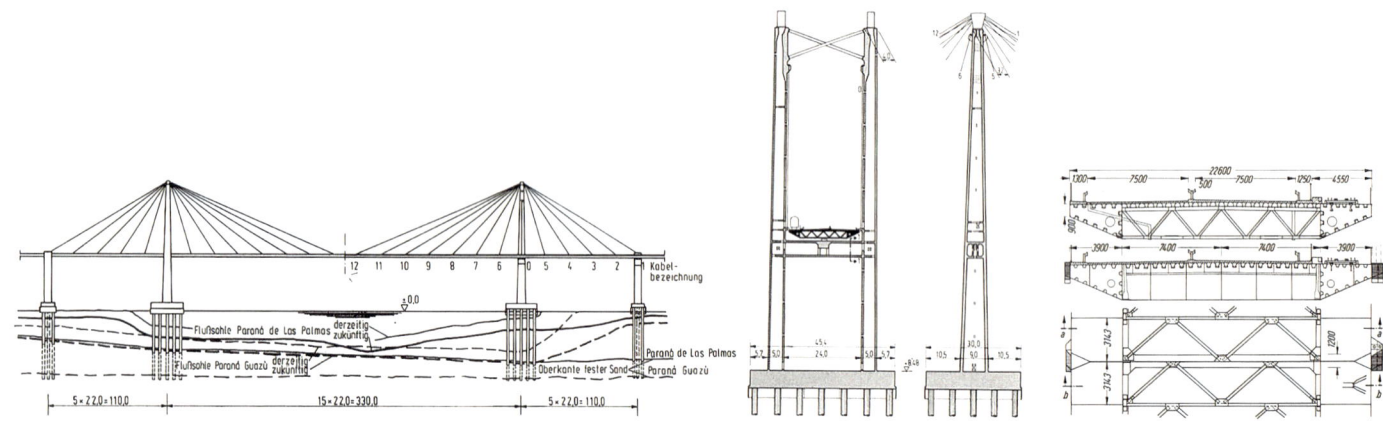

Figure 2.89 Zárate-Brazo Largo Bridge, Argentina, 1977, 330 m

Figure 2.90 Reiner Saul

Born 1938 in Lünen, Westfalen, Germany
1963 Dipl.-Ing. TH Hannover
1964–68 Hein, Lehmann AG
1968–2003 Leonhardt, Andrä and Partners
since 1995 Licensed Checking Engineer for Road and Rail Bridges
2003 Dr.-Ing. E. h. Braunschweig University
since 2004 Consultant Leonhardt, Andrä and Partners
Bonn North Bridge
Mannheim North Bridge
Zárate-Brazo Largo Bridge, Argentina
Double composite beams
Heinola Bridge, Finland
Stonecutters Bridge, Hong Kong
Kehl-Straßburg Pedestrian Bridge
Rosario-Victoria Bridge
Kap Shui Mun Bridge
Geo Geum Bridge, Korea

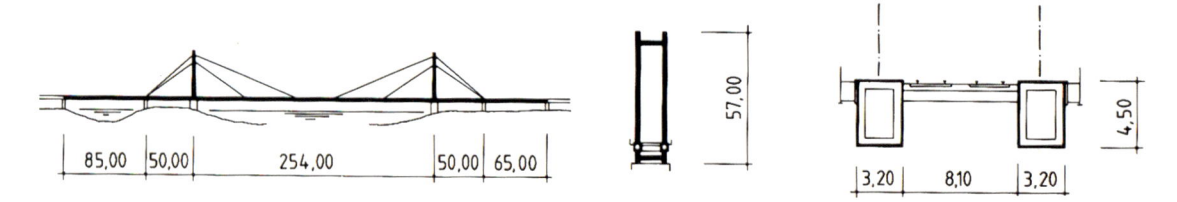

Figure 2.91 Save Railway Bridge, Belgrade, Serbia, 1977, 254 m

2.2 Steel cable-stayed bridges

In the same spirit the beam cross-section was developed from the frontal view of a bull's head, Fig. 2.93. The head became central box girder, the horns developed into the outside cantilevers.

The construction of the bridge, especially of the tower with its continuously changing cross-section, required high skills of the design engineers and artisans on site. The eventual success proved their outstanding ability. Even taking into account the inefficient flow of forces, Calatrava succeeded once more in designing a sculpture fulfilling high aesthetic requirements instead of a straightforward rationally engineered bridge.

The centerpiece of the 4100 m long second Strelasund Crossing is formed by the 600 m long cable-stayed steel bridge with a main span of 198 m and a single tower, Fig. 2.95 [2.65]. Parallel strand cables were used for the first time in Germany.

The bridge across the Sava River at the island of Ada Ciganlija in Belgrade, Fig. 2.96, has a main span of 376 m with a single tower. This 200 m high A-tower has partially spread legs through which the two central railway tracks pass, while three roadway lanes are located on each side, so that the beam has the remarkable width of 45 m. The two cable planes are partially spread as well, in line with the tower legs.

The main span has a steel beam while the side span is of concrete, but not supported on piers as had always been done before, for example for the Mannheim-Ludwigshafen Bridge, the Normandy Bridge and the Stonecutters Bridge. The dead weight of the concrete side span thus exactly balances the weight of the steel main span, while live loads in the main span are only carried back by the backstays.

Born 1951 in Valencia, Spain
1974 Architect Escula Tecnica Superior de Arquitectura
1979 Dipl.-Ing. ETH Zurich
1981 Dr.-Ing. ETH Zurich
since 1981 Independent Consulting Engineer and Architect, Zurich
Alamillo Bridge

Figure 2.92 Santiago Calatrava Valls

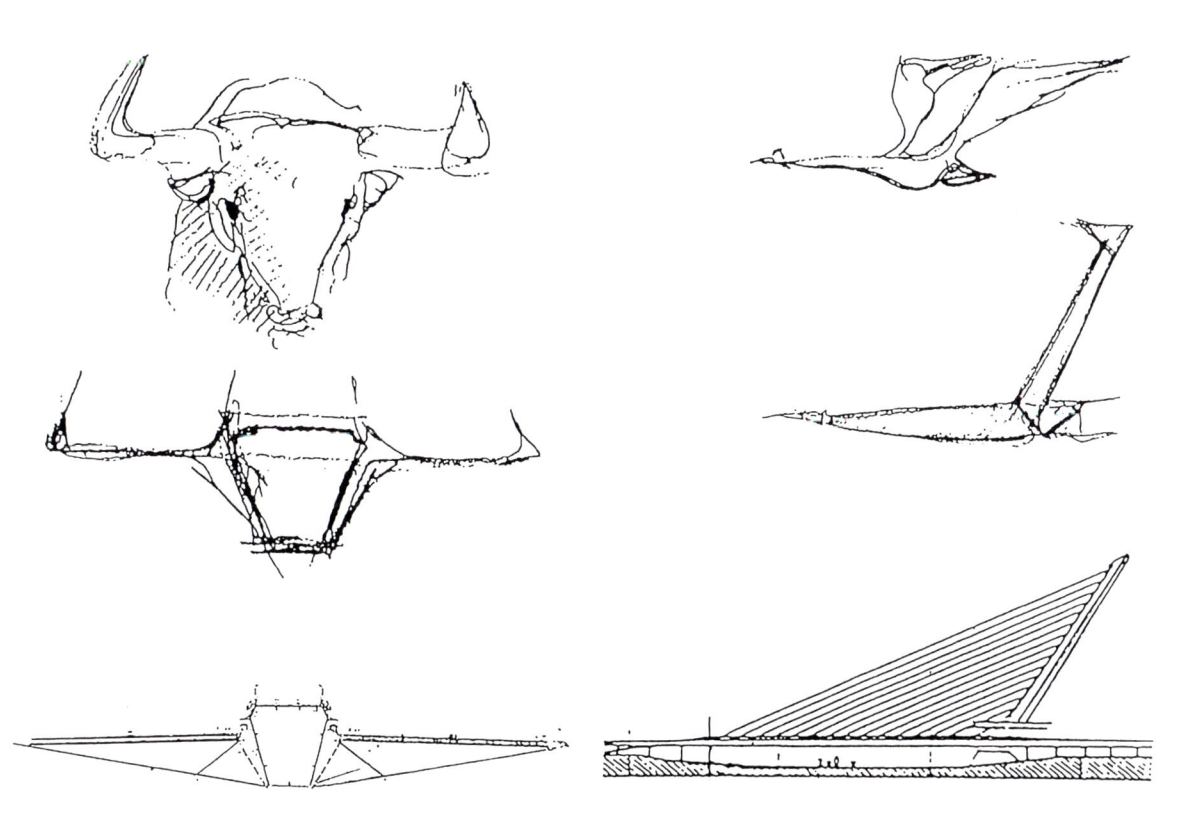

Figure 2.93 Development of bridge and beam shapes

74　　2 The development of cable-stayed bridges

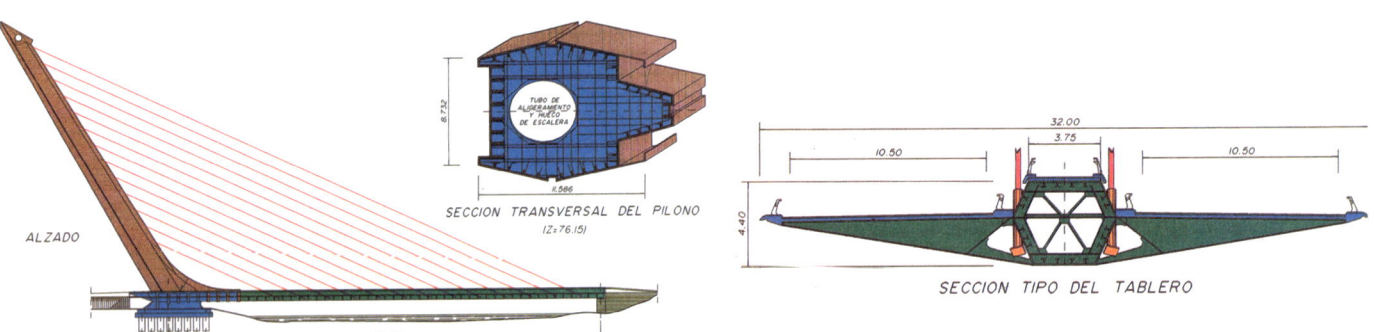

Figure 2.94 Alamillo Bridge, Spain, 1992, 200 m

2.2 Steel cable-stayed bridges

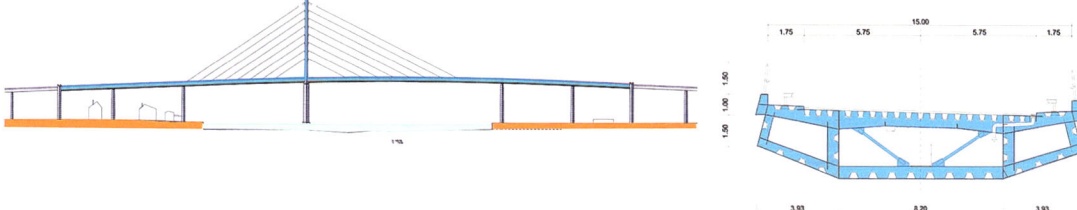

Figure 2.95 Strelasund Bridge, Germany, 2007, 198 m

Figure 2.96 Sava Bridge at Ada Ciganlija, Belgrade, Serbia, 2011, 376 m

2.2.6 Cable-stayed bridges with record spans

The record span lengths for steel cable-stayed bridges belonged to Rhine river bridges until 1975, Fig. 2.97, but from then on international bridges dominated.

The Rhine River Bridge in Duisburg-Neuenkamp with a record span of 350 m in 1907, Fig. 2.98, has a continuous beam between abutments which is supported by a central cable plane [2.66]. The central box of the beam has a width of 12.7 m in order to achieve the required torsional stiffness for the limitations of transverse inclination under live load, completed on each side by 11.8 m wide cantilevers.

The steel towers are fixed to the beam. In addition to compression they have to carry longitudinal restraint moments from cable elongations. Consequently, the towers were designed as slender as possible in order to keep the restraint moments small, Fig. 4.22. This resulted in the use of the special steel, N-A-XTRA 70, with a yield strength of 700 N/mm2 [2.67].

In 1975 the span record of 404 m went to the St Nazaire Bridge across the Loire in France [2.68]. The required vertical navigational clearance of 61 m led to a total bridge length of 3356 m in the flat terrain. Due to the nearby airport striped hazard painting appeared necessary, which gives the bridge its characteristic appearance, Fig. 2.99.

The approach bridges with lengths of 50.7 m were built from prestressed concrete girders with cast-in roadway slabs. The central steel box girder is continuous over the three stayed spans and has an aerodynamically shaped cross-section. The legs of the A-shaped steel towers with 2.0 m · 2.5 m boxes straddle the beam. The cables are anchored in 80 mm thick plates.

After six years of construction the Normandy Bridge across the Seine between Honfleur and Le Havre was opened for traffic in 1995, Fig. 2.100. It captured the record span with 856 m after the 404 m record had been held by the St Nazaire Bridge for 20 years [1.29].

The main span has a central 624 m steel part, the outer 116 m on each side being an extension of the concrete approach bridges. In this region of high compression forces around the towers they are more economic than a steel beam and, in addition, stiffen the main span for improved aerodynamic stability.

The continuous concrete girders of the approach bridges on a 6% grade were built by a special incremental lifting-launching system. The beam is a flat aerodynamically shaped box girder with a depth of 3.00 m and slenderness 1 : 285. The engineer in charge was Michel Virlogeux, Fig. 2.101.

In 1999 the span record of 890 m went to the Tatara Bridge in Japan, Fig. 2.102 [2.69], which was planned by the Honshu-Shikoku Bridge Authority. The bridge was originally foreseen as a suspension bridge but the costs for the required large abutments for anchoring the cable forces were too high and the design was changed to cable-stayed. Steel towers were selected in order to minimize the foundations in the water. The slender steel beam, only 26 m above the water, with a total length of 1330 m and a width of only 25.4 m (length : width = 1 : 35) is supported from the two 220 m high towers. A beam depth of 2.7 m (length : depth = 1 : 326) proved to be sufficient.

The Stonecutters Bridge, Fig. 2.104, became the second longest cable-stayed bridge in 2009. It is the new landmark of Hong Kong [1.22]. The bridge crosses the Rambler Channel between Stonecutters Island in the west of Kowloon, adjacent to the city center and is thus prominently visible. The concrete towers are 298 m high and comprise an upwards tapering slender circular cross-section. The beam consists of two separate box girders, which straddle the central towers. The responsible engineer was Ian Firth, Fig. 2.103.

The two separate box girders of the superstructure are connected by cross girders at cable distance, creating a girder grid. The monolithic concrete side spans are 70 m long. The steel beam of the main

Table 2.3 Record spans

Steel cable-stayed bridges	Record span length	Year completed
Rhine River Bridge Düsseldorf North	260 m	1957
Severins Bridge	301 m	1959
Knie Bridge	319 m	1969
Rhine River Bridge Duisburg-Neuenkamp	350 m	1970
St. Nazaire Bridge	404 m	1975
Normandy Bridge	856 m	1995
Tatara Bridge	890 m	1999
Stonecutters Bridge	1.018 m	2009
Sutong Bridge	1.088 m	2008
Russki Bridge	1.104 m	2012

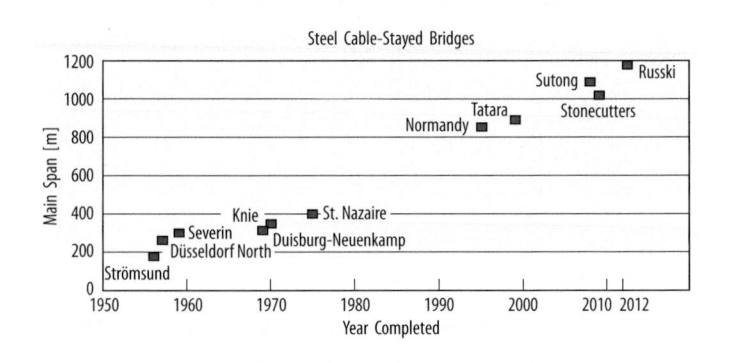

Figure 2.97 Record spans of steel cable-stayed bridges

2.2 Steel cable-stayed bridges

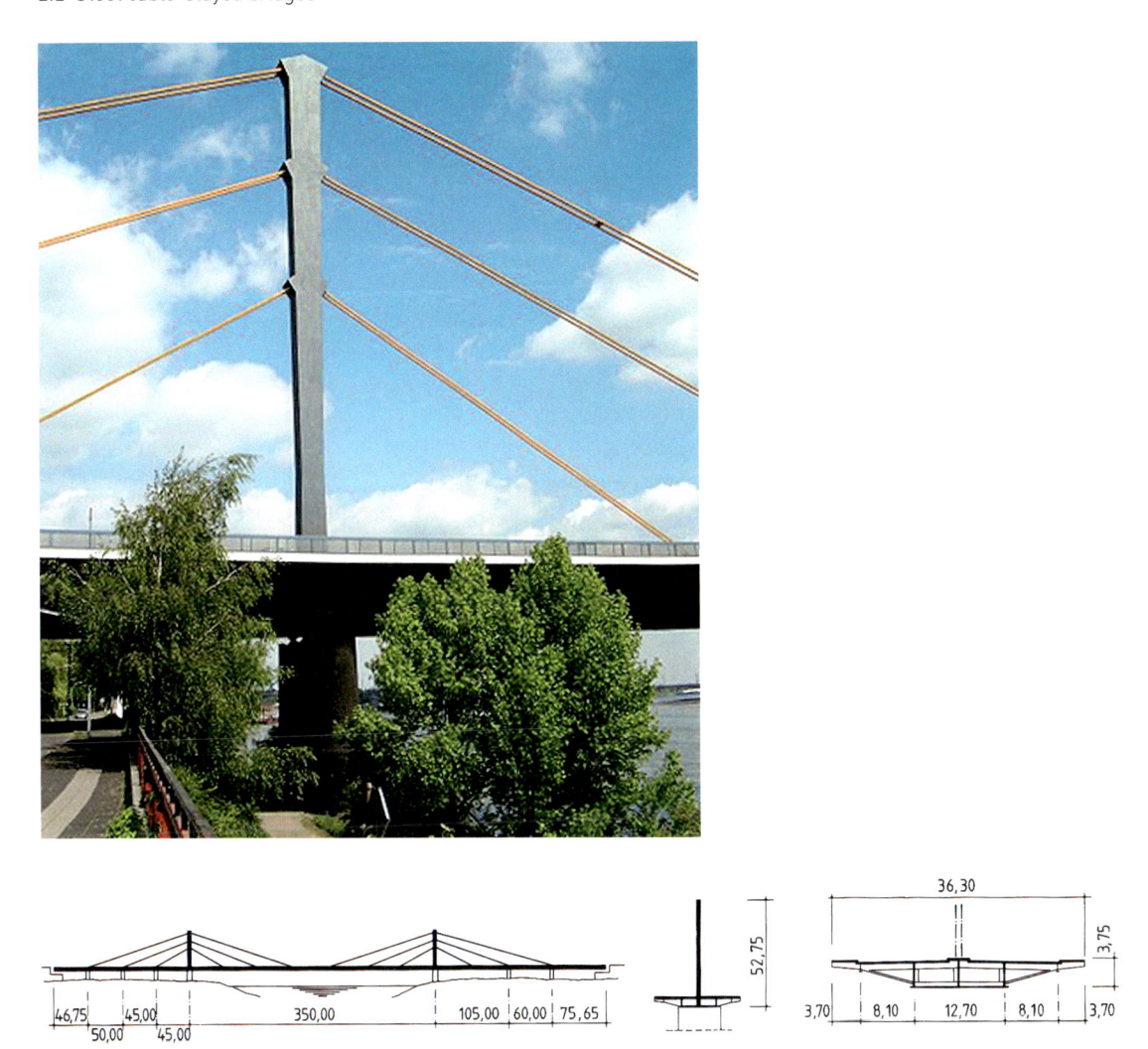

Figure 2.98 Rhine River Bridge Duisburg-Neuenkamp, Germany, 1970, 350 m

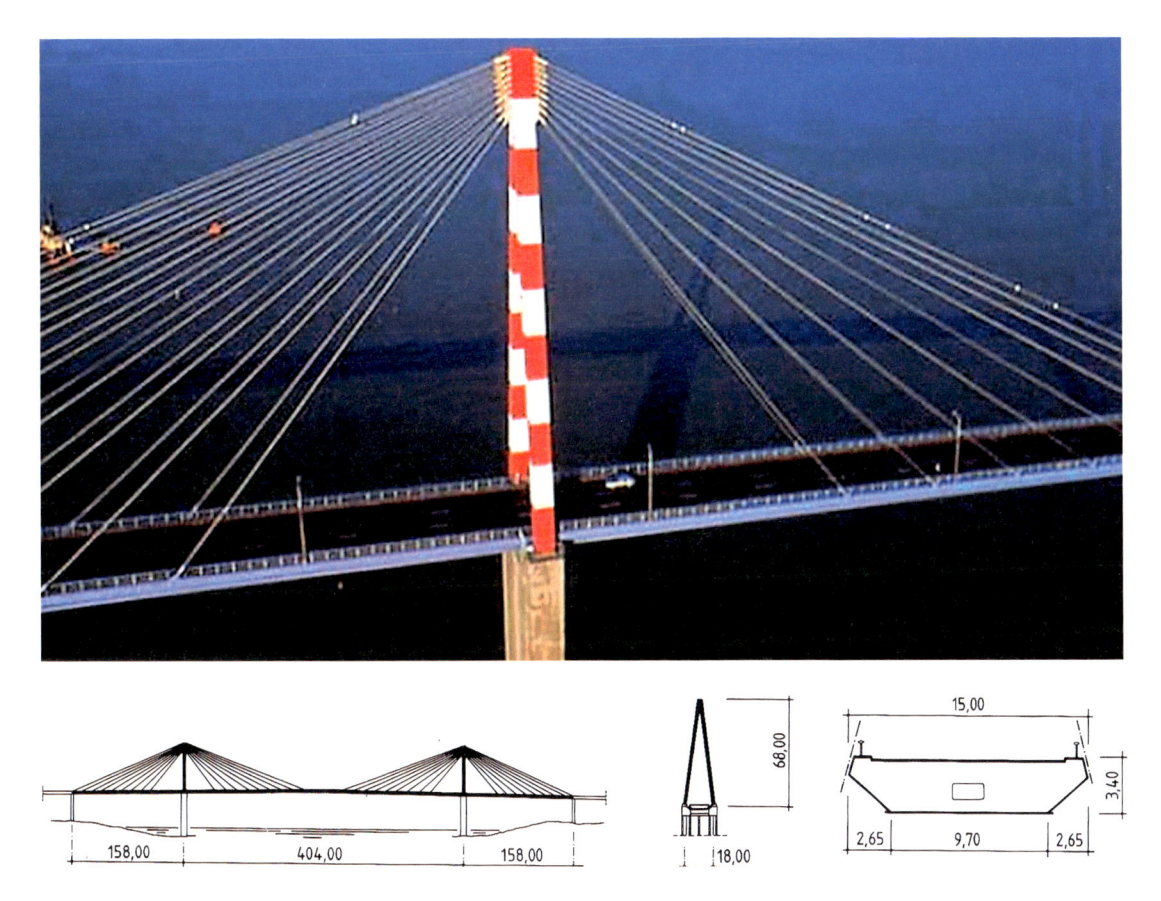

Figure 2.99 St. Nazaire Bridge, France, 1975, 404 m

Figure 2.100 Normandy Bridge, France, 1995, 856 m

Born 1946, La Flèche, France
1967 Ingénieur École Polytechnique
1970 Ingénieur École Nationale des Ponts et Chaussées
1970–73 Work in Tunisia
Doctor, Pierre and Marie Curie University (Sixth Arrondissement, Paris)
1974–94 SETRA, Director for major bridges
1977–91 Part time Professor for Structural Analysis at the École Nationale des Ponts et Chaussées
Independent Consulting Engineer
Normandy Bridge
Vasco da Gama Bridge
Millau Bridge

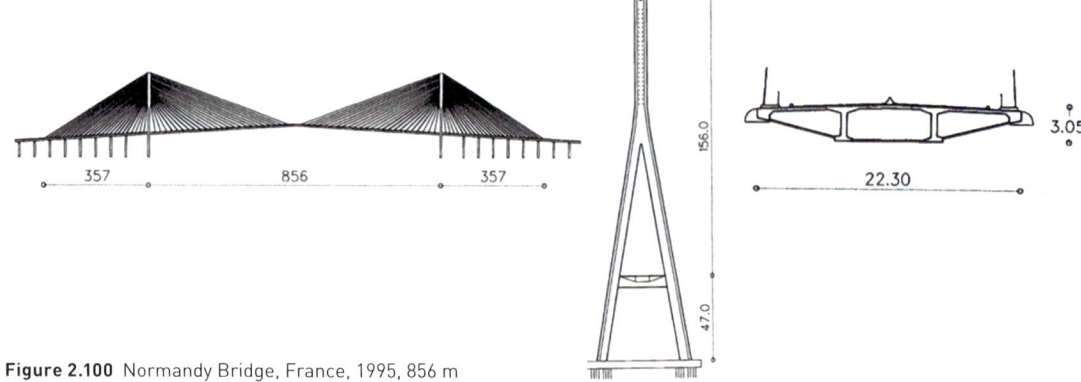

Figure 2.101 Michel Virlogeux

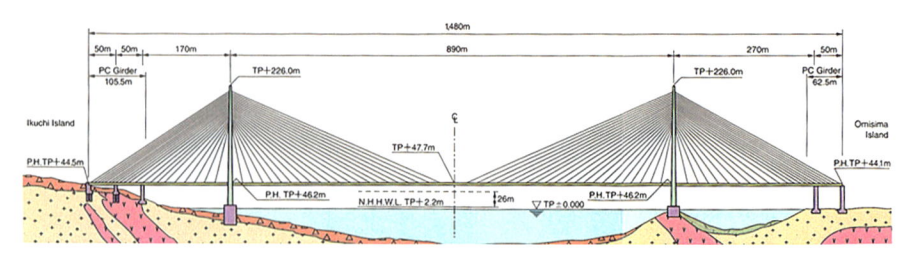

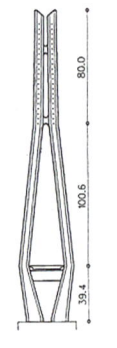

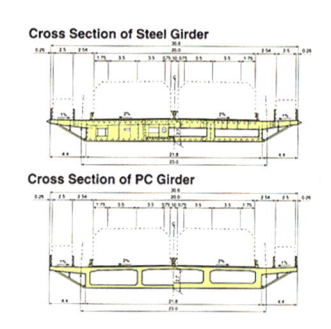

Figure 2.102 Tatara Bridge, Japan, 1999, 404 m

2.2 Steel cable-stayed bridges

Born 1956 in England
1979 BSc, University of Bristol
1982 MSc, DIC, Imperial College of Science
and Technology
1981–92 Wye Bridge, UK
1993–97 Kap Shui Mun Bridge, Hong Kong
1993–2000 Ting Kau Bridge, Hong Kong
1996–97 Poole Harbor Bridge, UK
since 1996 Erskine Bridge, Glasgow
1997–2000 Lockmeadow Bridge, Maidstone
1999–2000 Stonecutters Bridge, Hong Kong
2002–2003 Swansea Sail Bridge

Figure 2.103 Ian Firth

Figure 2.104 Stonecutters Bridge, Hong Kong, China, 2009, 1018 m

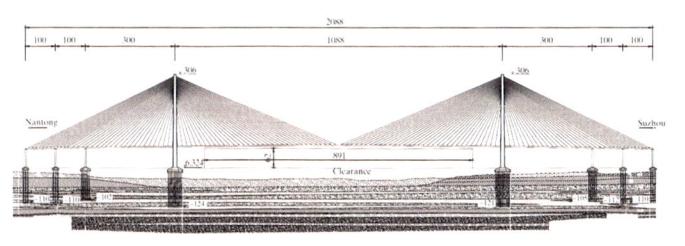

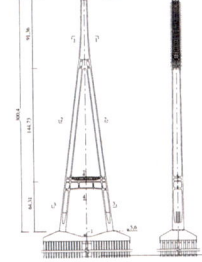

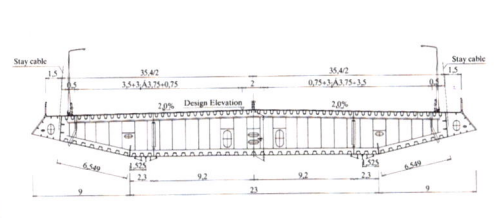

Figure 2.105 Sutong Bridge, China, 2008, 1088 m

Figure 2.105a Russki Bridge, Russia, 2012, 1104 m

span protrudes 50 m into the side spans where they join the concrete spans. The beam is supported by 8 · 28 parallel strand cables, which converge into steel boxes cast into the concrete shafts at the tips of the towers.

The Sutong Bridge in China, Fig. 2.105, with a main span of 1088 m and side spans of 300 m + 2 · 100 m is the current record-holder for cable-stayed bridges [1.21], crossing the Yangtze River about 100 km upstream of Shanghai. The 4 m deep steel box girder displays, together with the two outer cable planes, a high torsional resistance resulting in a good aerodynamic stability.

The two 300 m high concrete box towers have the simple shape of a straight A which, in the traditional Chinese culture, reflects the harmony between heaven and earth.

The Russki Bridge (Eastern Bosporus Stait crossing) with the new record span of 1104 m will connect the City of Vladivostok in Russia with Russki Island in 2012, Fig. 2.105a [2.138].

2.3 Concrete cable-stayed bridges
2.3.1 General

Since the construction of the Maracaibo Bridge in 1962, concrete cable-stayed bridges have achieved more and more acceptance. Today they form the main system for spans from about 200 m onwards. Their economic upper span length is constantly being pushed upwards. In 1973 a contractor's alternate with a lightweight concrete beam was proposed for the Rhine River Bridge Flehe with a main span of 365 m and one tower only – structurally comparable to a main span of about 730 m with two towers [1.16].

The cable-stayed bridges up to 1982 are shown in [2.70], and later examples can be found in [2.71] and [1.1].

For spans up to 400 m concrete beams were often economically superior to steel beams. One example is the Dame Point Bridge, FL, USA, with a main span of 396 m which was bid for $ 64.8 m in concrete against the $ 84.8 m in steel. Even the East Huntington Bridge, WV, USA, with a main span of 274 m (one tower only!) was bid for $ 23.5 m in concrete against $ 33.3 m in steel. Both bridges were tendered in steel and concrete. The steel alternates, however, used more steel than usual in Europe. The reason for the economic superiority of concrete is that towers and beam are mainly loaded in compression, and that a compression force can be carried less expensively in concrete than in steel. Also, sub- and superstructure can be built by the same contractor.

A portion of the cost advantage of the concrete itself is consumed by its higher weight which requires additional cable steel. Reasonable foundation conditions are required to carry the higher weight of concrete beam and towers into the ground without excessive additional costs for extra piles.

The forces and deformations of the beam are reduced advantageously due to its higher weight and corresponding stiffer cables. This is especially important for railway bridges, which otherwise show such large rotations at the bridge ends that extensive rail expansion joints are required.

The development proceeded from a few cables with stiff beams to multi-stay systems with slender beams for concrete cable-stayed bridges. Extreme examples are the Wadi Kuf Bridge in Libya, whose 282 m main span with a depth of 4.00 m to 5.90 m is only supported by two pairs of cables and the Helgeland Bridge in Norway with a main span of 425 m and a beam depth of 1.20 m, supported by many stays. The fan arrangement for the stay cables is more widespread than the harp arrangements.

In general, towers are fixed to their foundations but flexible in the longitudinal direction with their tips stabilized with backstays against the anchorage piers. Exceptions may be found in series of cable-stayed bridges for which the tower stiffness is achieved by the use of longitudinal A-frames for the towers, for example the Maracaibo Bridge. In the transverse direction, early cable-stayed bridges used mainly vertical tower legs, whereas currently A-towers dominate for larger spans.

2.3.2 Development of concrete cable-stayed bridges

The first modern cable-stayed bridge across the lock of Donzière in France, Fig. 2.106, nearly went unnoticed. It was designed by the French engineer for hydraulic concrete structures Albert Caquot (1881 – 1976), Fig. 2.107. He used high-strength steel tendons to support the stiff concrete trough roadway and anchored the stay cables at the beam in compression in order to save the heavy soil anchorages required for suspension bridges [2.72]. His grandson Kerisel [2.73] explains the ideas of Caquot to found modern stay cable bridge construction by stressing the high-strength stays. In 1995 Fritz Leonhardt in a letter to Kerisel accepted this view with his statement, 'I have to correct myself'.

It attracts attention in that, for the Donzière Bridge, two cross girders are placed adjacent, in the bridge center so that the front forestay cables are anchored with no free length between them. The end spans cantilever over the end supports. All stay cables are anchored in cast steel sleeves with adjustable support nuts. Caquot apparently followed Torroja's ideas by using a beam with balanced half main and side span. Instead of using hydraulic jacks to lift the tower heads to stress the stay cables, Caquot cast the beam in a super-elevated position and stressed the stay cables by lowering the beam into final position.

The structural system for the second Main River Bridge at Hoechst, Fig. 2.108, was chosen in accordance with the requirements of the client that the 148 m main span should only be extended by a 94 m approach bridge at the south side. Taking into account this rather short stay cable anchorage region, only a part of the main span was tied back to the tower. At the north side, the 26 m long approach bridge was extended into the main span to carry the loads back to the north approach but not across the river to the south approach. In order to carry the cantilever moments the beam was supported over the north pier by a so-called 'concrete sail' which extends the webs upwards by 6 m. The remaining part of the main span was hung into the two 50 m high tower legs with $4 \cdot 13$ stay cables within the plane of the webs, Fig. 2.108 [1.27].

The cross-section resulted from the client's request for a central area for one track of railway and utility pipes and 9 m wide roadways at the outside. In accordance with these three uses, the two planes of stay cables in a partly spread position between the roadways and the railway track were selected. The box underneath the bridge deck is continuous over the full length of the bridge. The webs together with the roadway slab and the bottom slab form a 2.7 m deep and 8 m wide box with sufficient stiffness in bending and torsion.

Cross girders at 3 m distance carry the roadway slab longitudinally. Secondary longitudinal girders at the outside of the cantilevers improve the distribution of concentrated loads. The responsible engineers for the Hoechst Bridge were Ulrich Finsterwalder, Fig. 2.109, supported by Herbert Schambeck, Fig. 2.133.

The first concrete cable-stayed bridges with stiff towers were designed by Morandi, Fig. 2.198. His influence on the design of the Chaco-Corrientes Bridges in Argentina, Fig. 2.110, is obvious [2.74].

The stiff A-towers support the beam together with inclined piers and two fore- and backstays on each side. Morandi's heavy cross girders at the cable anchorages are avoided by the use of two longitudinal box girders to which the stay cables are anchored directly. The towers and the inclined piers were CIP together with a 53 m long beam segment, and the remainder of the beam consists of precast elements. Drop-in girders in the center cater for the longitudinal beam deformations.

The bridge across the Danube canal in Vienna, Fig. 2.111, with spans of 55 m + 119 m + 55 m crosses a freeway and the canal with an angle of 45° [2.75]. Due to this skew arrangement, combined with the required navigational clearance and the walkways and feeder roads on both sides it was impossible to find an economic scaffolding in the final location of the bridge.

The many restraints with regard to location and elevation led to much reduced pier distances. By making use of the bridge ends which remained accessible during the full construction period a

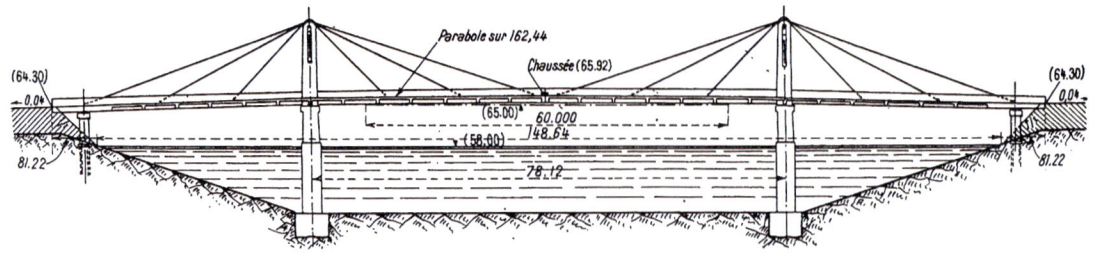

Figure 2.106 Donzière-Montragon Bridge, France, 1952, 78 m

Born 1881 in Vouziers, Ardennes, France
Diploma École des Ponts et Chaussées
Important hydraulic concrete structures
1952 Donzière Bridge
Died 1976 in Paris

Figure 2.107
Albert Caquot

Born 1897 in Munich, Germany
1923 Dipl.-Ing. Munich University
1923 Dyckerhoff & Widmann
1930 Dr.-Ing. Munich University
since 1933 Head of Design Office
Dyckerhoff & Widmann
Outstanding concrete structures
Prestressing bar system
Free cantilevering
Second Main River Bridge Hoechst
Died 1988 in Munich

Figure 2.109
Ulrich Finsterwalder

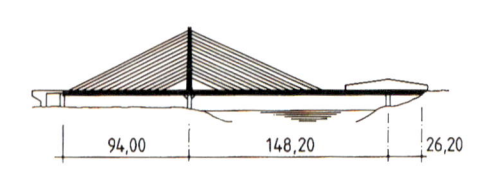

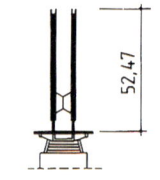

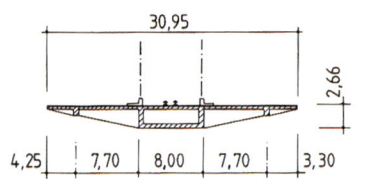

Figure 2.108 Hoechst Bridge, Germany, 1972, 148 m

2.3 Concrete cable-stayed bridges

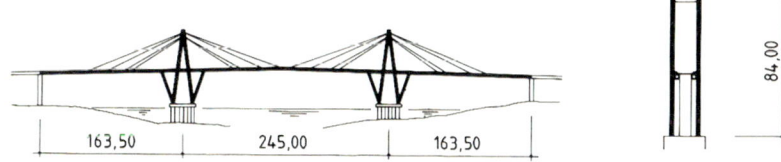

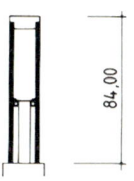

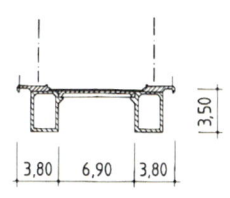

Figure 2.110 Paraná River Bridge between Chaco and Corrientes, Argentina, 1973, 245 m

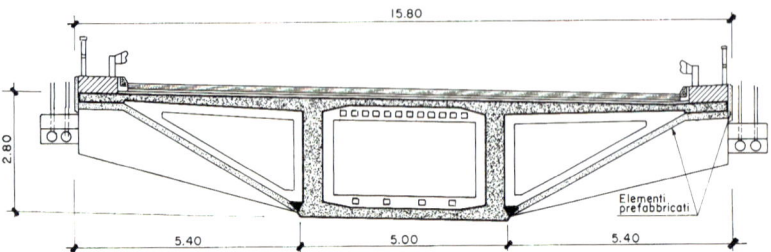

Figure 2.111 Danube Canal Bridge Vienna, Austria, 1975, 119 m

Figure 2.112 Danube canal bridge Vienna, Austria; casting on scaffolding

Figure 2.113 Beam rotation

Born 1930, Gmünd, Germany
1955 Dipl.-Ing.
since 1964 Independent Consultant
1975 Danube Canal Bridge Vienna
since 1982 Professor Vienna University

Figure 2.114 Alfred Pauser

solution was found to construct the bridge initially parallel to the Danube canal, Fig. 2.112, and then to rotate them into final position, Fig. 2.113. The gradient and the navigational clearance permitted a constant beam depth of just 2.8 m, resulting in a slenderness of 1:42. The responsible engineer was Alfred Pauser, Fig. 2.114.

The special characteristics of this bridge rotation are:
- the beam was supported by the final stay cables during construction
- the shifting of the whole bridge was not required and the rotational bearings could use large areas with low pressure.

The Pasco-Kennewick cable-stayed Bridge, Fig. 2.115, shows the following special characteristics [1.15]:
- main span about 300 m (a world record at the time of design) beam depth 2.13 m (1:140)
- beam built from precast elements with a weight of 270 t
- shop-fabricated parallel wire stay cables with HiAm-anchorages within cement-grouted PE-pipes
- design for earthquake
- construction by free cantilevering from the towers to both sides. For further details, see Section 6.1.2.

In 1981 the Sancho-El-Mayor Bridge across the Ebro River near Castéjon was completed, Fig. 2.116 [2.76]. The bridge comprises a main span of 146 m and a single tower with one cable plane in the center of the cross-section. The backstays, carrying the weight of the main span, are anchored in two counterweights on the two sides of the freeway. This arrangement creates three separate cable planes: the forestays in the center and the backstays in two outside inclined planes.

The horizontal components of the backstays are transferred by two buried concrete compression beams from the counterweights to the beam of the main span, meeting at the tower. The compression forces are thus balanced within the cable-stayed bridge. The concrete box girder of the main span has a depth of just 2.1 m and consists of precast elements with lengths equal to the cable distances of 6.4 m.

As with many other Spanish cable-stayed bridges the Sancho-El-Mayor Bridge gives the impression of an aesthetically pleasing sculpture.

The East Huntington Bridge, WV, USA, Fig. 2.117, crosses the Ohio River with a main span of 274 m and a single tower. This is a record span length; a structurally equal bridge with two towers would have a main span of about 450 m [2.77], which was actually exceeded only in 1991 by the Skarnsundet Bridge, Fig. 2.145.

The design of this concrete alternate was only commissioned after the design for a steel alternate was completed and the two main piers were already built. The problem to find a concrete alternate to fit the piers for the lighter steel alternate was solved by the use of steel cross girders and high-strength concrete B 56 for the precast beam elements and B 42 for the tower. (At that time only concrete B 35 was used for bridges in the USA.) This concrete alternate was bid 29% less expensive than the steel alternate with an orthotropic deck. The precast elements with a weight of 270 t were erected with a floating crane, free-cantilevering to both sides of the tower. This bridge is dealt with in detail in Section 6.1.3.

The Penang Bridge carries a six-lane highway to the island of Penang in the north-west of Malaysia. The concrete main span has a length of 225 m with two 101 m high towers, Fig. 2.118 [2.78]. Ship traffic required a navigational clearance of 33 m.

The bridge shows the characteristics of a design by Ulrich Finsterwalder, Fig. 2.109, and is similar to the second Main River Bridge Hoechst, Fig. 2.108. It uses symmetric side and main spans with lengths of 107.5 m and 2 · 112.5 m with stiff towers and parallel bar stay cables, see Section 2.3.2.

The bridge across the Paraná River between the cities of Posadas and Encarnación connects Argentina with Paraguay with a total length of 2550 m, Fig. 2.119 [1.24].

It comprises two long approach bridges with twin concrete box girders and spans of 55 m. The main cable-stayed bridge has spans of 115 m, 330 m and 115 m. The main girder is a three-cell concrete box beam, and the concrete towers have A-shape. The beam consists of precast elements with a weight of 300 t and was built by free cantilevering from the towers outwards to both sides, using a derrick located on the beam. One approach bridge was built by incremental launching, the other by self-launching formwork.

Unusual conditions were the eccentric one-track railway line and the variations in water level of up to 10 m. Special wind tunnel tests were executed for the high tornado wind speeds [2.79].

The Helgeland Bridge is a very slender concrete cable-stayed bridge with a main span of 425 m, Fig. 2.120 [2.80]. The aerodynamically shaped beam has a depth of only 1.2 m, resulting in a vertical slenderness of 1:355, and a width of only 12 m, giving a horizontal slenderness of 1:35. The towers are founded on rock in up to 30 m deep water. The bridge is located at the west coast of Norway near the Arctic Circle and is exposed to regularly occurring severe storms with gusts up to 77 m/s. This largely governing wind loading was investigated by a time-history analysis in the ultimate stage, taking into account the aerodynamic damping as well as geometric and material non-linearities. The bridge was built CIP by free cantilevering from the towers outwards to both sides and was opened in 1991 after a construction period of only two years. For further details see Section 6.2.

The Nordhordland Bridge across the Salhus Fjord in Norway is the only floating bridge with a cable-stayed span for the navigational channel. The concrete bridge has a main span of 163 m, Fig. 2.121 [2.81]. At the transition pier the main bridge is connected to the floating bridge, which has a steel box girder supported by buoys from lightweight concrete boxes at distances of 55 m.

The Leven River Bridge in Scotland near the city of Glenrothes, Fig. 2.122, is a slender, slightly curved cable-stayed concrete bridge with a main span of 115.2 m and a single tower. The beam is not post-tensioned, either longitudinally or transversely. The restraint

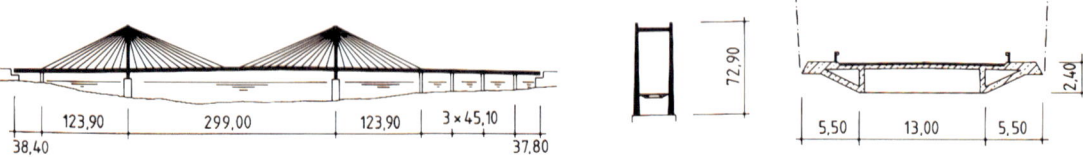

Figure 2.115 Pasco-Kennewick Bridge, USA, 1978, 300 m

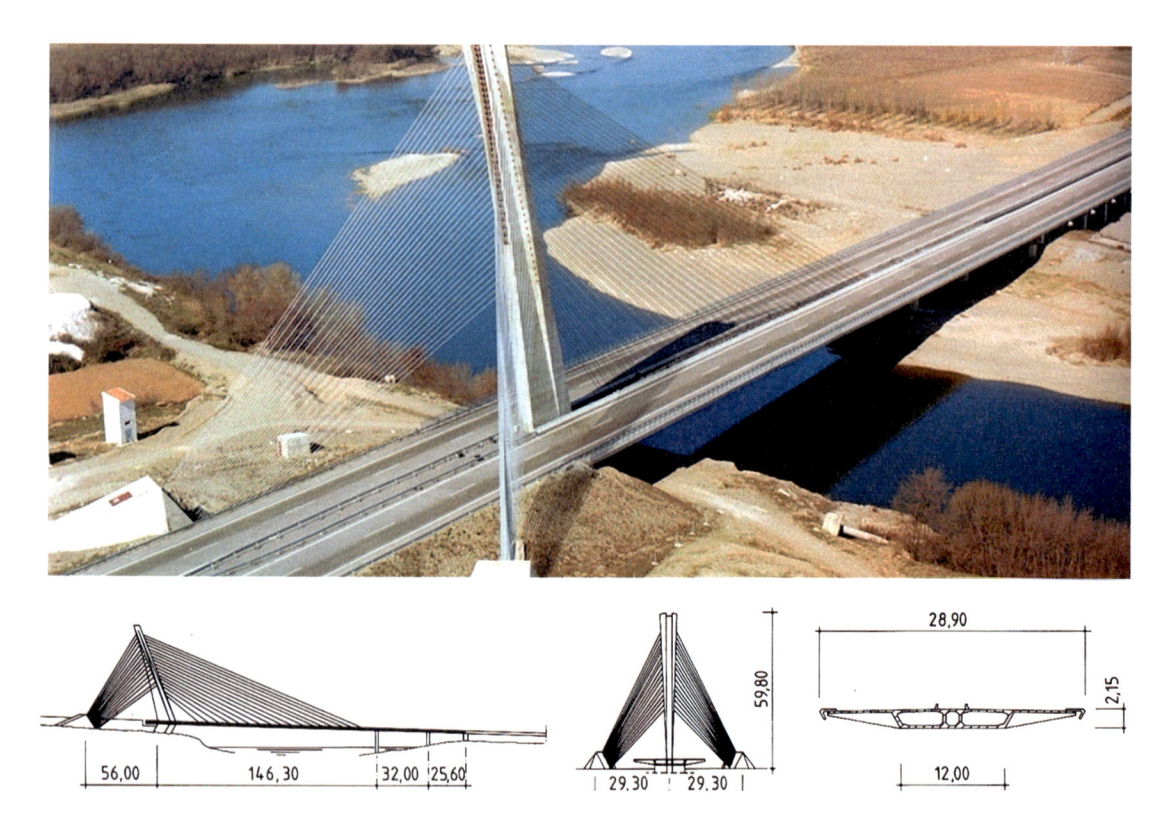

Figure 2.116 Sancho-El-Mayor Bridge, Spain, 1981, 146 m

moments from live load are reduced due to the cracking of concrete. The slender concrete A-tower is asymmetric transversely to reflect the beam curvature. The cables are anchored in a steel box between the two tower legs. The beam was CIP on scaffolding extending over the full length of the bridge, and lifted out of the forms by stressing the stay cables [2.82]. For further details see Section 5.1.3.2.

The second bridge across the Panama Canal, Fig. 2.123, has one central cable plane with a main span of 420 m and a total length of 1052 m [2.83]. The 24.1 m wide concrete box girder with cantilevers is located 80 m above the canal to ensure the passage of the largest ship's bridge passing the canal even after widening. The bridge carries three lanes in each direction.

The design and construction period was limited to two years for political reasons. In order to fulfill this extremely short time span, design and construction took place in a 'fast-track contract'. The tender documents were thus based on a preliminary design, the detailed design was prepared during construction. This arrangement is usual in Germany, but quite unusual in the Americas.

The main bridge has spans of 200 m, 420 m and 200 m and a tower with a height of 94 m above the deck. The fan-arranged stay cables have distances of 6 m at the beam and 1.5 m at the tower. The 4.5 m deep box girder with inclined webs and 9 m wide cantilevers supports the roadway slab on up to 1.55 m deep cross girders at 6 m distance. The towers taper in both directions. For aesthetic reasons the longitudinal walls are curved and structured by vertical grooves, and the front walls are distinctly set back. The cables are anchored within the box cross-section of the towers, the horizontal components being covered by loop tendons. The difficult foundation conditions are described in [2.84].

The Paraná River Bridge between the provinces of Rosario and Viktoria in Argentina has a total length of 3350 m required by the strong variation of the water level. The main bridge has a span of 330 m, the approach bridges have spans of 60 m, Fig. 2.124 [2.85]. The cross-section for three lanes in each direction of travel is a T-beam with a width of 22.8 m.

The piers in the river are provided with independent protection against the collision of 100 000 DWT ships. The piers and the protections rest on composite bore piles with 2.05 m diameter. For further details, see Section 4.4.3.5.

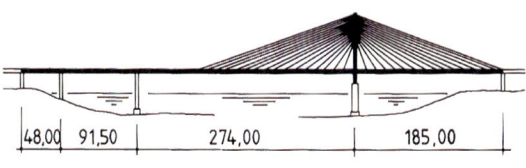

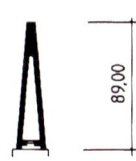

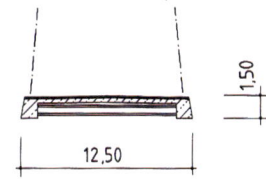

Figure 2.117 East Huntington Bridge, USA, 1985, 274 m

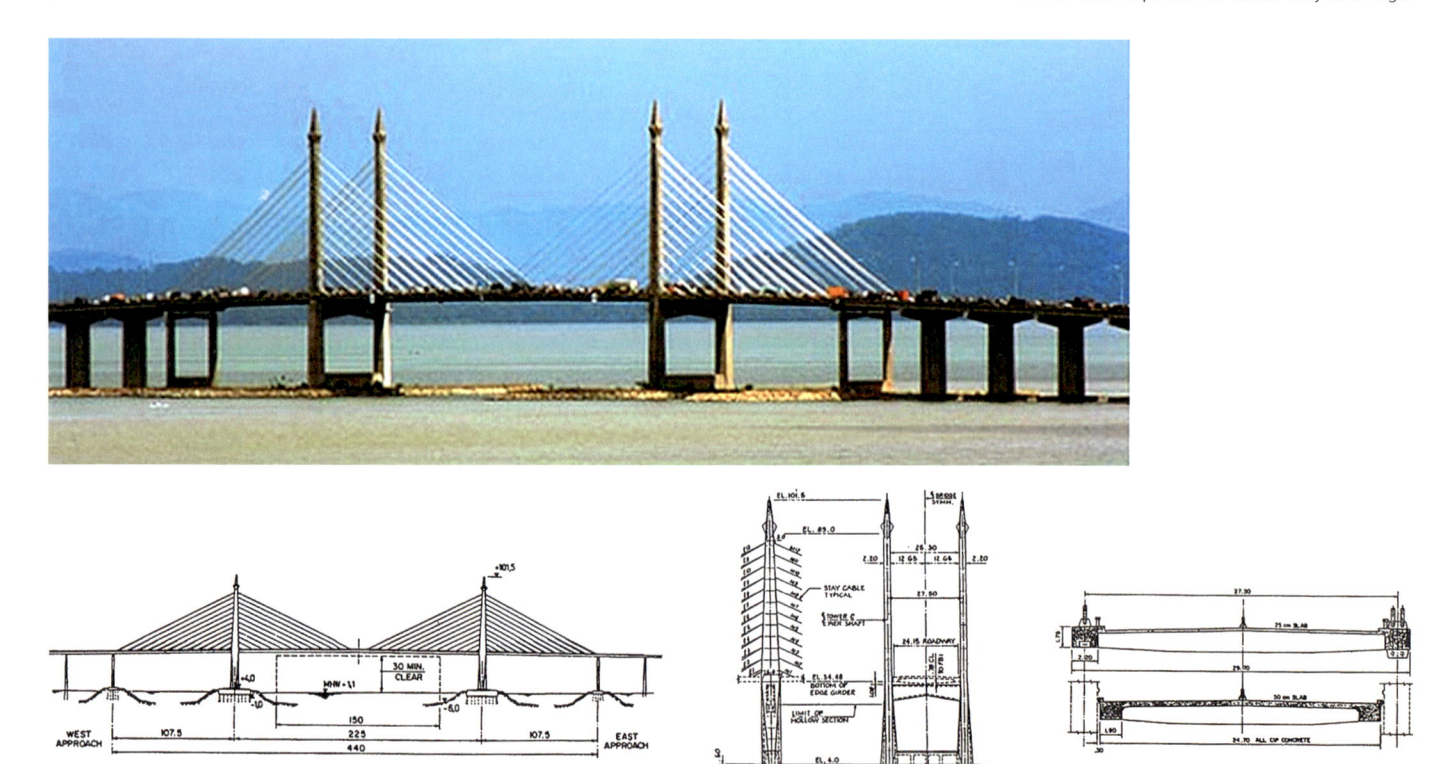

Figure 2.118 Penang Bridge, Malaysia, 1988, 225 m

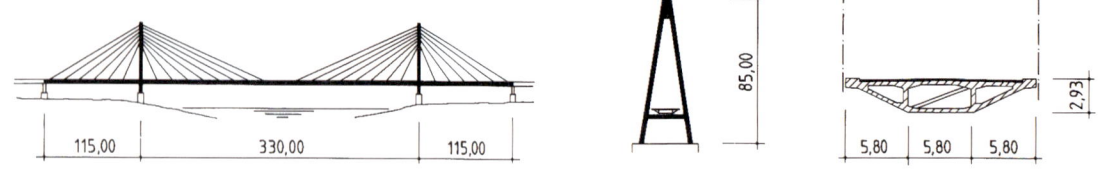

Figure 2.119 Paraná River Bridge Posadas-Encarnación, Argentina, 1987, 330 m

2.3 Concrete cable-stayed bridges

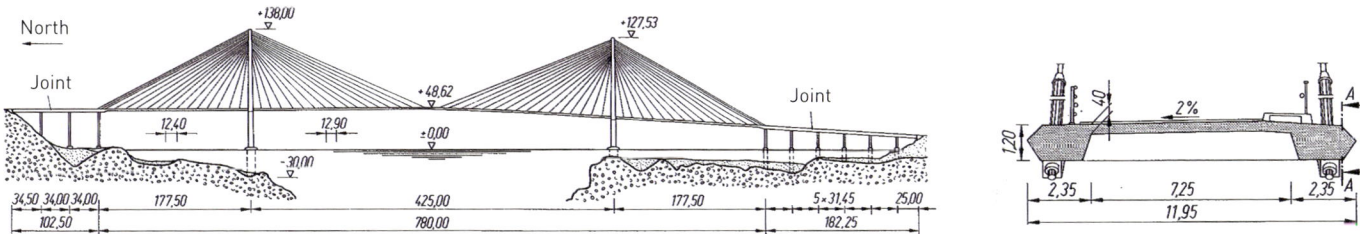

Figure 2.120 Helgeland Bridge, Norway, 1991, 425 m

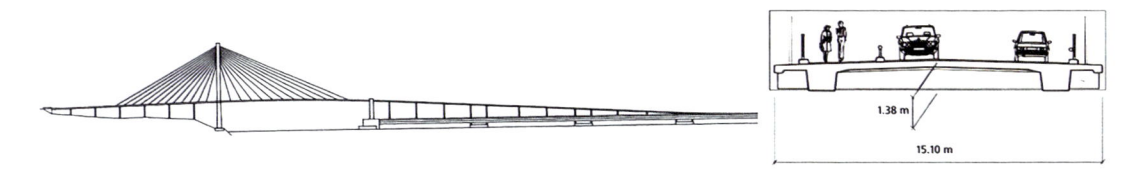

Figure 2.121 Nordhordland Floating Bridge, Norway, 1994, 163 m

90   2 The development of cable-stayed bridges

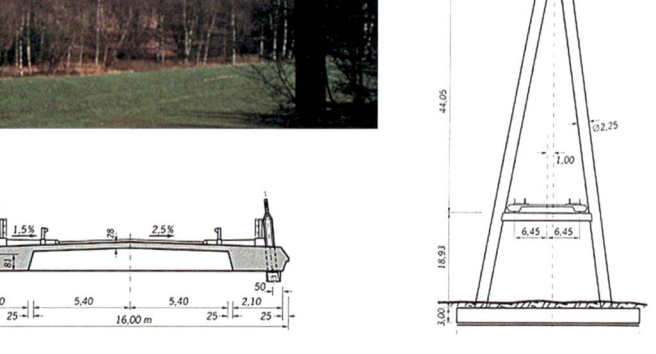

Figure 2.122 River Leven Bridge, Scotland, 1995, 115.20 m

Figure 2.123 Second Panama Canal Bridge, Panama, 2004, 420 m

2.3 Concrete cable-stayed bridges

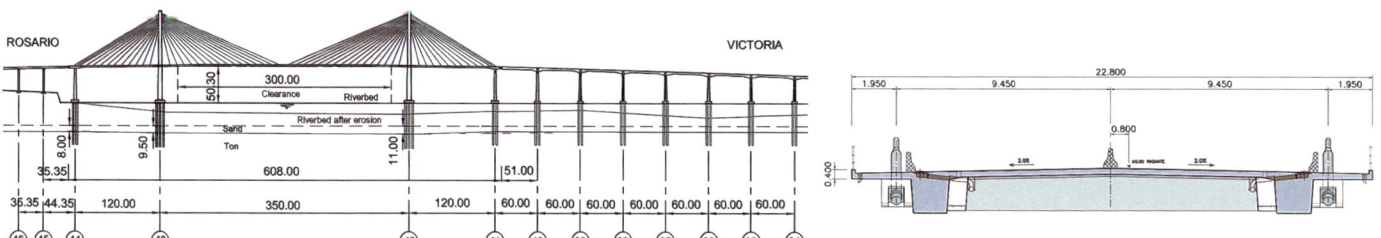

Figure 2.124 Rosario-Victoria Bridge, Argentina, 2000, 330 m

2.3.3 Bridges with concrete stays

The concrete stays of cable-stayed bridges consist of regular prestressing tendons cast into a beam as thick concrete cover for corrosion protection. After curing of the concrete the complete stays are post-tensioned in such a way that they remain in compression for all loadings. Only then are they grouted. In this way the stays have a reduced fatigue range and a corrosion protection similar to regular tendons and are thus designed as tendons and not as stay cables.

Since the complete cross-section participates in the load transfer these stays are very stiff. They are often used to replace missing piers in the main span. Their beam moments thus represent more those of a continuous girder than those of a beam elastically supported by stays at short distances.

2.3.3.1 Riccardo Morandi's bridges

The Italian engineer Riccardo Morandi, Fig. 2.125, was one of the most important pioneers of concrete cable-stayed bridges. Most famous is his Maracaibo Bridge with five spans of 235 m each which was completed in 1962 and which is treated in Section 2.5.1.7: 'Series of cable-stayed bridges'.

Morandi applied the design principles developed for the Maracaibo Bridge also for the Wadi Kuf Bridge in North Africa with a main span of 282 m, Fig. 2.126. Stiff A-towers with additional V-piers and single concrete fore- and backstays, beam monolithic with the towers and drop-in girder in the center of the main span to take up the shrinkage, creep and temperature deformations [2.86]. The bridge held the world record for concrete bridges between 1972 and 1977.

For his next bridges, the Magliana Bridge, Fig. 2.127 [2.87], the Bridge over the Rio Magdalena, Fig. 2.128 [2.88], and the Carpineto Bridge, Fig. 2.129 [2.89], Morandi used towers with single vertical legs instead of A-towers but kept the individual concrete fore- and backstays.

Morandi did not take up the development for concentrated individual stays to multi-stay systems which took place worldwide after 1972. This does not diminish his achievements as a pioneer of concrete cable-stayed bridges.

2.3.3.2 Later examples

In the search for new stay systems it was found, in Germany, that post-tensioned concrete ties instead of steel cables can be an efficient, robust, maintenance-free, cost-effective and visually impressive structural member, even if only individual fore- and backstays are required.

For the Danube Bridge Metten, Fig. 2.130, the individual fore- and backstays were, for the first time in Germany, sized not as a stay cable but as a prestressed concrete member in accordance with DIN 4227 [2.90], which permits higher steel stresses for fatigue and ultimate loads. The 600 m long and 30 m wide bridge has typical spans of 68 m, and only the river crossing proper has double spans of about 145 m. The box girder cross-section is continuous over the

Figure 2.125 Riccardo Morandi

Born 1902 in Rome, Italy
1927 Studies completed
1931 Independent Consultant
1959–69 Professor for Bridges, Florence University
1969–72 Professor for Bridges, Rome University
1962 Maracaibo Bridge
1967 Magliana Bridge
1968 Polcevera Bridge
1971 Wadi Kuf Bridge
1974 Rio Magdalena Bridge
1977 Carpineto Bridge
Died 1989 in Rome

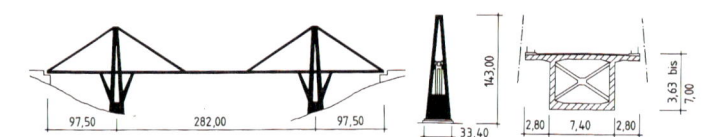

Figure 2.126 Wadi Kuf Bridge, Libya, 1971, 282 m

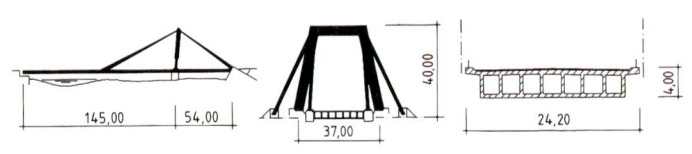

Figure 2.127 Magliana Bridge, Italy, 1967, 145 m

2.3 Concrete cable-stayed bridges

Figure 2.128 Rio Magdalena Bridge, Columbia, 1974, 140 m

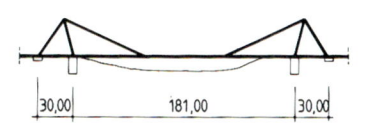

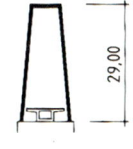

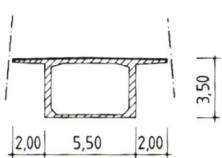

Figure 2.129 Carpineto Bridge, Italy, 1977, 181 m

full bridge length, the missing pier in the river being replaced by the fore- and backstays supported by a single concrete tower. It was, therefore, possible to use the same construction method over the full bridge length; the beam was incrementally launched from one abutment to the other.

The Blaubeurer Tor Bridge across a railway station in Ulm with a total length of 295 m, Fig. 2.131 [2.91], was similarly built by incremental launching with auxiliary piers at the location of the stays. The stays are again prestressed tendons encased in concrete.

The Flößer Bridge across the Main River in Frankfurt, Fig. 2.132 [2.92], also has concrete stays but is unique in its aesthetic appearance. The special local conditions for this bridge were a skew river crossing, an unsymmetric general layout with a single tower and a strong haunch of the beam at the east side of the river, very limited space for the counterweight to anchor the backstays and the requirement to build the main span without auxiliary piers in the river due to ship traffic, although the single forestays at great distances in the main span did not permit standard free cantilevering. All this had to be taken into account to find the appropriate layout, together with the structural details and the construction methods. The bold shape and coloring developed by the architect Egon Jux expresses the unique design task to the general public. The engineer in charge was Herbert Schambeck, Fig. 2.133.

2.3.3.3 Bridges with concrete walls

Concrete walls encase several stay cables and support the beam rather rigidly over a longer distance. They show similarities to the so-called 'extradosed' cable-stayed bridges which are a mixed system from haunched continuous girders and cable-stayed bridges with low towers, see Section 2.5.2.2.

Between the city of Brig and the Simplon ridge in Switzerland the deep Ganter valley with steep slopes had to be crossed. This required a 700 m long, up to 150 m high bridge, Fig. 2.134, with a double curvature in plan and gradient of 5 % in elevation [2.93]. The location of the piers was mainly governed by the requirement to avoid the steep slopes which resulted in a main span of 174 m.

The superstructure acts as a combination of continuous haunched girder and stays with towers reduced in height. The webs of the concrete cross-section protrude upwards outside the roadway and extend to the tower tips as continuous beam supports. These concrete walls follow the curvature of the beam and carry the transverse radial forces of the tendons inside, which were post-tensioned against the hardened concrete. Definitely a unique solution! The engineer in charge was Christian Menn, Fig. 2.135.

The Second Manama Crossing connects the islands of Manama and Muharraq in Bahrain. The main span of 102 m bridges a ship channel, Fig. 2.136 [2.94]. The beam supports are post-tensioned sails which comprise tendons cast into concrete walls.

Their design was governed by the aggressive atmosphere of great heat and salt water which resulted in the following characteristics:

– precast elements for the concrete walls
– use of Mikrosilika concrete with heat curing of the precast elements
– increased concrete cover
– cathodic protection for the reinforcement of the piles and pier caps
– reinforcement mesh from titanium as a skin reinforcement for the piers painting of all concrete surfaces with an anti-carbonization paint.

Monostrands within PE pipes, similar to modern parallel strand cables, inside the precast walls are post-tensioned in such a way that no tensile stresses are present under transient loads.

2.3.4 Cable-stayed bridges with thin concrete beams

The special characteristic of the Diepoldsau Bridge across the upper Rhine River, Fig. 2.137 [1.25], is the design of its beam which consists of a slender solid slab without additional main and cross girders. Although the main span of 97 m and the total length of 250 m would also be in the range of girder and arch bridges, the slender beam built by free cantilevering proved to be economically advantageous.

In order to realistically investigate the static and dynamic behavior of such a cable-stayed bridge with a very slender beam of $l:h = 79:0.55 = 1:176$, a 20 m long model in a scale of $1:20$ was built at the ETH Lausanne under Prof. René Walther. The 25 mm thick roadway slab was built from micro-concrete [2.95].

The tests confirmed that such bridges with very thin beams are not critical with regard to buckling, always provided that they are supported by closely spaced stays. For buckling calculations see Section 4.1.4.2.

The use of a very slender beam has been further developed for the Evripos Bridge, Fig. 2.138, which connects the island of Euböa with the mainland in Greece, the first cable-stayed bridge for roadway traffic in Greece [2.96], with a main span of 215 m, a transverse cable distance of 13.5 m and a longitudinal cable distance of 5.9 m.

The solid concrete beam has a constant thickness of 45 cm over its full width, resulting in a transverse slenderness of $0.45:13.5 = 1:30$ and a longitudinal slenderness of $0.45:215 = 1:478$ (a world record!). The concentrated loads on the thin slab elastically supported by the cables are distributed over a large area as confirmed by the tests for the Diepoldsau Bridge.

The concrete towers in H-shape have box cross-sections with 40–50 cm wall thickness, monolithically connected with the beam. The bad soil conditions required piles with lengths up to 50 m. The towers were consequently flexible enough to follow the changes in beam length by elastic deformations.

The slender concrete slab for this typical two-lane bridge proved to be very economic due to its superior load distribution, its simple CIP construction by free cantilevering, its aerodynamically advantageous shape and its robustness.

2.3 Concrete cable-stayed bridges

Figure 2.130 Danube Bridge Metten, Germany, 1980, 145 m

Figure 2.131 Blaubeurer Tor Bridge, Germany, 1989, 74 m

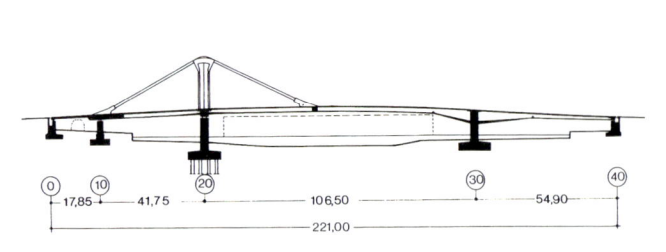

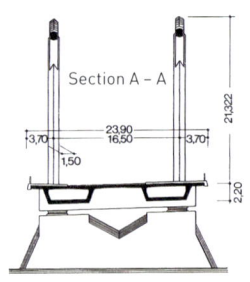

Figure 2.132 Flößer Bridge, Germany, 1985, 106.5 m

Born 1927
1949 Dipl.-Ing. Munich University
1950–90 Dyckerhoff & Widmann
1972 Second Main River Bridge Hoechst
1977 Düsseldorf Flehe Bridge
1980 Danube Bridge Metten
1985 Flößer Bridge
2001 Dubrovnik Bridge

Figure 2.133 Herbert Schambeck

Born 1927 in Meiringen, Switzerland
1957 Independent Consultant
1970 Felsenau Bridge
1971 Professor ETH Zurich
1980 Ganter Bridge
1998 Sunniberg Bridge
Honours:
1982 Fritz Schumacher Prize
1990 Freysinnet Medal
1996 Honorary PhD Stuttgart University

Figure 2.135 Christian Menn

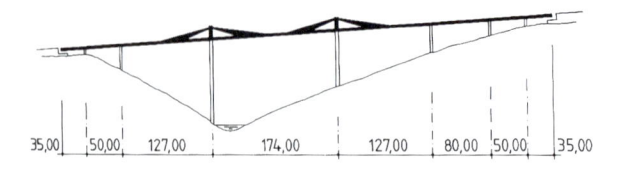

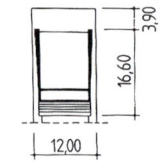

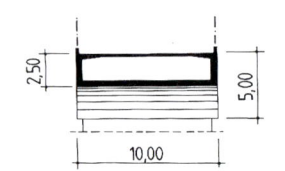

Figure 2.134 Ganter Bridge, Switzerland, 1980, 174 m

2.3 Concrete cable-stayed bridges

Figure 2.136 Second Manama-Muharraq Crossing, Bahrain, 1997, 102 m

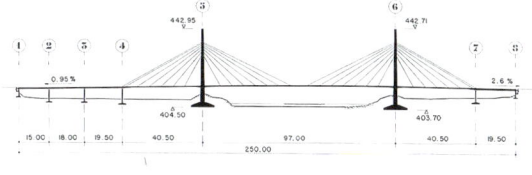

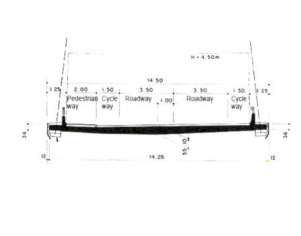

Figure 2.137 Diepoldsau Bridge, Switzerland, 1985, 97 m

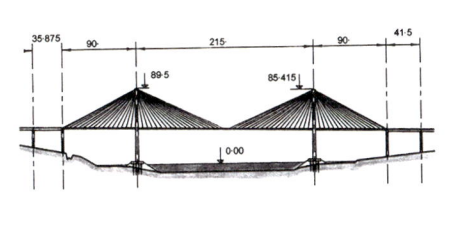

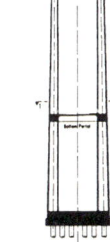

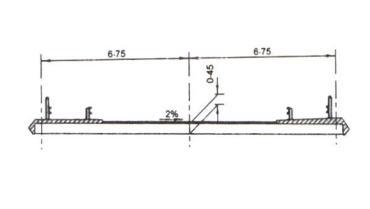

Figure 2.138 Evripos Bridge, Greece, 1993, 215 m

2.3.5 Record spans

The Brotonne Bridge across the Seine downstream of Paris, Fig. 2.140 [1.26], with a total length of 1280 m, is a concrete cable-stayed bridge with a main span of 320 m.

The beam comprises a single-cell post-tensioned box girder with constant wall thickness and inclined webs from precast elements. All members are post-tensioned in three dimensions, the box longitudinally, the roadway slab transversely, the webs vertically and the internal diagonals centrically. In this way, 'full prestress' in accordance with Freyssinet's philosophy of post-tensioning is achieved, Fig. 2.141.

The cables are arranged in one central plane. At the beam the vertical cable components are transferred into the webs by inclined ties. The torsional moments due to eccentric live load are carried by the box girder.

Construction took place by free cantilevering. The precast webs with protruding reinforcement were placed inside the form and the CIP concrete was cast around them. The engineer in charge, Jean Muller, Fig. 2.142, later remarked to the author that, for practical reasons, he would never again use a combination of CIP with precast concrete.

With a main span of 440 m the concrete cable-stayed bridge Barrios de Luna held the span record for all types of cable-stayed bridges from 1983 to 1991, Fig. 2.143 [2.97]. The bridge crosses a dam in the Spanish province of Leon. A large number of alternate systems were investigated. The selected solution takes into account the difficult geological conditions and the large changes of 15 m to 48 m in water level of the dam.

The span ratio of 99 m to 440 m comes only to 0.23 as compared with the regular ratio of about 0.4, see Section 4.1.2. Thus unusually heavy 35 m long concrete abutments were required on both sides as counterweights. In addition, the beam runs monolithically into the abutments, which requires a longitudinal joint in the center of the main span to account for changes in lengths due to temperature and shrinkage and creep. The location of the towers was chosen in such a way that they are never in the water. The engineer in charge was the Spaniard, Carlos Fernández Casado, Fig. 2.144.

Since 1991 the span record for concrete cable-stayed bridges has belonged to the Skarnsundet Bridge at 530 m, Fig. 2.145 [2.98]. The bridge crosses the Trondheim Fjord, north of Trondheim. The concrete towers have box girder cross-section, the beam has an aerodynamically advantageous triangular cross-section which is solid in the rather short side spans (ratio 0.36) to provide sufficient counterweight. The stay cables are fully galvanized locked coil ropes with diameters up to 85 mm.

The towers were built with sliding forms as usual in Norway, and the beam built CIP by free cantilevering in sections similar to the cable distance of 10 m.

Table 2.4 Record spans

Concrete cable-stayed bridges	Main span	Completed
Maracaibo Bridge	235 m	1962
Wadi Kuf Bridge	282 m	1972
Brotonne Bridge	320 m	1977
Barrios de Luna Bridge	440 m	1983
Skarnsundet Bridge	530 m	1991

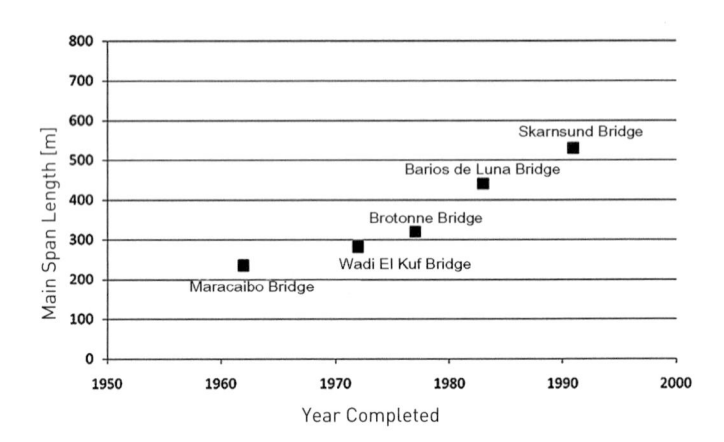

Figure 2.139 Concrete cable-stayed bridges with record spans

2.3 Concrete cable-stayed bridges

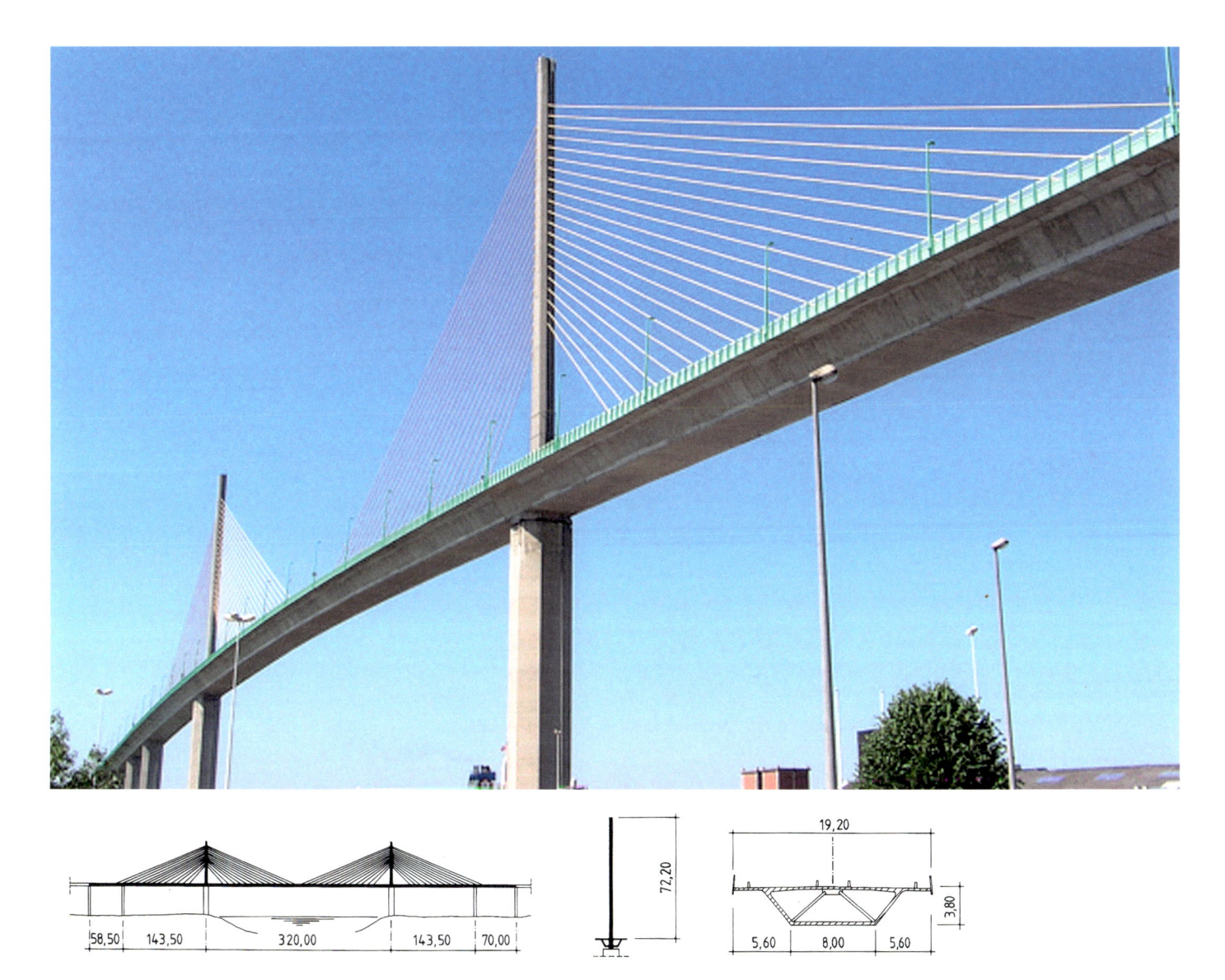

Figure 2.140 Brotonne Bridge, France, 1977, 320 m

Figure 2.141 Box beam with cable anchorage

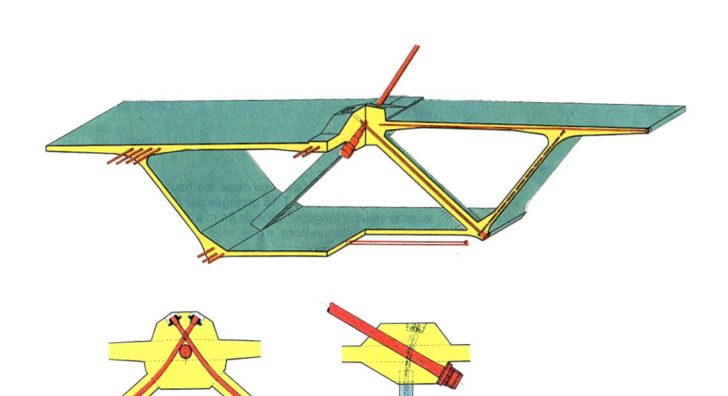

Figure 2.142 Jean Muller

Born 1925 in Levallois-Perret, France
1947 Diploma from the École Centrale in Paris
Start of Work with Campenon-Bernard and STUP
1951 Chief Engineer with Freyssinet, New York
1955 Technical Director with Campenon-Bernard
and Technical Advisor with Europe-Étude
1978 Co-founder of Figg & Muller Engineers Inc.
with Eugene Figg, Technical Director
1986 Return to France, Founder of Jean Muller
International and Chief Engineer of Scetauroute
1977 Brotonne Bridge
1987 Sunshine Skyway Bridge
1991 Isère-Viaduct
Died 2005 in Paris

Born 1905 in Logroño, Spain
1924 Diploma
1949–63 Spanish Building Ministry
1963–68 Department of Bridges and Roads
1966 Co-founder Consultancy Carlos Fernando Casados SL with Javier Manterola Armisén in Madrid
1983 Cable-Stayed Bridge Barrios de Luna
Died 1988

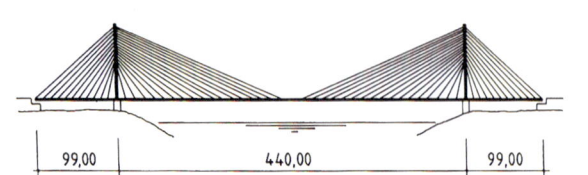

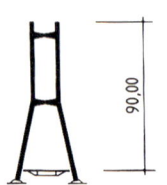

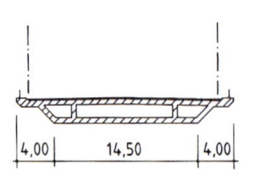

Figure 2.144 Carlos Fernández Casado

Figure 2.143 Barrios de Luna Bridge, Spain, 1983, 440 m

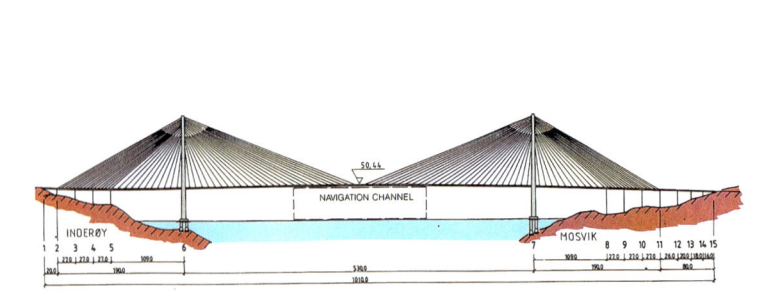

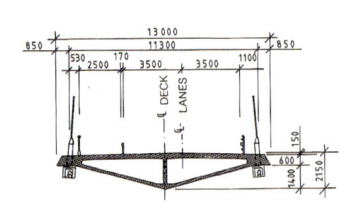

Figure 2.145 Skarnsundet Bridge, Norway, 1991, 530 m

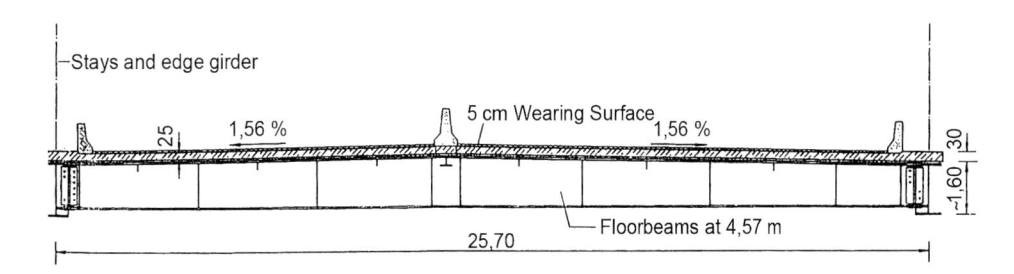

2.4 Composite cable-stayed bridges

2.4.1 General

Lately cable-stayed composite bridges have been used increasingly due to their economy and constructability. A typical composite cross-section is shown in Fig. 2.146.

Savings are achieved as against steel bridges because concrete can carry compression forces more economically than steel, and because a concrete roadway slab is more economic than an orthotropic deck. Compared with concrete bridges less cable steel and smaller foundations are required.

A composite beam can be erected in small units with steel main and cross girders and precast slabs, Fig. 2.147, for which light lifting equipment is sufficient. Even the weight of a completely pre-assembled girder grid remains low. If precast roadway slabs are used they can be connected with CIP joints and overlapping reinforcement. The low weight and the absence of joints between precast segments are the advantages of a composite girder against concrete girders. This tendency was pointed out by the author in 1984 [1.7, 2.99].

2.4.2 Cross-sections

The main task when designing a composite beam is to avoid tensile stresses in the roadway slab as much as possible. In the transverse direction this is achieved because the roadway slab acts in compression as the top flange of a beam simply supported by the two outer cable planes. For this reason one central cable plane is not efficient for composite cross-sections.

In the longitudinal direction compression forces in the beam are created by the inclined cables. In the center of the main span, where these compression forces are small and where substantial negative moments from live loads occur, positive moments under permanent loads can be created by a hogging shop form which causes compression in the roadway slab.

The concrete roadway slab can be built from precast elements or CIP. Precast elements have a higher age at installation and thus show less shrinkage and creep. Therefore, the transfer of compression forces due to creep from the concrete roadway slab into the supporting steel structure is minimized. The joints between precast elements, however, have to be designed and built so that they do not create any weak spots. The same is true for any pockets into which shear studs from the top flanges protrude.

Fig. 2.148 shows a typical precast roadway slab. The shear studs on top of the cross girders are accommodated in pockets and are also cast-in. At the edges, the precast slabs are extended by CIP edge strips which include the shear studs on top of the main girders. The joints with overlapping reinforcement are also closed with CIP concrete. All joints are under compression from permanent loads. The construction of CIP roadway slabs is outlined in Section 5.1.3.21.

The composite cable-stayed bridges built until 2001 are listed in Table 2.5 [2.99].

Figure 2.146 Typical composite cross-section

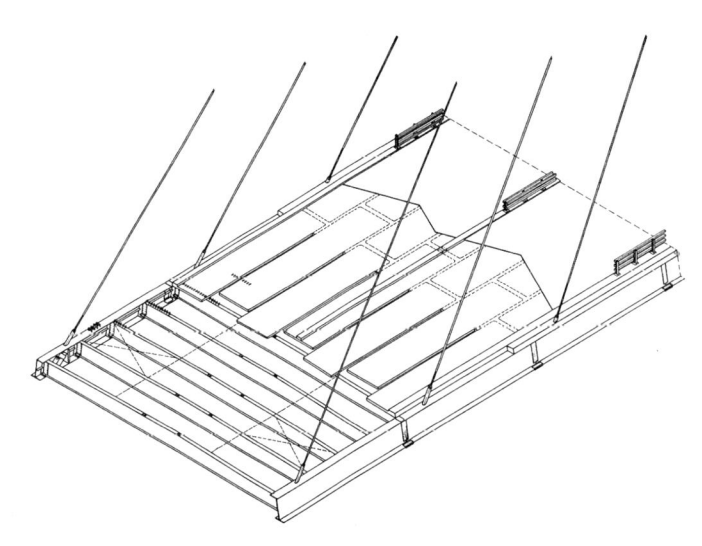

Figure 2.147 Erection of a composite beam with precast slabs

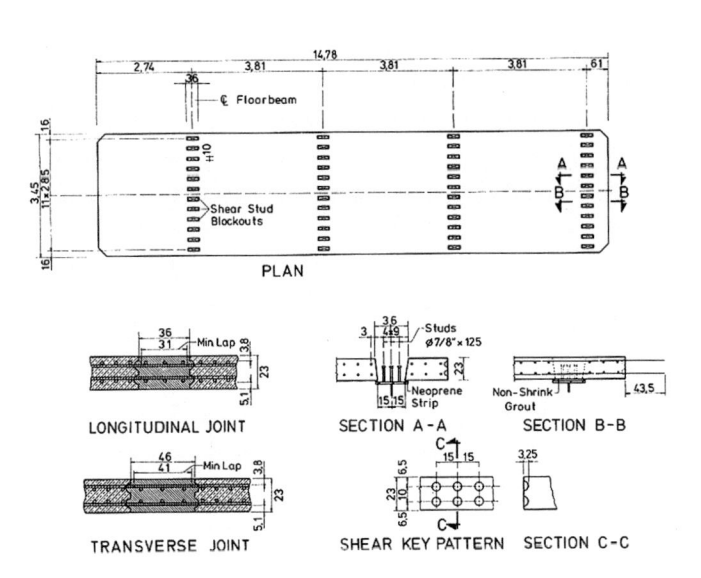

Figure 2.148 Precast slabs for the roadway of a composite beam

Table 2.5 Composite cable-stayed bridges

No.	Name Location	Year completed	Main span	Deck width	Deck depth	Slab type	Post-tensioning	Tower type	Cable planes
1	Strömsund Bridge Sweden (not composite)	1955	183 m	14.30 m	3.00 m	CIP, 0.20 m	none	2 H inclined	2
2	Büchenau Bridge Germany	1956	58.80 m	20.80 m	1.40 m	CIP, 0.25 m	longitudinal	2 H w/o struts	2
3	Pont des Iles, Expo '67 Canada	1967	2 · 105 m	28.65 m	2.82 m	CIP, 0.19 m	none	1 H w/ struts	2
4	Second Hooghly River Bridge, India	(1980) 1992	457 m	25.00 m	2.33 m	CIP, 0.23 m	none	2 H inclined	2
5	Sitka Harbor Bridge Alaska, USA	1972	157 m	11.00 m	1.80 m	CIP, 0.20 m	none	2 H w/o struts	2
6	Heer-Agimont Bridge Belgium	1975	123 m	4.50 m	2.05 m	CIP, 0.19 m	none	2 H w/ 2 struts	2
7	Steyregger Danube Bridge Linz, Austria	1979	161.2 m	24.86 m	4.07 m	CIP, 0.20 m	longitudinal	2 A asymmetric	2
8	Sunshine Skyway Bridge Florida, USA; Design	(1982)	366 m	27.50 m	2.34 m	PC, 0.23 m	none	2 Diamond	2
9	Annacis Bridge Vancouver, Canada	1986	465 m	28.00 m	2.10 m	PC, 0.27 m	none	2 H inclined	2
10	Saint Maurice Switzerland	1986	105 m	2 · 11.75 m	1.01 m	CIP, 0.22 m	none	1 A inclined longitudinal	2
11	Quincy Bridge Mississippi River, USA	1987	274 m	13.80 m	2.10 m	PC, 0.23 m	longitudinal	2 H inclined	2
12	Kemjoki Bridge Finland	1989	126 m	25.50 m		CIP, varies	transverse	1 centre mast	2
13	Weirton-Steubenville Bridge Ohio River, USA	1990	250 m	28.00 m	2.74 m	CIP, 0.22 m	none	2 A asymmetric	2
14	Nan Pu Bridge Shanghai, China	1991	423 m	25.00 m	2.10 m	PC, 0.26 m	longitudinal	2 H inclined	2
15	Burlington Bridge Ohio River, USA	1993	195 m	25.70 m	1.85 m	PC, 0.25 m	longitudinal	2 H inclined	2
16	Tähtiniemi (Heinola) Bridge Finland	1993	165 m	22.00 m	3.20 m	CIP, varies	transverse	2 H, 1 strut	2
17	Utsjoki Bridge Finland	1993	155 m	12.00 m	1.76 m	PC, 0.26 m	none	2 H inclined	2
18	Karnali River Bridge Nepal	1993	325 m	11.30 m	3.00 m	PC, 0.23 m	none	1 H asymmetric	2

2.4 Composite cable-stayed bridges

No.	Name Location	Year completed	Main span	Deck width	Deck depth	Slab type	Post-tensioning	Tower type	Cable planes
19	Mezcala Bridge Mexico	1993	300/311 m	18.10 m	2.79 m	CIP, 0.20 m	none	3 H towers	2
20	El Canon Mexico	1993	166 m	21.00 m	2.11 m	0.20 m		1 H inclined	2
21	El Zapote Mexico	1993	176 m	21.00 m	2.11 m	0.20 m		1 H inclined	2
22	Yang Pu Bridge Shanghai, China	1993	602 m	32.50 m	3.00 m	PC, 0.26–0.40 m	longitudinal + transverse	2 inverted Y	2
23	Clark Bridge Mississippi River, USA	1994	230 m	30.50 m	1.90 m	PC, 0.27 m	longitudinal	2 centre masts	2
24	Baytown Bridge Texas, USA	1995	381 m	2 · 23.83 m	1.83 m	PC, 0.20 m	none	2 twin diamond	4
25	Second Severn Bridge UK	1996	456 m	34.60 m	2.70 m	PC onto grid, 0.25 m	none	2 H w/2 struts	2
26	Kap Shui Mun Bridge Hong Kong	1996	430 m	35.20 m	7.46 m	PC onto grid, 0.25 m	none	2 H inclined	2
27	Ting Kau Bridge Hong Kong	1997	448/475 m	43.00 m	1.75 m	PC, 0.23 m	none	3 centre masts	4
28	Raippaluoto Bridge Finland	1998	250 m	15.00 m	2.80 m	PC, 0.27 m	none	2 diamond	2
29	Öresund Bridge Sweden – Denmark	2000	490 m	30.50 m	10.20 m	CIP	transverse	2 H w/o strut	2
30	Sunningesund Bridge Uddevalla, Sweden	2000	414 m	26.50 m	2.20 m	PC, 0.24 m	none	2 diamond	2
31	Kolbäcks Bridge Ume Älv, Sweden	2001	130 m	17.65 m	2.59 m	CIP, 0.27 m	none	1 H inclined	2

Table 2.6 Cable-stayed bridges with composite cross girders

No.	Name Location	Year completed	Main span	Deck width	Deck depth	Cross-Girder, depth	Cross-Girder, distance	Tower type	Cable planes
32	East Huntigton Bridge Ohio River, USA	1985	274 m	12.20 m	1.52 m	I-deck, 0.91 m	2.73 m	1 A asymmetric	2
33	Vasco da Gama Bridge Lisbon, Portugal	1998	420 m	31.20 m	2.50 m	I-Section, 0.20 m	4.41 m	2 H inclined	2

2.4.3 Special details

Composite cross girders

In order to build the concrete beam of the East Huntington Bridge to be as light as possible, the concrete cross girders were replaced by composite cross girders – see Fig. 2.149 and Section 6.1.3. The Vasco da Gama Bridge across the Tejo River in Lisbon also uses composite girders, Table 2.6 [2.100].

Composite roadway slab

In addition to high costs, orthotropic steel decks can also have further disadvantages. The asphalt wearing surface may disconnect from the steel plate and slippery ice may form faster than on a more inert concrete deck. These problems may be partly avoided by using a CIP concrete surface composite on top of the orthotropic deck. Today, this is called 'ortho-composite', Fig. 2.150.

The distances between the stringers may be increased from the usual 300 mm and the thickness of the top plate may be reduced down to 10 mm, all depending on the local stiffnesses.

Homberg used this system in 1969 for the Massená Bridge in Paris [2.101] and in 1991 for the Dartford Bridge near London [2.102].

The two Zárate Bridges [1.14] use a 10 cm thick concrete top slab, Fig. 2.150. Its stiffness allowed the stringer distance to be increased to 400 mm. The concrete was not considered to carry any overall main girder forces, but the shear studs and reinforcement were sized in such a way that they could carry all overall forces in composite action in addition to local wheel loads.

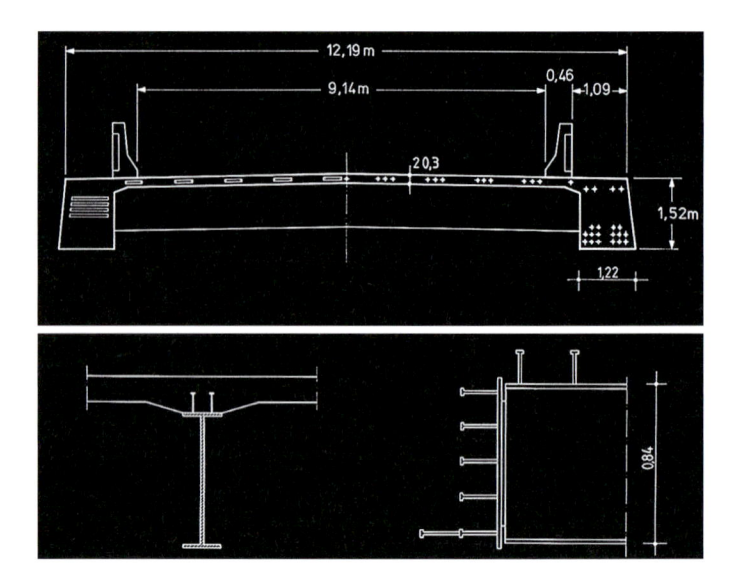

Figure 2.149 Composite cross girders of the East Huntington Bridge, USA

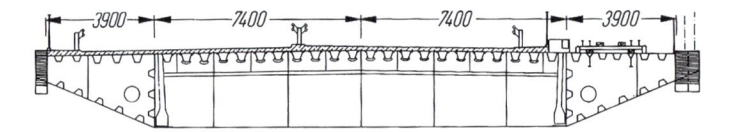

Figure 2.150 Ortho-composite deck of the Zárate-Brazo Largo Bridges, Argentina

2.4.4 Economic span lengths

To carry a normal force it costs generally only two-thirds as much in concrete as in steel. An orthotropic deck costs at least four times as much as a 25 cm thick concrete roadway slab. The general costs per m² for a bridge beam are given in Fig. 2.151 in relation to main span lengths and the beam material.

It follows that concrete beams should be most economic for spans up to 400 m and composite beams for spans between 400 and 900 m. The very long spans are dominated by steel beams. A further refinement can be achieved with hybrid systems which have a steel or composite main span and concrete side spans.

These hybrid systems reach their optimum if the concrete cross-section extends about 100 m into the main span due to the high compression forces around the towers. This advantage is, however, compensated if an additional construction system is required for the concrete beam in the main span, for example if the side spans can be built by incremental launching, and free cantilevering would have to be used for the main span part, as done for the Normandy Bridge, see Section 6.5.

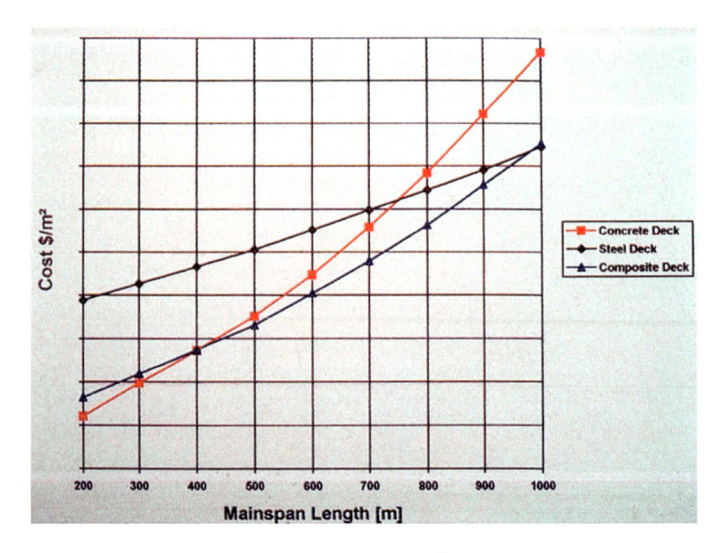

Figure 2.151 Comparison of beam costs per m²

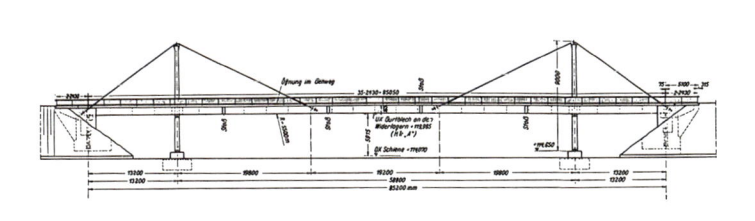

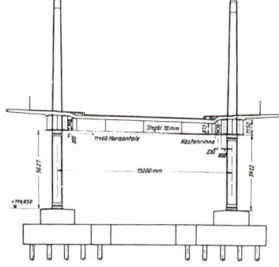

Figure 2.152 Büchenau Bridge, Germany, 1956, 59 m

2.4.5 Beginnings

Nearly overlooked, even by bridge engineers, the small Büchenau Bridge across the railway in the City of Bruchsal was the first modern cable-stayed bridge in Germany with a composite beam, Fig. 2.152 [2.103]. The locked coil ropes were stressed by lowering the beam from the scaffolding.

2.4.6 Record spans

The seagoing ships on the Hooghly River into the Calcutta harbor required a main span of 457 m (1500 feet) and a navigational clearance of 38 m, Fig. 2.154 [1.18]. The most economic bridge system for these conditions was a cable-stayed bridge. A suspension bridge would have been more expensive due to the high costs of the anchorage abutments in the poor alluvial soils. The ratio of beam dead weight to live load in the range of 3:1 led to side spans of 183 m (600 feet), equal to 40 % of the main span, Fig. 4.2.

An important condition for the design of foundations, towers, beam and stay cables was the request of the Indian client to use local materials as well as fabrication and construction techniques which were used in India or which could be adapted. The aim was to build the most economic bridge system with local labor and equipment.

This meant the use of local steel which, however, was not weldable as welding equipment was not available. Rivets had to be used for all connections. Thus an orthotropic deck could not be used and a composite deck had to be used – the first time for a major cable-stayed bridge.

The beam required a width of 35 m for two-directional carriageways and 2.5 m wide walkways on each side. The composite cross-section comprises the concrete slab supported on a steel girder grid with three longitudinal main girders and cross girders at 4.1 m distance, Fig. 2.155. The 23 cm thick concrete slabs cantilever beyond the outer main girders and taper to a thickness of only 15 cm at the outside. The slab is under compression due to the cable forces only. It is not post-tensioned but reinforced with thin bars at close distances for crack control. The shear connection between the concrete slab and the girder grid is provided by block studs with loops on top of the main and cross girders.

Due to the poor soil conditions, a steel tower was used which is distinctly lighter than an otherwise more economic concrete tower. The steel boxes taper from 4.0 · 4.0 m on top of the foundations to 3.0 · 4.0 m at the top. The tower legs are stiffened by two transverse cross girders.

The available sizes and thicknesses of Indian steel required double plates for all vertical tower walls with a longitudinal joint to halve the width of 4.0 m, as well as transverse joints at a distance of 6.0 m. All joints were connected with cover plates, corner angles and stiffeners by rivets! The cables support the beam at distances of 12.3 m. Parallel wire cables were selected, using up to 220 wires of 7 mm diameter made from high-strength Indian steel 1460/1570. Their corrosion protection comprises an outer thick-walled PE-pipe, filled with permanently elastic polyurethane.

The wires are anchored in steel sleeves by button heads and steel balls in accordance with the BBR-HiAm-System which could be fabricated in license in India. The engineer in charge was Jörg Schlaich, Fig. 2.156, assisted by Rudolf Bergermann, Fig. 2.157.

The Sunshine Skyway Bridge crosses the Tampa Bay in Florida with a main span of 366 m. The steel alternate uses a composite beam with a steel grid made from open steel sections, and a roadway slab of precast elements with CIP joints on top of the cross and main girders, Figs 2.158 and 2.159 [2.104]. For the towers, a diamond shape was developed which combines the advantages of an A-tower above deck with a single foundation. In competitive bidding against the concrete alternate, this design lost by a small margin, the design ideas for the composite beam and the towers have later been used successfully for several cable-stayed bridges.

The Annacis Bridge, Fig. 2.160, held the span record of 465 m from 1986 onwards [1.33]. The design follows the design principles developed for the Sunshine Skyway Bridge. A new cable anchorage at the beam was used. The cables are anchored to a web protrusion above the roadway, Figs 2.161 and 2.162. The engineers in charge for the Annacis Bridge were Peter Buckland, Fig. 2.163, and Peter Taylor, Fig. 2.164

The Yang Pu Bridge in Shanghai conquered the record in 1993 with a main span of 602 m, still unsurpassed, Fig. 2.165a [2.105]. The beam comprises two box girder main beams connected by cross girders on which the concrete roadway slab rests. The main span length resulted from the request to place both towers outside the Yang Pu river in order to make them unreachable by ships. The approaches have spans of 144 – 99 – 44/40 m, the outer 44 m being built from precast elements which serve as ballast for the anchor piers.

The Yang Pu Bridge was designed by the Shanghai Municipal Design Institute (SMEDI) under the direction of Lin Yuan Pei, Fig. 2.165b. The design and construction was reviewed by an international team of experts engaged by the Asian Development Bank, Fig. 2.165c. No exceptions were taken.

Table 2.7 Record spans

Composite bridges	Main span	Completed
Hooghly River Bridge	457 m	1970 – 1994
Annacis Bridge	465 m	1986
Yang Pu Bridge	602 m	1993

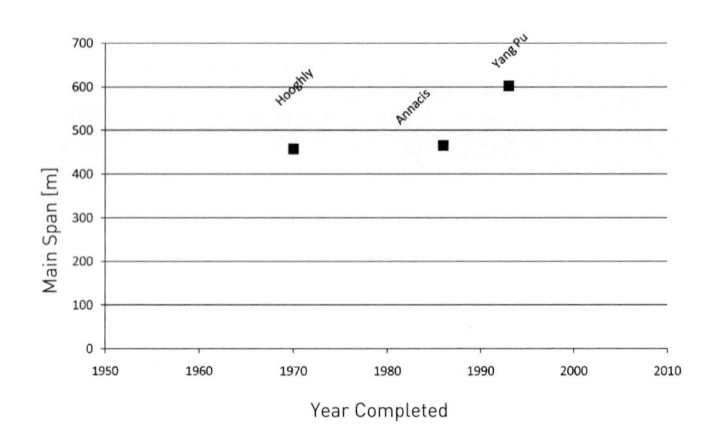

Figure 2.153 Record spans of composite bridges

2.4 Composite cable-stayed bridges

Figure 2.154 Hooghly River Bridge, Calcutta, India, 1974–1994, 457 m

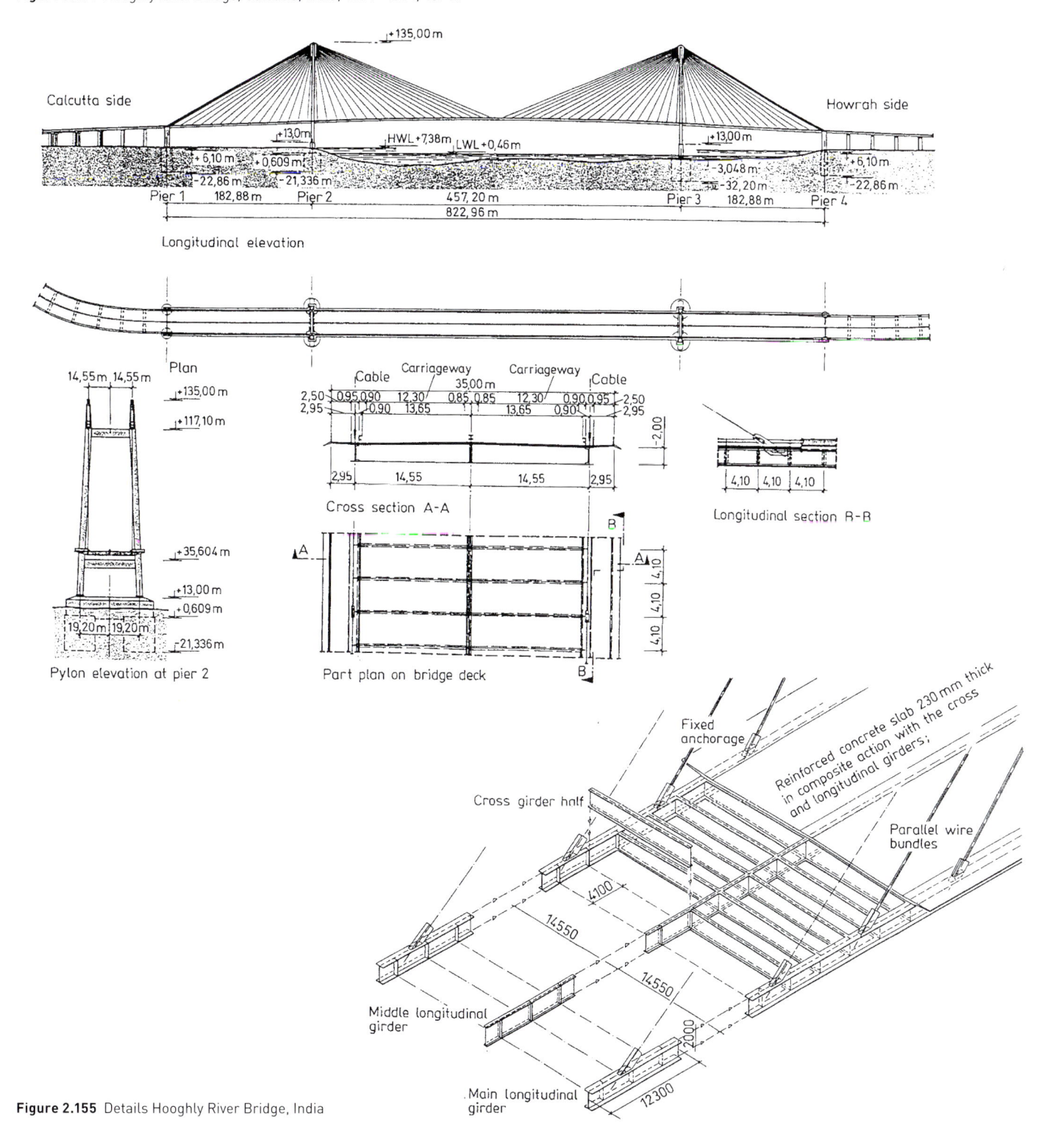

Figure 2.155 Details Hooghly River Bridge, India

108   2 The development of cable-stayed bridges

Born 1934 in Stetten, Germany
1959 Dipl.-Ing. University Berlin
1960 MSc, Case Institute
1962 Dr.-Ing. University Stuttgart
1974 Prof. TU Stuttgart
1963–79 Leonhardt, Andrä and Partners
Since 1980 Schlaich, Bergermann and Partners
Hooghly River Bridge Calcutta
Evripos Bridge
Many further pedestrian cable-stayed bridges
Obere Argen Bridge

Figure 2.156
Jörg Schlaich

Born 1941 in Düsseldorf, Germany
1961–66 Dipl.-Ing. University Stuttgart
1966–67 Ed. Züblin AG, Stuttgart
1968–79 Leonhardt, Andrä and Partners
since 1980 Partner in Schlaich, Bergermann and Partners
Hooghly River Bridge Calcutta
Al Bridge
Obere Argen Bridge
Evripos Bridge
Ting Kau Bridge

Figure 2.157
Rudolf Bergermann

Figure 2.158 Composite alternate for the Sunshine Skyway Bridge, USA, 1982, 366 m

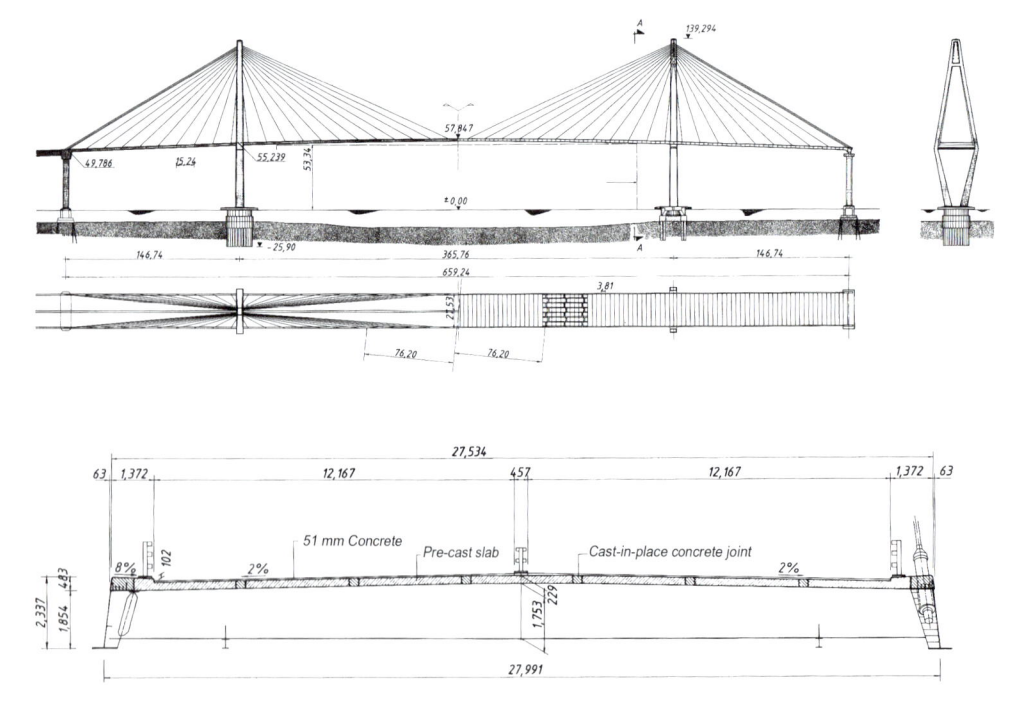

Figure 2.159 Details Sunshine Skyway Bridge, steel alternate

2.4 Composite cable-stayed bridges

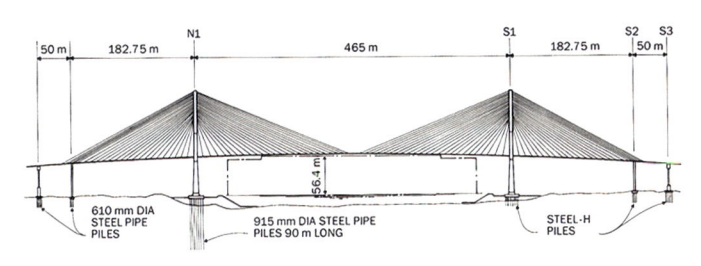

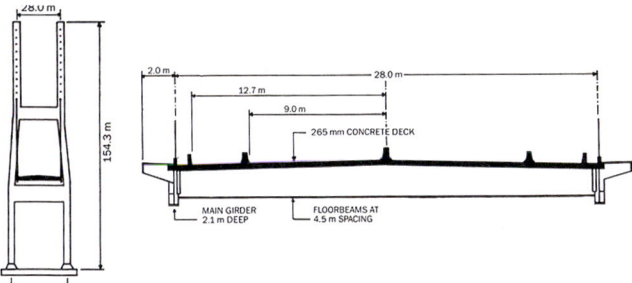

Figure 2.160 Annacis Bridge, Canada, 1986, 465 m

Figure 2.161 Cable anchorage for Annacis Bridge

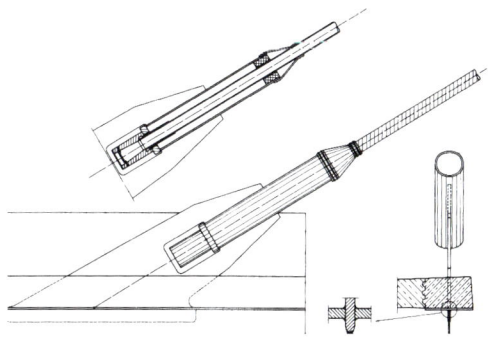

Figure 2.162 Anchorage detail

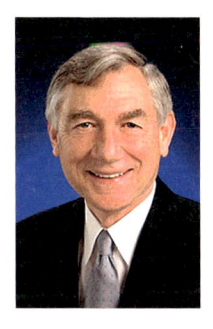

Figure 2.163 Peter Buckland

1960 Graduate Cambridge University
1960–65 Work with consultants in the UK and Canada
1966–70 Work with steel contractors
1970 Founder Buckland & Associates
1972 Buckland & Taylor Ltd, Vancouver BC
Work on various suspension and cable-stayed bridges
Alex Fraser (Annacis Bridge)
Lions Gate Bridge
Confederation Bridge
Golden Gate Bridge Rehabilitation

Figure 2.164 Peter Taylor

1965 Dominion Bridge Co., Canada
1972 Buckland & Taylor Ltd, Vancouver BC
Papineau Bridge
Alex Fraser (Annacis Bridge)
Cooper River Bridge
Second Severn Bridge

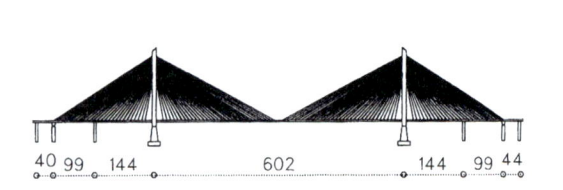

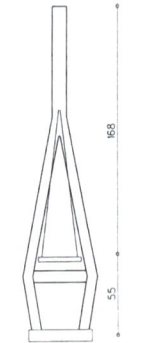

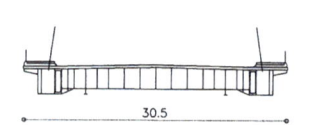

Figure 2.165 a Yang Pu Bridge, Shanghai, China, 1993, 602 m

Chief Engineer SMEDI
State Design Master
Nan Pu Bridge
Yang Pu Bridge
Lupu Bridge

Figure 2.165 b Lin Yuan Pei

Figure 2.165 c International team of experts (from left): Lin Yuan Pei (SMEDI); Walter Podolny (FHWA); Commander Zhu Zhi Hao, Site Manager; the author; Man Chung Tang (T.Y. Lin); Prof. Fumio Nishino, Tokyo University; Bruce Murray (ADB)

2.4 Composite cable-stayed bridges

2.4.7 Latest examples

The composite cable-stayed bridge across the Mississippi at Burlington, Iowa, has a main span of 201 m with a single tower, Fig. 2.166 [1.28]. The main characteristics are a steel girder grid with precast slabs, a tower post-tensioned in the cable anchorage area, stay cables with parallel epoxy-coated strands and beam erection by free cantilevering to both sides by geometry.

The Tähtiniemi Bridge crosses the Ruotsalainen Lake near Heinola in Finland with a length of 924 m, curved in plan with R = 4000 m, Fig. 2.167 [2.106]. The main spans of 165 m and 127 m are supported by stay cables, and the approach bridges have spans of up to 70 m. The composite beam uses an open steel grid with a CIP concrete slab, supported by a single concrete tower in H-shape.

The steel grid of the beam was launched from both sides, in the later stayed regions on auxiliary piers. After the concrete slab was cast intermittently, the stay cables were installed and stressed, thereby relieving the auxiliary piers. The engineer in charge was Esko Järvenpää, Fig. 2.168.

The Kap Shui Mun Bridge has a main span of 431 m and side spans of 2 · 80 m each, Fig. 2.169 [2.107]. The roadway traffic runs on six lanes on the upper deck, two light rapid transit tracks plus four emergency lanes for traffic run on the lower deck, Fig. 2.170. The central 387 m of the beam are composite, and the remainder has a prestressed concrete cross-section. The concrete parts of the beam were built by incremental launching, using the first 26 m of the composite central section as launching nose. The remainder of the central span was built by free cantilevering, installing complete sections with up to 500 t weight. Extreme typhoon winds had to be accounted for, including during construction. The engineer in charge of the detailed design and construction engineering was Reiner Saul, Fig. 2.90, assisted by Siegfried Hopf.

The twin Baytown Bridge across the Houston Ship Channel in Texas, with an area of 32.800 m², is one of the largest of its kind, Fig. 2.171 [1.17]. The cable-stayed main bridge has a span of 381 m with a navigational clearance of 53 m. Each bridge beam is 24 m wide and comprises a steel girder grid with open steel main girders at the outsides, and cross girders, supporting a concrete roadway of precast slabs. The double-diamond-shaped concrete towers are 140 m high and transfer the transverse wind loads by truss action to the foundations.

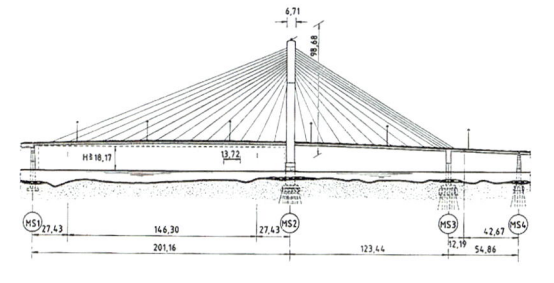

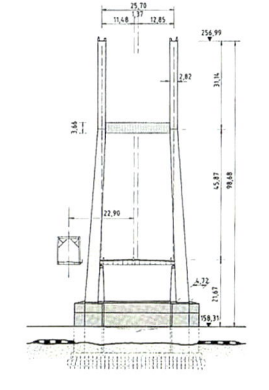

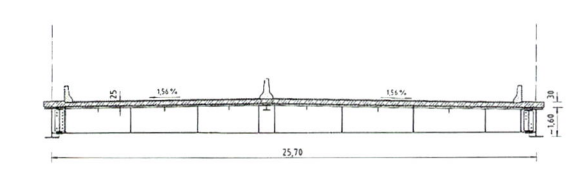

Figure 2.166 Burlington Bridge, Iowa, USA, 1993, 201 m

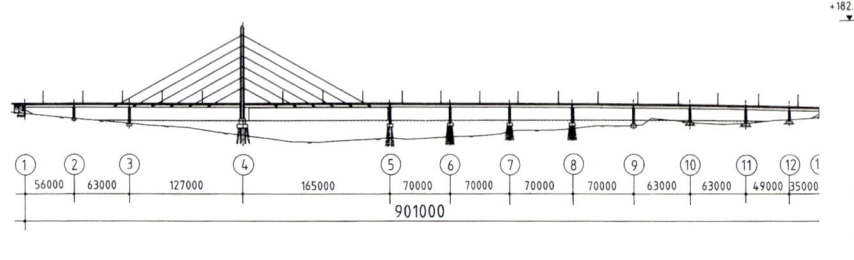

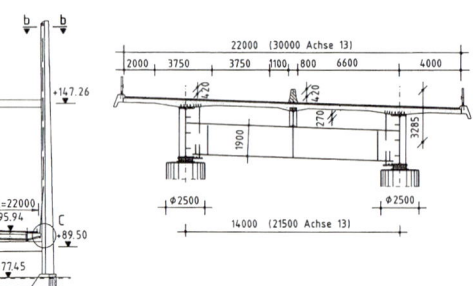

Figure 2.167 Heinola Bridge, Finland, 1994, 165 m

Figure 2.168
Esko Järvenpää

Born 1946
1972 MSc Civil Engineering, Oulu University Finland
1986 Lic. Tech. Civil Engineering, Oulu University Finland
1974–94 Lecturer in bridge design at Technical Institute, Oulu
1972–74 Bridge Engineer with WSP Consulting KORTES Ltd
1974–79 Chief Bridge Engineer WSP
1979–91 Technical Director and Bridge Specialist WSP
1992–1995 Deputy Director WSP
1995–2007 Managing Director WSP
since 2007 Director Bridges, Senior Consultant Bridges WSP
1989 Lumberjack's Candle Bridge
1997 Raippaluoto Bridge
1997 Thätiniemi (Heinola) Bridge
1998 Saame Bridge
2000 Stonecutters Bridge
2000 Swietokrzyski Bridge
2008/2009 Van Troi-Tran Thi Ly Bridge, Danang

2.4 Composite cable-stayed bridges

Figure 2.170 Double-stock composite beam of the Kap Shui Mun Bridge

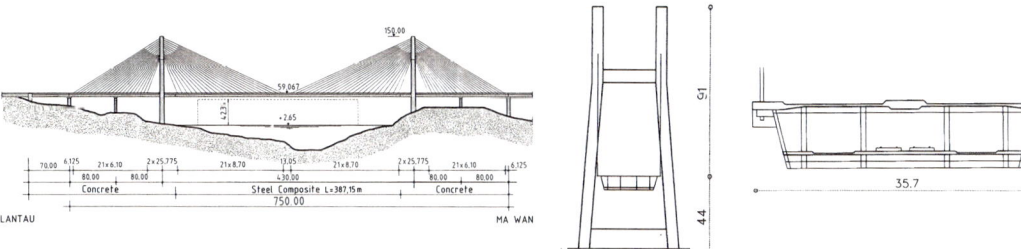

Figure 2.169 Kap Shui Mun Bridge, Hong Kong, China, 1995, 431 m

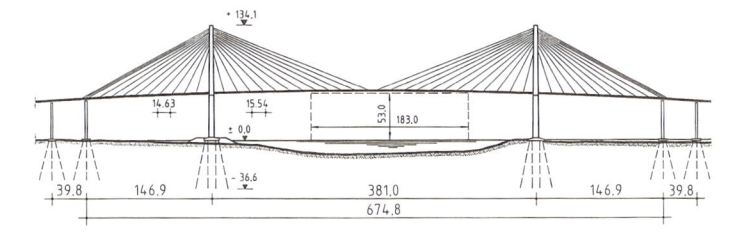

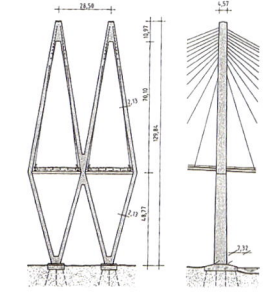

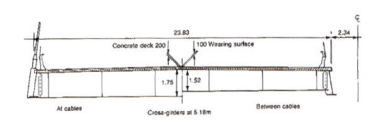

Figure 2.171 Baytown Bridge, USA, 1995, 381 m

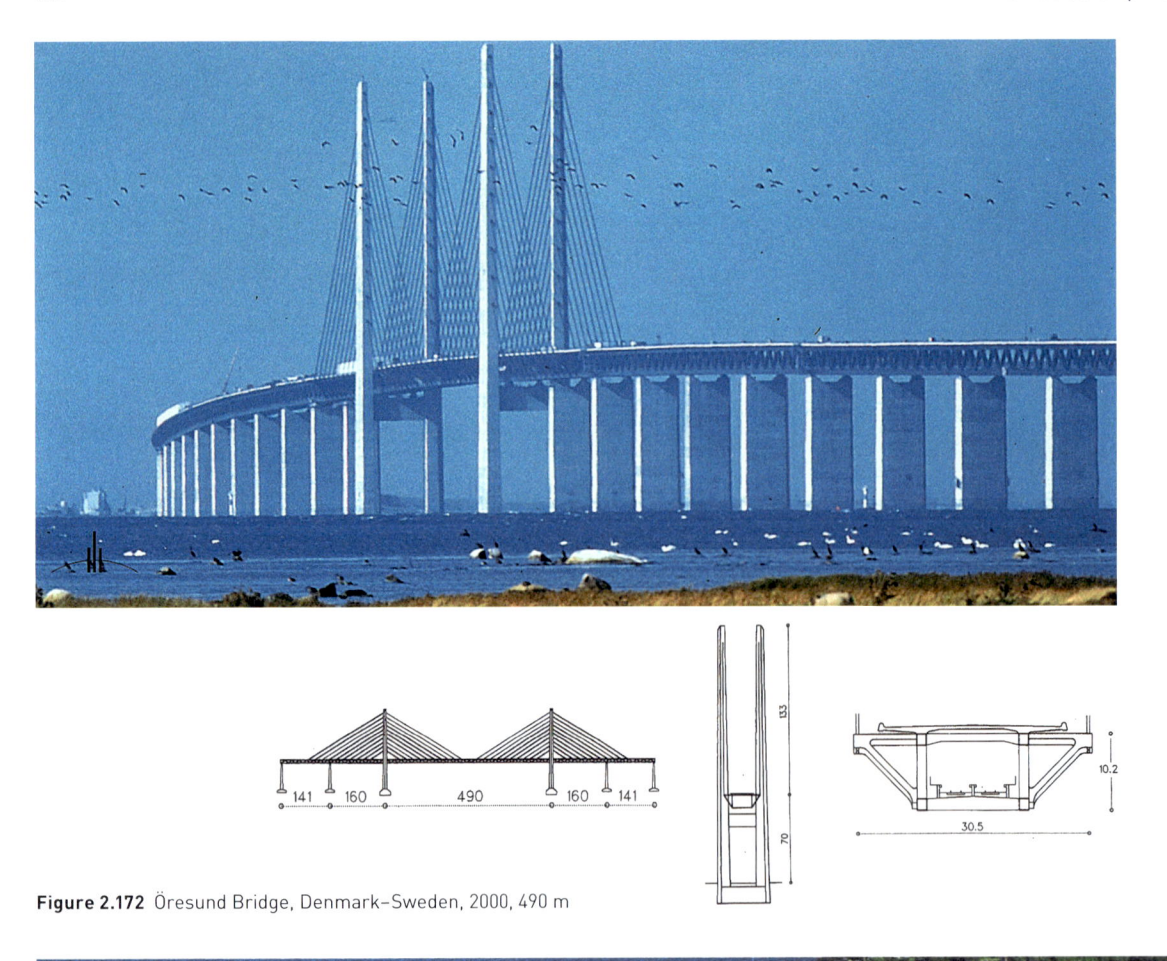

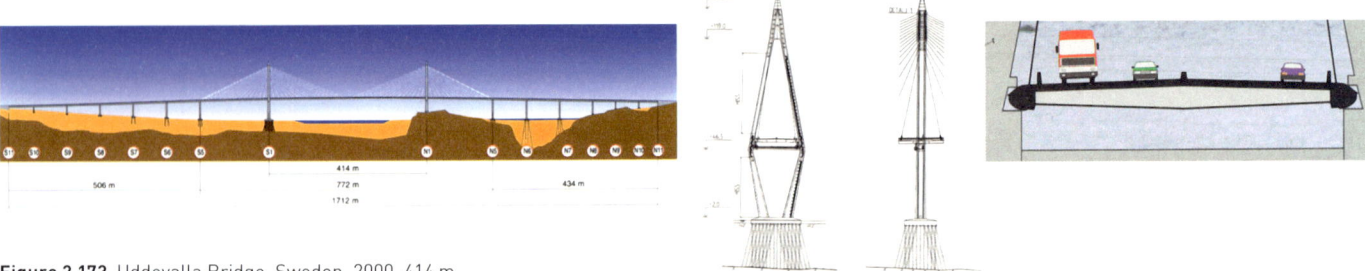

Figure 2.172 Öresund Bridge, Denmark–Sweden, 2000, 490 m

Figure 2.173 Uddevalla Bridge, Sweden, 2000, 414 m

2.4 Composite cable-stayed bridges

The bridge across the Öresund connects Copenhagen in Denmark and Malmö in Sweden, Fig. 2.172. It carries two lanes with full shoulders in each direction of travel on the top deck and two railway tracks on the bottom [1.30]. The box beam is a continuous truss over the full bridge length of 7845 m with a main span of 490 m (a world record for combined road and railway bridges) and 140 m spans in the approach bridges.

The Uddevalla Bridge on the west coast of Sweden carries six lanes of traffic with a main span of 414 m, Fig. 2.173 [2.108]. Its total length comes to 1712 m. The approach bridges have composite box girders, and the cable-stayed main bridge has an open grid cross-section with a precast concrete roadway slab. Parallel strand cables are used.

The Kolbäcks Bridge has a main span of 130 m with a single tower and a composite beam with a triangular cross-section, Fig. 2.174 [2.109]. The stay cables are connected to the steel cross girders. At the tower the stay cables are anchored to an inner steel box, cast into concrete. The steel grid of the beam was launched on auxiliary piers, then the roadway slab was CIP and the stay cables installed.

The new (2006) Berlin Bridge in Halle (Saale), Fig. 2.175, replaces a steel truss from 1916 and is Germany's second ever composite cable-stayed road bridge after the Büchenau Bridge of 1956 [2.110]. The reason for the unusual time gap of 50 years is due to the fact that the German codes for a long time limited concrete tensile stresses under service load. The longitudinal post-tensioning of the concrete roadway slab thus required would have been transferred by creep into the stiffer steel parts which would have produced an uneconomic structure. Only after the German codes were changed to sizing in the ultimate limit state and crack control in the service stage did composite cable-stayed bridges become economic in Germany as well.

The Berlin Bridge carries a road and a tramway across a railway yard. The steel tower in the shape of an inverted Y is 73.5 m high and rests on piles. Six pairs of locked coil ropes in both spans in fan arrangement support the composite beam. The total length of the bridge comes to 171 m with main spans of 87 m and 84 m.

The Elbe River Bridge at Niederwartha near Dresden has a main span of 192 m with a single A-tower, 77 m high, Fig. 2.176 [2.111], with a total length of 366 m and a roadway width of 9 m. The tower is built in concrete, and for the cable anchorages at the top a steel box is cast-in. The main bridge spans 192 m wide across the Elbe River with a side span of 82.5 m. The main span has a composite cross-section, the side span is built in concrete as a counterweight to the main span. This is one of the first hybrid bridges with a composite main span instead of the usual steel span. The tower legs were built with jumping forms above deck, the side span was built span-

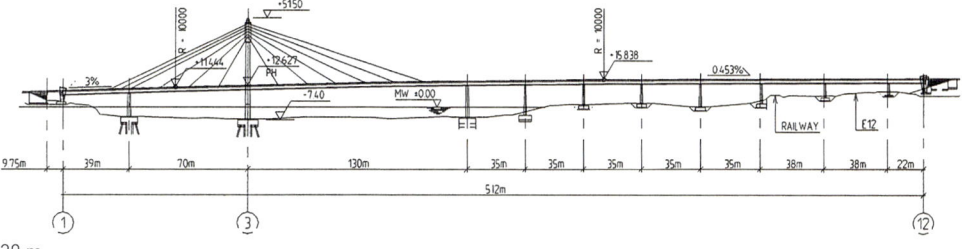

Figure 2.174 Kolbäcks Bridge, Sweden, 2001, 130 m

wise on stationary formwork and the main span was free-cantilevered as usual.

The third bridge across the Orinoco River for road and rail crosses the Orinoco and the adjacent flood plains with a total length of 11.1 km, Fig. 2.177 [1.31]. This great length is required to accommodate the changes in river level of up to 12 m. The main span with a length of 360 m comprises a composite steel truss with four traffic lanes on the top flange on the top deck and one railway track on the bottom deck. Unusually, a central cable plane anchored to the deep top cross girders is used as the most economic solution for the given conditions.

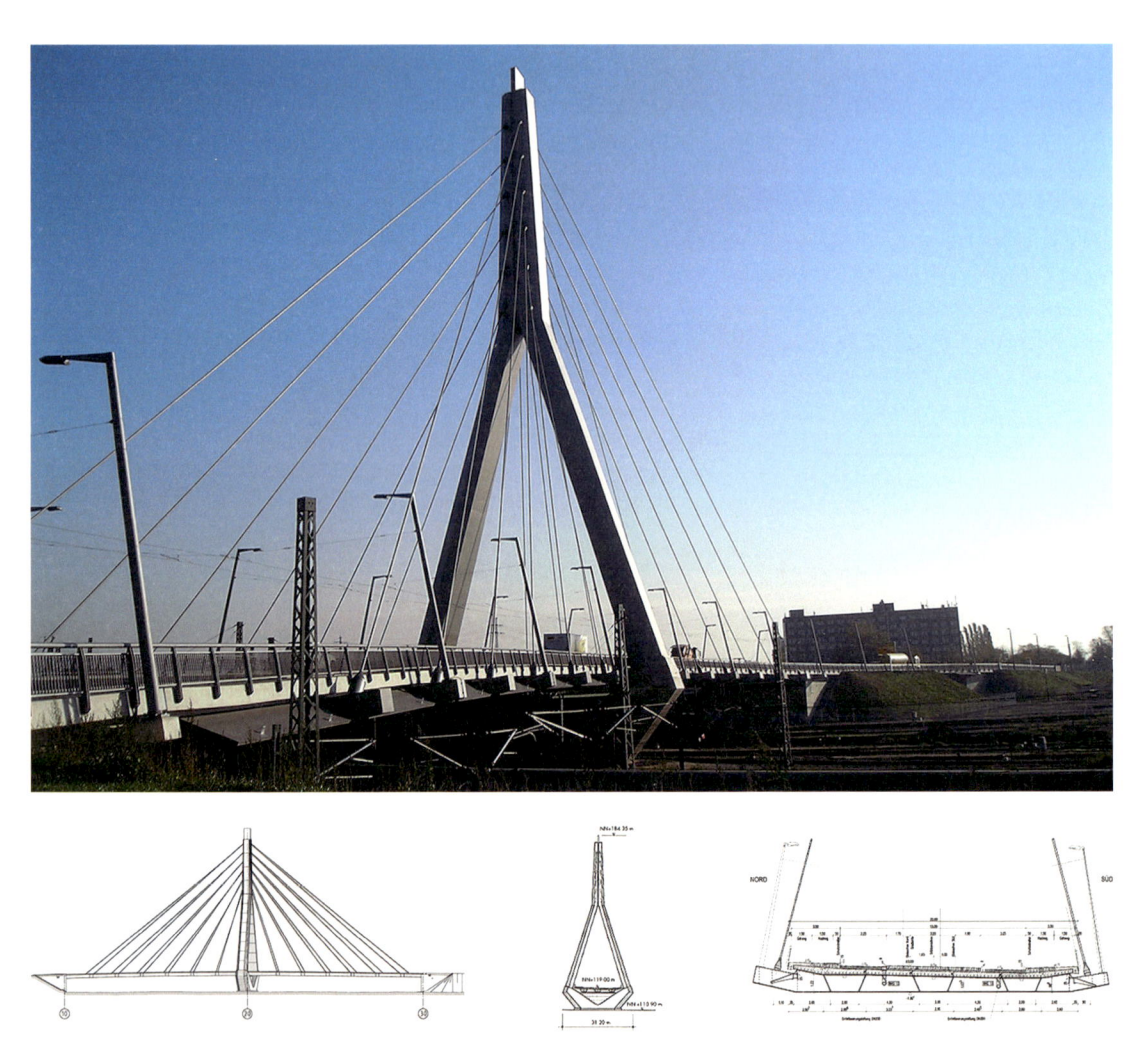

Figure 2.175 Berlin Bridge, Halle, Germany, 2006, 87 m

2.4 Composite cable-stayed bridges

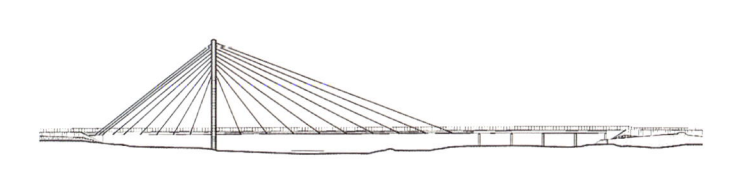

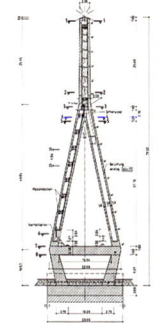

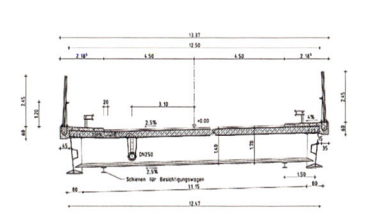

Figure 2.176 Elbe River Bridge Niederwartha, Germany, 2008, 192 m

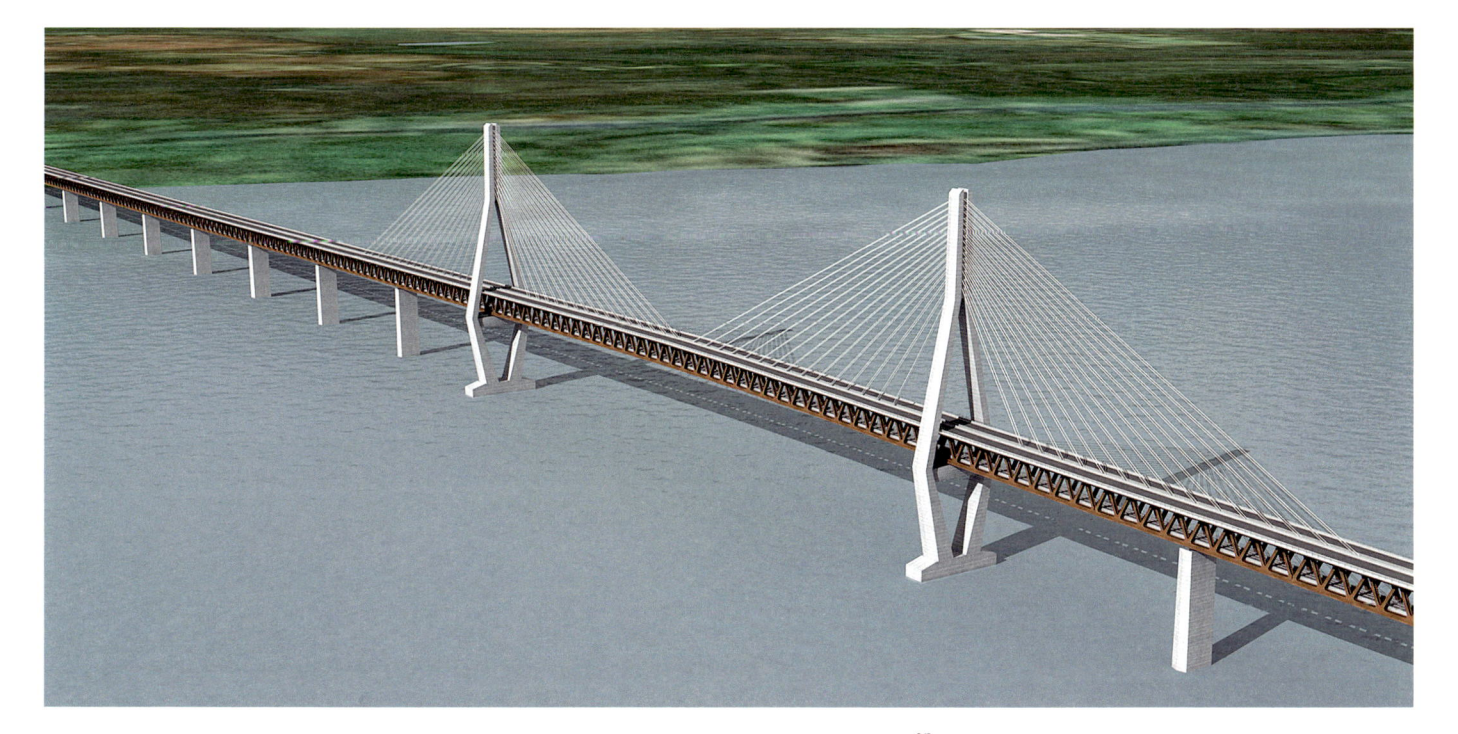

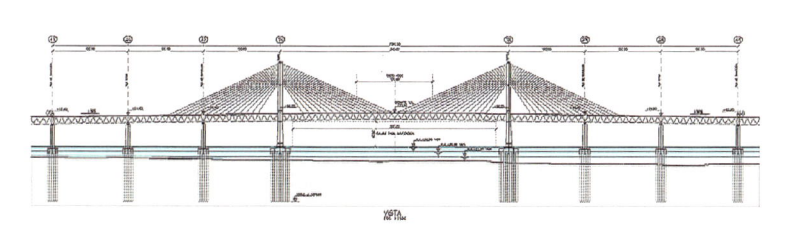

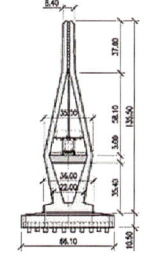

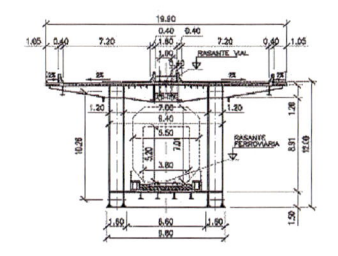

Figure 2.177 Third Orinoco River Bridge, Venezuela, 2011, 360 m

2.5 Special systems of cable-stayed bridges
2.5.1 Series of cable-stayed bridges

When planning the design for the Millau Bridge, the French engineer Michel Virlogeux investigated in detail the design basis for a series of cable-stayed bridges. We follow here his fundamental publication [2.112].

2.5.1.1 Load transfer

The load transfer of a three-span cable-stayed bridge from the main span to the abutments always runs from the forestay cables via the tower head into the backstay cables which are anchored in hold-down piers, Fig. 2.178a and b. Compression is created in the main span and side span beam and uplift forces in the side span pier. Live loads in the side spans are also carried in tension by the side span stays to the tower head, but from there they run in compression through the backstays to the anchor piers, creating tension in the side span beam and compression in the hold-down pier, Fig. 2.178c.

For a series of cable-stayed bridges, the backstays, which retain the tower head, are missing. The main problem, therefore, is how to retain the tower head by other means in order to limit the deflections of the bridge from live loads, Fig. 2.179. The second problem is how to accommodate the changes in lengths – due to shrinkage, creep and change of temperature – for the long beams of a series of cable-stayed bridges, while at the same time carrying longitudinal forces in the beam for example due to braking and longitudinal wind over the bridge length.

2.5.1.2 Intermediate piers

If piers are introduced in every second span of a series of cable-stayed bridges, these can be used as hold-downs for the backstays, Fig. 2.180. Such a system can be designed as a series of three-span cable-stayed bridges as done for the second Orinoco Bridge, Fig. 2.181. In this case of a combination of two three-span cable-stayed bridges for road and rail the important railway braking forces are carried by the central pier which is, therefore, built as an A-frame in the longitudinal direction.

2.5.1.3 Stiff towers

If intermediate piers are not possible the tower itself can be stiffened. An obvious choice is the use of longitudinal A-frames for the towers. Leonhardt proposed this solution in 1968 for his design of the Ala Habat Bridge across the Ganges River in India, Figs 2.282 and 2.283. In accordance with Indian practice the bridge dimensions were selected in such a way that each tower leg could be supported by one well foundation.

2.5.1.4 Stayed towers

Another possibility for retaining the tower heads is to stay them: Fig. 2.184 shows three possibilities. The tower heads can be directly connected to one another with a horizontal cable which itself is an-

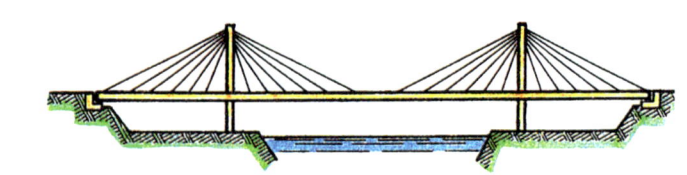

a) Static configuration

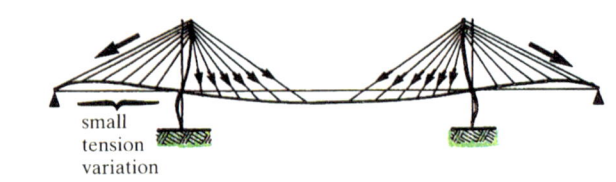

b) Loading the main span
small tension variation

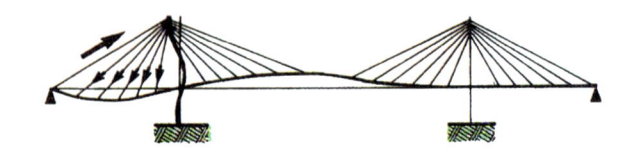

c) Loading a side span

Figure 2.178 Load transfer in a typical three-span cable-stayed bridge

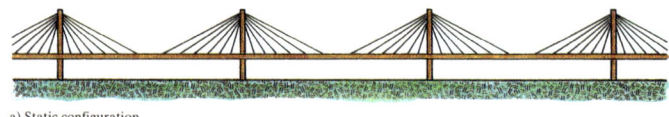

a) Static configuration

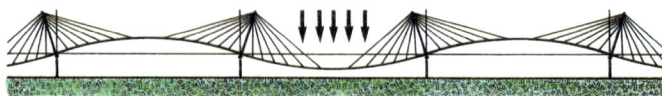

b) Effect on a span loading with a simple line of bearings on the piers to support the system of deck and pylons, no participation of the piers, and pylons turning almost freely

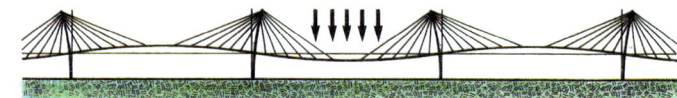

c) Effect of a span loading with bending piers taking part in the limitation of deflections

Figure 2.179 Load transfer in an unstiffened series of cable-stayed bridges

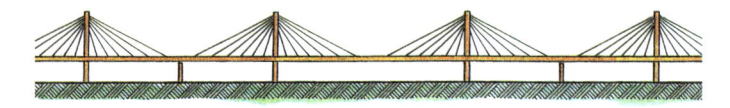

Figure 2.180 Series with intermediate piers

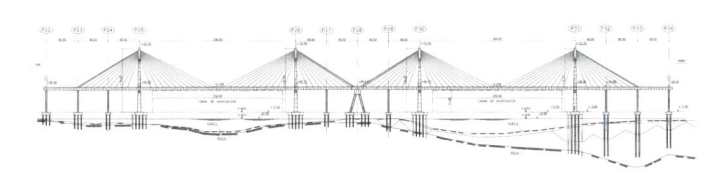

Figure 2.181 Second Orinoco Bridge with intermediate A-frame pier

2.5 Special systems of cable-stayed bridges

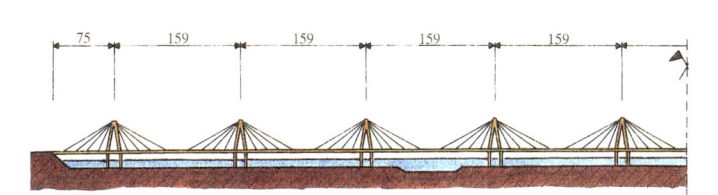

Figure 2.182 Ala Habat Bridge across the Ganges River, India

Figure 2.183 Ala Habat Bridge, artist's rendering

a) Intermediate support every second span

b) Head-cables

c) Long cables from a pylon head to an adjacent pylon at the deck level

d) Cable-stays coming from both adjacent pylons to support the central part of each span

Figure 2.184 Possibilities for staying the towers

Figure 2.185 Poole Harbor Bridge, GB, first prize in a design competition 1996

Figure 2.186 Munksjö Bridge, Jönköping, Sweden

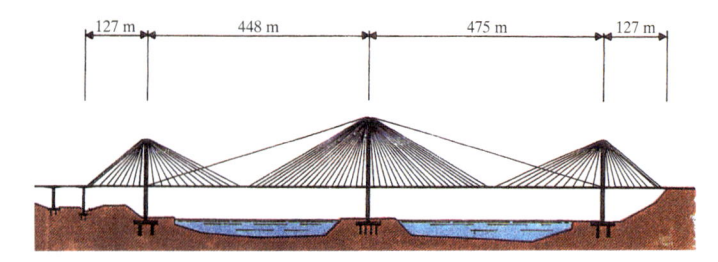

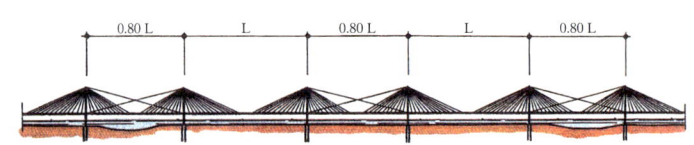

Figure 2.187 Ting Kau Bridge, Hong Kong, China, 1998, 448 m + 475 m

Figure 2.188 Design for the Patna Bridge across the Ganges River, India, 1971

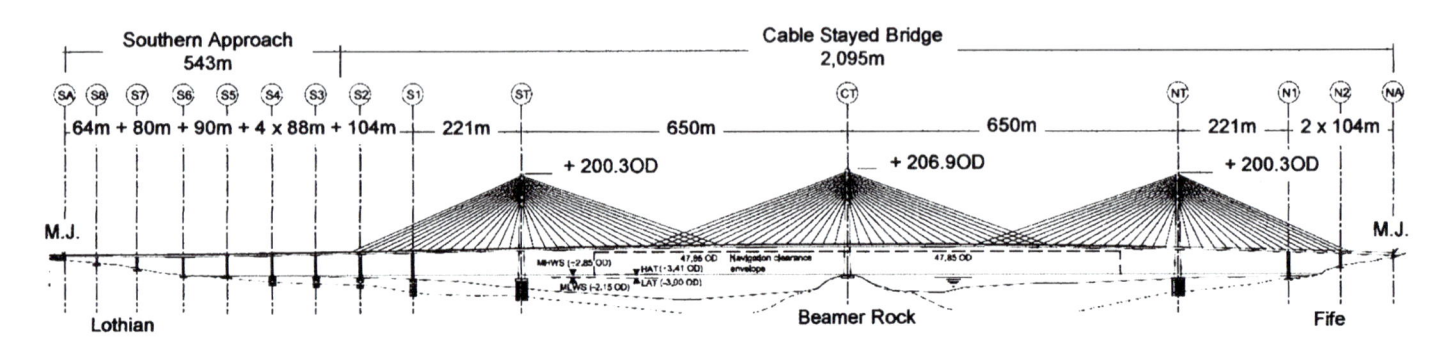

Figure 2.189 Design for the new Forth Bridge, Scotland, under design 2011

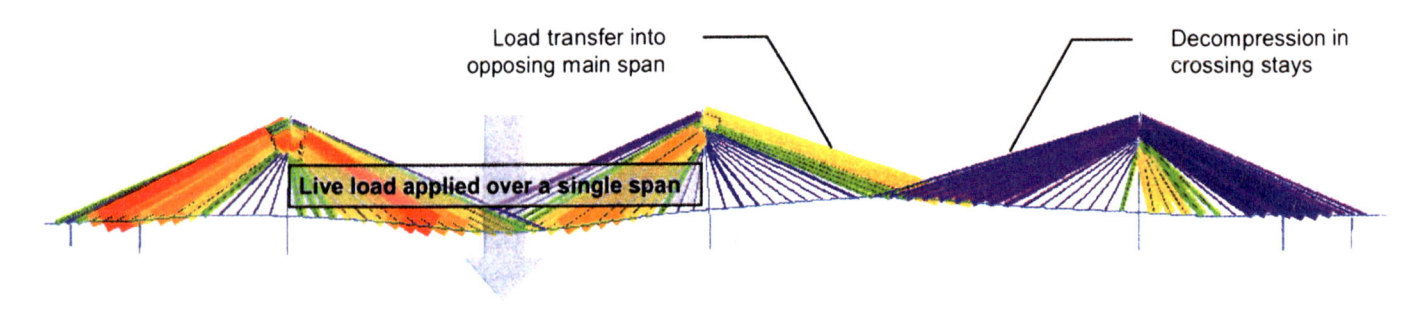

Figure 2.190 Cable forces from live load in a single span

2.5 Special systems of cable-stayed bridges

chored at the outer tie-down piers, Fig. 2.184 b. The disadvantage of this solution is the sag of the cables which increases with a square of the tower distances and which can reduce the effective stiffness of the horizontal cables significantly. If the horizontal cables are provided with higher tensile forces, then these have to be carried in compression over the full length of the beam. Examples for this solution are the design of the Poole Harbor Bridge, Fig. 2.185, and the realized Munksjö Bridge, Fig. 2.186.

Fig. 2.184 c shows how the tower heads can be tied back to the left and right towers at beam level. An example is the Ting Kau Bridge, Fig. 2.187. For the Patna Bridge, Fig. 2.188, every second span has only 0.8 times the length of the main spans like a series of three-span cable-stayed bridges back-to-back, which helps to stiffen the tower heads with crossing ties over the shorter spans. But for this configuration there is also the disadvantage that the tie cables become very long and the corresponding sag reduces their effective stiffness.

In Fig. 2.184 d the final possibility is shown of how to stiffen two adjacent main spans by overlapping the stays at their centers. An example is the design for the new Forth Bridge, Fig. 2.189, which is currently under construction. Live loads in one of the main spans produces compression (unloading) in the stays of the adjacent spans, Fig. 2.190 [2.113]. This system requires an increased stiffness of the beam in the region of overlapping cables.

2.5.1.5 Frames
Finally, we can try to achieve a continuous frame system by connecting beams and towers monolithically, Figs 2.191 and 2.192. In order to achieve the required stiffness, either the beam (Fig. 2.191 a) or the towers (Fig. 2.191 c) can be strengthened.

Examples of frame systems are the Kuang Fu Bridge, Fig. 2.193, and the Arenas Viaduct, Fig. 2.194.

2.5.1.6 Accommodation of longitudinal deformations
For the extra-long beam of a series of cable-stayed bridges the accommodation of the longitudinal beam deformations due to shrinkage, creep and change of temperature have to be carefully considered.

If intermediate movement joints are placed at suitable distances, they can accommodate the beam changes in length. Their disadvantage, a kink in the deformation line, can theoretically be avoided by installing some type of sleeve joint which can slide to absorb the longitudinal movements, but which transmits moments and shear, Fig. 2.195.

Figs 2.196 and 2.197 show that the towers can be made flexible below deck level by splitting the pier into two halves, so that they can follow horizontal movements but can take moments by a force couple in the two tower legs. This system is well known for continuous girders built by free cantilevering, when unbalanced loads are carried in a force couple in the split pier legs.

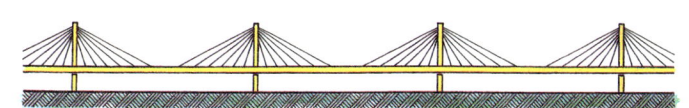

a) Rigid deck, with different possible types of connection between deck and piers *(rigid connection, two lines of bearings or one)*

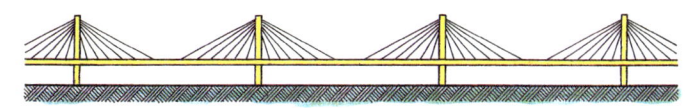

b) Intermediate solutions, with rigidity distributed between piers deck and pylons

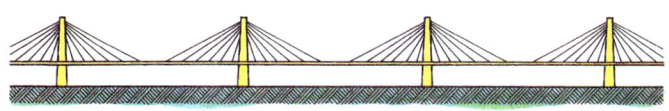

c) Rigid pylons and flexible deck, with a transmission of moments between pylons and piers *(rigid connection or two lines of bearings)*

Figure 2.191 Stiffness distribution between beam and towers

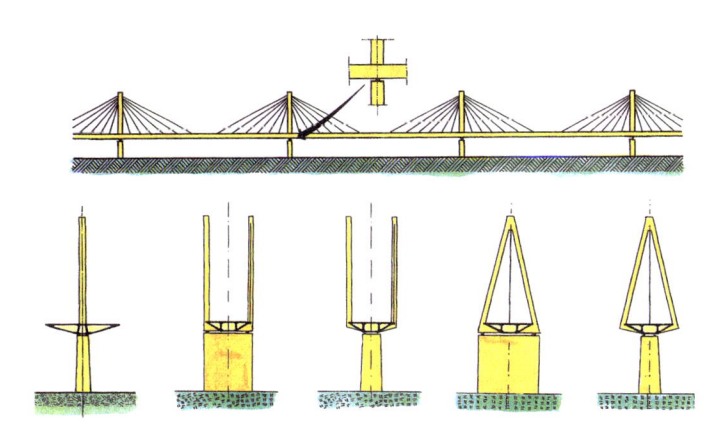

Figure 2.192 Monolithical connection of beam and towers

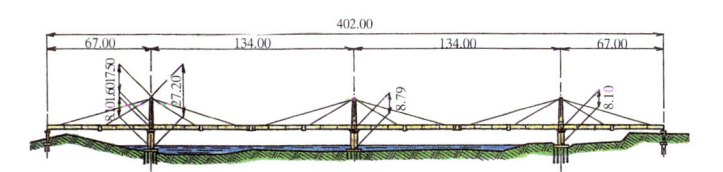

Figure 2.193 Kuang Fu Bridge, Taiwan, 1977

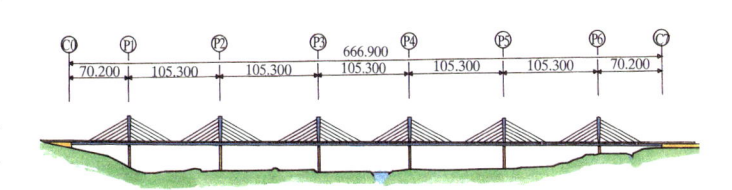

Figure 2.194 Arenas Viaduct, Mexico, 1992

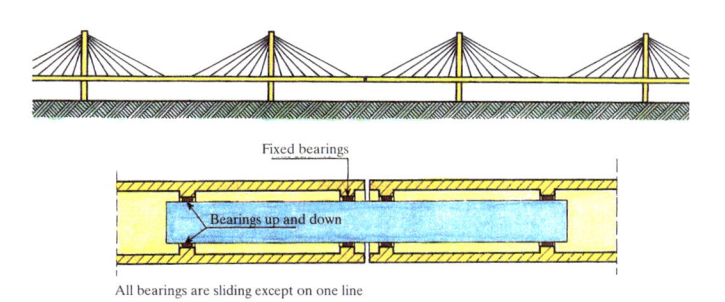

Figure 2.195 Longitudinal movement joints carrying moments and shear

122   2 The development of cable-stayed bridges

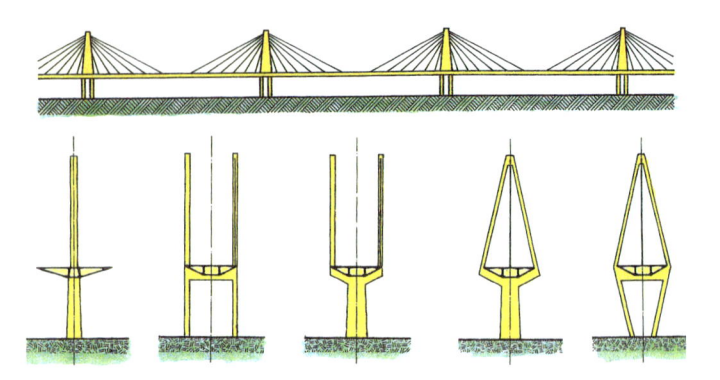

Figure 2.196 Horizontally flexible but moment-stiff towers

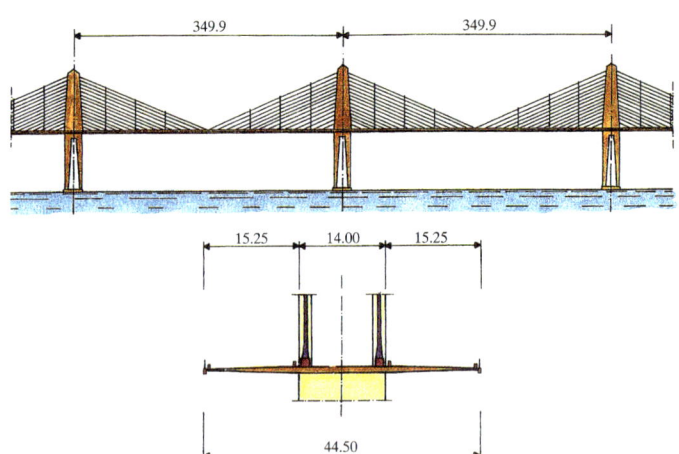

Figure 2.197 Great Belt Bridge, Design by Finsterwalder, 1967

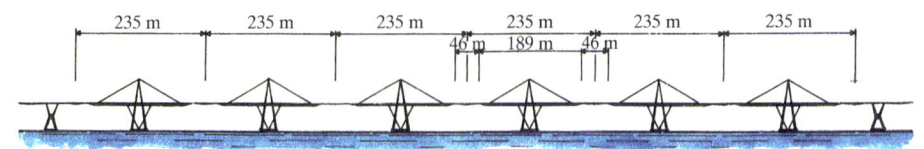

Figure 2.198 Stiff A-towers with V-piers of the Lake Maracaibo Bridge, Venezuela, 1962, 6 x 235 m

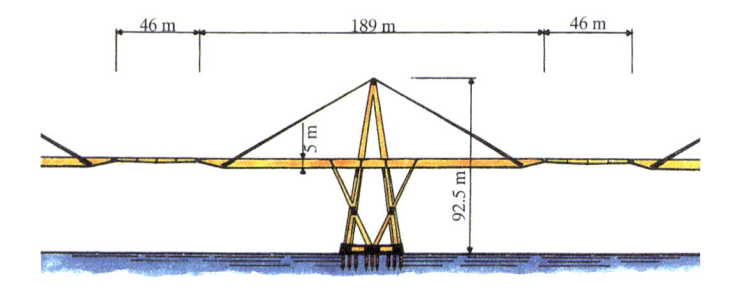

Figure 2.199 Individual span

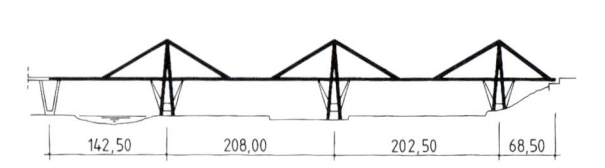

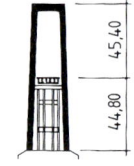

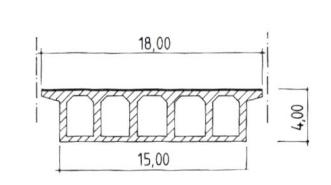

Figure 2.200 Polcevera Viaduct, Italy, 1964, 208 m + 203 m

2.5.1.7 Examples

The bridge across Lake Maracaibo in Venezuela by Morandi, Fig. 2.125, is the outstanding first example for a series of concrete bridges, Fig. 2.198 [2.114]. The additional challenge for this nearly 9 km long bridge was the required 30 to 45 m deep foundations through a thick overlaying silt. The large size of the bridge permitted the use of special equipment for this purpose.

Each of the six towers, Fig. 2.199, supports the adjacent concrete box girders with a depth of 5 m by single pairs of concrete forestays and backstays at distances of 79 m. The beam cantilevers a further 15 m to support a 46 m long drop-in girder in the center of each main span. The A-shaped towers support the beam additionally by two V-piers, and the long approach bridges use X-shaped piers.

The A-frames transmit eccentric loads in tension and compression directly into the foundations. The changes in beam lengths are taken up by the joints of the drop-in girders.

Since the main spans of 235 m are only supported by two concrete stays, the beam requires the additional support of the V-piers. The run of moments in the beam with stiff supports by the concrete stays and V-piers thus resembles more a rigidly supported continuous girder than a modern multi-stay cable bridge continuously elastically supported. The disadvantage from today's viewpoint is the obviously large amount of material used and the susceptibility to wear and tear and the driving discomfort of the many movement joints at the drop-in girders.

The bridge across Lake Maracaibo was an outstanding success of a new design under difficult conditions at the time of its completion in 1962. Its design and construction methods were appropriate at that time.

Shortly after the Maracaibo Bridge, Morandi designed a similar system for a freeway viaduct at Polcevera in Italy, Fig. 2.200 [2.115]. Its piers, towers and beams were again CIP monolithically. The stays were surrounded by CIP concrete beams which were post-tensioned in such a way that the concrete remains under compression for all loadings.

The Kuang Fu Bridge in Taiwan, Fig. 2.201 [2.116], is a series of cable-stayed concrete bridges with four spans built from precast concrete girders. The outer two steel towers are restrained by regular backstays. The center tower is relatively flexible, supported by the stays to the adjacent spans of 134 m each. This is a relatively overall flexible system. The bridge was designed by T. Y. Lin, Fig. 2.202.

The Arenas Viaduct in Spain is an elegant-looking bridge, Fig. 2.203, with an S-shaped roadway. This series of cable-stayed bridges comprises six towers with five main spans of 105 m each. These relatively short spans reduce the deflection problems, the stiff beam mainly carries asymmetric live loads together with the towers. The beam is not monolithic with the towers but rests on two neoprene bearings in each axis. This floating system accommodates the changes in beam lengths by shear deformations of the neoprene bearings and tower deflections.

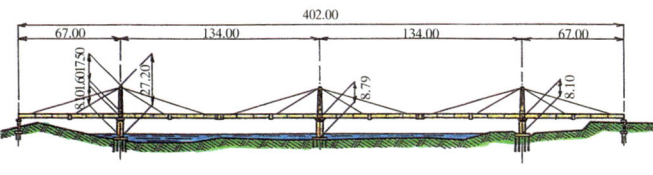

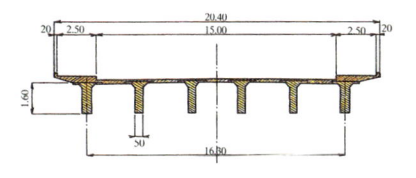

Figure 2.201 Kuang Fu Bridge, Taiwan, 1977, 2 x 134 m

Born 1912 in Fuzhou, China
1931 BSc Tangshan Jiaotong University
1933 MSc University of California, Berkeley
1946 Returns to the USA
1954 Founder of T. Y. Lin International
1992 Leaves T. Y. Lin to found Consultancy in China
Died 2003 in El Cerrito, USA

Figure 2.202 T. Y. Lin

124 2 The development of cable-stayed bridges

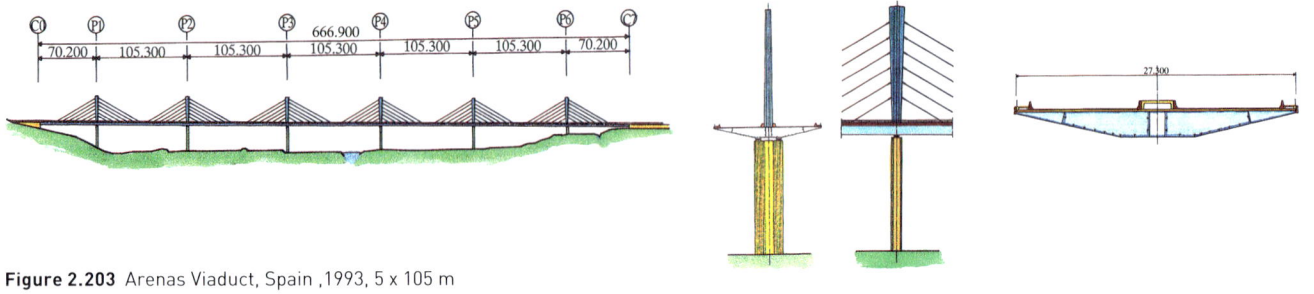

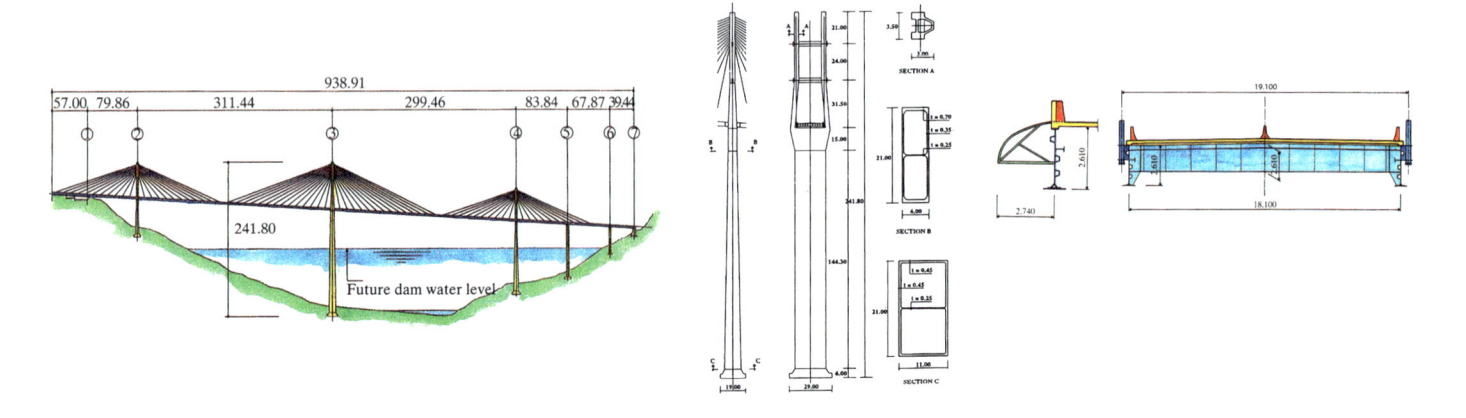

Figure 2.203 Arenas Viaduct, Spain, 1993, 5 x 105 m

Figure 2.204 Mezcala Bridge, Mexico, 1993, 300 m + 311 m

2.5 Special systems of cable-stayed bridges

The Mezcala Bridge in Mexico, Fig. 2.204 [2.117], comprises three towers with two main spans of 300 m and 311 m. The two shorter outer towers are directly stabilized by backstays to the hold-down piers. The composite beam runs between the tower legs and is supported by the solid lower part of the towers.

The Macau Bridge, Fig. 2.205 [2.118], has two main spans of 112 m each. Since drop-in girders are located in the main span centers the individual bridge units behave like independent cable-stayed bridges with single fore and backstays.

The Ting Kau Bridge in Hong Kong, Fig. 2.206 [2.119, 2.120], has two main spans of 448 m and 475 m with two outer shortened towers, similar to the Mezcala Bridge. The composite beam straddles the tower. The tip of the inner tower is stayed to the two outer towers at beam level. The large sag of the nearly 500 m long stiffening cables is obvious from Fig. 2.206. The slender central single towers are also uniquely stayed transversely, as the shrouds of sailing ships, Fig. 2.2. The engineers in charge were Jörg Schlaich and Rudolf Bergermann, Figs 2.156 and 2.157.

The basic design concept of the Sunniberg Bridge in Switzerland, Fig. 2.207 [2.121], is a series of extradosed cable-stayed beams with low towers and a slender beam. Five spans with four towers fit the landscape and the road alignment. The location of the 2 · 140 m main spans are adjusted to the course of the river in the valley below.

The towers above the road were to be as low as possible because the piers underneath are already high. One tenth of the spans equal to a quarter of the piers was finally chosen. Flatter inclinations of the stays were not possible with respect to the beam deflections. Beam and piers were to form a unit because live loads in individual spans cannot be carried by the flexible adjacent spans but have to be transmitted directly into the piers. The tower heads were stabilized horizontally by the continuous beam curved in plan over the full bridge length. For the beam width of 10 m between the cable planes a thickness of 40 cm, with 80 cm edge thickenings, was selected in which the cables were anchored at corbels. The longitudinal distance between the cable anchorages was chosen as 6 m, taking into account the permissible beam moments during free cantilevering, the cable sizes and the beam stability. The engineer in charge of the Sunniberg Bridge was Christian Menn, Fig. 2.135.

The Rion-Antirion Bridge across the Corinthian Straights in Greece is a composite cable-stayed bridge with concrete towers and three main spans of 560 m each, Fig. 2.208 [2.122]. The main design problem was the danger of large earthquakes. The towers are supported on 60 m long steel piles in 55 m deep water. The octagonal piers extend up to the beam where they widen to accept the width of the beam. On top of each pier a four-legged tower is placed, A-shaped in longitudinal and transverse direction. The four legs unite at the top for a composite cable anchorage.

The 2252 m long beam is continuous over its full length and is only supported by the stay cables, even at the towers. At both ends, roadway joints for ± 2.5 m movement can take the changes in lengths

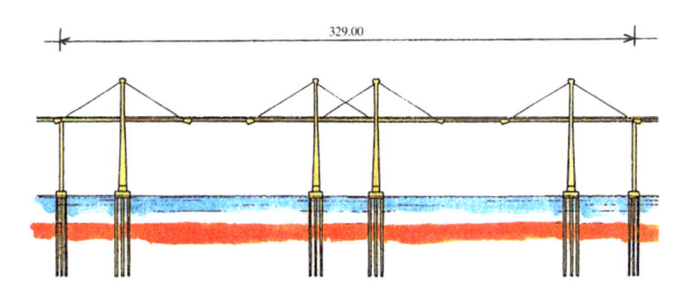

Figure 2.205 Macau Bridge, China, 1994, 2 x 112 m

126　　　　　　　　　　　　　　　　　　　　　　　　　　　　　　　　　　2 The development of cable-stayed bridges

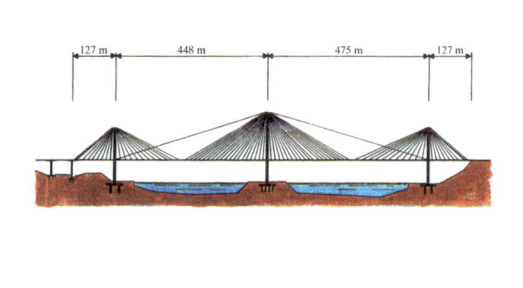

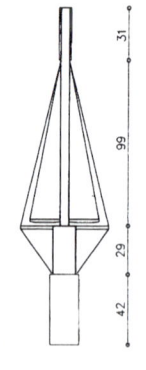

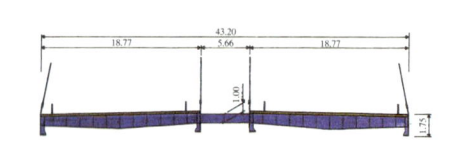

Figure 2.206 Ting Kau Bridge, Hong Kong, China, 1998, 448 m and 475 m

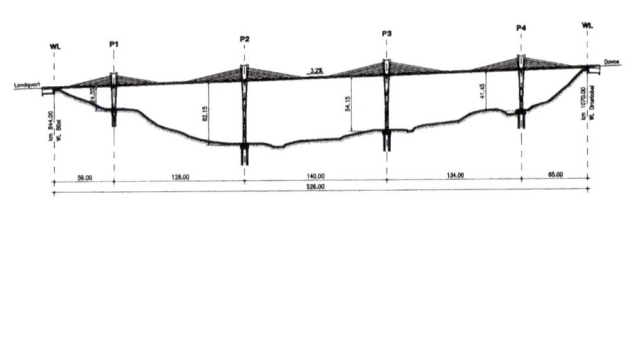

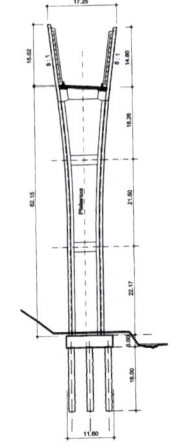

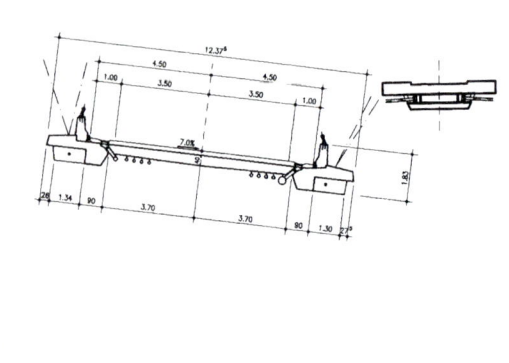

Figure 2.207 Sunniberg Bridge, Switzerland, 1998, 140 m

2.5 Special systems of cable-stayed bridges

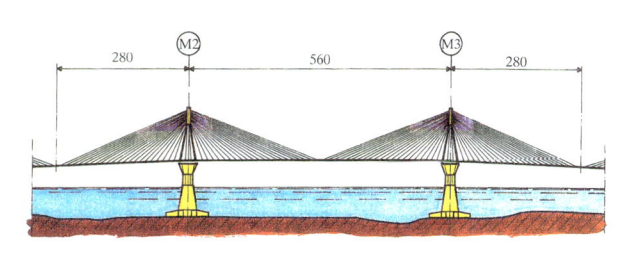

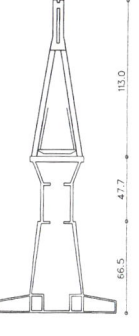

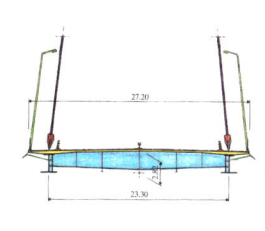

Figure 2.208 Rion-Antirion Bridge, Greece, 2004, 3 x 560 m

Born 1943 in Houilles, France
1965 MSc École Centrale de Lyon
1970–93 Campenon Bernard
1993–2001 GTM
since 2001 Advisor to GTM
since 2005 Finley Engineering Group, USA
since 1996 Professor École National
des Ponts et Chaussées
1977 Brotonne Bridge
2004 Rion-Antirion Bridge

Figure 2.209 Jacques Combault

due to temperature and earthquake. The transverse earthquake movements are restrained by dampers at the towers. The tower legs are provided with a high ductility in case of earthquake. The design and construction of this exceptional bridge is described in detail in Section 6.6.2. The engineer in charge of the Rion-Antirion Bridge was Jacques Combault, Fig. 2.209. Michel Virlogeux, Fig. 2.101, was an important advisor.

The shape of the Millau Bridge across the Tarne valley in France, Fig. 2.210 [2.123], as a series of cable-stayed bridges, was selected in 1996 by a jury with decisive input from the engineer Michel Virlogeux, Fig. 2.101, and architectural advice from the architect Sir Norman Foster. Two alternates were developed, one with a post-tensioned concrete beam and the other with an orthotropic steel deck. Both solutions looked similar and fitted the topography and the high wind speeds at up to 350 m above the valley floor. The live loads are distributed by the box girders and stay cables to the tips of the A-towers and from there as a force couple into the bifurcated pier heads. The piers can follow the longitudinal beam deformations due to their bifurcations in the upper part. Great emphasis was placed on an elegant appearance in which all bridge elements are integrated into a visual unity. The simple form makes the flow of forces apparent. In the end the steel alternate was selected, not least to permit beam construction by launching.

Its 2460 m long orthotropic deck has a trapezoidal, almost triangular shape with a narrow bottom flange, a depth of 4.5 m and a slenderness of 1:76, Fig. 2.210. The towers are 90 m high above the beam with an A-shape in the longitudinal direction in order to achieve the required stiffness and, at the same time, to appear light and transparent. For aesthetic reasons a central cable plane was selected so that the cables in the shape of a modified fan do not visually intersect. The piers below the beam with a height of up to 270 m above the Tarn river have to withstand high transverse winds, but they also have to be flexible enough to follow the beam changes in lengths due to temperature. All piers have the same shape for aesthetic reasons. The design and construction of this outstanding bridge is explained in detail in Section 6.6.1.

The second Orinoco River Bridge at Ciudad Guayana, Venezuela, for road and rail, has a composite beam with a total length of 3.6 km [2.124]. It consists of two three-span cable-stayed bridges back-to-back with main spans of 300 m, Fig. 2.211. The common hold-down pier between the two bridges creates the usual flow of forces as in regular three-span cable-stayed bridges. In order to carry the heavy braking loads from railways the hold-down pier is A-shaped in the longitudinal direction. Due to the alluvial soils all piers and towers are founded on large diameter bore piles.

A 17.6 km long permanent connection across the Femer Belt of the Baltic Sea will replace the current ferry connection between Puttgarden on the isle of Femarn in Germany and Rødby in Denmark [2.125]. In addition to a tunnel, the bridge shown in Fig. 2.212 has been preliminarily designed. The central part of the alternate bridge

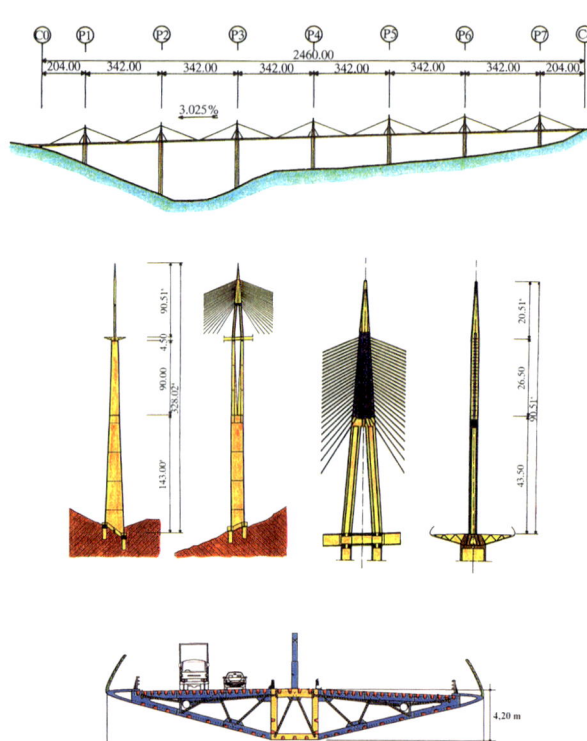

Figure 2.210 Millau Bridge, France, 2004, 6 x 342 m

2.5 Special systems of cable-stayed bridges

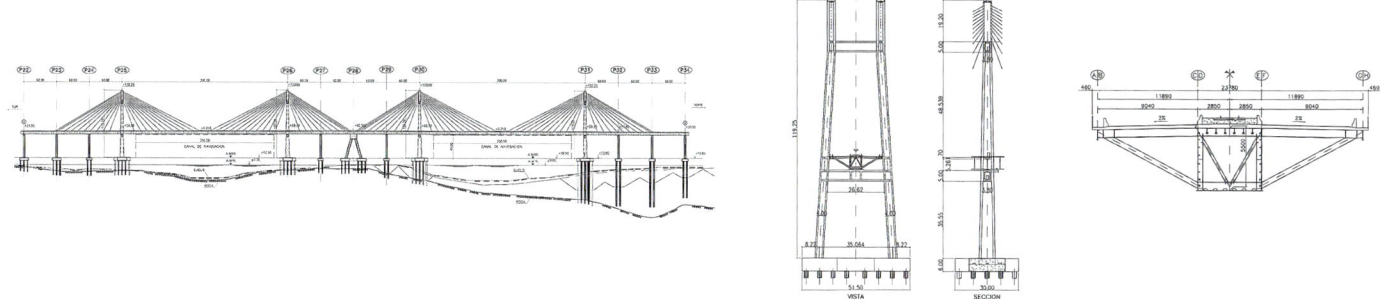

Figure 2.211 Second Orinoco Bridge, Venezuela, 2006, 2 x 300 m

Figure 2.212 Femer Bridge between Germany and Denmark, design 1999, 3 x 724 m

comprises three large spans of 724 m each. The two outer spans provide navigational clearance for the ships in each direction of sailing while the central span divides the two outer shipping lanes safely. The total length of the cable-stayed bridge comes to 2414 m. The truss cross-section with road traffic on top and two railway tracks on the bottom runs continuously over the full length of the bridge, which is fixed longitudinally at the central tower.

2.5.2 Stayed beams

Some bridge beams have been provided with spans stayed from above or from underneath in order to increase their structural depth and thus their moment capacity, because a stayed slender beam is less obtrusive than a continuous beam with considerably greater depth.

2.5.2.1 Stayed from underneath

Beams stayed from underneath (underslung supports with kingpost) have been used for a long time in timber construction for strengthening longer spans, for example for the roof structure of houses. For bridges they are used only as an exception for individual large spans in order to keep their depth as low as required for the adjacent side spans and, in this way, to provide a slender overall appearance to the bridge. These stays can reduce the moments under permanent loads considerably, but they are less effective for live loads due to their small additional stiffness.

Examples

The six-span Neckar Valley bridge near Weitingen with span lengths of 234 – 134 – 134 – 134 – 264 m reaches up to 124 m above ground, Fig. 2.213 [2.126]. Due to bad soil conditions the slopes were expected to creep with time and thus no foundations could be placed there. This led to the very long side spans which were stayed from underneath with locked coil ropes supporting a central strut. For permanent loads the side spans were thus in effect nearly halved and a constant beam depth of 6 m over the full length of the bridge was possible.

The roadway with a width of 31.5 m is supported by a central box girder with cantilevers supported by inclined struts. They are supported by twin piers at the footings of the slopes and two single piers in the middle part. These inner piers act like pendulum piers and carry only vertical loads, the wind loads on the bridge beam being taken by the outer piers and abutments. The construction of the bridge took place by free cantilevering on the final piers and auxiliary piers in between.

The steel beam of the Obere Argen Valley Bridge is stayed from above and from underneath, Fig. 2.214 [2.127], in order to overcome the 344 m on top of a creeping slope. The beam is supported by stays and by struts at six distances of about 43 m each.

The cross-section is a single-cell box, 3.7 m deep and 9.4 m wide with cantilevers supported by inclined struts. The three struts underneath the beam are supported by six parabola-shaped locked coil ropes ø 126 mm. The horizontal curvature of the beam causes deviations of the struts in the horizontal direction also. The two outer V-shaped struts are designed as pendulum piers in the longitudinal direction and thus transfer only vertical loads. The engineer in charge was Jörg Schlaich, Fig. 2.156.

2.5.2.2 Stayed from above (extradosed)

Stays above the beam are chosen in order to increase the effective beam depth above the piers. If the stays are arranged as in a cable-stayed bridge but with low towers, they are called 'extradosed'. They differ from usual cable-stayed bridges by having towers with a height of distinctly less than the usual 20 % of the adjacent spans.

The shear transfer of live loads from the beam to the piers does not take place exclusively via the stay cables. The model of a beam articulated at the cable anchor points as outlined in Section 4.1.3 would not work. The beams often have haunches. There is a sliding transition in the shear transfer from loads in the main spans to the piers. Continuous girders proper transfer 100 % of the loads via shear to the piers, whereas cable-stayed bridges with very slender beams carry nearly all the loads via the cables to the towers and the piers. Extradosed cable-stayed bridges lie in between with something like 50 % of the loads being carried by the beam and the cables each [2.128].

Examples

The Golden Ears Bridge across the Fraser River in Vancouver, BC, has three main spans of 242 m each, Fig. 2.215 [2.129]. The 968 m long composite beam comprises a simple steel girder covered by concrete slabs with a low weight in order to load the foundations as little as possible in very difficult soil conditions. The beam is connected to the piers, the bridge acting like a continuous frame with its fixed point in the center of the bridge. The towers are bifurcated underneath the beam to be flexible in the horizontal direction in order to follow the beam movements, but stiff for the bending from the towers above deck. The stay cables are arranged in harp shape with flat inclinations which results in short, slender towers above deck. The beam depth increases from 3.07 m in mid span to 4.87 m above the piers. The shear transfer is thus split between the stays and the beam.

The Tisza River Bridge near Szeged, Hungary, has spans of 95 m, 180 m and 95 m, Fig. 2.216. The composite cross-section with upper and lower concrete flanges and steel webs has haunches above the main piers. The flat stay cables enhance the load-bearing capacity of the beam.

2.5 Special systems of cable-stayed bridges

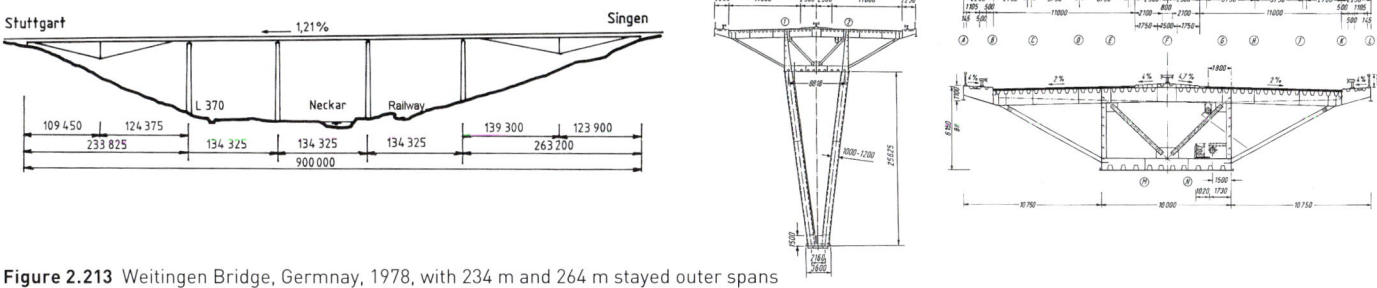

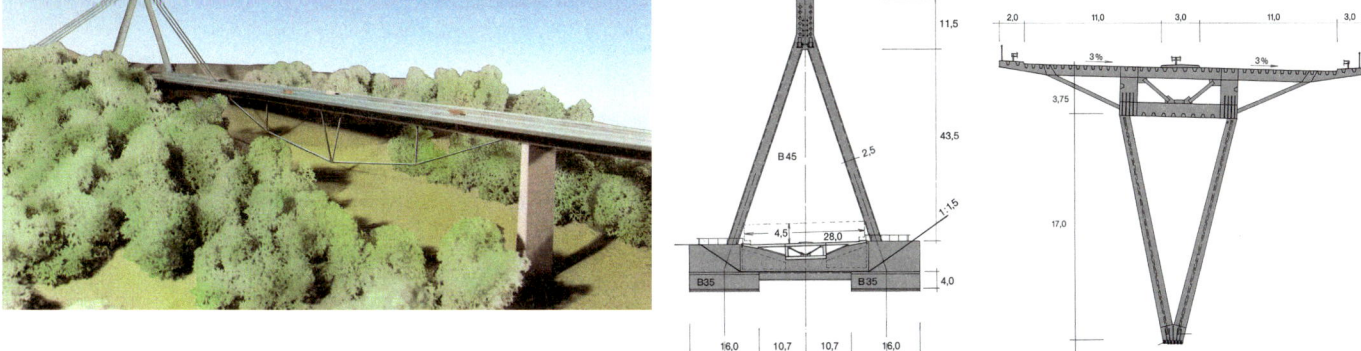

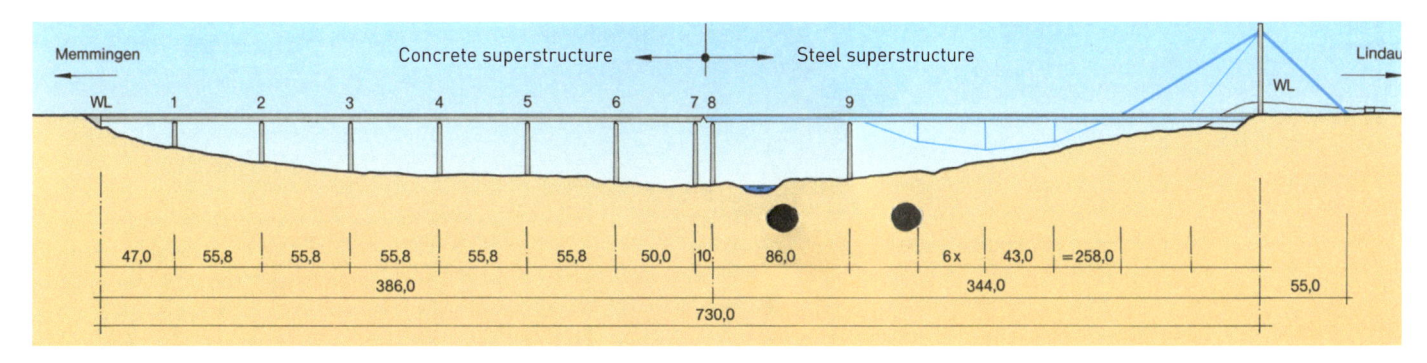

Figure 2.213 Weitingen Bridge, Germnay, 1978, with 234 m and 264 m stayed outer spans

Figure 2.214 Obere Argen Valley Bridge, Germany, 1990, 344 m

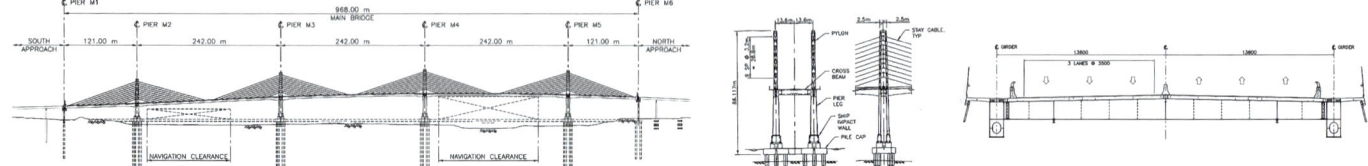

Figure 2.215 Golden Ears Bridge, Vancouver, Canada, 2009, 3 x 242 m

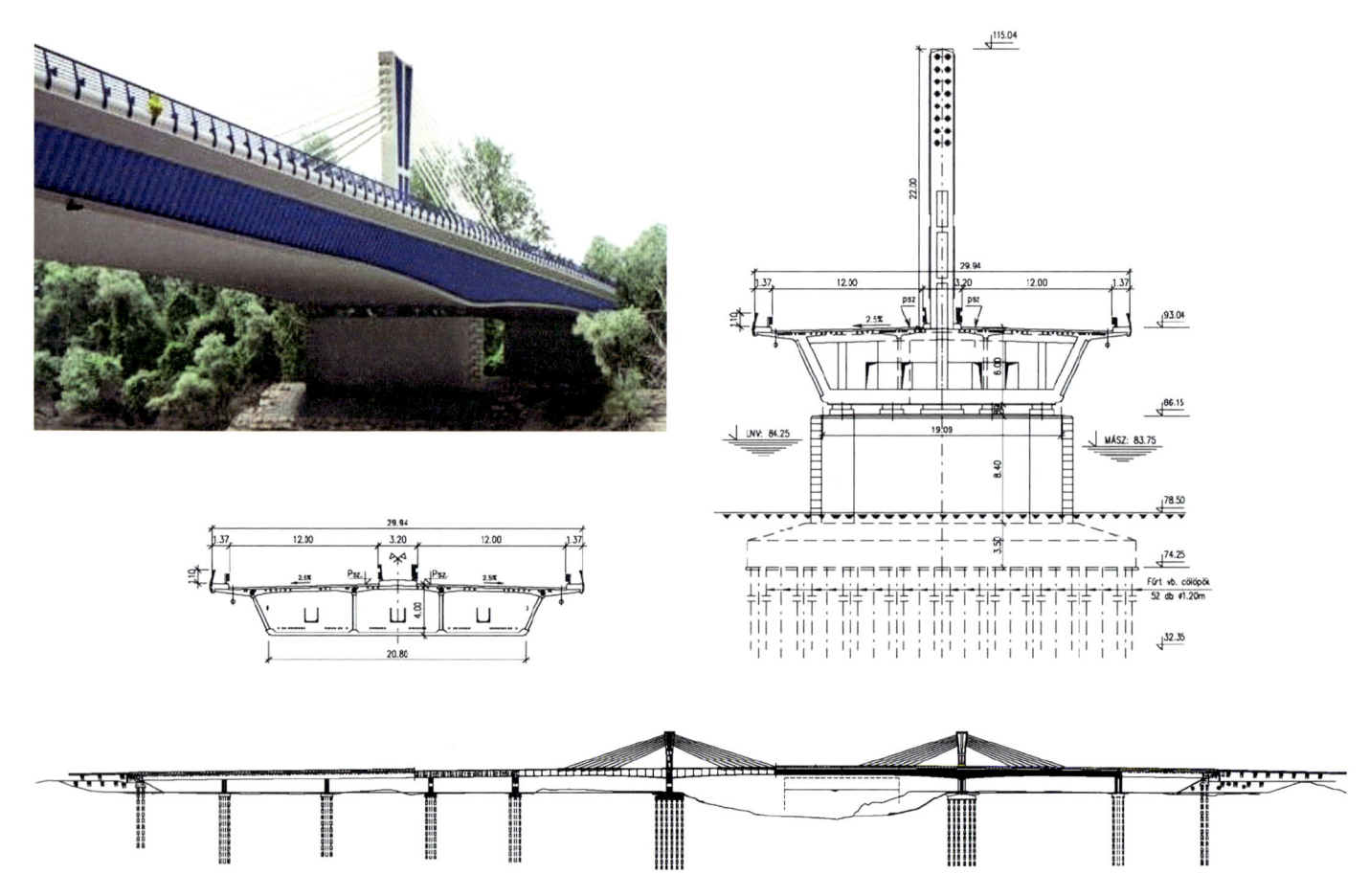

Figure 2.216 Tisza River Bridge at Szeged, Hungary, 2010, 180 m

2.5.3 Cable-stayed pedestrian bridges

Pedestrian bridges are the playground of bridge engineers. Their variety of shapes is almost unlimited. This results in the use of all possible materials with structural systems which do not always use the most effective flow of forces.

Comprehensive descriptions of pedestrian cable-stayed bridges can be found in the books of Jörg Schlaich and Rudolf Bergermann [2.130] as well as Ursula Baus and Mike Schlaich [2.131], both entitled *Fußgängerbrücken (Pedestrian Bridges)*. Latest examples are given by Idelberger in [2.131a].

For our purposes we only outline some cable-stayed bridges designed by Leonhardt, Andrä and Partners which show the basic principles of pedestrian cable-stayed bridges, especially their lightness, but have a rational flow of forces and use traditional materials of steel and concrete.

Schillersteg, Stuttgart

The Schillersteg crosses a main street in front of the railway station in Stuttgart, Fig. 2.217 [2.132]. Built in 1960, it is one of the early cable-stayed bridges, using parallel wire cables in PE-pipes with cement grout for the first time. It is distinguished by its lightness.

The beam depth of the Schillersteg had to be kept as low as possible in order to save the pedestrians ascending unnecessarily. The bridge connects two parks, and on one side follows the natural split of the pedestrians' destinations by bifurcating the superstructure. The bridge extends beyond the street on both sides in order to avoid large abutments so that the bridge seems to grow out of the park area. This resulted in a span of 90 m which had to be supported by stays to permit the low beam depth of 1:180.

The tower is placed just outside the street. A light steel box from thin plates and a thin asphaltic wearing surface was selected for the beam.

The bridge beam can easily be excited by pedestrians, which made its users initially question its structural safety. With time, however, people have become used to slight vibrations and hardly pay any attention to them anymore.

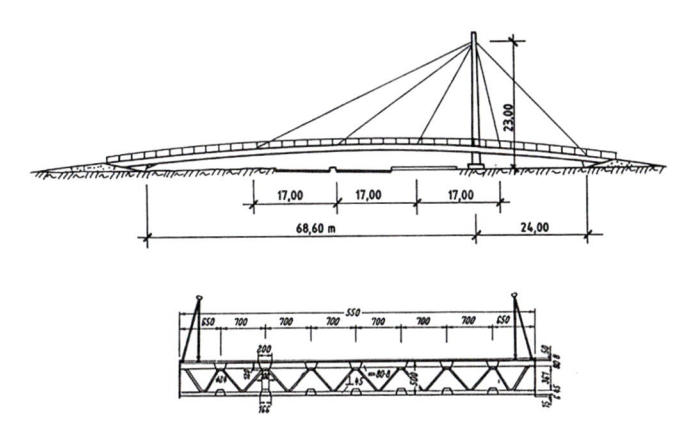

Figure 2.217 Schillersteg, Stuttgart, Germany, 1960, 90 m

Bickensteg, Villingen

The pedestrian Bickensteg in Villingen, Black Forest, connects the western part of the City of Villingen across a railway yard with the eastern district, Fig. 2.218 [2.133]. A main span of 66.5 m was required which had to be crossed by a beam as slender as possible.

This task was accomplished with a cable-stayed bridge with only one tower and a central cable plane. For the 5 m wide solid concrete beam a depth of only 0.6 m was sufficient to provide the necessary torsional stiffness. In the area across the railway yard five up to 14.10 m long precast elements from lightweight concrete were supported on auxiliary piers. Their reinforced joints were then closed with normal-weight concrete. For further reduction of the dead weight, the beam was provided with internal cylindrical block-outs.

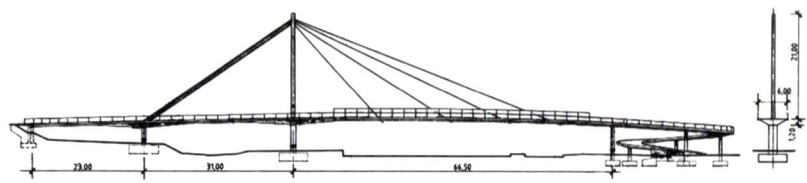

Figure 2.218 Bickensteg, Villingen, Germany, 1972, 66.5 m

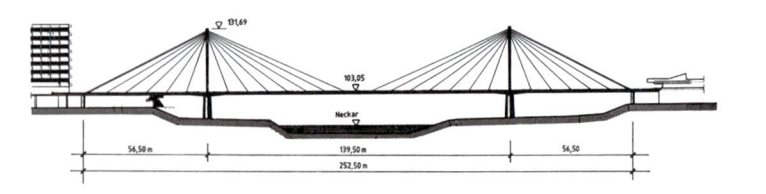

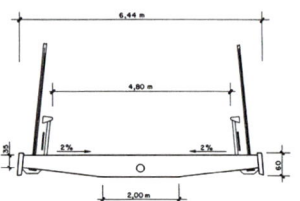

Figure 2.219 Neckarcenter, Mannheim, Germany, 1974, 139.5 m

2.5 Special systems of cable-stayed bridges

The end span was cast with normal-weight concrete without block-outs, in order to serve as a counterweight for the main span.

The bridge is fixed at the tower, where a concrete hinge connects the beam and the pier. At the anchor pier the beam is tied down to the foundations with tendons. The steel tower is cast into the beam, the stay cables using parallel wire cables with HiAm-anchorages.

Neckarcenter Bridge, Mannheim
The Neckarcenter Bridge, Mannheim, comprises a slender concrete beam supported by two outer cable planes, Fig. 2.219 [2.134]. The stay cables are anchored separately at the top of the steel towers.

At the beam the cables are anchored at 9–10 m distances so that the trapezoidal cross-section requires a depth of only 60 cm. At the towers the beam is haunched in both transverse and longitudinal directions so that its depth rises to 1.2 m.

Since the beam is fixed to the piers at both towers but can rotate, Fig. 2.220, an expansion joint is required in the bridge center. This joint permits rotations and longitudinal movements of the beam due to shrinkage and creep and change of temperature, but it transmits shear and torsional moments, Fig. 2.221.

Both steel towers are fixed to the concrete beam. Their cross-section increases from 1.0 m · 1.0 m at the bottom to 1.4 m in the longitudinal direction at the tower tip in order to accommodate the anchorages of the parallel wire cables. The engineer in charge was Wilhelm Zellner, Fig. 2.222.

Pedestrian bridge, Pforzheim
The unique characteristic of this pedestrian bridge in Pforzheim is the two vertical kinks in the beam at the tower and at the beam end, required by the local run of the pedestrian promenade, Fig. 2.223, which provides access to the town hall by crossing a busy inner city arterial road with a total length of 72.4 m. The cross-section of the superstructure is a concrete trough. The parallel wire stay cables in harp configuration are anchored to the single steel tower. For the

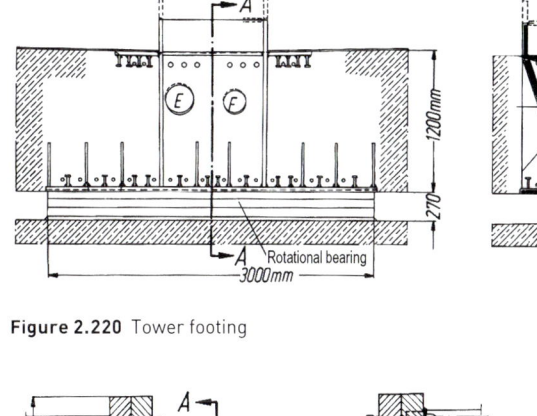

Figure 2.220 Tower footing

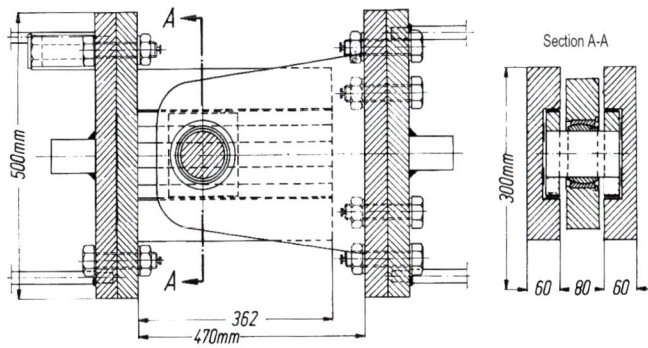

Figure 2.221 Shear hinge

Born 1932 in Frankenmarkt, Austria
1960 Dipl.-Ing. Vienna University
1962–97 Leonhardt, Andrä and Partners
1970–97 Partner/Executive Director
Pedestrian bridge, Villingen-Schwenningen
Pedestrian bridge, Neckarcenter Mannheim
Pasco-Kennewick Bridge
East Huntington Bridge

Figure 2.222 Wilhelm Zellner

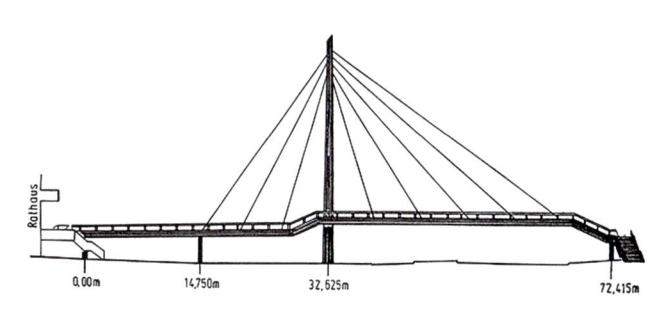

Figure 2.223 Pforzheim pedestrian bridge, Germany, 1987

first time the PE-pipe was factory-filled with an elastic corrosion protection which permits coiling and uncoiling of the complete cables for transportation. The anti-corrosion filler comprises a mixture of ground cork and bitumen.

The tower is cast into the beam and extends down to the ground and forms the fixed point of the bridge. Stairways are located at both bridge ends.

Pedestrian bridge, Ravensburg

The two pedestrian bridges near Ravensburg comprise a cable-stayed bridge, Fig. 2.224a, and a curved steel pipe truss with a concrete slab, Fig. 2.224b. Together with the truss of the sideways open tunnel, Fig. 2.224a, they form a group visually connected by the bright red color of the steel parts.

Pedestrian bridge, Kehl-Straßburg

In preparation of the cross-border horticultural exhibition 'Le jardin des deux rives' 2004 between Germany and France a competition for a pedestrian and cycle bridge across the Rhine was tendered. This was won by the French architect Marc Mimram together with LAP engineers with a very slender cable-stayed bridge, Figs 2.225 – 2.227 [2.135].

The cable-stayed main span has a length of 183 m with approaches on both sides. The composite beam is split in two parts, connected by cross beams and composite girders. The lower beam connects the two river promenades, Fig. 2.226, straight in plan and curved in elevation with a gradient of up to 18%. The upper beam is curved in plan and connects the levees on both sides so that the bridge can also be used during high water. The distance between the two beams reaches its maximum in the center of the bridge. There they are interconnected by a 300 m^2 platform which provides a look-out over the Rhine valley and invites people to rest.

For protection against wind-induced oscillations of this extraordinarily light bridge a special solution was developed, consisting of a transversely acting compensator in the shape of a see-saw that is tuned to counter the modes of rotational vibration, Fig. 2.227.

Special concrete guide structures protect the piers against ship collision, and are founded on piles with 1.2 m diameter, Figs 4.159 and 4.160.

Pedestrian bridge, Weil der Stadt

For the crossing of the Weil der Stadt bypass a cable-stayed bridge for pedestrians and cyclists was built, Fig. 2.228 [2.136]. The slender 39 m long concrete slab with a depth of only 25 m is monolithic with the abutment.

Steel was chosen as the material for the tower to permit rapid construction. The diameter of the cigar-shaped steel pipe comes to 40 cm at top and bottom and 65 cm in the middle to ensure safety against buckling. In order to permit rotation of the tower footing during installation and stressing of the cables the tower rests on a sphere from cast steel.

The distinctive bridge appears delicate and preserves an uninterrupted view of the unspoiled valley, so a cable-stayed bridge is the best structure in spite of the short span.

2.5 Special systems of cable-stayed bridges

Figure 2.224a Ravensburg cable-stayed pedestrian bridge and open tunnel, Germany, 1995

Figure 2.224b Ravensburg curved pedestrian bridge

Figure 2.225 Kehl-Straßburg pedestrian bridge, Germany–France, 2004, 183 m

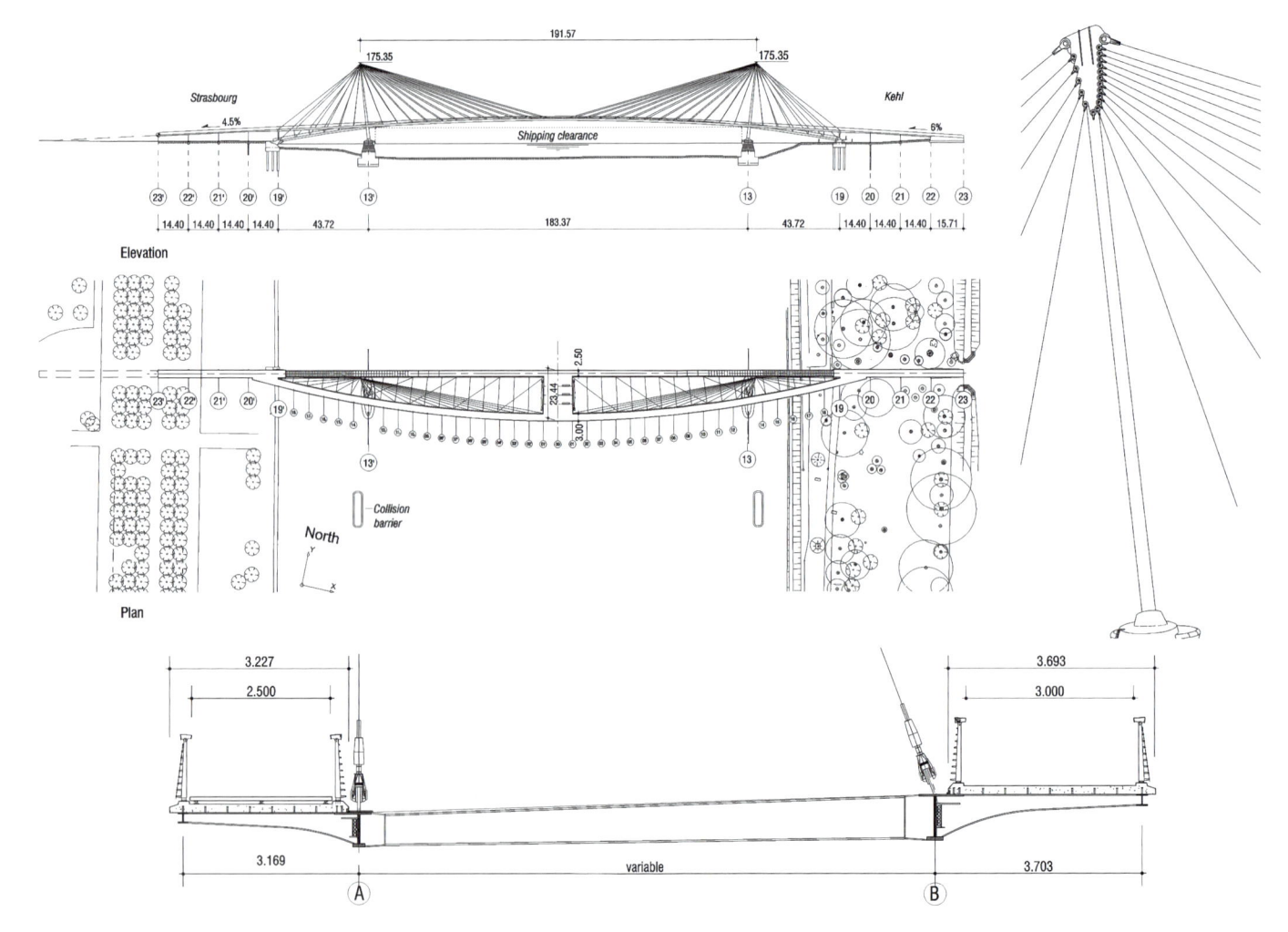

Figure 2.226 General layout Rhine River Bridge Kehl-Straßburg

2.5 Special systems of cable-stayed bridges

Figure 2.227 See-saw compensator

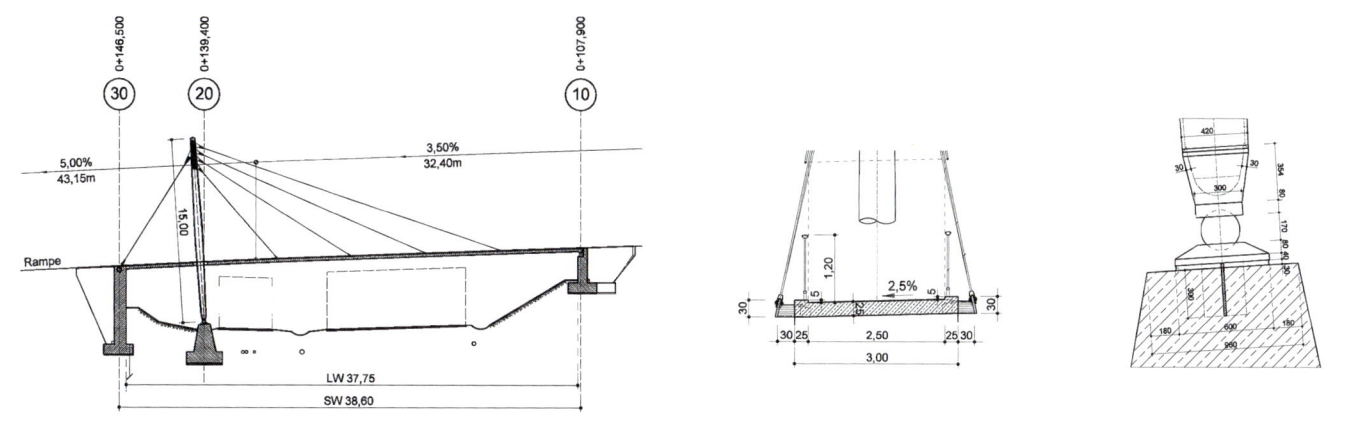

Figure 2.228 Pedestrian bridge, Weil der Stadt, Germany, 2006

3 Stay cables

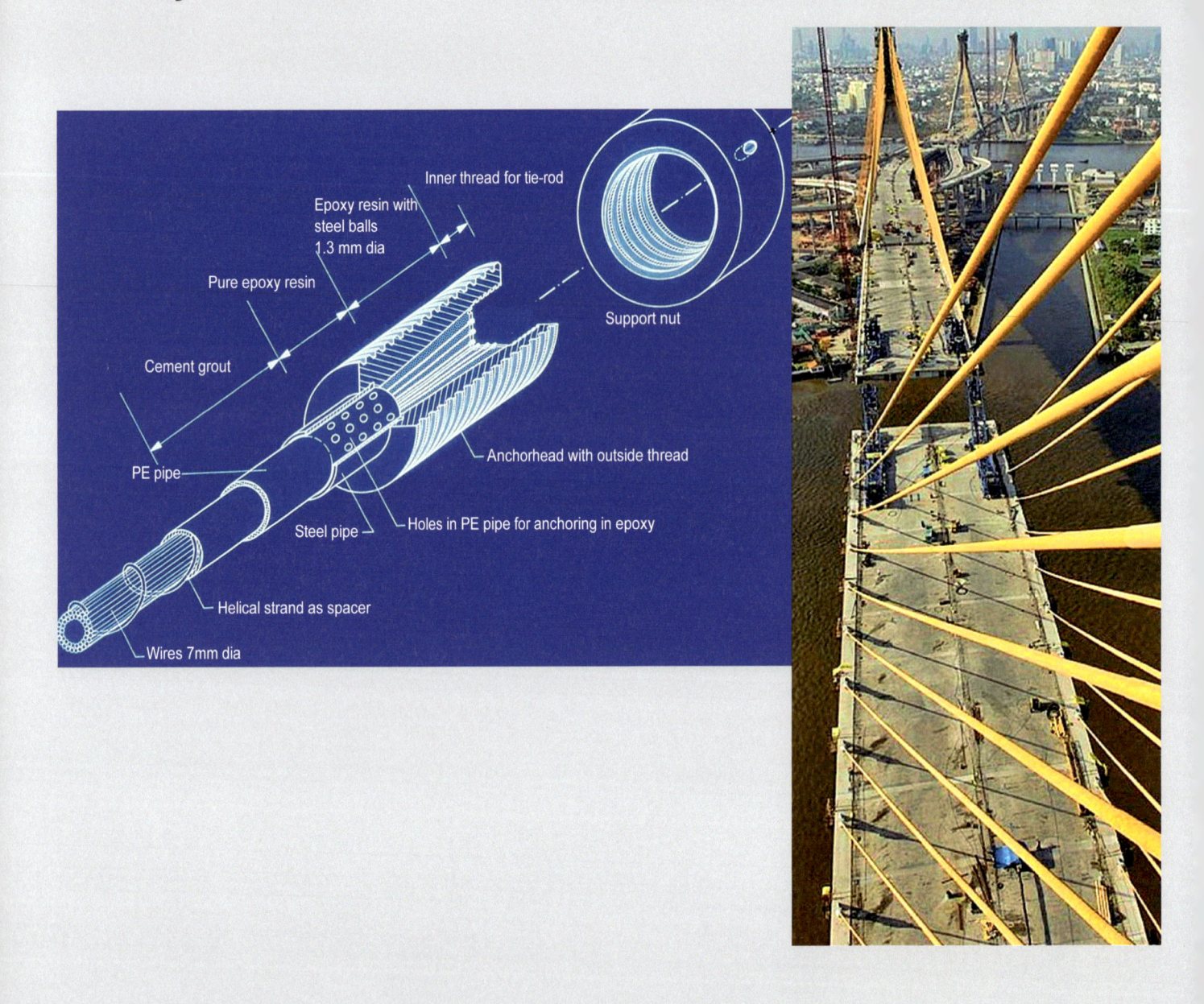

Cable-Stayed Bridges. 40 Years of Experience Worldwide. First Edition. Holger Svensson.
© 2012 Ernst & Sohn GmbH & Co. KG. Published 2012 by Ernst & Sohn GmbH & Co. KG.

Characteristics	Modern locked coil rope	Parallel wire cable	Parallel strand cable
$E \cdot 10^{-6}$ [N/mm²]	0.170	0.205	0.195
f_u [N/mm²]	1470	1670	1870
$\Delta\sigma$ [N/mm²]	150	200	200

Figure 3.1 Currently used stay cable systems

3.1 General

The stay cable systems in use today are listed below and shown in Fig. 3.1.

- *Locked coil ropes:* Traditionally used in Germany, completely shop fabricated, permit construction by geometry.
  Advantages: good corrosion protection, simple maintenance
  Disadvantages: reduced stiffness, subject to creep, reduced tensile strength and fatigue strength
- *Parallel wire cables:* Developed by LAP in the 1960s from BBR post-tensioning system in order to overcome the disadvantages of locked coil ropes, almost exclusively used in Germany since then, also completely shop-fabricated
  Advantages: high stiffness, no creep, high tensile and fatigue strength with HiAm-anchorage
  Disadvantage: complex corrosion protection with several components
- *Parallel strand cables:* Developed from strand tendons in order to exploit higher tensile strength and better availability of strands
  Advantages: cost-effective, fabrication on site from components, exchange of individual strands
  Disadvantage: slightly reduced stiffness

All these systems are continuously being further developed. The latest development is cables from synthetic materials such as glass fibers and especially carbon fibers which are still being developed.

Advantage: not corrosion sensitive
Disadvantage: expensive, fiber anchorages are difficult.

In addition, there is a parallel bar system, developed from DSI stress bars, which is hardly used any more.

3.2 Locked coil ropes

3.2.1 System

All early cable-stayed systems in Germany used locked coil ropes for stay cables. They were originally developed for heavy lifting equipment with winding machines in mining and for high lines.

Locked coil ropes consist of internal round wires with a diameter of 5 mm and outer layers of Z-shaped wires with a depth of 6–7 mm, Fig. 3.2. Their modern corrosion protection comprises galvanizing of all wires, filling the interstices with a corrosion inhibitor and painting the outside in several layers.

The different wire layers rotate in opposite directions in order to achieve twist-free ropes, Fig. 3.3. The most important manufacturer today is the British company Bridon [3.1, 3.2].

When stressing the cables, the Z-shaped outer wires are pressed against one another by lateral contraction and which 'locks' the rope surface against intrusion of water, hence the name 'locked coil ropes'.

The individual wires today reach tensile strengths of 1570 N/mm² by cold drawing and tempering. Their E-modulus remains unchanged 21 000 kN/cm². Depending on their composition, the complete ropes suffer a spinning loss, i.e. a reduction of their ultimate

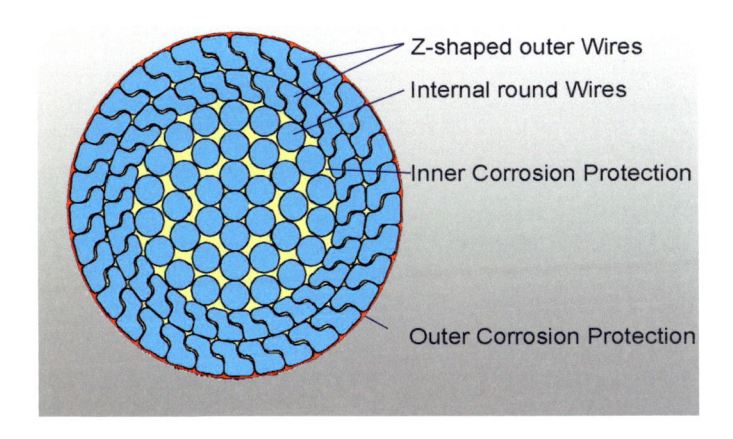

Figure 3.2 Cross-section of locked coil ropes

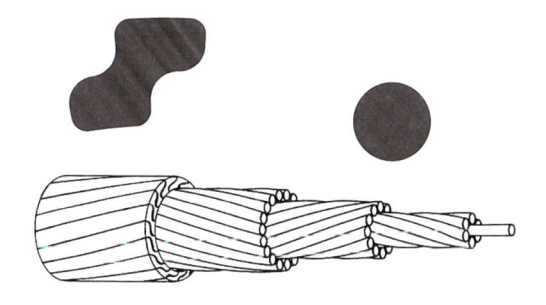

Figure 3.3 Composition of locked coil ropes

load compared with the sum of the wires' ultimate loads. The results of tensile stresses on the complete ropes serves as a design basis.

The modulus of elasticity of the complete rope depends on its composition. For completely locked coil ropes $E = 12 - 18$ MN/cm^2 may be assumed. For lower loads, the individual wires 'settle', resulting in a lower E-modulus; for higher loads the E-modulus rises to about $E = 18$ MN/cm^2. Sometimes the stress–strain behavior is determined by tests with a full cable under varying loading conditions, see Fig. 3.4.

Locked coil ropes are cast into steel anchor heads with a tensile strength of 700 N/mm^2. Their typical shapes are given in Fig. 3.5. Basic tests on their strengths on which their dimensioning rests were executed during the 1950s and 1960s [3.4] and [3.5]. Today three-dimensional FEM-calculations are done to optimize the anchor head dimensions.

A recent example of the use of locked coil ropes for a cable-stayed bridge is the Rhine River Bridge at Ilverich, Fig. 3.6 [3.6].

3.2.2 Fabrication

The locked coil ropes are fabricated in accordance with codes EN 10264 [3.7] and EN 12385 [3.8]. Fig. 3.7 shows one of the huge stranding machines for big diameter locked coil ropes.

Connection between the individual wires and the anchor head is done by hot casting, Fig. 3.8. The wires are broomed out and cast in at 425 °C with a so-called Zamak alloy, consisting of 93 % zinc, 6 % aluminum and 1 % copper.

This casting temperature is lower than used originally, when the alloys were not of such good quality so that now the microstructure of the wires is very little changed and thus the fatigue strength at the anchor head is almost that of the free length. This was formerly a disadvantage of locked coil ropes.

3.2.3 Modern corrosion protection systems
3.2.3.1 General
As a result of the failures outlined in Section 3.2.5 several improvements of the corrosion protection for locked coil ropes have been developed, based on extensive tests [3.9].

3.2.3.2 Galvanizing of the wires
All wires are hot-dip galvanized with 280 g/m^2 or a thickness of 45 µm. The galvanizing improves not only the corrosion protection of the wires but also their fatigue strengths as it acts as a lubricant [3.10]. Today Galfan is used for galvanizing [3.1]. Its aluminum oxide, which develops with time, adheres better to the steel surface and thus provides a better basis for additional coatings than the formerly used zinc oxide.

3.2.3.3 Filling
Initially red lead was used to fill the interstices between the wires. Due to its toxicity its replacement has to be done under enclosed

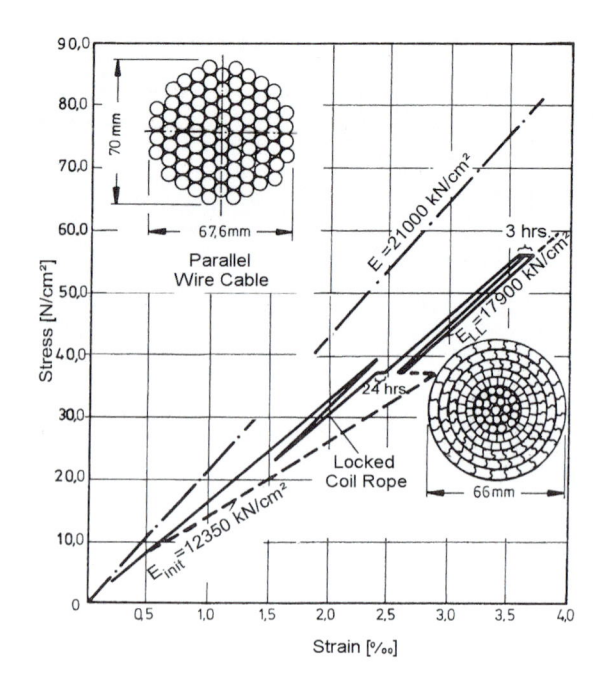

Figure 3.4 Stress-strain diagram of a locked coil rope in comparison to a parallel wire cable [3.3]

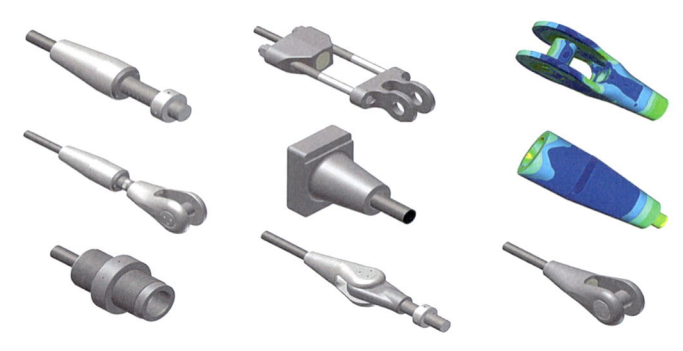

Figure 3.5 Anchor heads

Figure 3.6 Locked coil ropes for the Rhine River Bridge Ilverich, Germany

Figure 3.7 Closing machine for locked coil ropes

Order online: www.ernst-und-sohn.de

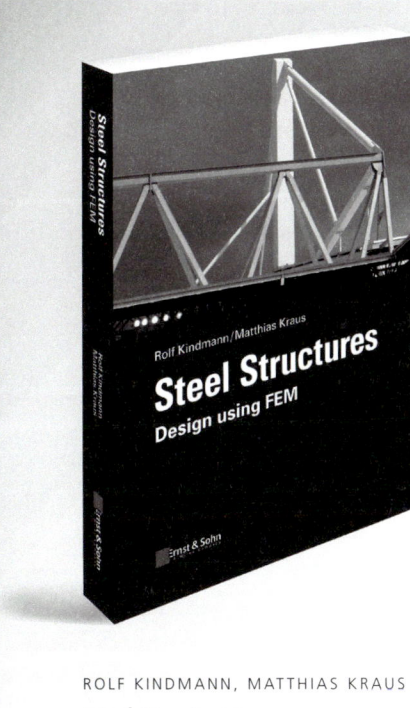

*€ Prices are valid in Germany, exclusively, and subject to alterations. Prices incl. VAT. Books excl. shipping. Journals incl. shipping. 0200200006_dp

ROLF KINDMANN, MATTHIAS KRAUS
Steel Structures
Design using FEM
April 2011. 542 pages. 365 fig.
90 tab. Softcover.
€ 59,– *
ISBN 978-3-433-02978-7

■ The Finite Element Method (FEM) has become a standard tool used in everyday work by structural engineers having to analyse virtually any type of structure.

After a short introduction into the methodolgy, the book concentrates on the calculation of internal forces, deformations, ideal buckling loads and vibration modes of steel structures. Beyond linear structural analysis, the authors focus on various important stability cases such as flexural buckling, lateral torsional buckling and plate buckling along with determining ideal buckling loads and second-order theory analysis. Also, investigating cross-sections using FEM will become more and more important in the future.

For practicing engineers and students in engineering alike all necessary calculations for the design of structures are presented clearly.

Author information:
■ Univ.-Prof. Dr.-Ing. Rolf Kindmann teaches steel and composite design at the Ruhr University in Bochum and is a partner of the Ingenieursozietät Schürmann-Kindmann und Partner in Dortmund.
■ Dr.-Ing. Matthias Kraus is a research assistant at the same chair.

Ernst & Sohn
Verlag für Architektur und
technische Wissenschaften
GmbH & Co. KG

Customer Service: Wiley-VCH
Boschstraße 12
D-69469 Weinheim

Tel. +49 (0)6201 606-400
Fax +49 (0)6201 606-184
service@wiley-vch.de

Ernst & Sohn
A Wiley Company

Inspected –
and declared fit for the job
Choosing the right testing method is critical

Whenever you need to decide on how suspension cables, and their termination connectors, are to be tested and assessed, that's when you will want to turn to DMT – a partner that has the right testing equipment even for those hard-to-reach spots.

Founded in 1903, our Rope Testing Centre offers more than 100 years of experience in wire rope testing. We introduced electromagnetic inspection equipment in the 1930s. In the many years since those early days, we have continuously refined and developed the testing equipment required. That's what sets us apart from the rest!

- Visual inspections
- Partially automated visual testing
- Magnetic inductive testing
- Ultrasonic testing
- Other non-destructive testing methods

DMT GmbH & Co. KG
DMT Laboratory for Non-Destructive and
Destructive Testing – Rope Testing Centre –

Dinnendahlstr. 9   Phone +49 234 957157-51
44809 Bochum,   bs@dmt.de
Germany   www.dmt.de
                                   Member of TÜV NORD Group

DIN EN ISO
9001
certified

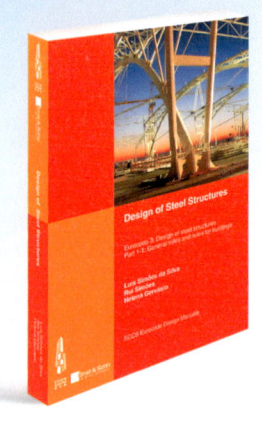

ED.: ECCS – EUROPEAN CONVENTION FOR CONSTRUCTIONAL STEELWORK

Design of Steel Structures
EC3: Design of Steel Structures.
Part 1-1: General Rules and Rules for Buildings.
April 2010.
446 pages, 295 fig., 105 tab., Softcover.
€ 70,–*
ISBN 978-3-433-02973-2

■ This book introduces the basis design concept of Eurocode 3 for current steel structures, and their practical application. Numerous worked examples will facilitate the acceptance of the code and provide for a smooth transition from earlier national codes to the Eurocode.

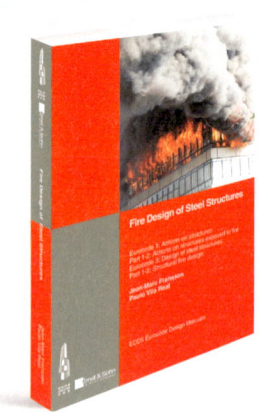

ED.: ECCS

Fire Design of Steel Structures
EC1: Actions on Structures.
Part 1-2: Actions on Structures exposed to Fire.
EC3: Design of Steel Structures.
Part 1-2: Structural Fire Design.
May 2010.
428 pages, 134 fig., 21 tab., Softcover.
€ 70,–*
ISBN 978-3-433-02974-9

■ This publication sets out the design process in a logical manner giving practical and helpful advice and easy to follow worked examples that will allow designers to exploit the benefits of the new approach given in the Eurocodes to fire design.

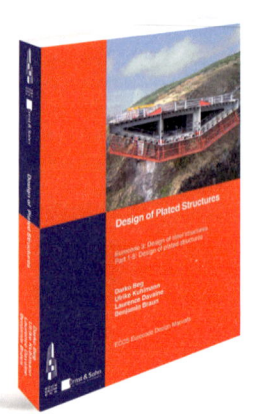

ED.: ECCS

Design of Plated Structures
EC3: Design of Steel Structures.
Part 1-5: Design of Plated Structures.
January 2011.
272 pages, 139 fig., 19 tab., Softcover.
€ 55,–*
ISBN 978-3-433-02980-0

■ This design manual provides practical advice to designers of plated structures for correct and efficient application of EN 1993-1-5 design rules and includes numerous design examples.

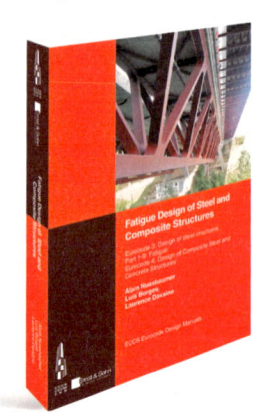

ED.: ECCS

Fatigue Design of Steel and Composite Structures
EC3: Design of Steel Structures.
Part 1-9: Fatigue.
EC4: Design of Composite Steel and Concrete Structures.
October 2011.
311 pages.
250 fig. Softcover.
€ 55,–*
ISBN 978-3-433-02981-7

■ This volume addresses the specific subject of fatigue, a subject not familiar to many engineers, but still relevant for proper and good design of numerous steel structures. It explains all issues related to the subject: Basis of fatigue design, reliability and various verification formats, determination of stresses and stress ranges, fatigue strength, application range and limitations. It contains detailed examples of application of the concepts, computation methods and verifications.

ED.: ECCS

Design of Cold-formed Steel Structures
EC3: Design of Steel Structures.
Part 1-3: Design of Cold-formed Steel Structures.
II. Quarter 2012.
approx. 512 pages.
approx. 300 fig. Softcover.
approx. € 70,–*
ISBN 978-3-433-02979-4

■ The book is concerned with design of cold-formed steel structures in building based on the Eurocode 3 package, particularly on EN 1993-1-3. On this purpose, the book contains the essentials of theoretical background and design rules for cold-formed steel sections and sheeting, members and connections.

ED.: ECCS

Design of Connections in Steel and Composite Structures
III. Quarter 2012.
approx. 500 pages, approx. 300 fig. Softcover.
approx. € 70,–*
ISBN: 978-3-433-02985-5

■ This volume elucidates the design rules for connections in steel and composite structures which are set out in Eurocode 3 and 4. Numerous examples illustrate the application of the respective design rule.

Ernst & Sohn
Verlag für Architektur und technische Wissenschaften GmbH & Co. KG

Customer Service: Wiley-VCH
Boschstraße 12
D-69469 Weinheim

Tel. +49 (0)6201 606-400
Fax +49 (0)6201 606-184
service@wiley-vch.de

Ernst & Sohn
A Wiley Company

Online-Order: www.ernst-und-sohn.de

3.2 Locked coil ropes

conditions which is very expensive. Therefore, modern corrosion protection systems consist of a filler with polyurethane (PU) and zinc or aluminum dust.

3.2.3.4 Paint
After installation, stressing and introduction of all permanent loads, the rope surface is cleaned thoroughly. Then a base layer, two intermediate layers and an outer layer with a total thickness of 410 µm is applied, Fig. 3.9. The individual ropes of the initially used concentrated stay cables, consisting of several ropes, had to be painted before installation and the joints between the individual ropes had to be sealed with putty, which gave rise to uncontrolled corrosion inside the concentrated stay cables.

Since zinc chromate is considered to be carcinogenic the use of new painting systems without active pigments is being investigated and these are composed using the barrier principle [3.11]. For this purpose, fillers and paints called Metalcoat are used which consist of phenol formaldehyde resin with aluminum flakes as the active pigment [3.12].

3.2.4 Inspection and maintenance
In Germany, stay cables are inspected visually from carriages running on the cables or by self-propelled video cameras. In addition, the inner wires of stay cables are inspected by magnetic induction, Fig. 3.10.

In this way wire breaks and their location within the rope cross-section can be determined, Fig. 3.11.

Where the ropes of existing bridges cannot be inspected visually and/or by magnetic induction, for example inside of anchorage pipes, they should be inspected using an endoscope and the result recorded on video.

Such an intensive inspection enables the detection of damage early on and thus permits repair, for example by touching up the paint.

3.2.5 Damage
3.2.5.1 Köhlbrand Bridge
Typical damage to locked coil ropes took place at the Köhlbrand Bridge [3.9, 3.13], and it was caused by the characteristics of the old type of locked coil ropes used and so these are outlined here in some detail. The Köhlbrand Bridge, Fig. 3.12, was built between 1969 and 1974. The 88 locked coil ropes with diameters of 54 mm to 104 mm had the following corrosion protection:
- bare wires, because the engineers at that time were afraid of hydrogen embrittlement due to galvanizing,
- inner filling of linseed-oil-based red lead,
- two base layers of phthalate-resin-based red lead paint with a thickness of 50 µm,
- two outer layers of linseed-oil-based iron mica paint with a thickness of 50 µm each, the total thickness thus reaching just 200 µm.

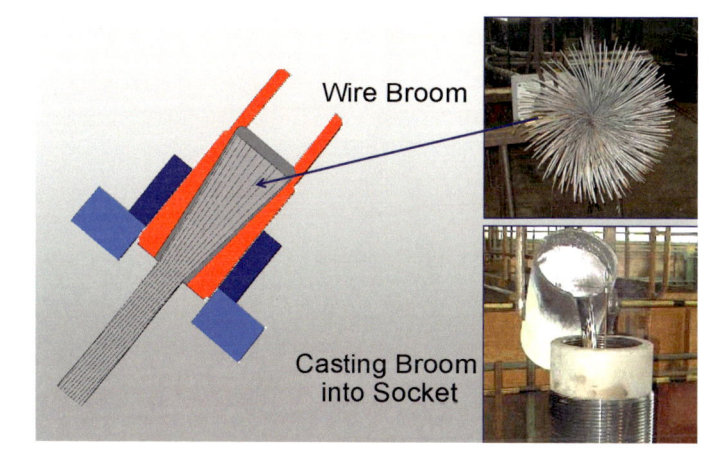

Figure 3.8 Casting of the rope into an anchor head

Figure 3.9 Painting of ropes

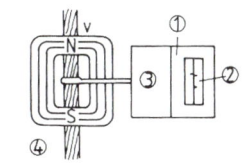

Figure 3.10 Magnetic inductive investigation

① Recorder
② Recording paper
③ Amplifier
④ Registering equipment

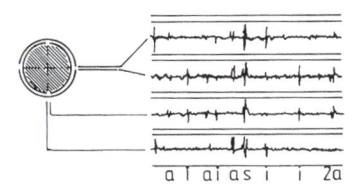

a Break of outer wire
i Break of inner wire
s Break of steel core

Figure 3.11 Registration of wire breaks

144   3 Stay cables

Figure 3.12 Köhlbrand Bridge, Hamburg, Germany

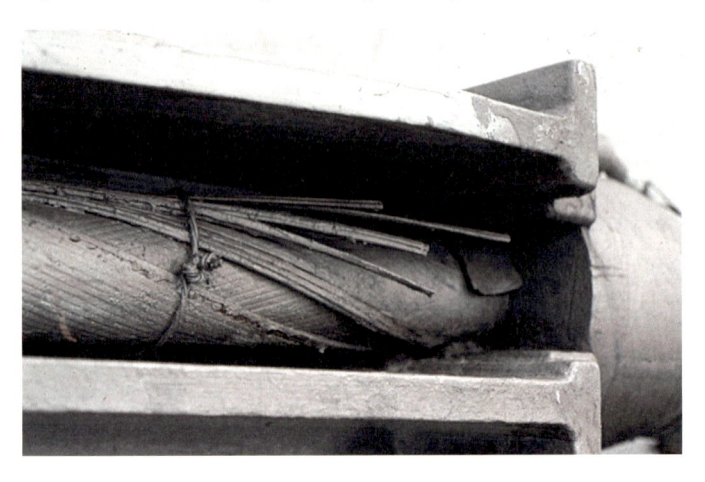

Figure 3.13 Lower cable anchorage with wire breaks

Figure 3.14 Cable exchange

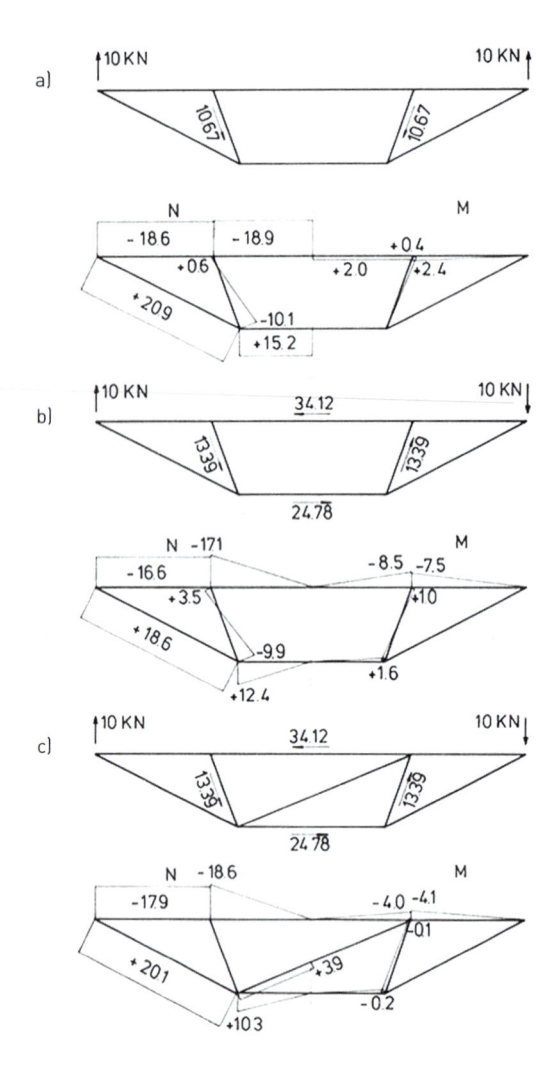

Figure 3.15 Torsional forces in beam with and without auxiliary diagonal

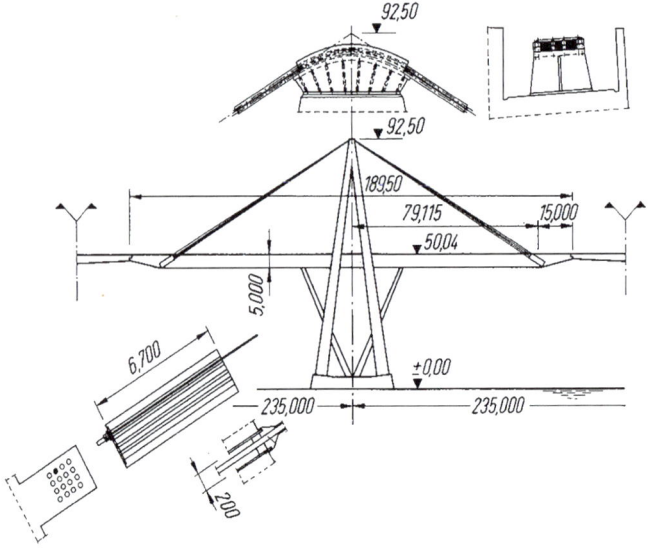

Figure 3.16 Maracaibo Bridge, Venezuela

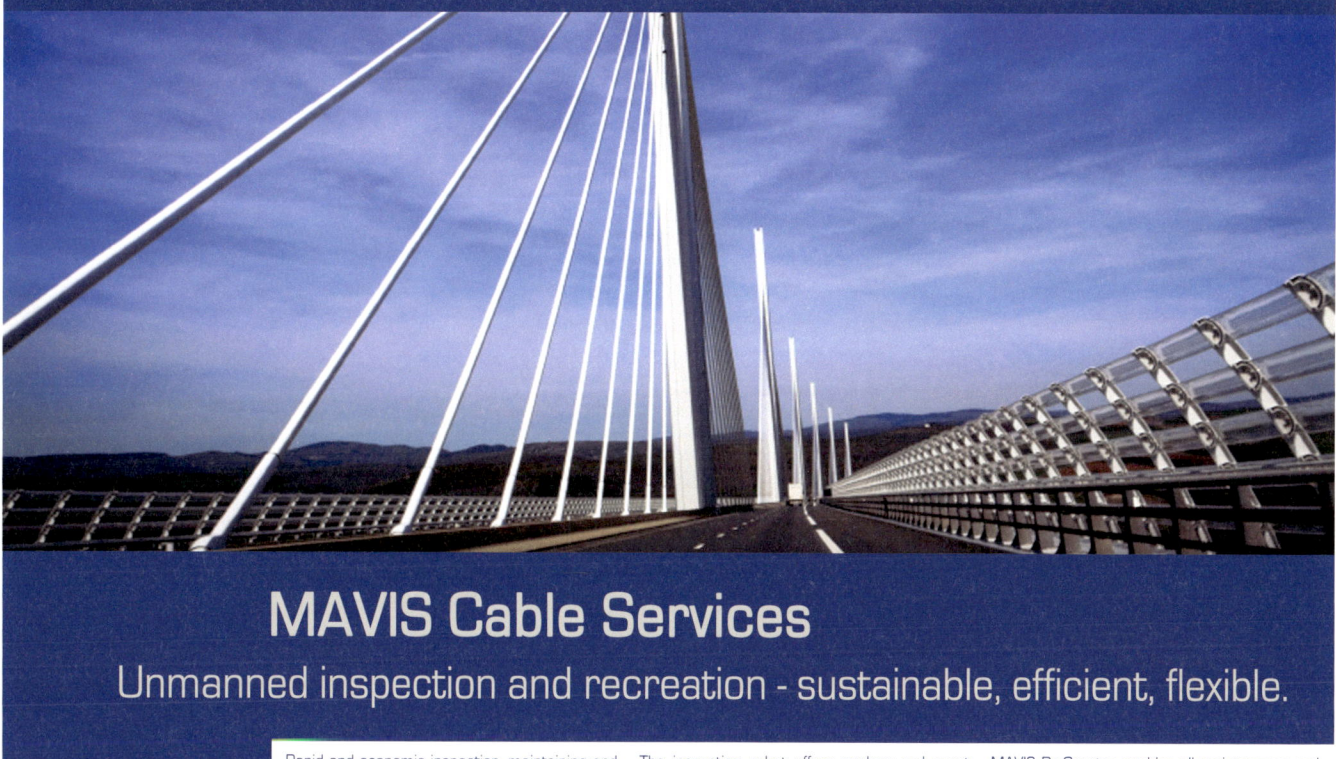

mavis-inspect.com
MAVIS Cable Services GmbH, Aachen

MAVIS Cable Services
Unmanned inspection and recreation - sustainable, efficient, flexible.

Rapid and economic inspection, maintaining and coating of bridge cables with MAVIS not only reduces costs by making scuffolding obsolete, it improves the quality and speed of work as well as offering a digital data basis for long-term analysis.

The inspection robot offers gapless and exact magnetic field-, oscillation- and coating thickness-tests as well as HD video surface documentation. Collected data is archived digitally by position for comfortable evaluation with MAVIS Software.

MAVIS Re-Creator enables all maintenance and corrosion protection procedures like cleaning, de-varnishing or painting on mounted cables, even if access is difficult. Automated application and full workflow documentation create optimal results.

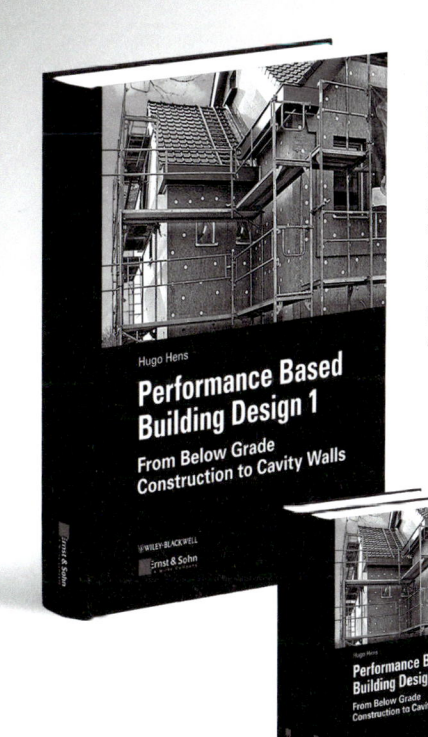

HUGO S.L.C. HENS

Performance Based Building Design 1
From Below Grade Construction to Cavity Walls

approx. 260 pages,
approx. 172 figures, Softcover.

approx. € 59,–*
ISBN 978-3-433-03022-6

Date of Publication:
April 2012

**Package:
Performance Based Building Design 1 and 2**
approx. € 99,–*
ISBN: 978-3-433-03024-0
Date of Publication:
September 2012

■ Just like building physics, performance based building design was hardly an issue before the energy crisis of the 1970ies. With the need to upgrade energy efficiency, the interest in overall building performance grew. As the first of two volumes, this book applies the performance rationale, advanced in applied building physics, to the design and construction of buildings. After an overview of materials for thermal insulation, water proofing, air tightening and vapour tightening and a discussion on joints, building construction is analysed, starting with the excavations. Then foundations, below and on grade constructions, typical load bearing systems and floors pass the review to end with massive outer walls insulated at the inside and the outside and cavity walls. Most chapters build on a same scheme: overview, overall performance evaluation, design and construction.

The book should be usable by undergraduates and graduates in architectural and building engineering, though also building engineers, who want to refresh their knowledge, may benefit. The level of discussion assumes the reader has a sound knowledge of building physics, along with a background in structural engineering, building materials and building construction.

Order online! www.ernst-und-sohn.de

Ernst & Sohn
Verlag für Architektur und technische Wissenschaften GmbH & Co. KG

Customer Service: Wiley-VCH
Boschstraße 12
D-69469 Weinheim

Tel. +49 (0)6201 606-400
Fax +49 (0)6201 606-184
service@wiley-vch.de

Ernst & Sohn
A Wiley Company

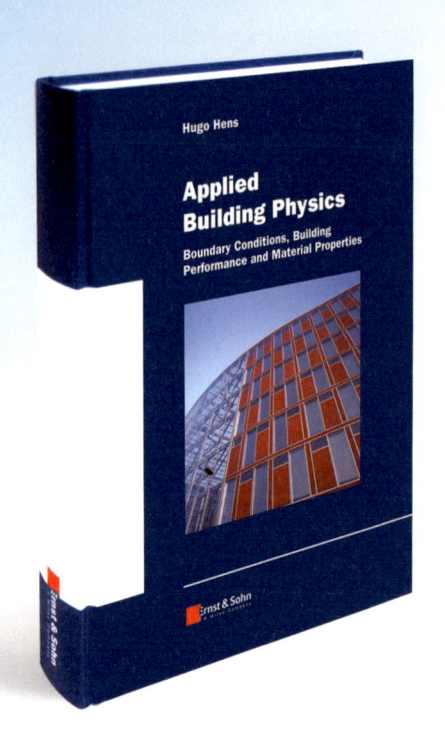

Applied Building Physics

■ The energy crises of the 1970ies, the persisting moisture problems, the complaints about sick buildings, thermal, visual and olfactory discomfort, and the move towards more sustainability in building construction pushed Building Physics to the frontline of building innovation. The societal pressure to diminish energy consumption in buildings without degrading usability acted as a trigger that activated the whole notion of performance based design and construction. As all engineering sciences, Building Physics is oriented towards application, reason why, after a first book on fundamentals, this second tome looks at the performance rationale and performance requirements.

HUGO S.L.C. HENS
Applied Building Physics
Boundary Conditions, Building Performance and Material Properties
2010. 316 pages,
100 figures,
Softcover.

€ 59,–*
ISBN 978-3-433-02962-6

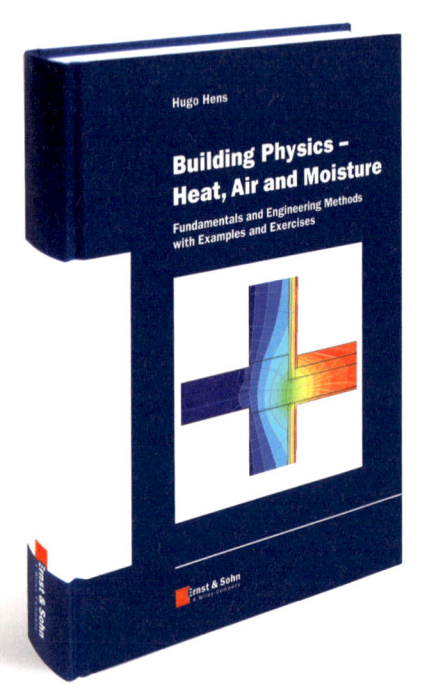

The outdoor and indoor climate conditions are described and calculation values are discussed, the performance concept is specified at the building level, at the building envelope level and at the materials' level. Definability in an engineering way, predictability at the design stage and controllability are the measures of concepts' quality. Thus, the author gives a practical guide of the performance approach which helps consulting engineers, architects and contractors guaranteeing building quality.

This book is the result of 35 years of teaching architectural, building and civil engineers, coupled to 40 years of experience, research and consultancy.

Building Physics – Heat, Air and Moisture

■ The book discusses the theory behind the heat and mass transport in and through building components. Steady and non steady state heat conduction, heat convection and thermal radiation are discussed in depth, followed by typical building-related thermal concepts such as reference temperatures, surface film coefficients, the thermal transmissivity, the solar transmissivity, thermal bridging and the periodic thermal properties. Water vapour and water vapour flow and moisture flow in and through building materials and building components is analyzed in depth, mixed up with several engineering concepts which allow a first order analysis of phenomena such as the vapour

HUGO S.L.C. HENS
Building Physics – Heat, Air and Moisture
Fundamentals and Engineering Methods with Examples and Exercises
2007.
270 pages, 133 figures,
16 tables, Softcover.

€ 59,–*
ISBN 978-3-433-01841-5

balance, the mold, mildew and dust mites risk, surface condensation, sorption, capillary suction, rain absorption and drying. In a last section, heat and mass transfer are combined into one overall model staying closest to the real hygrothermal response of building components, as observed in field experiments.

The book combines the theory of heat and mass transfer with typical building engineering applications. The line from theory to application is dressed in a correct and clear way. In the theory, oversimplification is avoided.

This book is the result of thirty years teaching, research and consultancy activity of the author.

Package-Price
€ 99,–*
instead of € 118,–*
ISBN 978-3-433-02963-3

■ Hugo S. L. C. Hens was professor at the University of Louvain, Belgium. After four years of activity as a structural engineer and site supervisor, he returned to the university to receive a PhD in Building Physics. He taught Building Physics from 1975 to 2003 and Performance Based Building Design from 1970 to 2005 and still teaches building services. Hugo Hens has authored and co-authored over 150 articles and conference papers, wrote hundreds of reports on building damage cases and their solution, introduced upgraded, research-based concepts for highly insulated roof and wall construction and directed several programs on building-energy related topics.

For ten years he has been coordinator of the international working group CIB W40 on Heat and Mass Transfer in Buildings. At present he is operating agent of the International Energy Agency's EXCO on Energy Conservation in Buildings and Community Systems, Annex 41 Whole Building Heat, Air and Moisture Response.

Ernst & Sohn
Verlag für Architektur und technische Wissenschaften GmbH & Co. KG

Customer Service: Wiley-VCH
Boschstraße 12
D-69469 Weinheim

Tel. +49 (0)6201 606-400
Fax +49 (0)6201 606-184
service@wiley-vch.de

Ernst & Sohn
A Wiley Company

During an inspection of the cables in 1976, after two years of service 25 wire breaks were detected, Fig. 3.13. A detailed investigation indicated the following causes:

a) Structural details
- Alignment of the cable anchorages in the chord but not the tangent of the deflected stay cables caused additional bending stresses in the wires.
- The stay cables were protected by collars of lead, which is located unfavorably in the electrochemical series and thus enhanced corrosion.

b) Damage of inner filler
- The inner corrosion protection filler in front of the anchor heads was burned away by the hot casting, which was especially damaging for the ungalvanized wires.

Since it could not be ruled out that all the ropes had been similarly damaged they all had to be exchanged, one after another, Fig. 3.14.

Since the loading 'cable exchange' was not accounted for in the design, the torsional moments could not be taken by the steel box beam, but an auxiliary diagonal had to be installed at the location of the actual cable exchange, Fig. 3.15.

3.2.5.2 Maracaibo Bridge, Venezuela

The bridge across Lake Maracaibo in Venezuela was built between 1959 and 1962 [3.9, 3.13, 3.14]. Its five identical main spans of 235 m each are supported by single concentrated stay cables, each comprising 16 locked coil ropes of 74 mm diameter, Fig. 3.16. Their corrosion protection was similar to that of the ropes for the Köhlbrand Bridge. At the beam, the ropes run through steel pipes and are anchored behind concrete corbels. The steel pipes were filled with bitumen and their upper ends sealed with wooden plugs and neoprene boots, Fig. 3.17.

Between 1974 and 1978 several wire breaks were detected. An in-depth inspection at the end of 1978 disclosed more than 500 broken wires, Fig. 3.18. In the beginning of 1979 three complete ropes had failed. The damage had two main causes:
- complete failure to maintain the paint in the hot sea climate for a period of over 16 years,
- the neoprene boots were not replaced after an earlier inspection and thus a permanent wet microclimate was present underneath the wooden plugs.

Due to the rapid progress of the breaks, the cables were initially strengthened with tendons and then all the ropes were replaced. For this purpose, new cable anchorages had to be installed on top of the existing cable saddles at the tops of the towers, Fig. 3.19.

This lack of inspection can only be overcome by strictly enforced rules like in Germany:
- all three years a complete visual inspection,
- all six years an in-depth inspection, including magnetic induction.

Cable damages were caused worldwide by not adhering to such inspection guidelines.

Figure 3.17 Rope anchorage at beam

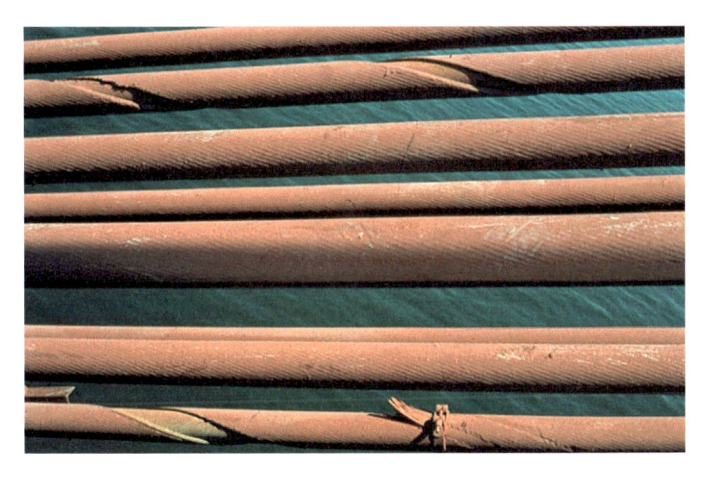

Figure 3.18 Wire breaks on the free length

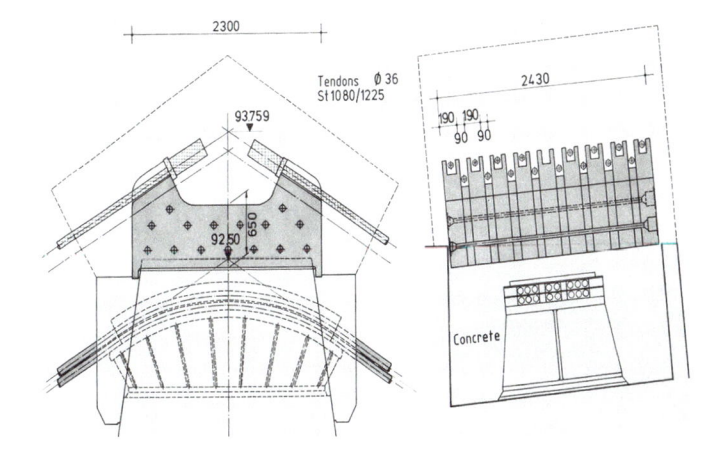

Figure 3.19 New cable anchorages on top of the existing cable saddles

3.2.5.3 Flehe Rhine River Bridge

Several locked coil stay cables had to be exchanged on the Flehe Rhine River Bridge in 2009 because many wires were broken. The traffic was maintained on the bridge during cable exchange. In [3.15] the damage, the cable exchange and the new corrosion protection are outlined in detail.

3.2.5.4 Lessons from the damage

The corrosion protection of the ropes was radically improved and is regularly inspected, cable oscillations are prevented and the loading 'cable exchange' is taken into account in the bridge design.

3.3 Parallel bar cables

Parallel bar cables were developed as stay cables by Ulrich Finsterwalder, Fig. 2.109, by using DSI stress bars on the design for the second Main River Bridge Hoechst in 1972, Fig. 3.20 [3.16].

These stay cables comprise DSI stress bars St 135/150 with 16 mm diameter and rolled-on threads. Due to the harp arrangement, all cables receive about the same load and have, consequently, the same cross-section. The 25 stress bars are installed in a thick-walled steel pipe made from St 37, fixed in place by spacers, Fig. 3.21. After installation, the steel pipes were grouted. In this way, the stress bars are permanently protected against corrosion.

A special structural detail ensures that the steel pipes receive, almost exclusively, the stresses from the live load. They are locally thickened where they are cast into the concrete at the beam and at the tower to avoid stress concentrations due to rotations and wind

Figure 3.20 Second Main River Bridge Hoechst, Germany

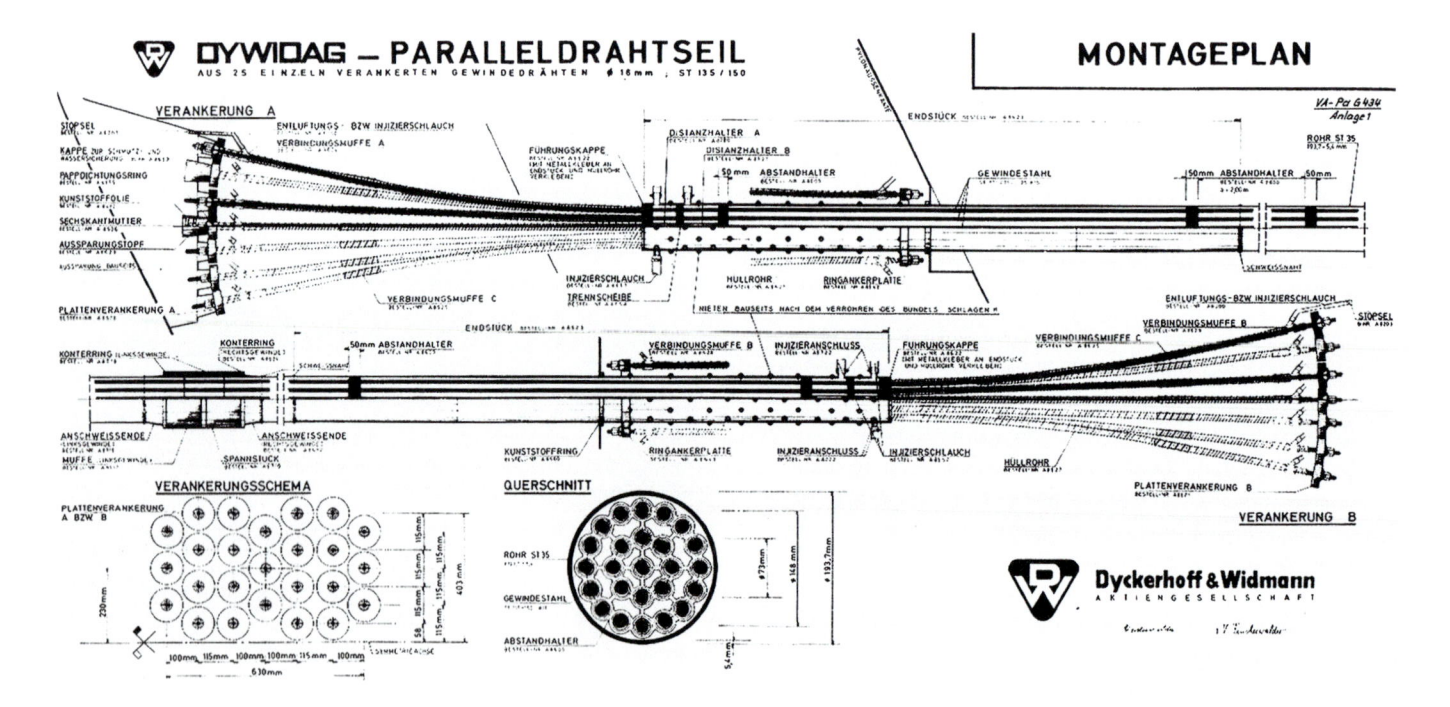

Figure 3.21 Installation drawing

3.4 Parallel wire cables

loads. The cable forces from permanent loads are introduced into the concrete at the ends of the stress bars (Verankerung A), the cable forces from live loads are introduced by shear into the concrete (Verankerung B) which is also post-tensioned at this location. In this way the fatigue strength of the cast-in stay cable is strongly reduced.

Later applications are the Penang Bridge in Malaysia in 1984 [3.17] and the Dame Point Bridge in Jacksonville, Florida, 1987, Fig. 3.22 [3.18]. This type of stay cable has been used for about 25 bridges all over the world. Today, however, they have virtually disappeared from the market for economic reasons.

3.4 Parallel wire cables
3.4.1 System

Parallel wire cables were developed by Leonhardt and Andrä, Figs 2.62 and 2.63, during the 1960s, in order to achieve a higher stiffness and a higher fatigue strength than locked coil ropes which were predominantly used at that time [3.19].

Parallel wire cables comprise a bundle of straight wires with 7 mm or ¼ inch diameter which are anchored with button heads in a retainer plate, Fig. 3.23, similar to BBR tendons.

The first use of such BBR tendons as stay cables was for the Schillersteg pedestrian bridge in Stuttgart in 1960 where a PE pipe with cement grout filling was used for corrosion protection, Fig. 3.24 [2.132]. Such stay cables are still made today by BBR under the trade name DINA, and are mainly used for pedestrian bridges, Fig. 3.25 [3.20].

In order to further improve the fatigue strength, the so-called 'HiAm' anchorage was developed and investigated in many tests [3.21]. The basic design idea is to anchor the individual wires gradually by lateral pressure exerted by small steel balls, Fig. 3.26, into which the wires are broomed-out inside a conical anchor head.

The wires are additionally secured in a retainer plate, Fig. 3.27. A mixture of epoxy resin and zinc dust is used as a filler to fix the steel balls in place and to connect the PE pipe to the anchor head. In this way there is no change in the microstructure of the steel of the wires as with hot casting material and so the fatigue strength of the complete HiAm cables is similar to that of the free length of the wires.

HiAm cables are fabricated today by BBR in Switzerland, Fig. 3.28, by Shinko in Japan, by CableTek in South Korea and China, and they are still used all over the world.

Fig. 3.29 shows the distribution of wire breaks after a tensile strength test subsequent to a fatigue test. For a fatigue range of 200 N/mm² all wire breaks of a locked coil rope are concentrated at the exit of the anchor heads and thus indicate a weakness there, which is caused by the change in microstructure due to hot casting, Fig. 3.29 a.

For a HiAm parallel wire cable with a fatigue range of 250 N/mm² the wire breaks are statistically distributed over the full free length, which means that the HiAm anchorage does not reduce the tensile strength and fatigue strength of the wires, Fig. 3.29 b [3.19].

Figure 3.22 Dame Point Bridge, Florida, USA

Figure 3.23 BBR Button head

Figure 3.24 Schillersteg, Stuttgart, Germany, 1960

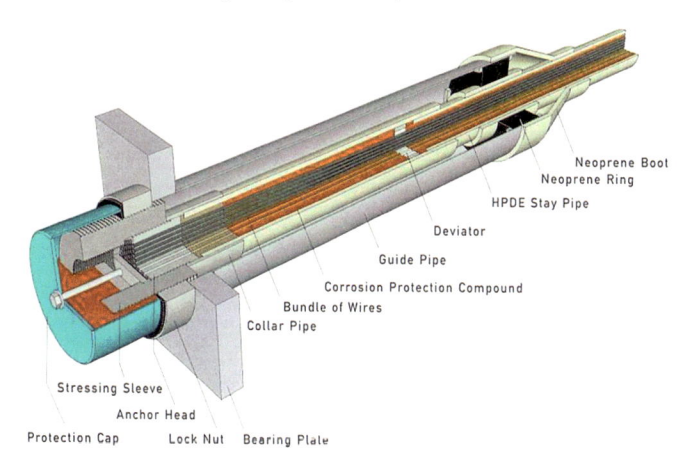

Figure 3.25 BBR DINA anchorage

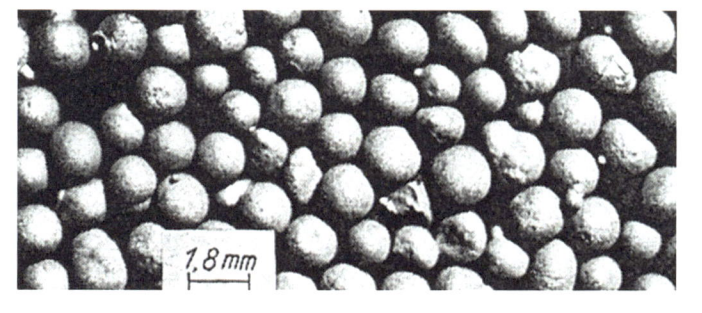

Figure 3.26 HiAm steel balls

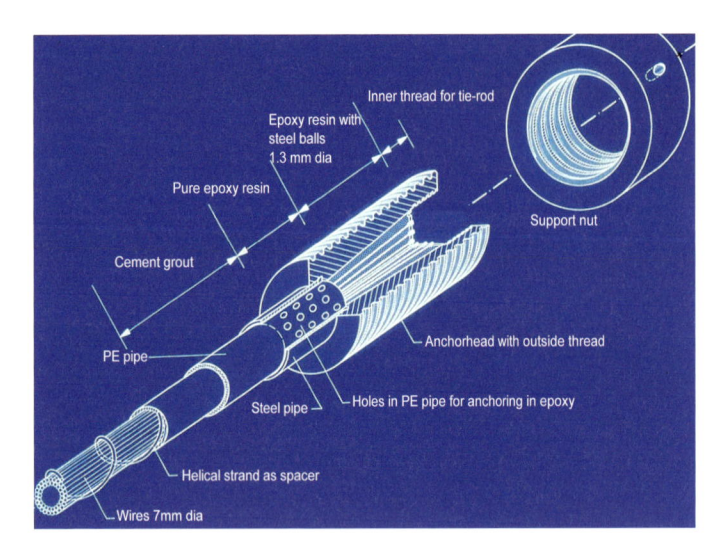

Figure 3.27 HiAm anchorage

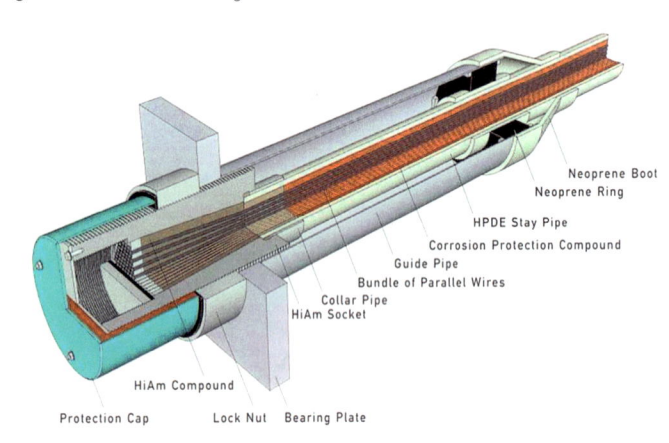

Figure 3.28 BBR HiAm anchorage

In another test, the anchor head of a HiAm cable was heated in steps up to 125 °C, Fig. 3.30. At this temperature the epoxy resin softens, and thus a failure of the epoxy resin is simulated. Nevertheless, the wires elongated at their ends only by an amount corresponding to the elastic strain of the steel balls, which means that the epoxy resin is not needed for anchoring the wires, but only serves to fix the location of the steel balls [3.21].

3.4.2 Corrosion protection

The corrosion protection of parallel wire cables comprised initially an outer PE pipe filled with cement grout, Fig. 3.31. Variations include the wrapping of the PE pipe with a colored self-adhesive, UV-resistant tape and to replace the cement grout by some type of flexible wax. The wires can additionally be protected by galvanizing or painting with epoxy. The general requirements for parallel wire cables are given in the 'Recommendations for Stay Cable Design and Testing' by the Post Tensioning Institute (PTI) [3.22].

3.4.2.1 Polyethylene (PE) pipes

Weather resistance

The durability of PE against UV-radiation is achieved by adding 2 percent of finely dispersed soot with a particle diameter of less than 0.1 µm. A low melting point and a reduced specific viscosity of about 3.5 also improve the durability of PE. Many tests with PE specimens have been performed worldwide since early on [3.23, 3.24, 3.25].

Bursting strength

At the end of the installation the PE pipes of parallel wire cables were filled with cement grout in the early applications. The fill pressure and height have to be determined in such a way that the PE pipes do not burst, either during filling or later. During filling, the weight of the grout exerts an internal pressure on the PE pipes, causing hoop stresses which widen them. After hardening of the grout the PE strain is frozen-in, Fig. 3.32. The corresponding hoop and tensile stresses are reduced by creep of the PE with time, Fig. 3.33 [3.26].

The strain of the PE pipe at the end of filling should not exceed 1.5 percent. For a pipe with 150 mm diameter, this means an increase of circumference from 471 mm to 481 mm, which can easily be controlled on site. The corresponding vertical filling height can reach up to 40 m, depending on the inner friction.

Temperature due to direct solar radiation

The temperature of the PE pipe plays an important role for its longevity, Fig. 3.34.

As the temperature coefficient of PE ($\alpha_T = 2 \cdot 10^{-4}/°C$) is distinctly bigger than that of the grout ($\alpha_T = 2 \cdot 10^{-5}/°C$), a change in the PE temperature causes a change in the hoop stresses. The surface temperature of black PE pipes may reach 65 °C or more from direct solar radiation, whereas a white surface reaches only about 40 °C under the same conditions.

3.4 Parallel wire cables

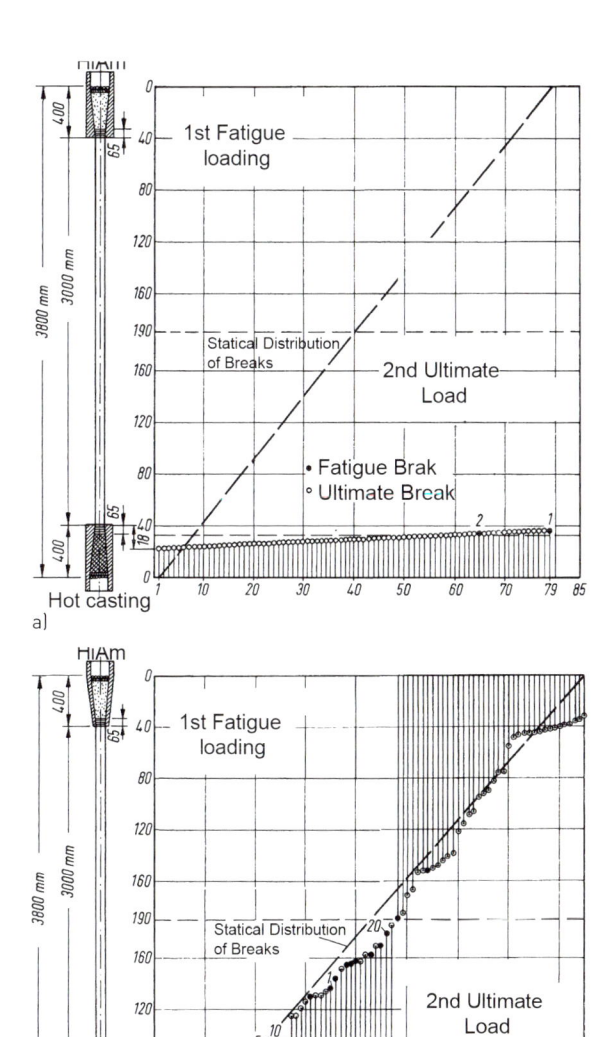

Figure 3.29 Distribution of wire breaks
a) Locked coil rope
b) HiAm cable

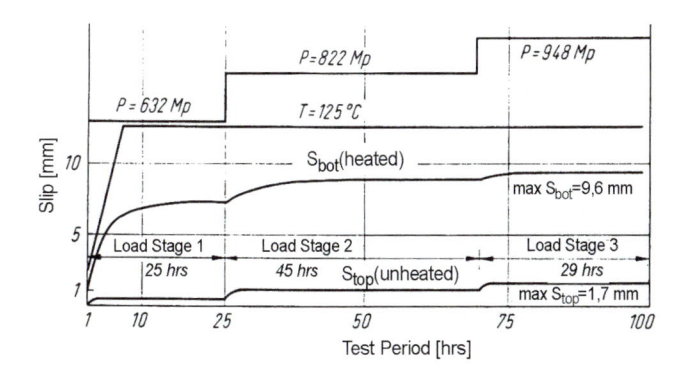

Figure 3.30 Temperature movements

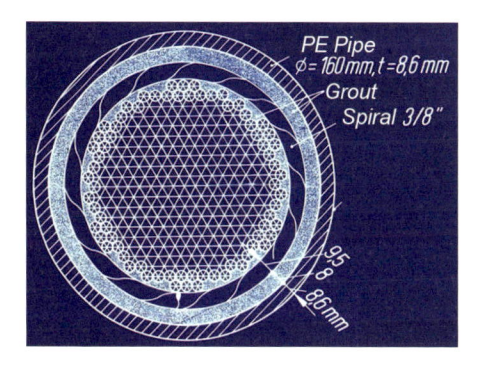

Figure 3.31 Cross-section of a parallel wire cable

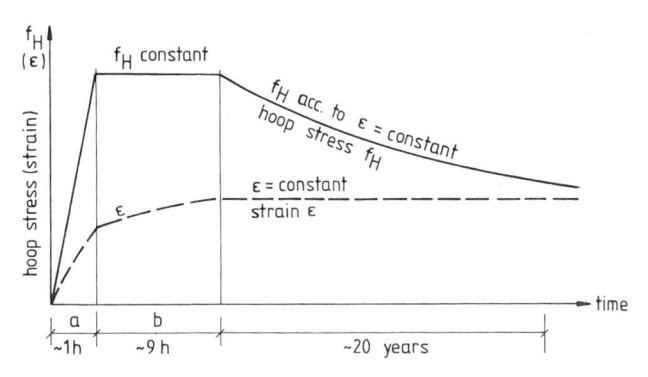

Figure 3.32 Hoop stress/strain curves for PE as a function of time

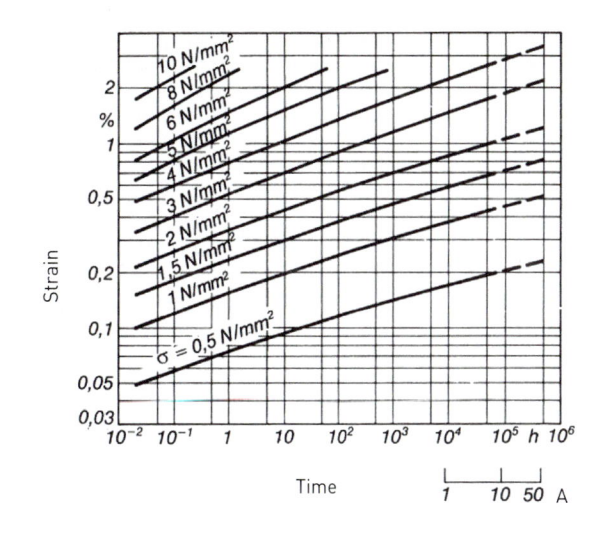

Figure 3.33 Stress/strain curves of PE specimens from Hostalen GM 5010 T2 at 20°C as a function of time

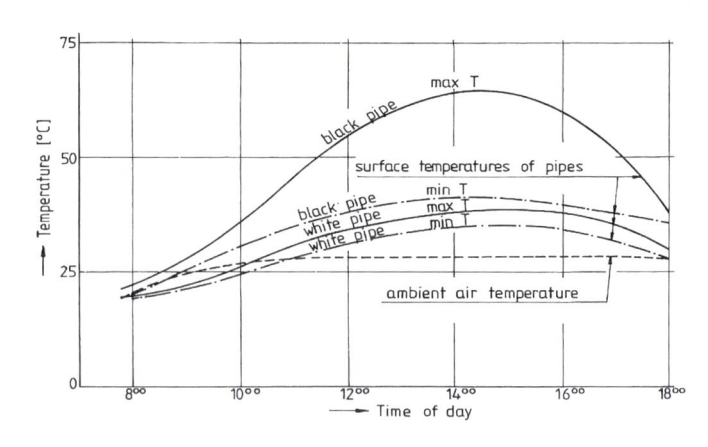

Figure 3.34 Surface temperature of black and white PE pipes from direct sun radiation

Coiling of PE pipes

The shop-fabricated parallel wire cables are coiled onto reels for transportation. On site, they are uncoiled, installed and grouted. During coiling, longitudinal bending stresses are created in the PE pipes due to bending around the reel and transverse bending stresses due to ovalizing of the pipes, Fig. 3.35 a, b. The strains remain unchanged during the coiled stage, but the bending stresses decrease due to creep. During uncoiling, only bending stresses are created, but they have the opposite sign and are again subject to creep, Fig. 3.35 c.

A detailed investigation of the coiling of PE pipes [3.9, 3.27], led to the recommendation that the reel diameter should be at least 18 times that of the PE pipe. At low temperatures the PE pipes may have to be heated, Fig. 6.44.

3.4.2.2 Wrappings

In order to improve the appearance of cable-stayed bridges and to reduce the temperature of the PE pipes a white color is advantageous. Paint does not adhere well to the chemically inert PE, so overlapping self-adhesive tapes are used, which stick to themselves. These tapes have to be weather-resistant and also should have a high tensile strength in order to improve the bursting resistance of PE pipes.

UV-resistance

During 1983/84 time accelerating tests on the UV-resistance of the polyvinyl fluoride (PVF) tapes were performed [3.28]. The composite white-grey PVF tape comprises transparent Tedlar tape from PVF, an adhesive and a PVC tape used for coloring and to provide additional stiffness, and an additional adhesive.

As a result it can be assumed that a PVF tape has a maintenance-free service life of more than 25 years. This assumption is supported by tests in which wrapped PE pipes were exposed to the natural climate. After four years, the PVC tape was delaminated, whereas the PVF tape did not show any deterioration [3.29].

As a new development an additional outer layer of PE with any color is used, which is directly extruded onto the inner black PE pipe. The UV-resistance may be less.

3.4.2.3 Grouting

At the end of the installation, when all permanent loads were present, the stay cables were cement-grouted, Fig. 3.36. Due to its alkalinity this injection material provided active corrosion protection for the wires, which remains unchanged over the service life of the bridge because the high-density PE pipe has such an extremely low water permeability that no carbonization of the cement grout takes place [3.30].

Stay cables have to undergo a fatigue range of up to 200 N/mm^2. The wires and the grout are uniformly strained by the live loads as long as the bond between them exists. The biggest stress changes, between +150 N/mm^2 and −50 N/mm^2, cause strains between

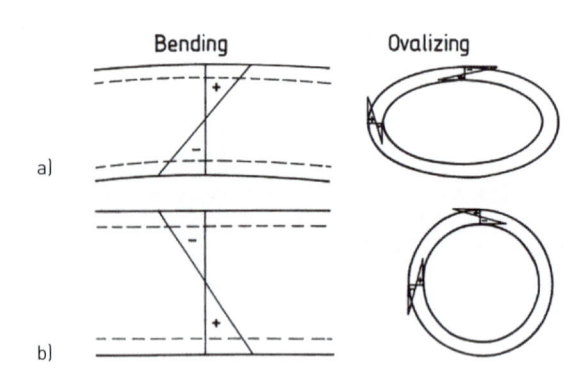

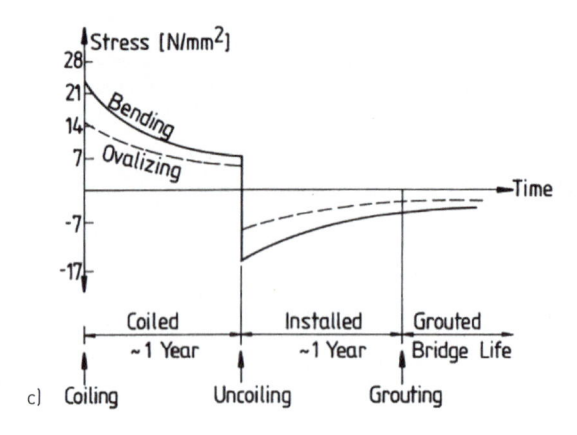

Figure 3.35 PE pipe stresses from bending and ovalizing
a) coiled on a reel
b) during uncoiling
c) change of stresses as a function of time

Figure 3.36 Cement-grouting

$+750 \cdot 10^{-6}$ to $-250 \cdot 10^{-6}$. The ultimate strain before cracking in regular cement grout lies in the range of $100 \cdot 10^{-6}$ which means that the injected cement grout within the PE pipe will crack.

In order to study this behavior in detail, test cables with 19 wires of 7 mm diameter within a PE pipe were tested. The crack widths for the maximum load were smaller than 0.1 mm. The cracks closed completely after the load was removed and could not be detected with the naked eye. The widths and distribution of the cracks are shown in Fig. 3.37 [3.31].

In 1970, after 10 years of service, two 'windows' 115 mm · 50 mm, Fig. 3.38, were cut into the top regions of a PE pipe of the very flexible Schillersteg pedestrian bridge, Fig. 3.24. At both locations the PE pipes were completely filled with cement grout, cracks could not be detected. The pH value of the cement grout was in excess of 11, which means that the original alkalinity was still present and no carbonization had taken place. At several locations the cement grout was removed so that the wires became visible. Their surfaces were still smooth and uncorroded [3.32].

In 1974 and 1979 additional windows were cut into the PE pipes and identical results were obtained. The strain characteristics of PE specimens, which were cut out of the upper side of PE pipes after 14 years of service, were compared with those of virgin PE material Hostalen GM 5010. The yield stress reached was in both cases about 24 N/mm², with a corresponding strain of about 20%. The necking in the strained area came to 17 percent of the original cross-section and the strain at failure reached more than 400 percent, Fig. 3.39. This indicates that the PE had not deteriorated after 13 years of solar radiation in an aggressive inner-city atmosphere.

3.4.2.4 Damage

The steel cable-stayed bridge that crosses the Mississippi River near Luling has a main span of 377 m [3.33]. The 72 parallel wire cables have a corrosion protection of PE pipes filled with cement grout. The injection of the cables was completed in September 1983, and in April 1985 cracks were detected in two backstay cables, Fig. 3.40. During the following winter more cracks occurred. A detailed investigation into the origins of the cracks indicated that the main cause was an over-straining of the PE pipes of up to 8.9% during grouting [3.34].

The cracks were repaired by filling them with a polyurethane paste and wrapping them with a filament tape for additional strength and an additional Tedlar tape for UV protection.

3.4.2.5 Petroleum wax

For the Helgeland Bridge at the Polar Circle, Section 6.2, cement grouting of the PE pipes was not possible because it would have had to take place at too low a temperature. The PE pipes were filled with a petroleum wax as an anti-corrosive medium, Fig. 3.41. This was done in the shop, and the completely shop-fabricated units were then transported to site. The petroleum wax remains flexible enough

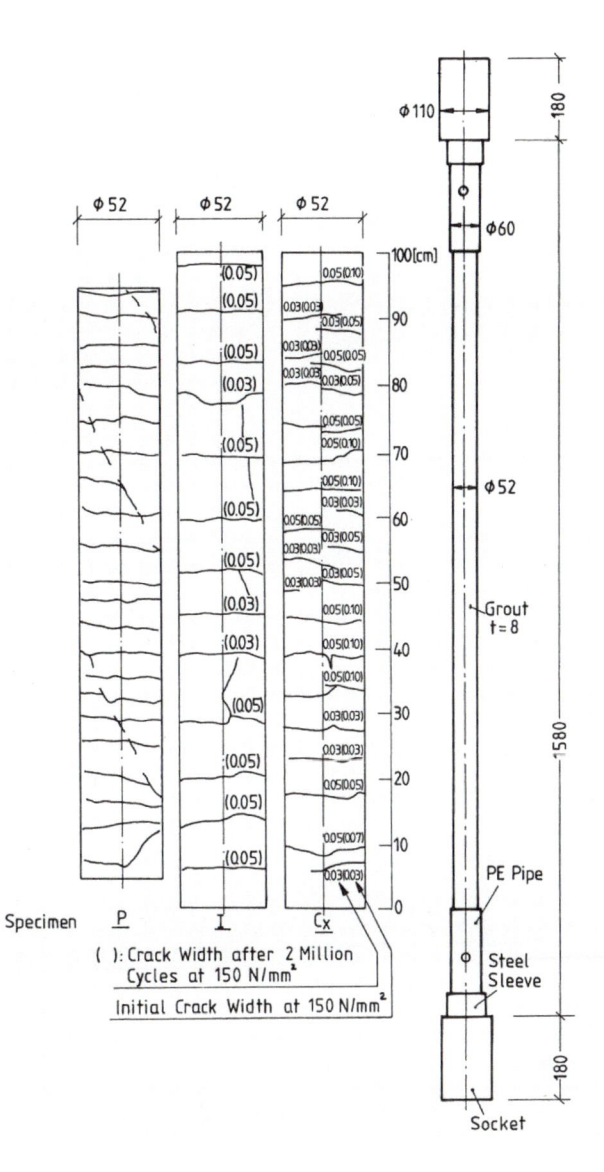

Figure 3.37 Crack pattern of the cement grout after loading

Figure 3.38 Investigation of the Schillersteg stay cables

Figure 3.39 Tensile specimens after 13 years' exposure

to permit reeling and unreeling of the cables, and yet it is stiff enough even under direct solar radiation to prevent a hydrostatic pressure build-up which might cause bursting of the pipes. By applying petroleum wax in advance, the somewhat messy and expensive cement grouting on site can be avoided.

3.4.3 Fabrication

Parallel wire stay cables were originally fabricated by BBR in Switzerland. In the US Prescon fabricated the cables for the Pasco-Kennewick Bridge, Section 6.1, assisted by Reiner Saul and the author. Later Shinko Wire also fabricated them in Japan. In 1992 the author inspected a fabrication plant near Shanghai, where the parallel wire cables for the Yang Pu Bridge were fabricated.

In the following, the typical fabrication process for parallel wire cables by the Shinko Wire Company, Japan, is outlined in Figs 3.42 – 3.45. The button-headed wires are introduced into the retainer plates at both cable ends, Fig. 3.43.

Figure 3.40 Cracks in PE pipes of the Luling Bridge

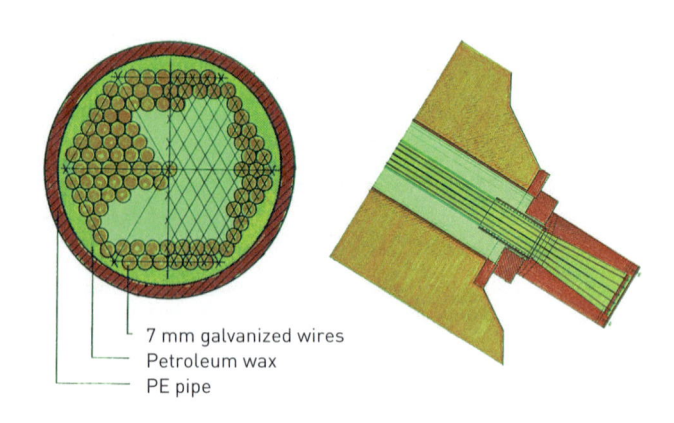

Figure 3.41 Petroleum wax filling

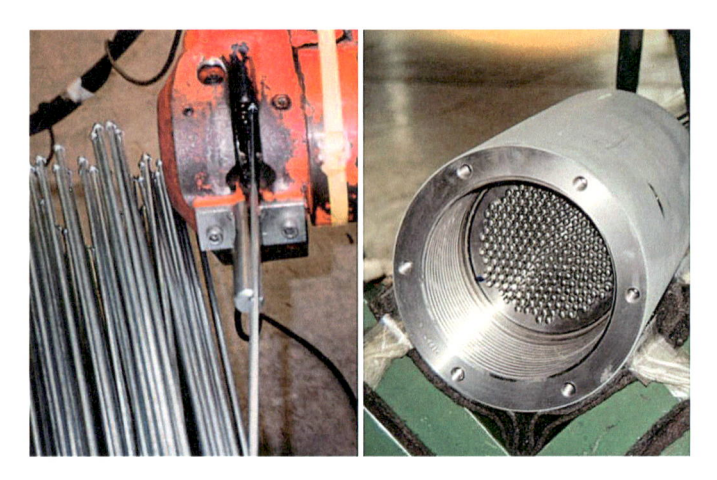

Figure 3.42 Button heads

Figure 3.43 Retainer plate for securing button heads

Figure 3.44 Welding of PE pipes

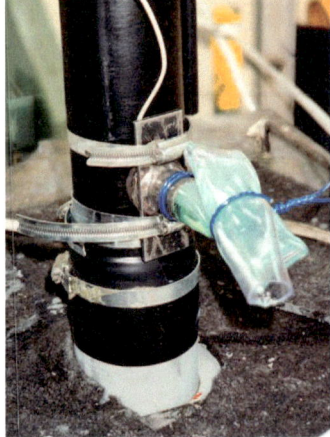

Figure 3.45 Filling of the anchor head with steel balls and epoxy resin

3.5 Parallel strand cables
3.5.1 General

The cables comprise 7-wire strands with 0.6" (15 mm) diameter. The individual wires with 5 mm diameter are cold-drawn and thus have an increased tensile strength of 1870 N/mm² compared with wires of 6.4 mm with a tensile strength of only 1670 N/mm². Strands are thus more economical and currently govern the market. Additional advantages of modern parallel strand cables are as follows:
- Their fabrication takes place from individual strands on site, the transportation weights are thus much smaller.
- Individual strands can be exchanged.
- The triple corrosion protection of individual mono strands does not require additional cement grout, wax or similar.

Parallel strand cables are today provided by all cable fabricators worldwide, BBR [3.35], DSI [3.36], Freyssinet [3.37], VSL [3.38], Shinko [3.25, 3.28, 3.31] and CableTec [3.34]. All systems have about the same quality and are permanently improved.

3.5.2 System

The system of all parallel strand cables is shown in Figs 3.46 and 3.47. The strands run parallel and are tightly packed over their free lengths. Near the anchorages they spread out, the corresponding deviation forces retained by a tension ring.

At their ends the strands are anchored with wedges, Fig. 3.48 [3.18]. The wedges have specially formed teeth which prevent sliding. By increasing the depth of the teeth in the direction of the cable ends a nearly uniform force introduction is achieved which hardly reduces the fatigue strength in the anchorage region as compared with the free length. The teeth cut through any Epoxy coating or galvanizing of the strands.

3.5.3 Corrosion protection
3.5.3.1 Traditional

The corrosion protection of individual monostrands comprises three components, Fig. 3.49 top:
- galvanizing of individual wires (or epoxy coating),
- wax filling of interstices between wires,
- extruded PE-sheath.

All monostrands are placed into an outer PE pipe, Fig. 3.49 bottom. Sometimes these PE pipes comprise two half shells that are clipped together after installation, Fig. 3.47.

For coloring today an additional outside PE-layer is extruded, Fig. 3.50, instead of the former wrapping with a PVF-tape, Section 3.4.2. Examples of such coloring are given in Fig. 3.50.

3.5.3.2 With dry air

The cable system VSL shows the latest development for protecting stay cables with dry air against corrosion, Fig. 3.51 [3.39]. This cable system comprises strands which are only galvanized and lie inside a PE pipe. The air around the strands inside the PE pipe is dried to a

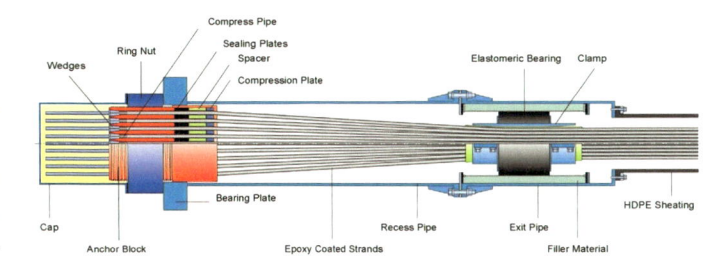

Figure 3.46 Typical elevation of a parallel strand cable (DSI)

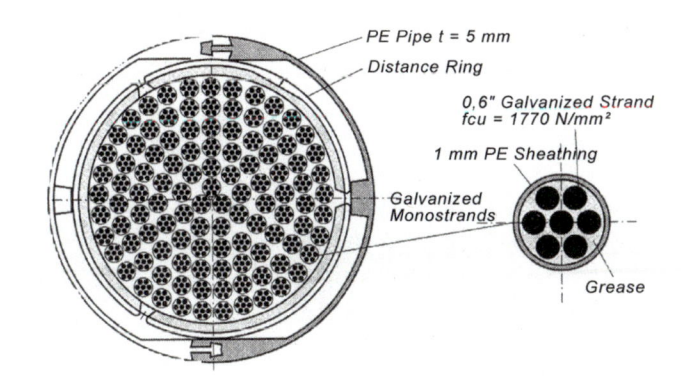

Figure 3.47 Typical cross-section of a parallel strand cable

Figure 3.48 Wedge anchorage of epoxy-coated strands

humidity of less than 60% which excludes corrosion of the strands [3.38].

The airtight welded PE pipes are provided with dry air from a unit located at the tower, Fig. 3.52. The dry air flows with slightly increased pressure from the cable anchorages at the tower via the PE pipes to the anchorages at the beam.

Today the inside of many steel boxes is protected against corrosion by dry air, Fig. 3.53. A closed dry air circuit can be installed by using the PE pipes. (The main cables of some suspension bridges are also corrosion protected with dry air, for example the Akashi Bridge in Japan from the beginning and the Forth Road Bridge in Scotland some time after construction.)

3.5.4 Fabrication

The fabrication of parallel strand cables takes place on site by installing one individual strand after another. The strands are pulled in and stressed individually. An advantage to completely shop-fabricated cables is the low weight of the individual strands and the mono jacks for stressing individual strands. Details of the fabrication of parallel strand stay cables are outlined in detail in Section 3.9.4.

3.5.5 Durability tests

3.5.5.1 Tensile strength and fatigue strength

The tensile strength and fatigue strength are tested with large hydraulic equipment, Fig. 3.54.

A standard test condition is that each cable has to reach 95% of the guaranteed ultimate tensile strength (GUTS) after a fatigue test with 2 million cycles and a range of 200 N/mm^2 with an upper stress of 45% GUTS.

3.5.5.2 Water tightness

The water tightness of the individual mono strands is tested with a pressure of 1 m of water head, Fig. 3.55. The water tightness of the complete cable is tested over a period of 6 weeks in accordance with Fig. 3.56 [3.37].

3.5.5.3 Sustainability

The durability of black PE pipes has already been discussed for parallel wire cables, Section 3.4.2, for which originally a wrapping with colored PVF-tapes was used. Today the black PE-cables are colored with an additional directly extruded PE-layer, Fig. 3.57.

The test results indicate that the tensile strength and the color brightness of the outer PE-layers remains unchanged after extended exposure to UV radiation.

3.5.6 Monitoring

The stay cables are the most vulnerable structural element of cable-stayed bridges. In order to achieve their desired 100 years of use they have to be inspected regularly. In addition to inspecting their corrosion protection the continuous supervision of cable forces and cable

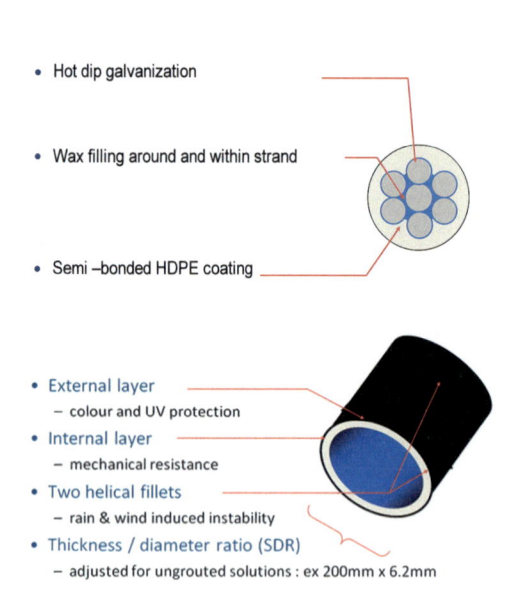

Figure 3.49 Monostrands and external PE pipe

Figure 3.50 Coloring

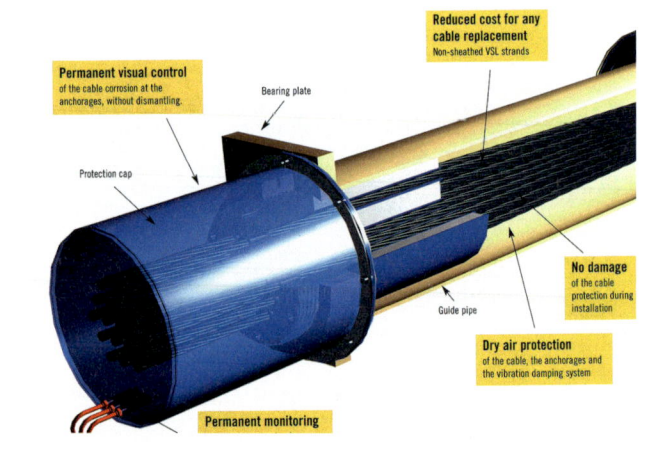

Figure 3.51 Cable anchorage for a corrosion protection with dry air

3.5 Parallel strand cables

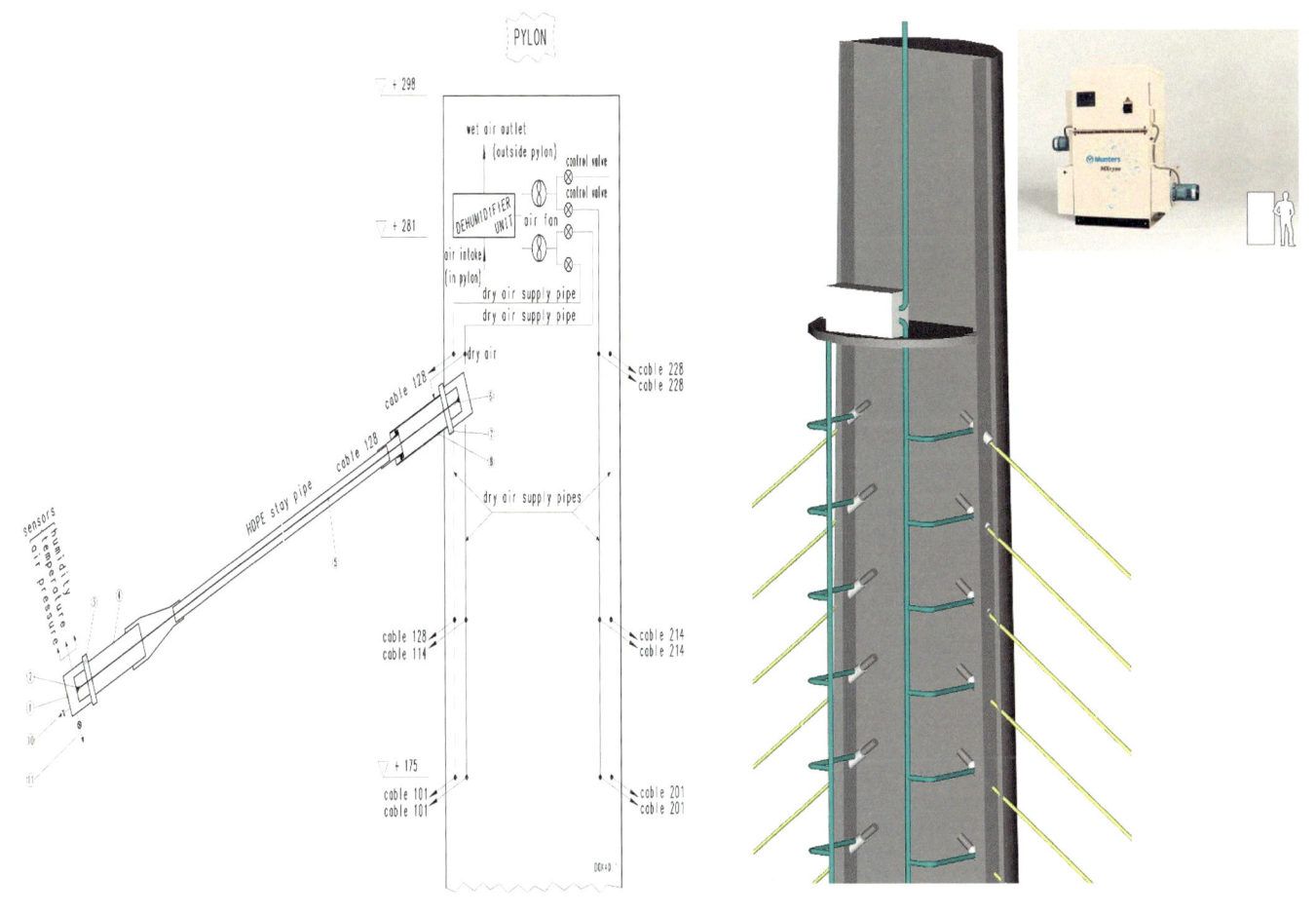

Figure 3.52 Dry air supply unit

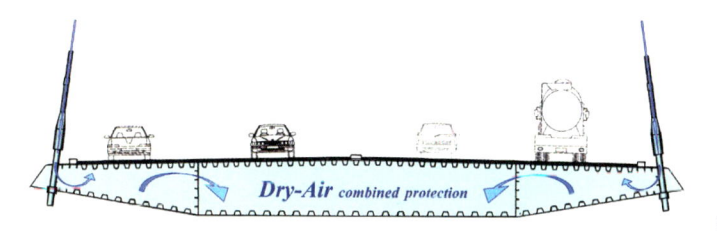

Figure 3.53 Steel beam and stay cables with corrosion protection by a dry air circuit

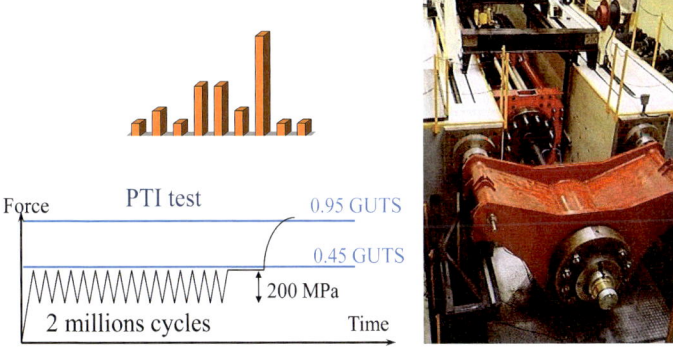

Figure 3.54 Cable tests

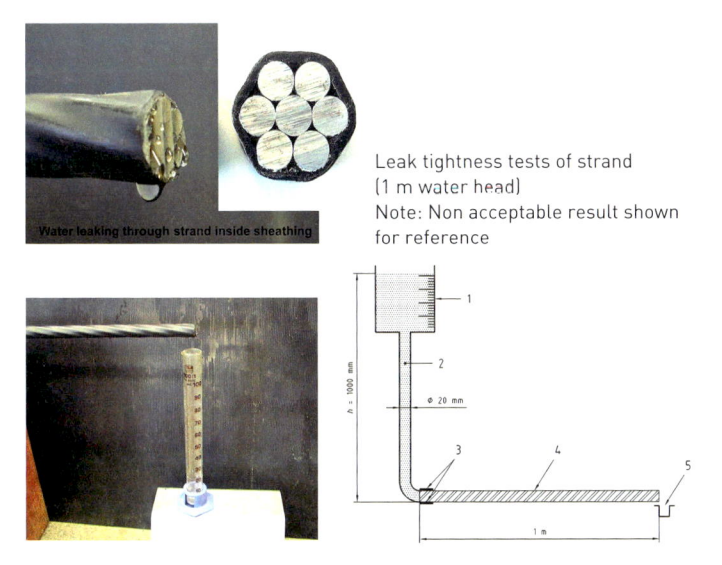

Leak tightness tests of strand (1 m water head)
Note: Non acceptable result shown for reference

Figure 3.55 Water tightness of mono strands

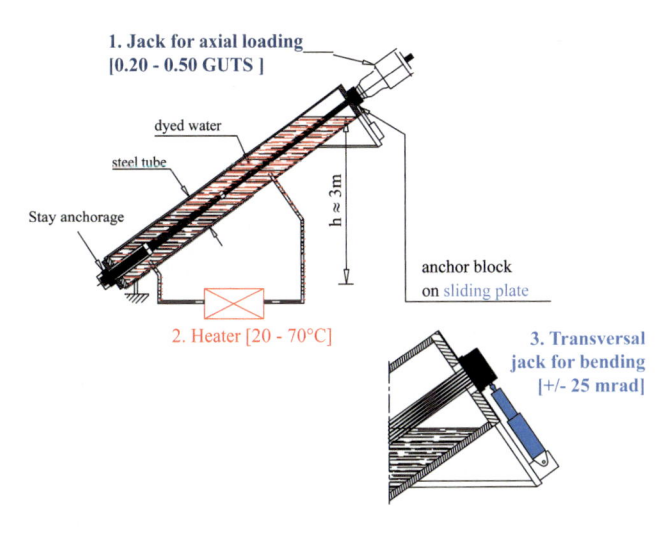

Figure 3.56 Water tightness of complete cables

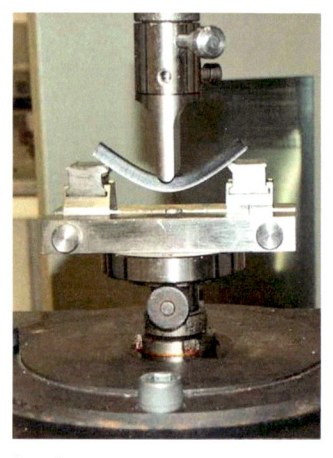

Bending test   Tensile test   Durability

Figure 3.57 Tests on the durability of colored PE pipes

oscillations can be of importance. Fig. 3.58 shows an accelerometer attached to a stay cable. The results are registered electronically and can be analyzed, Fig. 3.59.

In order to directly measure cable forces, permanently load cells can be installed into the cable anchorages, which register all changes, Fig. 3.60.

All registered data are combined in a computer on the bridge which can be accessed via the internet, Fig. 3.61.

3.6 Cable anchorages
3.6.1 General

The cable anchorages induce the cable forces into the structure. Traditional anchorages for concrete towers are shown in Fig. 3.62 and for steel towers in Fig. 3.63. A typical anchorage at a concrete beam is shown in Fig. 3.64, and an anchorage at the web extension of a steel beam is shown in Fig. 3.65.

The cable anchorages have already been treated in general in Section 1.1.5.2. In the following, further details of cable anchorages are outlined.

3.6.2 Support of anchor heads

The anchor heads of stay cables can be supported in three main ways:
– Cylindrical anchor heads, Fig. 3.66, are supported by split shims against a pressure plate, Fig. 3.67. The effective length of a stay cable can be adjusted by changing the amount of the shims.
– Cylindrical anchor heads can also be retained by an outside support nut, Fig. 3.68. The effective cable length can be adjusted by changing the position of the support nut.
– Open sockets are anchored with pins, Fig. 3.69. Changes in cable length can be achieved by activating the turnbuckle.

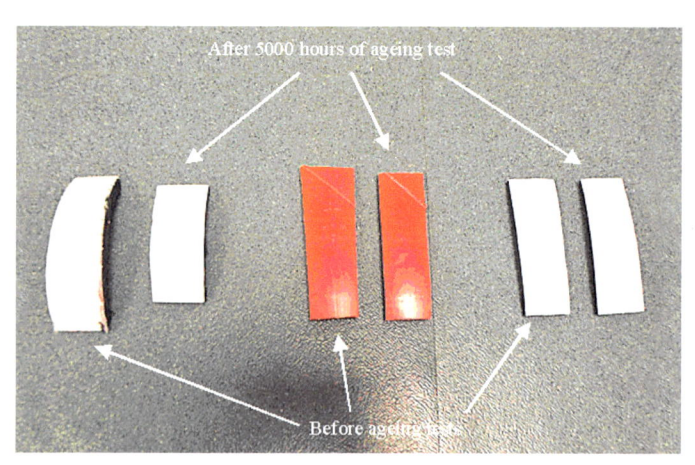

Figure 3.58 Cable Accelerometer

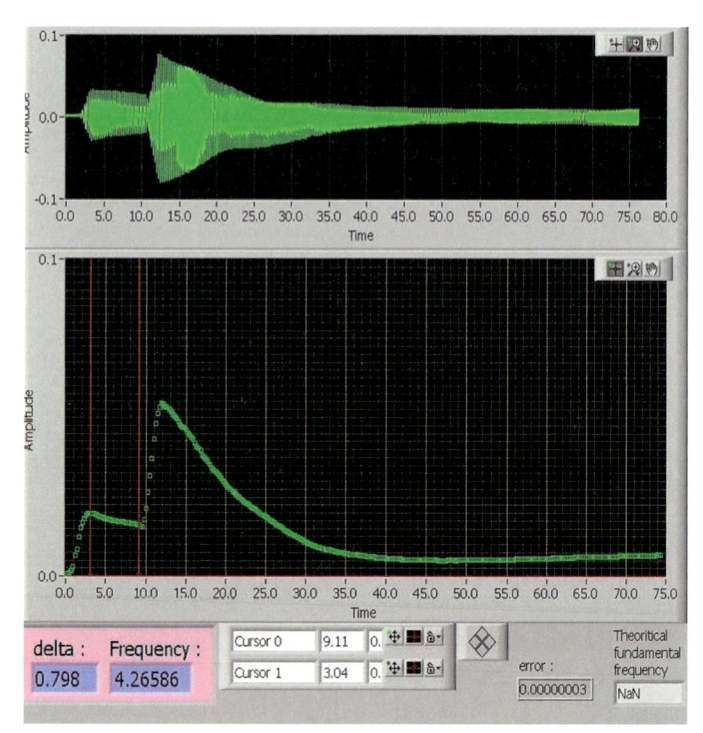

Figure 3.59 Record of cable oscillations

Figure 3.60 Load cell

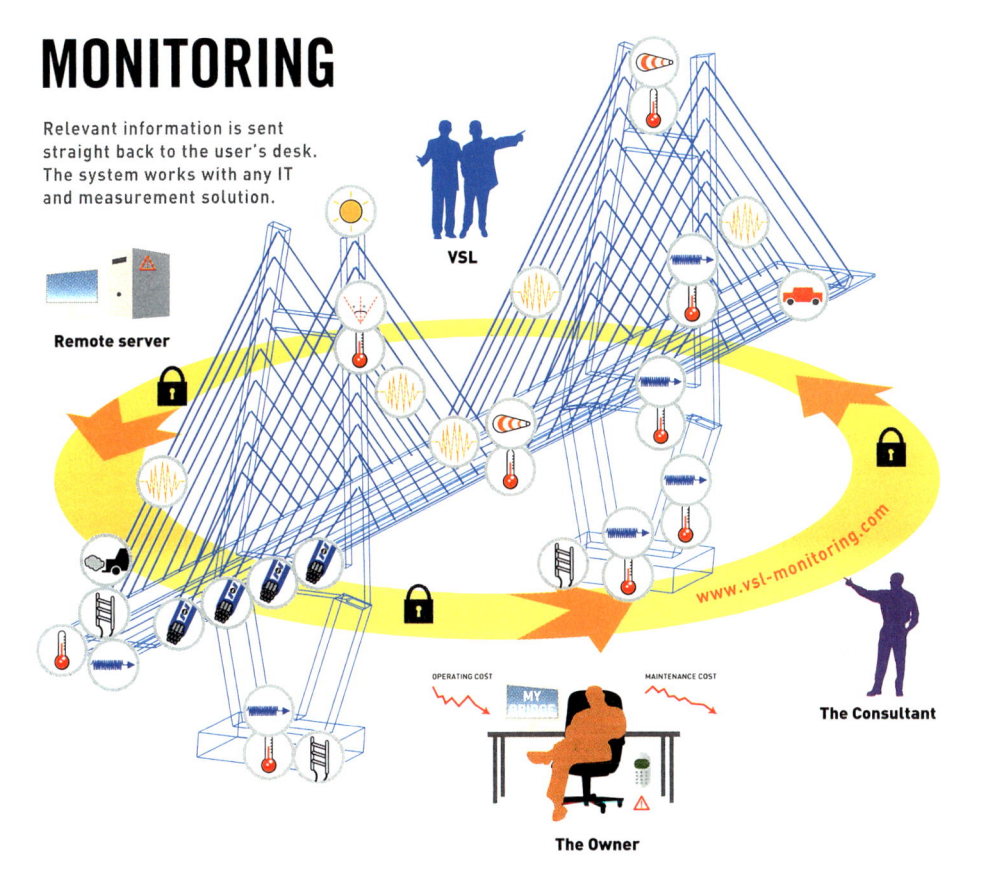

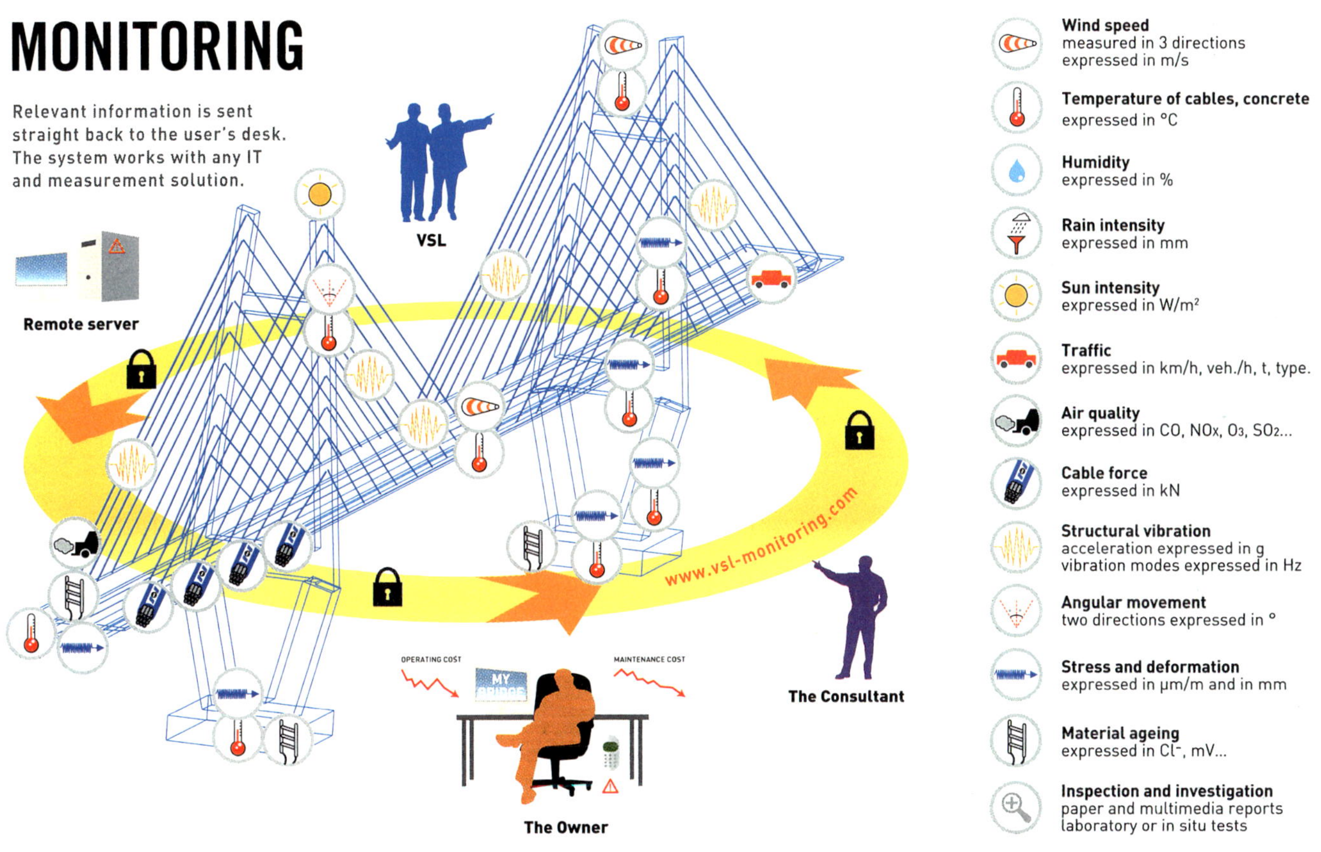

3.6 Cable anchorages

MONITORING

Relevant information is sent straight back to the user's desk. The system works with any IT and measurement solution.

	Wind speed — measured in 3 directions expressed in m/s
	Temperature of cables, concrete expressed in °C
	Humidity expressed in %
	Rain intensity expressed in mm
	Sun intensity expressed in W/m²
	Traffic expressed in km/h, veh./h, t, type.
	Air quality expressed in CO, NOx, O₃, SO₂...
	Cable force expressed in kN
	Structural vibration acceleration expressed in g vibration modes expressed in Hz
	Angular movement two directions expressed in °
	Stress and deformation expressed in µm/m and in mm
	Material ageing expressed in Cl⁻, mV...
	Inspection and investigation paper and multimedia reports laboratory or in situ tests

Figure 3.61 Central registration of control data

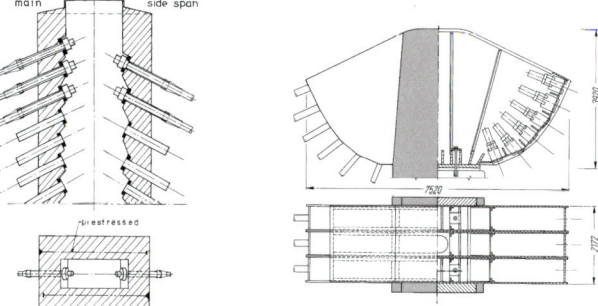

Figure 3.62 Modified fan anchorage at concrete tower head

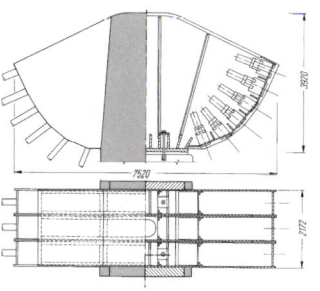

Figure 3.63 Fan anchorage at steel tower head

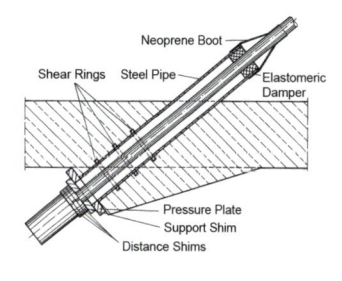

Figure 3.64 Typical cable anchorage at a concrete beam

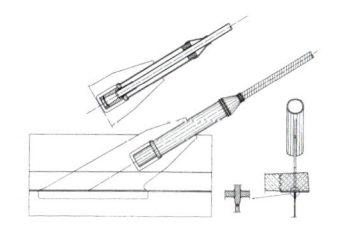

Figure 3.65 Anchorage by web extension at a steel beam

Figure 3.66 Cylindrical anchor head

Figure 3.67 Support with shims

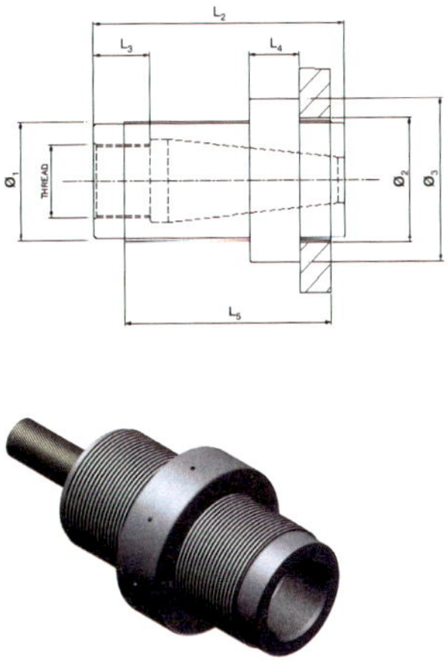

Figure 3.68 Support nut

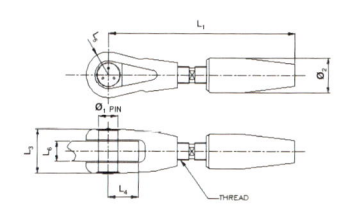

Figure 3.69 Open socket with turnbuckle

3.6.3 Anchorage at the tower
3.6.3.1 Continuous
Saddles are advantageous against anchorages of discontinuous cables because they require less space (and may be more economic). They are thus often used for central towers which have to be as slender as possible, Fig. 3.70.

Early cable-stayed bridges sometimes used locked coil ropes continuous through the tower, Fig. 3.71. The ropes were placed between two covers with fitting grooves, which were pressed against one another by bolts to achieve the required resistance against sliding by transverse pressure and friction.

The use of saddles for parallel strand cables was questioned for a long time due to uncertainty about their fatigue strength and exchangeability. The latest VSL systems claim the following advantages for their new development shown in Fig. 3.72 [3.38].

The saddles comprise a rectangular curved steel box filled with a special 'Ductal' cement which provides good corrosion protection and high friction. Additional friction is achieved by the V-shaped openings in transverse plates which clamp the strands additionally, Fig. 3.73. In total a friction factor of up to 0.7 is achieved, which can carry the differential cable forces between main span and side span.

Tests with this system showed no reduction in tensile strength or fatigue strength.

3.6.3.2 Composite cable anchorages at tower head
In the majority of cable-stayed bridges, the stay cables are anchored individually in the tower head. In this way they can individually be exchanged and remain straight. Stressing of the stay cables inside the tower head is possible.

Lately composite stay cable anchorages have been preferred for which a steel anchorage box is cast into concrete. A typical anchorage is shown for the Normandy Bridge in Figs 3.74 and 3.75. Inside a concrete box the horizontal tension components between forestays and backstays are directly connected by a steel tie. The steel anchorage units are stacked directly onto one another and in this way they ascertain the required exact placement of each anchor head during construction.

The tower cable anchorage of the Rion-Antirion Bridge is designed similarly, Fig. 3.76.

3.6.3.3 Cable anchorage in concrete
If cables are individually anchored inside the box cross-section of a concrete tower the tension forces between the forestays and the backstays are appropriately carried by tendons. If loop tendons are used as shown in Fig. 3.77 the force introduction can be explained by applying a strut and tie model. One half of the cable force is introduced via virtual struts directly into the tendon anchorages. The other half is balanced by the radial forces in the tendon loops.

Figure 3.70 Central tower with cable saddles for C and D Canal Bridge, USA

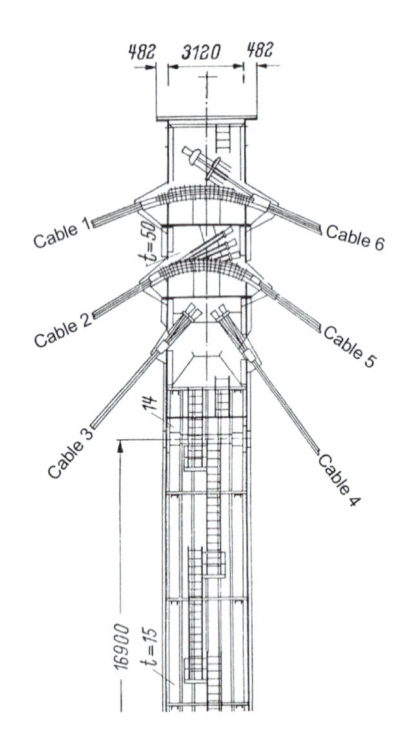

Figure 3.71 Cable anchorages and saddles for locked coil ropes at steel tower

Figure 3.72 VSL cable saddle

3.6 Cable anchorages

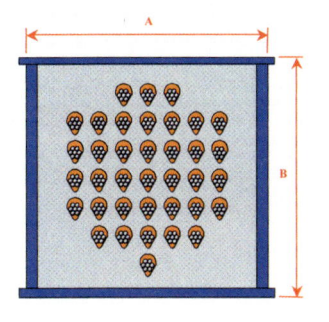

Figure 3.73 V-shaped openings for clamping strands in the saddle

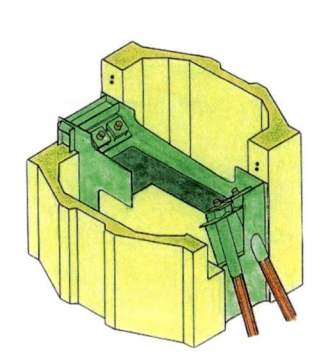

Figure 3.74 Composite cable anchorage of the Normandy Bridge

Figure 3.75 Fabrication of steel anchorages

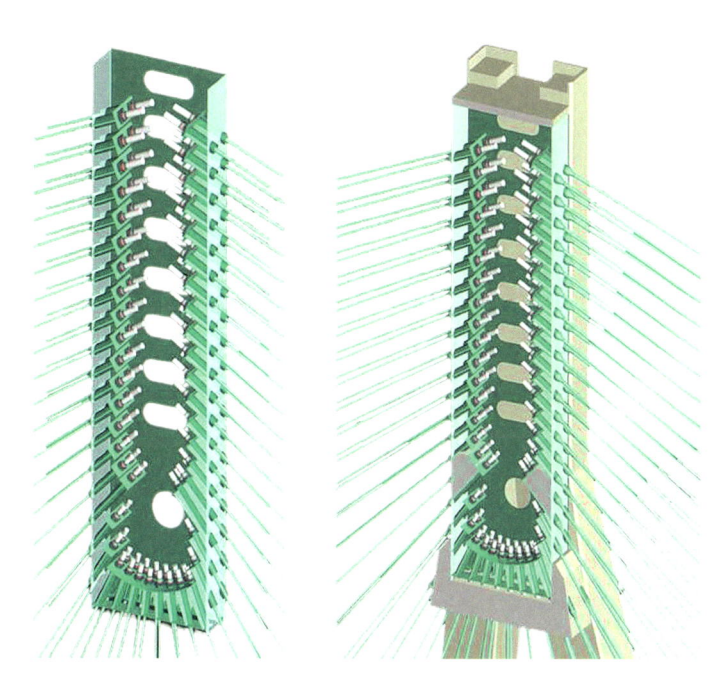

Figure 3.76 Composite cable anchorage of the Rion-Antirion Bridge

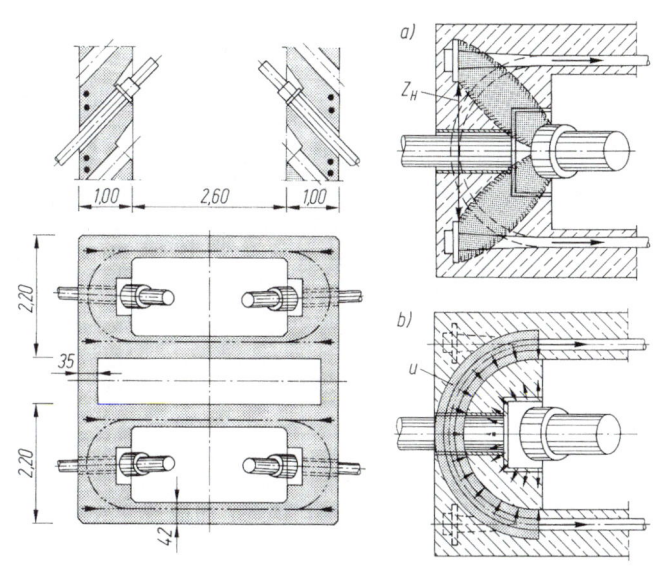

Figure 3.77 Cable anchorage with tendons

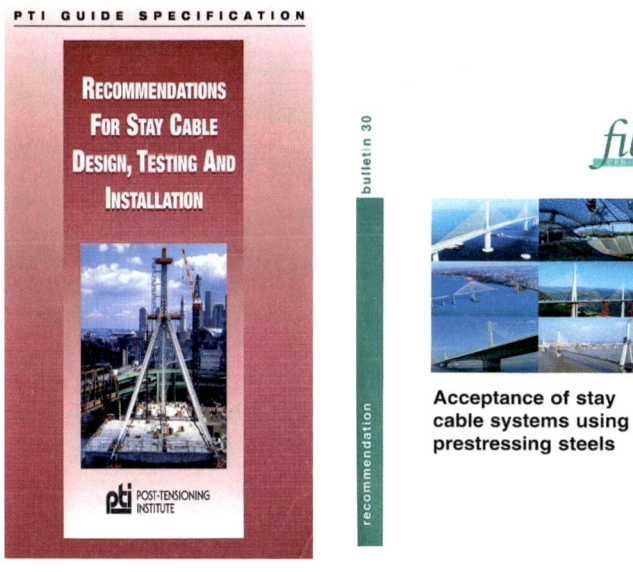

Figure 3.78 PTI Recommendations

Figure 3.79 The fib bulletin 30

3.7 Cable sizing
3.7.1 General

The following rules for sizing of stay cables only refer to the free length between the cable anchor heads and are based on a multitude of international tests.

The anchor heads themselves are today either sized with the help of three-dimensional FE-models or have a general certification.

The sizing of stay cables in Germany was done until about the year 2000 by the permissible stress concept, including the proof of fatigue stresses for 2 million cycles. Today, however, more and more countries adopt proof in the ultimate limit state, including the proof of fatigue endurance stresses for 100 million cycles.
- In Germany   [3.40] – [3.45]
- International   [3.46] – [3.48], Figs 3.78 and 3.79.

National codes [3.40], [3.41] and [3.46] are based on the permissible stress concept, national codes [3.43], [3.46] and [3.47] are based on the ultimate limit state. Extensive reports on tests with locked coil ropes and cables are given in [3.49 – 3.51].

The author is very grateful to his colleague of many years, Dr.-Ing. E. h. Reiner Saul, Fig. 2.90, for writing the following section on cable sizing. Dr. Saul is very experienced on ropes and cables; he was a member of the German DIN code commissions and was checking engineer for the Strelasund Bridge.

3.7.2 Sizing by permissible stresses
3.7.2.1 Permissible stresses for static loads
Locked coil ropes

When the permissible stresses for locked coil ropes were determined for German cable-stayed bridges the following two considerations were taken into account:
- The yield point lies relatively low and is – contrary to the individual wires – not well defined because the stress–strain curve for locked coil ropes is time dependent, Fig. 3.80 and Table 3.1.
- By the use of the hot casting material Zamak Z610 the actual tensile strength is about 8 % smaller than the calculated guaranteed ultimate tensile strength (GUTS).

The permissible stresses were thus determined lower in comparison to post-tensioning steel.

permissible $f_S = 0.42 \cdot f_{GUTS}$
→ safety $\gamma_M = \dfrac{1}{0.42} = 2.38$ for main loads

permissible $f_S = 0.46 \cdot f_{GUTS}$
→ safety $\gamma_{Add} = \dfrac{1}{0.46} = 2.17$ for main plus additional loads

with GUTS: guaranteed ultimate tensile strength
- In this way the usually applied safety against yield with $f_{Yield} = 0.55\ f_{GUTS}$ comes to
  safety against yield $\gamma_M = 2.38 \cdot 0.55 = 1.31$ for main loads
  safety against yield $\gamma_{Add} = 2.17 \cdot 0.55 = 1.20$ for main plus additional loads

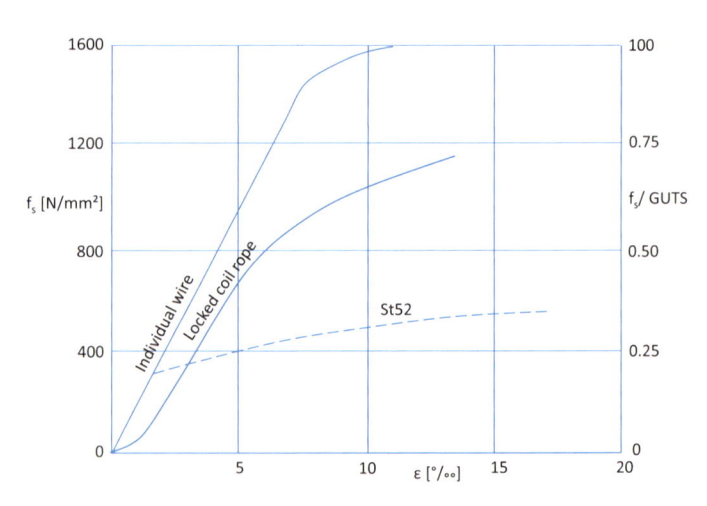

Figure 3.80 Stress-strain diagram of high-strength steel St 52 and locked coil ropes

Table 3.1 Relation of yield stress to tensile strength

	Structural steel St 52	Wire/Strand/Cables	Locked coil ropes
Yield stress [N/mm²]	355	1570	865
Tensile strength [N/mm²]	510	1770	1570
Yield stress / Tensile strength	0.70	0.89	0.55

Parallel wire and strand cables

The ultimate load of these cables equals the sum of the ultimate loads for the individual wires or strands, which means that there is no reduction in strength from the cold casting material or the wedge anchors. The permissible stresses are, therefore, assumed higher:

permissible $f_S = 0.45 \cdot f_{GUTS}$
→ safety $\gamma_M = \dfrac{1}{0.45} = 2.22$ for main loads

permissible $f_S = 0.50 \cdot f_{GUTS}$
→ safety $\gamma_{Add} = \dfrac{1}{0.50} = 2.00$ for main and additional loads

The safety against yield for a strand St 1570/1770 thus comes to:

safety against yield $\gamma_M = 2.22 \cdot \dfrac{1570}{1770} = 2.11$ for main loads

safety against yield $\gamma_M = 2.00 \cdot \dfrac{1570}{1770} = 1.77$ for main and additional loads

3.7.2.2 Permissible fatigue range
The permissible fatigue range for the locked coil ropes of many early cable-stayed bridges in Germany was assumed as $\Delta f_S \leq 150\ \text{N/mm}^2$ for 60 % of live load.

German DIN 18809 [3.41], valid until 2003, later stated the following: 'Railway loads have to be applied with their full weight, live loads with 50% of their full weight for the determination of the fatigue range Δf_S. For the fatigue range determined in this way the locked coil ropes have to be tested in sufficient numbers to prove that the ropes including their anchorages can withstand 2 million load cycles with amplitudes increased by 1.15.'

The fatigue range – not determined in the codes – was usually applied in the tests with test range $\Delta f_S = 150$ N/mm² for locked coil ropes: $\Delta f_S = 200$ N/mm² for parallel wire and strand cables.

Only the normal force was considered relevant for the fatigue strength.

3.7.2.3 Permissible stresses during cable exchange
For this loading, it is not the permissible stresses but the required safety that is determined in [3.41].
Ultimate safety $\gamma_{GUTS} = 1.6$
Safety against yield $\gamma_{yield} = 1.1$
This corresponds to the permissible stresses:
permissible $f_S = 0.50\ f_{GUTS}$ for locked coil ropes
permissible $f_S = 0.63\ f_{GUTS}$ for parallel wire and strand cables

3.7.3 Sizing in ultimate limit state
3.7.3.1 Ultimate limit state
The partial safety factors γ_M are assumed as:
– for locked coil ropes
   against GUTS $\gamma_M = 1.65$
   against yield $\gamma_M = 1.10$
– for parallel wire and strand cables
   against GUTS $\gamma_M = 1.65 \cdot 0.42 / 0.45 = 1.54$
   against yield $\gamma_M = 1.10$

3.7.3.2 Fatigue
General
The new codes require consideration of axial loads as well as bending in the cables due to live load, wind, etc. The governing loads and load cycles are, for example, determined in DIN FB 101, Section IV-4.6.

Locked coil ropes
If no applicable tests are available, the ropes may be classified as category 112. For a material factor of safety of 1.15 in accordance with DIN FB 103-II-9.3 thus the fatigue range in a virtual test comes to: $\Delta f_S = 1.15 \cdot 112 \approx 130$ N/mm².

The actual cable tests are usually executed with $\Delta f_S = 150$ N/mm² which classifies them as category $1.50 / 1.15 = 130$ N/mm².

Parallel wire and strand cables
The approach is demonstrated for the example of the parallel strand cables of the Strelasund Bridge [3.52].

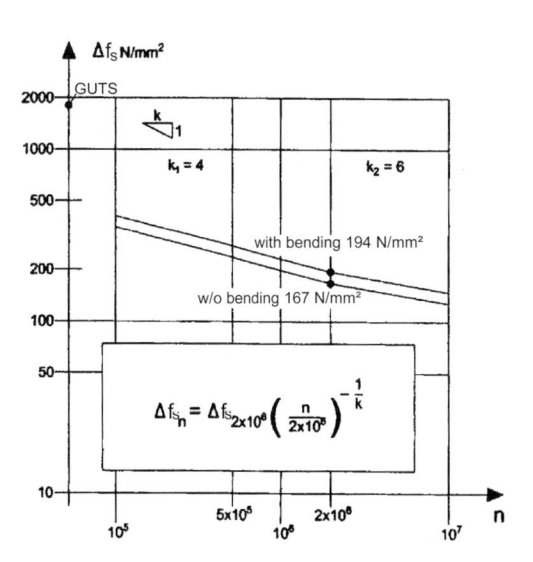

Figure 3.81 Wöhler curve [2.137] for an upper stress of $f_0 = 0.45$ GUTS

Permissible fatigue range
The tests executed for the licensing were executed with a base stress $f_S = 797$ N/mm², a fatigue range of $\Delta f_S = 200$ N/mm², a deviation angle of $\Delta \alpha = 0.6 = 0.01$ mrad and $n = 2 \cdot 10^6$ cycles. Taking into account the additional bending stresses due to the deviation angle the effective fatigue range comes to 233 N/mm². Other values of n result in:

$$\Delta f_S = \Delta f_{S\,2\cdot 10^6} \left(\frac{n}{2 \cdot 10^6} \right)^{-1/k} \text{ with } k = 6, \text{ Fig. 3.81.}$$

In accordance with DIN FB 103-II-9.3 the safety factor for main load-bearing members is $\gamma_{Mf} = 1.15$. As a consequence the permissible stresses for normal forces and bending moments come to:
– live load, $n = 2 \cdot 10^6$
   $\Delta f_0 / \gamma_{Mf} = 233 / 1.15 = 203$ N/mm²
– wind vibrations, $n = 10^8$
   $\Delta f_W / \gamma_{Mf} = 233 \cdot \left(\frac{10^8}{2 \cdot 10^6} \right)^{-1/6} / 1.15 = 121 / 1.15 = 105$ N/mm²

The elastic support of the stay cables at the centralizer reduces the bending stresses of the strands significantly compared with the fatigue tests, in which the deviation angle is located directly at the anchorage wedges, Fig. 3.82.

The actual bending stresses reach only about 25% of those at a rigidly fixed support, depending on the actual normal stresses and the stiffness of the centralizer support, Fig. 3.83.

Fatigue due to traffic
In accordance with a fatigue load model in DIN FB 101, Section 6.4, the fatigue range of the normal forces comes to max $\Delta N = 114$ kN and the deviation angle to max $\Delta \alpha = 0.17 = 0.003$ m between beam and stay cables. Taking into account the adaption value of $\lambda = 2.0$ and the partial safety factor $\gamma = 1.0$, we arrive at:

$\Delta f_N = 45$ N/mm²
$\Delta f_B = 15$ N/mm² at the anchor head
$\Delta f_N = 29$ N/mm² at the support
thus an actual total fatigue range of
$\Delta f_S = 45 + 29 = 74$ N/mm² ≪ 203 N/mm² permissible

Fatigue due to wind vibration
Oscillations and strains due to wind result from a multitude of exciting mechanisms which are not completely determined even today, see Section 3.8. Modern parallel strand cables are less sensitive to wind excitations because their elastic supports act like dampers and an outer helix on the PE pipes reduces wind-rain excitations.

On the basis of the fib criteria, oscillations with an amplitude of A ≤ L/1700 were investigated, which means 10 cm for the longest stay cable.

With $\Delta\alpha = 2 \cdot \dfrac{\pi \cdot A}{L} = 3.70 \cdot 10^{-3}$ rad

the bending stresses come to $\Delta f_B = 9$ N/mm² at the anchor head and 18 N/mm² at the support, and are thus significantly less than the base fatigue strength.

Summary
As a result of these theoretical considerations the fatigue range ($n = 2 \cdot 10^6$) comes to only about 35% of the permissible values, and the fatigue stresses due to assumed oscillations with A = L/1700 ($n = 10^8$) come to only 17%, Fig. 3.84.

3.7.3.3 Cable exchange
The increased permissible stresses under service conditions are taken into account at ultimate state by modifying the partial safety and combination factors for loads.

3.7.3.4 Service limit state
This limit state considers, for example, the correct functioning of the corrosion protection system. For parallel strand cables it is requested to perform the proof of stresses in the service limit state for deviation angles due to structure deformations, which do not exceed 0.3°, without taking into account the transverse bending stresses. The permissible stresses are given in Table 3.2.

Table 3.2 Permissible service limit state stresses

	Construction stage, cable exchange	Service stage
Deviation angle ≤ 0.3°	0.55 GUTS	0.45 GUTS
Deviation angle > 0.3°	0.60 GUTS	0.50 GUTS

3.7.4 Summary
A simplified summary of the sizing regulations is given in Tables 3.3 and 3.4.

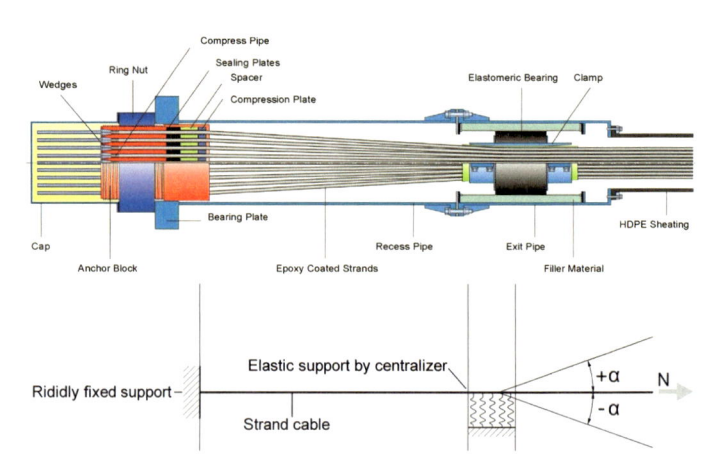

Figure 3.82 Elastic support of the cable at the centralizer

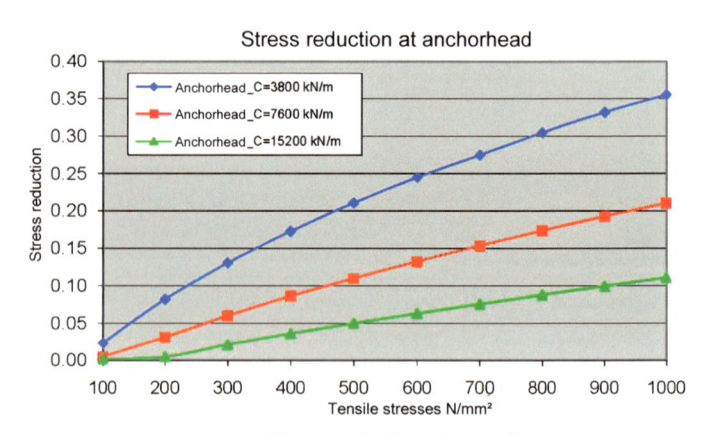

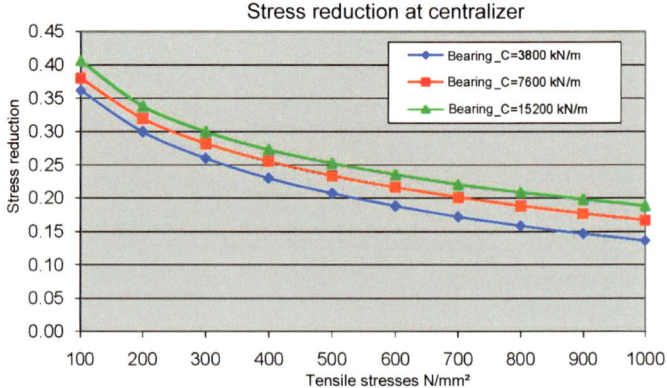

Figure 3.83 Influence of normal forces and support rigidity on the bending stresses of the strands due to a deviation angle

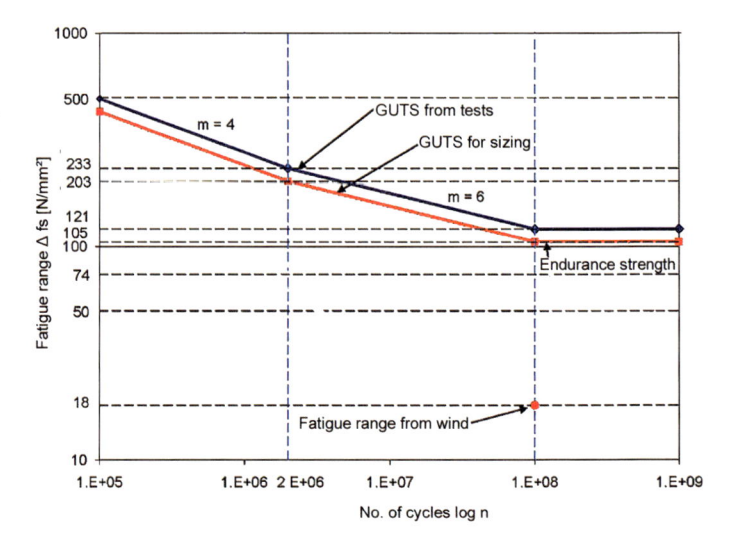

Figure 3.84 Wöhler curve ($n = 2 \cdot 10^6$) and endurance strength curve ($n = 10^8$) for parallel strand cables

Table 3.3 Permissible cable stresses in service limit state

	perm. f_S	perm. Δf_S	Cable exchange
Locked coil ropes	0.42 β_N	150 N/mm²	0.50 β_N
Parallel wire Parallel strand	0.45 β_N	200 N/mm²	0.63 β_N

Table 3.4 Permissible cable load factors at ultimate limit state

	Against tensile strength	Against 0.2 % yield strength
Locked coil ropes	1.65	1.10
Parallel wire Parallel strand	1.54	1.10

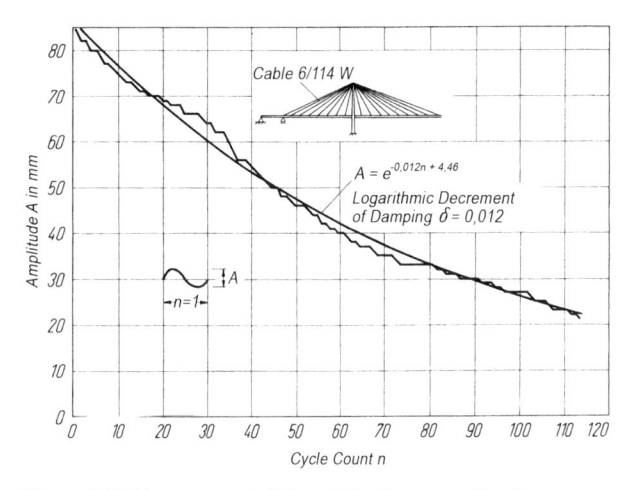

Figure 3.85 Measurement of decay behavior on a cable of the Pasco-Kennewick Bridge, USA

3.8 Cable dynamics

3.8.1 General

A straight cable can be idealized as a string: it behaves like a prestressed chain. Its special properties originate from the fact that its stiffness arises almost entirely from the longitudinal prestress. This stiffness is termed geometrical stiffness because on a transversally deflected cable the retracting forces are generated by the axial tension acting on the curvature of the deformed shape.

Cables are very easily excited into oscillations, mostly because their inherent structural damping is extremely low. Fig. 3.85 shows a convincing example of the decaying response of a bridge cable: the logarithmic decrement is on average 0.012, sometimes even less [1.15]. The following values can be used as a guideline:

$\delta \approx 0.005 - 0.006$ minimum
$\delta \approx 0.010 - 0.012$ expected value

The structural damping of a cable is about three to four times lower than that of a welded steel structure. The main reason for the low damping is the fact that, as explained before, the stiffness is of geometrical nature.

Because during an oscillation the longitudinal stresses remain constant, there is no energy-dissipating hysteresis, for example damping. It is because of the low damping that a number of excitation phenomena could be discovered at all, e.g. rain–wind-induced galloping cannot occur on other members of a steel structure, such as a girder. The susceptibility to oscillations can be further increased because of:
– the variability of the natural frequencies of the cables: because of variations in the cable tension (loading condition of the structure) the frequencies cover a certain range. There are therefore various different resonant frequencies and they can assume unexpected values.
– the low mass of the cables. If a cable is excited by oscillations of the attached members (deck or pylon), the cable amplitudes will be significantly larger than those of the primary member. This will be further discussed later on.

Cable vibrations can be dangerous because of the bending stresses that are induced near the anchorage. Although the idealized cable system does not feature a fixing, in reality a fixation does occur at this location because of the bending stiffness of the cable. This results in friction corrosion and fatigue at the effective anchorage point. A paramount role is played by the axial cable force, which acts unfavorably, see Fig. 3.86. Comparing two cable oscillations of equal amplitude, a higher cable force leads to a sharper kink and thus a higher fixation moment. The result can be dangerous cable bending moments even for small cable amplitudes, see also Section 4.1.6, Fig. 4.26.

Strong cable vibrations are usually also deemed unacceptable in terms of human perception: they are unaesthetic and induce uneasiness in observers. Often this becomes a key consideration.

The excitation phenomena are manifold:
– direct turbulent excitation of the cable – the static and quasi-static component of the wind load is small; only the resonant component is of interest. The intensity remains weak, and this type of excitation has not been observed as being unfavorable.
– several types of self-excitation – these are important as they lead to a dynamic instability. A characteristic feature of this phenomenon is that usually groups of cables are affected, but not single cables. Usually large amplitudes are observed. The most common form is that of rain–wind-induced galloping oscillations. Countermeasures to suppress the unstable oscillations may be required.
– parametric excitation – a relatively rare type of excitation, which only occurs on single cables and also results in large amplitudes. Countermeasures may be required.
– coupled vibrations (also termed anchorage excitation) – the cable and the structural member oscillate together, i.e. the oscillation is being transferred onto the cable from adjacent members.

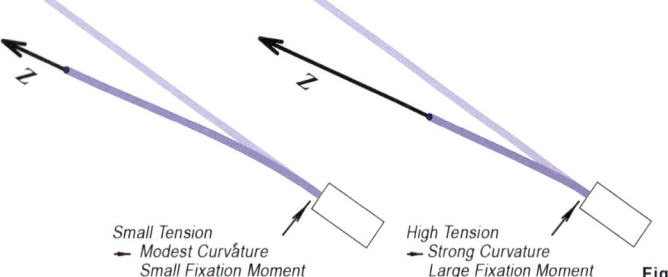

Figure 3.86 Fixation moments after second order theory

– vortex induced oscillations – a minor form of excitation without practical relevance, whose importance is often over-rated.

Cable dynamics have become more important with the increase of span lengths because longer cables are more prone to vibrations. Specialist sub-consultants are often employed to ensure cable stability, but the design engineer in charge should be able to understand and evaluate the findings of the specialists.

On the problem of cable vibrations the author has cooperated with Dr. Imre Kovacs, Fig. 3.87, for many years. Dr. Kovacs is one of the most successful engineers in this field. The author would like to thank him for contributing this chapter about cable dynamics.

3.8.2 Fundamental parameters
3.8.2.1 Static wind load

In later sections it will be shown that the susceptibility of cables to galloping excitation can be reduced by a profiling of the cable sheething. This, however, has an effect on the cable drag loads.

As explained above, the static wind load acting on the cable is of minor importance for the cable itself. However, it can be significant as an indirect load transferred to the main structural elements such as girder and pylon.

For large span lengths the cable drag is critical for the design of the main elements; see Fig. 3.88 for the cable-stayed bridge alternative of the Great Belt East Bridge with a 1205 m main span, [7.1]. Here, 65 % of the lateral wind load acting on the girder originates from the cable drag. The large indirect contribution of cable loads is important for cable dynamics because such drag must be reduced at the same time as trying to reduce susceptibility to galloping instabilities.

Of major technically significance for the economical design has been the reduction of cable diameters, the perimeter of corrosion-protected cables has been modified substantially over the past two decades. The efficiency of corrosion protection has been increased, cable diameters of up to 300 mm common 20 years ago are have become extremely rare now. Fig. 3.89 shows various options for reducing the cable cross-section. Diameters of over 200 mm are rarely found today.

The dependence of the drag coefficient of a cylindrical section on the Reynolds number is shown in Fig. 3.90. The drag changes drastically in the Reynolds number range between 2E5 and 5E5, the same range where real stay cables can be found. As can be seen in Fig. 3.91, in this region the drag actually reduces with the introduction of a surface profiling and hence a higher friction. This phenomenon varies depending on the geometry of the surface pattern and can be found in many other applications such as golf ball aerodynamics. It can be explained by the fact that in this Reynolds number regime the profiling leads to a turbulent boundary layer, which tends to remain attached for longer. The point of separation of the boundary layer thus moves leeward and the drag reduces. Modern cable design tries to exploit this effect to the maximum. Methods of achieving a favorable galloping behavior will be discussed later.

Figure 3.87 Imre Kovacs

Born 1943 in Bekes, Hungary
1966 Dipl.-Ing. Budapest University
1973 Dr.-Ing. Stuttgart University
1975–91 Leonhardt, Andrä and Partners
since 1997 Independent Consultant, Company Dynamik Consulting
Bridges, for which cable dynamics was investigated:
1986 Faroe Bridge, Denmark
1991 Helgeland Bridge, Norway
1995 Burlington Bridge, USA
1996 Penang Bridge, Malaysia
1996 Badajoz Bridge, Spain
1998 Uddevalla Bridge, Sweden
2001 Gdansk Bridge, Poland
2002 Monterrey Bridge, Mexico
2004 Stonecutters Bridge, Hong Kong
2004 Ma-Chang Bridge, South Korea
2006 M0 Bridge Budapest, Hungary
2008 Anzac Bridge Sidney, Australia

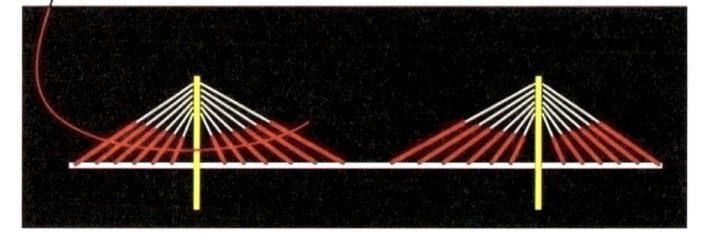

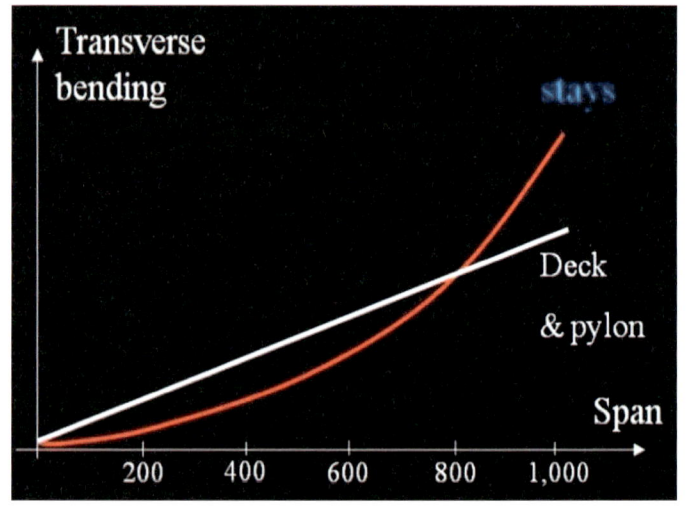

red = Cable Drag transfered to the Girder

Figure 3.88 Indirect cable drag load contribution on deck and pylon

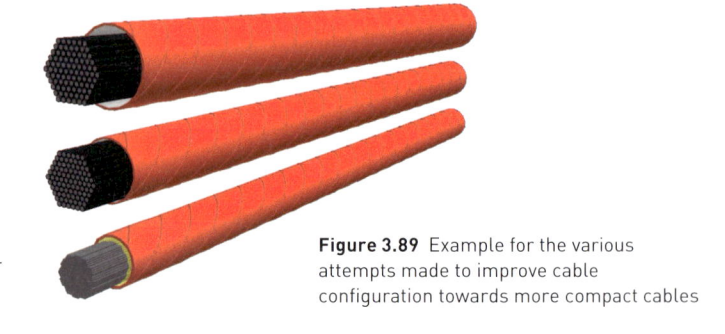

Figure 3.89 Example for the various attempts made to improve cable configuration towards more compact cables

3.8.2.2 Natural frequencies

A cable's natural frequencies follow an arithmetic series, Fig. 3.92, see Section 3.9.5.2. This is typical of a string, the reason being the purely geometric stiffness: the restoring force is proportional to the second rather than the fourth derivative of the displacements. Vertical and lateral eigenfrequencies are nearly identical. The cable frequencies are usually of interest if they are close to other natural frequencies of the bridge system. Short cables are usually unproblematic.

3.8.3 Dynamic excitation
3.8.3.1 Galloping oscillation
Ice galloping

This is an excitation mechanism which originates from an unfavorably altered cross-sectional shape, see Fig. 3.93.

Ice galloping leads to a dynamic instability. The circular cross-section itself is rather tame and usually unproblematic except for a slight susceptibility to skew angles of incident flow (if the flow has a component along the cable then the effective cross-section becomes an ellipsoid). A circular section can move toward an angular shape through the accumulation of material, which leads to a destabilization if the accumulation is strong enough.

The typical case of ice galloping with its lift force characteristic is shown in Fig. 3.93. For the critical cross-section $a_G \approx -4$ to -5 (after Davenport), but that is only realistic for about 20–25% of the cable length, so a sensible criterion is $a_G = 1.0 \cdot$ stability value after DIN [3.53] for ice galloping.

The excitation mechanism can be explained as follows when random ambient oscillations of the cable occur, an inclined apparent angle of incidence of the flow occurs when the oscillation passes the static cable position, because of the vector sum of the (horizontal) flow speed and the velocity of the moving cable. When the cable is moving upwards, the apparent incident flow onto an equivalent static section is approach downwards, which in turn corresponds to the section rotated in a counter-clockwise direction against a horizontal wind.

As can be seen in Fig. 3.93, the negative gradient of the angle–lift relationship thus leads to a lift force which acts in the same direction as the sectional velocity, thus feeding energy into the system. This energy can be thought of as a negative damping component. If the overall damping becomes negative, instability occurs. The intensity of galloping excitation can be calculated using the methodology of the DIN [3.53].

Ice galloping only becomes an issue with appropriate climatic conditions; this needs to be checked on a case-by-case basis. To assess the probability of ice formation and high wind speeds occurring concurrently is challenging and should be tackled together with the client. As a side note, in northern Germany wind speeds that can cause ice galloping of cables longer than 200 m occur several times each year.

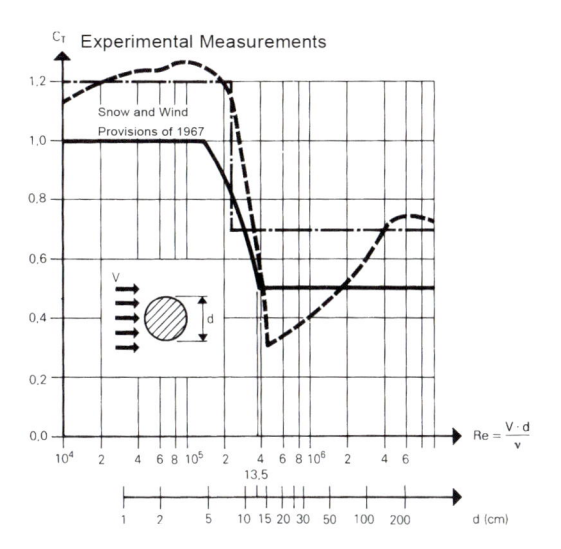

Figure 3.90 Examples of measured and codified drag coefficient of an infinitely long cylinder over a range of Reynolds numbers.

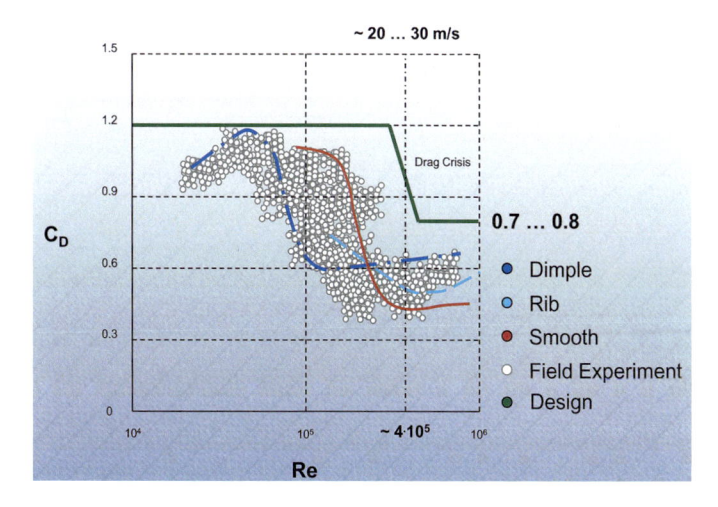

Figure 3.91 Measurements taken in the BBR lab for different cable surface patterns

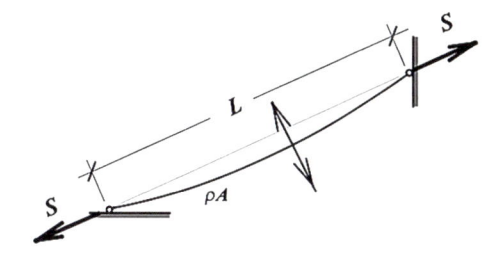

Figure 3.92 Stay cable – boundary conditions and cable mode of vibration

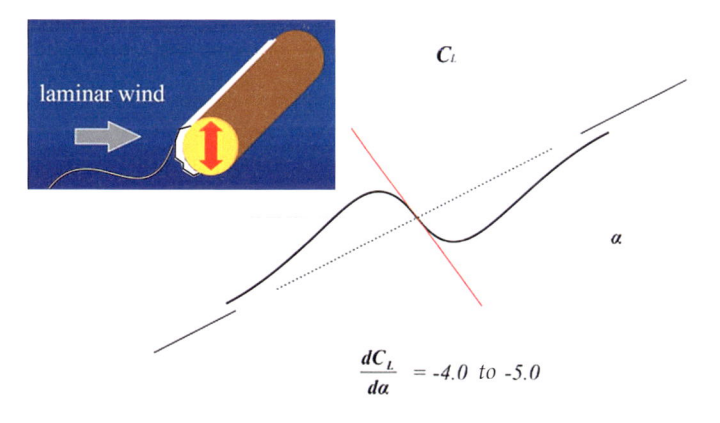

Figure 3.93 Ice galloping, change in lift force gradient typical of galloping susceptibility

Rain–wind-induced galloping

Rain–wind-induced galloping oscillations are significantly more frequent than ice galloping. The cross-section modification originates from rivulets of water forming along the cable. The pair of rivulets, one at the top and one at the bottom lead again to an unstable system with a negative lift gradient. The upper rivulet appears because the wind prevents the water from flowing down sideways, which only occurs in certain wind speed and wind direction conditions.

The cable is critical if the wind speed is above the galloping limit of the modified complex section.

Fig. 3.94 shows for which wind directions β and cable inclinations α the cable is susceptible to rain–wind-galloping. Practically all cables subject to flow from below can become critical. (During the lecture a video of galloping oscillations on the Meikon Nishi Bridge in Japan is shown. This is explained by: (a) the violent nature of the excitation, (b) that there is always a group of cables oscillating and (c) the strong demand acting on the cable anchorage.)

It can be seen on video recordings that cable oscillations are initiated very rapidly, practically with the first raindrops start arriving and in fact before real rivulets form, i.e. even without a fully established galloping-critical geometry.

This leads to the conclusion, that the root cause may not be fully understood. Perhaps it is related to the humidity and the surface roughness thereby altered thus influencing the flow? This problem is still subject to much debate. Fig. 3.94 shows one of further possible effects which has been studied by Prof. U. Peil: here the upper rivulet exerts a tangential oscillation along the perimeter and leads to an instability.

The list of Fig. 3.94 shows which parameters contribute to the excitation. It is clear, that a complete analysis accounting for all parameters is very complex; practical checks are, irrespective of other parameters, therefore based on the curve of the excitation constant

$$V \cdot dC_L / d\phi$$

shown in Fig. 3.94 on the bottom right. The diagram was derived from wind tunnel measurements in Japan [3.54 – 3.56], and shows the intensity in relation to the wind speed. It is obvious that rain–wind galloping does not occur above wind speeds of 20 m/s, and this is because the upper rivulet ceases to exist. The diagram shows that, for example, a 200 m long cable with 200 mm diameter gallops at about $\delta_{gall} = 0.02 - 0.025$, while for longer bigger cables (e.g. Stonecutters Bridge, Fig. 2.104) this value may rise to 0.04 – 0.05. Below the diagram of excitation intensity the graphical representation of amplitudes for damped and undamped cables is plotted, as determined in laboratory experiments by Freyssinet.

It should be mentioned, that the literature offers different stability criteria, some of which are not based on galloping considerations. For example, the well-known PTI condition [15.3] defines the stability in terms of the Scruton number as follows:

$$Sc = m \cdot \xi / \rho \cdot D^2 \geq 10 \quad \text{Stability criterion after PTI,}$$
$$\text{with critical damping } \xi = \delta / 2\pi.$$

In reality, the Scruton number of any bridge stay cable is between 2 and 5. The PTI condition then basically predicts any cable to be unstable and does not consider effects such as the cable length and/or the cable natural frequency. For example, cables stabilized by auxiliary cross tie cables – an excepted method – would remain unstable as per the formula, which shows the limitations of the criterion.

In some cases bridge cables remained stable even though conditions would have led us to expect unstable oscillations. One example the author has worked on is the Penang Bridge in Malaysia, which features cables with steel rather than a PE surface – could this be the reason?

Rain–wind-induced vibrations basically have to be expected for any bridge with undamped cables. The resulting amplitudes are large, often around 0.5 m, sometimes up to 2 m. Nowadays checks for galloping instability are a standard requirement in cable-stayed bridge design.

Wake galloping

A less important type of galloping is wake galloping, see Fig. 3.95: when two cables are aligned parallel and close to each other, the downstream cable can be forced into unstable oscillations. Ruscheweyh [3.57] found that the condition does not occur for cable spacings larger than 5 d. In a cable-stayed bridge, critical conditions can thus only occur if cables are combined in bundles, as is sometimes done with backstays.

3.8.3.2 Anchorage excitation

Internal resonance, also termed anchorage excitation or coupled oscillation, is sometimes responsible for cable oscillations when bridges have heavy girders or pylons. It occurs, Fig. 3.96, when a cable frequency equals that of the heavy bridge member. In that case combined modes of vibration occur, where the cable amplitudes are large compared to those of the adjacent member.

The external excitation, usually caused by wind, acts directly on the main member and the cable participates through coupling at its anchorage point. The internal resonance slightly reduces the amplitudes of the main member, but cable amplitudes in turn are very large. The intensity of such indirect cable oscillations is a function of the mass ratio, after [3.58].

$$A_{Cable} \approx A_{Member} \cdot \sqrt{(m_{Member} / m_{Cable})}$$

The mass ratio deck : cable or pylon : cable can be of the order of 200 to 400 for large bridges, i.e. even a modest member vibration leads to violent cable oscillations that are 15 to 20 times larger.

Internal resonance does not occur often, because usually the lowest natural frequencies of deck and pylon are below the lowest cable frequency. However, the dampers required to control the oscillations can be very large.

3.8 Cable dynamics

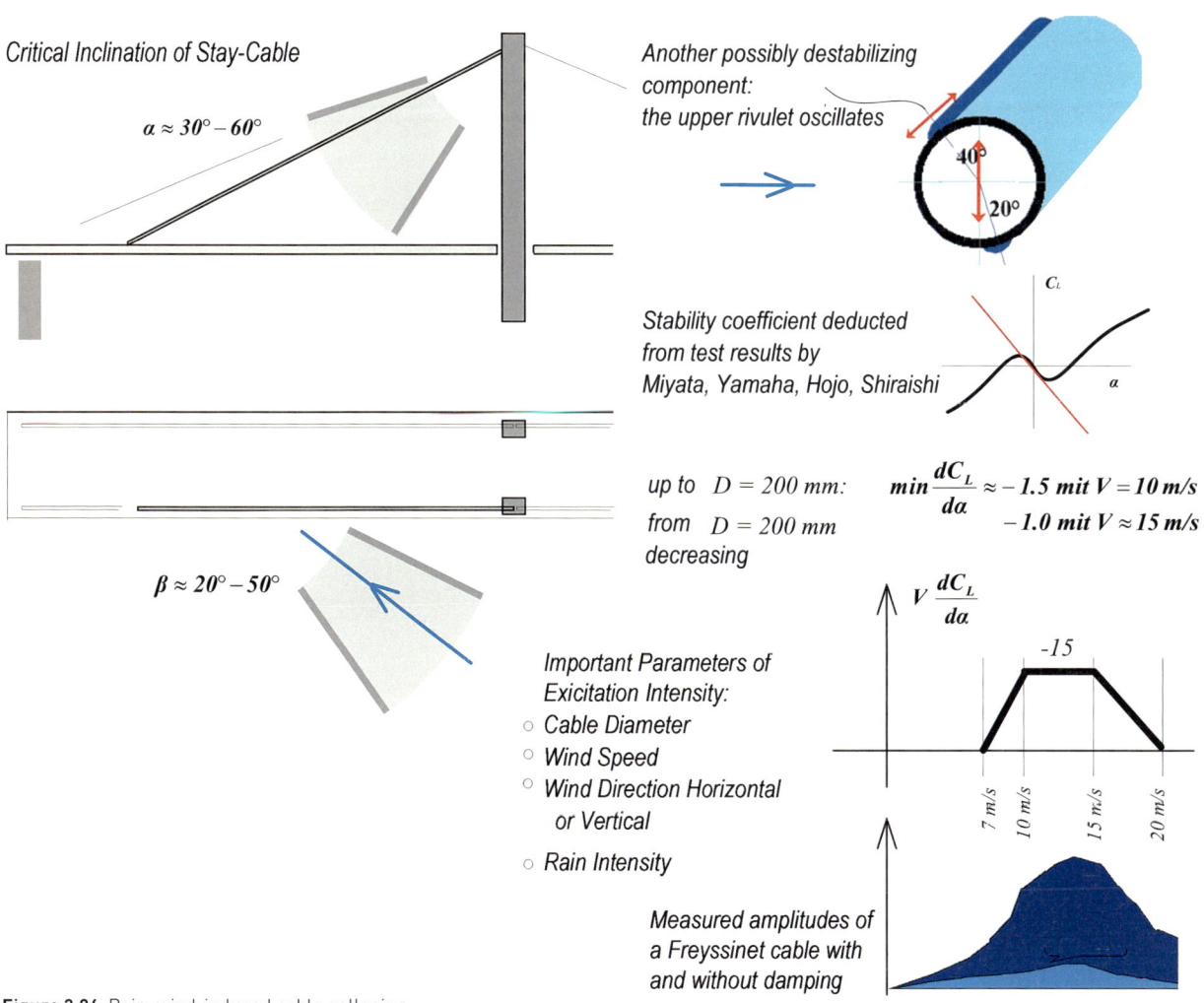

Critical Inclination of Stay-Cable

$\alpha \approx 30° - 60°$

$\beta \approx 20° - 50°$

Another possibly destabilizing component:
the upper rivulet oscillates

Stability coefficient deducted from test results by
Miyata, Yamaha, Hojo, Shiraishi

up to $D = 200\ mm$: $\quad \min \dfrac{dC_L}{d\alpha} \approx -1.5\ mit\ V = 10\ m/s$
from $D = 200\ mm$ $\qquad\qquad\qquad -1.0\ mit\ V \approx 15\ m/s$
decreasing

Important Parameters of Exicitation Intensity:
- Cable Diameter
- Wind Speed
- Wind Direction Horizontal or Vertical
- Rain Intensity

Measured amplitudes of a Freyssinet cable with and without damping

Figure 3.94 Rain–wind-induced cable galloping

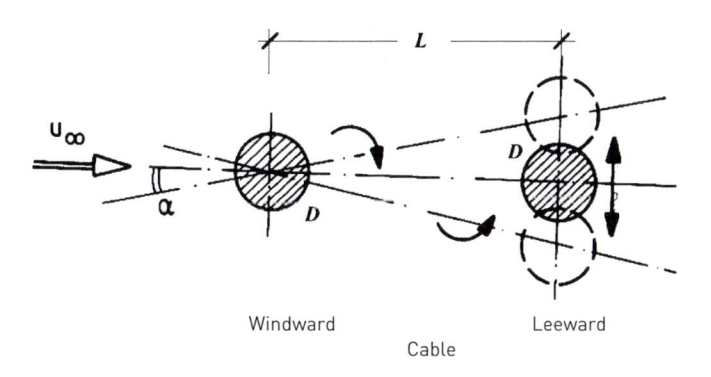

Windward   Leeward
Cable

Figure 3.95 Wake galloping

3.8.3.3 Parametric resonance

The reason for parametric resonance is the temporal variation of a natural harmonic parameter, leading to the variation of the natural frequency. This is only of practical relevance if the variation is rhythmic in nature. If the rhythm happens to be exactly $\eta = 2$ times the varying natural frequency, the fundamental mode of parametric resonance occurs.

Cables are particularly susceptible to parametric resonance, because the cable force – one such natural oscillation parameter – can be easily forced into natural vibrations through rhythmic movements of deck or pylon, Fig. 3.97. A more detailed explanation of the phenomenon can be found in [3.58, 3.59].

Parametric resonance can have many characteristics. Fig. 3.98, taken from [3.59], shows the possible resonance cases (the peaks of the shaded excitation areas). The following relationships are defined:
S_0 = cable force
X_0 = static elongation under S0
Ω_1 = first transverse circular natural frequency
$\xi \cdot X_0 \cdot \sin(\Omega \cdot t)$ = end point (anchorage) displacements

The diagram gives the critical excitation amplitudes $\xi \cdot X_0$ at the end point leading to instability in relation to the ratio $\eta = \Omega_1/\Omega$ (the frequency ratio between parametric variation and natural frequency). In the shaded areas the condition is unstable.

The fundamental and most important resonant case M_1, where $\eta = 2$ (i.e. the exciting oscillation has double the frequency of the cable natural frequency) only rarely occurs in completed structures, because the cable natural frequency is too high compared to the dominant structural frequencies. There are however a few configurations under which case M_1 may occur:

a) various erection stages, for example segment lifting or where cables have very low forces during free cantilevering;
b) in the final state if the span ratio is nearly 0.5:1:0.5, because backstays may encounter very low forces for ll in the side spans.

Of further relevance is case R particular longitudinal cable frequencies may lead to high-frequent beating transverse oscillations with large amplitudes. This was, for example, observed with the Badajoz Bridge. Other parametric resonance phenomena are of no practical relevance.

3.8.3.4 Buffeting

Buffeting is an excitation through the turbulent nature of the wind, Fig. 3.99, and affects cables as it does other structural members [3.60]. Only the resonant component of the pressure is relevant because of the low inherent damping. The intensity of the oscillations remains well below those of the galloping type and neither the peak nor the fatigue component are important for cable design.

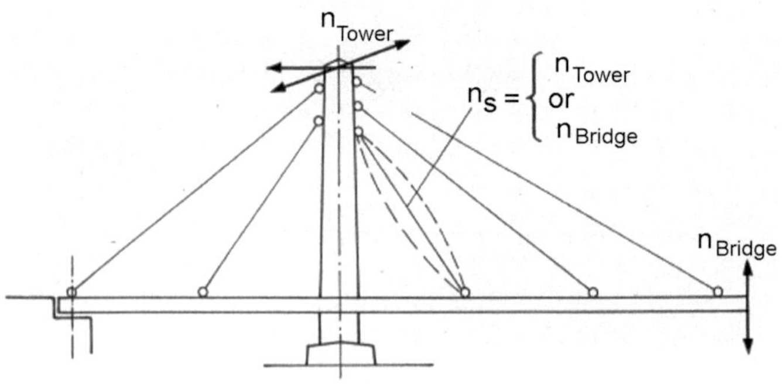

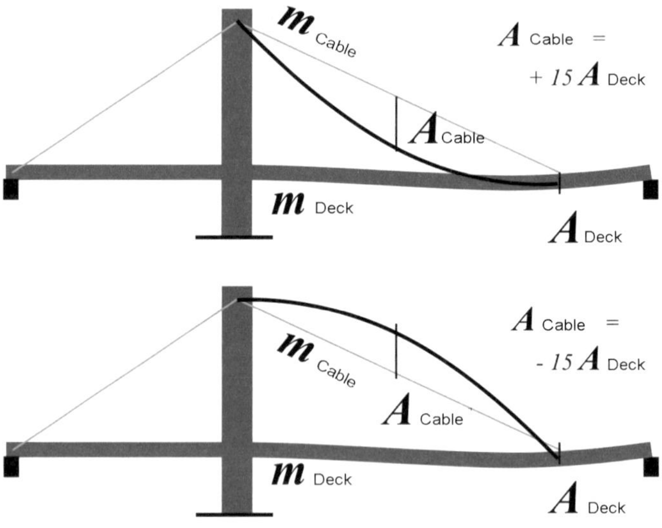

Figure 3.96 Internal resonance

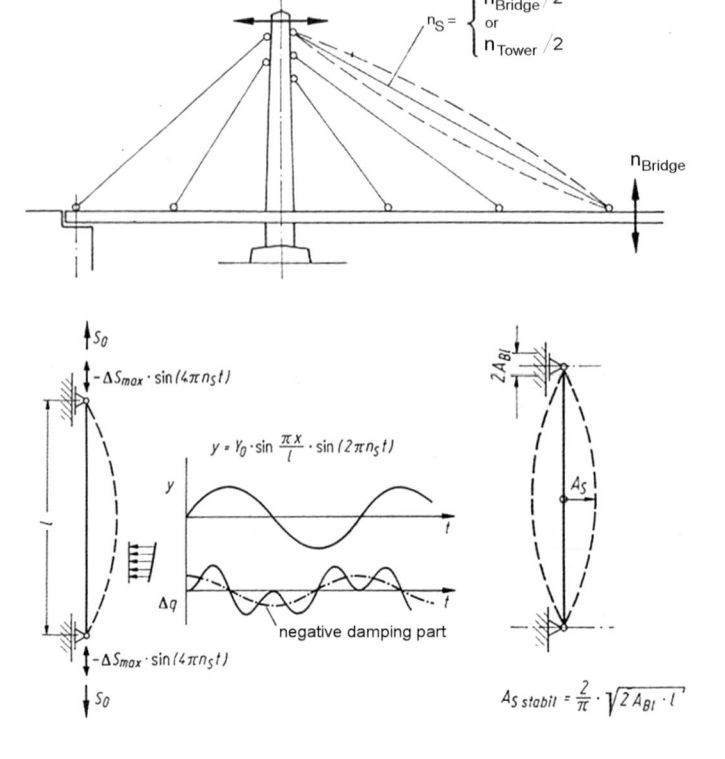

Figure 3.97 Parametric resonance

3.8 Cable dynamics

3.8.3.5 Vortex-induced vibrations

The same is true for vortex-induced vibrations, Fig. 3.100. The correlated length of the vortices along the cable is about 6 · D and hence small. When synchronization (locking-in) between oscillatory movement and vortex shedding occurs, the correlation length increases, but the cable amplitudes still remain of low magnitude.

3.8.4 Countermeasures

Measures for suppressing cable oscillations are geared towards rain–wind oscillations because of its high significance. The designer usually considers this phenomenon first. When galloping is controlled, it needs to be checked whether the measures are also sufficient for suppressing any of the other relevant excitation mechanisms. Usually that will be the case.

3.8.4.1 Dampers
Artificial damping

The best way of controlling aerodynamic cable instabilities is by supplying artificial damping, particularly in the vertical plane. The most common technique is to apply dampers fixed to a point on the cable near the anchorage at the deck or, less commonly, at the pylon, Fig. 3.101.

Theoretically it would also be possible to use a tuned mass solution, but that has aesthetic and technical problems. It has not been done on a cable-stayed bridge so far.

Rain–wind-induced galloping occurs at wind speeds between 7 and 20 m/s. The design of the damper considers a wind speed of V = 15 m/s, following

$$\gamma \cdot \delta_{gall}^{(-)} + \delta_{eige,min} + \delta_{arti} \geq 0$$

with

$\gamma = 2.0$ safety factor, common assumption
$\gamma = 1.5$ possibly for full load or rare load combination
$\delta_{eige,min} \approx 0.006$ assumed minimum internal damping
$\delta_{eige,exp} \approx 0.012$ probable, expected internal damping

The example for the design of cable dampers for the M0 Bridge in Budapest is shown in Fig. 3.102. The natural frequencies follow a common pattern the fundamental frequency of the shortest cable is at about 1.5 Hz, that of the backstays is from 0.6 Hz (bridge under self-weight only) down to 0.35 Hz (side spans fully loaded). The susceptibility to galloping is then distributed accordingly the backstay has the highest negative damping values of $\delta_{gall}^{(-)} \approx 0.025$ (unloaded condition) and 0.035. The main span cables are somewhat less critical. When the expected internal damping and the required safety against galloping are considered, the required amount of additional damping according to the blue shaded diagram is about req $\delta_{arti} = 0.05$.

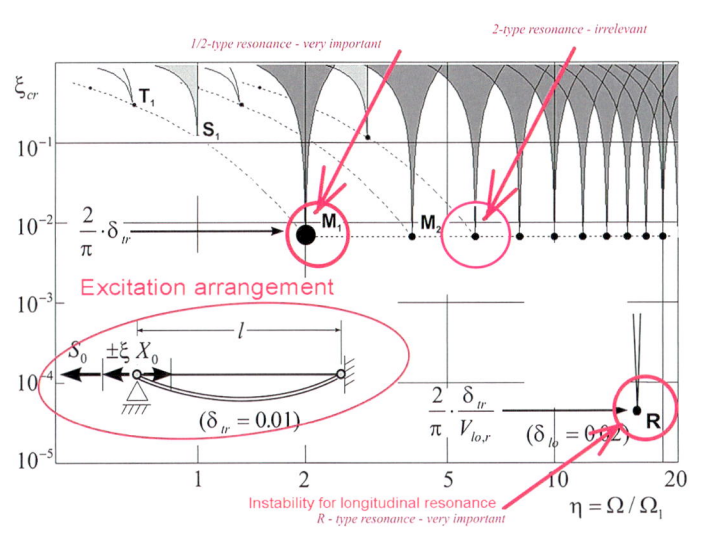

Figure 3.98 Overview of possible parametric resonance phenomena; cases M_1 and R are important

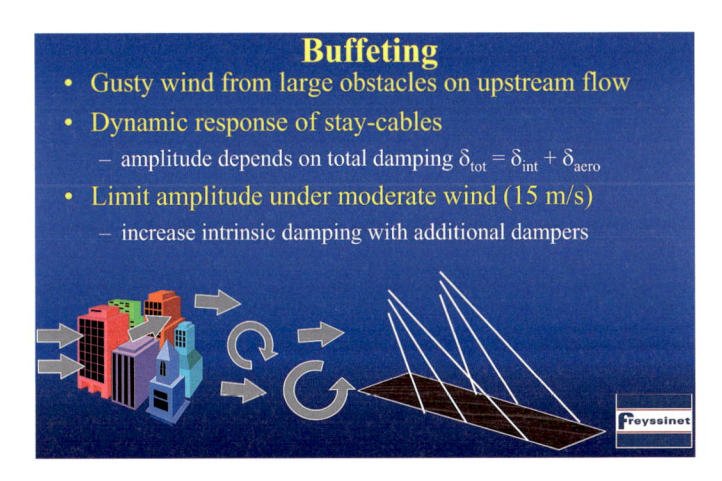

Figure 3.99 Explanation of buffeting excitation

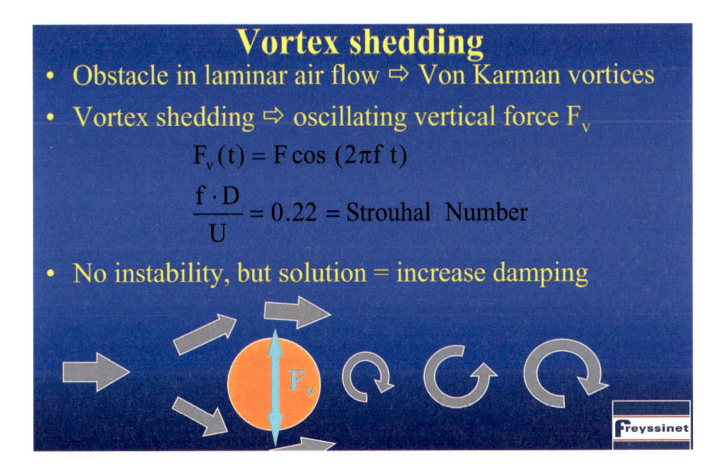

Figure 3.100 Explanation of vortex-induced vibrations

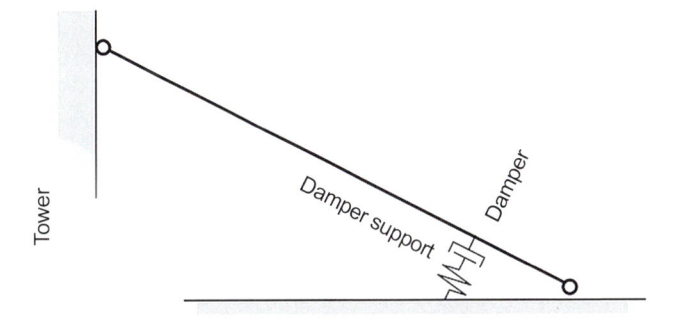

Figure 3.101 Simple damper arrangement in principle

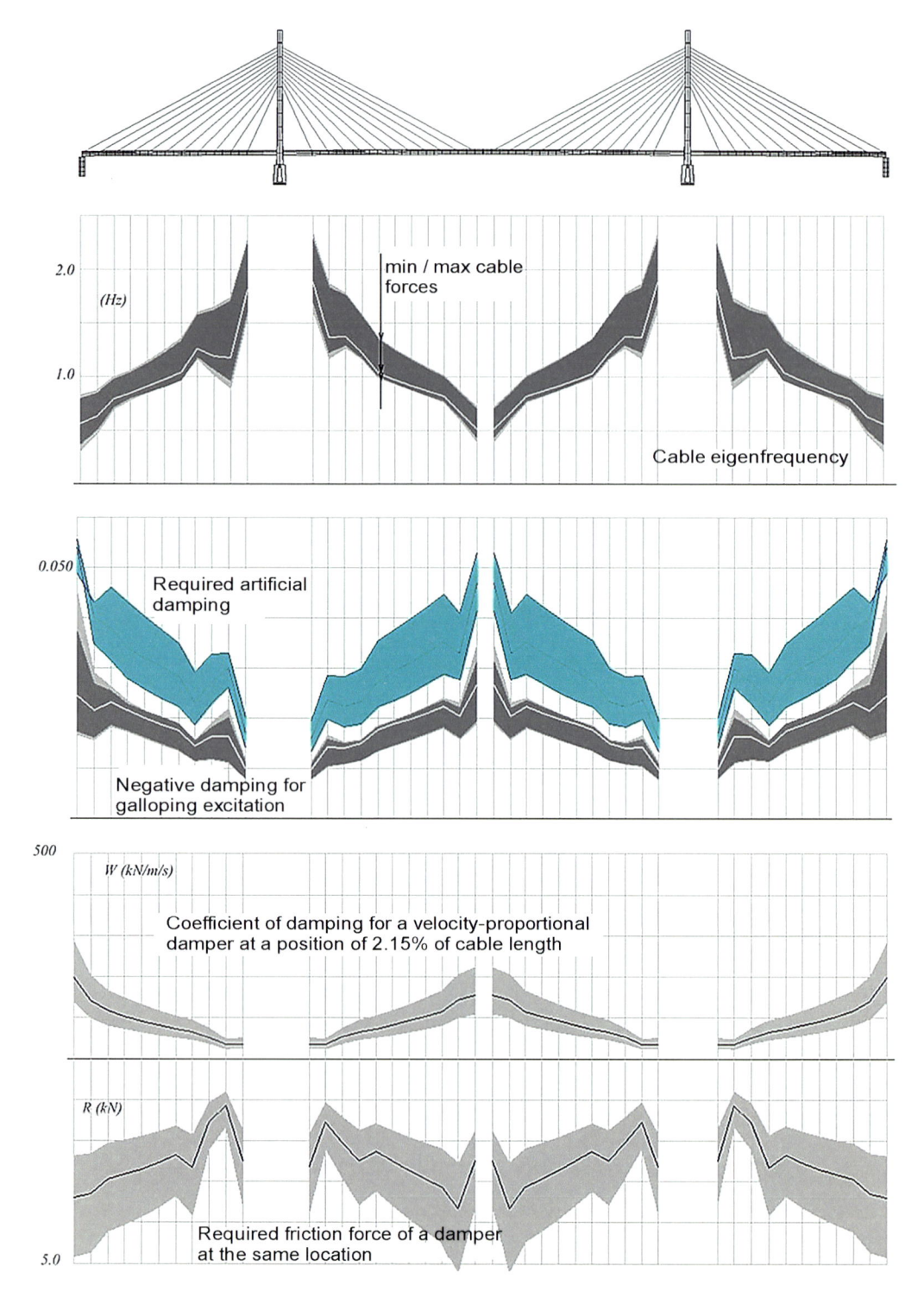

Figure 3.102 M0 Bridge Budapest; required additional damping to suppress galloping

Discrete damper

The discrete damper can either be a linear damping element with a telescopic arrangement or a non-linear friction damper. The achievable damping is, after [3.58], in both cases opt $\delta = \pi \, \Delta L / L$ optimum achievable damping resulting from a discrete damper.

The minimum required distance between attachment point and anchorage is 2.15 % of the cable length. For this position the parameters of the two damper (damping constant or friction force) options have been calculated.

It should be noted that cable oscillations are a special kind of loading which possibly lead to damage or failure only after a long period of time. This often allows us to use a strategy where no damping is used initially and additional dampers are fitted should oscillations be observed. Because of uncertainties, it is possible that cables may remain stable throughout.

Hydraulic dampers

Hydraulic dampers are a cylinder filled with a viscous liquid that passes between two chambers within the cylinder. The devices are telescopic pistons which generate reaction forces that are nearly proportional to the movement velocity. These solutions first had to deal with the problem of very small amplitudes of about 1 mm, which requires rather high damping constants and remains technically problematic to date.

The problem can be illustrated by looking at the early applications, Fig. 3.103: the constant of up to 300 kN/m/s can only be achieved by using three dampers, which is physically difficult to arrange. An alternative would be to mount the dampers at a higher position, as was done, for example, for the Erasmus Bridge in Rotterdam.

Fig. 3.104 shows examples of externally positioned hydraulic dampers, of which the damper bodies can be seen: the connection is realized through an appropriate articulation mechanism.

Somewhat more complex are so-called internal solutions, Fig. 3.105, for which the dampers are hidden inside the guide tube of the cable. The solution is fundamentally the same as for external dampers the guide tube must be stiff enough such that displacements at its mouth are identical to those of the deck. The HDPE version is, as shown in the sketch in Fig. 3.105, a single chamber damper, the IRD solution consists of radially arranged, normal telescopic systems. There are still problems with system tightness and with the way these systems can accommodate differential longitudinal displacements between cable and guide tube.

Friction dampers

Friction dampers are still relatively novel and operate on the basis of dry friction. Small displacement amplitudes are no problem with friction dampers, because the friction remains constant and the effective constant is high particularly for small amplitudes. The required friction force is in a manageable range of 2 to 6 kN for bridge cables [3.61].

Figure 3.103 An early example of a discrete telescopic cable damper

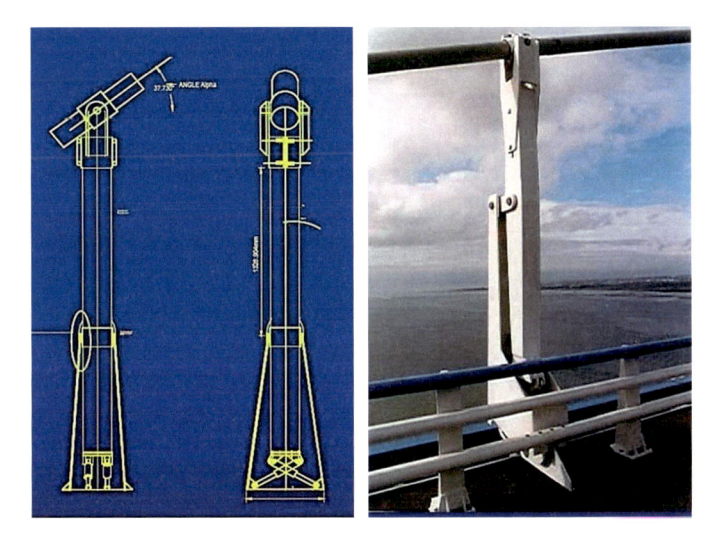

Figure 3.104 Damper by Freyssinet in external configuration (Elorn Bridge, France)

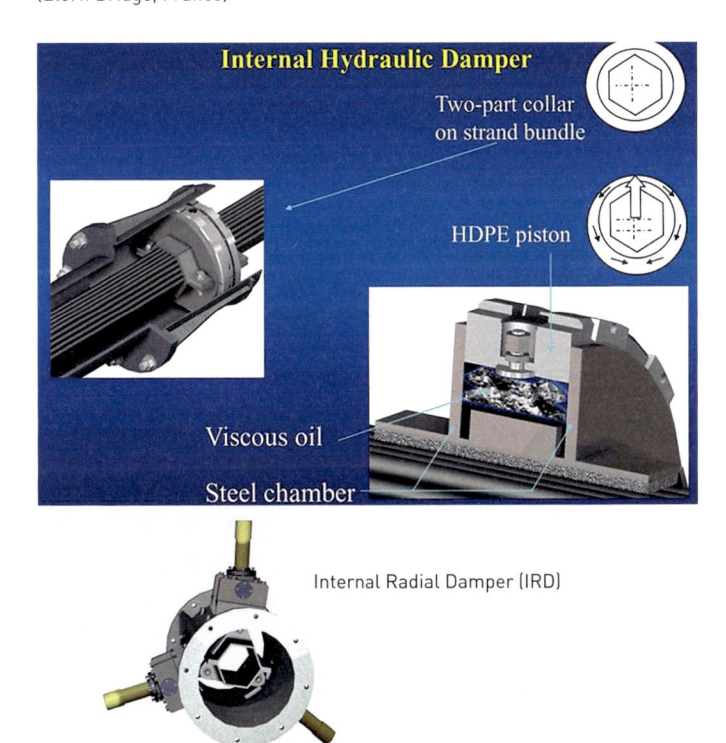

Figure 3.105 Examples of internal Freyssinet dampers

Friction dampers are inherently non-linear, i.e. amplitude-dependent according to the green line in Fig. 3.106. The amplitudes under galloping follow the principle

$\delta < \delta_{gal}$ → A grows

$\delta > \delta_{gal}$ → A reduces

such that any amplitude below P is pushed towards O. The starting amplitude A_0 is adjusted through the friction force. A sensible setting is $A_0 \approx L/3000$ recommended starting amplitude.

For medium to large cable lengths A_0 is 50–100 mm, so it is about ⅓–½ of the cable diameter, which is quite small. The intersection point P in Fig. 3.106 (green) is then 300–600 mm. If the amplitude were to exceed P, the damping would be insufficient to control the oscillation. This, however, is a theoretical case, because amplitudes of this magnitude cannot occur in a damped cable.

There are a few early friction damper designs (Köhlbrand Bridge, Fig. 3.12, Burlington Bridge, Fig. 2.166, Badajoz Bridge) which cannot satisfy today's requirements anymore. Fig. 3.107 shows a modern development of VSL and LAP, which has been installed on a number of bridges. This design is insensitive to the aforementioned longitudinal displacements and function perfectly.

The decay behavior of a cable with and without a friction damper is shown in Fig. 3.108. The right diagram shows the typical behavior of a friction damper: at first amplitudes decay linearly, then they remain nearly constant.

A major advantage of friction dampers is the simple installation and replacement. Fig. 3.109 shows some installation operations.

Finally, Fig. 3.110 depicts an internal friction damper model of BBR: in contrast to the VSL system which contains flexible leaf springs to deal with the longitudinal displacements, this design features a series of articulations.

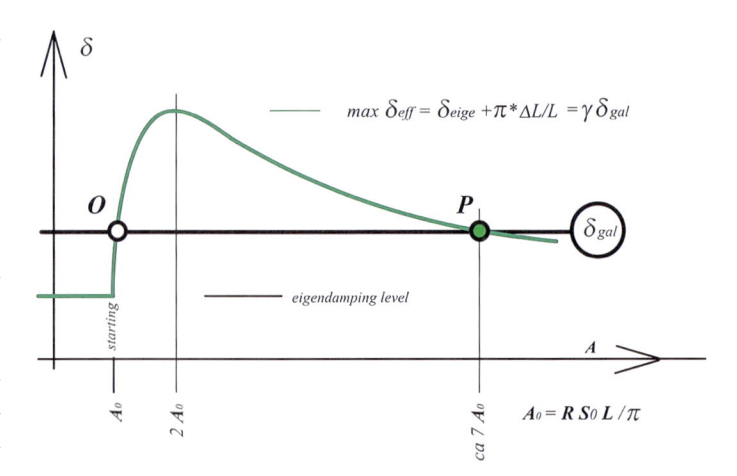

Figure 3.106 Friction damper – working principle

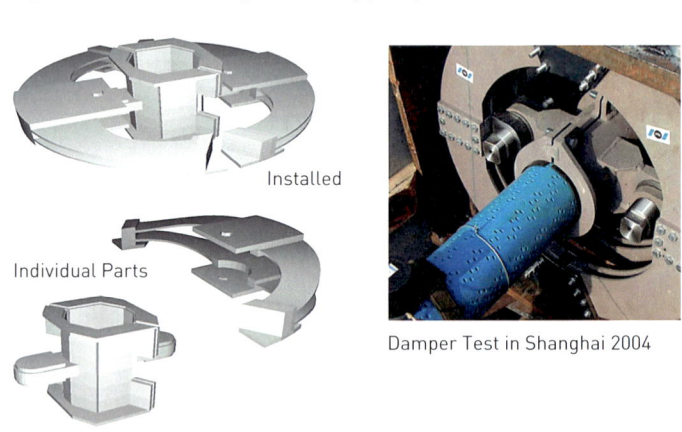

Figure 3.107 Monterrey Bridge Mexico; VSL friction damper of Type S

Elastomeric dampers

Among the recent developments are elastomeric materials with high plastic deformation capacity and with high fatigue resistance. The force-displacement relationship for an HDR-Gensui damper (HDR = high damping rubber) is shown in Fig. 3.111 in comparison to a friction damper. It can be seen that the energy dissipation during one cycle of oscillation reaches about 50%. The damping efficiency thus reaches approximately 70% of a friction damper.

Because of its unfavorable, partly elastic, behavior Gensui dampers cannot fully replace friction dampers. The spring effect reduces the efficiency such that an equivalent Gensui damper would need to be positioned about 50% higher than the friction damper. However, the simplicity of the Gensui damper is appealing and makes it easy to use, Fig. 3.111. VSL for example currently applies Gensui dampers to shorter cables.

An alternative rubber product by Freyssinet, which achieves similar efficiency, is shown in Fig. 3.112.

3.8 Cable dynamics

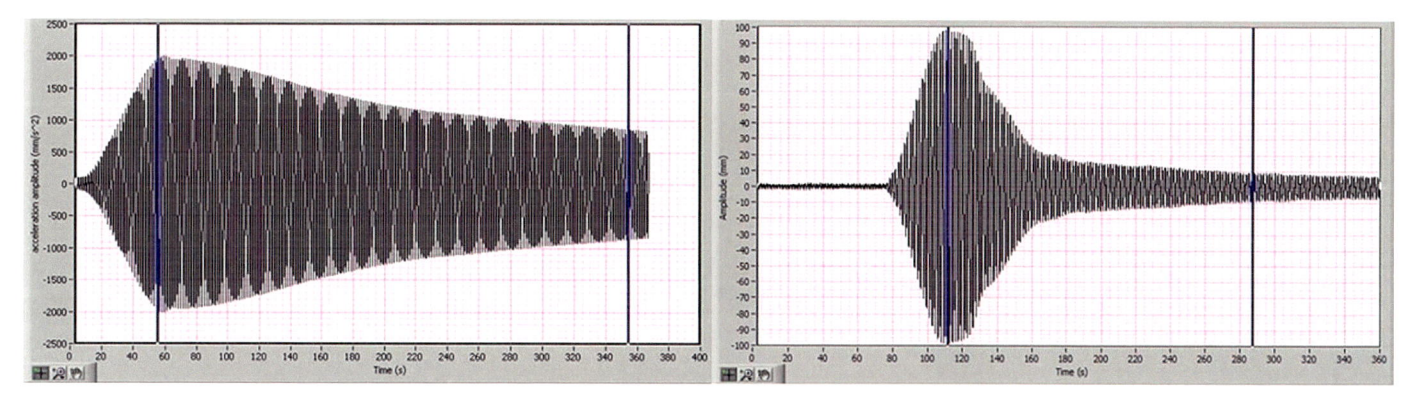

Figure 3.108 VSL cable with and without friction damper, from a test in Shanghai in 2004

Figure 3.109 Installation operations for a friction damper on Nuevo Leon Bridge, Mexico

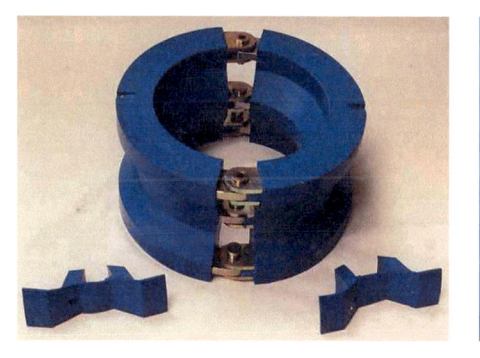

Figure 3.112 Elastomeric damper by Freyssinet

Figure 3.110 BBR friction damper

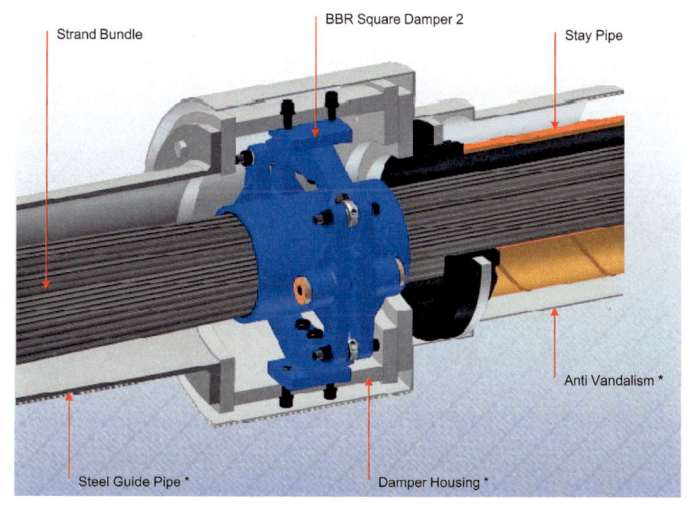

Figure 3.113 Some well-known applications of surface profiling

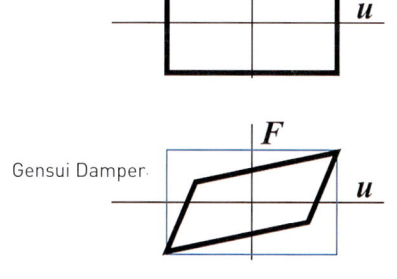

Figure 3.111 Gensui damper – application and characteristics

3.8.4.2 Surface profiling

The excitation mechanism of rain–wind-induced galloping can be disturbed by applying a special surface pattern to the cable tube which inhibits the formation of the critical water rivulets. The shape-based excitation mechanism is reduced, but does not vanish. Nowadays, a surface profiling is usually used as an additional countermeasure besides the application of a damper.

The effectiveness is not fully understood: for example, with Normandy Bridge [3.62] (see Section 6.5.1.2), a surprisingly low effectiveness – well below that expected from prior testing – was found. Ad hoc and debatable recommendations estimate that the required artificial damping can be reduced to around 50 % if profiling is used (Irwin).

For cables of large span bridges it is important to consider the drag loads on the cable planes which, as explained in Fig. 3.88, are being reduced through the surface pattern in most cases. It is important to ascertain this effect through wind tunnel tests as full scale Reynolds numbers [3.56].

Of course, surface profiling is, in contrast to additional damping, only effective against rain–wind galloping. Some solutions are shown in Fig. 3.113.

3.8.4.3 Cross ties

Cross ties are a common technique for stabilizing stay cables, see Fig. 3.114 for the Normandy Bridge (a further measure, additional to dampers and surface profiling). The cross ties suppress the most critical low frequency vertical oscillations of the cables [3.60]. Transverse oscillations perpendicular to the cable plane are not targeted directly, but are also somewhat reduced.

The function of the cross ties is in introducing additional fixation nodes which lead to higher order cable modes whose excitation intensity – according to the characteristics of galloping excitation – are lower. The cross ties need to be under tension in any loading state of the structure, otherwise high fatigue loads would arise in both the cross ties and the main cables. The ties are rather effective, but they usually spoil the aesthetics. The installation is always performed after completing the structure – often after the first photo shootings (see Figs 6.129 and 6.133)!

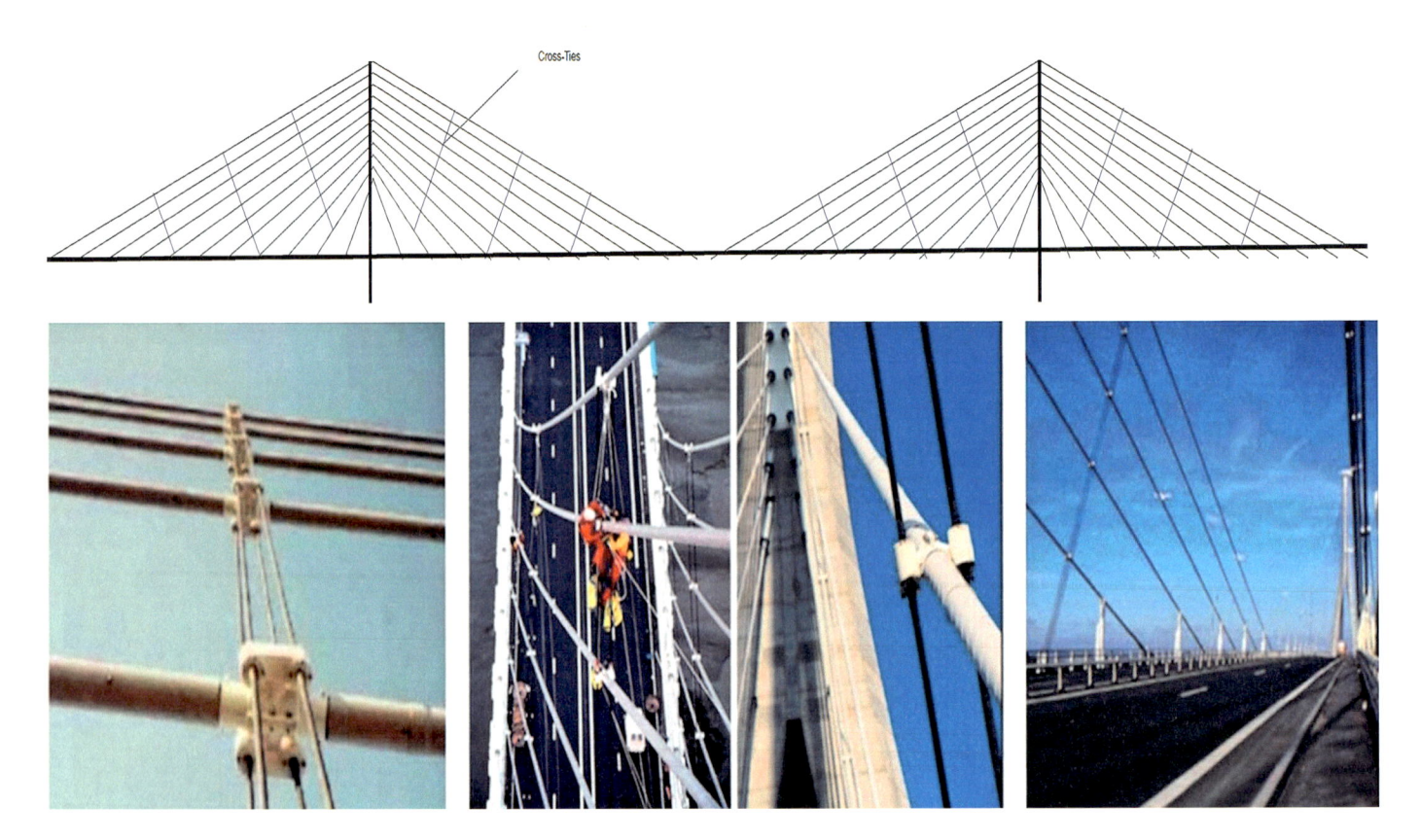

Figure 3.114 Details of the cross tie design for Normandy Bridge, France

3.9 Cable installation
3.9.1 General

An important part in the construction of cable-stayed bridges is the installation of the stay cables. The methods vary for the different types of cables which were outlined in Sections 3.2 – 3.5.

3.9.2 Locked coil ropes
3.9.2.1 General

Locked coil ropes consist of the outer helical layers of Z-wires and the inner round wires, Fig. 3.115 [3.2].

Locked coil ropes are completely shop fabricated and transported to the site on reels, Fig. 3.116. The bending radius should at least be 32 rope diameters. The size and weight of long, thick locked coil ropes thus can be very larger, and so transporting them can be a major issue.

Locked coil ropes for cable-stayed bridges are stressed with hydraulic jacks. These jacks are supported by jack chairs into which the cable anchor heads are pulled by means of a threaded tie rod and temporarily anchored, Fig. 3.117. For the permanent anchorage split shims, support nuts or hammerheads are used.

3.9.2.2 Example

The Elbe River Bridge Niederwartha near Dresden, Fig. 3.118, is a modern example for the use of locked coil ropes [2.111].

The passive anchorage of the ropes in the tower head uses composite action between an inner steel box cast into concrete on both longitudinal outsides, see Section 3.6.3.2.

The hammerhead anchorages are supported by crossbars which introduce the cable forces into the longitudinal outer plates of the inner box, Figs 3.119 and 3.120.

The active anchorage at the beam uses support nuts which introduce the cable forces via steel cantilevers into the main girder web, Figs 3.121 and 3.122. Fig. 3.123 shows the hammerheads for the cable anchorage inside the tower head.

A typical situation during installation of the ropes is shown in Fig. 3.124. The hammerheads are pulled up to the tower heads. They are supported in their free lengths by crawler cranes running on the bridge deck. Spreader beams prevent kinks which could open the lock between the outer Z-wires, the bending radius should not be smaller than 32 rope diameters also during installation.

The active rope anchorage is pulled with a come-along (grip hoist) near the final anchorage. At this stage the rope force is still small enough because of the big sag. When close enough to the anchorage, a pull-rod is threaded into the anchor head and guided through a hydraulic center hole jack underneath the beam.

Fig. 3.125 shows the pull-rod between the anchor head and the hydraulic jack; this rests on a jack chair which permits temporary anchorage of the tie rod with a support nut when the jack is retreated during strokes.

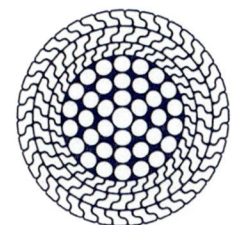

Figure 3.115 Locked coil rope

Figure 3.116 Locked coil ropes on reels

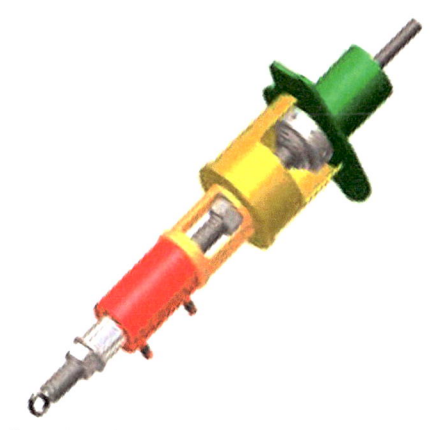

Typical tensioning scheme

Figure 3.117 Stressing of locked coil ropes

176　　3 Stay cables

Figure 3.118 Elbe River Bridge at Niederwartha, Germany, during construction

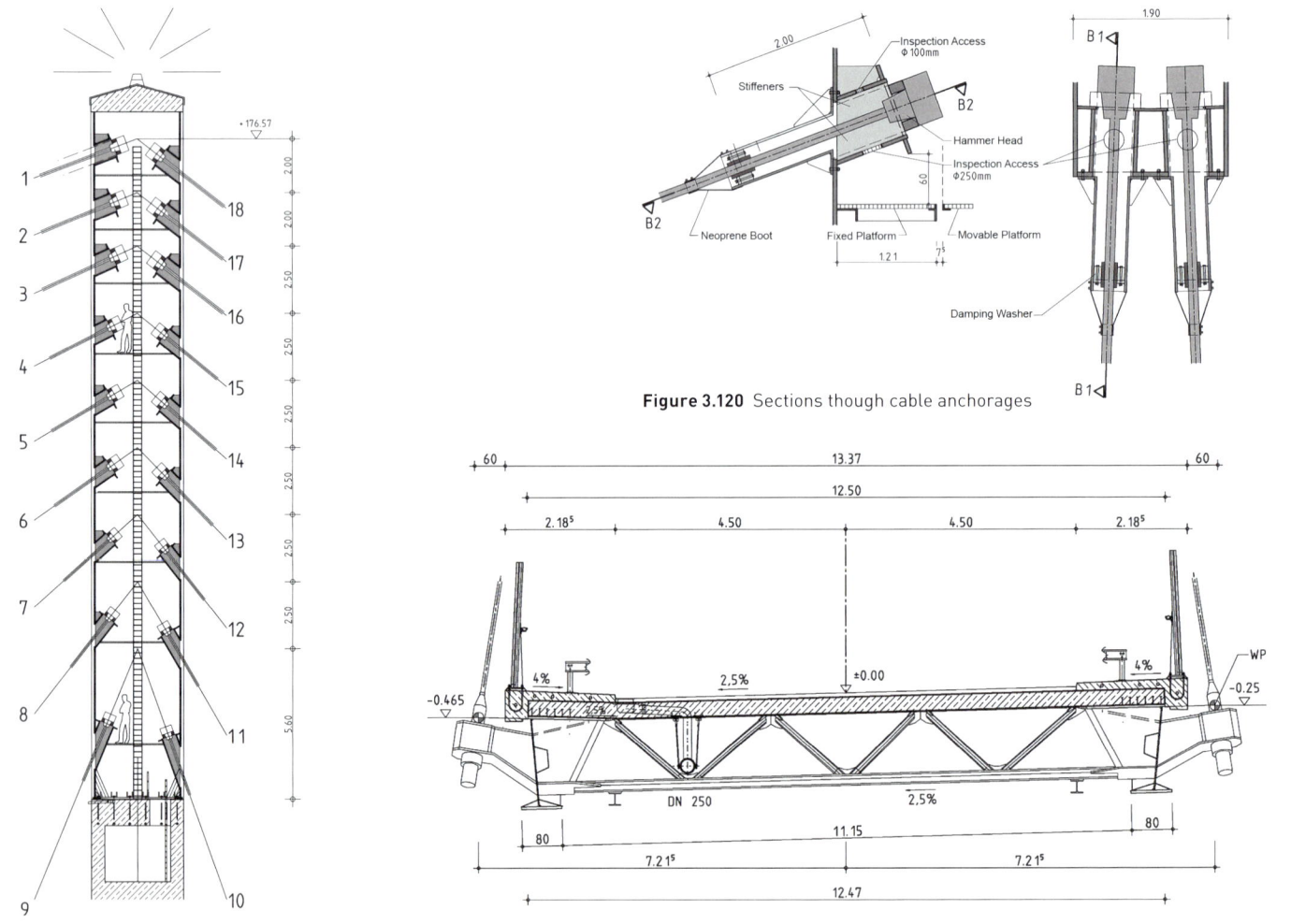

Figure 3.119 Cable anchorages inside tower head

Figure 3.120 Sections though cable anchorages

Figure 3.121 Cable anchorage at beam

3.9 Cable installation

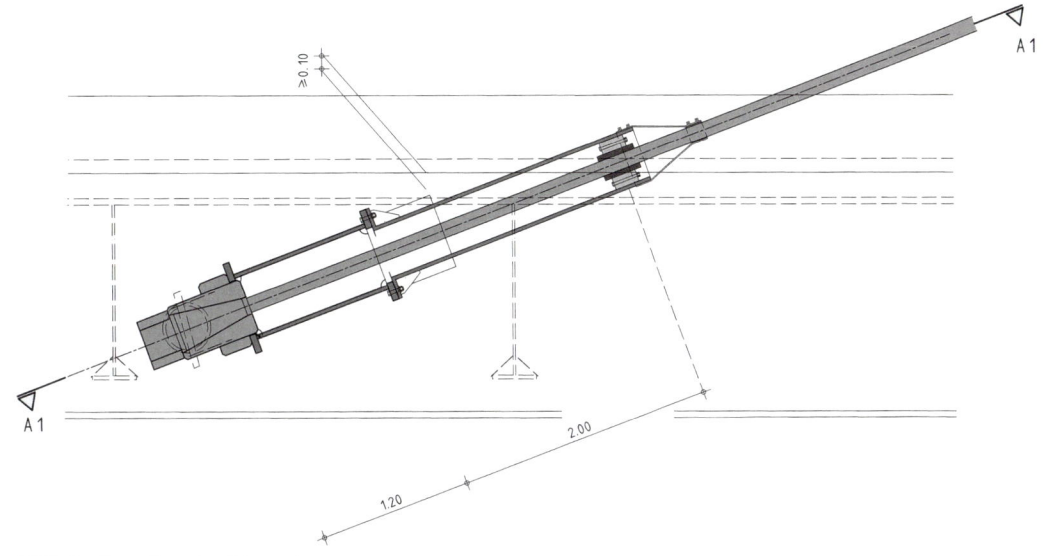

Figure 3.122 Sections though cable anchorages

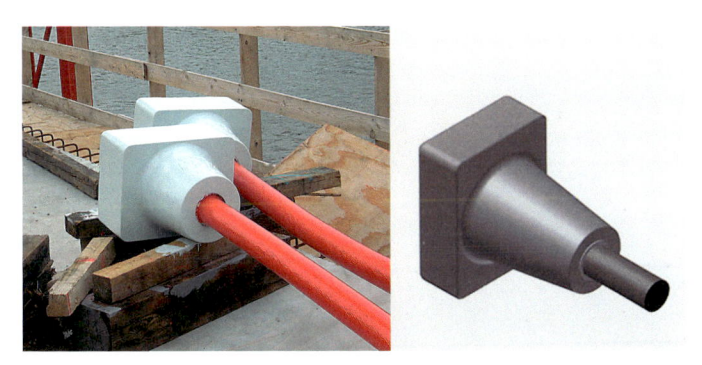

Figure 3.123 Locked coil ropes with hammerheads

Figure 3.126 Pull-rod in anchor head

Figure 3.124 Rope installation

Figure 3.127 Jack chair with support nut

Figure 3.125 Pulling the rope anchor head into the beam anchorage

Figure 3.128 Installed rope

The anchor head is then pulled through the final anchorage at the beam, Fig. 3.126.

A support nut is used to fix the rope anchor head in its final position, Fig. 3.127. The position of the support nut on the anchor head defines the total rope length. For restressing or relieving the rope, the position of the support nut is adjusted while the rope is restressed via the pull-rod to the hydraulic jack. It is not possible to turn the nut under full cable force because the friction in the thread would be too high. Fig. 3.128 shows a completely built-in rope with support nut.

3.9.3 Parallel wire cables
3.9.3.1 General
The development of parallel wire cables was outlined in Section 3.4. They consist of straight parallel round wires inside a PE-sheath and are also completely shop-fabricated and transported to site on reels. The surrounding PE pipe is sensitive to kinks and has to be supported during installation.

3.9.3.2 Example
An early example for the installation of parallel wire cables is the Zárate-Brazo Largo Bridge in Argentina [1.14].

The cables were transported on reels by ship to the site and pulled up to the beam. A curved guide ('banana') at the tip of the beam ensured a minimum radius while pulling the cable up to the tower head, Fig. 3.129. The free lengths of the cables were supported by hangers from a highline and were connected by a pull rope, Fig. 3.130.

The following pictures show a modern installation of parallel wire cables in Japan [3.63]. The cables on reels are lifted onto the already completed bridge deck and are unreeled on auxiliary carriages, Fig. 3.131 and 3.132.

At the active cable anchorage a pull-rod is screwed into the inner thread of the cable anchorage, Fig. 3.133.

The passive cable anchor head is pulled up to the tower head with a winch or tower crane. A two-part spreader for cable support ensures a sufficient bending radius. Modern parallel wire cables, which were wax-filled before being brought to the site, are not as sensitive to kinks as the early parallel wire cables with a loose PE pipe which were cement grouted after installation. The anchor head is pulled into its final position with a tie rope running through the anchorage pipe, Fig. 3.134.

Once it is the anchorage pipe, two split shims are placed between the anchor head and the support to fix the cable in its final position, Fig. 3.135.

For anchorage at the beam, the anchor head is pulled with a grip hoist close to its final position. A crane and a spreader beam support the cable, thereby reducing the sag and preventing a kink. Then a pull-rod is used to haul the anchor head into its final position, Fig. 3.136.

Figure 3.129 Cable installation Zárate-Brazo Largo Bridge, Argentina

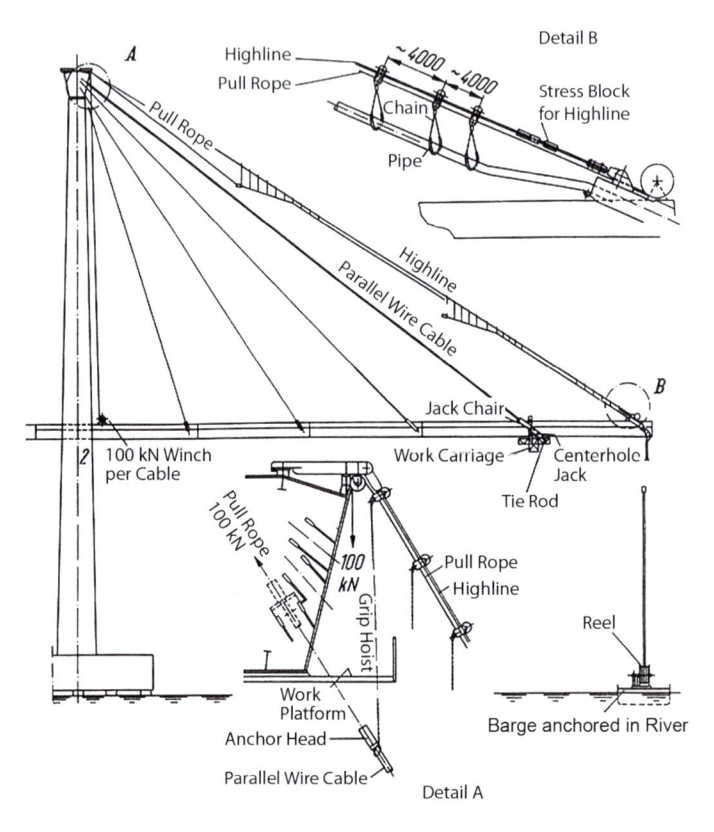

Figure 3.130 Pulling cable up to the tower head

Finally, the anchor head is pulled with a hydraulic jack into its final position underneath the beam and secured with shims. The lengths of the cables are defined by the thickness of the shims. In order to restress or relieve the cable, its anchor head is lifted from its anchorage again by a tie-rod and hydraulic jack, and shims are added or taken out as required, Fig. 3.137. Fig. 3.138 shows the final installed cable underneath the beam supported by shims.

3.9.4 Parallel strand cables
3.9.4.1 General
Modern parallel strand cables use seven wire mono strands as tension elements inside some sort of PE cover. Their main advantage lies in the fact that they are fabricated from their individual members on site, thereby avoiding transporting them on reels. This advantage increases with span length, i.e. increasing transportation weight.

The completely shop-fabricated locked coil ropes and parallel wire cables have a well-defined precise length and permit construction 'by geometry'. Parallel strand cables, on the other hand, are fabricated on site and thus can be installed 'by force' only.

The advantages of the installation of parallel wire cables are highlighted by their suppliers [3.28, 3.35 – 3.37]:
- reduced construction loads,
- greater mobility of the installation equipment,
- greater flexibility for site organization,
- easier maintenance and repairs.

3.9.4.2 Example
Site set-up

The individual components of parallel wire cables are separately delivered to the site. The strands arrive on reels, which are, however, smaller and less heavy than the reels of completely shop-fabricated cables, Fig. 3.139.

The protective PE pipes are supplied in long units by barge and stored on the already completed bridge beam, Fig. 3.140.

In order to install the cable components at the tower head, a light scaffolding is erected, Fig. 3.141.

Anchorage installation

The anchor heads are first temporarily fixed at the tower and the beam, in their final positions, Fig. 3.142.

HDPE stay pipe assembly

The PE pipes are aligned on the bridge deck, Fig. 3.143. The PE pipe sections are connected by fusion welding, Fig. 3.144. A heating plate between two pipe ends is used to melt the PE and then the two sections are pressed against one another. This type of fusion weld has the same strength as the free length and is also used for shop-fabricated cables.

Figure 3.131 Unreeling on carriages

Figure 3.132 Cable on auxiliary carriages

Figure 3.133 Connecting the tensioning rod

Figure 3.134 Pulling into the anchorage

Figure 3.135 Fixing the anchorage at tower head

Figure 3.136 Installation at girder

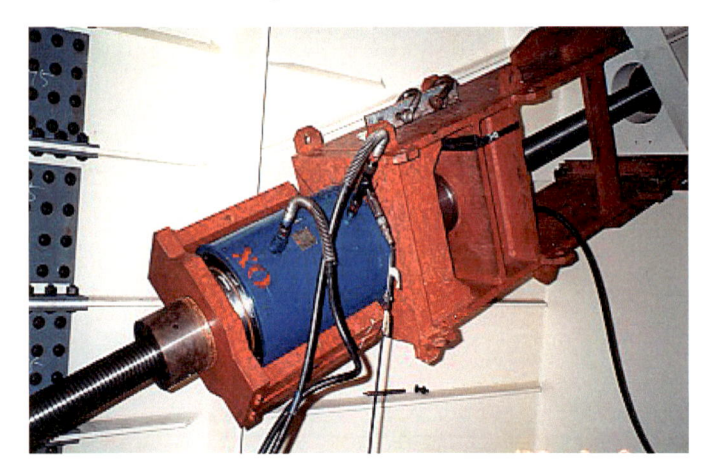

Figure 3.137 Cable tensioning

Figure 3.138 Final anchorage

Figure 3.139 Delivery of strands

Figure 3.140 Supply of PE pipes

Figure 3.141 Scaffolding at tower head

3.9 Cable installation

Preparation and lifting of the stay pipe

For lifting the cable, a temporary anchorage with a pull rope collar is connected to the upper end of the PE pipe, Fig. 3.145. The prepared PE pipe is moved with carriages to its lifting location, Fig. 3.146.

A pull rope from a winch at the tower head finally lifts the upper end of the PE pipe from the carriages, Fig. 3.147, and pulls it to the anchorage at the tower head, Fig. 3.148.

The PE pipe is hanging from the tower head under its own weight with a big sag, Fig. 3.149. The first strand is installed and supports the PE pipe so that its sag is strongly reduced, Fig. 3.150.

Strand installation

The individual installation steps for the strands are listed in Fig. 3.151. For lifting the PE pipe and strands, no heavy lifting equipment is required at the tower head, just simple winches placed at the towerhead, Fig. 3.152.

In the following, the individual steps for strand installation are shown, Figs 3.153–3.156.

When installing the cable by force (instead of geometry) several influences have to be accounted for, including the deformations of the tower and beam, the actual load condition on the beam (e.g. derrick weight and location) and the temperature conditions, Fig. 3.157.

The individual strands are separately stressed with a mono-jack, Figs 3.158 and 3.159, which can be handled much more easily than the multi-strand hydraulic jack required to stress the complete cable. A carefully determined stressing system is required to ensure that, at the end of stressing, all strands have the same force ('isotensioning').

This means basically that the first strand is temporarily overstressed. With the installation of the next strand this initial force is slightly reduced, and so on, until at the end all strands have the same predetermined force.

Practically this is achieved by stressing each new strand exactly to the same stress which all formerly installed strands have at that point of time. In other words, all strands are immediately installed to their final lengths. But because these lengths cannot be measured, the detour over measuring temporary strand forces has to be taken.

After stressing, the monostrands are closely packed. The PE pipe is reinforced against vandalism at the lower end near the deck, e.g. by a steel pipe, Fig. 3.160.

The strands protrude from the lower anchorage so that the complete cable can be restressed, Fig. 3.161.

Restressing is done with a multi-strand jack for the complete cable, Figs 3.162 and 3.163.

Final tuning of the cable tension

At the end of stressing, the steel wedges are forced hydraulically into their anchorages in order to completely avoid any later slipping of individual strands, Fig. 3.164.

Figure 3.142 Anchor heads at the tower

Figure 3.143 Aligning PE pipes on bridge deck

Figure 3.144 Fusion welding of PE pipes

Figure 3.145 PE pipe with lifting collar

Figure 3.146 Transportation of PE pipe on carriages

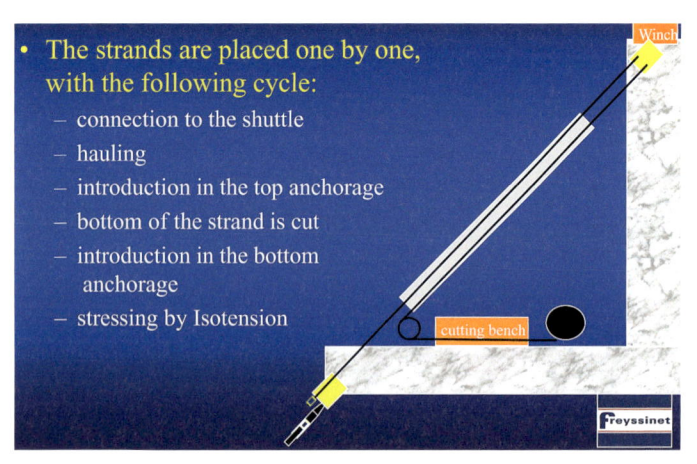

Figure 3.151 Installation of strands

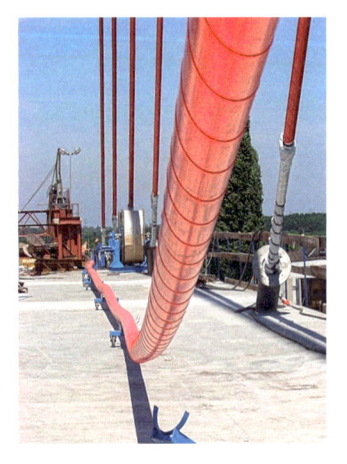

- The strands are placed one by one, with the following cycle:
  - connection to the shuttle
  - hauling
  - introduction in the top anchorage
  - bottom of the strand is cut
  - introduction in the bottom anchorage
  - stressing by Isotension

Figure 3.147 Lifting from carriages

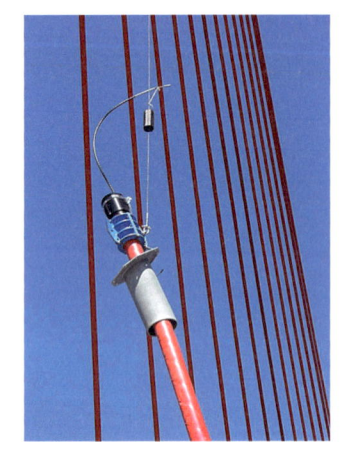

Figure 3.148 Pulling up

Figure 3.152 Winch at towerhead

Figure 3.149 Slack PE pipe

Figure 3.153 Preparing a new strand

Figure 3.150 PE pipe supported by first strand

Figure 3.154 Shuttle to connect the winch rope to the strands

3.9 Cable installation

Figure 3.155 Strand pulled through the stay pipe

Figure 3.156 Temporary connection of the stay pipe at the tower

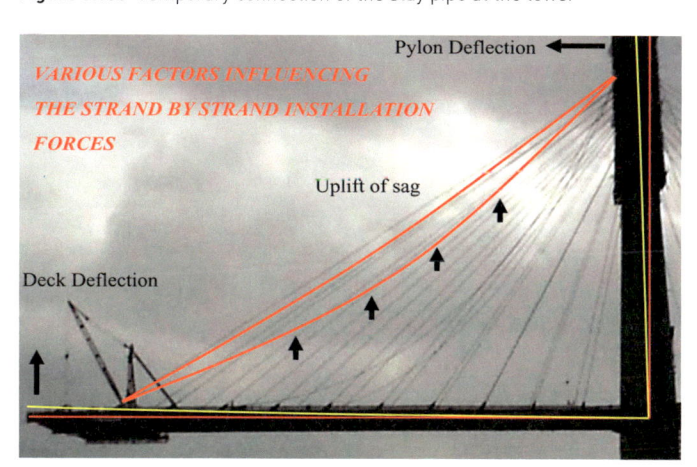

Figure 3.157 Deformations influencing cable lengths

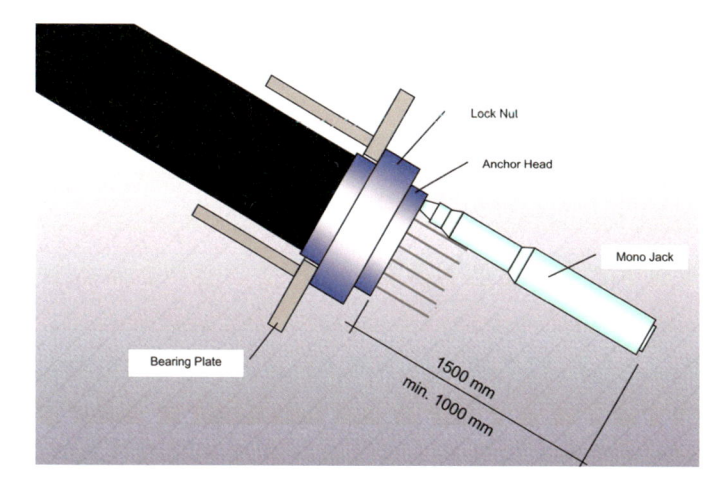

Figure 3.158 Mono-Jack for stressing individual strands

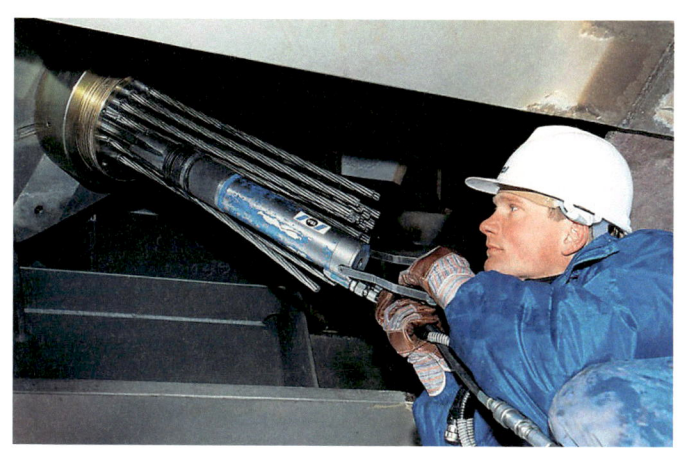

Figure 3.159 Stressing of individual strands

Figure 3.160 Stay cable without deviator and anti-vandalism pipe

Figure 3.161 Protruding strands for restressing

Stay pipe connections

Finally, dampers are built in, as required, at the tip of the steel pipe, hidden from view by boots, Fig. 3.165.

Finishing

At the end of the installation the cable anchorage system and the anchor heads are permanently protected against corrosion by grease inside the strand cap and by applying several layers of corrosion-protective paint, Fig. 3.166.

3.9.5 Cable calculations
3.9.5.1 Cable deformations

The knowledge of cable deformations (strains and sag) is important for determination of the exact cable length for shop fabrication, and calculation of the angle of the cable anchorage at the beam.

The cable length comprises the chord distance between the theoretical cable anchorage points plus the additional length due to sag plus the total shim thickness (or support nut positions) minus the elastic strain under dead weight after shrinkage and creep.

The cable force under varying sag conditions during installation is required for shop-fabricated cables. The force required to pull the tie rod into the jack underneath the final anchorage determines the grip-hoist capacity for a given tie rod length.

The use of the temporary effective modulus of elasticity accounts for elongation of the cable under additional loads due to change of sag. The basic relations of cable geometry and corresponding forces are given in handbooks, e.g. [3.3, pp. 246–254].

3.9.5.2 Measuring of cable forces

During bridge construction the cable forces have to be measured at each new erection stage for comparison with the theoretical values given in the construction manual.

In the following, the three practical methods for measuring cable forces are described and evaluated as used for the Mannheim-North Bridge [1.19].

Comprehensive control measurements were carried out after completing the bridge in order to find out if the actual forces and geometry correspond to those calculated. Most important was the control of cable forces, which took place with three different methods:
- direct measuring with a hydraulic Jack,
- determining the cable eigenfrequencies,
- surveying the cable sag.

Using a hydraulic jack required the biggest effort, because the voluminous jack with a weight of 3 T had to be carried together with the tie-rod and a jack chair into the cable anchorage chamber. The cables were stressed until the anchor heads lifted off the bearing plates and the corresponding force was read from the gauge.

In order to determine the eigenfrequency, a hemp rope was tied to each cable about 2 m above deck and by suitably pulling at it, the cables were excited in their first eigenfrequency. The time for 100 oscillations was measured by a stop watch and from the time, the cable force was determined in accordance with the following formulas:

$$S = 4 \cdot L_2 \cdot f_1^2 \cdot \mu$$

with
S   cable force
L   cable length between hinged anchorages
f_1 first eigenfrequency
$\mu = G/g$ = mass distribution

The influence of fixing the anchor heads to the structure as well as the bending stiffness $E \cdot I$ of the cable was accounted for with the following formula:

$$S = 4 \cdot L^2 \cdot f_1^2 \cdot \mu \cdot \left(1 - \frac{2}{f_1 \cdot L^2} \cdot \sqrt{\frac{E \cdot I}{\mu}}\right)$$

whereby $E \cdot I \ll S \cdot L^2$ was used.

The cable sag was surveyed between the anchor points and at two cable points with a theodolite. The corresponding cable force was calculated in accordance with

$$S = \frac{G \cdot L^2}{8 \cdot f_m}$$

with G the weight per m of the cable.

The accuracy of the different methods vary, Table 3.5. When using hydraulic jacks, it is difficult to determine the precise moment at which the anchor head lifts off. The result also depends on the accuracy of the gauge, which has to be specifically calibrated. The most accurate results were achieved by measuring the eigenfrequencies. The measured 100 oscillations gave the precise frequency. Measuring the sag gave slightly unreliable results because of the practical difficulties of obtaining a precise value by surveying.

The differences between the various measured values and between those and the theoretical values were small. The measured values varied by about ±5%. They were generally above the theoretical values. The actual cable difference was applied as an additional load case to the bridge in the final stage. The permissible cable forces were exceeded only by 1.7% at t=0 for main loads, which was considered acceptable.

After opening for traffic the bridge was loaded with 20 tanks on flatbeds with a total weight of 1564 T equivalent to 8.10 T/m.

The actual deflection of the beam was measured as 542 mm, which differed only by 8 mm from the theoretical value. The cable forces measured by the eigenfrequency method also tallied well.

As a result, it can be stated that the desired run of forces together with the corresponding geometry was achieved at the end of erection.

3.9 Cable installation

Table 3.5 Result of the cable force measurements with different methods

Cable no.	Cable force [T]						
	Theoretical		Actual				
	Without taking into account S + C	Taking into account S + C	A	B_1	B_2	$(B_1 + B_2)/2$	C
1 Aui	591	579	606	626	620	623	606
3 Aui	347	300	349	315	303	309	337
5 Aui	510	494	512	527	518	523	493

Theoretical values A: Using hydraulic jacks
Theoretical values B: Measuring eigenfrequencies (B_1 assuming hinged supports and J = 0
  B_2 assuming fixed supports and actual J)
Theoreticical values C: Measuring sag

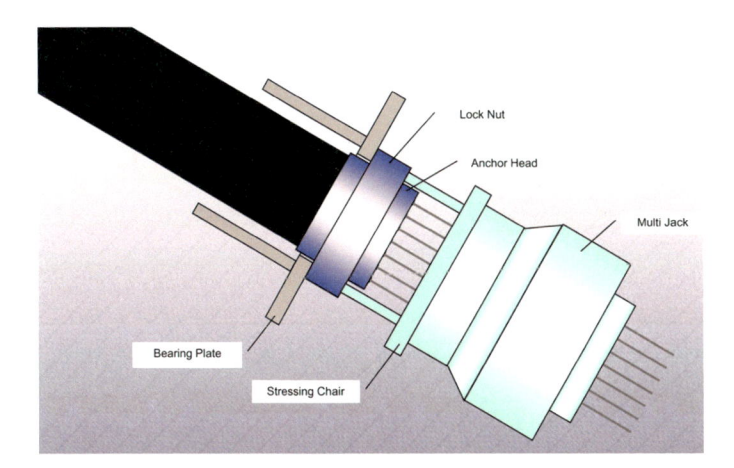

Figure 3.162 Multi-strand jack for stressing complete cable

Figure 3.163 Stressing of complete cable

Figure 3.164 Power seating of the wedges

Figure 3.165 Dampers inside boots and anti-vandalism pipes

Figure 3.166 Corrosion-protection paint and strand cap

4 Preliminary design of cable-stayed bridges

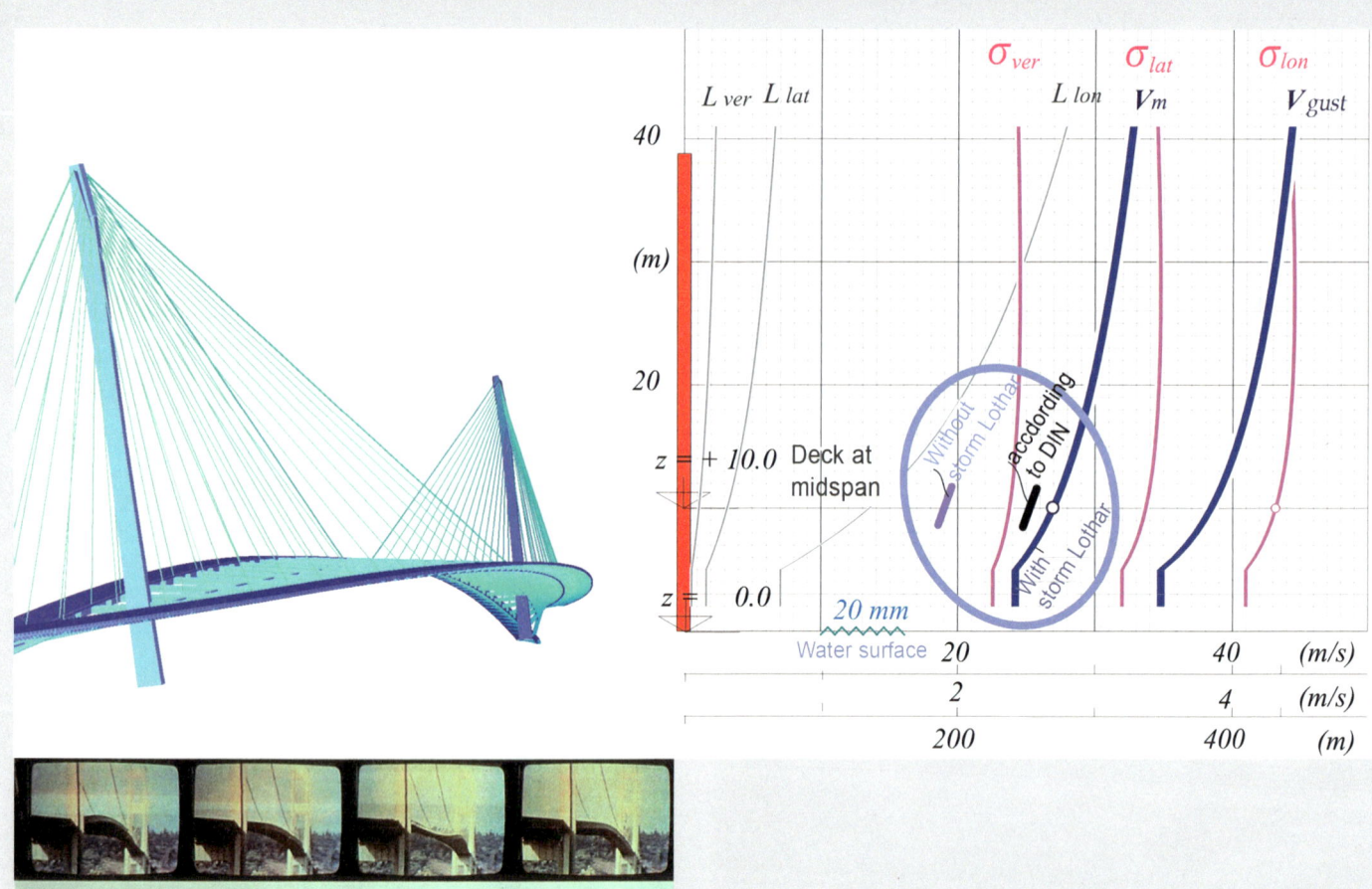

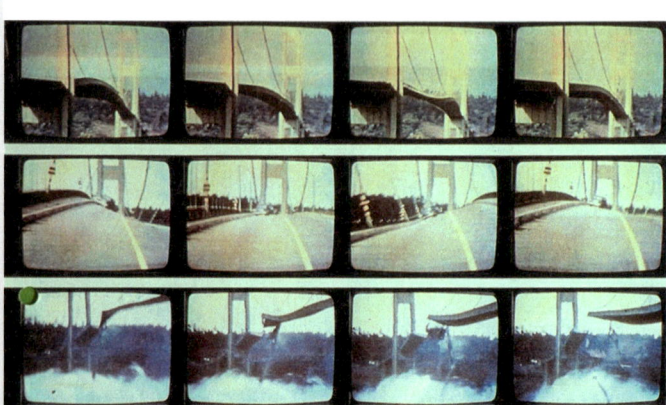

4.1 Action forces for equivalent systems
4.1.1 General

The purpose of this section is to explain the load transfer and flow of forces of cable-stayed bridges. This is best achieved by outlining approximate methods for the preliminary design which leads to initial dimensions of all structural parts which can then be used for the comparison of the quantities for different systems, and as input values for a detailed computer calculation.

Furthermore, such preliminary designs permit completely independent checking of computer calculations, in order to avoid the computer being used as 'black box', the results of which are accepted without questioning them.

These preliminary design principles have been used and partly developed by the author in order to size his first cable-stayed bridges during a time when high capacity computers were hardly available and were only used for the final design of systems based on preliminary calculations.

4.1.2 System geometry

The geometry of cable-stayed bridges is determined by the ratio of side spans to main span and by tower heights to main span. An important criterion in this regard is the required amount of cable steel. The total costs comprise, in addition, those for beam, towers and hold-downs at the bridge ends. Often the local conditions enforce deviations from the ideal geometry. In the following, these considerations are outlined for typical three-span cable-stayed bridges.

The basic load-bearing behavior of cable-stayed bridges is outlined in Fig. 4.1 from [2.112]. Loads in the main span are carried by the forestays to the tower heads and from there anchored by tension via the concentrated backstays in the anchor piers. The inner stay cables of the side spans receive virtually no forces at all from this loading. The horizontal cable components act in compression in the beam and equal one another out – look forward to Fig. 4.4 b, c.

Loads in a side span are transmitted by the side span cables to the tower head and from there via compression (meaning reduction of tensile forces from permanent loads) in the backstays to the anchor piers where they cause compression. The horizontal components of the side spans are balanced by those of the backstays by tension in the side spans, Fig. 4.4 d, e.

The backstays are thus governing the stiffness of a cable-stayed bridge and receive important load changes.

The ratio between side spans I_1 and main spans I, Fig. 4.2, influences strongly the stress changes in the backstay cables. Live loads in the main span increase the stresses from permanent loads, while live loads in the side spans decrease them. These stress changes must not exceed the permissible fatigue range of the actual stay cable system. These fatigue stresses increase with increasing span ratios.

The span range also influences the size of the vertical uplift forces in the outer anchor piers. These anchorage forces decrease with increasing span ratios. The compression created in the backstays by live loads in the side spans decreases their tensile force from perma-

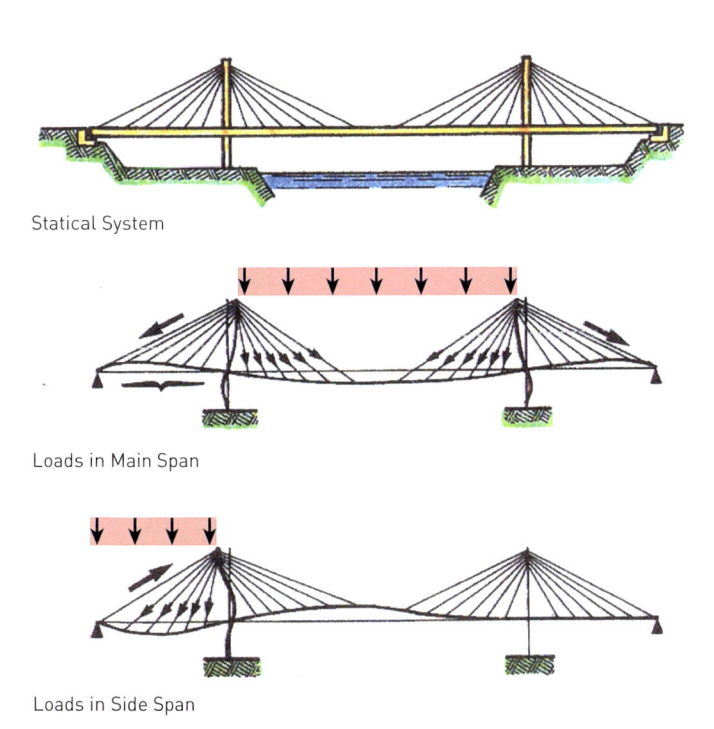

Figure 4.1 Load transfer in cable-stayed bridges

nent loads increasingly with increasing span ratios. In this way the effective cable stiffness decreases.

A good ratio of span length is thus important for an economic design of cable-stayed bridges [1.6]. Fig. 4.2 outlines the above considerations for the backstay cable stresses. Vertically the ratio of live load to dead load is given, horizontally the main span length.

To the right of the nearly vertical straight line the tensile forces are governing for an assumed minimum cable stiffness of E_{eff} = 180 000 N/mm², to its left a fatigue range of Δf_S = 200 N/mm² for 40 % of live load is governing. The curves within the diagram give the span ratio.

For concrete road bridges the ratio between live load to dead load often comes to about 0.25. The corresponding span ratio of $l_1 : l$ = 0.4 for a typical main span of 400 m gives the same amount of cable steel required for maximum cable load and fatigue range.

Steel railway bridges may have a ratio of live load to dead load of up to 0.6 which results in a span ratio of $l_1 : l$ = 0.3 for similar stress conditions.

The diagram indicates that steel bridges will have a smaller ratio of side span to main span than concrete bridges in order to obtain the same optimized stresses.

The ratio of tower height above deck also influences the required amount of cable steel; Fig. 4.3 indicates that the optimum tower height comes to about 1/5 to 1/6 of the main span, taking into account the additional costs for higher towers [1.3].

Fig. 4.3 also gives formulas for an approximate determination of the total amount of cable steel required. This can be very helpful when comparing the costs of different preliminarily bridge systems.

4.1.3 Normal forces of articulated system

The deformation characteristics of a cable-stayed bridge with the usual slender beam are dominated by the changes of lengths in the stay cables. The influence of beam and towers is comparatively small, often in the range of only 5 % of the total deformations.

The geometric conditions and the corresponding simple formulas for the normal force distribution are given in Fig. 4.4.

The loads of cable-stayed bridges are predominately carried by normal forces by exploiting the large structural depth between beam and tower tips. The moments in the beam and the towers are mainly caused by restraint, which means that they are not necessary for load balancing.

The size of the normal forces can be well approximated for an articulated system in which the beam is provided with hinges at the cable anchor points of the beam and the tower footing, Fig. 4.4a. In this way a statically determined equivalent system is created which represents closely the actual run of normal forces and deformations.

As mentioned before, live loads in the main span, Fig. 4.4b, are transferred in tension from the forestays into the backstays which are anchored in tension to the anchor or hold-down piers, where the uplift forces are tied down to the foundations.

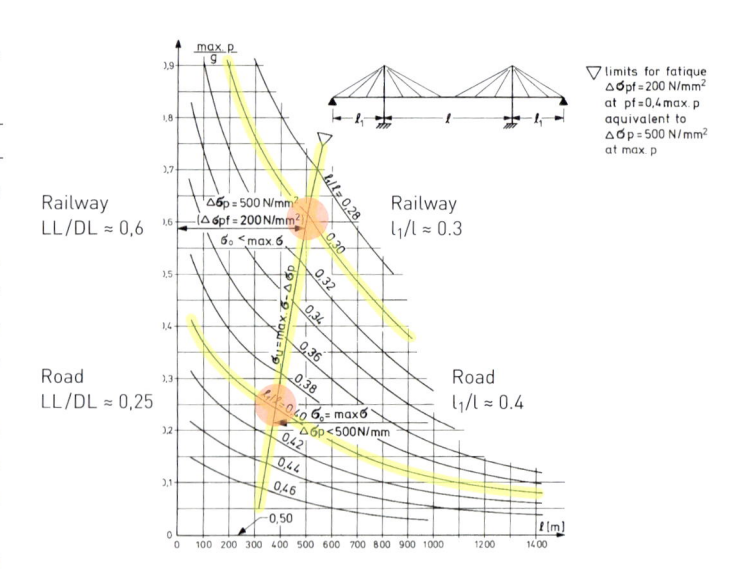

Figure 4.2 Relationship between span ratios and cable stresses

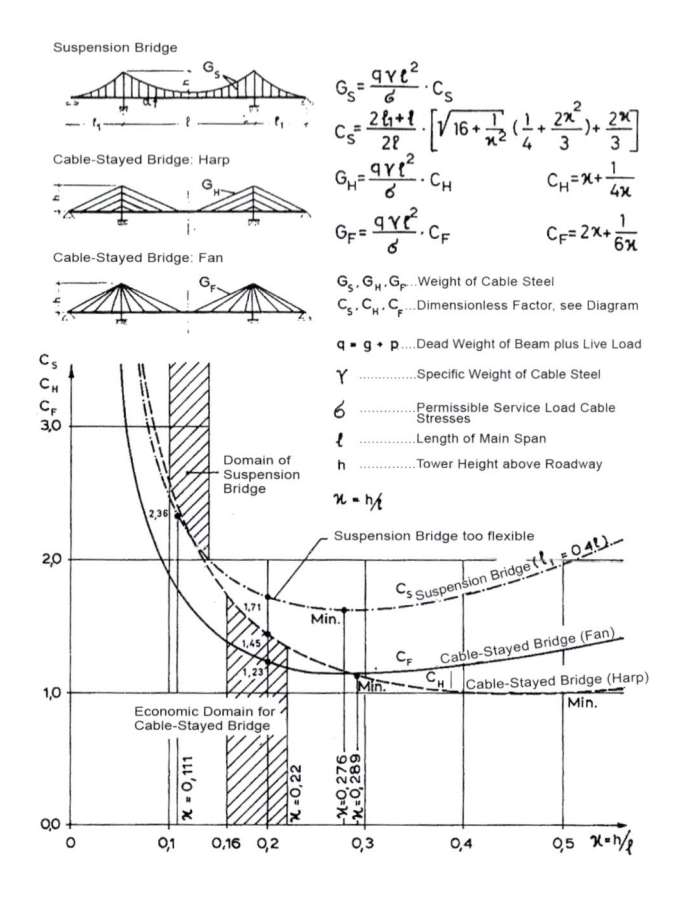

Figure 4.3 Optimum ratio of tower height above deck to main span and total quantity of cable steel

4.1 Action forces for equivalent systems

The cable forces can be calculated from the various cable inclination angles. The corresponding normal forces in the beam and the towers are constant from one point load, Fig. 4.4c.

Loads in the side span are also transferred in tension by the corresponding cable to the tower head and from there in compression in the backstay cable to the anchor pier, Fig. 4.4d. A cable cannot, of course, carry compression, so compression in this connection means a reduction of the tensile forces in the backstays from permanent loads. The shortening of a cable under compression is calculated linearly as an elongation from tension. The corresponding normal forces are given in Fig. 4.4e: compression in the tower and tension in the bridge beam between the load and the anchor pier.

It is important to know the different flows of forces from loads in the main span and in the side span because they form the basis for the load-bearing behavior of a cable-stayed bridge.

The support reactions at the tower can be calculated from $\Sigma M = 0$ about the anchor pier, and those at the anchor pier can be calculated from $\Sigma M = 0$ about the tower footing.

The loads for the governing forces in the articulated system are placed in the main span or in the side span, Fig. 4.5. The uplift force in the anchor pier and thus the tensile forces in the backstay cable reach their maximum for load in the main span, and they become minimal for load in the side span.

The biggest compression in the tower results from full load in the main and side span.

For uniform loads and small cable distances the normal force in the beam can be approximated by a parabola, Fig. 4.6. The break in the normal force at the anchor pier is due to the horizontal component of the backstay cable anchored here.

$$N = \frac{DL + LL}{2h} \cdot (l^2 - x^2) \cong 1.25 \cdot (DD + LL) \cdot \left(1 - \frac{x^2}{l^2}\right) \cdot l$$

At tower for full load: $x = 0$

$$\min N = \frac{DL + LL}{2h} \cdot l^2$$

4.1.4 Live loads on elastic foundation

For the approximate calculation of the beam moments the beam with its actual bending stiffness is isolated from the complete structural system.

4.1.4.1 Beam on elastic foundation [4.1]

The approximate system for the beam is a beam elastically supported at the cable anchor points, Fig. 4.7a. In a first step the cables are replaced by springs, Fig. 4.7b, in a second step these individual springs are spread out to a continuous elastic support, Fig. 4.7c.

The governing characteristic for a beam on elastic foundations is the elastic length L, which is proportional to the 4th root of the ratio of beam stiffness to the elastic support, Fig. 4.7.

The bedding factor c is determined from the vertical displacement due to a unit load on the articulated system divided by the cable dis-

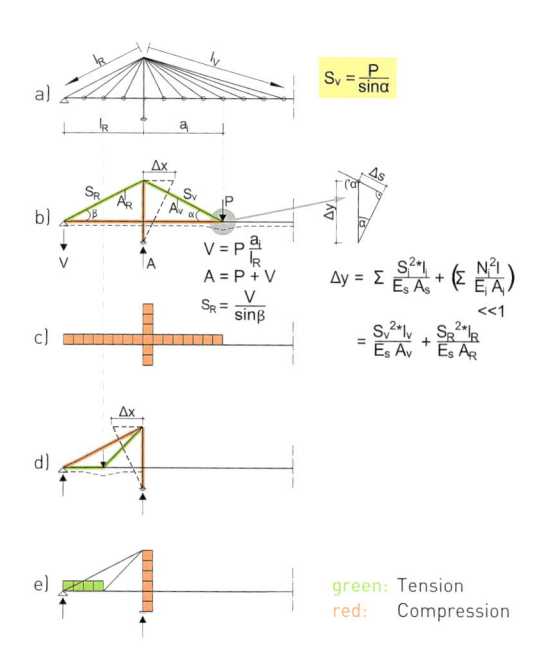

$S_v = \frac{P}{\sin\alpha}$

$V = P \frac{a_i}{l_R}$
$A = P + V$
$S_R = \frac{V}{\sin\beta}$

$\Delta y = \Sigma \frac{S_i^{2} \cdot l_i}{E_s A_s} + \left(\Sigma \frac{N_i^2 l_i}{E_i A_i}\right)_{\ll 1}$

$= \frac{S_v^{2} \cdot l_v}{E_s A_v} + \frac{S_R^{2} \cdot l_R}{E_s A_R}$

green: Tension
red: Compression

Figure 4.4 Normal forces in an articulated system

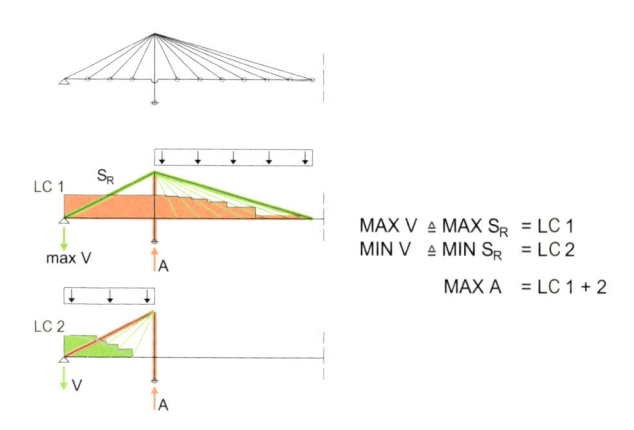

MAX V ≙ MAX S_R = LC 1
MIN V ≙ MIN S_R = LC 2
MAX A = LC 1 + 2

Figure 4.5 Biggest normal forces in tower and beam

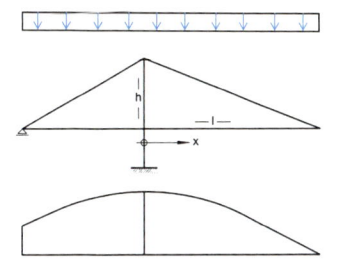

Figure 4.6 Approximation for normal forces in beam

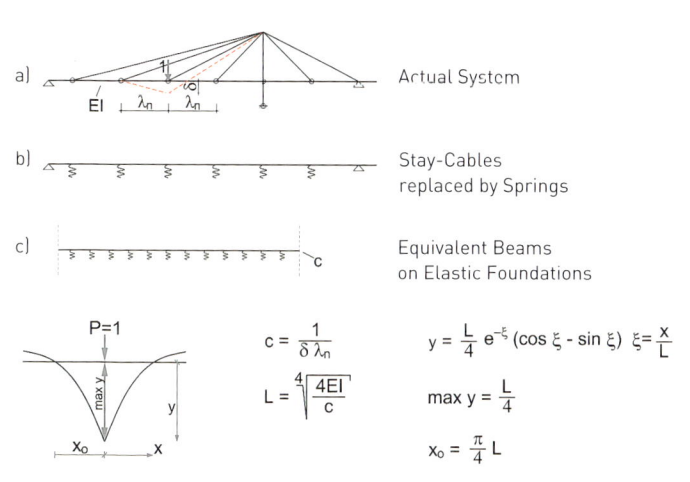

a) Actual System
b) Stay-Cables replaced by Springs
c) Equivalent Beams on Elastic Foundations

$c = \frac{1}{\delta \lambda_n}$

$L = \sqrt[4]{\frac{4EI}{c}}$

$y = \frac{L}{4} e^{-\xi}(\cos\xi - \sin\xi)$ $\xi = \frac{x}{L}$

$\max y = \frac{L}{4}$

$x_0 = \frac{\pi}{4} L$

Figure 4.7 Bending moments of a beam on elastic foundation

Figure 4.8 Bending moments due to a) concentrated load, b) uniformly distributed load

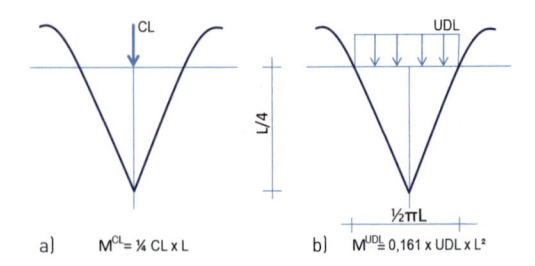

tance λ_n. For a concentrated load P the bending moment of a beam on elastic foundations comes to $M^P = \frac{1}{4} \cdot P L$, Fig. 4.8a, for a uniformly distributed load p the bending moment is $M^{UDL} = 0.161 \cdot p \cdot L^2$, Fig. 4.8b. With these formulas the governing bending moments over the length of the bridge can be approximated.

A comparison of the positive moment influence line for a beam on elastic foundation with the moment influence line of the actual bridge calculated by computer shows in general good agreement, Fig. 4.9 [1.3]. The approximation is better the more slender the beam is. For very stiff beams, for example double-deck bridges, the agreement is not as good. The negative moment influence lines do not show good agreement.

An overview of typical loads and action forces of a cable-stayed concrete bridge (Pasco-Kennewick Bridge) is shown in Fig. 4.10 [1.15]. The envelopes for moments and normal forces from the live load are given in 4.10f. The effective widths for bending can be determined in accordance with national codes. The appropriate effective lengths l are load dependent:
- For permanent loads which are approximately determined for a beam rigidly supported at the cable anchor points the effective length is the cable distance. It can, however, also be argued that the loading 'permanent loads' consists of the superposition of the two loadings 'dead load acting on the elastic system' and 'cable shortening', which each have long influence lengths. As a compromise the action forces can be determined for the long effective lengths (dead load, cable shortening), but the corresponding stresses under permanent loads from the short effective lengths between cable anchor points.
- For live loads the effective length is $\pi/2 \cdot L$ for a beam on elastic foundation, Fig. 4.9.

The moments from permanent loads are generally much smaller than those from live loads, see Fig. 4.10b, f. The effective width for normal forces is always the full beam area after a certain force introduction length.

To understand the inner workings of a cable-stayed bridge it is interesting to know in which way – for a given bridge system under given loads – the stresses are related to the beam depth. Fig. 4.11 left explains that for an idealized cross-section comprising only top and bottom chord the bending stresses from transient loads are indirectly proportional to the beam area only, but are independent of the beam depth:

$$f = \frac{const}{A}.$$

This result is at first glance perplexing. It can be explained by the fact that the moments from transient loads are primarily restraint loads which solely depend on the curvature of the beam. For a stay system with a given stiffness, the radius of curvature becomes bigger the greater the depth and thus the greater the stiffness of the beam becomes.

For a solid cross-section, Fig. 4.11 right, the bending stresses are indirectly proportional to the square root of the beam depth [4.57]:

$$f = \frac{const}{\sqrt{h}}.$$

4.1.4.2 Buckling – non-linear theory

The slenderness of a beam finds its lower bound by the required safety against buckling. The safety against buckling for a long beam on elastic foundation with end supports is shown in Fig. 4.12. This approximation assumes a constant elastic support and a constant normal force [4.2]. For the beam of an actual cable-stayed bridge,

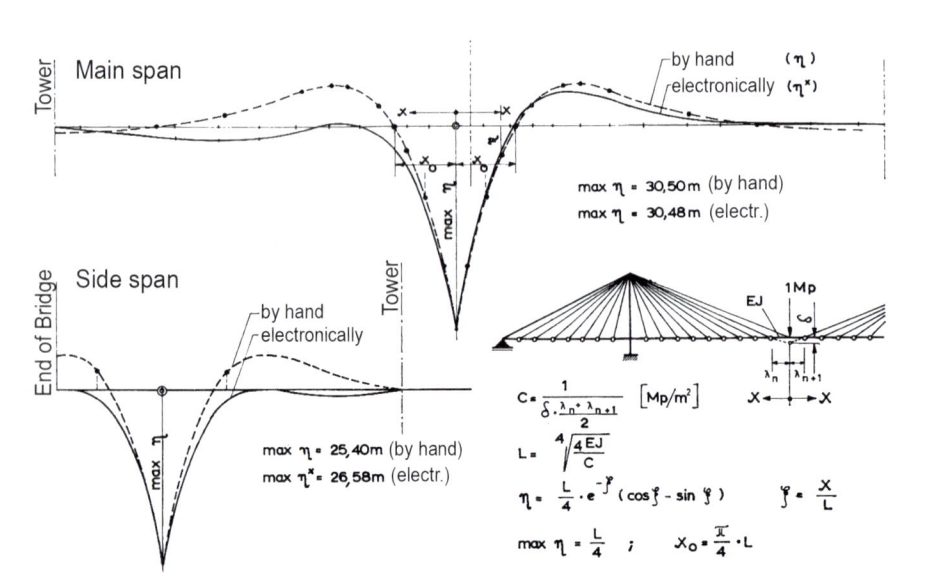

Figure 4.9 Comparison of moment influence lines for a beam on elastic foundation with the actual bridge

4.1 Action forces for equivalent systems

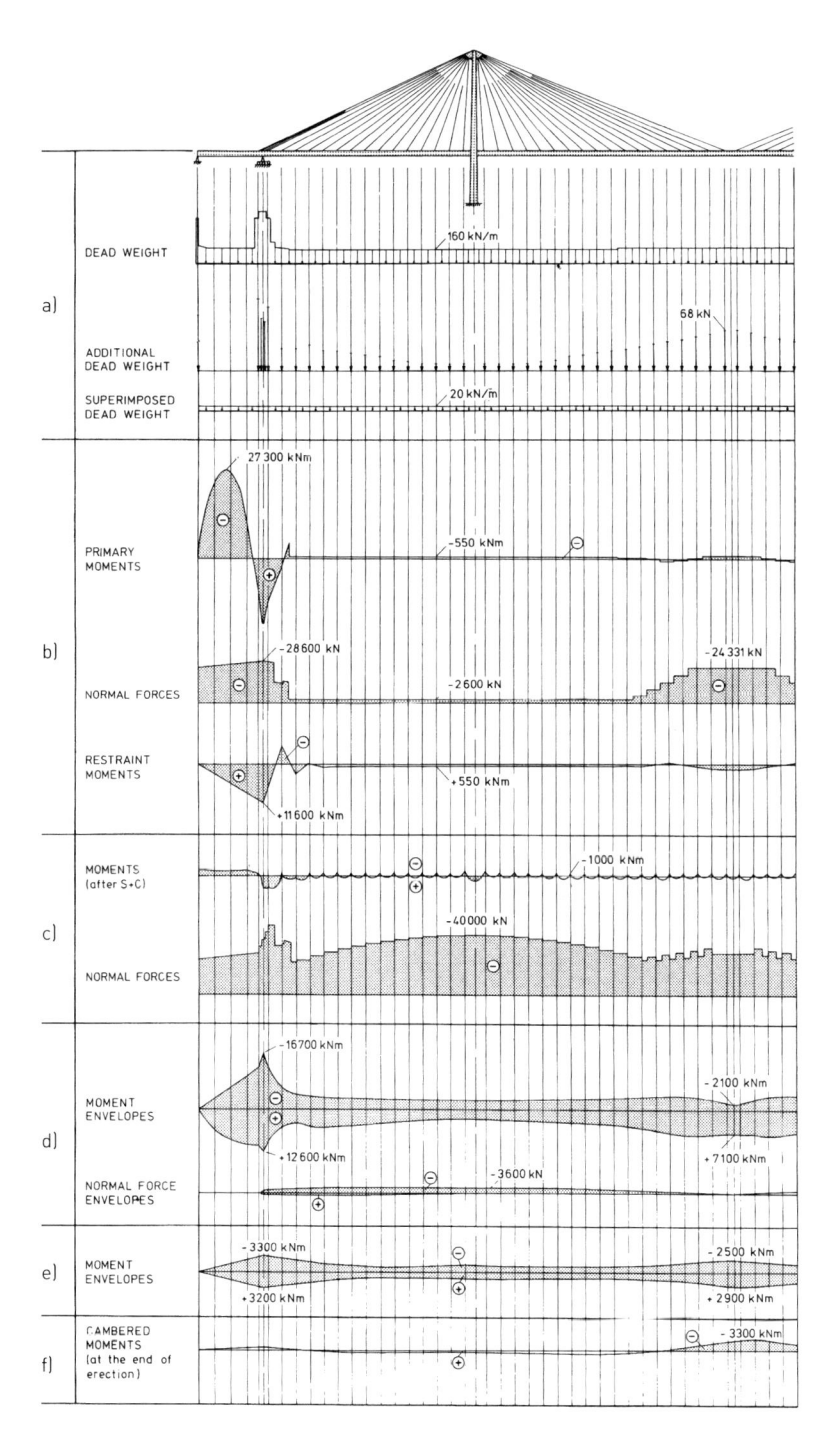

Figure 4.10 Typical loads and action forces of a cable-stayed concrete bridge (Pasco-Kennewick Bridge)
a) dead weight
b) prestress
c) permanent loads
d) live load + impact
e) temp. combinat.
f) S + C final stage

Moments

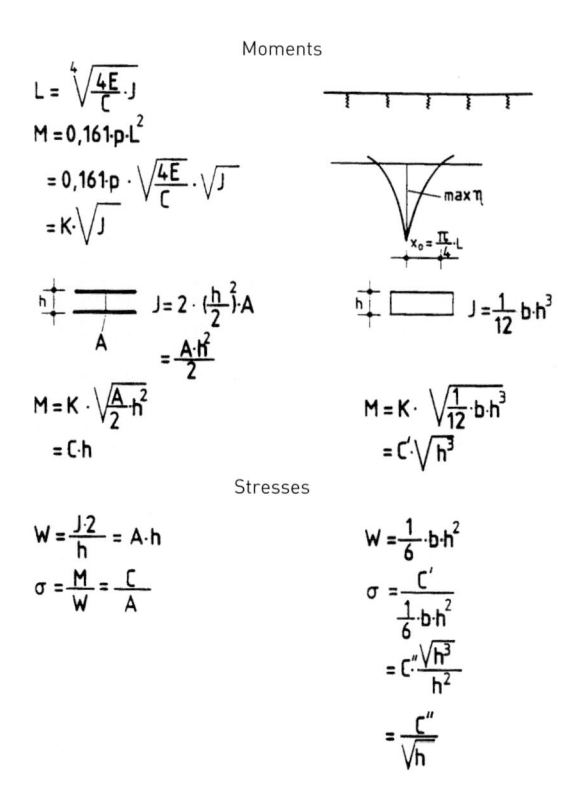

$$L = \sqrt[4]{\frac{4E \cdot J}{C}}$$
$$M = 0{,}161 \cdot p \cdot L^2$$
$$= 0{,}161 \cdot p \cdot \sqrt[4]{\frac{4E}{C}} \cdot \sqrt[4]{J}$$
$$= K \cdot \sqrt[4]{J}$$

$$J = 2 \cdot \left(\frac{h}{2}\right)^2 \cdot A \qquad J = \frac{1}{12} b \cdot h^3$$
$$= \frac{A \cdot h^2}{2}$$

$$M = K \cdot \sqrt[4]{\frac{A}{2} \cdot h^2} \qquad M = K \cdot \sqrt[4]{\frac{1}{12} \cdot b \cdot h^3}$$
$$= C \cdot \sqrt{h} \qquad = C' \cdot \sqrt[4]{h^3}$$

Stresses

$$W = \frac{J \cdot 2}{h} = A \cdot h \qquad W = \frac{1}{6} \cdot b \cdot h^2$$
$$\sigma = \frac{M}{W} = \frac{C}{A} \qquad \sigma = \frac{C'}{\frac{1}{6} \cdot b \cdot h^2}$$
$$= C'' \cdot \frac{\sqrt{h^3}}{h^2}$$
$$= \frac{C''}{\sqrt{h}}$$

Figure 4.11 Bending stresses in a beam in relation to the beam area and depth

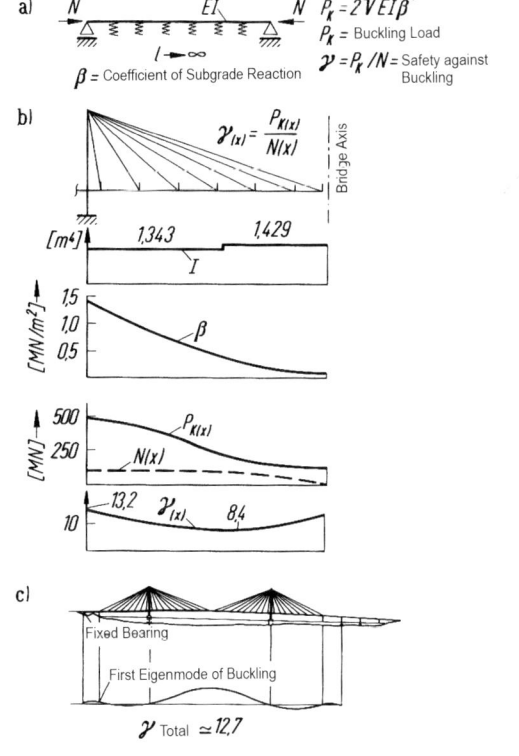

a) $N \longrightarrow EI \longleftarrow N \quad P_K = 2\sqrt{EI\beta}$
P_K = Buckling Load
β = Coefficient of Subgrade Reaction
$\gamma = P_K / N$ = Safety against Buckling

b) $\gamma_{(x)} = \frac{P_{K(x)}}{N(x)}$

c) First Eigenmode of Buckling
$\gamma_{Total} \simeq 12{,}7$

Figure 4.12 Safety against buckling
a) beam on elastic supports
b) local safeties against buckling
c) overall safety against buckling

$$M^{II} = \frac{M^I}{1 - \dfrac{N_i}{P_k}} = \frac{M^I}{1 - \dfrac{1}{\gamma_i}}$$

M^{II} Non-Linear Moment of 2nd Order
M^I Linear Moment of 1st Order
N_i Normal Force at Location i
P_k Buckling Load of actual System

Example Pasco Bridge

$\max N_i = 38 \text{ MN} \quad P_k = 480 \text{ MN}$

$M^{II} = M^I \cdot \dfrac{1}{1 - \dfrac{38}{480}} = 1.09 \, M^I$

Figure 4.13 Non-linear increase of moments

however, the elastic support and the normal force increase towards the tower.

When investigating this problem Man-Chung Tang [4.3] has shown that the governing location for the determination of the overall safety against buckling is near the tower, which means that the elastic support and the normal force of the beam near the tower are representative of the whole beam.

Fig. 4.12 shows, for the Pasco-Kennewick Bridge [1.15], the run of the normal force and the elastic supports over the length of the bridge which give local safeties against buckling between 8.4 and 13.2, Fig. 4.12a and b. The exactly determined overall safety against buckling comes to 12.7, Fig. 4.12c, near the tower as predicted by Man-Chung Tang.

Important for the bridge design is not so much the safety against buckling, which is a more theoretical value, but the non-linear increase of moments (P-δ effect, second order theory). In a first approximation this increase may be calculated from the local safeties against buckling. The example in Fig. 4.13 [1.15] for a cable-stayed concrete bridge shows a non-linear increase of moments of 9%. This very approximate method gives the designer early on a feel of the susceptibility of a design.

4.1.5 Permanent loads on rigid supports
4.1.5.1 Dead load

For the preliminary calculation of the action forces under permanent loads rigid supports of the beam are introduced at its cable anchor points. For this continuous beam on rigid supports the moments are determined. The cable forces are determined from the support reactions.

This run of moments on rigid supports is generally chosen for a concrete beam in the final stage because only these moments are not subject to creep, Fig. 4.14 and Section 5.3.

For steel beams, that run of moments under permanent loads in the final stage is selected which requires the minimum of steel in combination with live loads. Since the positive live load are governing, a negative run of moments under permanent loads is often selected.

Since the concrete deck of a composite beam is also subject to creep, the moments of a beam rigidly supported at the anchor points are approximately selected. In the bridge center, where the compression forces in the beam from the cables are small, a composite beam may be cambered so that a positive moment occurs which gives compression to the roadway slab as top chord of the beam cross-section. This camber moment, however, creeps partly away.

The action forces from permanent loads for a rigidly supported system result in the moments and normal forces shown in Fig. 4.10d for the actual system.

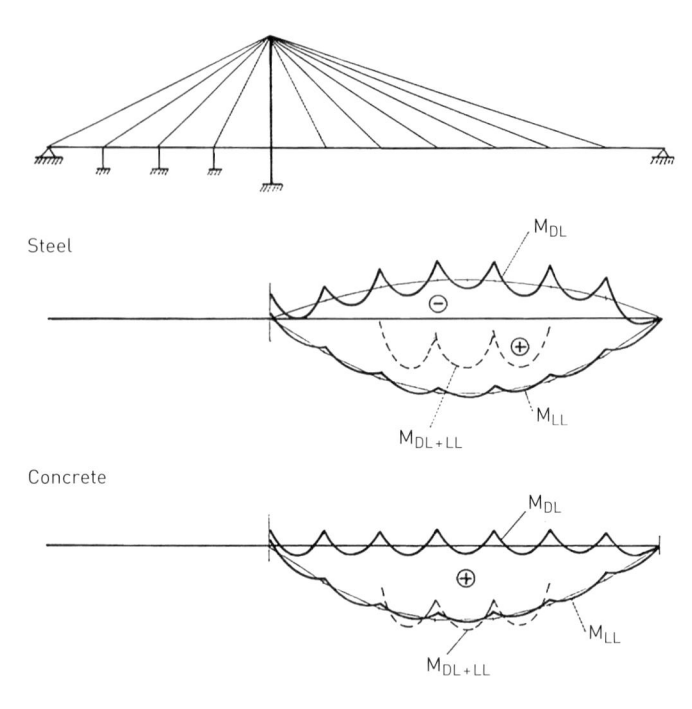

Figure 4.14 Chosen run of moments under permanent loads for steel and concrete beams

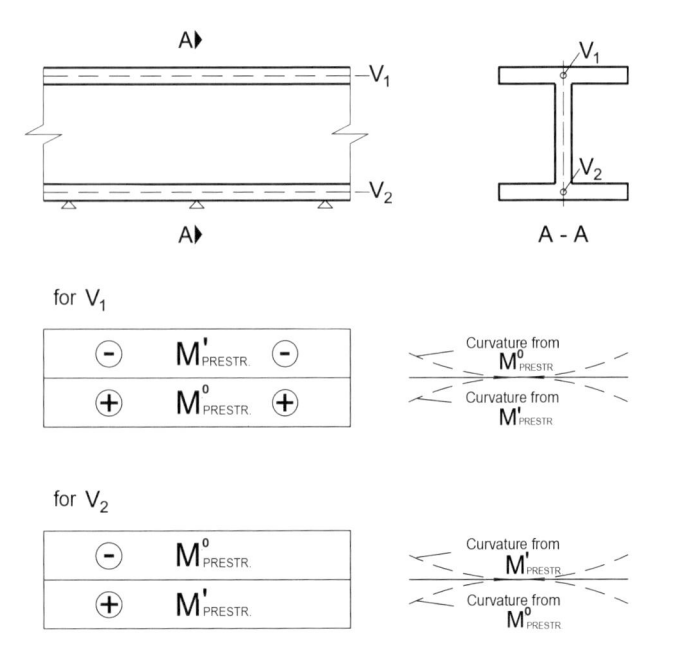

Figure 4.15 Beam moments on rigid supports from straight tendons

4.1.5.2 Post-tensioning

The service load moments from post-tensioning in a cable-stayed concrete beam are also perplexing at first glance. If straight tendons are used – which is usual for the small cable distances – the beam moments from eccentricity and restraint exactly balance one another so that only a centric compression force remains.

The reason is that any curvature of the beam from eccentrically positioned tendons is flattened out by the rigid supports and this flattening produces exactly the same (restraint) moments as the original moments which caused the curvature, Figs 4.15 and 4.10 c.

Fig. 4.16 shows for the example of the Helgeland Bridge [2.80], the arrangement of tendons for a cast-in-place concrete beam. In the center of the bridge and near the anchor piers, where the normal forces from cables are small and the bending moments large, straight tendons are arranged which in some way replace the missing normal force from stay cables.

Precast concrete beams, for example for the Pasco-Kennewick Bridge, require an increased construction post-tensioning over their full length in order to keep the joints under compression during construction also.

Fig. 4.10 gives the distribution of moments and normal forces due to post-tensioning. The primary moments from eccentricity and the secondary moments from restraint balance one another and only a normal force from post-tensioning remains.

4.1.5.3 Shrinkage and creep

Concrete beams carry action forces due to shrinkage and creep. These have two origins. Creep of moments takes place:
- if a run of moment is chosen which is not equivalent to that of a beam of rigid foundations, and
- due to the shortening of the beam from shrinkage and creep which causes a sagging of the beam due to the cable inclination with the corresponding elastic moments which themselves are subject to moment creep.

The creep of moments is known from a rigidly supported continuous concrete girder for which a freely selected run of moments always creeps in the direction of a system that occurs at the removal of the forms on which the girder is cast continuously. The change of moments depends on the creep value and the time sequence.

Depending on the time sequence the creep of moment reaches different values:
- like originating from a sudden pier settlement
- like originating from a slow pier settlement.

These phenomena are outlined for a typical two-span concrete beam on rigid piers in Fig. 4.17 [1.15].

If a concrete beam is elastically supported on steel springs which are not subject to creep, the moments of a rigidly supported beam are also not subject to creep of moments.

This means that the beam moments of a cable-stayed concrete bridge are not changed due to creep of moments if the vertical com-

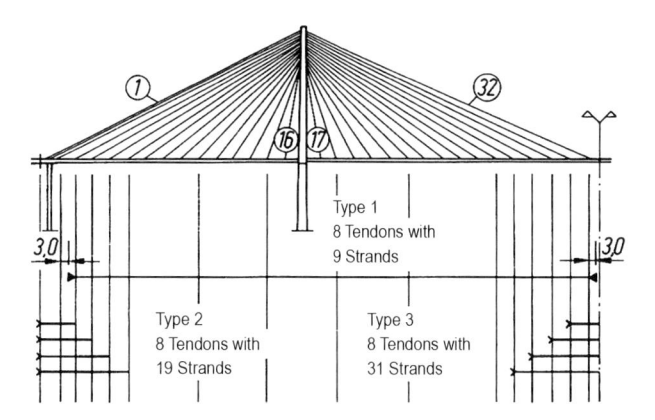

Figure 4.16 Longitudinal post-tensioning of a CIP concrete beam

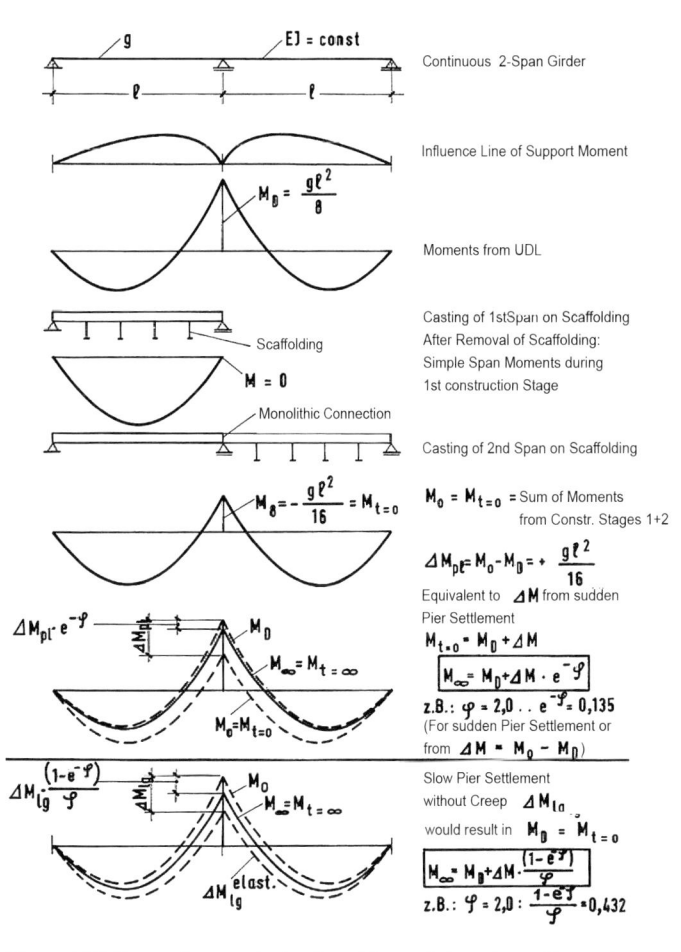

Figure 4.17 Creep of moments due to sudden and slow pier settlement for a rigidly supported beam

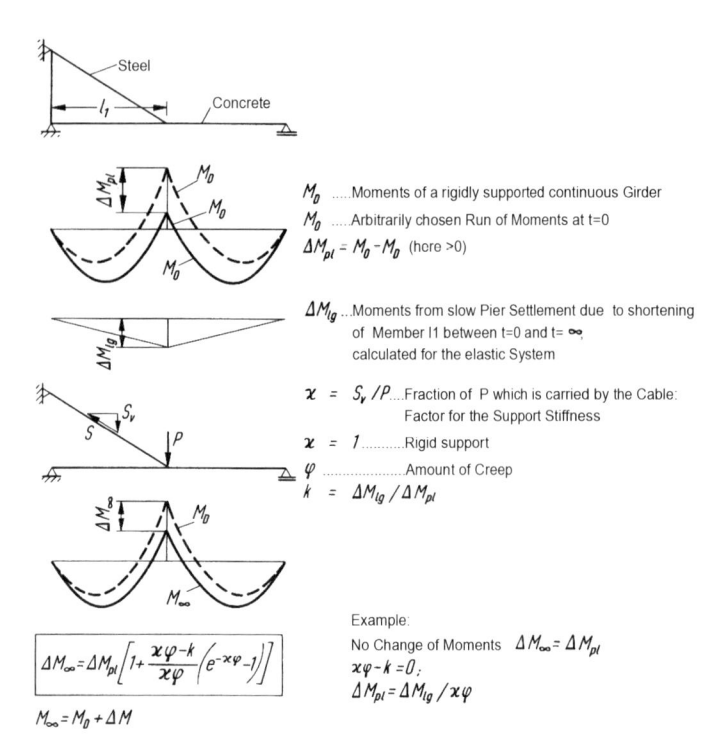

Figure 4.18 Simultaneous creep of moments due to sudden and slow pier settlement for a cable-stayed bridge

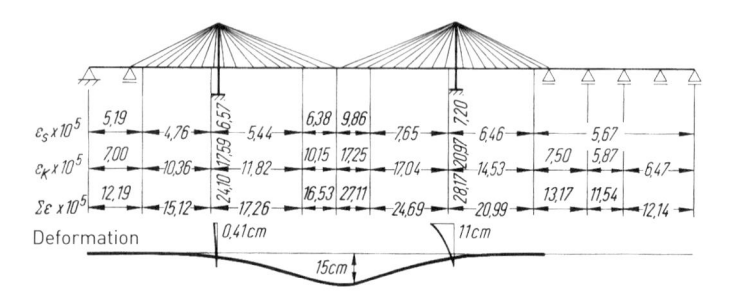

Figure 4.19 Shrinkage and creep deformations of the Pasco-Kennewick Bridge

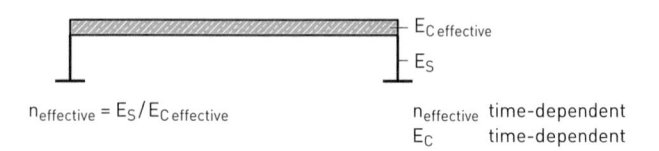

Figure 4.20 Shrinkage and creep of a composite cross-section

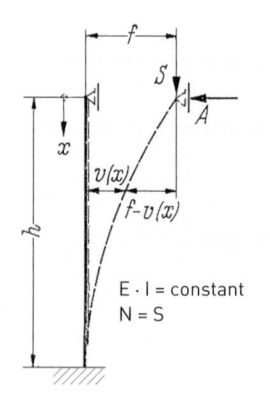

Figure 4.21 Statical tower system

ponents of the cable forces from permanent loads are selected in such a way that they are equal to the support reactions of a beam rigidly supported at the cable anchor points.

Due to the shrinkage and creep shortenings of the beam (and the concrete towers) moments are building up in the beam similar to those due to slow pier settlement of a continuous girder. Theoretically a run of moments different from that of a rigidly supported beam may be chosen in such a way that these two creeps of moments cancel one another.

This collaboration of the different types of moment creep is shown for a system with only one cable in Fig. 4.18.

If the shortening of the beam and the towers due to shrinkage is neglected, only creep due to normal force is present. It is very helpful to know that no changes of beam moments due to creep take place if at the end of erection a run of moments is chosen which is equal to that of an elastic system in which the creeping concrete members (beam and towers) are provided with their true stiffnesses E_C, A_C and I_C, and the non-creeping steel parts (stay cables) with an infinite stiffness $E_S \cdot A_S = \infty$. These considerations clarify the meaning of a 'rigidly supported beam'.

Fig. 4.10e shows the moments due to shrinkage and creep for the Pasco-Kennewick Bridge. In Fig. 4.19 the corresponding shrinkage and creep shortenings of the beam at the end of construction (t = 1) and after shrinkage and creep has taken place (t = ∞). During that time the beam deflects downwards by 15 cm and the tower tips move towards the fixed bearing at the left-hand side. The shop form of the beam was chosen in such a way that the beam gradient was too high at the end of construction and the intended gradient was only reached after shrinkage and creep. The towers were constructed vertically, but at the end of construction they were pulled with the back-stay cables by half the amount of the expected shrinkage and creep movements away from the fixed point, so that they ran through the vertical position while shrinkage and creep took place.

A composite beam also undergoes a change of action forces due to shrinkage and creep in its concrete roadway slab. If the cross-section properties are determined for an equivalent steel beam in which the concrete slab is replaced by a steel plate with a thickness reduced by the factor $n_{effective} = E_S / E_{C\,effective}$, then the time-dependent concrete characteristics can be accounted for in a simplified way by adjusting the n-values, Fig. 4.20.

For short-term loadings (live load) an $n_{effective}$ = 6 to 7 may be used. For permanent loads, the effective n-values depend on the age of the concrete. If a cast-in-place concrete slab is used,

the effective n-values at opening for traffic come to about 10 and after shrinkage and creep to about 20.

If precast concrete slabs are used for the roadway slab, the n-values are advantageously reduced. For the Sunshine Skyway Bridge [2.104], for which the precast slabs were supposed to be one year old at installation, the n-values at opening for traffic were estimated to be 9 and after shrinkage and creep 11, Fig. 4.20.

4.1 Action forces for equivalent systems

The stiffness-dependent beam moments can be approximately calculated by using its time-dependent stiffnesses corresponding to the $n_{\text{effective}}$-values.

4.1.6 Towers
4.1.6.1 In the longitudinal direction
For the preliminary design of the towers an equivalent system of a vertical cantilever fixed at the bottom and connected to the cables, Fig. 4.21 is used. The deflection of the tower head results from the elastic change in length of the backstay cable, resulting in linear tower moments with the maximum at the fixation.

In the seminal publication by Klöppel, Esslinger and Kollmeyer [4.4], the system of Fig. 4.22 was solved generally for non-linear effects (second order theory). It is apparent, that the non-linear tower moments are subject to significant change if the vertical cable components are taken into account as loads. The fixation moments may decrease while the moments in the lower third of the tower may increase with decreasing bending stiffness of the tower.

For a large tower stiffness – on the left-hand side of Fig. 4.22 – the non-linear effects are small. With decreasing stiffness, the fixation moments decrease, run through zero and may even change sign. Such a small stiffness can effectively only be reached for steel towers, for example the Duisburg-Neuenkamp Bridge with a high-strength steel, N-A-XTRA 70, see Fig. 2.98.

Concrete towers are generally rather stiff as shown for the example of the Pasco-Kennewick Bridge, Fig. 4.23, but even in this case the non-linear fixation moments decrease by about 16 % [1.15].

4.1.6.2 In the transverse direction
For towers with vertical legs the transverse moments can be calculated in accordance with [4.11] by taking into account the tower head deflection. The transverse restoring action of the stay cables may also be taken into account as elastic tower support.

For straight A-towers the transverse moments due to wind on cables and beam remain small because the loads are carried in tension and compression in the inclined tower legs and moments are only created by restraint. The legs of straight A-towers for bridges with beams high above water often spread too wide at water level, thus creating a danger for shipping. In these cases, the tower legs are pulled together underneath the beam in a diamond shape, Fig. 4.24. In this case the strut and tie action is lost below deck and significant moments are created in the lower tower legs which require an increase in their transverse width [1.17].

A unique type of tower was designed for the Baytown Bridge, Fig. 4.25. For the two separate beams two A-towers above deck were chosen which were both A-shaped underneath the roadway also, resulting in two foundations. Both diamond-shaped towers are connected at roadway level, forming a truss in the transverse direction with a so-called double-diamond shape. The transverse loads from wind on cables and beam thus create only tension and compression

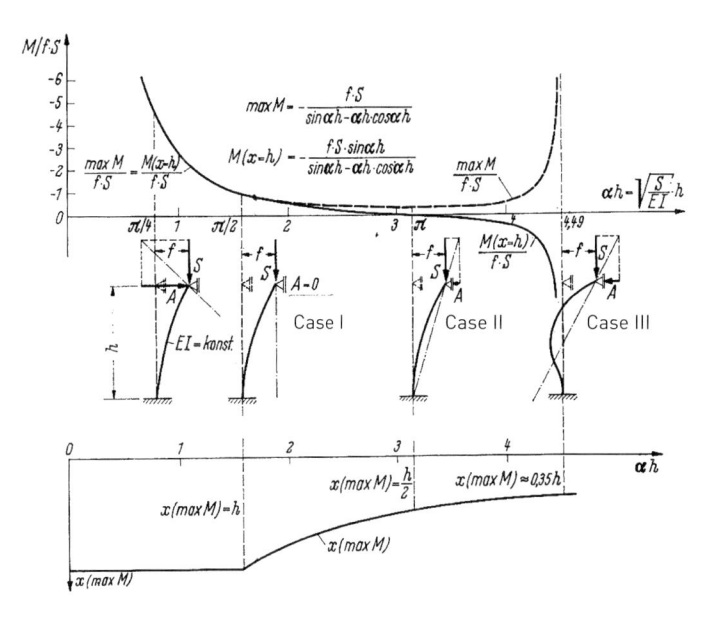

Figure 4.22 Tower moments M as a function of the tower stiffness $E \cdot I$

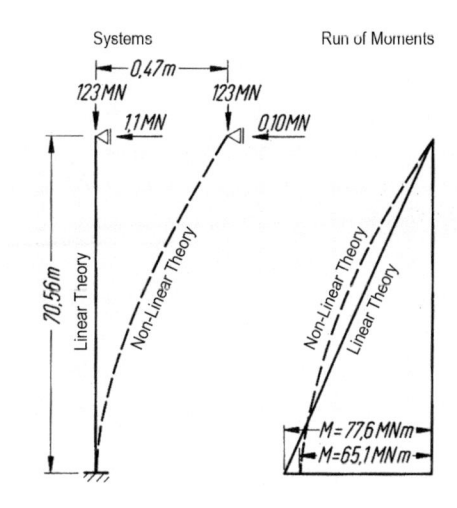

Figure 4.23 Tower moments due to linear and non-linear theory for the Pasco-Kennewick Bridge

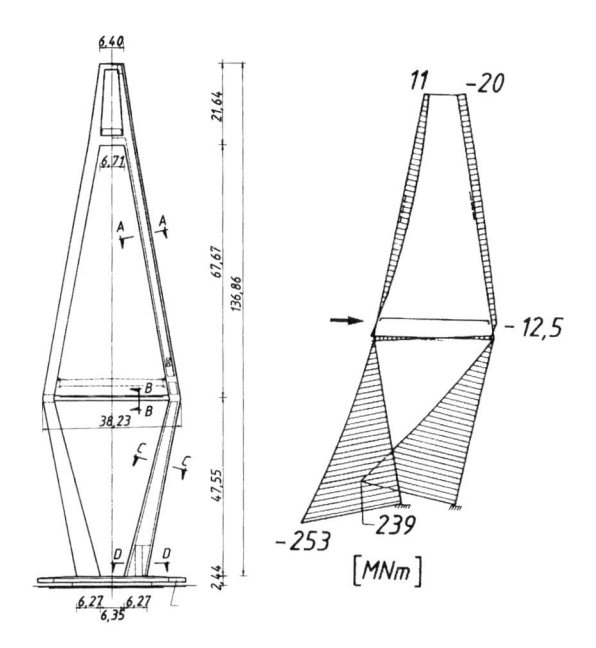

Figure 4.24 Transverse moment of a diamond-shaped tower

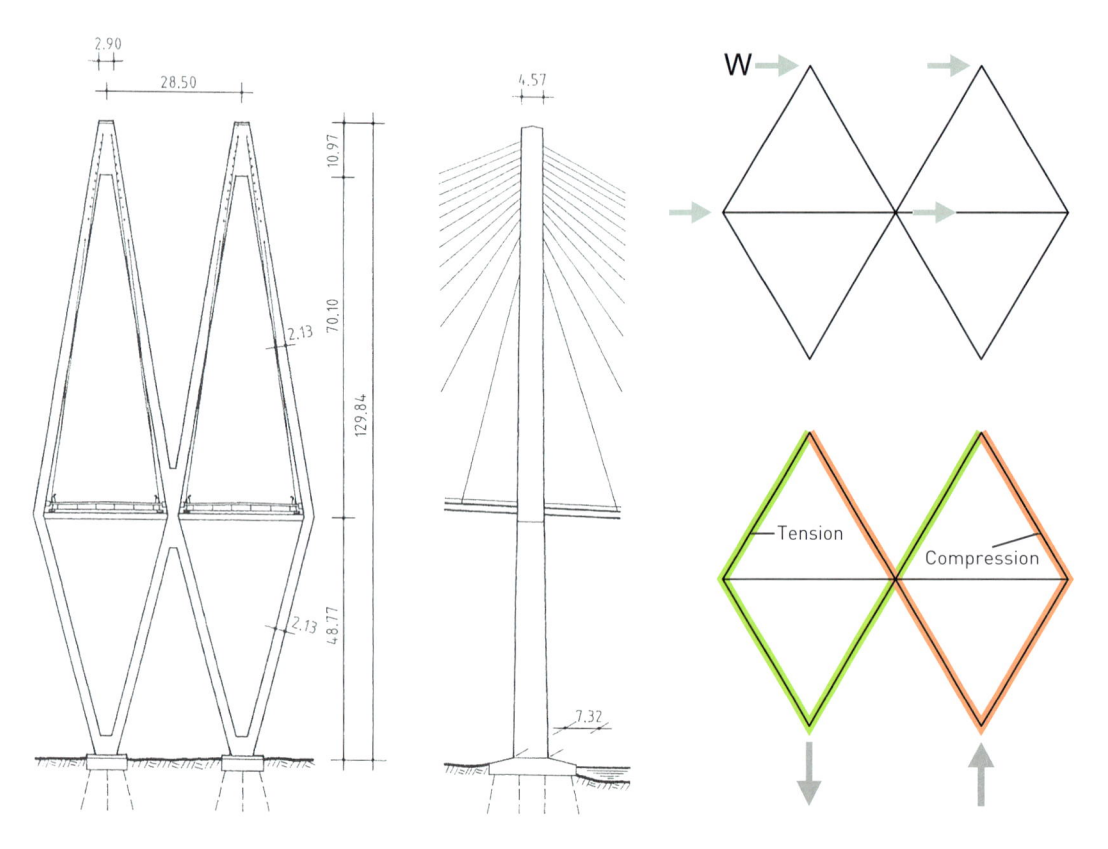

Figure 4.25 Transverse strut and tie action of a double-diamond tower

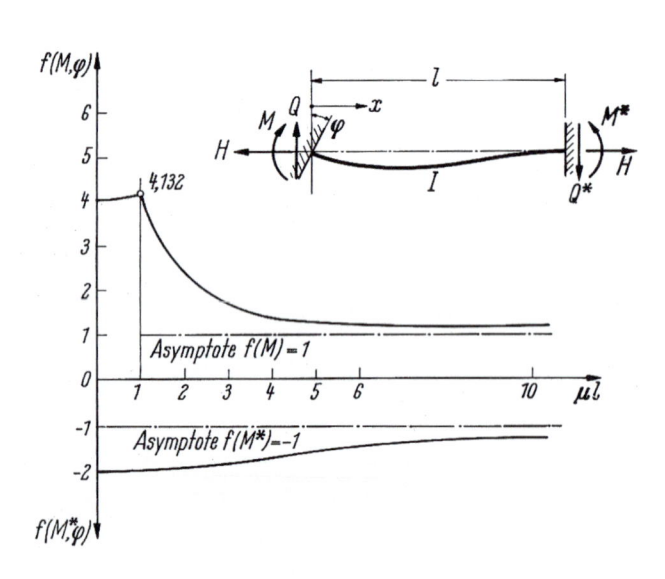

$$M = \varphi \cdot EJ \cdot \mu \cdot \frac{\mu l - \tanh \mu l}{\mu l \cdot \tanh \mu l - 2 + \dfrac{2}{\cosh \mu l}}$$

$$0 \leqq \mu l \leqq 1 \qquad \frac{M}{\varphi \cdot \dfrac{EJ}{l}} = f(M, \varphi)$$

$$1 \leqq \mu l \leqq 6 \qquad \frac{M}{\varphi \cdot \sqrt{HEJ}} = f(M, \varphi)$$

$$\mu l \geqq 6 \qquad \frac{M}{\varphi \cdot \sqrt{HEJ}} \approx \frac{\mu l - 1}{\mu l - 2}$$

$$\mu^2 = \frac{H}{EJ}$$

$$M^* = -\varphi \cdot EJ \cdot \mu \cdot \frac{\tanh \mu l - \dfrac{\mu l}{\cosh \mu l}}{\mu l \cdot \tanh \mu l - 2 + \dfrac{2}{\cosh \mu l}}$$

$$0 \leqq \mu l \leqq 6 \qquad \frac{M^*}{\varphi \cdot \dfrac{EJ}{l}} = f(M^*, \varphi)$$

$$\mu l \geqq 6 \qquad \frac{M^*}{\varphi \cdot \dfrac{EJ}{l}} \approx \frac{\mu l}{\mu l - 2}$$

$$Q = -\varphi \cdot EJ \cdot \mu^2 \cdot \frac{1 - \dfrac{1}{\cosh \mu l}}{\mu l \cdot \tanh \mu l - 2 + \dfrac{2}{\cosh \mu l}} = Q^*$$

Figure 4.26 Non-linear fixation moment of tension members for rotation of the end anchorage

in the legs – see right-hand side of Fig. 4.25 – accompanied by small restraint moments. Therefore, the transverse beam depth can be kept constant 2.13 m (7 feet) over the full height of about 130 m and thickening for transverse bending in the lower part, as used for a single diamond-shaped tower, is not required.

4.1.7 Stay cables

The main load for stay cables is direct tension. Deflections and rotations of the beam from transient loads produce a kink in the stay cables at the anchor heads which causes bending in the wires/strands of the stay cables. The combination of anchor head rotations with tension in the cables causes a non-linear effect (second order theory) by which the bending radius of the wires is strongly reduced and the bending stresses are accordingly increased. The non-linear increase of moments for anchor head rotations under tension is shown in Fig. 4.26 [4.5].

The increase of moments is governed by the bending stiffness of the cables. Solid post-tensioning bars are especially sensitive. Cables with individual wires or strands are less sensitive if the bending stiffness of the complete cable is assumed to be the sum of the bending stiffnesses of the individual wires/strands. For locked coil ropes, however, there is a certain shear friction between the Z-shaped outer wires due to transverse contraction under load, so that their true stiffness lies between that of the sum of the individual wires and a solid bar with the diameter of the rope. For parallel wire cables it may be assumed that the individual wires do not act together, because there is no transverse pressure, so that the cable stiffness is equal to the sum of the wire stiffnesses. For parallel strand cables it is generally assumed that the seven wires of each strand act together and have a common strand stiffness. The stiffness of the strand cable is then equal to the sum of the strand stiffnesses.

4.2 Action forces of actual systems
4.2.1 Permanent loads
4.2.1.1 General

The action forces for permanent loads are calculated for the true elastic system by superpositioning the two loadings dead loads DL and cable shortenings CS in such a way that the predetermined run of beam and tower moments is achieved. Fig. 4.27 shows, for a simplified system with only one cable, the moments from dead loads and the freely chosen moments from cable shortenings which in sum give the desired run of moments for permanent loads [4.6]. The determination of the desired run of moments for permanent loads is outlined in Fig. 4.14.

For an n-time statically indeterminate system, n section forces can be chosen freely. For a cable-stayed bridge the numbers of statically indeterminate parameters may be determined by releasing all internal beam supports so that a simply supported beam results.

For the example in Fig. 4.28 the statically indeterminate parameters are the six stay cables and the tower bearing. If they are released,

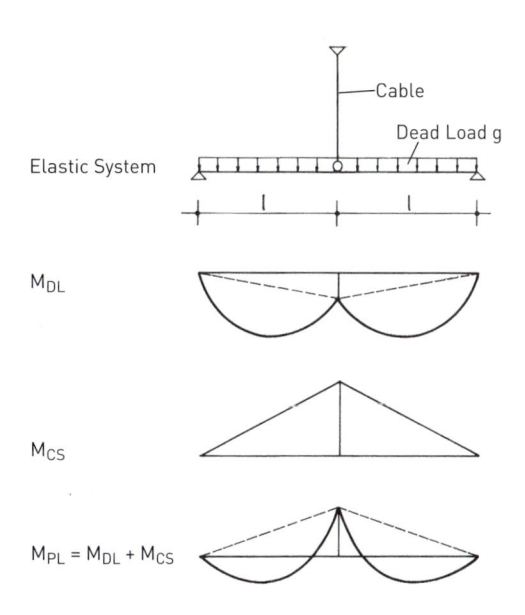

Figure 4.27 Superposition of the two loadings resulting in the permanent load stage

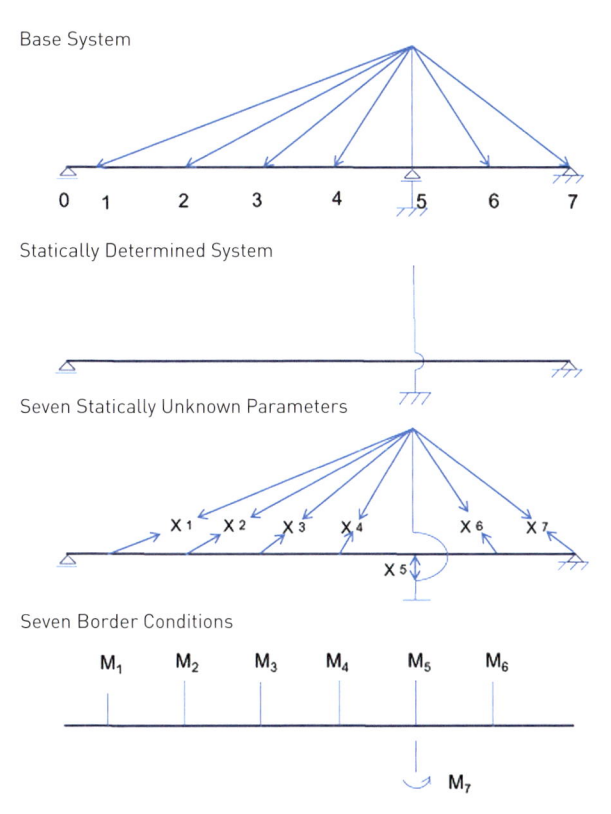

Figure 4.28 Statically indeterminate parameters and border conditions

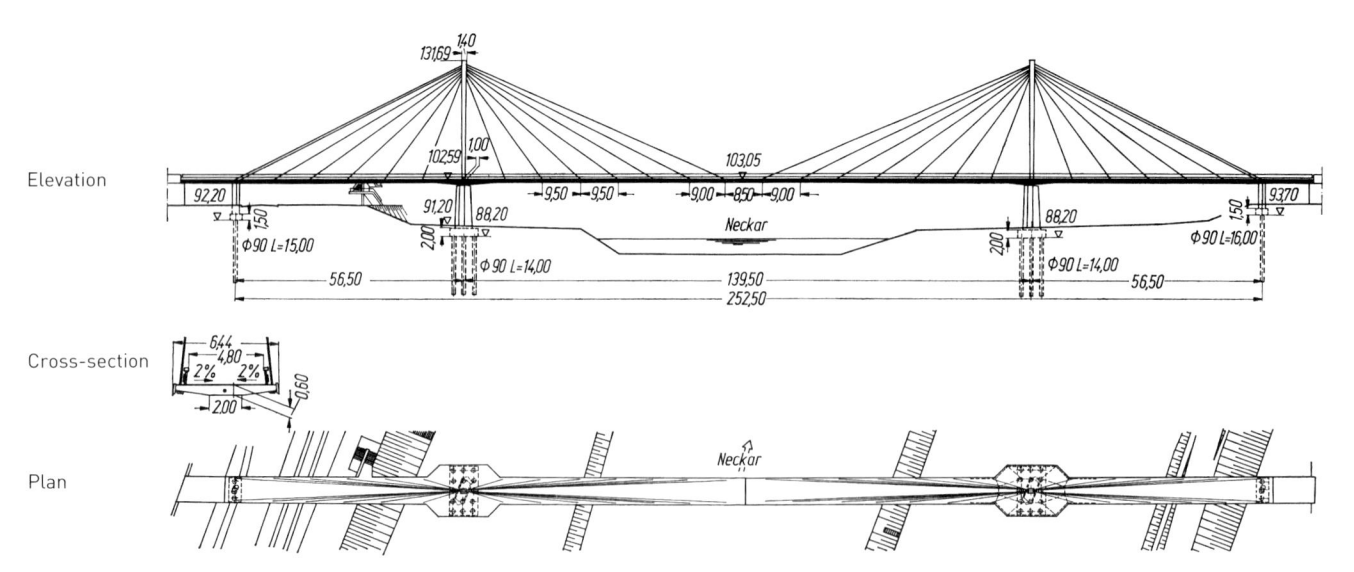

Figure 4.29 Neckarcenter Bridge, with steel towers and concrete beam

a statically determinate beam supported at both ends remains. The released supports are introduced as statically indeterminates X_i. The corresponding seven border conditions X_{0i} may also be chosen freely. It is, however, expedient to select them in such a way that the individual indeterminates X_i directly correspond to the border conditions X_{0i}. The beam moments at the anchor points of cables 1 to 4 and 6 appear to be suitable as well as a pier settlement at tower 5. The cable shortenings of the backstay cable 7 may be combined with the tower fixation moment.

The desired run of beam moments under permanent loads may be selected for a concrete beam as those for rigid supports at the cable anchor points and tower bearing as outlined in Fig. 4.14. The fixation moment of the tower may be selected as zero.

Alternatively the vertical beam deflections as well as the tower head movement may be selected as zero. (A certain inaccuracy in this regard is created by the normal force shortening of an inclined beam.)

4.2.1.2 Concrete bridges

As an example for a concrete cable-stayed bridge the Neckarcenter Bridge in Mannheim, Germany, is selected, Fig. 4.29 [4.7].

For structures consisting of a uniformly creeping material (concrete) the run of moments from permanent loads does not change with time if a run of moments is selected such that it is inherently compatible with the system, for example the moments of a continuous girder cast on scaffolding [4.7]. For further details see Section 4.1.5.3. In order to minimize the amount of material (reinforcement/post-tensioning) required for the beam, the moment envelopes from permanent loads plus live loads have to be minimized, which means that the live loads need only be superimposed onto constant permanent load moments, and not onto the different permanent load moments before and after shrinkage and creep.

The towers and stay cables of the Neckarcenter Bridge consist of non-creeping steel, whereas the bridge beam comprises concrete, subject to creep. The beam deformations under live load are 97% created from cable elongations and tower shortenings and only 3% from beam shortenings. These values indicate that the vertical beam deformations are by far governed by the non-creeping steel parts. The run of moments for the concrete beam which is not subject to creep is thus equal to that of a beam 'nearly rigidly supported' at the anchor points. For further details see Section 4.1.5.3.

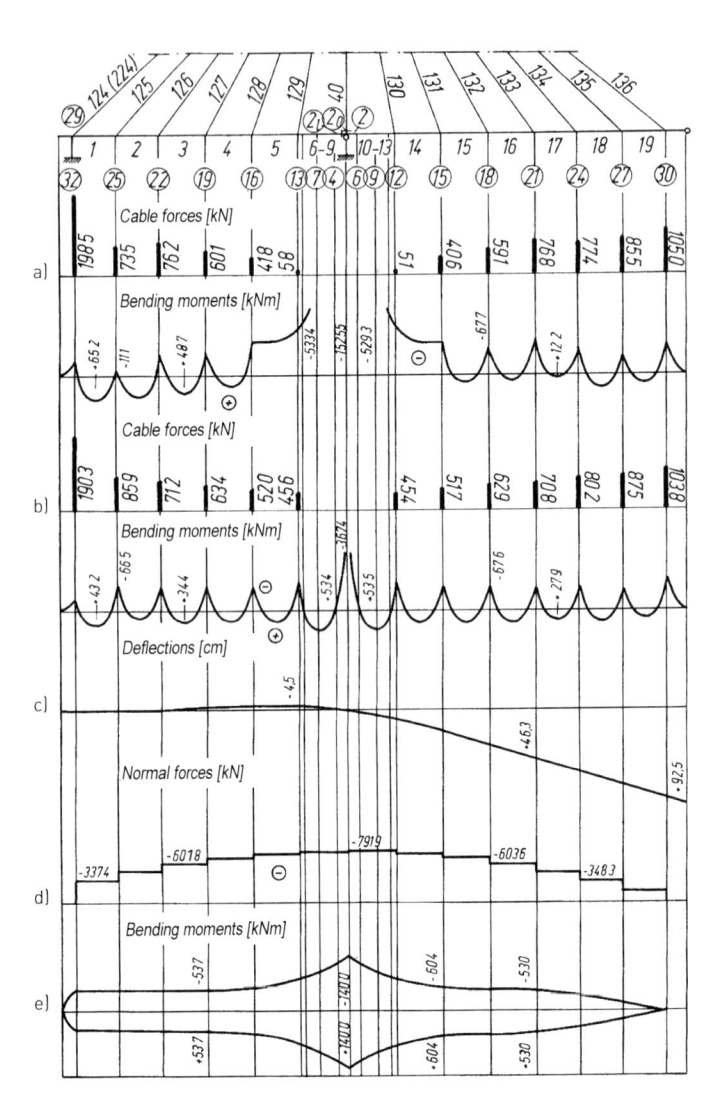

Figure 4.30 Bending moments, normal forces and deflection of beam

This run of moments has to be determined from the superposition of the loadings dead load on the elastic system plus cable shortenings.

If moment changes due to creep were completely to be avoided, a run of moments would have to be selected that lies between the rigidly and elastically supported girder, but only 3 % away from the rigidly supported.

For simplicity the cable shortenings for the Neckarcenter Bridge were selected in such a way that the permanent load moments are those of a rigidly supported beam. In this way the beam deflections at the cable anchor points become zero and the action forces in the beam and the stay cables are favorable, Fig. 4.30. The creep of moments for permanent loads is practically negligible. The bending moments and cable forces with and without cable shortenings are shown in Fig. 4.30 a–d, together with the deflections without cable shortenings and the run of normal forces.

Since the steel tower does not creep, its moments under permanent loads were chosen in such a way that a minimum amount of material is required if combined with the live load moments:
a) bending moments and cable forces from dead loads on elastic system,
b) bending moments and cable forces from permanent loads including cable shortenings (equivalent to the moments of a rigidly supported beam),
c) deflections of beam from dead load without cable shortenings,
d) normal forces in beam from permanent loads.

An important change of action forces due to cable shortenings takes place at the beam around the tower. In that location the cable shortenings unload the tower support and the dead load is transported more by the stay cables into the tower. This rearrangement of forces is shown schematically in Fig. 4.31.

For the Posadas-Encarnación Bridge, Fig. 4.32 shows the moments from dead loads acting on the elastic system, the corresponding moments due to cable shortening and their superposition to the permanent load stage. The large negative bending moments of the elastic system around the tower indicate that the beam is not supported by a tower bearing but is continuously supported by the stay cables over the full length of the bridge. This means that there is no support reaction of a tower bearing. For this bridge the tower consists also of concrete, the tower moments from dead load and cable shortening are thus selected in such a way that they also balance one another, similarly as for the beam. Calculation examples are given Sections 4.5.2 and 4.5.7.

4.2.1.3 Steel bridges

Since steel bridges are not subject to creep, their run of moments for permanent loads is selected in such a way that after superpositioning with all transient loads a minimum amount of material is required. A calculation example for the dead load moment calculation of a steel bridge is given in Section 4.5.3.

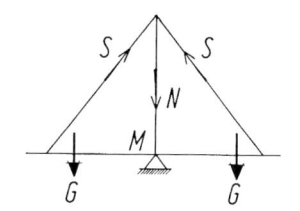

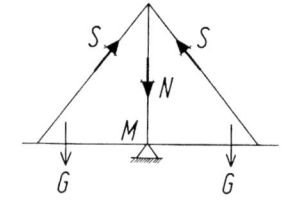

Without additional cable shortenings
S small – M large

With additional cable shortenings
S large – M small

Figure 4.31 Rearrangement of dead loads by cable shortening

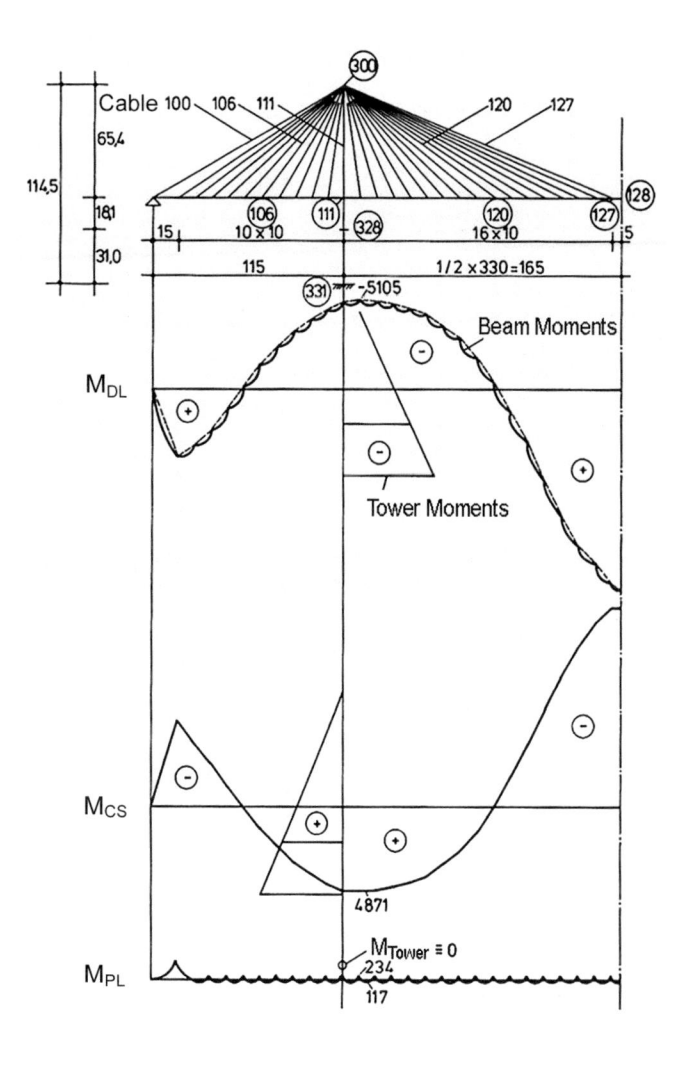

Figure 4.32 Superposition of beam and tower moments from dead load on elastic system and cable shortenings to permanent load moments

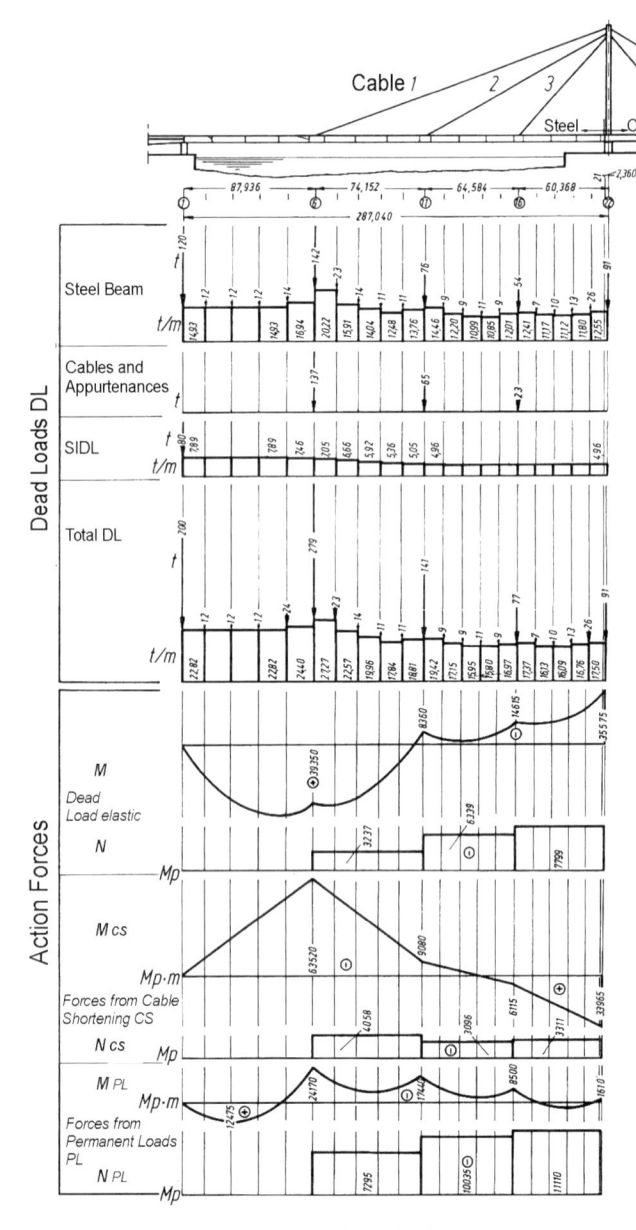

Figure 4.33 Dead loads, moments of the elastic system, moments due to cable shortenings and moments for permanent loads

The basic relations are shown in Fig. 4.33 for the steel main span of the Rhine River Bridge at Mannheim-Ludwigshafen [1.19]. The moments from cable shortening were selected in such a way that the sum of moments from dead load plus cable shortenings result in a run of moments from permanent loads that result in a minimum of material required if the live load and other transient loads are superimposed.

Another example is given in [2.71] for the steel main span of the Rhine River Bridge at Flehe, Fig. 4.34 gives the individual moments together with the superimposed moment envelope from dead load on the elastic system plus cable shortenings, Fig. 4.34b, together with the live load moments, Fig. 4.34d.

For the determination of the cable shortenings the following considerations are made:

– By a vertical movement of the support at abutment 14 and the shortening or releasing of the seven ropes the run of moment shown with a dotted line in Fig. 4.34d is created. These moments from cable shortening counteract the moments from permanent loads on the elastic system.

– The superposition of the permanent load moments with the live load moments result in the shaded area between the upper and lower moment envelopes, Fig. 4.34d, for which the beam has to be sized.

– The selection of the cable shortenings was governed by the live load moment envelopes. Near the tower (axis 13) negative live load moments dominate, while near the abutment (axis 14) positive ones are governing. The moments due to cable shortening divide these two moment areas in such a way that the extreme positive and negative moments have similar amounts. In the center of the main span the moments due to cable shortenings result in similar stresses in the top and bottom flanges.

– The cable shortenings result in a change of cable forces due to unloading of the tower support which require an increase in cable areas. Their influence on the normal forces and the shear in the beam is comparatively small.

As a result, it can be stated that by choosing a suitable pier settlement in axis 14 and suitable cable shortenings in all cables V 100 to V 700 the run of moments in the main span for permanent loads can be chosen freely. The large main span is governing the costs for such a major bridge. It is, therefore, expedient to choose the support reaction and the seven cable forces in such a way that the sizing of the main span beam results in a minimum of material. By incorporating the resulting superelevation of the beam into the shop form it can be ensured that the desired run of moments is indeed present at the end of construction when the beam has reached its predetermined gradient.

The two loadings – dead load on the elastic system and cable shortenings – which constitute the permanent loads do not appear separately on site. Both partial load components should thus receive the same load factor in the ultimate limit state.

4.2 Action forces of actual systems

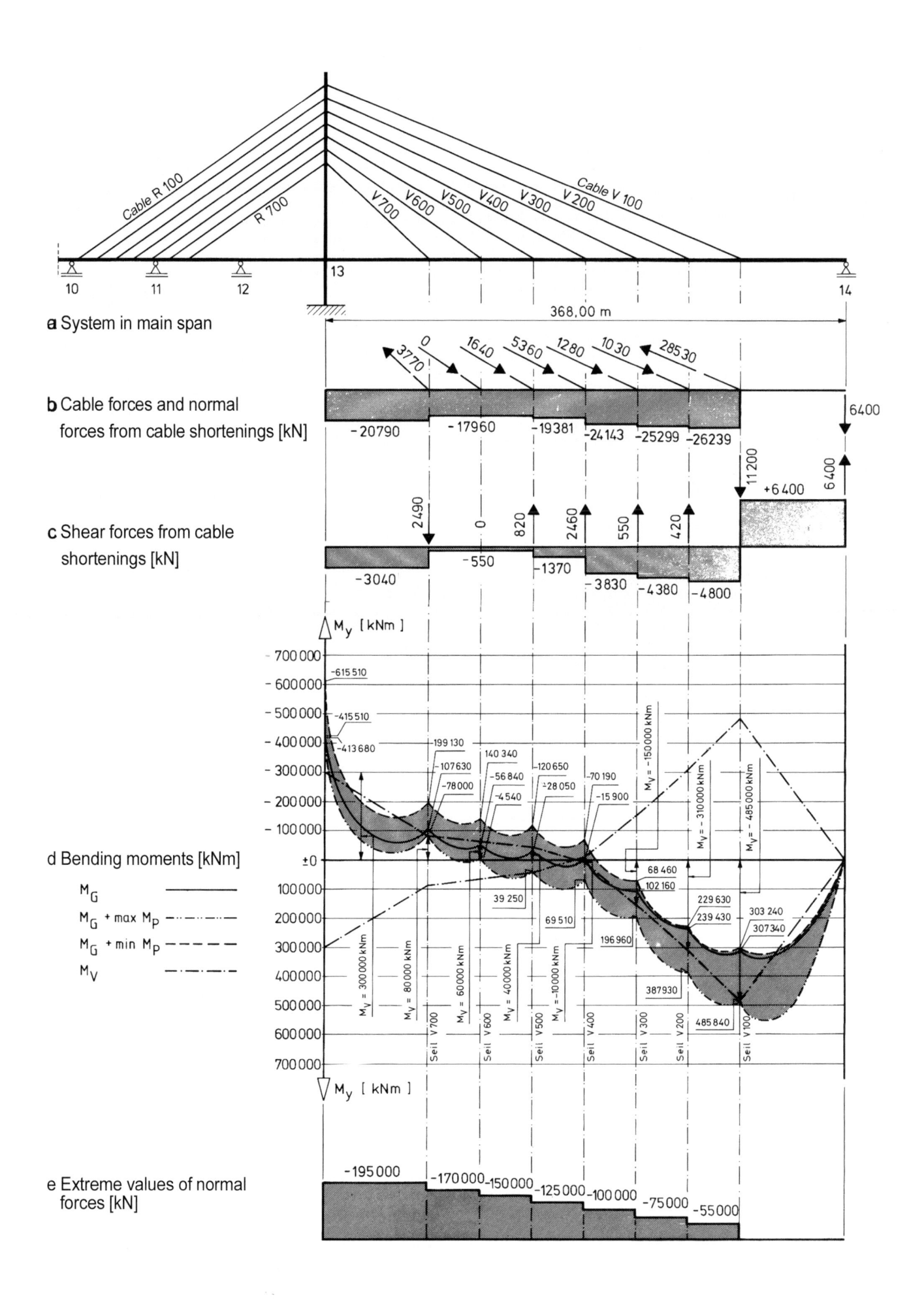

Figure 4.34 Superposition of moments for the Rhine River Bridge at Flehe, Germany

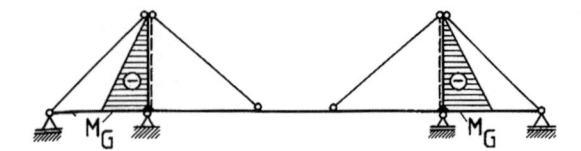

a) Tower moments from dead load

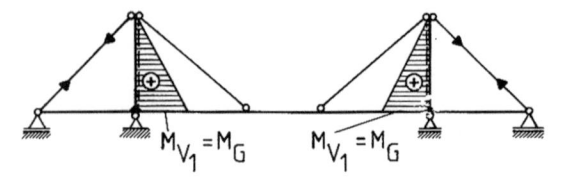

b) Tower moments from cable shortenings

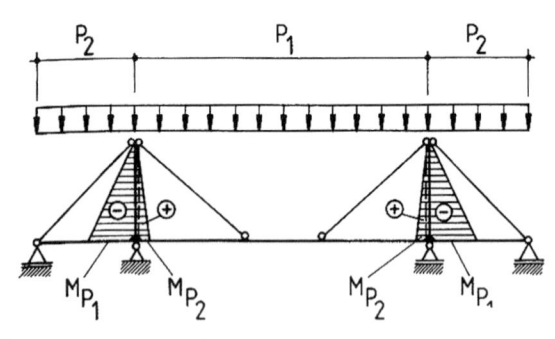

c) Tower moments from live load

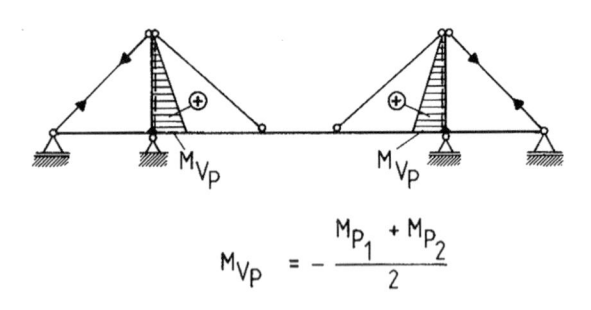

$$M_{V_P} = -\frac{M_{P_1} + M_{P_2}}{2}$$

d) Tower moments from additional cable schortenings

Figure 4.35 Determination of tower moments

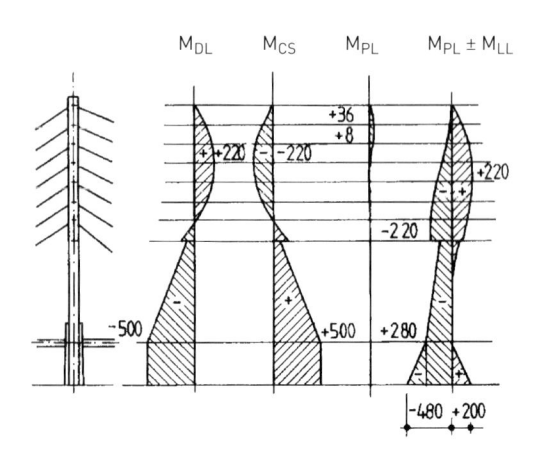

Figure 4.36 Tower moments of the Rhine River Bridge at Flehe

4.2.1.4 Towers

Cable shortenings can also be applied to the towers. They are anchored to the nearly rigid concrete beam of the approach bridge by the backstay cables. The tower moments for permanent loads are chosen for optimum sizing by suitable cable shortenings. The basic approach is demonstrated for the simplified tower as shown in Fig. 4.35.

- The dead loads on the elastic system create a fixation moment M_G, Fig. 4.35a, which can be counteracted by cable shortenings V_G, Fig. 4.35b. In this way, the tower does not carry any moments from permanent loads.
- The live loads P_1 in the main span create fixation moments M_{P1}, and the live loads in the side span P_2 create fixation moments M_{P2}, Fig. 4.35c. In general $M_{P1} \gg M_{P2}$.
- It is now possible to introduce additional cable shortenings V_P, Fig. 4.35d, in the direction of the smaller live load moment M_{P2} with the amount of $1/2\,(M_{P1} + M_{P2})$.

The choice of moments outlined above leads to the minimum amount of material for steel towers. The moments of concrete towers, however, are strongly reduced due to creep. For them, cable shortenings V_G in accordance with Fig. 4.35b are efficient, which can reduce the permanent load moments to nearly zero. The different live load moments, however, have to be accepted.

Fig. 4.36 shows this method for the concrete tower of the Düsseldorf Flehe Bridge. After the selection of the forestay cable forces V 100 to V 700, the backstay cable forces R 100 to R 700 are selected in such a way that the tower remains nearly free of moments under permanent loads, and the different live load moments remain.

The nearly free selection of moments under permanent loads is an excellent tool for the design engineer to achieve an economical structure. Obviously a detailed investigation is required for every bridge, taking into account the specific statical system and cross-sections. A mathematical formulation of this optimization problem appears not to be useful in view of the complex interdependent structure of cable-stayed bridges. It is recommended to investigate the influence of various combinations of cable shortenings on the overall sizing and to arrive at the final solution.

4.2.2 Live loads

The permanent load moments are reduced to a minimum by appropriate cable shortenings. This is not possible for transient loads, for example the live loads. Live load moments are thus the governing moments. Their corresponding normal forces are small in comparison with those from permanent loads.

An overview of the moments and of the normal forces due to spanwise uniformly distributed loads for a typical cable-stayed bridge are given in Fig. 4.37 in accordance with Homberg [4.8].

4.2 Action forces of actual systems

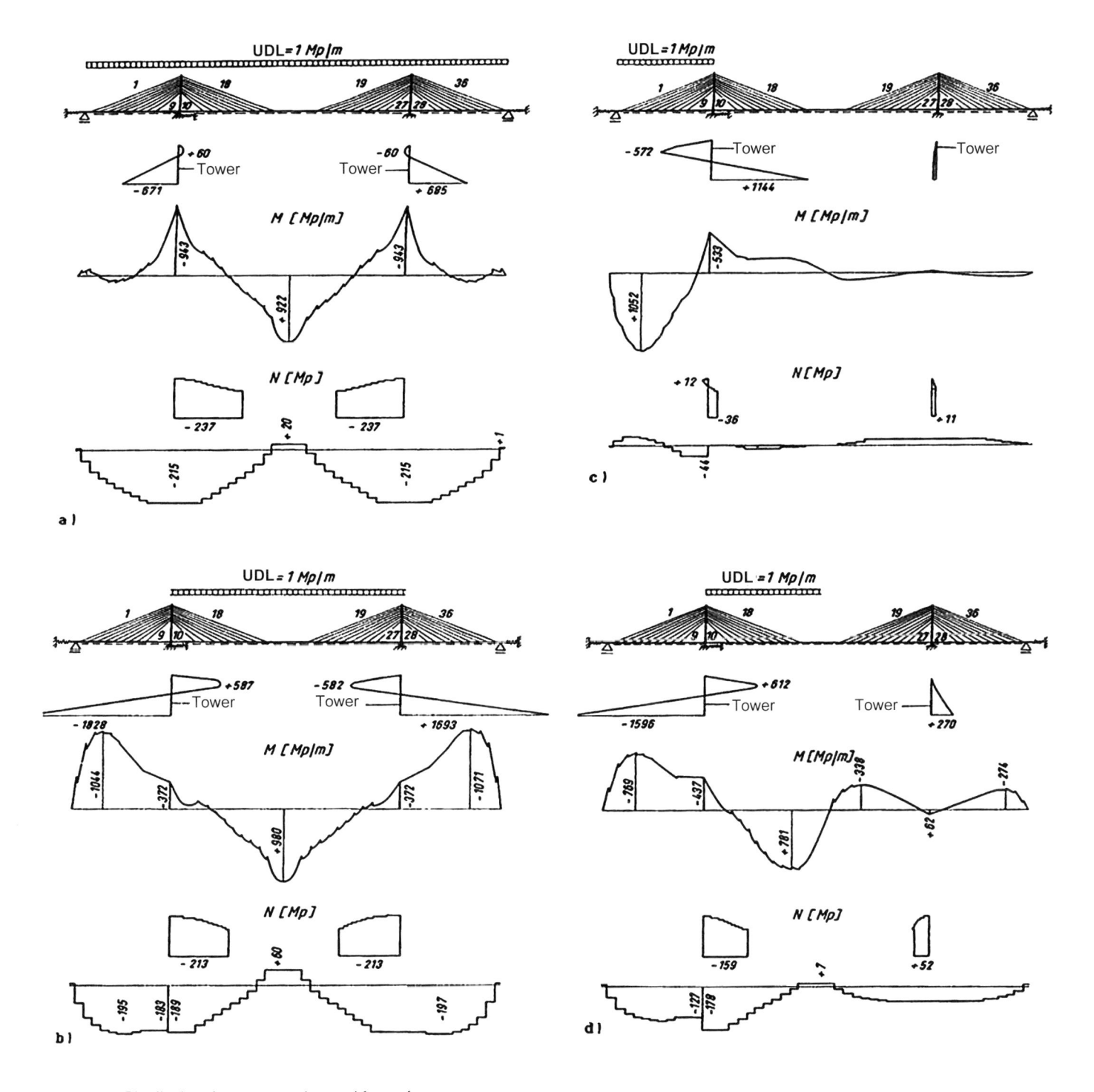

Figure 4.37 Distribution of moments and normal forces for
various types of uniformly distributed loads UDL (1 Mp = 10 kN)
a) full load
b) main span loaded
c) side span loaded
d) half main span loaded

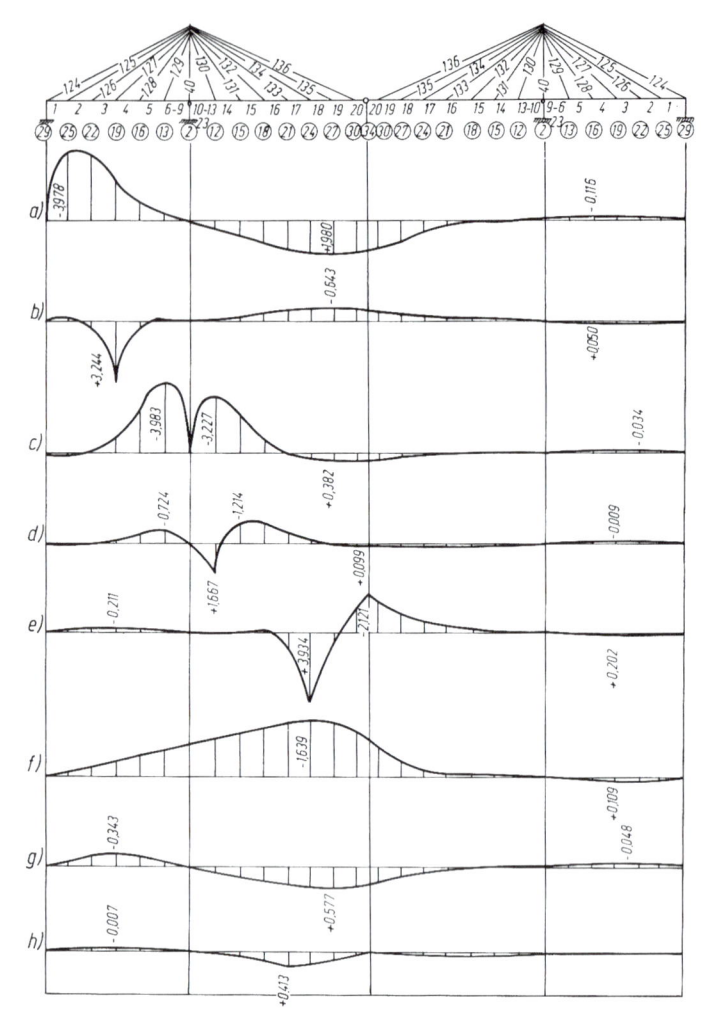

Figure 4.38 Influence lines
a) a moment at member 1, node 29
b) a moment at member 4, node 19
c) a moment at member 9, node 2 left
d) a moment at member 14, node 12
e) a moment at member 18, node 24
f) normal force in tower pier, member 23, node 2
g) normal force in backstay cable 124
h) normal force in stay cable 133

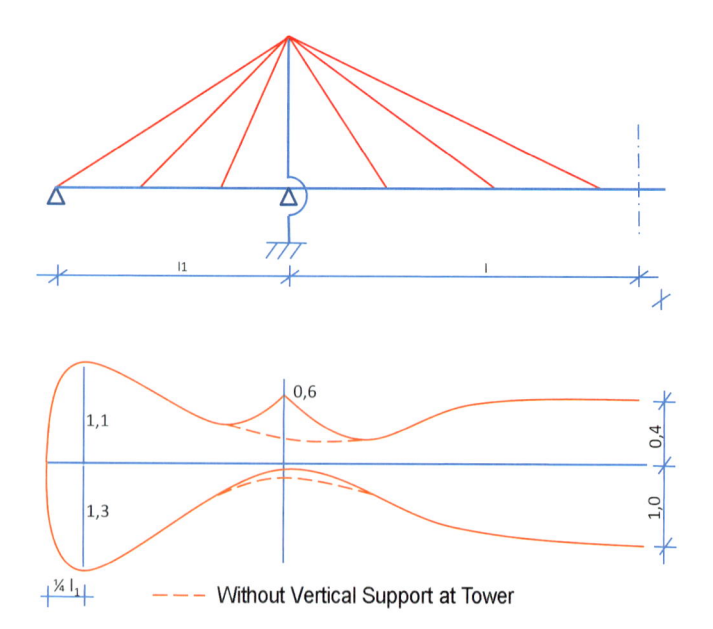

Figure 4.39 Typical run of live load envelopes

Live loads generally comprise a uniformly distributed load UDL and a concentrated load CL. They create the biggest moments in a given cross-section if arranged in accordance with influence lines. Typical influence lines are given in Fig. 4.38 [4.7].

A comprehensive overview on influence lines for different systems of cable-stayed bridges is provided by Homberg in [4.9].

As a first approach for preliminary designs Fig. 4.39 gives the live load envelopes for a typical cable-stayed bridge.

The largest live load moments occur in the side spans near the anchor piers because there the otherwise elastically supported beam is rigidly supported.

If the beam is supported at the tower by a bearing, a negative moment peak is created there, which disappears if the beam is only supported by cables. For a concrete beam the corresponding stresses, however, can easily be overcome by the high normal forces in this region. For a steel beam only a little additional material for compression is required. The costs for the bearings, on the other hand, are smaller than the additional costs for the cable steel required to carry the beam weight by a detour via the tower head into the foundations. Homberg used his 'window' for this economical purpose, Fig. 2.79.

4.2.3 Kern point moments

If a load creates moments M and normal forces N at the same time, and if both forces have different influence lines, it is not immediately clear what the governing load position for the extreme stresses in the top and bottom flanges are.

The combination of max/min M with the corresponding N from associated loadings often provides too small stresses. The combination of max/min M with max/min N from non-associated loadings may be far on the safe side and thus uneconomical. The governing loadings can, therefore, not be determined from the M and N influence lines alone, but from stress influence lines. These are calculated with the help of so-called 'kern point moments' M_K. In Table 4.1 the general definition of kern point moments from Roik 'Lectures on steel structures' is used [3.3].

Since the moments from live load and the normal forces from permanent loads are governing for the beams of cable-stayed bridges, the stress influence lines are often very close to the influence lines of the live load moments. This is especially true for concrete beams, but steel beams with smaller dead weight are more sensitive in this regard.

The situation at the footings of towers, however, is different. As outlined in Fig. 4.40 the influence lines for moments and normal forces differ there strongly, and the normal forces from the live load may be considerable because spanwise loading is governing.

The application of kern point moments is shown for the footing of the Neckarcenter Bridge tower [4.7]. For the steel tower footing the kern point moment influence lines were determined and the governing loadings applied, which were then used as load packages in the final computer calculations, Fig. 4.41.

4.2 Action forces of actual systems

The stresses σ due to bending moments M and normal forces N are with W = section modulus.

$$\sigma_o = \frac{N}{F} - \frac{M}{W_o}$$

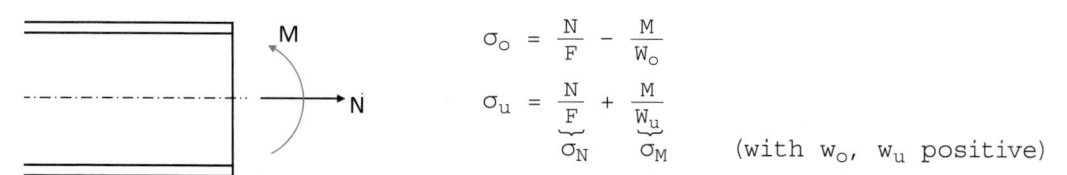

$$\sigma_u = \underbrace{\frac{N}{F}}_{\sigma_N} + \underbrace{\frac{M}{W_u}}_{\sigma_M} \quad \text{(with } w_o, w_u \text{ positive)}$$

Fig. K1 Moments and normal forces in center of gravity

For the stress calculations, especially for the proof of fatigue, those load positions are required which render the largest edge stresses as the sum of both stresses due to moments and normal forces.

$(\sigma_N + \sigma_M) \longrightarrow$ extreme values

For this purpose the so-called 'kern point moments' and their influence lines are used.

The kern points

If a cross-section is loaded by the eccentric normal force with a lever arm z, the stresses become: $\sigma_N = \frac{N}{F}$; $\sigma_M = \frac{N \cdot z}{W} = \frac{M}{W}$

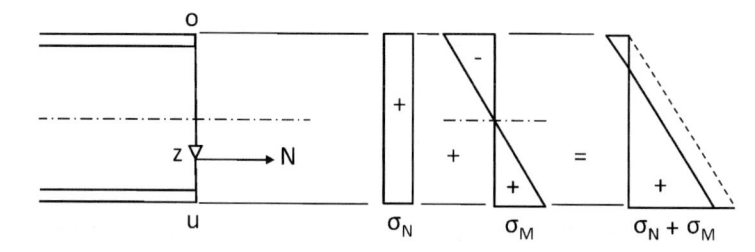

Fig. K2 Stress distribution

If z is chosen in such way that the stress $\sigma_o = 0$, then this $z = k_u$ the lower kern width or the lower kern point.

$$\sigma_0 = \frac{N}{F} - \frac{N \cdot z}{W_o} = \frac{N}{F} - \frac{N \cdot k_u}{W_o} \stackrel{!}{=} 0$$

Renders the kern widths

$$k_u = \frac{W_o}{F}; \quad k_o = \frac{W_u}{F} \quad \text{(analogous)} \quad \text{(with } w_o, w_u \text{ positive)} \tag{3.20}$$

The kern point moments

If the moments M and normal forces N in the cross-section are not referred to the center of gravity but to the kern points, then the corresponding moments become the kern point moments.

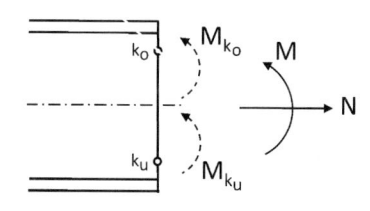

$$M_{ko} = M + N\, k_o$$
$$M_{ku} = M - N\, k_u \tag{3.21}$$

The extreme values of the edge stresses can now be calculated from the extreme values of kern point moments (with w_o, w_u positive):

$$\sigma_u = \frac{M_{ko}}{W_u} \quad \text{Beweis: } \sigma_u = \frac{M + N\, k_o}{W_u} = \frac{M}{W_u} + \frac{N}{F} \tag{3.22}$$

$$\sigma_o = \frac{M_{ku}}{W_o}$$

Fig. K3 Kern point moments

Kern point moment influence lines

These influence lines are combined as with IL = influence line.

$$\text{IL ``}M_k\text{''} = \text{IL ``}M\text{''} + k_o\, \text{IL ``}N\text{''} \quad \text{and} \quad \text{IL ``}M_{ku}\text{''} = \text{IL ``}M\text{''} + k_u\, \text{IL ``}N\text{''} \tag{3.23}$$

Table 4.1 Definition of kern point moments

For the concrete bridge beam the influence lines for the moments and the kern point moments were so close that the determination of its kern point moment influence lines was omitted. For the governing load combinations max M with corresponding N or max N with corresponding M were used.

Modern computer programs deliver max/min stresses by re-arranging the loadings internally as outlined for the kern point moments.

4.2.4 Non-linear theory (second order theory)

As was shown in Fig. 4.22, the moments of a beam under compression increase due to its deformations under live loads. Thereby two influences work against each another, Fig. 4.42 [1.19]. The beam moments increase due to the shifting of the normal force caused by beam deflections from the live load. At the same time, the beam moments decrease due to the unloading of the beam caused by the increase of cable inclination.

Due to the relieving effect of stay cable change of inclination the non-linear moment increase is reduced.

The amount of non-linear moment increase depends on the system stiffness, i.e. the stay cables stiffness in combination with the beam bending stiffness.

Traditionally the non-linear increases ΔS^{II} of action forces have been calculated linearly for the action forces ΔS^{I} of the deflected system by iteration. The iterated increases $\Delta_i S^I$ can be considered as terms of a geometrical sequence and, so the sum of the sequence can be determined from the first increased $\Delta_i S^I$, see Fig. 4.43.

For modern computer calculations the non-linear increase of moments is determined by loading the system with load packages comprising dead load and cable shortenings plus the governing live load for the beam node under consideration, Fig. 4.44.

The non-linear increase for the kern point moments of several Rhine river bridges reach only 2–5% of the linear moments, Fig. 4.45. Their corresponding beam slenderness depth:span length comes to 1:60 (Mannheim) and 1:80 (Oberkassel).

All the Rhine river bridges have comparatively stiff box girders. For the Baytown Bridge with an open composite cross-section with a slenderness of 1:180 the non-linear increase under service loads reaches up to 20%, Fig. 4.46 [1.17].

For the Helgeland Bridge with an open concrete cross-section and a slenderness of 1:354 the non-linear increase under service loads reaches nearly 30%, Fig. 4.47 [2.80].

The beam slenderness is closely related to the non-linear increase of beam moments under live load.

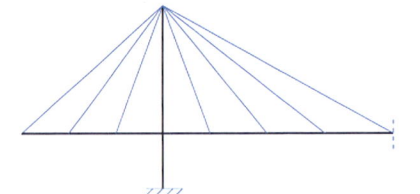

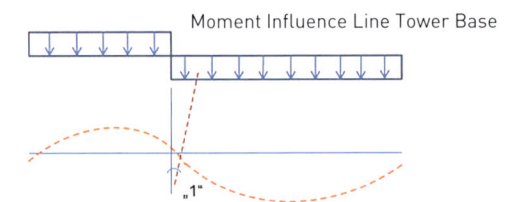

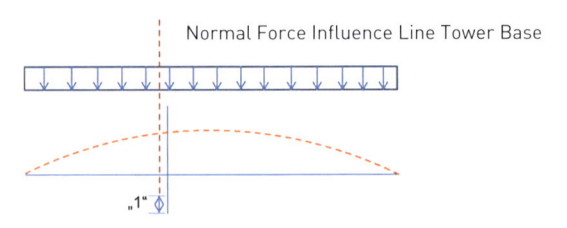

Figure 4.40 Influence lines for a tower footing

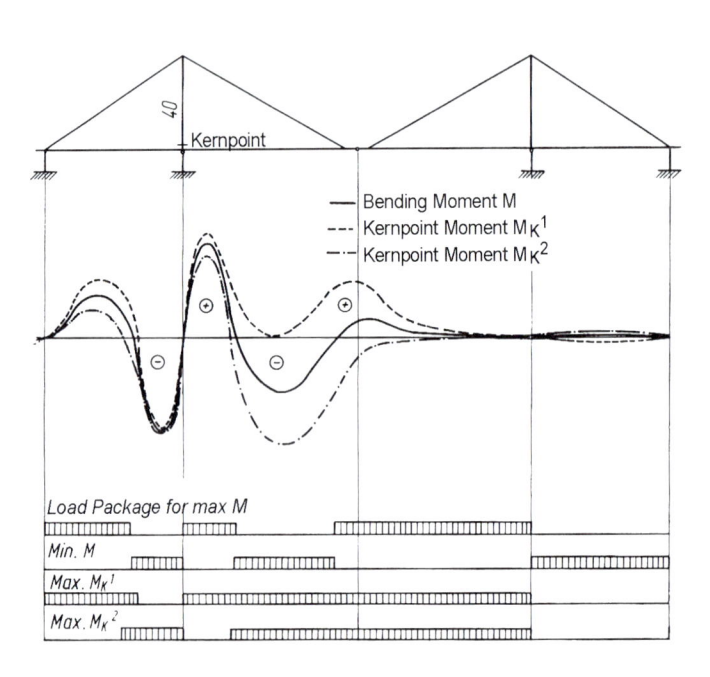

Figure 4.41 Governing loaded lengths for the bending moments and the kern point moments at the tower fixation to the beam (member 40, node 2)

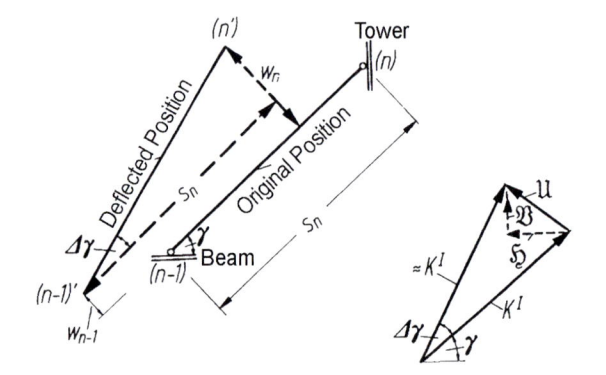

Figure 4.42 Non-linear change in stay cable inclination

4.2 Action forces of actual systems

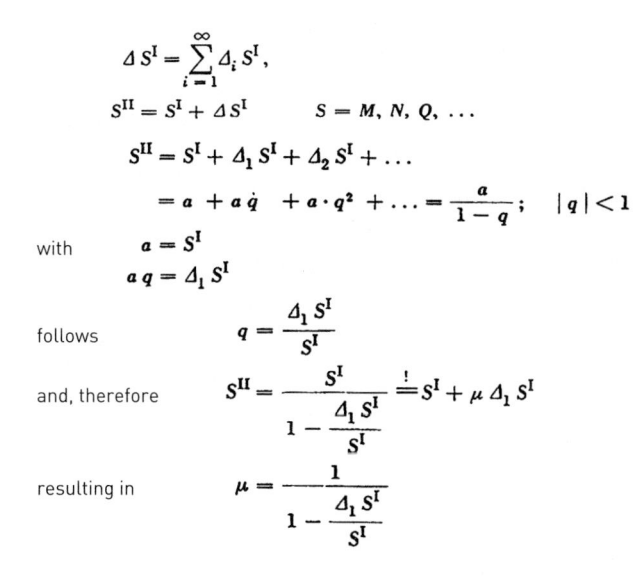

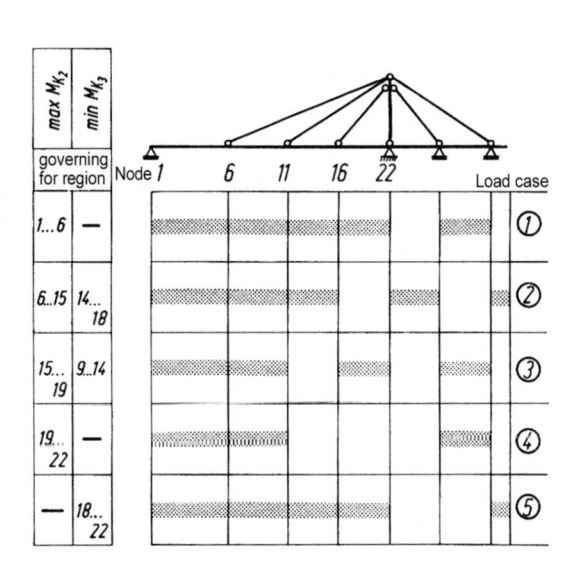

$$\Delta S^{\mathrm{I}} = \sum_{i=1}^{\infty} \Delta_i S^{\mathrm{I}},$$

$$S^{\mathrm{II}} = S^{\mathrm{I}} + \Delta S^{\mathrm{I}} \qquad S = M, N, Q, \dots$$

$$S^{\mathrm{II}} = S^{\mathrm{I}} + \Delta_1 S^{\mathrm{I}} + \Delta_2 S^{\mathrm{I}} + \dots$$

$$= a + a \dot{q} + a \cdot q^2 + \dots = \frac{a}{1-q}; \quad |q| < 1$$

with $\quad a = S^{\mathrm{I}}$
$\quad a\,q = \Delta_1 S^{\mathrm{I}}$

follows $\quad q = \dfrac{\Delta_1 S^{\mathrm{I}}}{S^{\mathrm{I}}}$

and, therefore $\quad S^{\mathrm{II}} = \dfrac{S^{\mathrm{I}}}{1 - \dfrac{\Delta_1 S^{\mathrm{I}}}{S^{\mathrm{I}}}} \stackrel{!}{=} S^{\mathrm{I}} + \mu\,\Delta_1 S^{\mathrm{I}}$

resulting in $\quad \mu = \dfrac{1}{1 - \dfrac{\Delta_1 S^{\mathrm{I}}}{S^{\mathrm{I}}}}$

Figure 4.43 Iterative determination of non-linear moment increase from a geometrical sequence

Figure 4.44 Governing load packages for non-linear live load moments

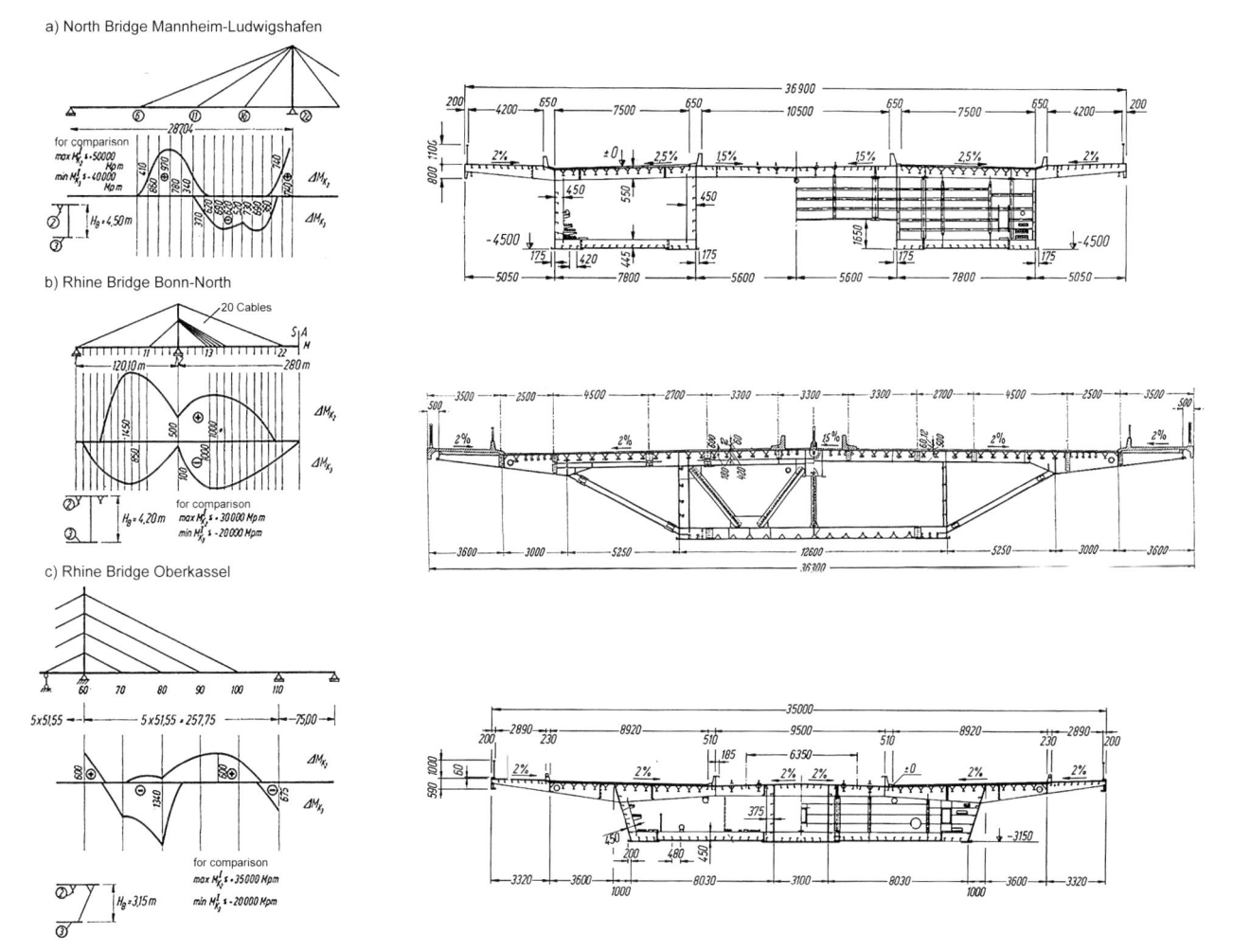

a) North Bridge Mannheim-Ludwigshafen

b) Rhine Bridge Bonn-North

c) Rhine Bridge Oberkassel

Figure 4.45 Non-linear increase of kern point moments for Rhine river bridges, slenderness 1 : 60 to 1 : 80

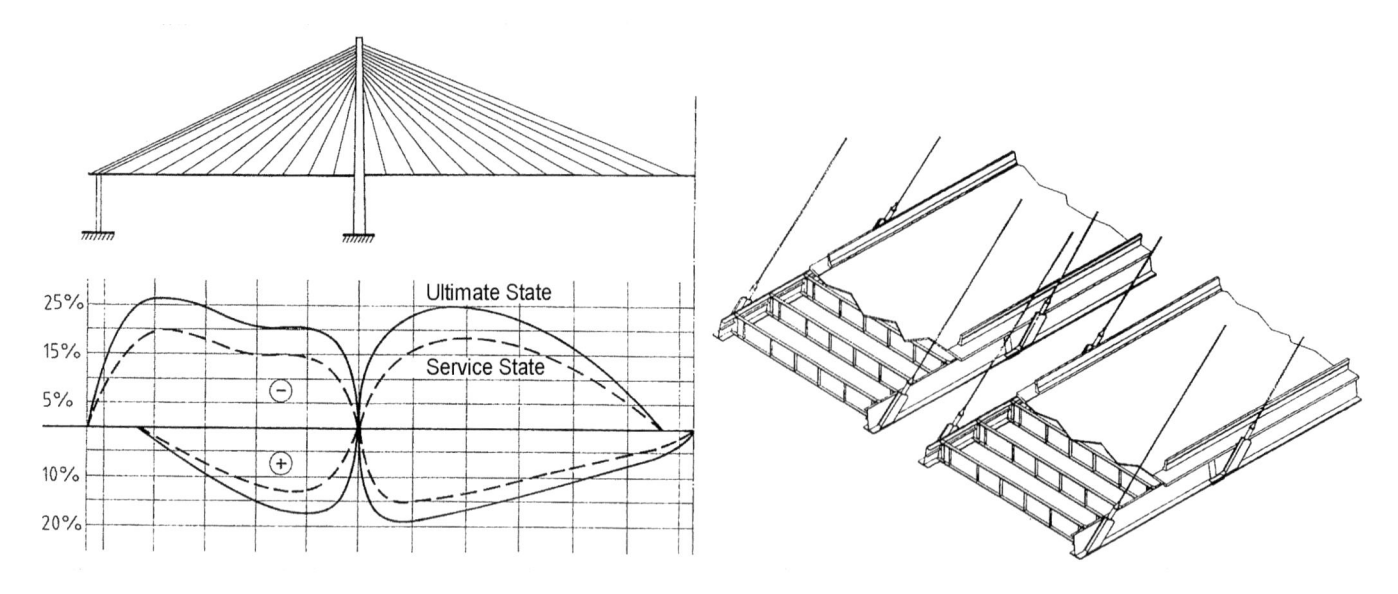

Figure 4.46 Non-linear moment increase for the Baytown Bridge, slenderness 1 : 180

4.2.5 Superposition

Homberg published the results of his preliminary design for the Brotonne Bridge with a steel beam in [4.8]. (The actual Brotonne Bridge has a concrete beam.) Fig. 4.48 left shows the run of moments for permanent loads. The large positive moments at the tower due to cable shortening act against the large negative moments from the dead load. These are caused by the beam bearing at the tower which is heavily loaded because of the 'Homberg window', Fig. 2.79. The cable distance to the towers comes to 30 m whereas the regular cables have lengths of 12 m. In this way the beam moments at the tower bearings increase strongly.

Fig. 4.48 right shows the superposition of the permanent load moments with the live load moment envelope due to linear and non-linear theory. The non-linear increase remains small for a beam slenderness 1:81.

The steel towers received a fixation moment +1140 Mpm under permanent loads. The total moments including live loads reach 6700 Mpm to the side span and 7110 Mpm to the main span, Fig. 4.49. Their amounts are close to one another and thus permit symmetrical sizing of the towers.

4.2.6 Temperature

For the determination of the bridge forces from change of temperature, a linear and non-linear increase has to be taken into account. Their amount depends on the climatic conditions of the site and the material: steel adjusts faster to temperature changes than concrete.

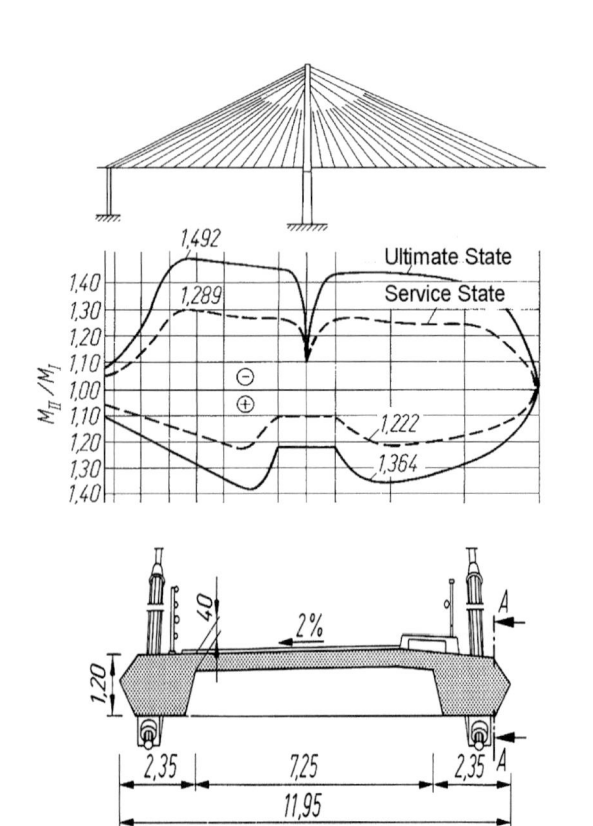

Figure 4.47 Non-linear moment increase for the Helgeland Bridge, slenderness 1 : 354

4.2 Action forces of actual systems

Permanent load moments

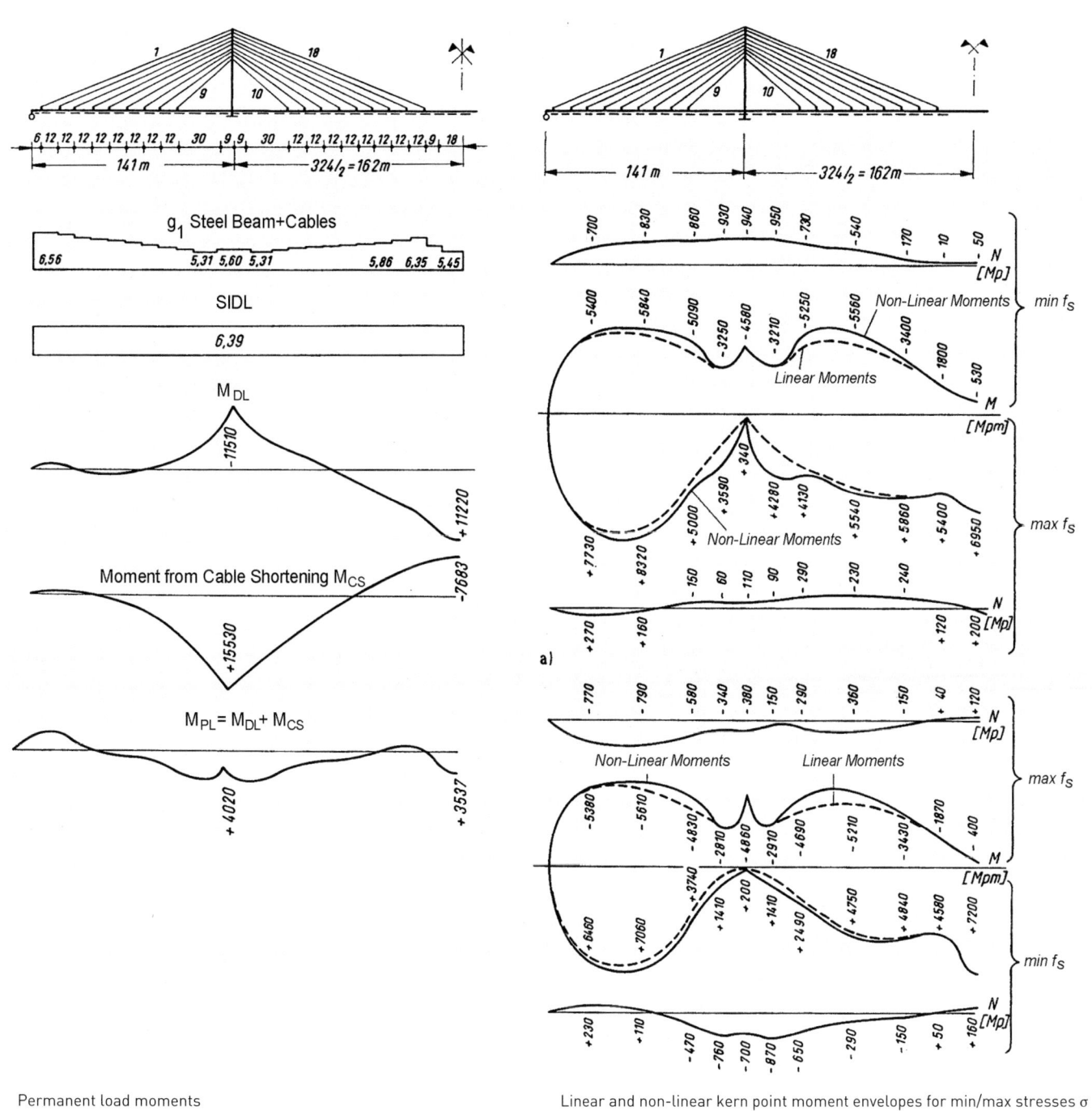

Linear and non-linear kern point moment envelopes for min/max stresses σ

Figure 4.48 Moments and normal forces of the steel beam for the Brotonne Bridge, slenderness 1 : 81 [1 Mp = 10 kN]

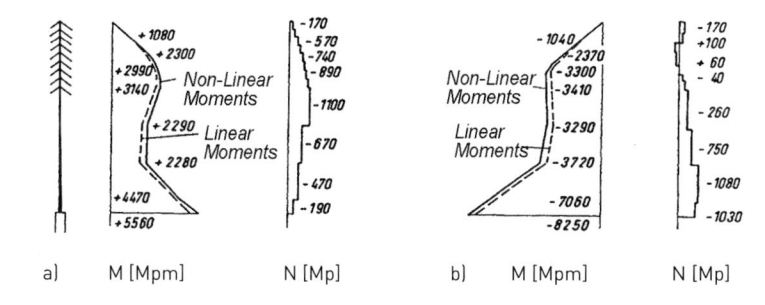

Figure 4.49 Linear and non-linear tower moment envelopes

For the hybrid beam (steel in main span, concrete in side span) and the steel tower of the Mannheim-Ludwigshafen Bridge the assumed changes in temperature in comparison with an average overall temperature of 10 °C are given in Fig. 4.50 [1.19]. The resulting moment envelope is shown in Fig. 4.51.

For the concrete beam of the Pasco-Kennewick cable-stayed bridge in the north-east of the USA, the following temperature load cases were taken into account, Fig. 4.25 [1.15]. From these temperature load cases come the moment envelopes in Fig. 4.10 e.

4.2.7 Eigenfrequencies

The governing first bending eigenfrequency can approximately be determined for the main-span of a cable-stayed bridge as for a corresponding simply supported beam [4.10, 4.11].

$$f_b = \frac{1}{2\pi} \cdot \sqrt{\frac{K_B}{m}} = \frac{0.58}{\sqrt{\delta}}$$

with

K_B spring stiffness
δ deflection in center span from dead load

For the calculation of the deflection in the center of the main span the bridge has to be loaded with the dead load, but applied spanwise in the local direction of the first eigenmode, Fig. 4.53.

The corresponding first eigenfrequency in torsion for the main span comes approximately to:

a) for an open cross-section

$$f_t = \frac{b_s}{2\,i_y} \cdot f_b \quad \text{(neglecting the torsional stiffness of the beam)}$$

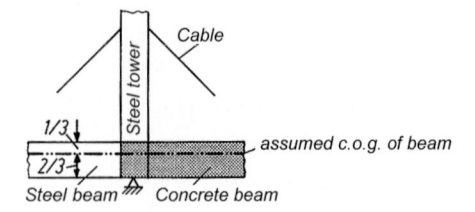

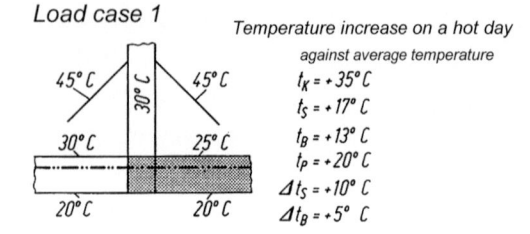

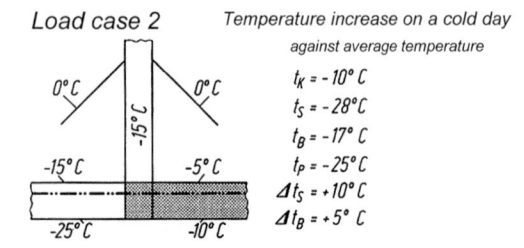

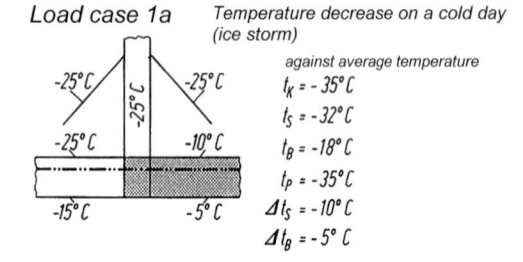

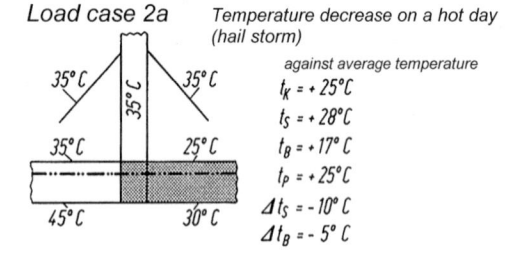

Figure 4.50 Temperature load cases for the hybrid beam and steel tower of the Mannheim-Ludwigshafen Bridge

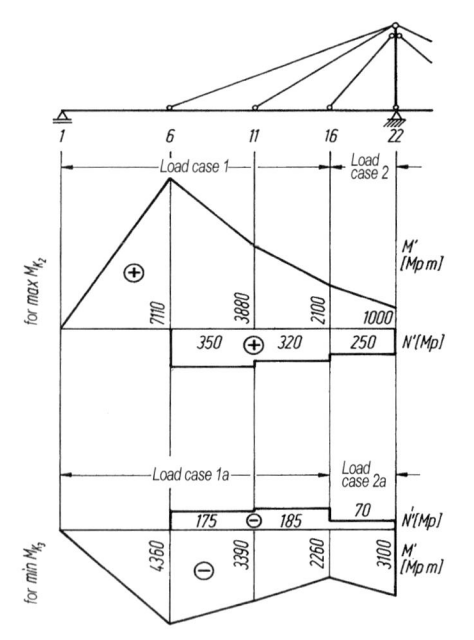

Figure 4.51 Forces from temperature loadings for steel beam

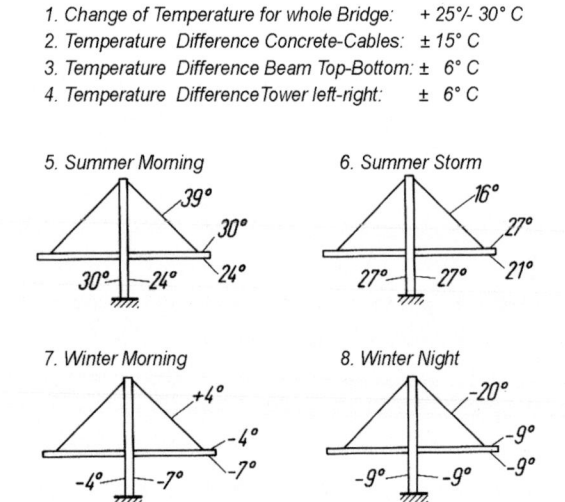

Figure 4.52 Temperature load cases for concrete beam (Pasco-Kennewick Bridge)

If an A-tower is used, the two tower legs cannot deflect independently in the longitudinal direction but remain stationary. The deflection caused by the elongation of the back stays are thus excluded. In this way, the torsional eigenfrequency of a cable-stayed bridge with an A-tower is strongly increased against that with an H-tower with independently moving legs – see the calculation example in Section 4.5.7.4.

b) for a closed box girder

$$\vartheta = m \cdot \frac{I_P}{A} = m \cdot i_P^2$$

$$f_T = \frac{1}{2\pi} \cdot \sqrt{\frac{K_T}{\vartheta}} = \frac{1}{2L} \cdot \sqrt{\frac{G \cdot I_T}{\vartheta}} \quad \text{(neglecting the torsional stiffness of transversely spread cables)}$$

with
- b_s distance of cable planes
- i_P polar mass moment of inertia
- i_y horizontal radius of mass moment of inertia
- K_T torsional spring stiffness
- I_P polar moment of inertia
- I_T torsional moment of inertia
- I_Y horizontal moment of inertia
- G shear modulus
- ϑ torsional mass per unit length of beam.

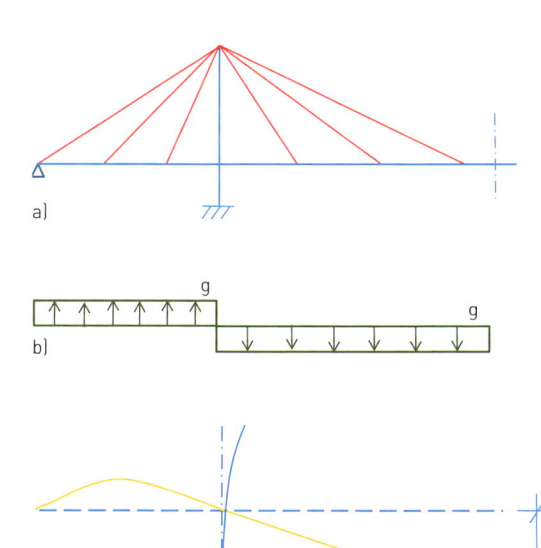

Figure 4.53 Load arrangement for first eigenmode
a) system, b) dead load arrangement, c) deformation line

4.3 Bridge dynamics
4.3.1 General

At the beginning of building major bridges, particularly suspension bridges, many of these early bridges were destroyed by wind effects. The list in Table 4.2 shows well-known bridge collapses due to wind since 1818 [2.71].

The nature of the various dynamic phenomena was not understood for a long time. John Röbling was one of the first engineers, who felt that stiffening the girder of suspension bridges through stay cables should increase the aerodynamic stability of bridges, see Section 2.1.2.

The destruction of Tacoma Narrows Bridge by wind effects in 1940 attracted immense attention. The bridge had exhibited significant wind-induced oscillations since its opening – even at low wind speeds – and was thus being monitored by the University of Washington. The dramatic collapse was then filmed by a local cameraman. These recordings are arguably the most famous movie about any bridge to date. They are still being shown to students. Fig. 4.54 shows a few snapshots from that movie.

This collapse was the origin of deep scientific research into the phenomenon of wind-induced bridge oscillations, led by Frederick Burt Farquharson of the University of Washington [4.12]. Previously Theodore Theodorsen had solved the flutter problem of airfoils mathematically [4.13]. These results were adapted to bridges by Friedrich Bleich [2.137] who thus also arrived at an analytical solution [4.14]. The attempts by practical bridge engineers to solve the problem with numerical formulae not substantiated by a sound mathematical theory, for example by David Steinman [4.15], were bound to fail.

The problem of wind excitation had not been studied much in Germany because the topography required no large span bridges. When Fritz Leonhardt in 1939 designed the then longest suspension bridge in Europe across the Rhine at Rodenkirchen with a main span of 478 m, he had a partial model of the cross-section built at 1:50 scale for testing in the wind tunnel. However, this was only for determining the aerodynamic force coefficients [4.16]. In the mid 1960's the phenomenon of wind excitation of bridges was studied more closely by Kurt Kloppel, [2.137] at the University of Göttingen. The findings were published in a fundamental paper by Klöppel and Thiele in 1967 [4.17], in which they solved the aerodynamic flutter stability problem by means of graphical representations based in the ideal flat plate solution. Reduction factors were also given which allowed a correction of the flutter prediction for real cross-sections. These had been determined from extensive wind tunnel testing. Such experimental testing is still being carried out today.

Later, further analysis methods were developed for gust-induced oscillations, particularly by Alan Davenport, who developed the first modern spectral method [4.18] and by Robert Scanlan [4.19] who introduced aerodynamic derivatives as new parameters. Both these methods are generally used today.

Table 4.2 Bridges destroyed by wind effects

Bridge	Location	Designer	Span length [m]	Year of failure
Dryburgh Abbey	Scotland	John and William Smith	80	1818
Union	England	Sir Samuel Brown	140	1821
Nassau	Germany	Lossen and Wolf	75	1834
Brighton Chain Pier	England	Sir Samuel Brown	80	1836
Montrose	Scotland	Sir Samuel Brown	130	1838
Menai Straits	Wales	Thomas Telford	180	1839
Roche-Bernard	France	Le Blanc	195	1852
Wheeling	USA	Charles Ellet	310	1854
Niagara-Lewiston	USA	Edward Serrell	320	1864
Tay	Scotland	Sir Thomas Bouch	85 · 38	1879
Niagara-Clifton	USA	Samuel Keefer	380	1889
Tacoma Narrows	USA	Leon Moisseiff	850	1940

Today the behavior of bridges in all kinds of dynamic excitation scenarios is being more and more frequently studied by sophisticated computer simulations [4.20].

The characteristic aerodynamic properties of bridge cross-sections can today also be determined numerically through fluid flow simulations. However, for non-standard sections it remains the standard to test partial models in the wind tunnel to determine characteristic parameters as input for further computer analyses. Further, very large wind tunnels have been built in order to test whole bridge models with model dimensions of up to 40 m length, for example the suspension bridge across the Great Belt in Denmark (L = 1624 m) and the Akashi Straits Bridge (L = 1991 m). However, the usefulness of such tests is questionable, because the resulting scaled model dimensions are very small, which means that model laws cannot be satisfied because properties such as the viscosity of air is not being modified at the same time. The state of the art is rather to analytically analyze the whole bridge on the basis of experimental results obtained from partial model tests in the wind tunnel.

As part of his bridge design work the author has collaborated with many excellent aerodynamics experts. For his bridges in the USA these were:
– the Pasco-Kennewick Bridge with Bob Wardlaw [4.21],
– the Sunshine Skyway Bridge with Alan Davenport [4.22],
– the Baytown Bridge across the Houston Ship Channel with Robert Scanlan [4.23].

For all other cable-stayed bridges the author worked with Dr.-Ing. Imre Kovac (Fig. 3.87). Dr. Kovacs is one of the leading engineers in the field of structural dynamics with extensive experience. He has investigated aerodynamical the following cable-stayed bridges:
1989 Helgeland Bridge, Norway
1991 Storebelt Bridge, Denmark
1993 Höga Kusten Bridge, Sweden
1997 Sunningesund Bridge, Sweden
2000 Millau Bridge, France
2001 Stonecutters Bridge, Hong Kong
2002 Second Panama Bridge
2002 Geo Geum Bridge, South Korea
2003 Thuan Phuoc Bridge, Danang, Vietnam
2003 M0-Ring Budapest North–East, Hungary
2005 Bridge Kehl-Straßburg, Germany–France
2007 Brandanger Bridge, Norway
2008 M43 Bridge, Szeged, Hungary.

The author is grateful to Dr. Kovacs for drafting this chapter about aerodynamic behavior.

Figure 4.54 Tacoma Narrows Bridge, USA

4.3.2 Overview of wind effects

The overview of Fig. 4.55 is due to Professor Christian Petersen TU Munich [4.24], and their interrelations are commonly accepted. The load model of natural wind – previously only a velocity and stagnation pressure profile, possibly a stepped profile, with only static loads, has become successively more complex over the years. The reason for this development is the increased scale and slenderness of structures, which has led to lower natural frequencies and thus an increased susceptibility to wind effects. This has required a refined and more involved analysis of the structural behavior, explicitly including dynamic effects.

As for the definitions of Fig. 4.55: the *turbulence* is a random variation of the wind speed. The *turbulence-induced vibration* can only be described by statistical means. The vortex-induced oscillations are caused by a regular creation (shedding) of vortex structures from the body. The shedding itself depends on an interaction between structure and fluid flow. The *motion induction* is the feedback between the flow and the motion of the body. Its result, the motion-induced vibration can best be compared with static second order effects. Its most important property is the dynamic instability.

These three excitation mechanisms can also occur together and can amplify each other. It is a non-strict property of wind effects that they intensify with a reduction of the structural natural frequencies.

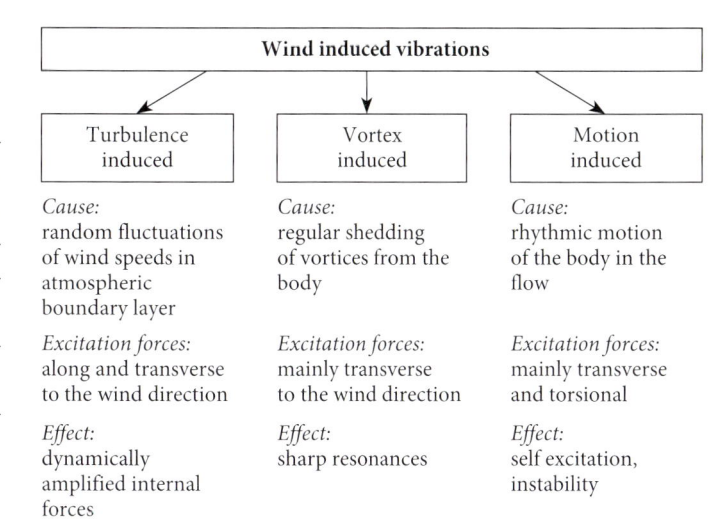

```
                  Wind induced vibrations
         ┌─────────────────┼─────────────────┐
   Turbulence           Vortex            Motion
    induced            induced           induced
```

Cause:
random fluctuations of wind speeds in atmospheric boundary layer

Excitation forces:
along and transverse to the wind direction

Effect:
dynamically amplified internal forces

Cause:
regular shedding of vortices from the body

Excitation forces:
mainly transverse to the wind direction

Effect:
sharp resonances

Cause:
rhythmic motion of the body in the flow

Excitation forces:
mainly transverse and torsional

Effect:
self excitation, instability

Figure 4.55 Overview of most important wind effects

4.3.3 Wind profile, turbulence and turbulence-induced oscillations

4.3.3.1 Wind parameters

Fig. 4.56 gives the most important terms with a focus on cable-stayed bridges.

The safety criterion of a structure against wind excitation is usually expressed today by regulations which require that within the life time of the structure – typically 100 years for bridges – the structure must retain sufficient resistance. This is usually analyzed through section forces scaled by safety factors.

The design check for the ultimate limit state is in practice reduced to a small time frame – in Europe: 10 minutes. Within this, the highest wind speed (the 100 year return period wind) is acting and there must be sufficient time for the structure to be subjected to all resonances. The wind speeds in these 10 minutes can thus be interpreted as the sum of

- a time-constant, spatially variable *mean wind speed*, and
- a superimposed, time and space variable *turbulence* part.

The V_m profile and the turbulence originate from the roughness of the earth or water surface. The V_m profile converges to an asymptotic value, the gradient wind, which is dependent on the global weather conditions. In the so-called boundary layer, i.e. near the surface, the V_m profile and the turbulence profile are influenced by the local roughness.

The $V_m(z)$ profile follows a logarithmic [4.19] or an exponential law [4.25]. Both are approximations derived from statistical analyses. They are typically defined by the $V_m(z = 10)$ value and the elevation profile which depends on the roughness. The turbulence is being described by its energy spectrum or its power spectral density, an idealized form of the Fourier integral of the time function. The amount of turbulence depends on the elevation and is described either by the turbulence intensity or sometimes by a 5 second gust speed profile.

Wind data available for Germany is extensive and has been determined from numerous long-term wind measurements. This is not necessarily the case for other countries.

In DIN 1055-4 [4.25] the wind properties are given in terms of standard profiles and mixed profiles: for the standard profiles the roughness is constant upstream and is described by roughness categories. Mixed profiles also take into account various artificial and unfavorable changes in upstream roughness conditions.

Further, DIN provides an idealized energy spectrum of the along-wind turbulence component: for cable-stayed bridges this is often not sufficient, they also require lateral and vertical components, which can be found in the literature.

The most frequently used definition of turbulence spectra with large cable-stayed bridges are the ESDU documents [4.26, 4.27]. They are contained in most tender documents of such projects. The spectrum, shown in Fig. 4.57, is arranged such that an identical shape is used and then scaled according to surface roughness and elevation through the integral length scales L. It is evident that the

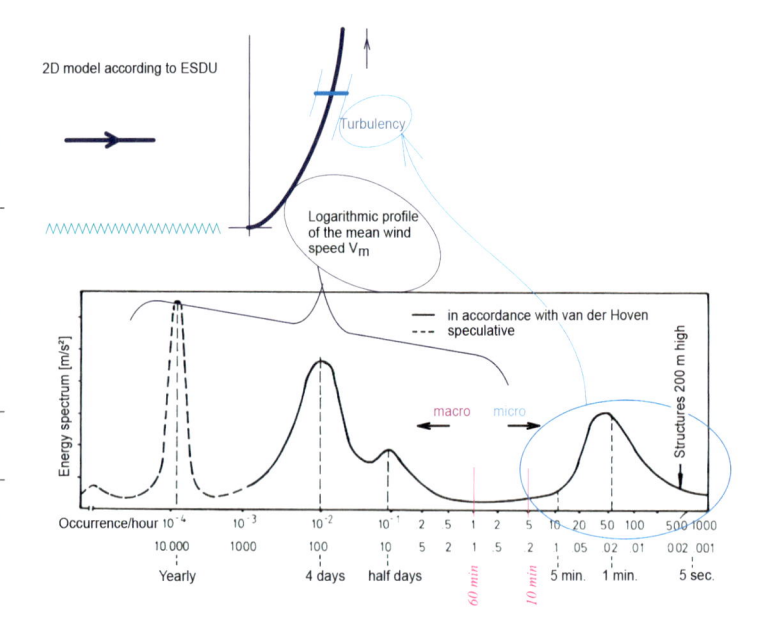

Figure 4.56 2D wind model and comprehensive spectrum

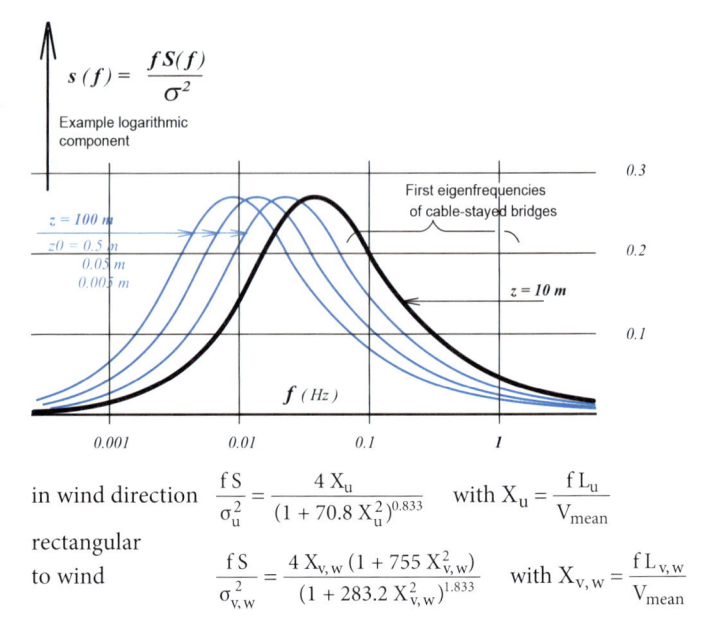

in wind direction
rectangular to wind

$$\frac{f S}{\sigma_u^2} = \frac{4 X_u}{(1 + 70.8 X_u^2)^{0.833}} \quad \text{with } X_u = \frac{f L_u}{V_{mean}}$$

$$\frac{f S}{\sigma_{v,w}^2} = \frac{4 X_{v,w}(1 + 755 X_{v,w}^2)}{(1 + 283.2 X_{v,w}^2)^{1.833}} \quad \text{with } X_{v,w} = \frac{f L_{v,w}}{V_{mean}}$$

Figure 4.57 Karman turbulencies

4.3 Bridge dynamics

excitation reduces for higher frequencies. The complete wind definition required for determining the wind effects on a cable-stayed bridge:

Comprehensive wind definition

$V_m(z)$ mean wind speed
$\sigma_u, \sigma_v, \sigma_w(z)$ scatter in wind speed process
         $\sigma = I \cdot V_m$ with I = intensity
                 u, v, w = index for the longitudinal,
                 lateral and vertical component
$L_u, L_v, L_w(z)$ integral length scale in [m], scale parameter for the turbulent spectrum

There is a definitive relationship between the gust velocity (5 second wind speed) and the gust duration in terms of L_u/V_m: Fig. 4.58 shows a sample relation for a parameter of 4.

The important height-dependent wind parameters for the footbridge Kehl-Straßburg, see Fig. 2.225, are shown in Fig. 4.59.

The longitudinal turbulence can be defined in two ways: through the turbulence component σ_{lon} or through the 5 second gust V_{5s}. They can be transformed into each other. The other parametric relations are also important and are given further down.

In Fig. 4.59 different approaches for V_m at z = 10 m are shown, the Kehl-Straßburg project is here a useful example: as is common in many projects an attempt was made to define the local profile on the basis of a limited number of measurements. Here, 20 measurements were used, which is insufficient and the values were low, apparently to achieve a cost efficient design.

The definition of wind properties on the basis of limited measurements is unsound. Initially V_m was fixed at a low value of 19 m/s = 75 % of the DIN specification. This had been done before the occurrence of the devastating storm, Lothar. Its inclusion led to a revision of the statistical analysis, now yielding much (possibly too much) increased V_m values. It is reiterated that the DIN values are founded on sound statistics and there is little room for justifying lower values. Similar approaches to wind definition were attempted elsewhere: Metsovo Bridge, Phuan-Thuoc Bridge. Engineers involved in such wind data processing and design parameter definitions must be aware of the significant responsibility associated with it.

It is of paramount importance that wind definitions be coherent. Relationships between individual parameters are not arbitrary but lie in narrow bands. It is for example not uncommon for the relationship between V_{gust} and V_m not to be suited to the surface roughness and hence irrational (e.g. the individual definition for Mecca: $V_{gust}/V_m = 21.5/15.3 = 1.40$ was much too low given an actual roughness of 0.05–0.10). The engineer needs to be aware of the local conditions and be able to recognize such discrepancies and correct where necessary.

The coherence – often not defined separately though important – signifies how in a turbulent wind field the instantaneous V value at one point is correlated to that at another point. Coherence is a stochastic property that can be determined from the cross-correlations

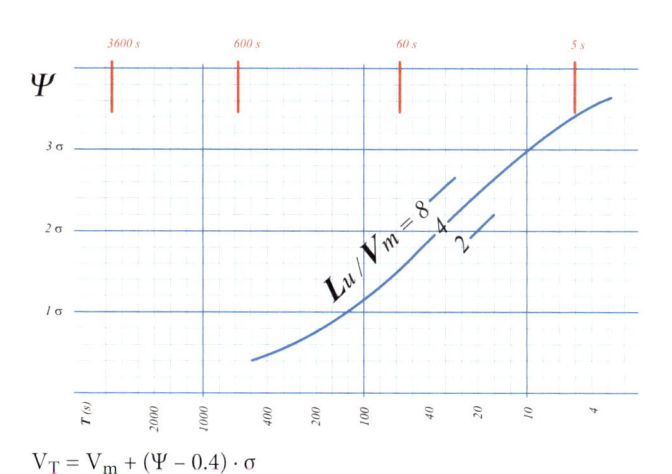

$V_T = V_m + (\Psi - 0.4) \cdot \sigma$

Figure 4.58 Gusts with different durations, based on the Karman spectrum

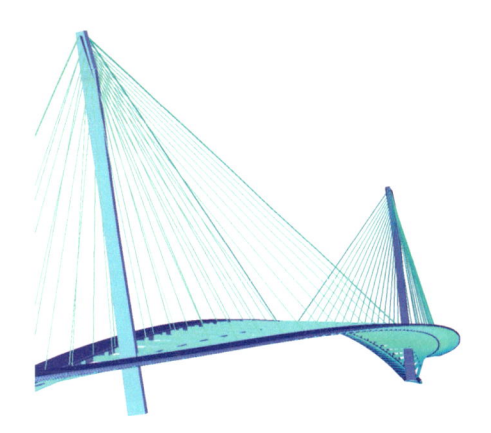

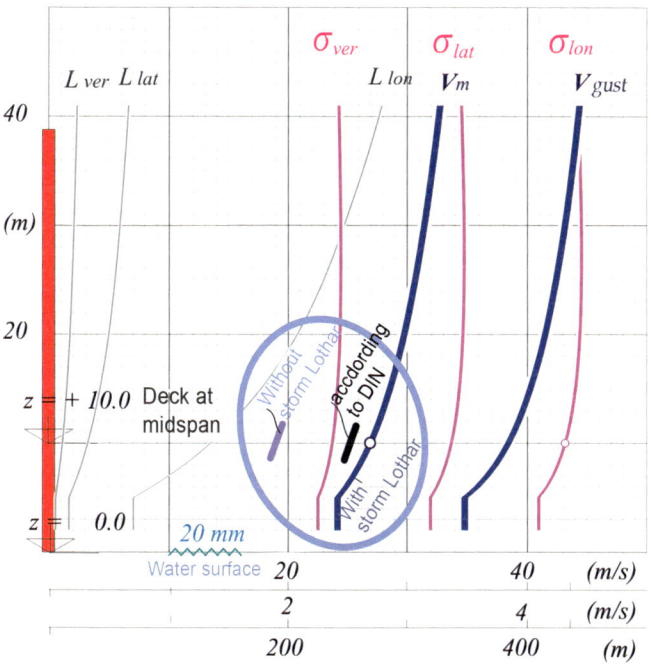

Figure 4.59 Pedestrian bridge Kehl-Straßburg, assumptions for wind speed and turbulence

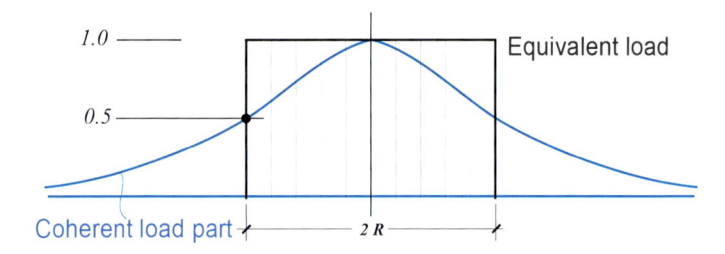

Figure 4.60 Correlated loads

between two measurements [4.29]. The coherence properties depend on surface roughness and elevation. They are relatively well known.

The coherence also defines the spatial dimension of a gust. It is frequency dependent and reduces with increasing frequency. Karman defines asymptotic values in the atmosphere, Fig. 4.60, as:

For a given frequency f the agreement between turbulent components (size and phase) reduces as one moves away from a point. This can be represented by an asymptotic correlated width = 2 R of a gust, with the half distance R being:

$R \approx 0.11 \cdot V_m / f$ half distance for longitudinal turbulent component
$R \approx 0.17 \cdot V_m / f$ for the lateral and vertical components.

Towards the ground, earth or water, R is reduced significantly.

DIN 1055-4 [4.25] specifies the wind speed of an aeroelastic instability limit. This will be covered later, but we note here that the required onset velocity of unstable oscillations is in itself a function of the size of the building and its natural frequencies. This is why an ad hoc definition is questionable, which is particularly true for bridges.

The data set of wind properties presented above can be used to generate natural random wind time-histories and to thus enable refined investigations. This will be discussed later.

4.3.3.2 Natural modes of vibration of structures

In cable-stayed bridges, as for any other structure, the natural modes of vibration are of major importance for the susceptibility to dynamic excitation. Low structural frequencies point to a high susceptibility. With growing structural dimensions or larger spans the frequencies decrease, so large bridges are particularly critical.

The frequency ratio between different modes can also be important in cable-stayed bridges. A classical example is the ratio f_T / f_B (torsional-to-vertical bending frequency), which is the critical parameter for the susceptibility against flutter instability.

Further of importance, though less well known, is that the ratio between the frequency of a massive structural member and that of an attached cable can be responsible for internal resonance through an indirect excitation (1:1) or parametric resonance (2:1), see Section 3.8.3.

A typical bridge system with two H-shaped pylons is shown in Fig. 4.61: it features a reinforced concrete cross-section and has a lateral natural frequency of around 0.3 Hz, being relatively low and thus sensitive to excitation through lateral wind (the sensitivity is also strongly dependent on the cross-sectional width). The deck and pylon natural frequencies fdeck and fpyl are very close in this case, which however has no significance for wind excitations.

Typically the vertical and torsional modes are dominated by the axial stiffness of the cables; the stiffness of the deck is of minor importance. The vertical modes are particularly susceptible to gust-induced response. Sometimes they form a series of closely spaced frequencies: here and with similar spans the series starts at around 0.2 – 0.25 Hz. In terms of flutter, the torsional frequency and its ratio to the vertical frequency are of interest.

With such classical bridge configurations the ratio is close to the quotient of half the distance between the cable planes and the radius of gyration. Typically this lies at 1.5 – 1.7, Section 4.2.7.

The structural damping is almost random, but is important for the design. The calculation assumptions are regulated relatively strictly through EC and DIN [4.25, 4.28] and [4.29, 4.30]. They are material dependent and generally lower than the values used in the past. The logarithmic decrement (= amplitude reduction in one cycle of oscillation) is given as follows:

Reinforced concrete $\delta = 0.04 + \ldots$
Steel $\delta = 0.02 + \ldots$
Deck vertical sensibly $\delta = 0.02$ (because supported by cables).

Some modes are damped relatively strongly by aeroelastic damping; this is an important wind effect critical for design as we shall see later.

For the design of major bridges the stability criterion is decisive. Here the use of an A-shaped pylon provides an improvement, see Fig. 4.62:
– the torsional stiffness – and hence the frequency ratio – is increased by the cable end points being fixed for longitudinal movements and
– additionally there are different bending and torsional modes along the bridge axis, which inhibit the dangerous coupling of modes.

For functional reasons a central cable plane is sometimes used instead of two planes at the girder edges. This is then dynamically unfavorable because the missing torsional stiffness through the cable system needs to be provided in other ways, typically by the (box) girder.

Cable modes are typically susceptible to excitations, Section 3.8, and can be problematic because of
– the low inherent damping and
– the large range of natural frequencies depending on the loading condition. Long, and thus low-frequency, cables are most critical.

Fig. 4.63 shows the example of cable vibrations of the M0 Bridge Budapest: particularly for span ratios of almost 0.5 – 1 – 0.5 as here, the backstay cable can be destressed significantly under some load conditions: the natural frequency becomes very low (here 0.37 Hz) and parametric resonance is possible [4.31].

Bridge cables are often exposed to internal resonance caused by an alignment of main structure and cable frequencies. From experience the most critical mechanism is through a deck or pylon excited by vortex shedding resonance.

The higher frequency modes of cables form an arithmetic series as $f_n = f_1 \cdot n$. The higher modes are less problematic than commonly considered as they are rarely observed and if so then they are typically caused by secondary effects only.

4.3 Bridge dynamics

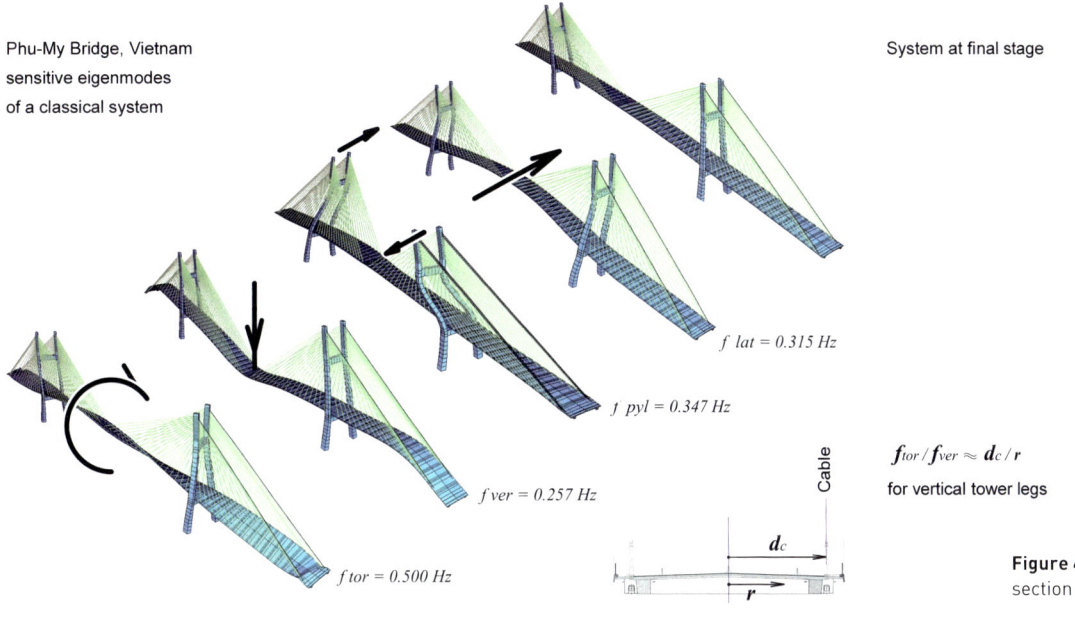

Phu-My Bridge, Vietnam
sensitive eigenmodes
of a classical system

System at final stage

$f\,lat = 0.315\ Hz$

$f\,pyl = 0.347\ Hz$

$f\,ver = 0.257\ Hz$

$f\,tor = 0.500\ Hz$

$f_{tor}/f_{ver} \approx d_c/r$
for vertical tower legs

Figure 4.61 H-Towers and open deck section

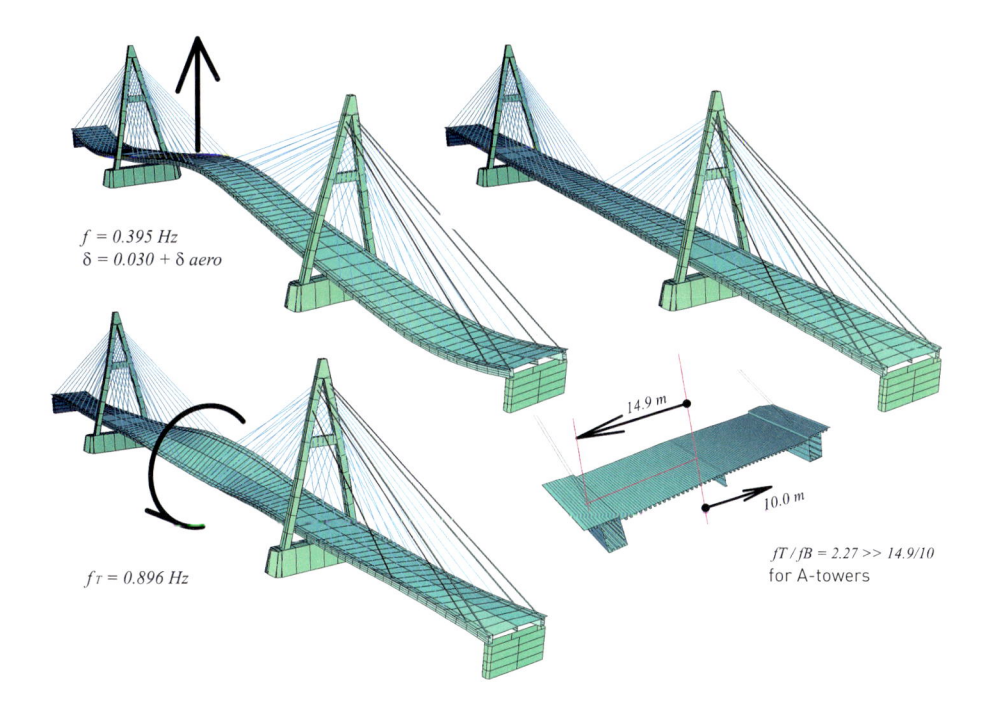

$f = 0.395\ Hz$
$\delta = 0.030 + \delta\,aero$

$f_T = 0.896\ Hz$

14.9 m
10.0 m

$f_T/f_B = 2.27 \gg 14.9/10$
for A-towers

Figure 4.62 Bridge with A-shaped pylon

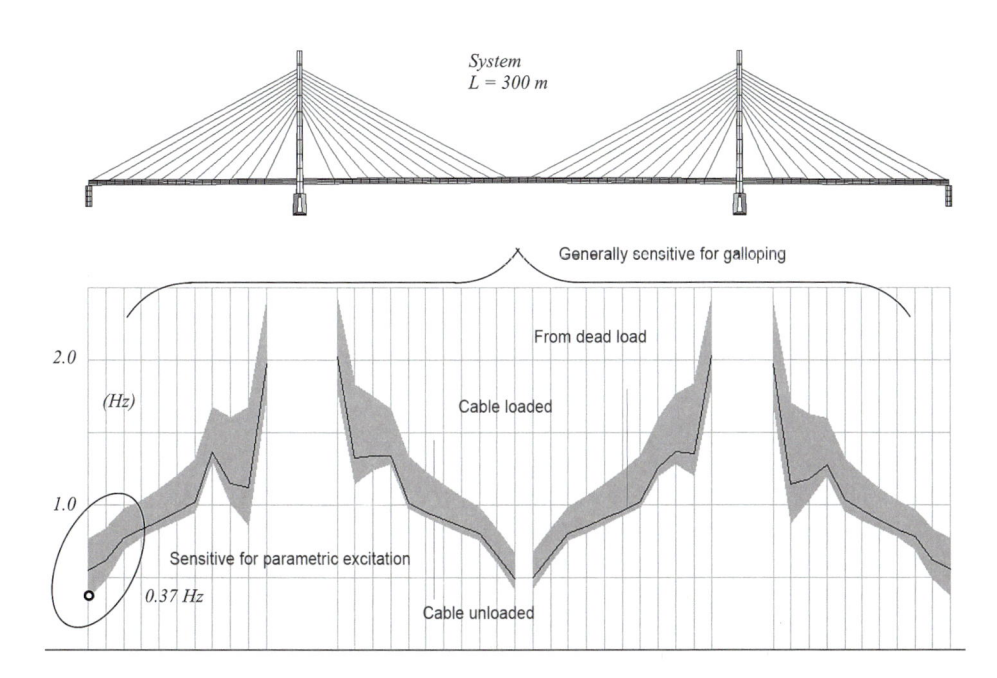

System
$L = 300\ m$

Generally sensitive for galloping

From dead load

Cable loaded

2.0

(Hz)

1.0

Sensitive for parametric excitation

0.37 Hz

Cable unloaded

Figure 4.63 Cable frequencies

4.3.3.3 Section forces under turbulent excitation

The turbulence excitation is of critical importance for the section force calculation. The forces exerted on the structure by the wind result from the product of stagnation (static) pressure and the so-called aerodynamic coefficient. Both components of the product vary with time, which leads to dynamic excitation.

Fig. 4.64, from [4.35], depicts the aerodynamic coefficients as a function of the angle of attack of the flow: they are measured in the wind tunnel on models of typical scales M = 1/~75–80. The measurement can be done in laminar flow – typically when stability properties are to be investigated – or in low turbulence flow. In the latter case the tunnel turbulence should be only about 50% of the real turbulence, because in the tunnel only the higher frequency turbulence components can be properly created. The asymptote of the spectrum towards the higher frequencies needs to be realistic, which is important to be verified.

Typically the aerodynamic coefficients in low turbulence flow are lower than those without turbulence and exhibit a flatter gradient.

According to [4.32, 4.33] the excitation arises (a) in the pressure direction through the *velocity fluctuations* resulting in a varying stagnation pressure. This leads to lateral excitation forces on deck and pylon, and also in the lift direction such that excitation contributions occur.

For the deck section there is further a very critical contribution (b) in the fluctuation of the *angle of attack* as a result of the vertical turbulence: with the angle of incidence, the lift coefficients also fluctuate. This fluctuation is more intensive the steeper the lift coefficient gradient in the diagram. Deck sections with low depth are particularly sensitive. Turbulence also produces a torsional excitation, but this is of little practical significance.

In cable-stayed bridges the contribution of the second of these phenomena to the deck vibrations is the more significant, i.e. the vertical excitation dominates.

This will also be a warning, since the effect is often underrated: neither EC nor DIN properly treat the vertical turbulent component, although its contribution can exceed that of live loads.

The vertical excitation can be so critical for the deck that hidden or typically neglected effects need to be utilized as far as possible. A significant such effect is *aeroelastic damping*, which will be treated together with the motion-induced effects further down. Of further importance is the effect of reduced coherence towards the ground.

Often – particularly during early design phases – approximations using *equivalent static forces* are used. The aim is a sensible concept for the estimation of section forces [4.35]. Let us analyze the 10-minute excitation process which needs to be considered for design:

a) The base contribution of the loading is the static component resulting from the mean wind profile, and the stagnation pressure as

$q_0(z) = 0.5 \rho \cdot V_m^2(z)$

b) This is followed by the time- and space-dependent fluctuations of wind speeds (pressures) q_q = quasi-static part or background part. This is best interpreted as a dynamic wind load acting on a structure without natural modes of vibration (equivalent to a structure without mass). In q_q the low-frequency parts of the spectrum dominate.

c) Finally the so-called q_r = resonant parts: the velocity fluctuations near the resonance points of the structure which lead to dynamically amplified inertial forces.

This results, considering the stochastic character of components b) and c), in:

$q = q_0 + \sqrt{(q_q^2 + q_r^2)}$

In this the summation of contributions b) and c) is performed through square root-sum-of-square (SRSS) because of the stochastic nature of the independent processes q_q and q_r (because their maxima do not occur simultaneously).

There is no hidden secret for finding and fixing components b) and c); this requires experience in order to find reasonable estimates. In any case, governing load combinations of type b) and c) need to be selected from an infinite number of possible load combinations and then combined using SRSS.

For the most important locations of the structure, however, the critical combinations can usually also be selected by visual inspection. Fig. 4.65 shows the procedure for the example of a transverse bending moment at midspan.

In case b) a sensible choice for the loaded length for fixing the quasi-static load magnitude is 0.75 of the span length, based on experience from statics (the shorter the selected length the higher is the built-in safety because the equivalent load is placed on the entire span). For the Budapest bridge of Fig. 4.65 the corresponding gust duration is then:

z = 25 m

$V_m = 28$ m/s → $t_{bö} \approx 0.5$ L/$0.22 V_m \approx 24$ s

From algorithms for gust duration and stagnation pressure, see above, the equivalent stagnation pressure for the gust is found as

$q_q \approx 1.36 \, q_0$

For c), at the resonant frequency of the transverse mode of $f_{lat} = 0.96$ Hz the ordinate of the Karman spectrum at z = 25 m is found from the equations of Fig. 4.57 as

$s(f_{lat}) = 0.032$

From this, assuming a structural damping of $\delta \approx 0.04$, the equivalent dynamic stagnation pressure and the resonant loaded length are computed as

$q_r = \pi \cdot 3.0 \cdot \sqrt{s(f)/2\delta} = 2.9 \quad q_0$ along a length of

$2 R = 0.22 \, V_m / f_{lat} = 6.7$ m

4.3 Bridge dynamics

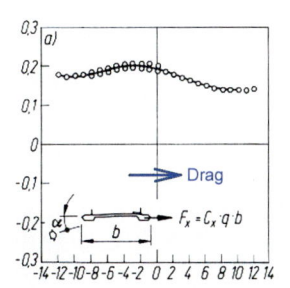

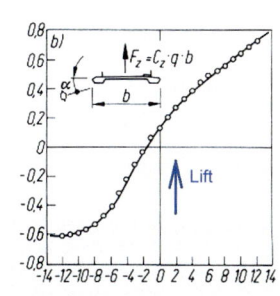

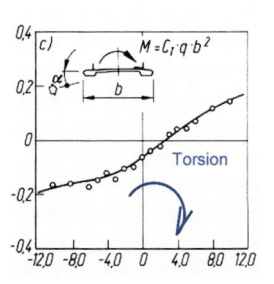

Figure 4.64 Aerodynamic coefficients

as shown in Fig. 4.65. The other load components are calculated by similar means.

These calculations can be further refined, for example through refining the coherence values, accounting for aeroelastic damping, considering further modes etc. But an exact calculation is not presented here as this would be the scope of more in-depth literature.

For a given section all considered equivalent loads need to be computed on the basis of one and the same wind direction. There can only be contributions from *one* horizontal and one vertical quasi-static component, but there are contributions from *all* resonant components that belong to the same wind direction, because they lie in different frequency ranges – unless the frequencies are very close. All selected q_q and q_r components are to be combined by SRSS and then added to q_0.

For the resonant components it should be noted that, strictly speaking, the load is not counted into the summation but into the modal contributions excited by it. The difference however is typically not big, one exception being the vertical bending of the bridge in the final condition. Here the modal contributions (and there are more than one!) should be included in the calculation because the superposition of the equivalent loads would greatly overestimate the demand on the structure.

For determining the section forces suitable for detailed design, several wind tunnel test strategies have been devised, which are still being used today, Section 4.3.7.

We are not convinced by these methods (there are better ones), but they will be mentioned here for the sake of completeness:
1) A method by Davenport [4.22]:
   turbulent testing on a section model, M = 1:70–80, which is supported elastically in three DOFs. Masses, stiffnesses and damping need to be correctly reflected according to similitude considerations.

   The turbulence in the tunnel is created through meshes, and the wind properties need to be measured and calibrated against the theoretical targets. The wind speeds in the tunnel are successively increased and the model displacements (max, min, rms) are measured. The displacements can then be converted to modal amplitudes and corresponding section forces.

   The procedure for processing the results is not very robust, because some parameters required (duration of testing, coherence relative to model length – it needs to be considered that the effect of the slow turbulence components is missing) are not well defined. Measurements also show scatter, Fig. 4.66, which leads the user – particularly the client – to use the most unfavorable values, rather than the mean or expected values as would be sensible. This then results in an over-design.

2) Measurements on a full model after [4.22], M = 1:150–200:
   very expensive and time consuming. Many natural frequencies need to be accounted for correctly and represented correctly at model scale, which is typically very difficult on a 3D model.

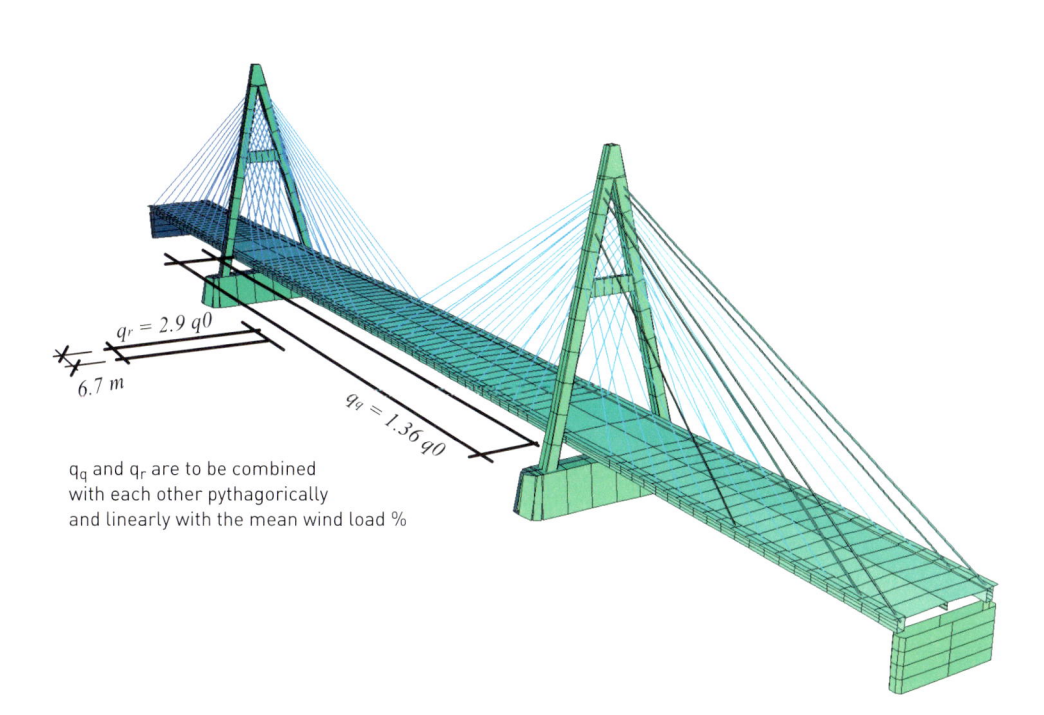

$q_r = 2.9\, q_0$
6.7 m
$q_q = 1.36\, q_0$

q_q and q_r are to be combined with each other pythagorically and linearly with the mean wind load %

Figure 4.65 Example of the computation of an equivalent static load for the transverse bending moment at midspan

The flow is being created with boundary layer turbulence through an artificial roughness in the tunnel. The turbulent properties including the spectra need to be measured and calibrated.

Successively increasing wind speeds are also used here, and the displacements are measured and interpreted as modal amplitudes. This procedure is not – or not much – more accurate than that of 1). The design checks of large cable-stayed bridges are today more frequently done through analytical and numerical calculations on the basis of 3) spectral methods or 4) time-history simulations.

3) Spectral methods

An automatic, complete spectral solution is possible, but the software known to us is not mature and sometimes incorrect. In principle a node-by-node optimized gust configuration needs to be calculated for the quasi-static and then again for the resonant part. The aerodynamic damping should then be, for safety sake, reduced somewhat.

4) Time-history simulations

The calculation using time histories is in many respects the most reliable solution to date, because for example it can account for non-linearities in force coefficients [4.32]. This is the only method capable of taking into account separation effects like those found on the East Huntington Bridge, Fig. 4.67.

A simulation using a linearized $M(\phi)$ diagram leads to the identification of a virtual torsional failure of the deck for large torsional deformation. However, when correctly accounting for the moment curve there should have been a relief in this condition. The spectral method is responsible for the ill-identified failure because it requires a linearization. A few important points on time-history methods:

The analysis is based on the natural wind field: based on a 100-year return period a 10 minute wind history is generated analytically [4.34]. Fig. 4.68 shows an instantaneous wind field snapshot. It is also possible to include an ultimate limit state analysis by using wind speeds scaled up by a factor $pf = \sqrt{\gamma}$.

An instantaneous snapshot of the resulting bridge vibrations is shown in Fig. 4.69. In the lectures the underlying results of Figs. 4.68 and 4.69 are shown as animations.

In Fig. 4.69: The dynamic model of the system is being subjected to the wind field and the response is being tracked. The interaction between the flow and the vibrations was being modeled in a somewhat simplified way during initial development of the method, in that each time step was studied as a stationary condition (Helgeland Bridge). The new simulation methods also include derivatives, such that stability effects can also be studied, as we shall see later; Fig. 4.70

The processing of the results is performed stochastically. Each constellation is typically analyzed by 11 time-histories of 10 minute duration. The expected collapse is derived in the form of complete section force diagrams.

As shown on Helgeland Bridge [4.35], any form of non-linear material behavior and section force interactions – M_{lat}, M_{ver}, M_{tor} – can be accounted for: here the axial force originating from the cable forces provided a large horizontal resistance of the deck at the pylon, such that span reinforcement and stiffness could remain limited.

4.3 Bridge dynamics

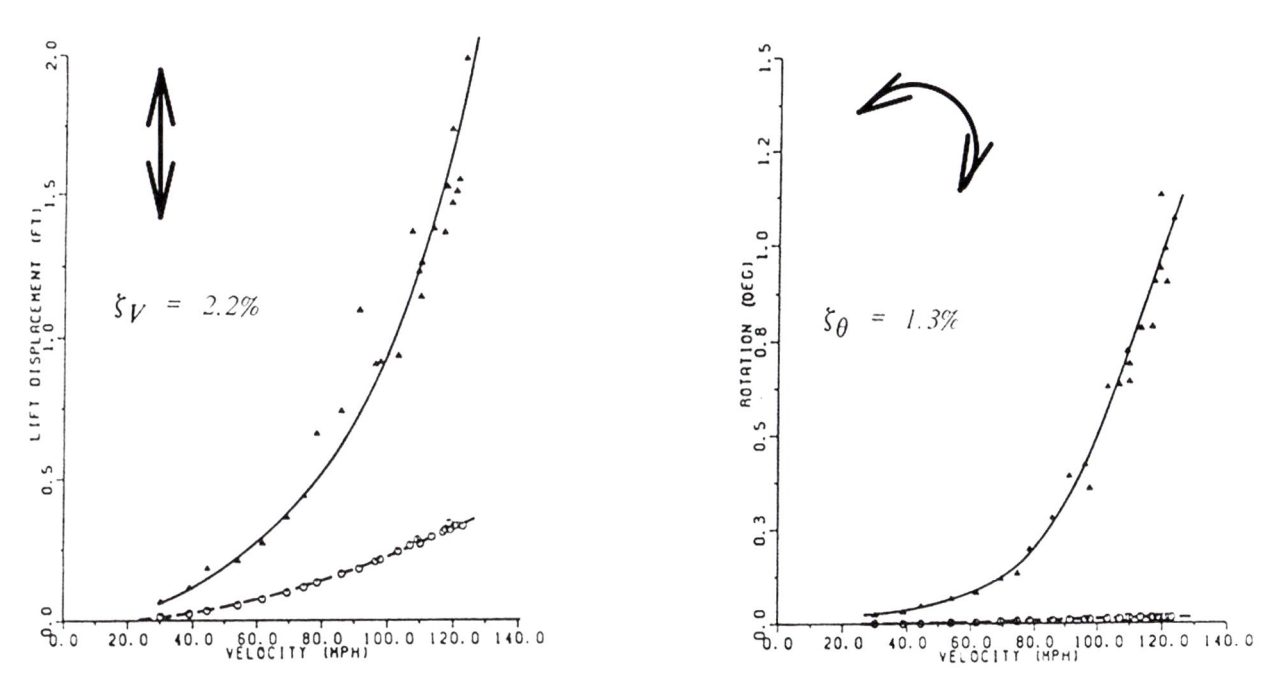

Figure 4.66 Example for amplitude measurements on section models in turbulent flow

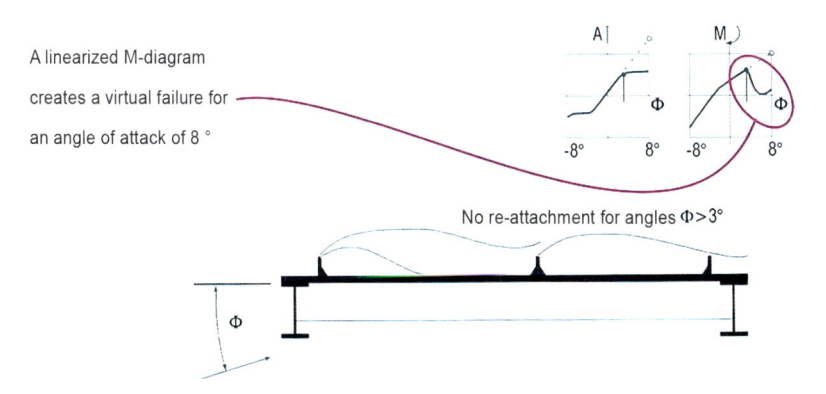

A linearized M-diagram creates a virtual failure for an angle of attack of 8 °

No re-attachment for angles Φ>3°

Figure 4.67 Cross- section with aerodynamic derivatives

Figure 4.68 Helgeland Bridge, instantaneous snapshot of a 800 x 250 m turbulent domain

Figure 4.69 Instantaneous snapshot of the dynamic response, 1989

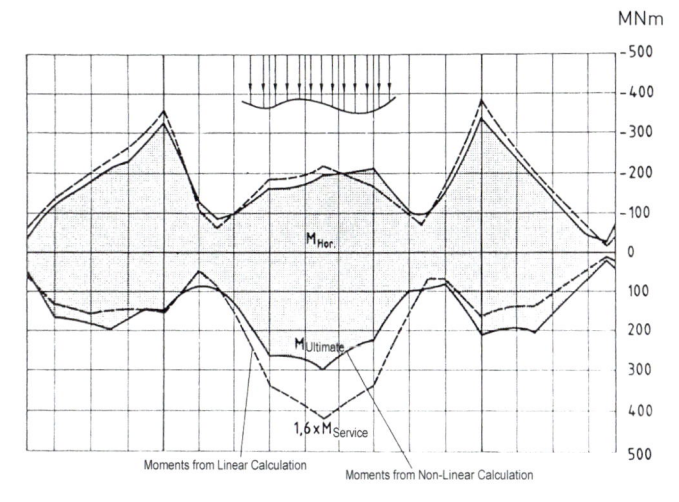

Figure 4.70 Design bending moments from the non-linear simulation

4.3.4 Vortex-induced vibrations

When a deck, pylon or cable section is being subjected to transverse flow, the separation of the boundary layer leads to the creation of eddies (vortices). These are shed alternatingly and, being then carried downstream, create a characteristic wake behind the structure. This is called the *Karman Vortex Street*, which is responsible for periodic excitation forces acting transversally to the flow direction.

Fig. 4.71 shows the geometric conditions during a vortex shedding process: a circular section creates an St = 0.2 (St = Strouhal number) geometric rhythm, which for a lift-type section becomes only St ≈ 0.07 – 0.12. The shed vortices extend only along part of the beam, i.e. the correlated shedding is only about 6 d long. This enables the reconstruction of the time-history of the excitation force as shown in the figure.

The vortex shedding behavior of a circular section depends on the Reynolds Number:

$Re = d \cdot V / \nu$ where viscosity $\nu = 1.5 \cdot 10^{-5}$ kg/(s · m).

There are three distinct Reynolds Number ranges [4.29], the sub-, super- and transcritical regions, with the boundaries lying at about $5 \cdot 10^5$ and $5 \cdot 10^6$. In contrast, sections with sharp edges such as typical bridge sections exhibit low Reynolds Number dependence.

$C_{lat,0}$ = lateral force coefficient, is not well known. DIN provides values, for example, for rectangles of up to 1.1, for thin-walled profiles up to 2.3. Such large values multiplied by the resonant factor can provide very large effective forces and highlight the fact that vortex-induced resonance is a dangerous phenomenon. This is particularly true if the resonance point lies at high wind speeds. Therefore: with vortex shedding resonance one should always check:

a) Is resonance actually possible?
b) What is the real magnitude of $C_{lat,0}$?
c) How regular is the vortex shedding?
d) Is self excitation possible?
e) Can the oscillation be damped if necessary?

There are a number of effects which may cause attenuation in the analytically computed excitation forces, see a) – e). Therefore the items in this list should be considered carefully.

a) A full resonance is not usually possible if the corresponding wind speed is too high. In DIN 1055, [4.25] therefore correctly limits the range required to be checked: at very high wind speeds, the 50-year wind ± 25 %, there is only a reduced excitation with reference to a shorter duration, which means that the shedding excitation cannot develop fully. Above these wind speeds no vortex excitation needs to be considered.

b) The code values for the lateral force coefficient are rather exaggerated, further these peak values would only be applicable for a very limited range of angles of incidence. It is thus beneficial to study the effects carefully.

Example rectangle: the DIN value of 1.1 in reality only holds for a small angular range of 0° ± 5° and 90° ± 5°. Issue 1 therefore: is it possible for the wind to be acting in this dangerous direction (for the deck this is often not the case, because the wind can only be horizontal), and are wind direction and critical angle of incidence aligned, respectively? Issue 2: does the critical angle of incidence suit the plane of vibration?

For these reasons it can therefore be sensible to perform wind tunnel tests in order to determine the directional dependence of the lateral force coefficient. Care must then be exercised when using the test results (Reynolds number effects at model scale, length of the 2D model)!

c) The low frequency part of the turbulence leads to modulations of the shedding frequency, which may lead to drastic reductions (the oscillating structure moves out of resonance successively). As an example the pylon displacements of the Teresina Bridge are shown in Fig. 4.72. The maximum value is about 75 %, and the effective value about 45 % of the full resonance amplitude. This is important in terms of human perception as well as possible fatigue issues.

d) Lock-in at higher resonant amplitudes. This complex interaction effect is a synchronization between oscillation and shedding, such that, for example, frequency modulation does not occur.

Approximate values for the lock-in criterion according to DIN are amplitudes: below 10 % of H (= section depth) no lock-in, above 60 % of H perfect resonance with double the coherence length. This approach is definitely not on the safe side! Example test results of Bublitz Göttingen on the circular cylinder showed a synchronization at amplitudes as low as 1 % H.

We also observed significantly more critical synchronization cases, for example at the Nuremberg Tower in 1982 (which we retrofitted), and astonishing test results for Niederwartha, Section 3.9.2. In case of doubt, testing should be performed! Care should then be exercised with regards to possible Reynolds number effects in the test.

e) Artificial damping can be a last resort; benefits and feasibility must be checked carefully. A typical concept is a wait-and-see strategy: design for the addition of damping, but to first wait (measure, observe) and only install the damping if necessary.

Two further points should be considered: Often the vortex excitation is a nuisance, as on the Golden Ears Bridge Vancouver, Fig. A33, from our design. In such case a wait-and-see strategy is sensible. The acceptance criterion for a frequent resonant excitation may be chosen as effective acceleration ≤ 0.1g.

The excitation of deck or tower by vortex shedding can sometimes lead to indirect cable vibrations: this corresponds to the case of so-called internal resonance of the dynamic system. This issue is dealt with in the cable dynamics chapter, Section 3.8. As an example, for Stonecutters, Section 2.2.6, there were pylon dampers planned to avoid possible cable excitations. Similar issues exist with the Golden Horn Bridge Vladivostok, Fig. A37.

Oviously H-towers and even more so A-towers are inherently stiffer than single towers or V-towers and thus less sensitive for internal resonance.

4.3 Bridge dynamics

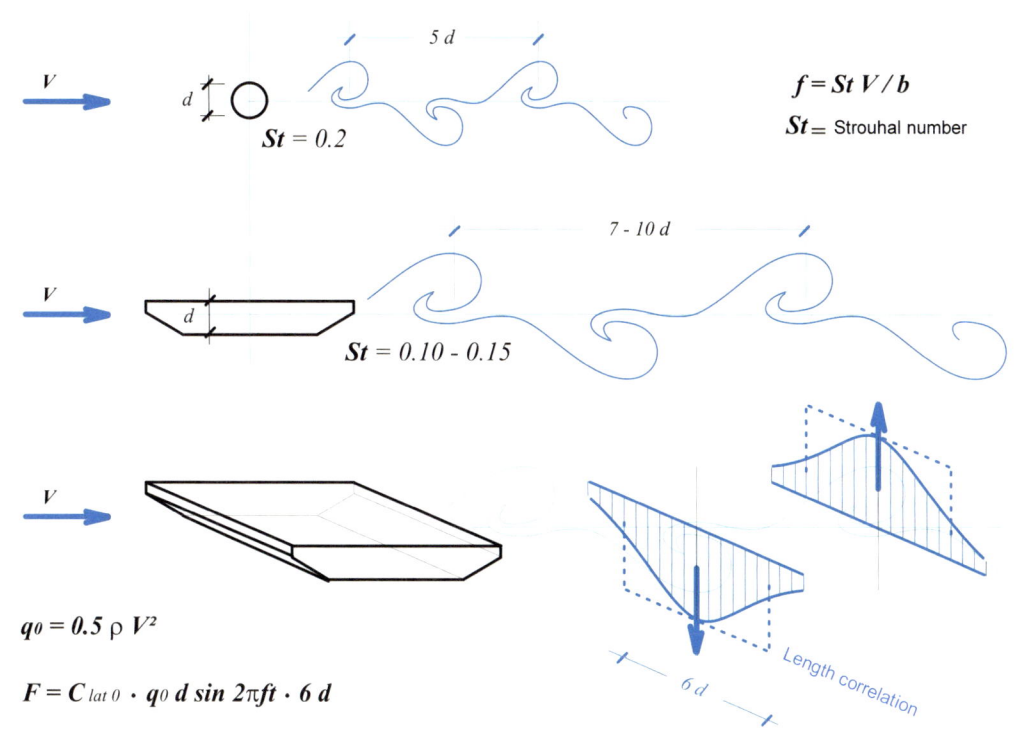

$f = St\,V/b$
$St =$ Strouhal number

$St = 0.2$

$St = 0.10 - 0.15$

$q_0 = 0.5\,\rho\,V^2$

$F = C_{lat\,0} \cdot q_0\,d\,\sin 2\pi ft \cdot 6d$

Figure 4.71 Karman Vortex Street – geometrical rhythm, excitation force

0.47 Hz

Response for perfect resonance

realistic response

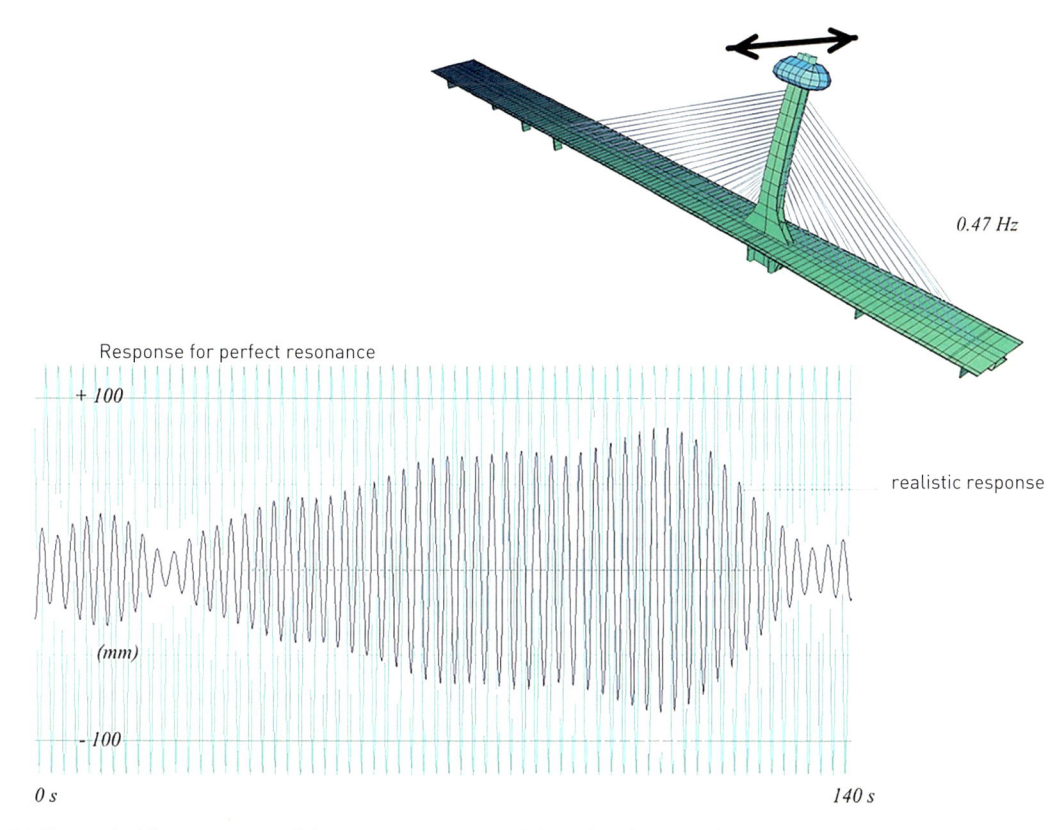

Figure 4.72 Vortex shedding resonance of the tower-comparisons of time histories assuming perfect resonance on wall as realistically varying frequencies

4.3.5 Self-excitation and other motion-induced effects
4.3.5.1 General description, background

In elastically vibrating structures the respective member – deck, pylon, cables – changes its position in, and/or its incidence towards, the flow periodically.

This creates additive periodical wind forces due to the motion, a motion which we will refer to as eigenmotion because this will be the component of major interest. The additive forces are either in phase with the instantaneous displacement or the instantaneous velocity of the motion.

The interaction between body and fluid hence creates forces that are solely due to the motion, i.e. practically the flow becomes part of the vibrating structure by acting like a (positive or negative) spring or damper, depending on the phase angle of the additive forces [4.32]. These effects can dominate in slender and large structures and govern their design. In the phenomenon discussed here, the amplitudes will, contrary to, for example, vortex-induced resonance, not be limited but practically increase without bound – a typical case of instability.

Some phenomena of the eigenresponse-flow interaction will be highlighted here, cf. Fig. 4.73. The terms introduced can be found in any long-span bridge project today.

The torsional moment acting on the beam subjected to the flow is typically proportional to the angle of incidence. A twisting beam would therefore be subject to a further-increasing torsional moment, the flow acting as a negative torsional spring. The corresponding spring constant is proportional to the square of the flow velocity, i.e. the negative stiffness increases rapidly with speed and can exceed the structural stiffness. This effect is termed torsional divergence, and the corresponding wind speed V_{div} = critical speed of divergence. The phenomenon is practically a static instability in which the total torsional stiffness of the system – and hence the natural frequency f_T – becomes zero.

Typical bridge decks of cable-stayed bridges have torsional frequencies above the vertical bending frequency. When the wind speed increases and the effective torsional stiffness decreases, there will be – before reaching the divergence state – a wind speed where the natural frequencies f_T and f_B match. These frequency-aligned modes can then couple to a combined oscillation. With lift-type cross-sections such as are typical of long-span cable-stayed bridge decks, the critical wind speed of *classical flutter* V_{CF} lies more or less exactly at this point. We shall return to this problem later.

The phenomenon of aeroelastic damping can be explained as follows: imagine the incidence angle of the flow to be rotated clockwise around the cross-section. The aerodynamic force usually – though not always – turns in the same direction. If we assume that the structure undergoes eigenoscillations transversally to the flow simultaneously, the relative (effective) flow turns – as calculated by vector summation – against this motion and creates an additional motion-inhibiting fluid force.

This force is proportional to the flow velocity and the phenomenon is called aeroelastic damping. The magnitude of the *aeroelastic damping* is often much larger that the structural damping and plays a significant role in slender bridge structures.

However, there are exceptions: sharp-edged 'bluff' sections may exhibit aerodynamic forces which for certain angles of incidence may act against the rotation of the flow. This usually originates from a change in flow pattern as the incidence angle changes, typically associated with a boundary layer separation or reattachment. This behavior is responsible for *galloping instability*, or bending galloping, and in the extreme case separation galloping. The phenomenon can be seen in the diagram of the lift coefficient; the curve which normally rises monotonically shows a local negative gradient, i.e. in this region *negative aeroelastic damping* occurs. This is the simplified, den-Hartog-type description of galloping sensitivity. The instability becomes more critical – i.e. the negative damping increases – for steeper negative lift gradients. The force itself rises with the wind speed, and the unstable condition is reached at a critical galloping speed, where the negative aeroelastic damping exceeds the structural damping. This simplified model is relatively easy to understand and allows the estimation of the main characteristics with good accuracy.

The mentioned destabilizing lift force effects may, for example, occur with triangular, rectangular and – to a lesser extent – polygonal, semicircular etc. sections, and at certain incidence angles. Cable cross-sections have a high technical relevance, where the body contour may be modified by water rivulets, Section 3.8.3.1. Other important cross-sections are typical box girder geometries.

There is a similar phenomenon of *aeroelastic torsional damping*, but this contribution is rather small and is of little practical significance.

For certain angled cross-sections and at particular angles of incidence a *torsional galloping* may occur, which again is mostly related to flow separation mechanisms. This is of practical importance usually with girder sections which feature edge beams located far from the bridge axis.

This phenomenon cannot be easily and reliably described, so for this the aerodynamic derivatives need to be studied [4.19].

Aeroelastic stability analyses of long-span bridges today are strongly related to the so-called *Scanlan aerodynamic derivatives*: they describe the additive aeroelastic forces acting on a body, for example a deck section, which oscillates in a fluid flow [4.19]. For a deck section usually two degrees of freedom are of interest: the aeroelastic forces can then be described through $2 \cdot 4 = 8$ specific diagrams. The Scanlan derivatives have a shape which depends on the sectional geometry and may be rather characteristic. They can today be measured with good accuracy by experiments, Fig. 4.74, see Section 4.3.7.2.

The theoretical background is not presented here. The equations given in the figure show that two generally independent types of motion – h(t) = vertical oscillation and a(t) = torsional oscillation – are

4.3 Bridge dynamics

Wind force as negative torsional spring

$C_{T\,aero} < 0$ reduces the torsional frequency

if $f_T = 0$ → Torsional divergence

if $f_T = f_B$ → Flutter

Wind force as translational damper

C_U-Gradient positive → aeroelastic damping

C_U-Gradient negative → galloping instability

small / deep Examples

Wind force as negative torsional damping

critical shape of derivative A_2^*

(Tacoma) → instability of torsinal galloping type

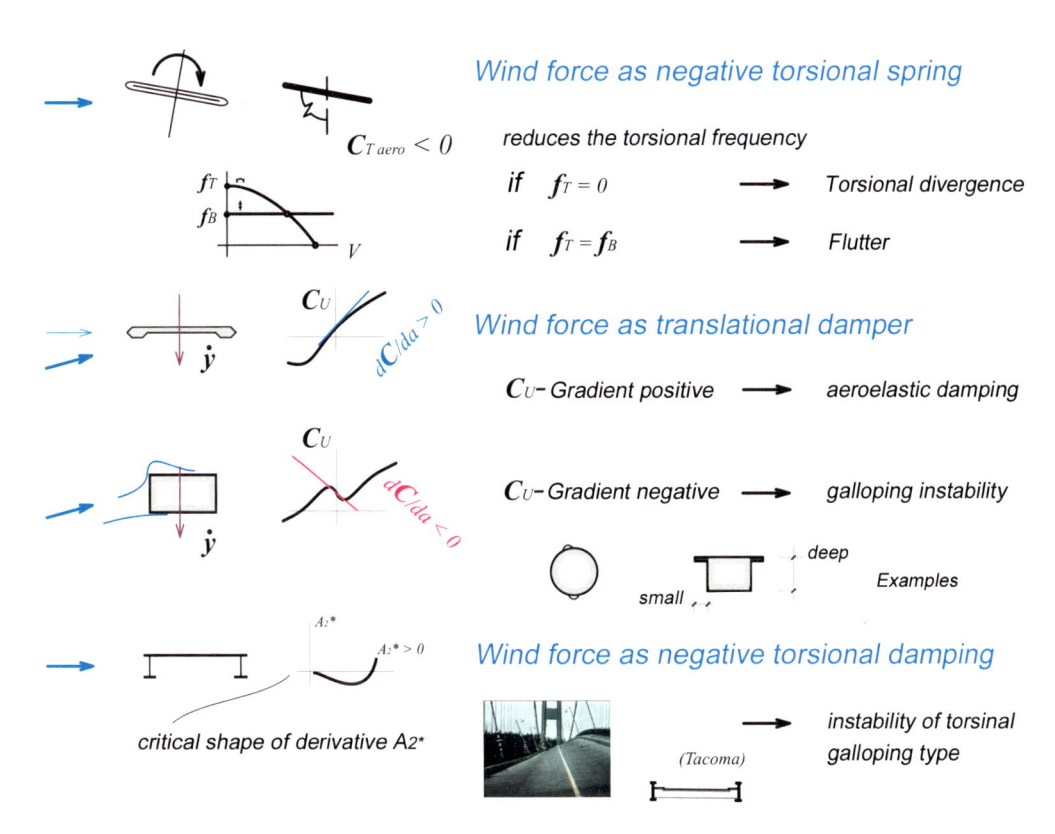

Figure 4.73 Overview of the motion-induced oscillations

coupled and, as a consequence, are forces arising from the flow (this means that a pure vertical oscillation also causes aeroelastic torsional forces and vice versa). The derivatives $H_1^* \div H_4^*$ and $A_1^* \div A_4^*$ are functions of K, or the so-called *reduced wind velocity* U_{red}. For large U_{red} – or very low frequency of motion – the problem is nearly stationary and the instantaneous wind forces correspond to the static wind forces.

This asymptotic behavior is being modified for low flow velocities because parts of the cross-section are shaded and subject to an already modified flow. The reduced velocity is the wind speed non-dimensionalized by the speed that would be required to pass the bridge width during one cycle of oscillation. It can also be thought of as the geometrical relationship between the wavelength projected into the flow and the section width.

Of the eight derivatives, the following have a particular significance: H_1^* (= lift force due to vertical oscillation) is important for bending-type galloping, which however – as discussed before – can be relatively reliably derived from a stationary consideration; A_2^* (= torsional moment due to torsional oscillation) is directly relevant for the mechanism of torsional galloping. For the latter, the zero-point of the curve is important, which can only be found through dynamic tests. This is discussed in detail in later examples.

The mechanism of the interactive excitation is summarized in Fig. 4.75: similar to a system loaded by a static compression force (example: Euler rod) there is a sharp stability boundary (equivalent to the rod under central compression) and an amplifying effect of second order (equivalent to the rod under eccentric compression) below the stability boundary. The engineer therefore has to solve a two-fold problem:

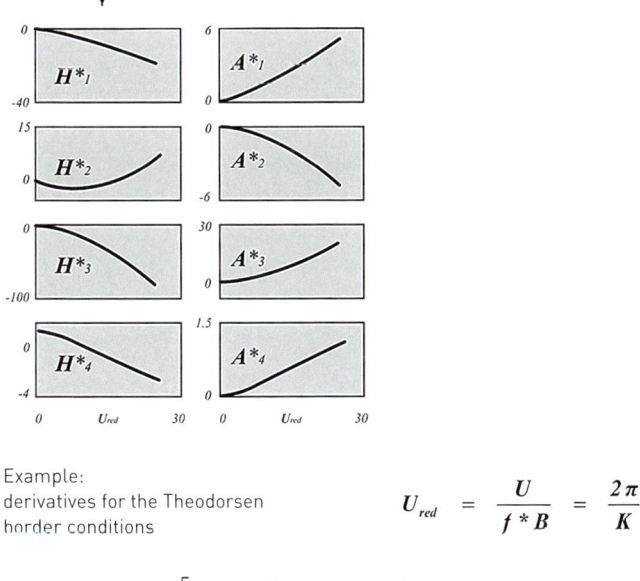

Example: derivatives for the Theodorsen border conditions

$$U_{red} = \frac{U}{f*B} = \frac{2\pi}{K}$$

$$F_L = \frac{1}{2}\rho U^2 B * \left[KH_1^*(K)\frac{\dot{h}}{U} + KH_2^*(K)\frac{B\dot{\alpha}}{U} + K^2 H_3^*(K)\alpha + K^2 H_4^*(K)\frac{h}{B} \right]$$

$$F_M = \frac{1}{2}\rho U^2 B^2 \left[KA_1^*(K)\frac{\dot{h}}{U} + KA_2^*(K)\frac{B\dot{\alpha}}{U} + K^2 A_3^*(K)\alpha + K^2 A_4^*(K)\frac{h}{B} \right]$$

Figure 4.74 Aeroelastic derivatives according to Scanlan

– He has to ensure, with a certain safety, that the 'dynamic equilibrium' does not bifurcate – i.e. avoid unstable, exponentially growing oscillations.
– He needs to adequately account for the unfavorable action of the negative aeroelastic damping on the response (section forces, displacements) below the stability limit.

The first task is a comparison of the actual wind speed n, site factored by a safety factor with the critical wind speed V_{div}, V_{CG} or V_{CF} derived from the system model. The safety factor should be at least $\sqrt{\gamma}$ with $\gamma = 1.5$. An important question in this context is, what actual wind speeds should be used.

The failure does not – in contrast to static bifurcation problems – occur abruptly, but the (over)-critical wind speed needs to act over a certain amount of time for the displacements and internal forces to rise to a catastrophic level. We should thus use a wind speed that is sustained for a longer time and will hence be lower than the known gust speed.

What wind speed should be assumed here: neither the codes nor the practitioners have reached a common understanding; indeed, they diverge significantly. DIN and EC are, from our point of view, relatively far on the safe side, but the criteria of the BD 49/01 Design Manual appear to be significantly on the unsafe side. We have typically used a gust duration of 30 to 50 oscillation periods (those of the critical mode of vibration). This implies acceptance of a duration of 30–50 oscillation periods of unstable behavior – but not 600 s or longer as required by EC and DIN at the moment. For a deck section one also needs to consider that an onset flow at inclination can be significantly more critical than a horizontal flow, but that an inclined gust of this duration has a lower corresponding speed. Fig. 4.76 shows in simplified form the achieved stability boundary as a function of incidence angle (green). The critical wind speeds also depend on α (red), such that the two curves can be compared. This procedure is recommended.

The secondary effect of instabilities needs to be added to the analyses described in Section 4.3.3.3. It can for example be added as a correction factor to the design checks, using approximation formulae. The derivatives are today implemented in the modern time-history software packages, e.g. in SOFiSTiK [4.32], such that the results already represent the non-linear, elevated section forces. The secondary effects are thus included, assuming that the derivatives are available. An example where the results are sensitive to the described effects will be presented further down.

4.3.5.2 Practical examples of bending-type galloping

The characteristic of a galloping instability is – as already discussed – the negative gradient of the lift curve of the cross-section under the acting incidence angle. The most important parameter expressing the susceptibility to galloping is the factor of galloping instability:
$a_G = -dC_L/d\alpha$ ref d (d = cross-sectional dimension across the wind).

Because of its importance, the rain-wind-induced galloping excitation of stay cables, Fig. 4.77, is covered Section 3.8.3.1.

The bridge in Fig. 4.78 is an interesting example of galloping excitation. This is a narrow, slender road bridge: narrow cross-sections with relatively deep girders are often sensitive to galloping.

The fundamental bending frequency of the unloaded bridge is at $f_B = 0.5$ Hz; from this and the assumed wind profile the stability limit for horizontal wind is

$$\text{erf} V_{CG} = \sqrt{\gamma} \cdot V_{60s} = \sqrt{1.5} \cdot 33.1 = 40.6 \text{ m/s} \quad \text{for } \alpha = 0°.$$

Fig. 4.78 shows the complete α-dependency of the criterion (green diagram).

The aerodynamic lift coefficient was obtained in the wind tunnel (red tangent in the diagram on the right) and shows that a flow approaching from below is more critical. The analysis showed that the damping of the system assumed as

$$\delta_{eig} = 0.03 \cdot \text{structural damping}$$

is not sufficient to ensure stability of the system: at an incidence angle of +2° galloping occurs at a low wind speed of

$$V_{CG} \approx 15.2 \ll \sim 35 \text{ m/s} = \text{erf} V_{CG}.$$

The red curve in the stability diagram shows that the onset wind speeds are below the criterion throughout. This was, of course, unacceptable and as a consequence, dampers were installed that would increase the critical wind speed to the double (not shown here).

Narrow deep sections (whether closed or not), and also box girders with short cantilevers, are typically prone to bending-type galloping. It should also be noted that high traffic envelopes increase the susceptibility to galloping.

This relationship can be seen in the example of the bridge in Fig. 4.79, which shows wind tunnel studies about the influence of the parameters, cantilever length and height of the traffic envelope.

The existence of the cantilever appears to only be important when the bridge is empty. With traffic, the stability value can reach very high values (a_G up to 5): the engineer then needs to establish whether high wind and large traffic envelope may occur simultaneously, i.e. whether a 4 m high traffic band can realistically coincide with a strong wind, when generally no traffic would be present but a traffic jam might actually be possible. Most bridges could not accommodate a stability value of 5, even at modest wind speeds.

Pylon sections are often prone to galloping. Oscillations transverse to the cable plane can occur in bridges with just one cable plane and a tower-like configuration of the pylon. Those pylons can become unstable during construction as well as in the final state of the bridge. Fig. 4.80 shows the results of wind tunnel tests undertaken for the second Panama Bridge, Fig. 5.17: even with the carefully rounded geometry the stability value reaches very high values (>5) in a 5–6° wide range of incidence angles. Extensive stability investiga-

4.3 Bridge dynamics

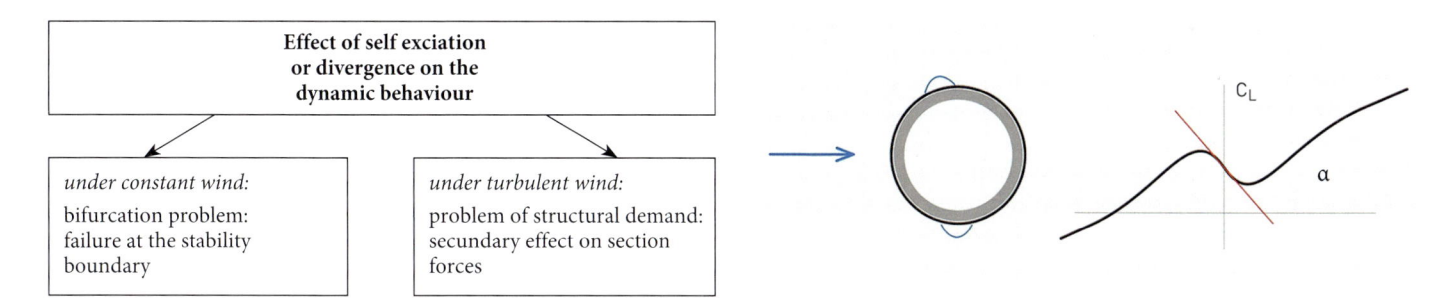

Figure 4.75 Characteristics of self-excited behavior

Figure 4.77 Rainwater-(rivulet)-induced galloping of cables

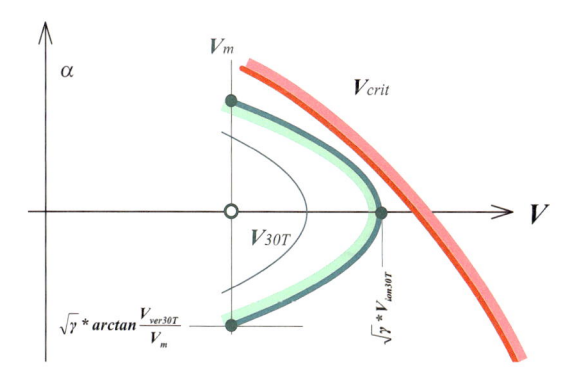

Figure 4.76 Comparison of required and actual critical wind speeds for a deck cross-section

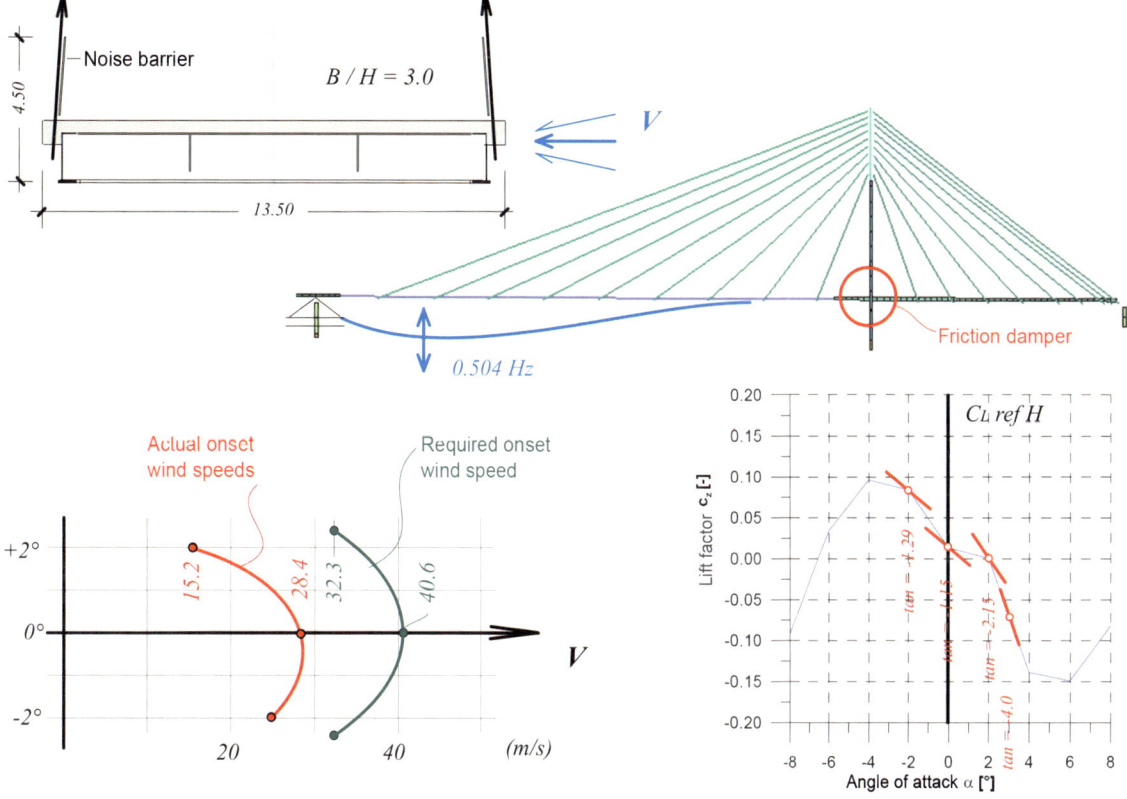

Figure 4.78 Galloping excitation in the first bending mode

tions have shown that the resulting damping just remains positive under the 100-year wind.

Finally we note that there are incorrect and unsafe values provided with respect to the galloping sensitivity of rectangular sections in DIN 1055-4 [4.25]. Unfortunately the diagrams of German as well as some international codes have been erroneous for years. In the example presented, there is a significant discrepancy between the stability value provided by the code and the classical test results of Försching and others (Fig. 4.81).

It can be seen that, for example, for a square section the DIN provisions lead to more than three times the galloping wind speed than what would be expected in reality. This can lead to a catastrophic misjudgment of the stability, reducing safety immensely. All tabulated values for rectangular sections are erroneous.

One interpretation may be that the values provided, correctly represent the results for flow exactly parallel to the body edges. However, then it would be imperative to mention that this is not the critical angle of incidence. At an angle of about ± 6° the sensitivity is – as shown in the diagram of Försching – two to three times higher. Therefore, if the angle of attack can vary, stability values of at least three times those provided in the code should be used.

4.3.5.3 Practical examples of torsional galloping

Torsional galloping is the most common type of instability with cable-stayed and suspension bridges of medium to large span lengths. Typical plate girder deck sections are excellent statically and yet very efficient and therefore used in about 80% of all cable-stayed bridges. However, these sections are all prone to torsional galloping. The deeper the girder and the further out the edge girders are located, the higher the susceptibility to aeroelastic effects. Only for the extremely long-span bridges which are equipped with aerodynamically shaped girders, like Vladivostok or Stonecutters, the classical flutter phenomenon is the governing mechanism.

The background to torsional instability is a negative effective damping of the torsional oscillations because of the motion-induced wind forces. It is rather difficult to identify the exact excitation mechanism: compared to vertical bending-type galloping the phenomenon is very complex and not easy to analyze. The oscillating cross-section strongly modifies the flow pattern, leading to complex pressure distributions along the section. From the distribution of these flow patterns it may be possible to deduce the mechanism. But reliable conclusions can only be drawn from wind tunnel tests. In the early stages of a design the derivative A_2^* needs to be measured at different reduced velocities, with a special focus on identifying the zero point of the diagram, beyond which the damping is negative.

It is certain that structural details at the upstream edge of the girder play an important role in the stability behavior of the section. It has, for example, been seen on the Ting Kau Bridge, shown in Fig. 4.82, how strong the influence of the edge detailing of the deck

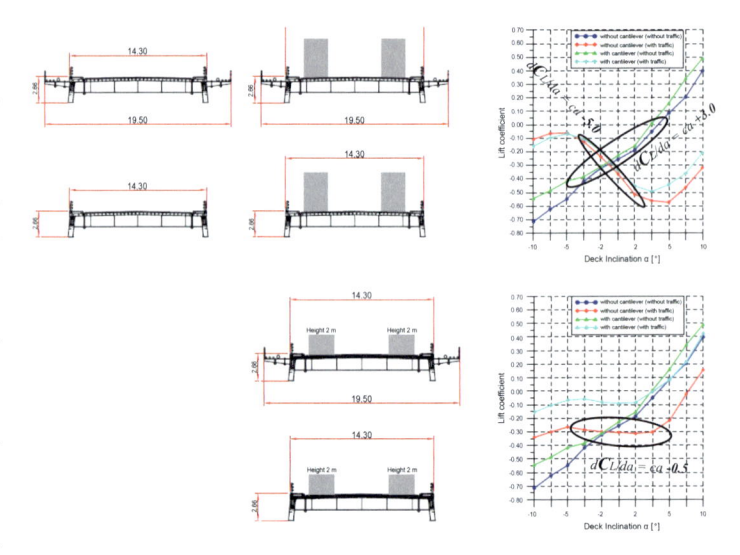

Figure 4.79 Relationship between lift coefficient and parameters, cantilever length and height of traffic band

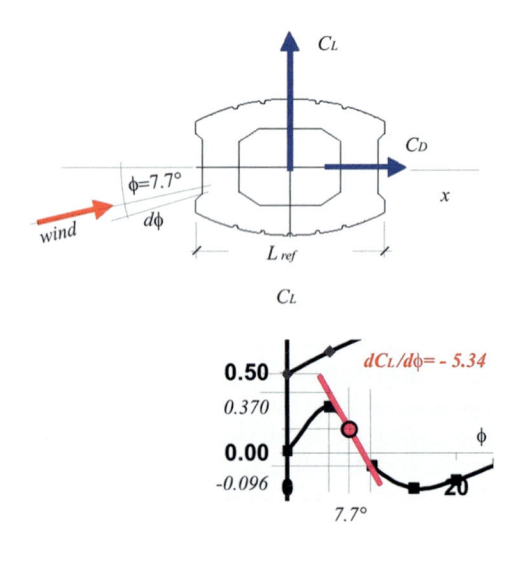

Figure 4.80 Second Panama Bridge, wind tunnel results on the lift behavior of the pylon section

4.3 Bridge dynamics

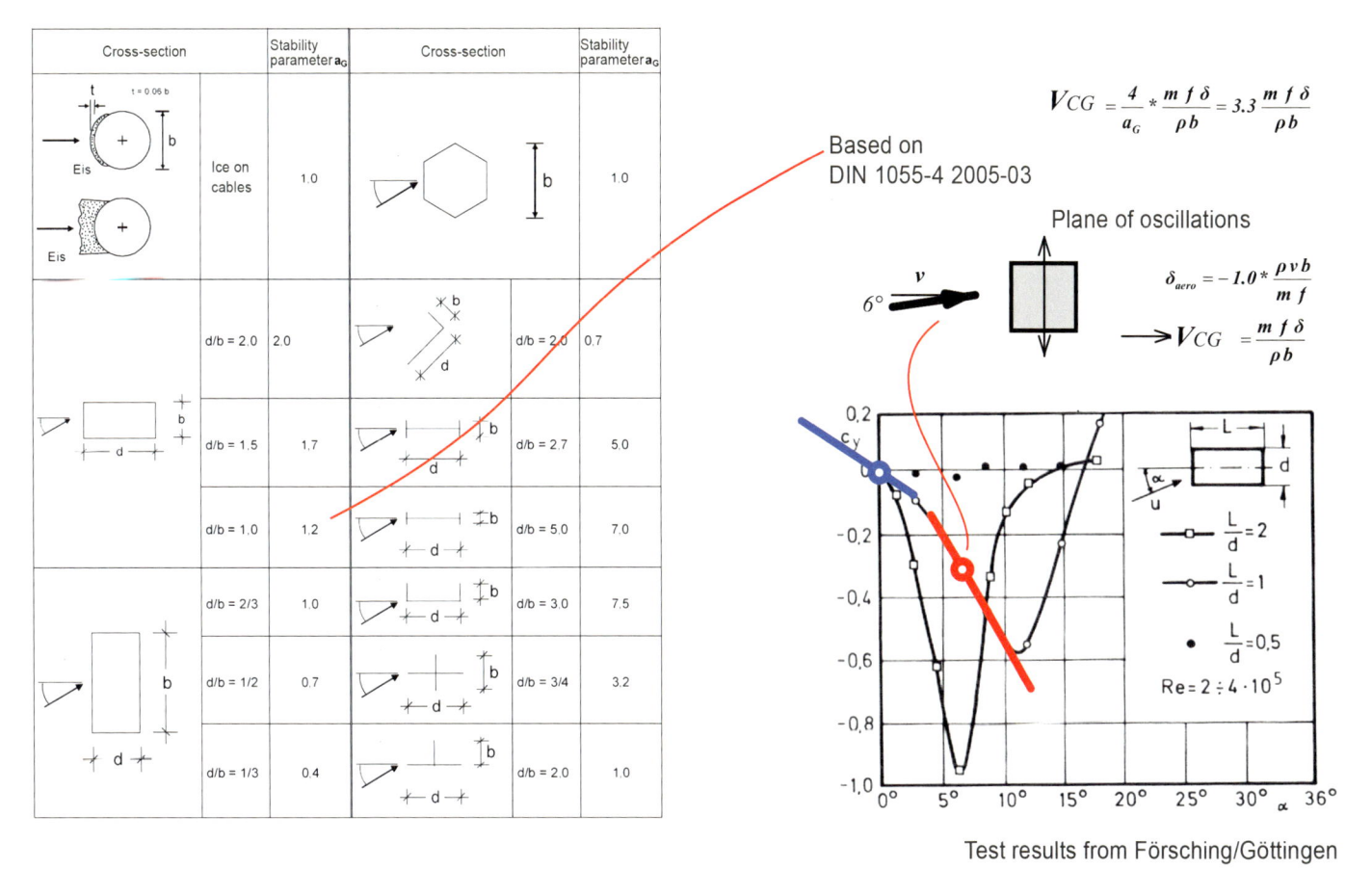

Figure 4.81 Contradictory stability parameters according to DIN 1055-4 : 2005-03

$$V_{CG} = \frac{4}{a_G} * \frac{m f \delta}{\rho b} = 3.3 \frac{m f \delta}{\rho b}$$

Based on DIN 1055-4 2005-03

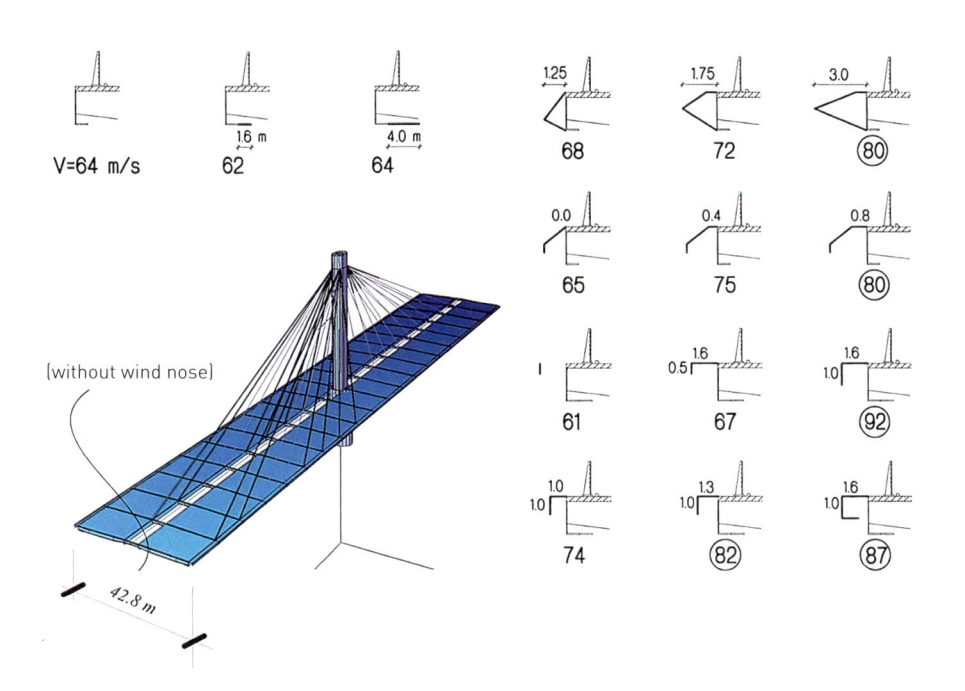

Plane of oscillations

$$\delta_{aero} = -1.0 * \frac{\rho v b}{m f}$$

$$V_{CG} = \frac{m f \delta}{\rho b}$$

Test results from Försching/Göttingen

Figure 4.82 Ting Kau Bridge, Hong Kong, stability boundary (m/s) for different wind nose concepts

can be. Davenport was involved in the aeroelastic design. Hong Kong is in a typhoon region and the stability requirements are thus very strict.

A number of edge beam geometries have been tested on a 2D section model of 1 ÷ 2 m size. The result is shown in Fig. 4.82. The critical wind speed varies between 61 and 92 m/s. From the figures it is difficult to deduce a logical explanation for which geometry behaves favorably and why, i.e. which section is more and which is less stable.

As a general rule, of course, modifications that increase the deck width tend to work favorably, but beyond this the results show no clear trend. Results almost appear contradictory; for example, when one compares the similar design providing 67 and 92 m/s.

Fig. 4.83 shows the ambitious design of the Bridge Kehl-Straßburg built for the International Garden Show of 2005, Fig. 2.225, [4.38, 4.39]. The main span length is 183 m. The structure is very complex and slender. From a technical perspective, wind effects are, despite relatively low wind speeds, the critical action.

There are two independent pedestrian walkways as shown in the plan. These form the structural skeleton: the southern girder is a horizontal, the northern a vertical arch, connected laterally through a light truss system to a combined cable-stayed girder. At midspan there is a platform connecting the two decks. This arrangement provides a global girder which changes significantly along the bridge length, also in terms of its aerodynamic characteristics.

Considering the bridge dimensions the natural frequencies are relatively low. The girder geometry is such that torsional instability may be expected to govern: therefore the torsional frequencies are of great importance.

Because of the varying cross-section, aeroelastic wind tunnel tests of three section models were performed at RWTH Aachen (Fig. 4.84): a) the centre section, b) the section at ¼ of the span and c) the section on land. All three sections were composed of two deck sections – in the case of (b + c) at different elevations – and a wide opening in between. It is this arrangement that gives rise to the torsional instability of the section.

The wind forces were measured as reactions on the testing rig during a harmonic excitation of the deck. These were then analyzed by decomposing the force signal for identification of components that are in and out of phase with the imposed velocity. These measurements were repeated in steps of $\Delta U_{red} \cong 0.25$. Fig. 4.86 shows the example of the thus computed derivatives for the centre section.

The curve of derivative A_2^* requires special attention because it is obviously strongly dependent on the spacing between the decks and the wave length of the flow (shaded area) induced by the upstream girder. This is the origin of the early onset of instability: where this wake hits the downstream deck exactly, the diagram of A_2^* is modified upwards (= in the direction of negative torsional damping, cf. red circular mark). For the section at ¼ span this effect occurs at a slightly shifted location in the diagram. The effect can actually be identified in all derivatives, but it is in A_2^* that it represents a critical problem.

In order to determine the onset wind speeds for the torsional instability, time-history analyses were performed using the derivatives and applying a constant wind speed. These show damped responses at low wind speeds but become negatively damped unstable responses with increasing wind speeds.

The thus-derived aeroelastic damping is shown at the top right of Fig. 4.86: the first torsional mode becomes critical at 13 m/s (the structural damping will be exceeded by the aeroelastic excitation) and the second at 19 m/s.

In order to ensure torsional stability, tuned mass dampers (TMDs) for torsion were provided. These were designed as rockers attached to springs and dampers. The TMD components were optimized as shown in Fig. 4.87 and tuned to the natural frequencies. Two pairs of rockers were required for damping the two torsional modes. This technical solution appears somewhat risky: large masses of up to 19 tonnes oscillate at a large height and are carried by a very slender structure (cross girder 13 300).

The TMDs made the system more complex as the TMDs are vulnerable and sensitive (failure, sabotage – e.g. through fixation, spring failure, maintenance works, temperature dependence of viscosity etc.). Therefore, extensive time-history analyses were required, for example, assuming various scenarios of failed TMD components. Fig. 4.88 shows the case in which one of the TMDs – at the 50-year return period wind – is stuck and thus fixed to the bridge, whilst the other is detuned by 10 %.

This scenario was considered with a safety factor of 1.1. An animation of the computed response with the moving TMDs is shown in the lecture.

4.3 Bridge dynamics

Figure 4.83 Rhine River pedestrian bridge between Straßburg and Kehl

Figure 4.84 Cross-section

Figure 4.85 Modes of vibration, wind properties

$f_{ver} = 0.556\ Hz$

$f_{tor1} = 0.648\ Hz$

$f_{tor2} = 0.805\ Hz$

$f_{lat} = 0.948\ Hz$

Negative aerolastic damping of the torsional eigenfrequencies of the undampened bridge as a function of the wind speed

Uniform wind $V = 13.5\ m/s$

Torsional eigenmode $0.65\ Hz$

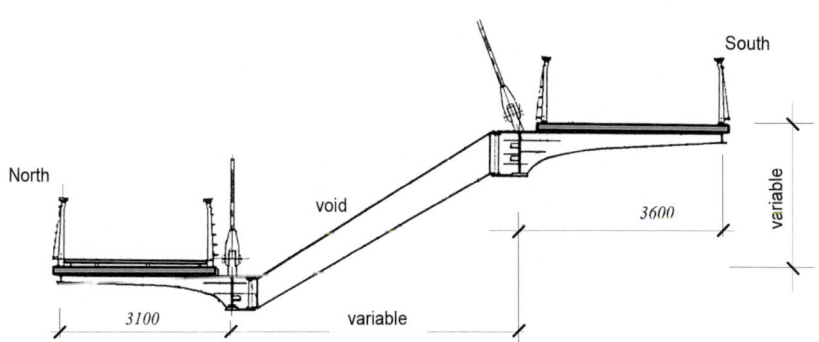

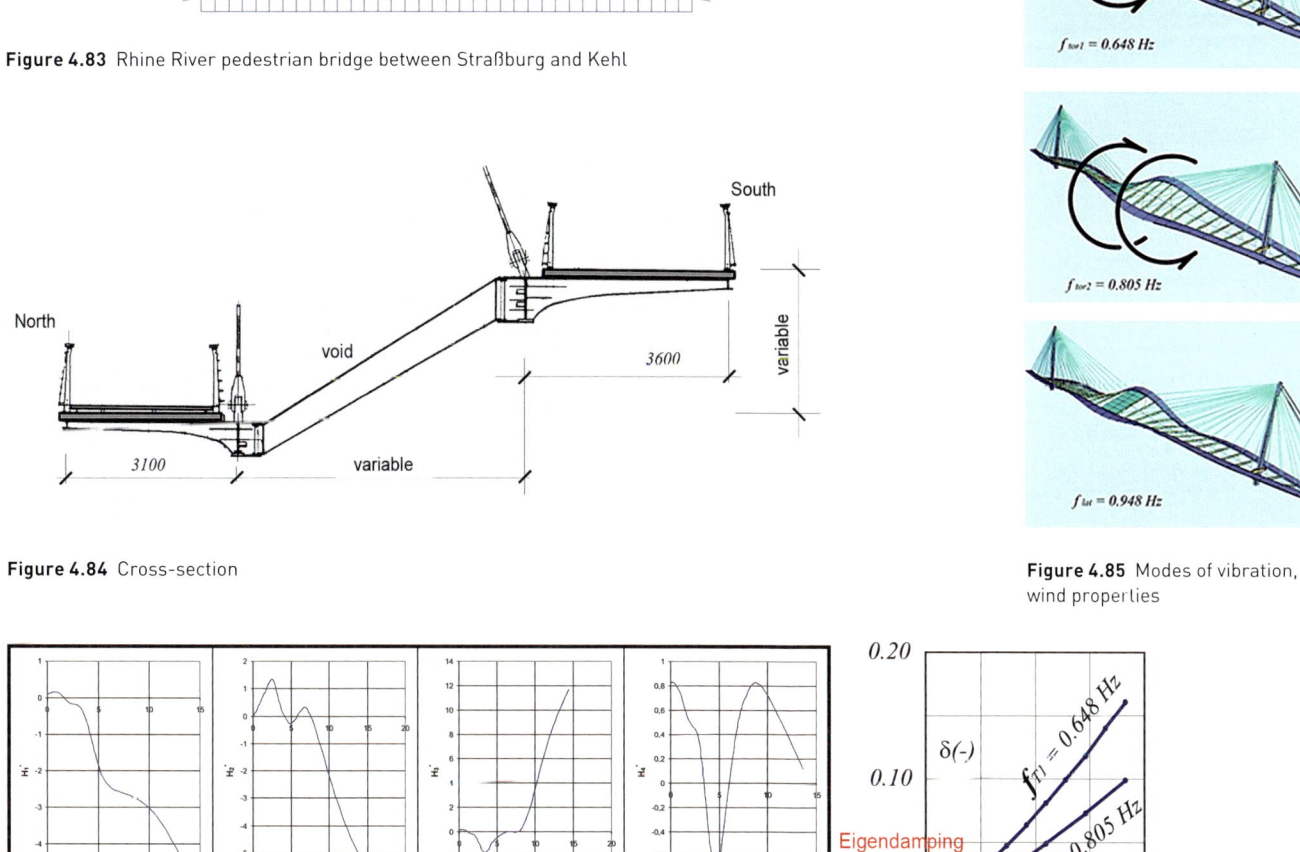

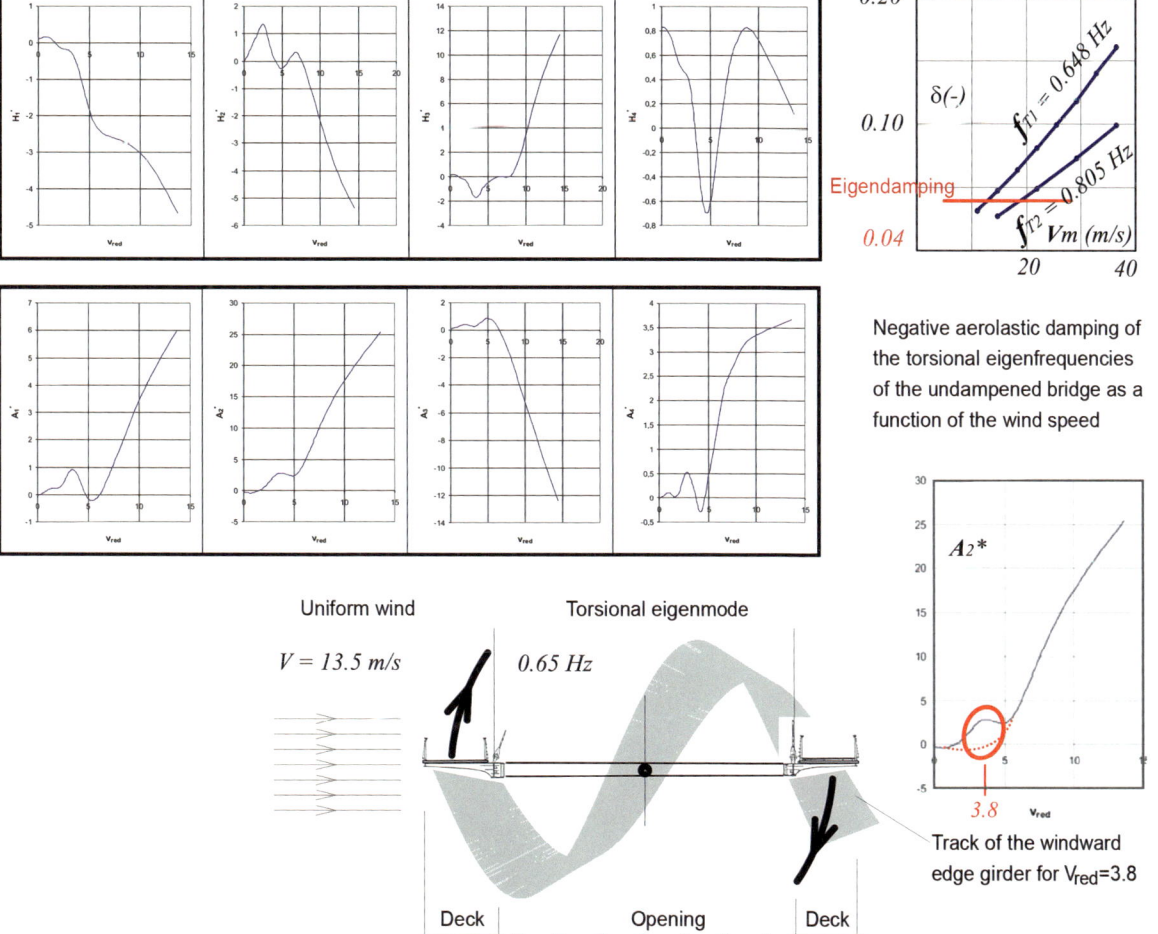

Track of the windward edge girder for $V_{red}=3.8$

Figure 4.86 Aerolastic derivatives, stability checks

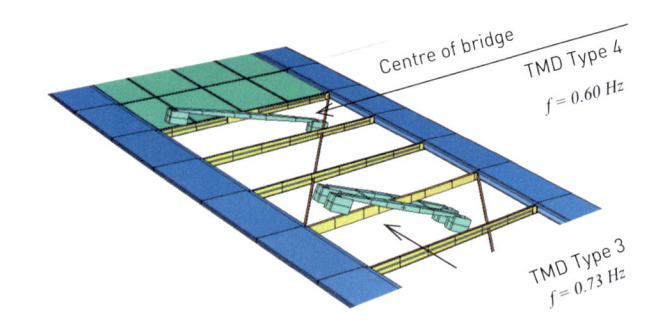

$x = 6.600$ m
$M = 3.451$ t
$x = 4.400$ m
$C = 121.1$ kN/m
$x = 1.480$ m
$W = 67.75$ kN/m/s

$x = 5.400$ m
$M = 2 \times 5.491$ t
$x = 2.830$ m
$C = 2 \times 484.3$ kN/m
$x = 1.480$ m
$W = 2 \times 89.95$ kN/m/s

Figure 4.87 Torsional TMDs of type 3 and 4, bridge Kehl-Straßburg

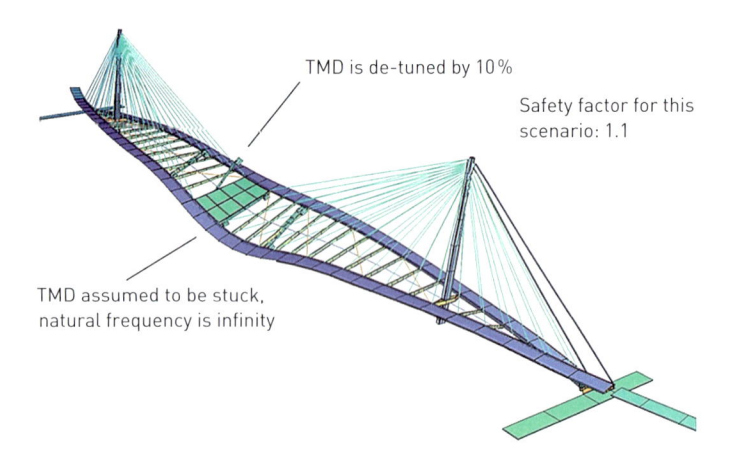

Figure 4.88 Simulation of an accidental load case spring is stuck, TMD de-tuned

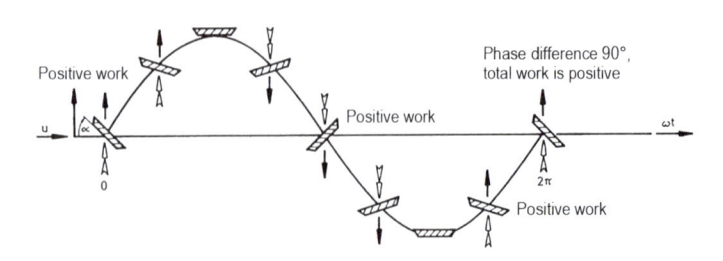

Figure 4.89 Excitation mechanism and energy effects of aerodynamic forces after Försching

4.3.5.4 Flutter

Classical or potential flutter strictly speaking is an instability of airfoil-type sections (slender, L ≫ h, smooth and with curved upstream edge) at which the flow remains attached to the top surface. Airfoil sections are not prone to bending or torsional galloping. Failure can only occur in a coupled response involving both vertical and torsional modes at f_B and f_T respectively.

In this case the vertical bending motion induces forces, which drive the torsional oscillation and vice versa; the flow thus couples the two response mechanisms which excite each other. According to Försching [4.40] the unstable oscillation becomes similar to a dolphin-like swimming motion, Fig. 4.89.

It is critical that the two initially different natural frequencies align – they move towards each other through the negative spring-type stiffness effect of the flow; and that the motion occurs with a certain phase shift between vertical and torsional response components. The two amplitudes A_B and A_T grow proportionally and exponentially.

An aeroelastic excitation of a general cross-section involving the coupling of the two modes is referred to as *flutter*, no matter what the flow pattern is. Thus, flutter is a general aeroelastic failure mode, the boundaries of which are the two galloping mechanisms described before. (Strictly speaking a slight mixture between motion components typically also occurs in galloping instabilities, but there it is quasi-steady and without practical significance.)

In practice, airfoil-type sections are rarely used for bridge sections, because their structural efficiency is not optimal and hence expensive. However, slender trapezoidal cross-sections are also aeroelastically stable, particularly if the edges are rounded for aerodynamic efficiency. Such deck sections are being used for bridges with large span lengths, particularly in typhoon and hurricane regions.

For Normandy Bridge a shallow closed box girder was used for the main span, in order to achieve the necessary torsional stiffness for a span of 856 m.

For this reason the concrete section in the side span is also designed as a box girder, Figs 4.90 and 4.91, Section 6.5.

Until recently the aeroelastic stability of bridge design was studied by hand calculations. Fig. 4.92 shows the well-known diagram of Klöppel/Thiele after [4.17]. The approximate solution solves the problem of the aeroelastic stability of a general bridge section through the behavior of an infinitely thin plate section. A flat plate is very stable and this fundamental case has been analytically solved by Theodorsen, Küssner and Bleich [4.13, 4.14]. The abscissa uses the ratio

$\varepsilon = f_T / f_B$ the ratio of natural frequencies,

whilst the ordinate shows the dimensionless onset wind speed of classical flutter. The authors thus define two groups of curves for two parameters

$\mu = m / \rho \cdot \pi \cdot b^2$ mass ratio 100, 50, 30, 10

r / b with r = radius of gyration 1.0, 0.75, 0.5

between which the user needs to interpolate twice. The use of the complex diagram requires some exercise. It allows the determination of the onset wind speed $V_{cr,0}$ of the flat plate using the parameters mentioned above.

The main diagram – only applicable to the idealized section – is accompanied by a series of further diagrams providing reduction factors $\eta(\varepsilon)$. These factors for various typical cross-sectional shapes (with edge beams, box-type with vertical and inclined webs, with various depth-to-width ratios, truss sections etc.) were determined by wind tunnel tests and the computation of the actual onset wind speed as

$$V_{cr} \approx \eta(\varepsilon) \cdot V_{cr,0}(\varepsilon, r/b, m) \quad \text{with } \varepsilon = f_T / f_B.$$

The sketches provided along the diagrams show a simplified mechanism for the phenomenon of frequency coupling. As mentioned above, the two frequencies (vertical and torsional) first need to align for coupled flutter to be possible. This mainly occurs through a downward shift of the torsional frequency through the negative torsional stiffness effect of the aeroelastic torsional moment (as explained before in the context of torsional divergence). With increasing wind speeds, on the way to torsional divergence the frequency meets the vertical frequency; at around this point there is potential for the coupled motion, and about here the flutter boundary of the ideal section is located. This point can be easily computed by hand.

This simplified model is the basis for the closed form solutions of Selberg and others, as for example:

$$V_{cr} = \eta \cdot 2\pi \cdot b \cdot f_B \left[1 + \left(\frac{f_T}{f_B} - 0.5 \right) \cdot \sqrt{\frac{0.72 \, m \cdot r}{\pi \cdot \rho \cdot b^3}} \right]$$

with

η reductionfactor compared to flat plate after [4.17]
b deck width
$m = \dfrac{G}{g}$ girder mass
$r = \sqrt{\dfrac{I_x + I_y}{A}}$ radius or gyration
$\rho = 1.225 \text{ kg/m}^3$ density of air
f_B 1st bending frequency
f_T 1st torsional frequency
$\varepsilon = f_T / f_B$ frequencyratio

The application of this equation is presented in Section 4.5.7.4.

A slight deficiency of the Klöppel/Thiele procedure is, as has been seen over the years, that the galloping critical sections can sometimes not be derived easily from the Theodorsen solution. The reduction factors provided by Klöppel/Thiele are relatively low, as they have resulted largely from the somewhat random damping of the tested models. This is the case for all low h factors. It is hence important to study deep sections and those with edge beam by other means.

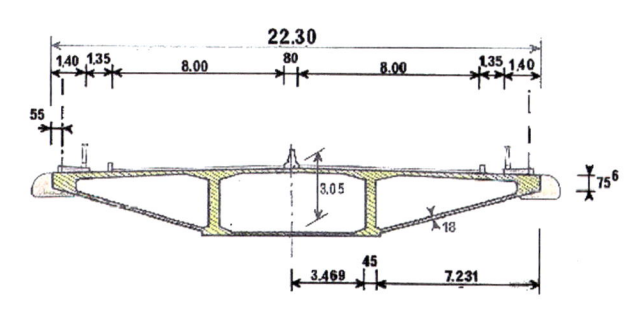

Figure 4.90 Concrete cross-section of the approach

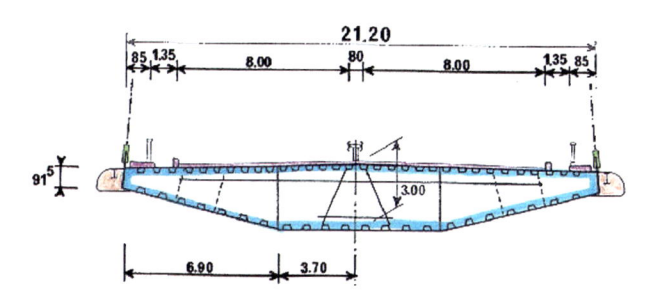

Figure 4.91 Steel cross-section of the main span

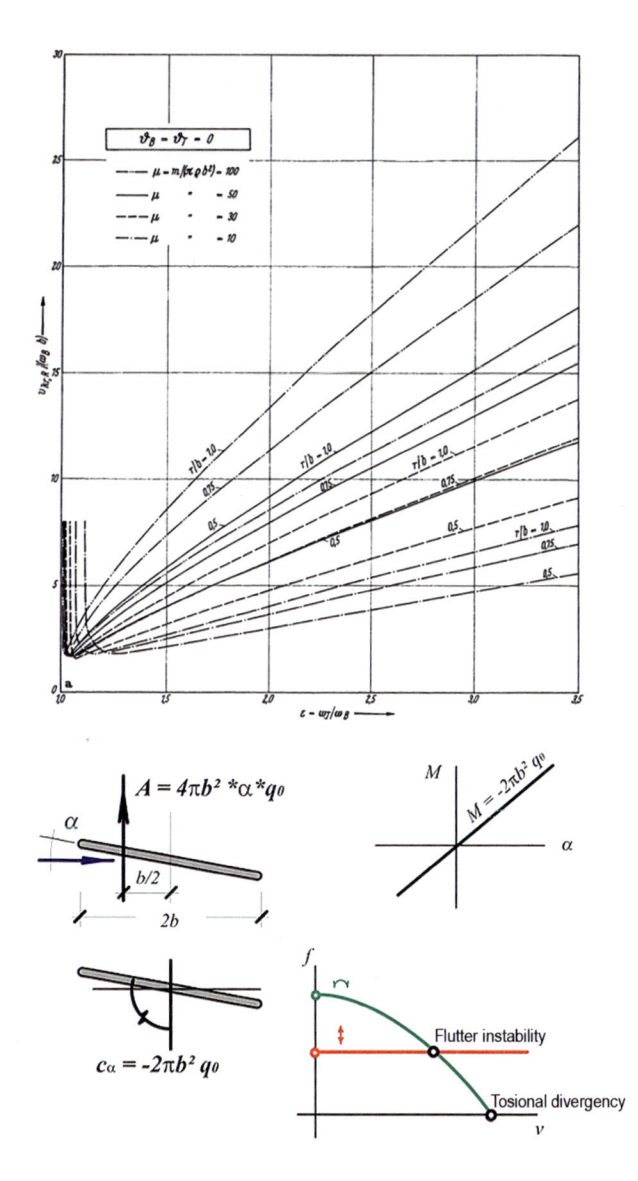

Figure 4.92 The classical Klöppel/Thiele flutter diagram for $\delta = 0$, based on the analytical solution of Theodorsen

In contrast to this, the accuracy of the method is impressive for aerodynamically shaped sections: we have in some cases performed back-up calculations using measured derivatives and the results have always been very good.

The Klöppel/Thiele procedure has been successful in general; it is an imaginative method that is still being used to date, at least in the early design stages, for smaller bridges but sometimes even in the detailed design.

It is actually still without competition; there is no comparable method and we still use the procedure in most cases, though in a slightly extended form: Fig. 4.93 presents a solution in which the influence of the affinity between the bending and the torsional modes is being accounted for [4.20] (ISALB '92). The two modes can only be coupled fully if the mode shapes are affine along the girder, or at least close to affine. In case the mode shapes are significantly different – as shown by the red and blue lines of the figure – the flutter boundary will be higher or even vanish. Thiele has already pointed out this possibility. It is the reason that modern cable-stayed bridges have traditionally been built with A- rather than H-shaped pylons whenever that proved necessary. Compared to a portal-type H-pylon the A-pylons cause very different mode shapes, thus increasing the stability limit which can be calculated from the diagrams of Fig. 4.93.

In order to describe the degree of affinity we have, in Fig. 4.93, defined an affinity parameter $\phi \leq 1.0$. Here $\phi \approx 1.0$ for H-pylons whilst it is about $0.8 - 0.9$ for A-pylons.

When the shapes differ, the mode-induced destabilizing excitation forces will be factored down by ϕ, whilst the remaining part would be responsible for higher-mode contributions which can be neglected.

According to the diagram instability can only occur above a limit value of ϕ^*; the corresponding wind speed is V_{cr}^*. The true critical wind speed corresponding to ϕ can be estimated by the parabolic interpolation shown in the figure.

A further refinement of the procedure is required for a case that is typical of very large spans: when the pylon top is relatively flexible but heavy [4.20]. In this case the pylon mass tends to oscillate with the girder and hence contributes as if it were part of the girder. The Klöppel/Thiele procedure, however, only accounts for the pure deck mass contributing to the dynamic properties, which leads to an understatement of the stiffness of the equivalent 2D oscillator. This may not be obvious to the user and whilst the error is on the safe side, this luxury may not be affordable in the actual project.

The Klöppel/Thiele procedure is hence strictly only correct when the pylon is weightless. In other cases the user should modify the system such that the pylon masses vanish, which is done by calculating equivalent deck masses and inertias whilst keeping the frequencies constant [4.20]. The thus modified system will then be used in the Klöppel/Thiele approximation.

A possible caveat should also be mentioned: it is possible to misinterpret the stability of a system critical to pure galloping when the flutter limit after Klöppel/Thiele is also low. In this case, the system may show sufficient stability for the pure flutter but exhibit a further reduced overall stability boundary. In the case of the Incheon Bridge in South Korea, Fig. 4.94, with

$$f_B = 0.16 \text{ Hz}; \quad f_T = 0.30 \text{ Hz}; \quad \text{req } V_{cr} = 74 \text{ m/s}$$

we determined a torsional stability limit of only $V_{cr} \approx 35$ m/s because the design had been driven by static considerations. This solution was not acceptable.

A modified cross-section with geometric corrections (wind nose) then provided a pure torsional stability limit of 76 m/s. However, it was found that the Theodorsen solution of the coupled oscillation of the ideal section was rather low at ~100 m/s and thus gave rise to concerns. In such cases it should always be expected that the ~35 m/s are reduced through the contribution of bending oscillations. It is possible to estimate the actual stability boundary using a Dunkerley-type approximation. In order to ensure the required stability, the torsional frequency had to be raised to 0.38 Hz as shown in Fig. 4.94.

This brings us to a simulation method that may be the future: Fig. 4.94 shows instantaneous images computed by the vortex particle method in a 2D simulation. This allowed the determination of the stability values to reasonable accuracy. It is our experience, however, that flow simulations of this or other kinds cannot be relied upon fully at the moment.

It may be of interest though, to study the effect of small section details – like the design of a wind nose, wind shield or similar – by such numerical simulations. It would also be expected to supplement such studies by wind tunnel tests at some stage of design implementation.

4.3 Bridge dynamics

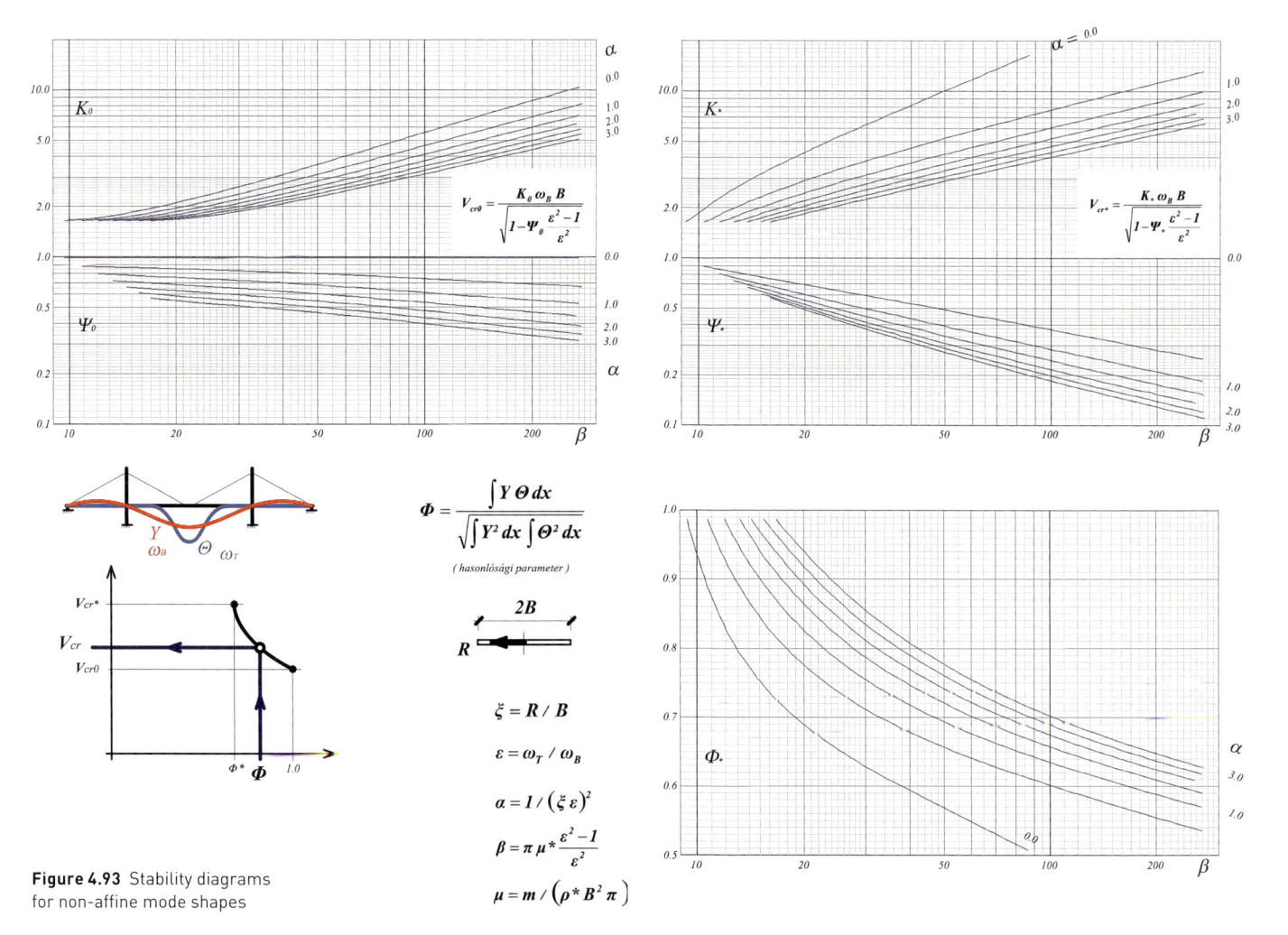

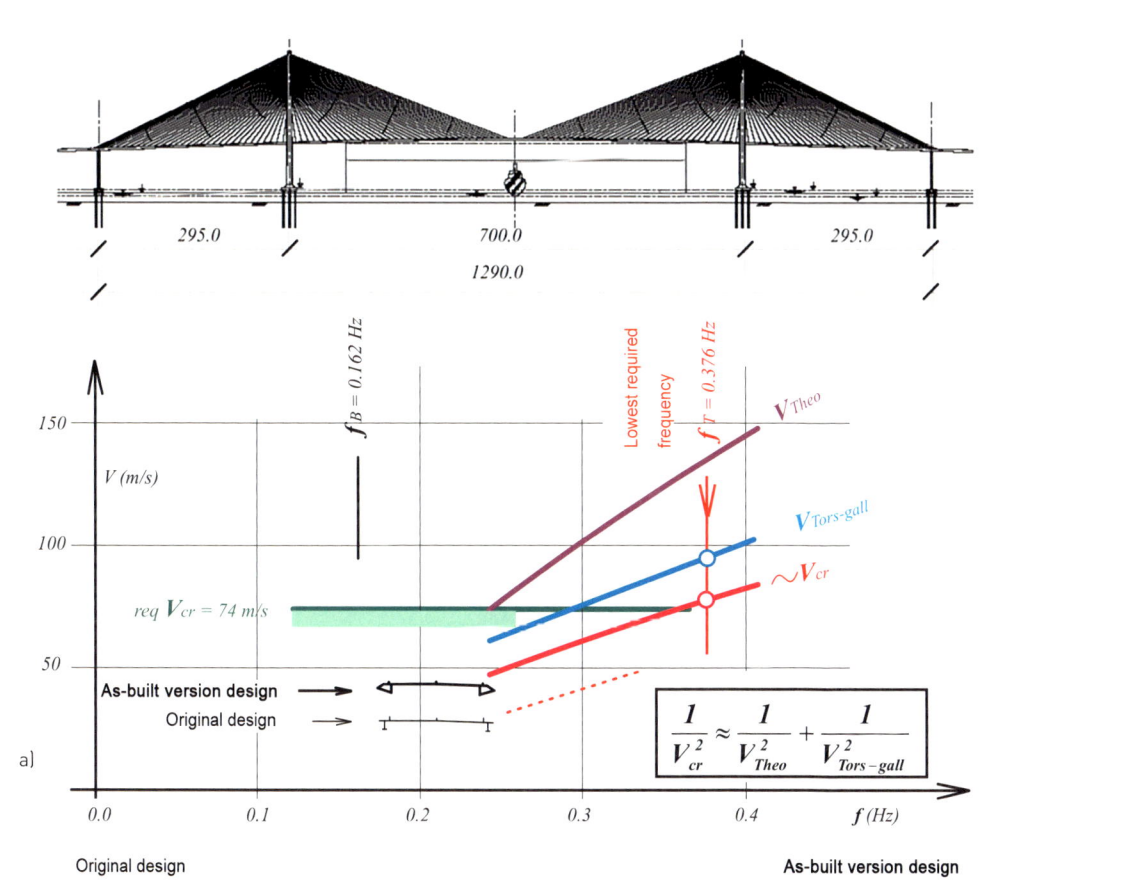

$$V_{cr0} = \frac{K_0 \omega_B B}{\sqrt{1-\Psi_0 \frac{\varepsilon^2-1}{\varepsilon^2}}}$$

$$V_{cr*} = \frac{K_* \omega_B B}{\sqrt{1-\Psi_* \frac{\varepsilon^2-1}{\varepsilon^2}}}$$

$$\Phi = \frac{\int Y \Theta \, dx}{\sqrt{\int Y^2 dx \int \Theta^2 dx}}$$

(hasonlósági parameter)

$\xi = R / B$

$\varepsilon = \omega_T / \omega_B$

$\alpha = 1 / (\xi \varepsilon)^2$

$\beta = \pi \mu^* \frac{\varepsilon^2 - 1}{\varepsilon^2}$

$\mu = m / (\rho^* B^2 \pi)$

Figure 4.93 Stability diagrams for non-affine mode shapes

$$\frac{1}{V_{cr}^2} \approx \frac{1}{V_{Theo}^2} + \frac{1}{V_{Tors-gall}^2}$$

Original design

As-built version design

Figure 4.94 Incheon Bridge, Korea
a) stability for the case galloping and classical flutter are at similar speed regions
b) simulation with Vortex Particle Method

4.3.6 Damping measures

Some applications of dampers have already been discussed along some previous examples: cable dampers, damping against galloping of bridge decks etc.

As mentioned before, the structural damping originating mostly from the materials in the structure is generally rather low: the logarithmic decrement for oscillations in the elastic range is about δ = 0.02 (pure welded steel structure) up to 0.04 (reinforced and post-tensioned concrete). The overall level of damping can be raised ten-fold by employing suitable damping measures. This renders damping measure capable of stabilizing an otherwise critical dynamic bridge system and/or reducing the dynamic section forces. The effectiveness of damping measures in instability scenarios is of course subject to the individual characteristics, because the aeroelastic properties can vary significantly. Their effectiveness can, however, be estimated – though only roughly. For bending-type galloping – and this includes the rain-wind-induced oscillations of stay cables – the damping linearly increases the onset wind speed of galloping, i.e. here the damping is very effective. Typically only little damping is required for stabilization. However, the case of torsional galloping is more complex: the effect of added damping is sub-linear when the natural structural damping level is low. Because the aeroelastic torsional excitation mechanism itself it highly non-linear, in practice the damping does still stabilize the system relatively effectively. Adding damping to prevent classical flutter, however, is normally not very effective. This should not rule the option out, as it is often the only possible measure that can be employed. The effectiveness of damping measures also varies for the case of excitation purely through wind forces. For vortex-induced vibrations the displacements and internal forces are indirectly proportional to the overall damping, where the effectiveness is highest. For buffeting, the damping only reduces the resonant component and only contributes with the square root of the damping value.

There are in general two types of artificial damping:
- *The application of standard damper elements*, Fig. 4.95 – such a device is installed either between two structural members or between a member and the ground. When relative movements between the members occur, resisting forces are created, which are linear to the velocity of motion in the case of a linear damper. Close to linear behavior is obtained using, for example, hydraulic, visco-elastic or pot dampers with tightly sealed oil filling. Rubber elements exhibit force components proportional to velocity as well as to displacements. Further there are friction dampers that react to relatively displacements by the activation of a constant friction force.
- *The application of tuned mass dampers (TMDs)*, more precisely of damped TMDs, Fig. 4.95 – these devices are oscillators themselves, composed of a concentrated mass, which is connected to the actual structure by a spring and a damper. When the structure oscillates, it causes TMD vibrations which are, if tuned correctly, in the opposite phase of the structural oscillations and hence calm the damped structure.

The efficiency of a standard damper depends strongly on the point that it is connected to, i.e. the influence that can be made to the targeted modes of the damped structure in this location. This efficiency can be best presented on a damped prestressed cable, Fig. 4.96.

The picture shows the two limit cases, from which the efficiency can be derived: the original undamped case (red, W = 0) on one side and the case where the damper is infinitely hard (blue, W = ∞, effectively a fixed support). The latter case produces an increase of the natural frequency of $\Delta f = \varepsilon \cdot f$. The optimum damper constant is somewhere between these two limits. The theoretically possible artificial damping that can be provided at a given point is

$$\text{opt}\,\delta = \pi \cdot \Delta f / f \quad \text{theoretically achievable damping with optimum tuning}$$

The corresponding damper constant of the damper can be calculated by FE software, which is somewhat cumbersome. Fig. 4.96 shows a schematic which allows the estimation of the efficiency of a standard damper on any system: it is proportional to the frequency difference of the blue and red equivalent systems. Figs 4.97 and 4.98 shows, besides the already discussed applications (Sections 3.8.4 and 4.3.5), a further selection of solutions with standard dampers as applied in actual projects.

The footbridge Kiel-Hörn is located at the entrance to the harbor at Kiel and is opened through an actuated cable mechanism several times a day in order to allow the passage of ships, Fig. 4.97. The deck in the main span is a chain of beams with three elements, moved and held by a system of tiltable masts. During opening the deck retracts like an accordion. The critical mode of vibration during service also has a zig-zag shape and can therefore most effectively be damped by rotational dampers located at the hinges.

The damper forces are being created through a friction coating and a stainless steel plate; both parts are being pressed together by a prestressed spring plate. The damper system is designed such that during bridge opening and closing it is simply be decoupled and engaged again.

A further example, the Nhattan Bridge in Vietnam, is shown in Fig. 4.98. The deck of this multi-span cable-stayed bridge has a total length of 1500 m. This girder is freely moveable in the longitudinal direction in order to prevent restraint effects between deck and pylons. The frequency of the longitudinal mode is 0.24 Hz. A damping is required to limit longitudinal movements under earthquake excitation. The strong hysteretic behavior of the arrangement can be seen on the F-w diagram: longitudinal displacements could be reduced to about 55 %.

A damped TMD is required where a standard damper is not feasible, i.e. no support against the ground and no other members that could be connected to effectively. A TMD is more complex and demanding than a standard damper solution, but it is also more effec-

4.3 Bridge dynamics

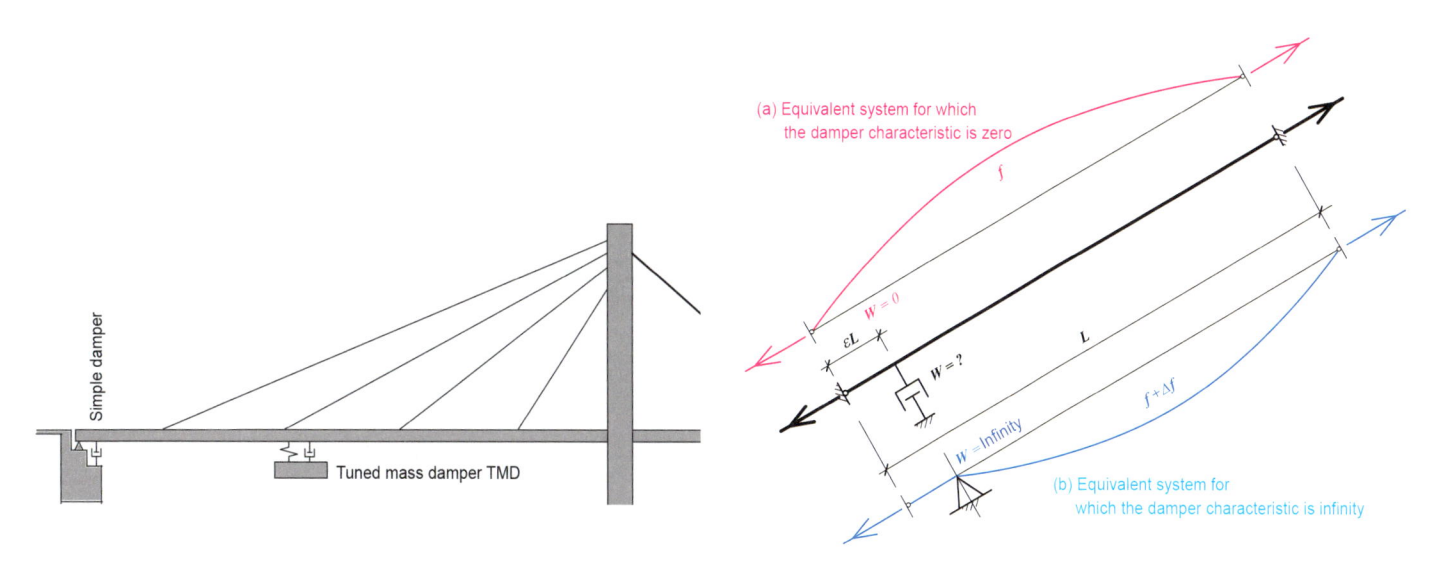

Figure 4.95 Important types of damping, schematically

Figure 4.96 Estimate of the best possible damping for a taut cable

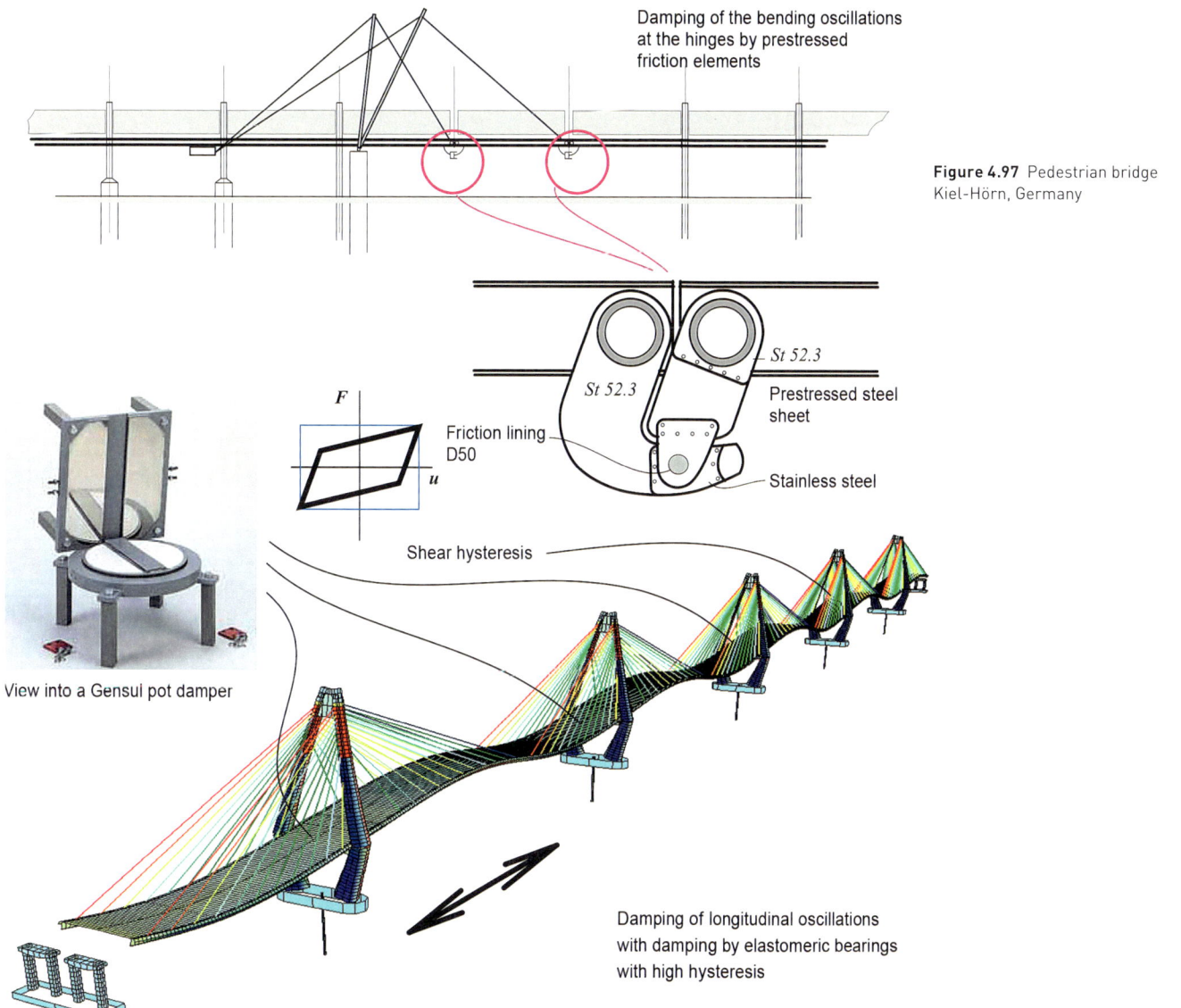

(a) Equivalent system for which the damper characteristic is zero

(b) Equivalent system for which the damper characteristic is infinity

Damping of the bending oscillations at the hinges by prestressed friction elements

Figure 4.97 Pedestrian bridge Kiel-Hörn, Germany

View into a Gensul pot damper

Damping of longitudinal oscillations with damping by elastomeric bearings with high hysteresis

Figure 4.98 Nhattan Bridge, Vietnam

tive. For maximum effectiveness the TMD should be located at the maximum of the shape of the mode to be damped. The most important element of the TMD is the mass – typically about 5% of the mass of the oscillating structure. If 5% mass is chosen and a parameters are optimized, the resulting damping achieved is δ = 0.30–0.35.

The effectiveness of the TMD can be deduced from Fig. 4.99: the same models as used for the standard damper can also be employed to estimate the effectiveness of the TMD, leading to an achievable damping of

opt δ = π · Δf / f   theoretical upper limit of damping by TMD, assuming optimum tuning.

The increase in natural frequency is, however, larger than for a standard damper, if optimum frequency tuning is assumed. The optimization of a TMD is somewhat complex, but there is clear guidance, particularly through the classical works of den Hartog.

The TMD mass of course also acts as a static mass, and accommodating it on the structure can pose problems, particularly if it is to be added as a retrofitting solution. There is hence a tendency to work with as low a mass as possible in order to limit the extra weight. Theoretically a mass of 1 to 2% may sometimes be sufficient. However, it should be kept in mind that the TMD tuning can vary as the structural frequencies shift with the loading conditions on the structure. The consequence is shown in Fig. 4.99: the resonance peak grows again, thus reducing the efficiency of a de-tuned TMD is lower. TMDs with low mass are particularly sensitive to de-tuning.

The design should also consider that TMDs, being somewhat like a machine, are more vulnerable than standard components. On the example of the Kehl-Straßburg pedestrian bridge, Section 4.3.5, we demonstrated how potential problems can be accounted for: the checks are performed assuming locked or imperfectly tuned TMDs, whilst at the same time reducing the required safety level.

Further examples of TMD applications are shown in Fig. 4.100. Lateral TMDs were required for the critical lateral oscillations during cantilevered erection stages of the Millau Viaduct (alternative design), Section 6.6.1.

The frequency of the critical lateral mode dropped below 0.2 Hz in some cases and there was no option of stiffening the deck at 300 m above ground.

The TMD shown in the figure travelled with the deck and was able to solve the problem: the frequency of the pendulum-type lateral TMD was adjustable through the curvature of the surface. This surface was a thin, flexible steel element the radius of which could be modified at its supports. The damper force was created through friction and was also adjustable. This alternative design of the bridge eventually was not chosen. Internal resonance between pylon or deck and cables is a typical problem with long-span cable-stayed bridges. Particularly during construction, pylon and cable frequencies often align near the tip of the free cantilever, as shown in the diagram.

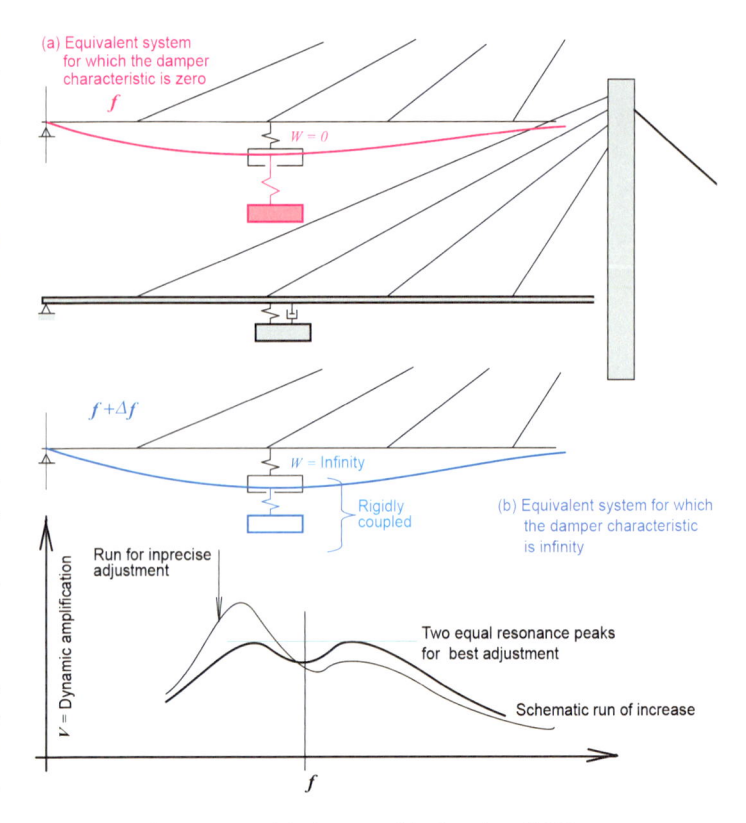

Figure 4.99 Assessment of the best possible damping of TMDs

4.3 Bridge dynamics

Millau Bridge, tip of free cantilevering

Friction surface

Friction mass-laterally fixed-1.5 t, adjustable for cantilever length

Roller on tracks

Roller on tracks

Steel box girder

Adjustable attachment

Laminated spring

35 t

lateral

Stonecutters Bridge, internal resonance during free cantilevering

The TMD eigenfrequency is adjustable by shifting the lateral guide

The friction damper is pressed by a laminated spring against the TMD, also adjustable via the friction force

1 Hz

Shortest cable: 0.85 Hz

Cable eigenfrequencies

Internal resonance possible

Lateral tower eigenfrequencies
0.19 Hz to 0.26 Hz

(longest) 0.23 Hz

Free cantilevering

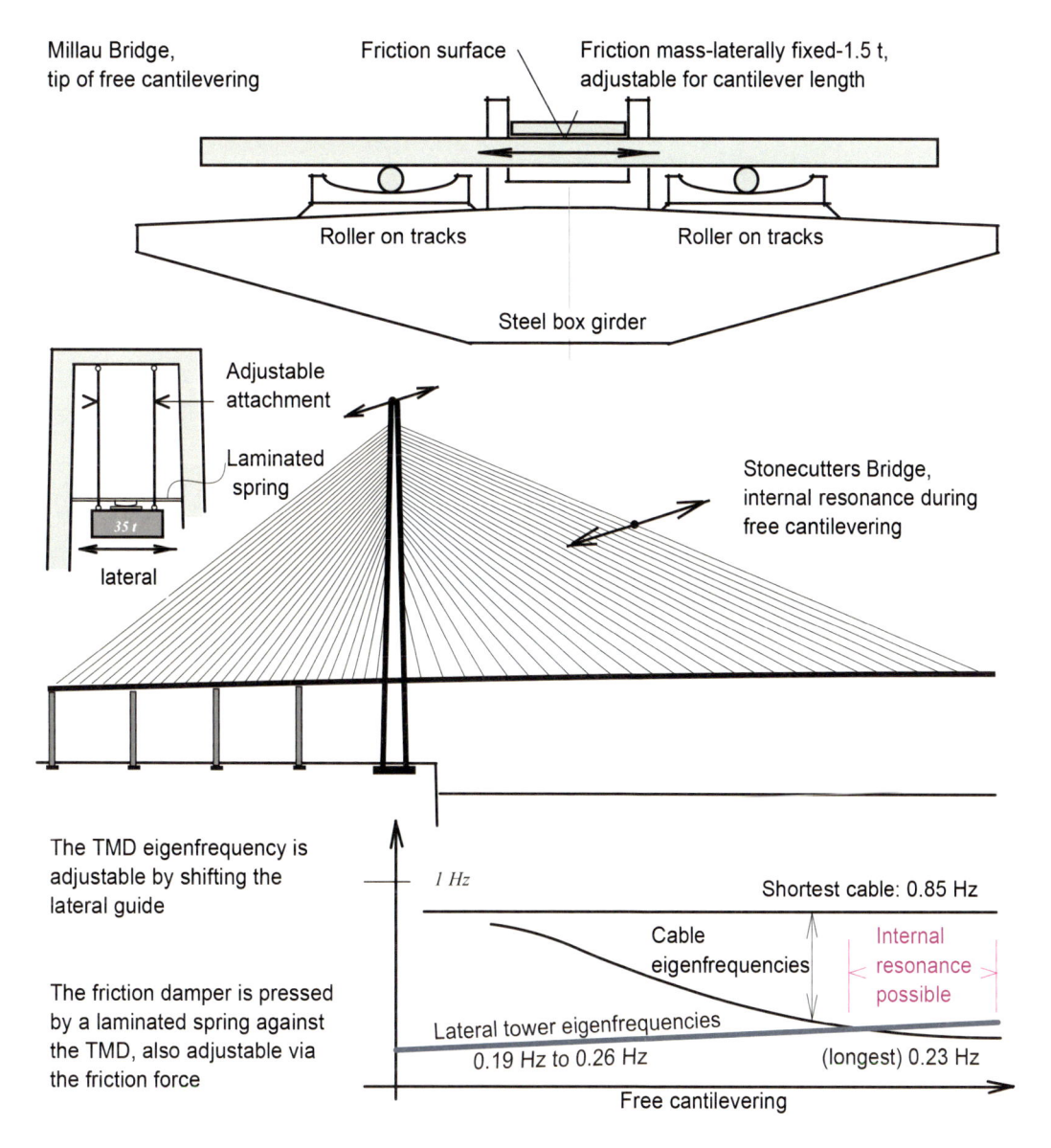

Figure 4.100 Examples of TMD applications

With the Stonecutters Bridge, Sections 2.2.6 and 5.1.3.3 it was found that the cylindrical pylon was prone to vortex-induced oscillations that could have amplitudes of 10–12 cm. The cables connected to the top of the pylon could then have developed amplitudes of several meters in the case of resonance. A standard cable damper typically used to prevent rain-wind-induced vibrations would not have been able to cope with this resonant excitation. Even with stronger cable damping solutions it would have been difficult to suppress the problem. The solution was therefore a TMD in the tower top, Fig. 4.100, which could reduce the tower oscillations to 1–2 cm amplitude. For the cables, a standard damping provided against galloping was then sufficient to also cope with the resonant case. This particular problem is quite often found in bridges with large span lengths. Sometimes, as was the case with the Stonecutters Bridge, the TMD is then initially omitted and kept as a possible remedial measure.

4.3.7 Wind tunnel testing
4.3.7.1 General
Wind tunnel tests are still the most important tool in the design of bridges with large spans. An important group of institutes, universities and private companies own large tunnels with up to 40 m widths. Well known are the following, though this list is not exhaustive:
- *Germany:*
  Aachen/Ruscheweyh
  Institutions with smaller wind tunnels: e.g. Wacker
- *Europe:*
  FORCE (previously Danish Maritime Institute)
  London, Nat. Phys. Lab., Novak + successor
  BMT Fluid Mechanics, UK
  Politecnico di Milano, Italy
  CSTB Nantes, France
- *worldwide:*
  BLWTL, Davenport (today King etc.), London, Ontario, Canada
  RWDI (Peter Irwin and Rowan Williams)
  Colorado University – successor of Cermak + Peterka
  Tongji University, China
  Akashi wind tunnel, Tokyo, Fujino, Shiraishi etc.

The institutions typically work according to the state of the art, based on information provided by the designing engineer. These should be as precise as possible because wind tunnel testing is not much regulated by codes or other guidelines. In the UK and, for example, Vietnam there are some guidance documents. Engineers need to thoroughly prepare the testing and measurements, such that a maximum of information can be obtained for the money invested. Typically wind tunnel tests deliver the following information:
- aerodynamic coefficients of the section (drag, lift, moment for the deck)
- wind forces on the freestanding pylon (typically given in terms of support reactions from the forces need to be back-calculated by the engineer)
- possibly measurements for the wind shields
- parameters of the vortex shedding: Strouhal number, force amplitude or amplitudes
- aeroelastic stability as a function of wind speed
- section forces.

4.3.7.2 Overview of important types of wind tunnel testing
Static tests
Tests on section models
Typical scope: section aerodynamics, C_D, C_L, C_M with
$F_D = C_D \cdot q \cdot H$, possibly $= C_D \cdot q \cdot B$ drag
$F_L = C_L \cdot q \cdot B$ lift
$M = C_M \cdot q \cdot B^2$ torsional moment

Measurements for α = ca. –12° to +12°, $\Delta\alpha$ = 1° or 2°
Provide aerodynamic coefficients
Of particular importance with bridge decks: $C_D(\alpha)$
$dC_L(\alpha)/d\alpha$ lift force gradient
$dC_M(\alpha)/d\alpha$
Model scale 1:50–100 particular attention to be paid to modeling the separation regions (structural details, handrails, steel sections, cable anchorages).

Where significant, the influence of the Reynolds number on the aerodynamic properties must be considered. Sections with sharp edges are more or less independent of Reynolds number. Problems may occur when the body is circular or oval or otherwise has a rather smooth shape or is composed partly of pipe sections. The distribution of the surface pressure is then Re-dependent, and consequently so are the aerodynamic force coefficients, particularly the drag. The Reynolds number is typically wrong by several orders of magnitude in wind tunnel tests, so the force coefficients may be erroneous.

For testing the aerodynamic force coefficients, the model is rigidly fixed, the forces are measured at the supports. The entire model is rotated to test at changing angles of attack. The forces are unsteady and therefore a time-averaging is performed.

Tests can, as shown in Fig. 4.101, be performed in smooth or low turbulent flow. The turbulence in the incident flow modifies the flow patterns (the location of separation points may change), resulting in different force coefficients. The most important difference is typically a reduction in the lift force gradient through the turbulence.

Typically the force coefficients in smooth flow are less favorable (higher drag, larger lift and lift gradient). Tests of aeroelastic stability are almost always performed in smooth flow.

The coefficients measured with some turbulence can be expected to be closer to reality and they normally lead to a more economic design. Turbulence is created in the tunnel by one or more grids and should be verified carefully if the results are sensitive to the level of turbulence. For a correct modeling, the turbulence intensity in the tunnel must be about 50% of that in reality, because the slow (low frequency) turbulent components cannot be properly reproduced in

4.3 Bridge dynamics

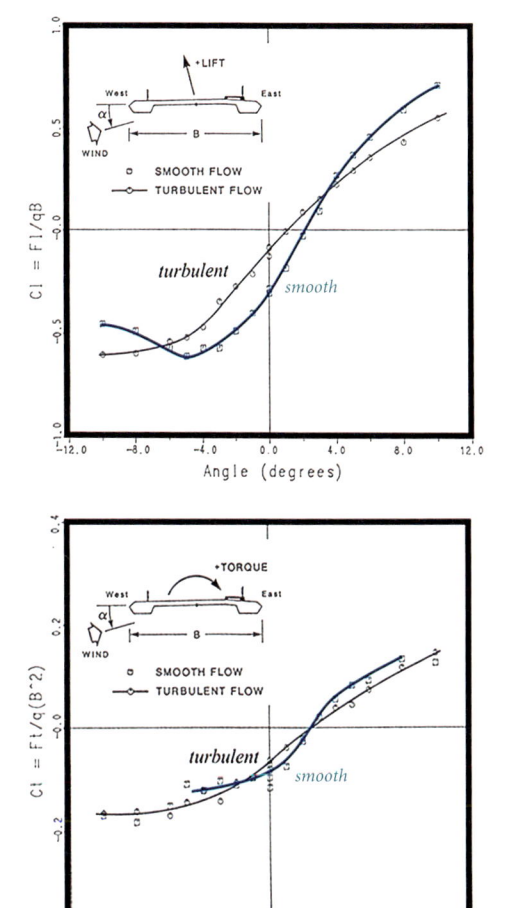

Figure 4.101 Helgeland Bridge, lift and moment coefficients in smooth and turbulent flow; measurements by Davenport

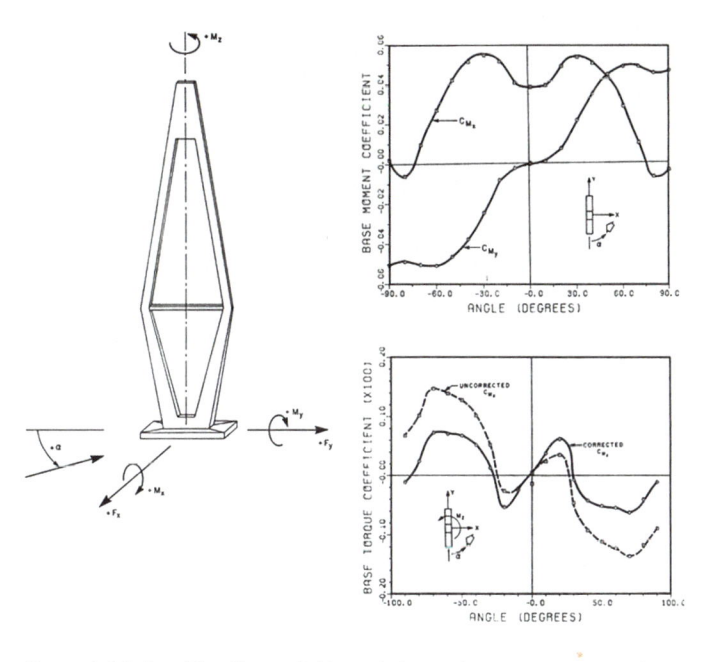

Figure 4.102 Sunshine Skyway Bridge, wind tunnel measurements on the tower, result plots

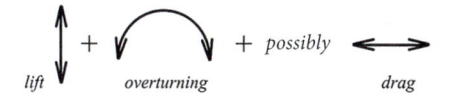

Figure 4.103 Active degrees of freedom

the tunnel due to the space limitations within the tunnel walls. We shall return to this issue later.

Sometimes mistakes are made in interpreting the wind tunnel results because the force measurements provide values in the wind coordinate system, see Fig. 4.101. These need to be transformed to the section coordinate system which is typically used in analyses.

Care should also be taken to properly represent cable anchorages in the tunnel; they can have a significant influence because they affect the section geometry of sometimes more than 10% of the length.

In the application of wind tunnel results, often the coefficients are linearized (C_D constant, C_L, C_M linearly dependent on α). The sign of the lift force gradient indicates – as discussed earlier – the susceptibility to galloping.

Tests on 3D structures with freestanding pylon
Such tests are performed in a so-called boundary layer wind tunnel in order to capture the effect of the turbulent wind speed distribution. The profile is being created using obstacles attached to the tunnel floor upstream of the test section.

Normally the test is performed on a rigid model. The support reactions are then measured and, assuming certain configurations for the coefficients, the coefficients are back-calculated for a range of angles of incidence (azimuth angle), Fig. 4.102. Recently, additional pressure taps have been being used, though their usefulness is questionable because the pressures vary very sharply along the structure.

Dynamic tests on 2D models

The main reason for such tests is to investigate the aeroelastic stability of the oscillating deck. Tests are performed on section models, i.e. on a cut out prismatic part of the deck. The model is supported elastically with a calibrated mass and stiffness, such that it can exert oscillation in the tunnel. The dynamic properties are calculated from the real structure according to similitude laws – see below.

The models are mounted with two or three degrees of freedom (lift and moment and sometimes drag direction), Fig. 4.103, while other directions are fixed. An example is shown in Fig. 4.104: this model has three degrees of freedom which are controlled by springs and dampers.

The model scale is typically 1:50 to 1:100, very small models are not considered reliable anymore because the separation regions cannot be properly replicated.

Fig. 4.105 shows for a 2D model the important model law: the geometric representation of the track of the oscillating body inside the flow needs to be correct. This is best expressed as follows: the reduced velocity $V_{red} = V/f \cdot B$ (ratio of track wavelength and deck width) needs to be identical in the model and on the full-scale structure.

In a flutter test the dynamic behavior of a freely oscillating model inside the flow is being tested: the flow velocity is increased in small steps and at each stage the model is being slightly displaced manually in order to measure the decay behavior of the torsional and vertical oscillation. In such test normally a smooth flow and very low damping is used, such that a sharp stability boundary exists. The test is being repeated for angles of incidence $\alpha = -6°$ to $+6°$, $\Delta\alpha = 2°$.

A characteristic response is shown in Fig. 4.106, which clearly identifies vortex shedding resonances for bending and torsion modes as well the aeroelastic stability limit. From such a diagram the flutter onset wind speed can be determined as well as the Strouhal number and an approximate value fo the lift coefficient.

The test results form a basis for determining the 3D response behavior of the complete structure, but it is not identical. A number of 3D effects such as coherences and mode shapes along the structure need to be taken into account.

Further information can be obtained from analyzing the decay behavior as well as the amplitude and phase relationships of the individual tests; the derivatives can be estimated from frequency and decay constant. It is useful to first perform the tests with only one degree of freedom, which simplifies the analysis. Scanlan himself still analyzed the tests this way. The information obtained, however, is not very accurate and also not complete because V can only be tested up to the stability limit, whereas for the analysis of the overall structural response derivatives beyond the stability limit are also required.

Fig. 4.107 shows that from 2D models under turbulent flow one can also calculate dynamic force reactions. This method was developed by Davenport. The turbulence is created by grids and must be verified accurately in terms of its spectrum. There are further uncertainties such as, for example, the missing low frequency turbulence components, and coherence-based corrections would also be required. The method is thus not very accurate and often leads to an over-dimensioning.

The reproduction of turbulence in the wind tunnel is shown in Fig. 4.108. As mentioned previously, the low frequency components are missing and are hence to be accounted for separately in the later analysis.

In terms of coefficients and derivatives it is most important to check that the reproduction is correct in the asymptotic region (the region of system resonances), which should be verified through additional validation measurements.

A safe and more accurate method for determining the derivatives is today the force measurement on harmonically excited 2D section models. This is a highly sensitive procedure which is standard nowadays. The wind speed or the frequency of oscillation is successively increased and the support reactions are measured simultaneously. The derivatives then follow directly through an analysis of phase and amplitude relationships between motion and forces.

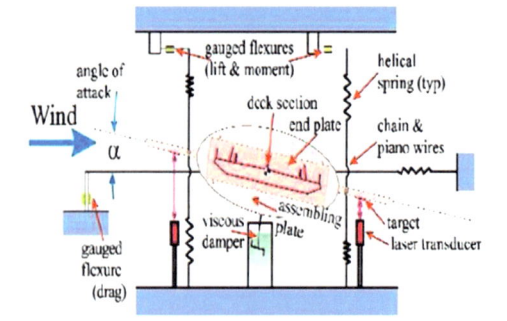

Figure 4.104 Audubon Bridge Louisiana, section model mounted in the wind tunnel of RWDI Consulting

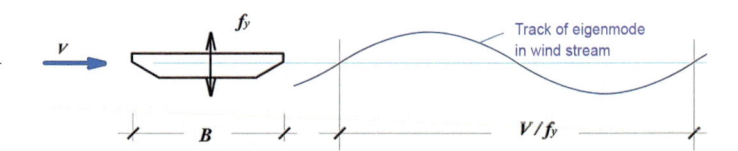

Figure 4.105 Model law reduced wind speed $v_{red} = v/f_v \cdot B$ to be identical

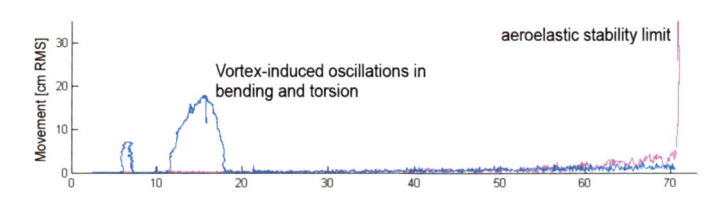

Figure 4.106 Phu My Bridge, model oscillations in the wind tunnel

Dynamic 3D full models in boundary layer wind tunnels
This is a very complex procedure: the static/dynamic bridge system needs to be as completely and accurately as possible replicated in the wind tunnel, following as well as possible the similitude laws. The main tasks of such 3D models are:
- the measurement of displacement amplitudes – typically performed directly by transducers or otherwise, or indirectly through accelerometers
- the measurement of stresses (curvatures) typically using strain gauges.

The conclusions to be drawn from such models are prone to inaccuracies. For studies of instabilities these models are too rough.

The tests are typically performed in boundary layer wind tunnels. Here the comments above, on the required accuracy of the spectra, also apply. Inaccurate spectra can induce large errors.

The similitude laws are only satisfied partially:
- Reynolds number not satisfied – but also not achieved in the section model tests
- Cauchy number not satisfied
- Froude similitude is typically satisfied – after long discussions, it is today accepted that this is the most important requirement for an accurate modeling.

The calibration of the relevant modal parameters is particularly challenging and expensive. This is typically only achieved approximately. In order to model the frequencies very light materials need to be used in the model such that the model can then be fine-tuned through the addition of masses.

The testing of a full model is typically required for major bridges in the USA, Canada and other countries in the Americas. The method is not popular in Europe because of the costs and its shortcomings. Typical model scales are M = 1:150 to 1:200. A special case was that of the Akashi Straits Bridge, where a comparatively large scale of M = 1:100 was chosen, achieved by a newly built tunnel of 42 m width.

As numerical models continue to improve, the focus will further shift towards section modeling and derivative measurements.

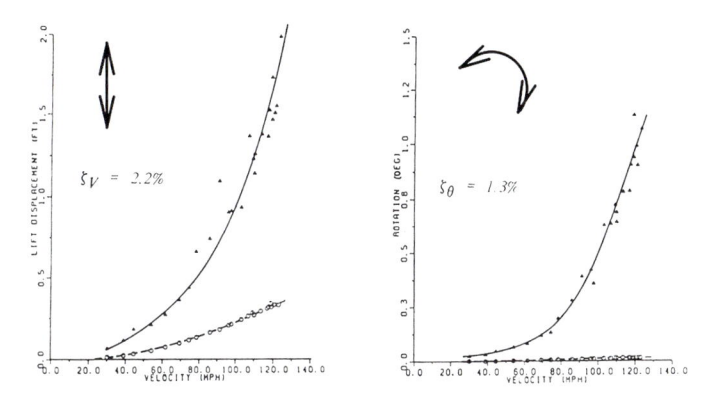

Figure 4.107 Example for amplitude measurements on section models in turbulent flow

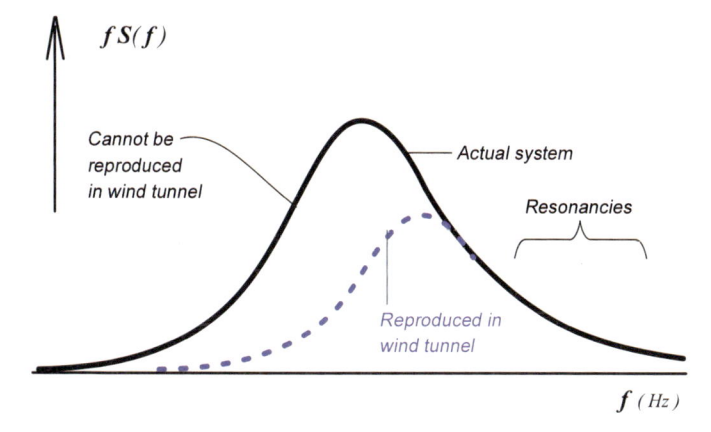

Figure 4.108 Tests with 50% turbulency: low frequency components missing

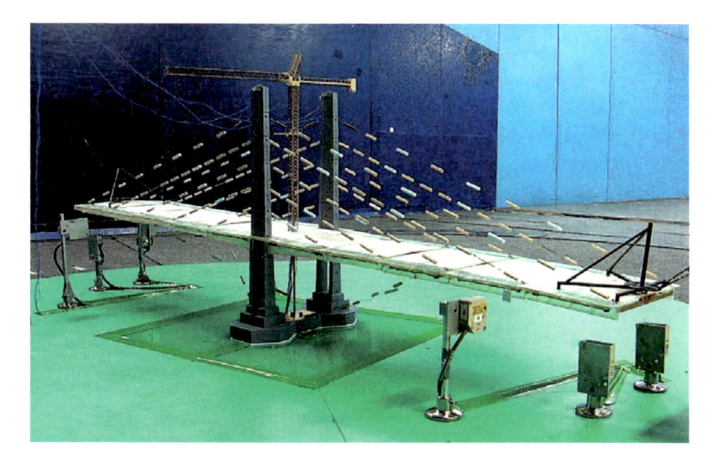

Figure 4.109 Golden Ears Bridge, Vancouver, B.C.: full model during construction in RWDI wind tunnel

4.3.8 Earthquake

Seismic loading is a dynamic support excitation through mostly lateral shaking of the ground. This causes inertial effects and thus time-dependent displacements in structures such as cable-stayed bridges. Because of the magnitude of the loading, the dimensioning is typically performed assuming plastic behavior, which is a significant difference to, for example, wind design or other dynamic effects. Earthquake loading of cable-stayed bridges is treated in a way similar to that of other bridges or buildings. Therefore the following will only provide a few general comments which are in line with numerous similar international and national specifications. Design for seismic effects should not violate some general principles:
- regularity, symmetry, compactness, such that forces can be resisted along short load paths and without major eccentricities,
- avoid coupled modes as far as possible,
- particular care with the design of joints (pounding or loss of bearing support),
- design for certain collapse mechanisms, e.g. formation of plastic hinges at equal elevations.

The design should satisfy two main requirements:
a) after the higher design earthquake there shall be a residual load capacity
b) for a lower intensity earthquake the degree of damage shall be limited to a defined level which may be defined on a member-by-member basis.

These requirements are defined in terms of limit states of ultimate or serviceability type. Under ultimate conditions plastic deformations are allowed to a certain degree, which is expressed through a behavior factor q which is determined by the provided ductility, i.e. the ratio between total and elastic deformations. Serviceability conditions are mostly defined through deformation limits.

For the design earthquake, a_g = design value of horizontal ground acceleration.

This is the maximum lateral acceleration of the rock surface as reference ground. Typically the magnitude of ground accelerations varies with soil conditions. The soil conditions also determine the spectral properties of the time-history of the shaking, through filtering effects originating from soil structure and damping properties. This complex behavior results in an elastic response spectrum which expresses the time-history in terms of the maximum acceleration response experienced by separated single-degree-of-freedom oscillators with a structural damping of critical value $\xi = 0.05$, Fig. 4.110.

Typically the vertical seismic component, which can also be expressed as response spectra, is smaller and less critical. The periods of vibration of cable-stayed bridges, as for other long-span bridges, are rather long (= low frequencies). For this reason, the hyperbolically decaying branch of the response spectrum is of major importance, Fig. 4.111. The values in this region are highly dependent on the soil conditions.

As the sketch of Fig. 4.111 from the book of Petersen shows, the response behavior for weak soft soil becomes less favorable for long vibration periods. This is reflected in all design specifications, though with some differences, Fig. 4.110. In design bases for major bridges often even more onerous assumptions are made, even though the soil properties are comparable all around the world. This fact is not satisfactory and has in the past often led to over- or under-designed structures.

There are a number of further definitions and regulations, for example on the basis of stiffnesses, damping, taking into account the artificial damping, system simplifications etc. These are to be found in design codes or, mostly for major bridges, project-specific design documents. The assumptions are similar around the world – see the relevant design codes.

Section forces under earthquake are usually calculated on linear systems, as if the system remained elastic. The design checks are then based on so-called design spectra, which account for the plastic behavior through the inclusion of the behavior factor. The analysis and design checks are typically performed using the multi-modal response spectrum method, in which the excitation and response are decomposed into modal components. The superposition of the modal contributions is done using probability theory. It is somewhat unsatisfactory that non-linear systems are treated by modal analysis, which is strictly only applicable to linear systems.

Also, the reductions for non-linear response are preempted through the design spectrum used, even though the amount of plastic deformation under the given earthquake input is not actually known until the response and the degree of ductility utilized is evaluated.

A correct, though more sophisticated and demanding, method is a non-linear time-history analysis, in which the force-deformation behavior of members can be defined. Using this method, however, a number of artificial seismic input series need to be derived. This requires a specific numerical methodology.

The seismic input typically also excites, unlike wind loading, the high frequency modes. The minimum limit of the total mass to be represented in the model (mass participation) is typically 80–95 % in each direction, which indirectly determines the number of modes to be taken into account. In cases where very heavy members, e.g. gravity foundations, are present this rule can often not be satisfied, because the high mass participation is then concentrated in very high modes. An example is the Geo Geum Bridge in South Korea, Fig. 4.112: the heavy caisson foundations account for about 75 % of the bridge mass. These are founded directly on rock and are also supported laterally, which gives them overturning modes with frequencies of 10–20 Hz. Critical mass contributions therefore occur for modes at around number 250, which renders the calculation rather expensive and inaccurate because a high number of mostly irrelevant modes have to be included. In such cases it is sensible and sufficiently accurate to perform the analysis in two stages: in the first

4.3 Bridge dynamics

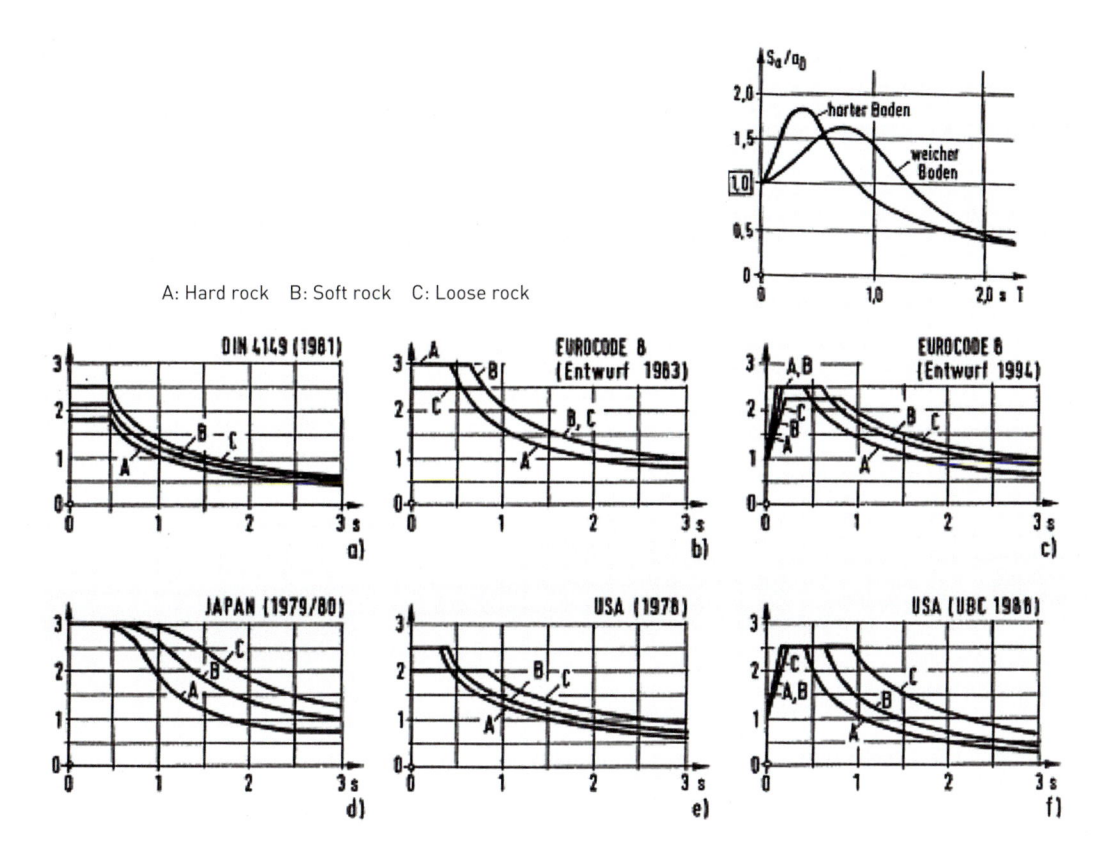

A: Hard rock   B: Soft rock   C: Loose rock

Figure 4.110 Different approaches for the shape of the response spectrum depending on soil conditions in the range of long periods

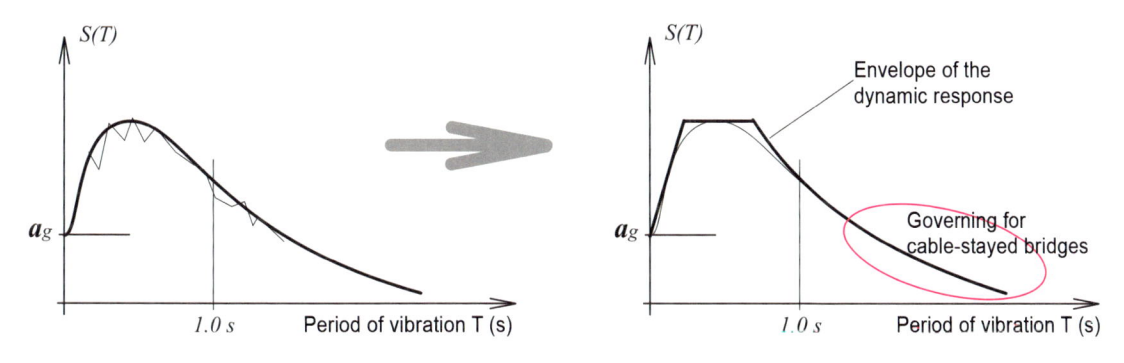

Figure 4.111 Period of vibration of a reference single-degree-of-freedom oscillator with $\xi = 0.05$

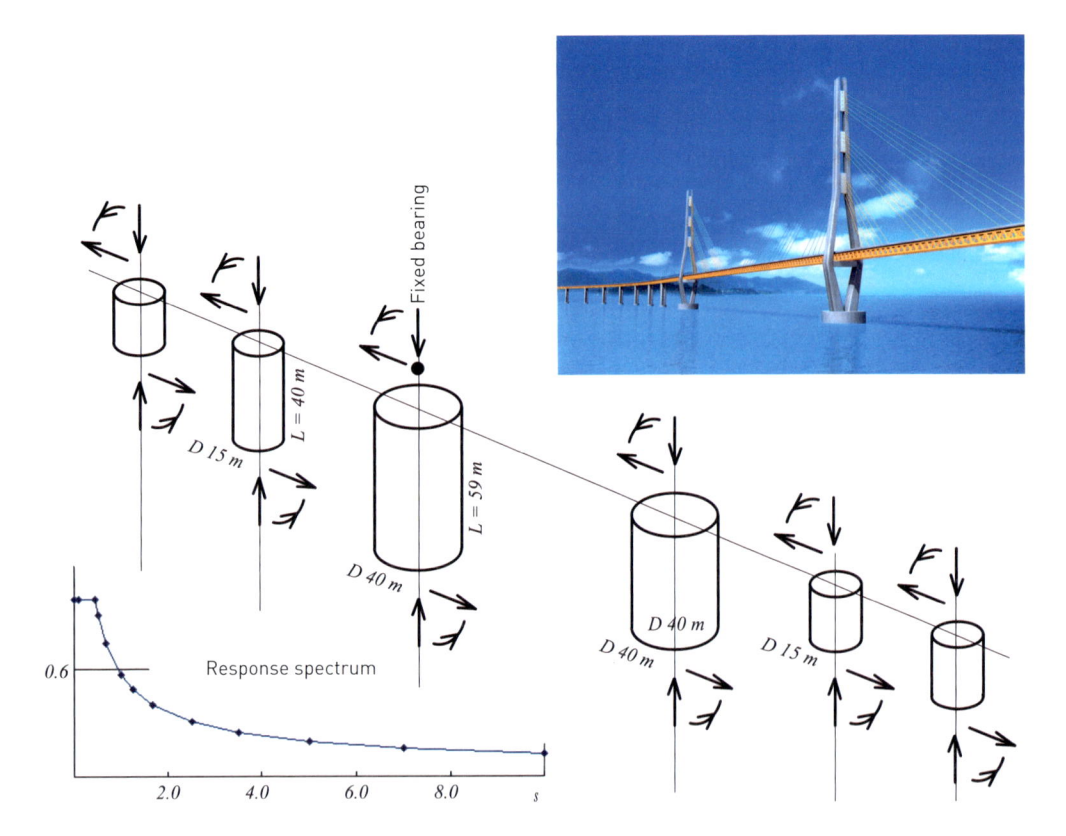

Figure 4.112 Caisson foundations of Geo Geum Main Bridge, South Korea, under longitudinal earthquake

stage the foundations are modeled as rigid elements with six degrees of freedom and subjected to linear seismic input. This simple model provides the design forces on the foundation itself. The calculation of section forces in the superstructure then proceeds by assuming the foundations as mass-less. Should this still provide significant loads on the foundations, these are to be superimposed with those determined in stage 1 by SRSS.

A major difference to the seismic behavior of buildings is that a potential difference of excitation between individual foundations should be taken into account. Major bridge structures have large longitudinal extensions, such that the excitation waves reach the foundations at different points in time. This leads to a phase shift in the direction of propagation which typically reduces the structural demand. On the other hand, coherences occur below the foundations because of the somewhat random nature of the ground motion and the local differences of soil layers or rock formation.

A synchronous acceleration history can only be expected directly in the epicentre of the earthquake; in this rare case the vertical S-waves reach the surface simultaneously. In most cases, however, a horizontal distance between site and epicentre, Fig. 4.113, leads to waves approaching the surface at an inclination and with time shifts along the structure. The superposition of approaching and reflected waves creates the so-called Love waves along the surface with propagation speeds between those of P- and S-waves, i.e. at around 500–1000 m/s. Disruptive shaking is also caused by Rayleigh waves, which are induced at singular surface configurations as secondary waves. Their speed equals that of S-waves, the slowest of all.

The phase shift as well as the coherences are important effects which tend to reduce the excitation. Unfortunately there are few methods to estimate the coherence even for homogenous soil conditions. This is of course related to the wave propagation near the surface. As a rough estimate from field measurements in Taiwan and Panama one may use:

$$\text{coh} \approx \exp(0.1 \cdot \omega \cdot x / V_S)$$

with V_S speed of Love waves.

For the 1500 m long Nhattan Bridge, Vietnam, Fig. 4.114 (bottom), this provided a factor 0.8 reduction in the longitudinal force in the girder through the phase shift and coherence decay. For similar reasons the reduction was 0.75 for the Bukhang Bridge, Fig. 4.114 (top).

It is also possible, particularly with large expansion lengths of a cable-stayed bridge that soil properties vary between different foundations, leading to different excitations. Fig. 4.115 shows as an example the 1052 m long Panama Bridge, Section 5.1.3; along the bridge axis there is a repeat change between stiff and soft foundations.

The soft foundations were the consequence of sand deposits above the rock. It was therefore to be expected that higher frequency surface oscillations could reach these points in antiphase. This could have led – particularly at piers 1 to 4 – to unwanted transverse deformations under lateral shaking, leading to large restraints in the superstructure. It was therefore decided to provide freely moveable bearings atop piers 1 to 3. Also at other points problematic effects were expected due to asynchronous excitation. The non-linearity, phase shifts and coherence decays cannot be properly accounted for by modal analysis. The analyses for Panama Bridge were therefore performed using complex time-history simulations. Considering carefully the amplitudes at stiff and soft response spectra an approximate distribution of the phase differences could be selected, Fig. 4.116. Based on this, a series of time-histories was created.

For the analysis the stiffness of dissipative members – foundations and pylons – was defined by a bilinear relationship. Further, the bearing friction was modeled realistically. A further example for seismic analysis is given in Section 6.6.2 for the Rion-Antirion Bridge.

4.3 Bridge dynamics

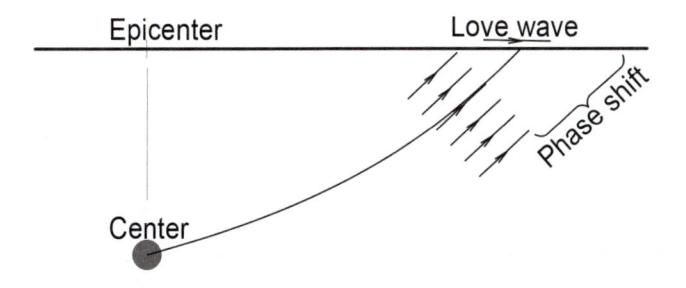

Figure 4.113 Origin of a Love wave

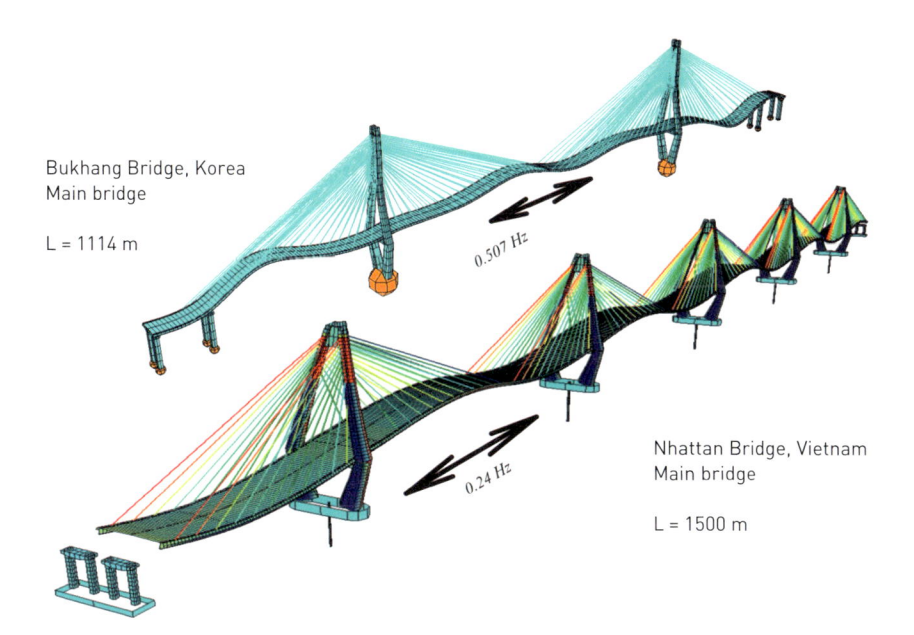

Bukhang Bridge, Korea
Main bridge
L = 1114 m

0.507 Hz

Nhattan Bridge, Vietnam
Main bridge

L = 1500 m

0.24 Hz

Figure 4.114 Earthquake cases with longitudinal excitation

1 – Laterally free
2
3 – Laterally free
4
hard
hard
soft
hard
soft
hard
hard
Laterally free

Love wave speed 750 m/s

Response spectra

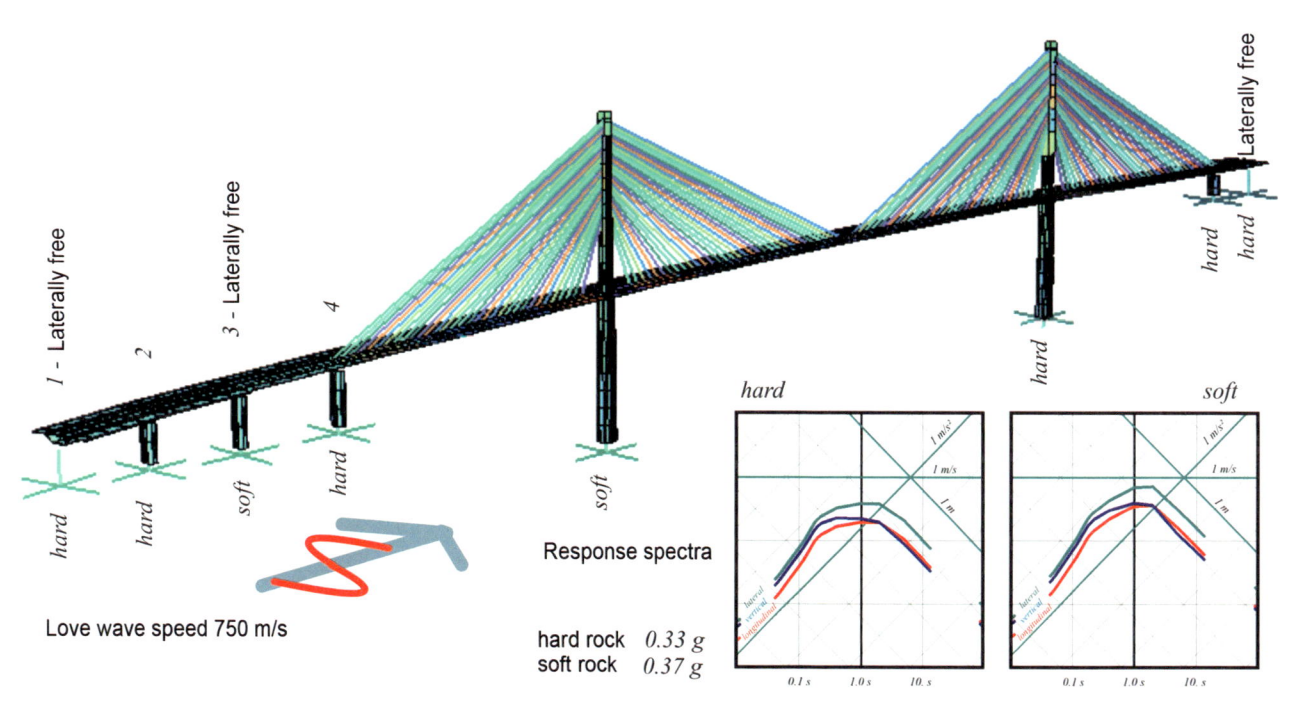

hard rock $0.33\ g$
soft rock $0.37\ g$

hard   *soft*

Figure 4.115 Panama Bridge, L = 1052 m

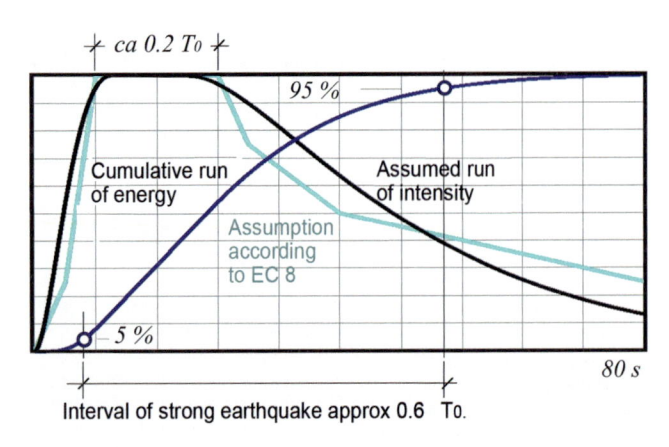

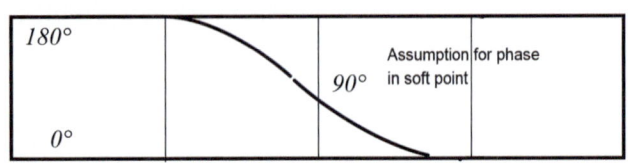

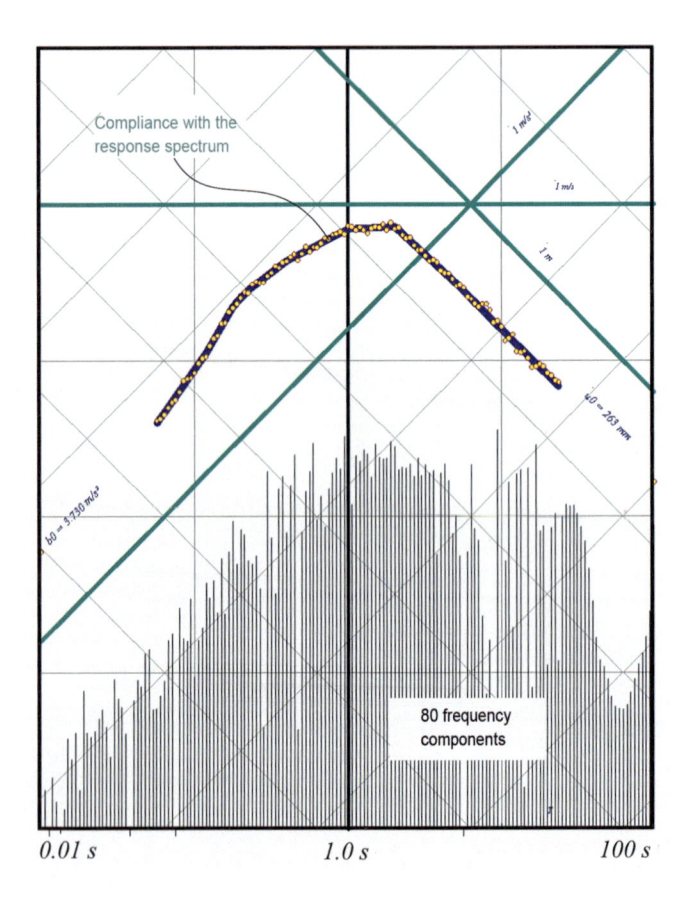

Figure 4.116 Panama Bridge, artificial earthquake input

4.4 Protection of bridges against ship collision
4.4.1 Introduction

The collision of ships with bridge piers represents a real danger for bridges over major waterways with large ships. Since long spans are required in these situations cable-stayed bridges are often an appropriate solution. They have, therefore, often to be protected against ship collisions.

Impact forces can be very high. Currently up to 800 MN is considered for seagoing ships against the main piers of the Femer Crossing over the Baltic Sea. Thus ship impact becomes an important design criterion. In order to resist large ship impact forces, the foundations may have to be strengthen considerably. Longer spans may thus be more cost effective than shorter spans because of savings in the foundations. Current worldwide codes show different impact forces up to a factor of 2. The local conditions for ship impact have thus carefully to be investigated. A list of ship collisions with bridges is outlined in Table 4.3.

Not only the main piers but also those in the side spans may be hit by ships, see Fig. 4.117.

The different probabilities of ship collision as a function of the distance from the navigational channel is given in national codes.

In 1980 the author had to design a protection against ship collision for the already completed Zárate-Brazo Largo Bridges across the Paraná River in Argentina. An investigation into the state of the art at that time showed that little systematic research and virtually no code recommendations existed. The author, therefore, had to make his own investigations.

The most important characteristics of any investigation on ship collision with bridges are the determination of the impact forces and the structural design of the protection against these forces. In the following, the author outlines his personal experience gained from more than 25 years of dealing with the protection of bridges against ship collision. The examples of the structural design of the different types of pier protection have mostly been designed or checked by the author's firm, Leonhardt, Andrä and Partners.

4.4.2 Collision forces

The only theoretical investigation available in 1980 was that by Minorsky [4.41], who had investigated the forces acting in head-on collisions between two ships. Fig. 4.118 shows the linear relation between the volume of the deformed steel in both ships and the absorbed energy. However, this determination of impact forces is only valid for the collision between two ships.

For the protection of bridge piers, a head-on collision of a ship against a rigid wall as shown in Figs 4.119 and 4.120 is important. For this purpose, the results of collision tests made in Germany under the direction of Gerhard Woisin were evaluated, Fig. 4.121 [4.42]. Woisin found a dynamic relation between the impact forces over time as shown in Fig. 4.122, whereby the maximum force at the beginning of the collision for about 0.1 to 0.2 seconds increases

4.4 Protection of bridges against ship collision

Table 4.3 Major ship collisions with bridges

Bridge location	Year	Lives lost
Severn River Railway Bridge, UK	1960	5
Lake Pontchartrain Bridge, USA	1964	6
Maracaibo Bridge, Venezuela	1964	0
Chesapeake Bay Bridge, USA	1970	0
Sidney Lanier Bridge, USA	1972	10
Chesapeake Bay Bridge, USA	1972	0
Lake Pontchartrain Bridge, USA	1974	3
Welland Canal Bridge, Canada	1974	0
Tasman Bridge, Australia	1975	15
Fraser Bridge, Canada	1975	0
Pass Manchac Bridge, USA	1976	1
Benjamin Harrison Memorial Bridge, USA	1977	0
Union Avenue Bridge, USA	1977	0
Tingstad Bridge, Sweden	1977	0
Berwick Railroad Bridge, USA	1978	0
Second Narrows Railway Bridge, British Columbia, Canada	1979	0
Tjorn Bridge, Sweden	1980	8
Sunshine Skyway, USA	1980	35
Lorraine Pipeline Bridge, France	1982	7
Hannibal Railroad Bridge, USA	1982	0
Sentosa Aerial Tramway, China	1983	7
Volga River Railway Bridge, Russia	1983	176
Lake Pontchartrain Bridge, USA (third major incident)	1984	0
St. Louis Bridge, Canada	1985	0
Bonner Bridge, USA	1990	0
Tostero Bridge, Sweden	1990	0
Hamburg Harbour Bridge, Germany	1991	0
Claiborne Avenue (Judge Seeber) Bridge, USA	1993	1
CSX/Amtrak Railroad Bridge, USA	1993	47

Figure 4.117 Benjamin Harrison Bridge, USA, 1977

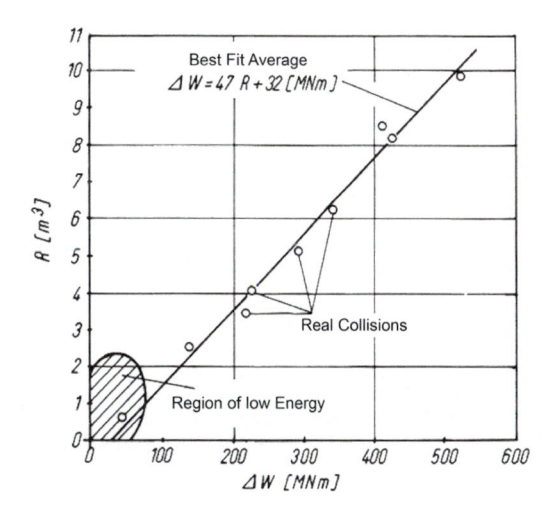

Figure 4.118 Collision energy by Minorsky

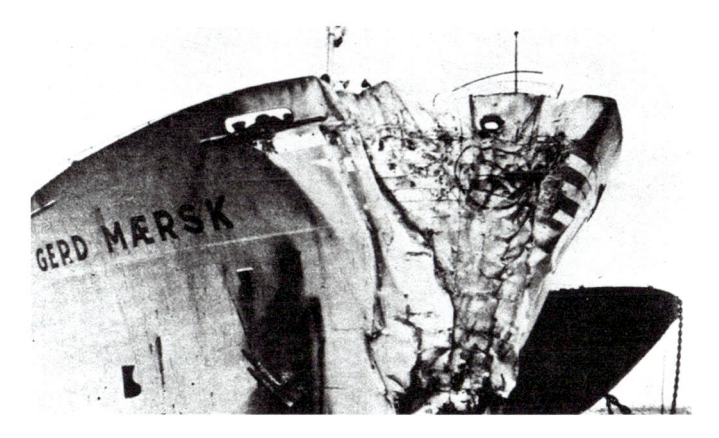

Figure 4.119 Collided ship, Newport Bridge, RI, USA

Figure 4.120 Damaged pier, Newport Bridge, RI, USA

up to double the value of the average collision force over several seconds.

From Woisin's test results the author concluded that equivalent static impact forces from a direct collision of a major ship against a rigid wall are, in a first approximation, proportional to the square root of the dead weight tonnage (DWT) of the ship, Fig. 4.123. However, within this relationship a wide scatter of impact forces exists for ships with the same DWT due to their widely varying hull structures and collision speeds. The author thus proposed the formula for the impact forces with a scatter of ±50%.

$$P = 0{,}88 \cdot \sqrt{DWT} \cdot \pm 50\% \quad \text{(Svensson, 1981 [4.43])},$$

where P is the equivalent average impact load in MN.

This formula was published by the author in Germany in 1981 [4.43], in the IABSE Periodica in 1982 [4.44], and in the Proceedings of the IABSE Colloquium Copenhagen on Ship Collision with Bridges in 1983 [4.45].

In 1980 a freighter collided with an unprotected pier of the Sunshine Skyway Bridge across Tampa Bay on the west coast of Florida, USA. Four hundred metres of bridge beam fell into the water and 38 people died, see Section 3.5. As a result of this tragedy, studies were initiated in the USA under the guidance of the Federal Highway Administration (FHWA) to develop a design code for use by bridge engineers in evaluating structures for vessel collision. This effort culminated in 1991 in the Guide Specification and Commentary for Vessel Collision Design of Highway Bridges [4.46]. The principal investigator was Michael A. Knott under the guidance of an AASHTO special ad-hoc committee of which the author was a member. Knott simplified the wide scatter from Woisin's impact tests as outlined in Svensson's ±50% relationship, Fig. 4.123, by using the 70% fractile of the average ship impact force to evaluate the bridge response to impact, and to size members to resist the impact forces, Fig. 4.124. The 70% fractile increased Svensson's 50% fractile factor 0.88 to 0.98. In addition, Woisin recommended to add a factor v/8 to account for varying velocities.

The AASHTO specification thus uses the relationship

$$F = 1{.}11 \cdot 0{.}88 \sqrt{DWT} \cdot \frac{v}{8} = 0{.}122 \cdot \sqrt{DTW} \cdot v \quad \text{Fig. 4.125 [4.46]}$$

where F is the equivalent average impact load in MN and v is the ship impact speed in m/sec.

The use of the maximum impact force, i.e., twice the average force, Fig. 4.122, was not recommended for use by the AASHTO specification since it was considered that its time duration at the beginning of the collision was too brief to cause major problems to most structures. The average impact force over time was, however, increased to a 70% fractile level to stay on the conservative side of the analysis.

Later other formulas were produced for the determination of impact forces. In the old Eurocode 1, Part 2.7, the design collision force was estimated as

$$F = \sqrt{K \cdot m} \cdot v \quad \text{(Eurocode)} \quad \text{whereby K is an equivalent stiffness,}$$

m is the impact mass and v the velocity. This old Eurocode equation gives slightly higher values than the AASHTO equation, Fig. 4.126.

In 1993, Prof. Pedersen developed an empirical formula through numerical computation which gives about twice the impact forces of the AASHTO formula [4.47].

Also in 1993 the International Association for Bridge and Structural Engineering (IABSE) published a state-of-the-art report on ship collision with bridges [4.48]. The Working Group, of which the author was a member, was headed by Ole Damgaard Larsen in 1997.

Recently, highly sophisticated numerical FEM-simulations of ship collisions using the interaction between master and slave surfaces to simulate a collision between a ship's bow and a concrete bridge pier have been executed by the Tongji University, Shanghai, China [4.49].

Fig. 4.126 gives the typical calculated dynamic impact forces for a head-on collision of a 50,000 DWT bulk carrier against a rigid wall vs time, compared with the estimated equivalent static loads from the various formulas mentioned above.

Please note that the indicated Eurocode is the old version. The current Eurocode uses Pedersen's formula.

Naturally the simplified code equations can only give rough estimations for impact loads. The AASHTO and Eurocode equivalent loads appear to lie on the safe side when compared with Tongji's calculated dynamic values, whereas the Pedersen equation gives about double those values.

The Tongji calculations have also shown that the collision against the elastic foundation on piles in Fig. 4.127, reduces the peak value of the collision force by about 50% against that against a rigid wall, Fig. 4.128.

The discussion of how to determine the impact forces from ship collisions against bridge piers has not yet been finally settled. If equivalent static loads on the bridge piers are used to simplify their sizing it is questionable what mean should be used: the absolute peak value, the local mean peak or the global mean value, Fig. 4.129. A comparison of these different peaks as calculated by Tongji University and the equivalent static loads from AASHTO and Eurocode is shown in Fig. 4.130. Please note that the AASHTO formula uses the global mean value over time corresponding to the average force from Fig. 4.122. Also, the different structural types of ships' hulls mean that single equations can only approximate the impact load. One formula cannot cover all types. The ±50% scatter of impact forces as indicated in Fig. 4.123 still gives a realistic scenario.

The best possible calculation of impact forces today would use dynamic numerical FEM-simulation similar to that outlined above by the Tongji University, taking into account the structure of the design ship's hull as well as the structural details of the collision structure.

4.4 Protection of bridges against ship collision

Figure 4.121 Collision tests by Woisin

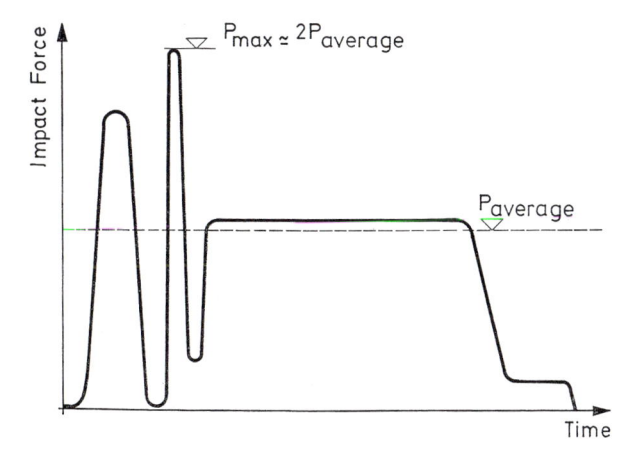

Figure 4.122 Dynamic impact forces vs time by Woisin

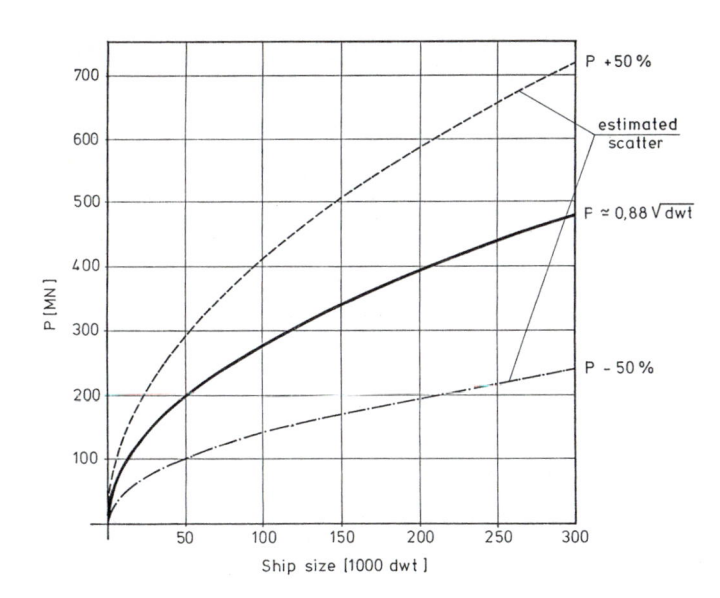

Figure 4.123 Equivalent static impact, forces vs ship size by Svensson

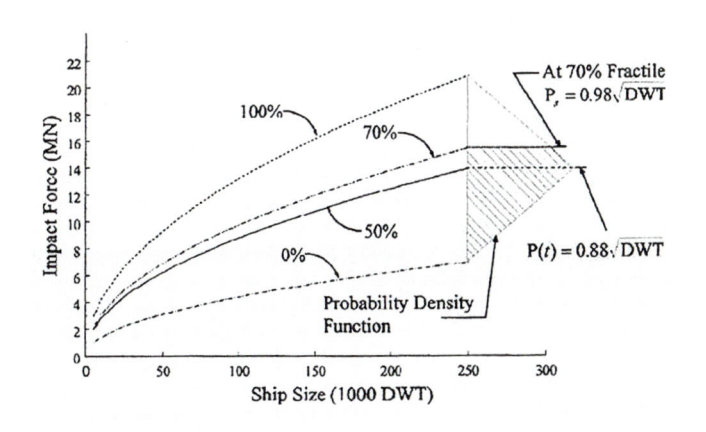

Figure 4.124 Impact forces vs ship size by Knott

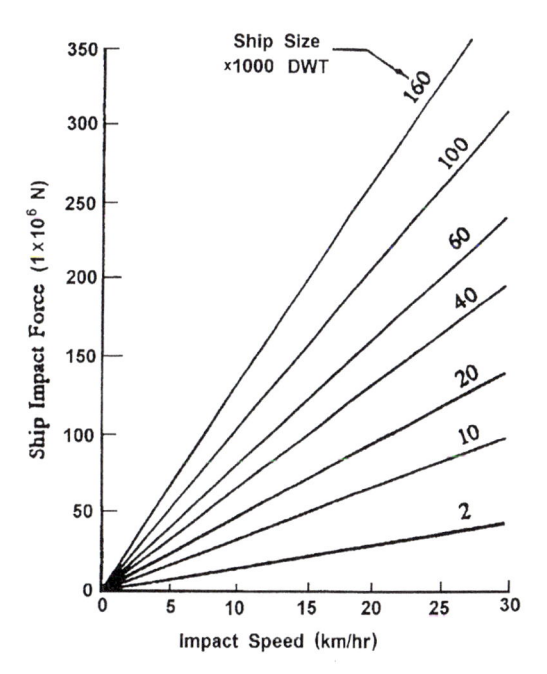

Figure 4.125 Impact forces by Knott

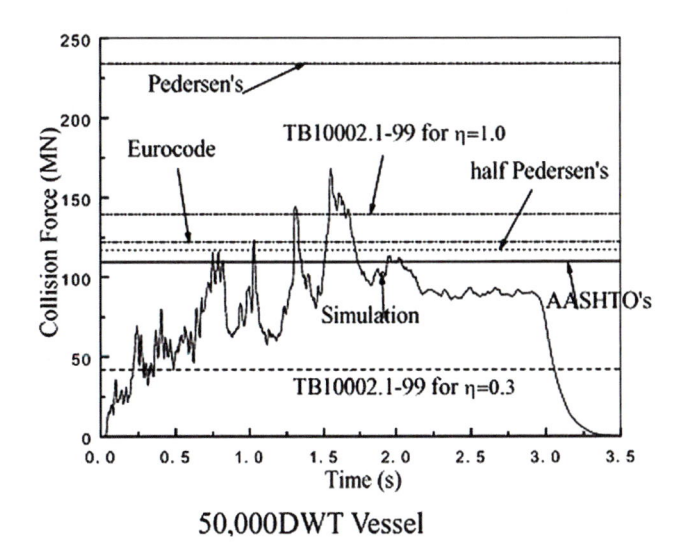

Figure 4.126 Collision forces vs time by Tongji University, Shanghai

In addition, the forces and stresses in the pier and the piles during the full impact time should be investigated. From the governing design stresses equivalent static loads could be determined.

4.4.3 Protective structures
4.4.3.1 General
The following possibilities exist for protecting bridges and other structures from ship collision:
- place out of reach on shore,
- deflect ship by artificial islands or by guide structures,
- make piers strong enough to withstand direct collisions.

In the following, executed examples of each these types of protections are given.

4.4.3.2 Out of reach
The safest method of protecting piers from ship collisions is to place them out of reach on land. The additional costs of an increased span length may well be offset by savings on pier protections. In the case of an arch bridge this means that the arch itself also has to be out of ship's reach as demonstrated by the collapse of the Tjörn Bridge in Sweden due to ship collision, Fig. 4.132. The new, cable-stayed Tjörn Bridge, see Figs 4.131 and 4.133, has a main span of 366 m, increased from the 217 m of the original arch, and a minimum navigational clearance over the whole length of the main span of 45.3 m [4.50]. The new Tjörn Bridge can thus not be reached by ships anymore.

The Panama Canal is the busiest artificial waterway in the world, Fig. 4.134. When a second crossing had to be built, the overriding design condition was that the bridge would not in any way hinder the ship traffic. Both piers were securely placed on land at a safe distance from the water, resulting in a main span length of 420 m. The navigational clearance permits the passage of the biggest ships which can pass through the current and future locks.

In order to place both tower foundations safely on the river banks, the Yang Pu Bridge across Huangpu River in Shanghai, China, was designed with the then world record main span of 602 m, Figs 4.135 and 4.136 [4.51].

Both towers for the Stonecutters Bridge in Hong Kong are placed on shore, which results in a new record main span of 1017 m, Fig. 4.137 [4.52]. The pile foundations are located close to the water, protected by sea walls, Fig. 4.138. If a ship were to collide with the sea wall in front of the foundations, the bow would exert lateral pressure onto the piles. The amount and distribution of this lateral load is shown in Fig. 4.139.

In the harbor of Portsmouth, UK, a viewing tower with a height of about 150 m has been built directly on the waterfront, Fig. 4.140. An accidental impact of berthing ships cannot be completely ruled out. Various scenarios including the approaches of a disabled aircraft carrier and disabled ferries have been investigated, Fig. 4.142, in order to prevent ship's bows to reach the tower legs [4.53].

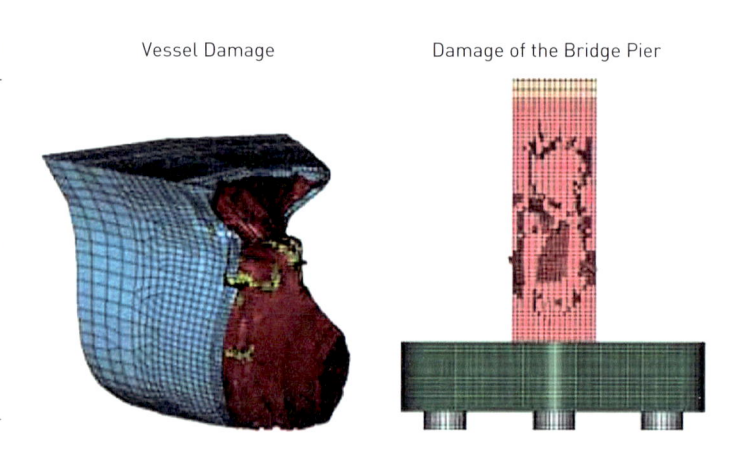

Figure 4.127 FEM model by Tongji University Shanghai

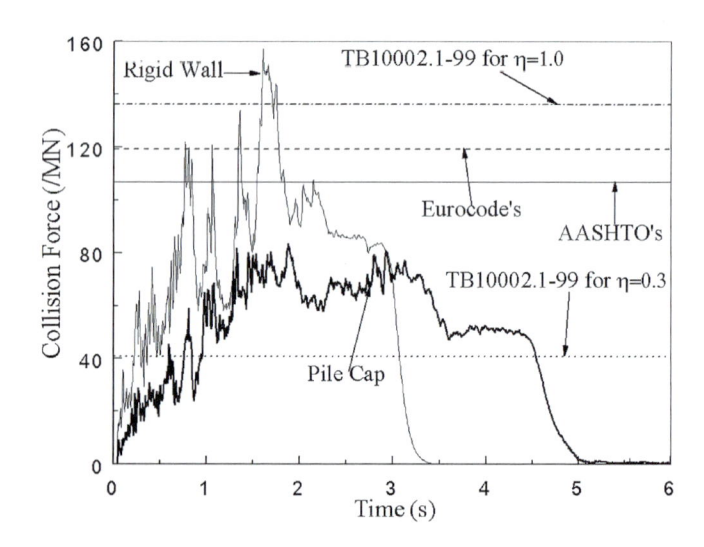

Figure 4.128 Collision force on rigid and elastic, pier by Tongji University

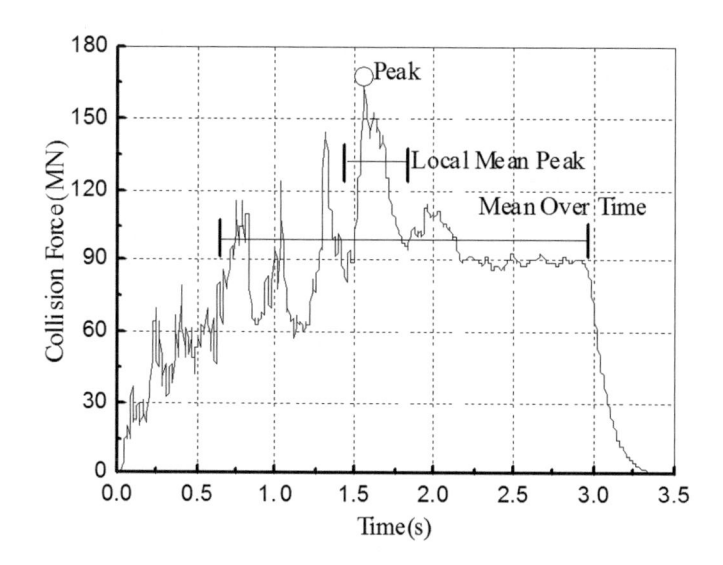

Figure 4.129 Definitions of time periods for equivalent loads by Tongji University

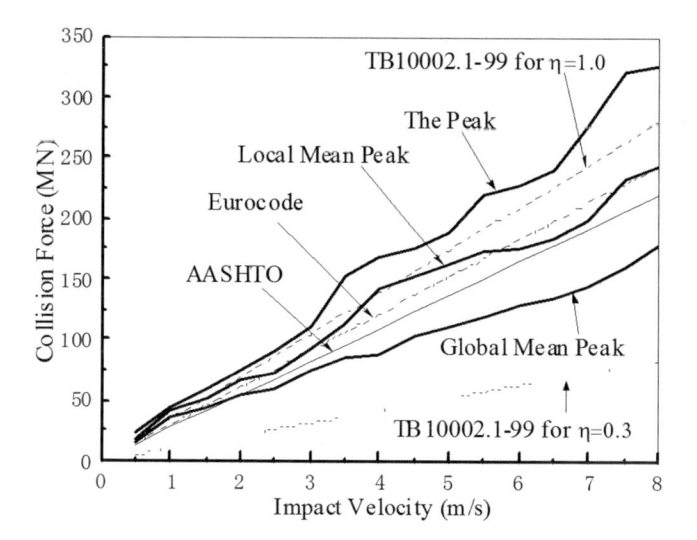

Figure 4.130 Comparison of equivalent loads by Tongji University

A crescent-shaped concrete fender has been designed which is connected to the pile cap of the tower by damper elements, Figs 4.141 and 4.142. The energy of a major impact should be absorbed by the plastic deformation of the vessel's hull, the local damage of the fender system, and the deformation of the damper elements. The tower foundation is designed to withstand the resulting impact forces.

4.4.3.3 Artificial islands

If the water is too wide to be crossed without piers in navigable water, it should be investigated whether artificial islands are possible. The advantages of such islands are that they provide a high degree of safety and that they stop a ship slowly, thereby limiting the extent of damage to the ship's hull. If provided with a proper lining against erosion, such islands are virtually maintenance free and require only minor additional fill after a collision. Their use is often limited in so far as they must not reduce the flow cross section so much that the speed of the current is dangerously increased.

For the Houston Ship Channel Crossing at Baytown, TX, USA, an artificial island was built for the foundation of the one tower which had to be placed in shallow water, Fig. 4.143 [1.17]. A similar situation exists for the Kap Shui Mun Bridge in Hong Kong: one tower is placed on shore, the other one in shallow water where an artificial island had to provide a minimum distance of 25 m to passing ships, Fig. 4.144.

4.4.3.4 Guide structures

Guide structures are designed to guide a ship away from the bridge piers or bridge superstructure. They are usually not designed for a head-on collision but for a glancing impact of a small or medium-size vessel.

The Crown Prince Arch Bridge spans the Spree River in the heart of Berlin, Germany, Fig. 4.145. In order to prevent ships reaching the arch supports near both river banks steel deflectors were installed. It is interesting to note that the amount of steel required for the deflectors is bigger than the amount of steel required for the bridge itself. It may have been more economic to widen the span of the bridge to place the supports on land and to use a shallow, constant depth girder completely out of the reach of ships.

The superstructure of the Neckar River Bridge near Kirchheim, Germany, had to be replaced with minimum disruption to traffic [4.54]. The new steel truss beam was therefore launched parallel to the existing bridge on auxiliary piers next to the existing piers, Fig. 4.146. In order to protect these auxiliary piers against ship collision guide structures were installed. The fender supports are steel pipes filled with gravel and drilled into the underlying rock.

The kinetic energy from a glancing impact is taken up by plastic deformation of the steel pipes. The corresponding large deformation of the guide structures were taken into account by placing them 10 m away from the auxiliary piers. This resulted in a reduction of navigation width from 40 to 20 m only. The permanent piers with

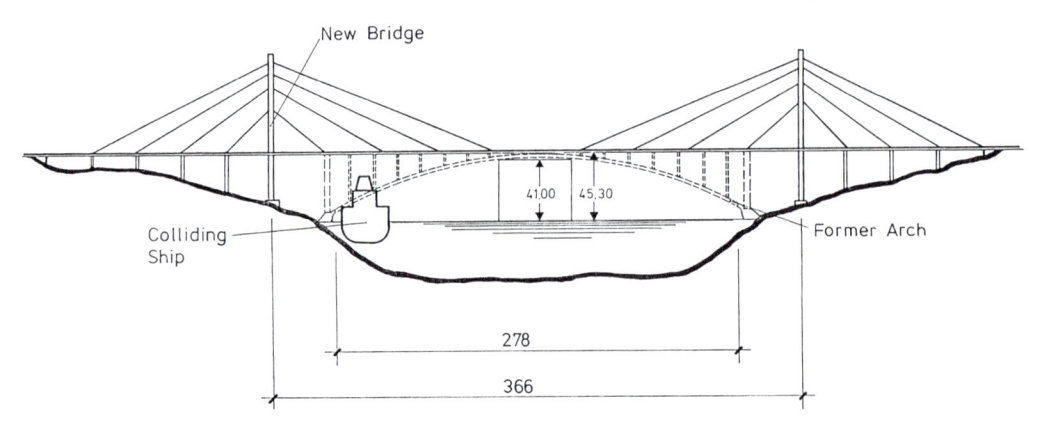

Figure 4.131 Old and new Tjörn Bridge, Sweden

Figure 4.132 Old Tjörn Bridge, ship collision

Figure 4.133 New Tjörn Bridge, Sweden

4.4 Protection of bridges against ship collision

Figure 4.134 Second Panama Canal Bridge

Figure 4.135 Yang Pu Bridge, Shanghai, China

Figure 4.136 Foundations for Yang Pu Bridge

256 4 Preliminary design of cable-stayed bridges

Figure 4.137 Stonecutters Bridge, Hong Kong, LAP's second prize design

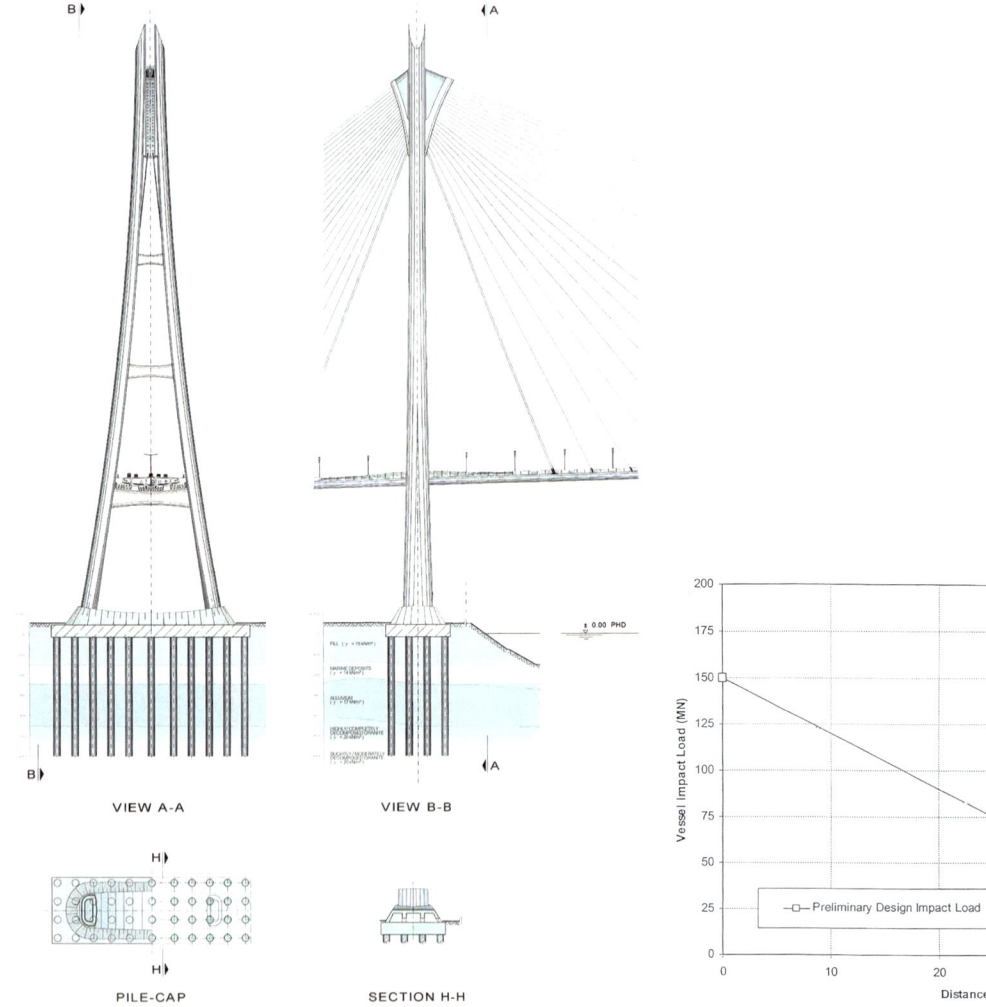

Figure 4.138 Pile foundations

Figure 4.139 Stonecutters Bridge, lateral collision pressure on piles

4.4 Protection of bridges against ship collision

Figure 4.140 Portsmouth Tower, UK

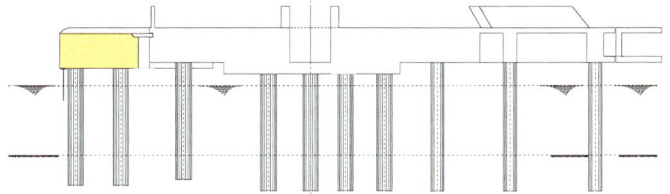

Figure 4.141 Portsmouth Tower, UK, elevation of foundation

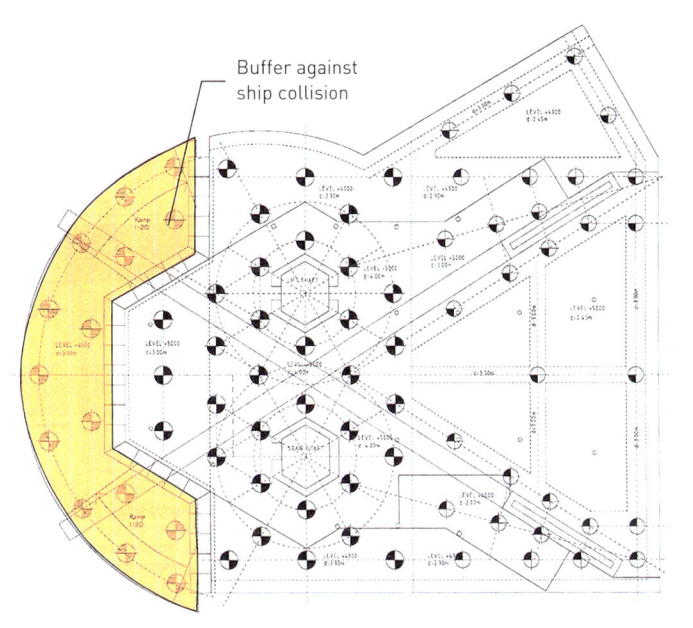

Figure 4.142 Portsmouth Tower, UK, plan of foundation

Figure 4.143 Houston Ship Channel Crossing, Texas, USA

Figure 4.144 Kap Shui Mun Bridge, Hong Kong, China

Figure 4.145 Crown Prince Bridge, Berlin, Germany

Figure 4.146 Neckar Bridge Kirchheim, Germany, during construction

Figure 4.147 Neckar Bridge Kirchheim, Germany, completed bridge

the new superstructure were designed to take a collision force by themselves, Fig. 4.147. For that purpose they were vertically post-tensioned into the underlying rock and provided with a 20 cm sacrificial concrete lining.

4.4.3.5 Independent protective structures

For the already mentioned Zárate-Brazo Largo Bridges in Argentina the center piers had to be protected independently of the bridge against collision with a 25 000 DWT ship, Fig. 4.148 [4.43]. Protective islands were not possible due to the extreme water depth, and since the bridge was already completed, the piers could not be made strong enough to withstand a collision due to the extremely bad soil conditions which required the bridge to be supported on up to 80 m long piles.

The author, therefore, proposed independent circular concrete dolphins founded on 70 m long piles with triangular platforms on top to withstand a collision or to guide ships away from the piers, Fig. 4.149.

However, for financial reasons the client decided to install floating protections, Fig. 4.150. The greatest risk of such floating protections lies in their vulnerability against being pushed under water by the ship's bow and then being run over, see Fig. 4.151. Furthermore, all their anchorage devices are subject to severe corrosion attack and are difficult to inspect.

The collapse of the Sunshine Skyway Bridge due to ship collision, see Fig. 4.152, was mentioned in Section 2.2. For the new Sunshine Skyway Bridge a pier protection system utilizing a combination of dolphins and island protections was developed, Fig. 4.153 [4.55]. The dolphins consist of steel sheet-piling with an 18 m diameter and a concrete cap, Fig. 4.154. The dolphins were designed to withstand the collision forces of either a loaded 23 000 DWT or an empty 87 000 DWT bulk carrier. The key to a dolphin's ability to absorb a major ship collision is to tie the top of the sheet-piling rigidly together with the concrete cap and to weld the sheet steel pile interlocks to one another. The dolphins are expected to act sacrificially, i.e. after a major collision the cell would be destroyed and would have to be replaced.

The main bridge of the Rosario-Victoria Crossing of the Paraná River in Argentina comprises a cable-stayed bridge with a main span of 350 m [2.85]. All piers in the Paraná rest on deep pile foundations of about 30 m length, Fig. 4.155. The governing design ship has 43 000 DWT, and a speed of 4.6 m/sec, resulting in an impact force of 118 MN in accordance with AASHTO [4.46] (228 MN in acc. with Pederson [4.47]).

Taking into account a variation of the water level of up to 6 m and local scour of up to 12 m the total free length of piles could reach up to 42 m. With this specific geotechnical and hydraulic situation the foundations themselves could not be designed to withstand the high impact forces economically within the elastic range. Independent defense structures, designed as sacrificial structures by

4.4 Protection of bridges against ship collision

Figure 4.148 Zárate-Brazo Largo Bridge, Argentina

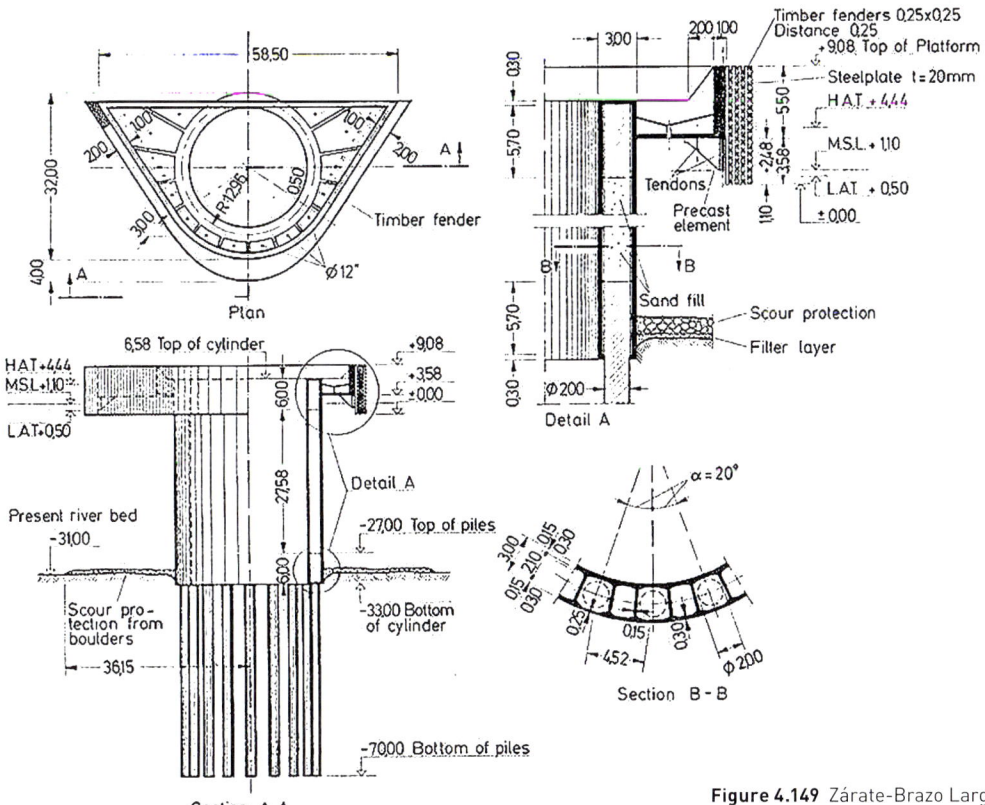

Figure 4.149 Zárate-Brazo Largo Bridge, proposed protection

Figure 4.150 Zárate-Brazo Largo Bridge, floating protection

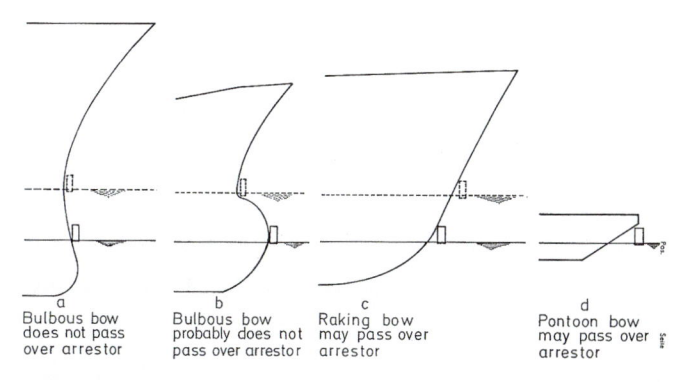

Figure 4.151 Risk of overrunning the floating protection

Figure 4.152 Sunshine Skyway Bridge, Tampa, Florida, USA, collapsed bridge

Figure 4.153 Sunshine Skyway Bridge, dolphin protection

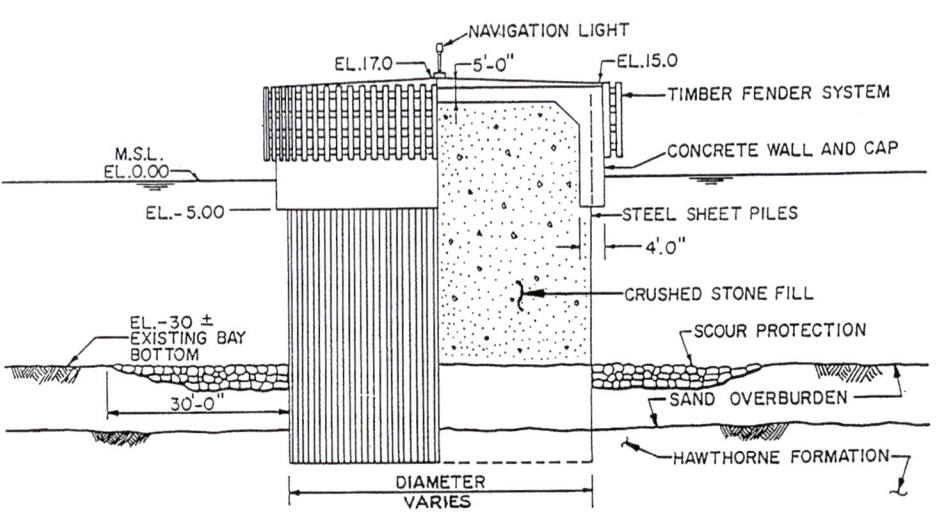

Figure 4.154 Sunshine Skyway Bridge, dolphin details

Figure 4.155 Rosario-Victoria Bridge, Argentina

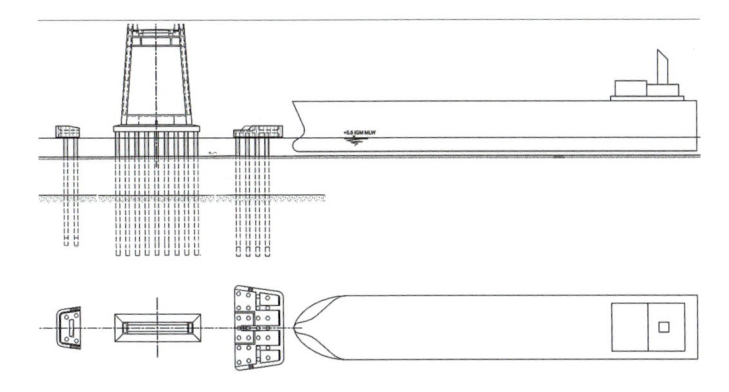

Figure 4.156 Rosario-Victoria Bridge, independent protection

Figure 4.157 Rosario-Victoria Bridge, construction of protection

Figure 4.158 Rosario-Victoria Bridge, pile test

exploiting their plastic capacities, were the appropriate solution, Fig. 4.156.

The adopted protective structures consist of a concrete platform supported on concrete-filled steel piles of 2 m dia. and a clear distance of 17.50 m to the bridge foundation in order to permit the high deflections resulting from plastic hinges, Fig. 4.157. The plastic rotational capacity of the concrete-filled piles was confirmed by tests, Fig. 4.158.

It is well known that not only the main span of a bridge but also the side spans have to be protected against ship collisions as shown for the Benjamin Harrison Bridge, USA, in Fig. 4.117.

This experience had to be taken into account for the cable-stayed steel composite pedestrian bridge across the Rhine River between Kehl, Germany, and Strasbourg, France, Fig. 4.159 [2.135].

The bridge can take the impact forces from ship collision by the main foundations themselves. However, since part of the bridge superstructure touches down to the river banks, the side spans had to be protected against ship collision by independent structures. For this purpose upstream independent impact collision protections supported on three piles were placed, Fig. 4.160. Downstream concrete-filled steel pipes were considered sufficient.

Two composite cable-stayed bridges for road and rail with main spans of 300 m each form the Second Crossing of the Orinoco River currently under construction in Venezuela, Fig. 4.161 [2.124]. The piers will be protected against ship collision by independent concrete beams supported on piles, arranged around the foundations proper, see Fig. 4.162.

4.4.3.6 Strong piers

Another possibility for protecting piers in deep water against ship collision is to make them strong enough to withstand the collision forces by themselves. The vertical loads on the foundations from the bridge weight assist considerably in designing the foundations against horizontal impact load. For foundations on solid rock not too deep down, this type of design can be more economical than designing independent protective structures without the benefit of the bridge dead weight. Since these foundations are rigid the ship's kinetic energy has mainly to be absorbed in the ship's hull and can, therefore, lead to more damage for the ship than sacrificial independent protections which absorb the ship's energy more gently by plastic deformations with large deflections.

In 1981 the Newport Suspension Bridge at Rhode Island, USA, was struck head on by a fully loaded 45,000 DWT tanker. The ship's bow was shortened by about 3.5 m, Fig. 4.119, but the pier suffered only superficial damage as shown in Fig. 4.120. The steel H-pile foundations were also strong enough to prevent any permanent displacement of the pier.

The center piece of the new Galata Bridge at Istanbul, Turkey, is a bascule bridge with a clear span of 80 m and a width of 42 m. It consists of four orthotropic deck flaps, Fig. 4.163. In the design of the bascule bridge piers, two contradictory requirements had to be fulfilled: for the absorption of the ship impact they had to be stiff, and for earthquake they had to be flexible. This was achieved by hollow concrete piers going down to the bottom of the Golden Horn, where they are founded on 32 m long piles, Fig. 4.164. The piers were designed for a head-on collision with a force of 40 MN.

The steel flaps of the bascule bridge themselves can not, of course, withstand a ship collision, should a ship run into them while closed. Two governing scenarios were investigated, Fig. 4.165:
– the formation of a plastic hinge in front of a pier in addition to a central plastic hinge,
– the loss of a flap between the two hinges.

Both scenarios did not lead to the loss of the another flap or the rear arm with the counterweight and were thus considered acceptable local failures.

The foundations of the towers for the Helgeland Bridge in Norway consist of solid rectangular piers up to 32 m down to solid rock, Figs 4.166 and 4.167 [2.80]. They were designed for an impact force of 54 MN.

The cable-stayed Uddevalla Bridge in Sweden has a main span of 414 m, Fig. 4.169. While the north tower could be founded on an island, unreachable for ships, the south tower had to be founded on piles in the high water flood plane, see Fig. 4.168. This exposed foundation was designed for a collision force of 70 MN in the direction of ship's travel and for 35 MN in the transverse direction, Fig. 4.169.

The cable-stayed My Thuan Bridge across the Mekong River in Vietnam is founded on up to 100 m deep piles, Fig. 4.170. The water depth may vary due to scour from 16 to 25 m. The foundations had to be designed in the downstream direction for a collision force of 32 MN, determined by the AASHTO guide specification, [4.46]. Under the very difficult design conditions the pile cap was enlarged upstream and downstream by deflectors. In case of a collision part of the ship's kinetic energy is taken by the hull's deformation and the crushing of the pile cap deflectors, Fig. 4.171.

The remaining part of the collision energy is carried elastically by the piles. Provisions were made in the design for the strengthening of the foundations for impact of larger vessels in future, which may cause a downstream collision force of up to 39 MN.

Figure 4.159 Rhine River Bridge Kehl, side span protection

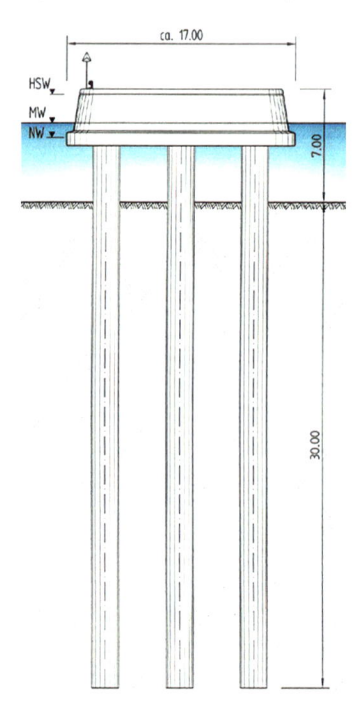

Figure 4.160 Rhine River Bridge Kehl, upstream protection

4.4 Protection of bridges against ship collision

Figure 4.161 Second Orinoco Bridge, Venezuela

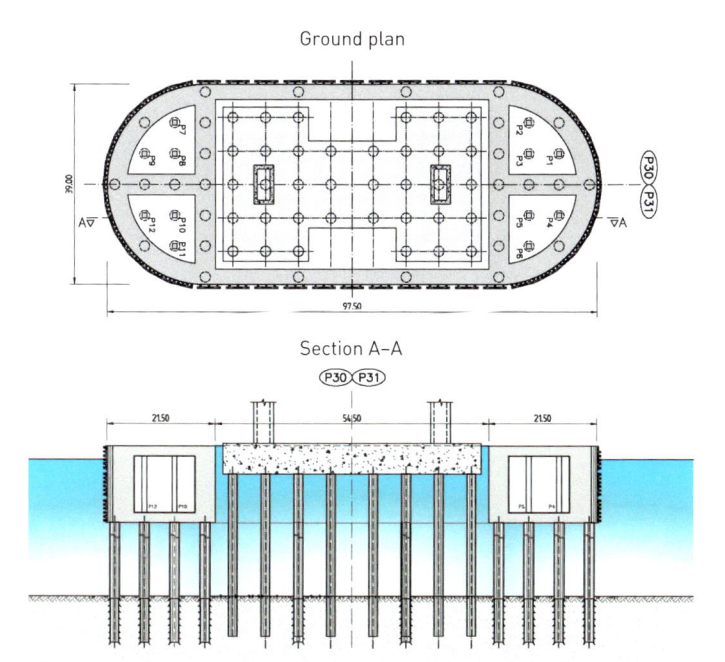

Figure 4.162 Second Orinoco Bridge, independent pier protection

264	4 Preliminary design of cable-stayed bridges

Figure 4.163 Galata Bridge, Istanbul, Turkey

Elevation

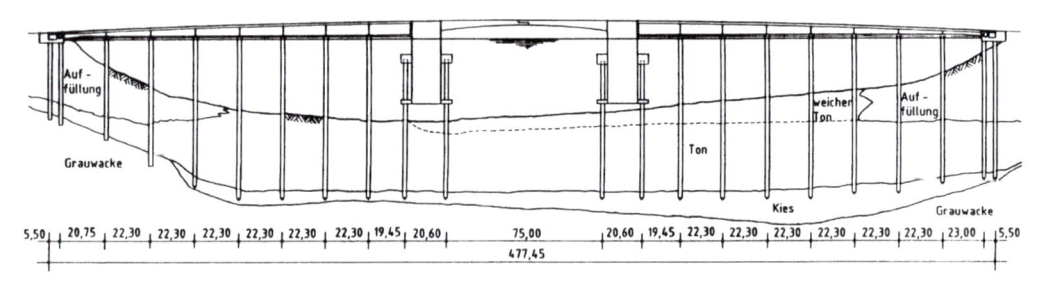

Plan

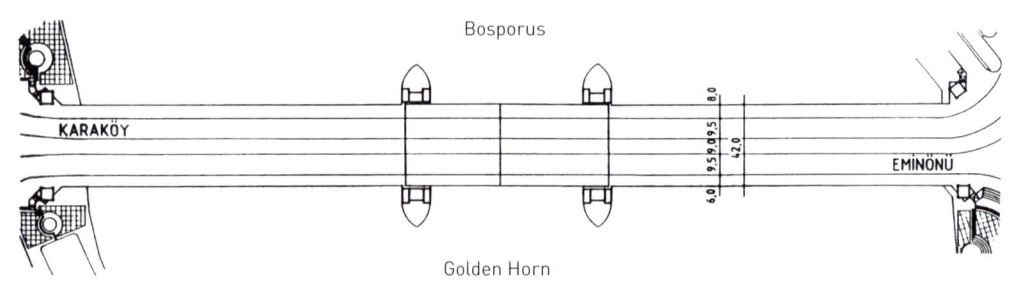

Figure 4.164 Galata Bridge, elevation

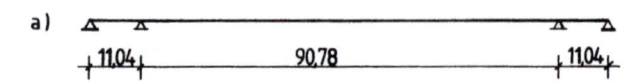

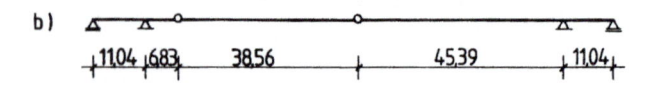

Figure 4.165 Galata Bridge, collision with flaps
a) base system, b) plastic hinges, c) collapsed beam

4.4 Protection of bridges against ship collision

Figure 4.166 Helgeland Bridge, Norway

Figure 4.167 Helgeland Bridge, construction of solid foundations

Figure 4.168 Uddevalla Bridge, Sweden

Figure 4.169 Uddevalla Bridge, elevation

Figure 4.170 My Thuan Bridge, Vietnam

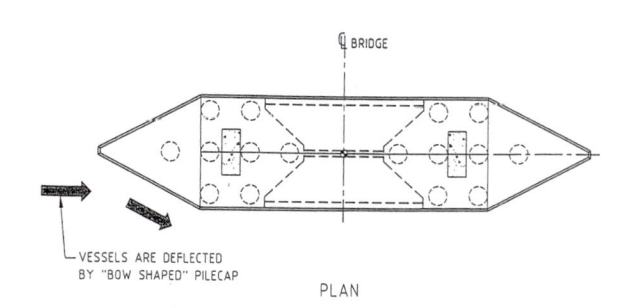

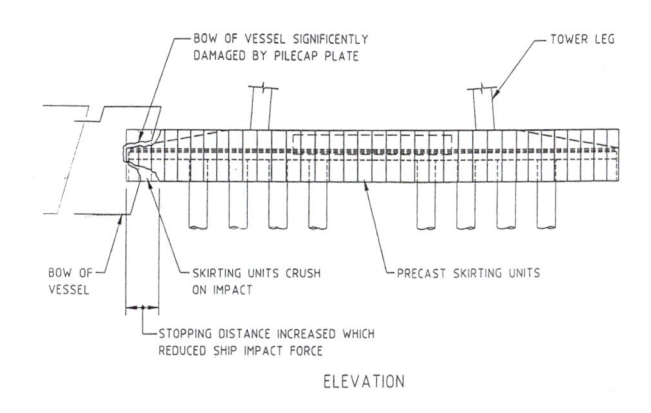

Figure 4.171 My Thuan Bridge, pier protection

4.5 Preliminary design calculations
4.5.1 General
The flow of forces in a cable-stayed bridge and their preliminary calculation have already been treated in Sections 4.1 and 4.2. The most important findings are the following:
a) The normal forces can be determined for an articulated system.
b) The moments for permanent loads at beam and towers can be freely selected. For concrete beams the moments of a beam rigidly supported at the cable anchor points are advantageously selected, because only this run of moment does not change with time, due to creep. For a steel beam that run of moments is selected which, in combination with the live load moments, requires the least amount of material.
c) The beam moments for live load can be approximated for a beam on elastic foundations. These live load moments have then to be superimposed onto those of a rigidly supported beam for dead load.
d) The non-linear increase of moments can be determined with the help of the overall safety against buckling of the beam.
e) The aerodynamic stability can be approximated for flutter by using the method of Klöppel/Thiele. The required eigenfrequencies can be calculated from the beam deflections for dead load.

With these approximate design calculations, any cable-stayed bridge can be preliminarily sized or independently checked. The precise computer calculations should not yield significant deviations.

4.5.2 Typical cable-stayed concrete bridge
4.5.2.1 System and loads
Consider a symmetric cable-stayed bridge with a total length of 243 m, Fig. 4.172. The bridge beam comprises a solid twin t-beam cross-section with two outer cable planes, Fig. 4.173. This system gives a clear indication of the particular forces in the backstays from live loads in main and side span.

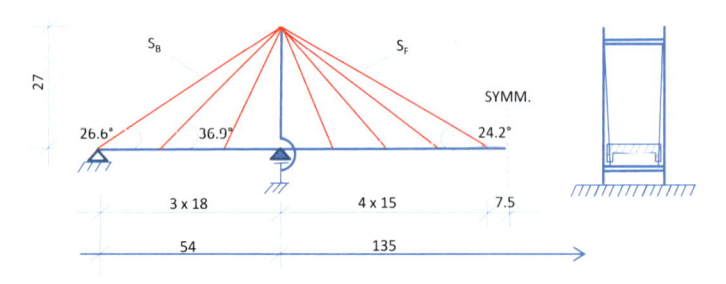

Figure 4.172 Elevation

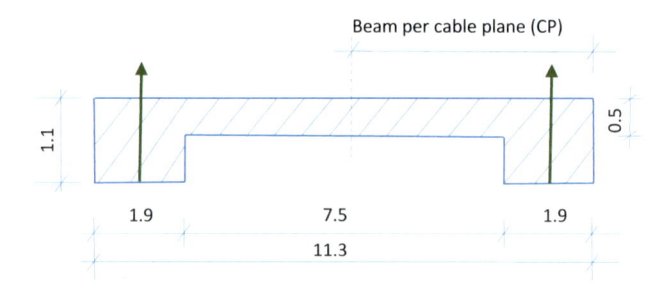

Figure 4.173 Cross-section

Design parameters
Cables
Parallel wire cables
ø 7 mm, A = 38.48 mm²
$E_{eff} = 2 \cdot 10^5$ N/mm²
Steel 1470/1670 (f_y/GUTS)
perm. f_S = 0.45 · 1.650 = 752 N/mm², see Section 3.7.4, Table 3.3
perm. Δf_S = 200 N/mm², see Section 3.7.4, Table 3.3

Alternatively: Parallel strand cables
Steel 1660/1860
$E_{eff} = 1.9 \cdot 10^5$ N/mm²
perm. f_S = 0.45 · 1.860 = 837 N/mm², see Section 3.7.4, Table 3.3
perm. Δf_S = 200 N/mm², see Section 3.7.4, Table 3.3
Strands ø 15 mm: A_S = 150 mm²

Beam
Concrete B 45
E_C = 37.000 MN/m
Central compression perm. f_C = 11.5 MN/m
Edge compression perm. f_C = -17.0 MN/m
A_C = 3.99 m²/CP, with CP = cable plane
Vertical moment of inertia I_V = 0.341 m⁴/CP
S_T = −0.83 m³/CP, with S = section modulus
S_B = +0.49 m³/CP

Dead load
Dead load DL, including super imposed dead load SIDL:
DL = 128 KN/m, CP
Live load LL
Uniformly distributed load: UDL = 11.5 kN/m³/CP
Concentrated load CL: CL = 375 kN/CP

4.5 Preliminary design calculations

4.5.2.2 Normal forces for articulated system
A) Stay-cables
Backstay cable S_B

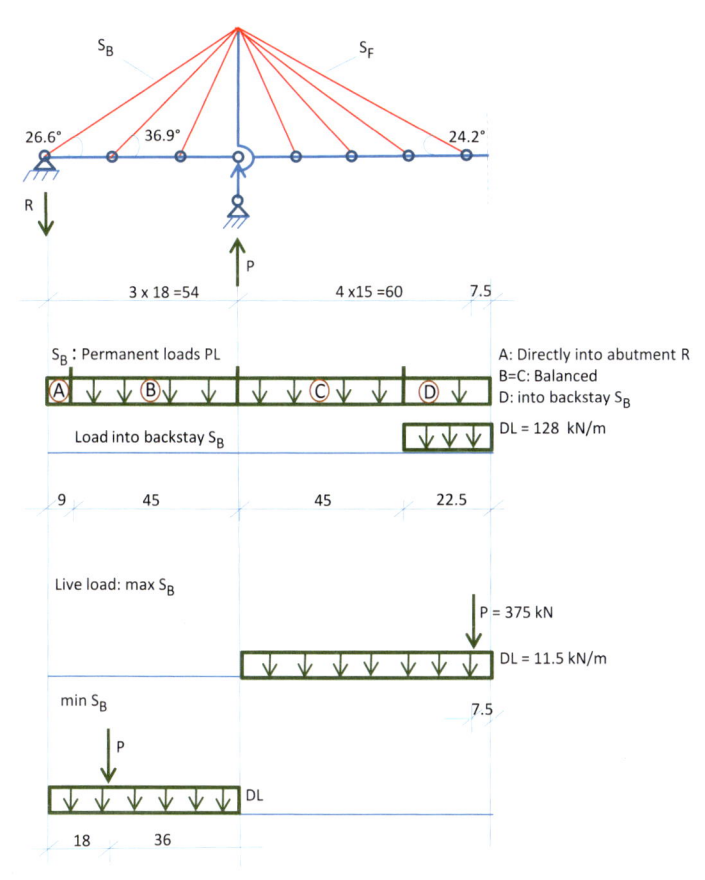

Figure 4.174 Load arrangement for backstay cable (S_B)

Permanent loads
As shown in Fig. 4.174: only DL, part D, goes directly into the backstay.

$\Sigma M_{Tower} = 0$

$\rightarrow R = 22.5 \cdot 128 \cdot \dfrac{56.25}{54} = 3000$ kN anchor force

$S_B = \dfrac{R}{\sin 26.6°} = \dfrac{3000}{\sin 26.6°} = 6700$ kN backstay cable force

Live load

$\max R = 67.5 \cdot 11.5 \cdot \dfrac{67.5/2}{54} + 375 \cdot \dfrac{60}{54} = 902$ kN uplift

$\min R = -54 \cdot 11.5 \cdot \dfrac{27}{54} - 375 \cdot \dfrac{36}{54} = -561$ kN compression

$\max S_B = 902 \cdot \dfrac{1}{\sin 26.6°} = 2014$ kN

$\min S_B = -1253$ kN

Parallel wire cables
Sizing for tension
DL + max. LL

req'd $A_S = \dfrac{6700 + 2014}{752} \cdot 10^3 = 11\,587$ mm^2

Sizing for fatigue for full live load

$\Delta S_B = |\min S^{LL} + \max S^{LL}|$

req'd $A_S = \dfrac{2014 + 1253}{200} \cdot 10^3 = 16\,335$ mm^2 for full LL

$\rightarrow$ governing

req'd $n = \dfrac{16\,335}{38.48} = 425$ wires

chosen: 2×217 wires (using hexagonal wire arrangement with $n_i = 6x + 1$)

act. $A_S = 16\,700$ mm^2

Note: depending on the corresponding codes, less than full LL has to be applied for Δf_S, for example only 50 %.

Alternatively: parallel strand cables
Backstay cable SB
max $S_B = 8714$ kN
max $\Delta S_B = 3267$ kN

req'd $A_B = \dfrac{8714}{837} \cdot 10^3 = 10\,411$ mm^2 governing

req'd $n = \dfrac{10\,411}{150} \cdot 10^3 = 69$

chosen: 2×37 strands in accordance with manufacturer's catalogue [3.35 – 3.38]

act. $A_S = 74 \cdot 150 = 11\,100$ mm^2

Fatigue stress for 50 % LL

$\Delta f_S = \dfrac{1}{2} \cdot \dfrac{3267}{11\,100} \cdot 10^3 = 147$ N/mm^2 < 200 N/mm^2

Central forestay cable S_F
In accordance with Fig. 4.175 for dead load

$S^{DL} = 15 \cdot 128 \cdot \dfrac{1}{\sin 24.2} = 4684$ kN

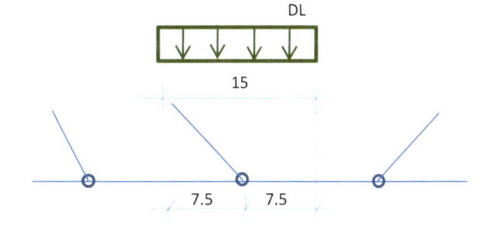

Figure 4.175 Dead load

In accordance with Fig. 4.176 for live load

max $S^{LL} = (15 \cdot 11.5 + 375) \cdot \dfrac{1}{\sin 24.2} = 1336$ kN

Fatigue from live load in main span and back span:
max $\Delta S_F = 1336$ kN for 100% live load
Sizing for strands
req'd $A_S = 7192$ N/mm² for tensior
chosen: n = 48 ø 0.62"
actual $A_S = 7200$ mm²
actual $\Delta f_S = \dfrac{1}{2} \cdot \dfrac{1336}{2700} \cdot 10^3 = 93$ N/mm² for 50% live load

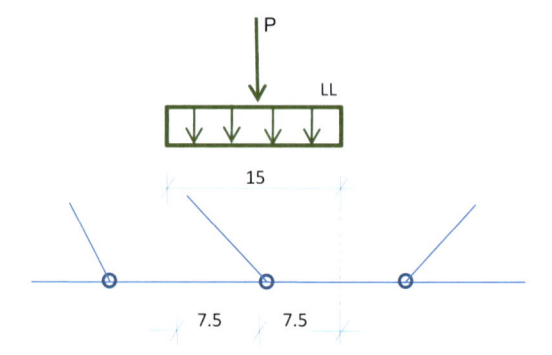

Figure 4.176 Live load

Alternatively, sizing for wires
max $S_F = 6020$ kN
req'd n $= \dfrac{6020 \cdot 10^3}{752 \cdot 38.48} = 209$ wires ø 7 mm
chosen with $n_i = 6x + 1$
chosen: n = 211 wires ø 7 mm
act. $A_S = 8119$ mm²

Note: The loading of each stay cable with the full concentrated load is on the safe side. Depending on the relation between cable stiffness and beam stiffness, the concentrated load is distributed over several cables. At the bridge center only about 25% may go directly into the affected cable, but at the tower, where the beam support from the cables is stiffer, up to about 50% may go into the affected cable. This slight oversizing is often required in the detailed design for additional effects such as local cable bending.

$\Delta f_S = \dfrac{1336 \cdot 10^3}{8119} = 164$ N/mm² < 200 N/mm² for 100% live load

B) Normal forces in beam
Approximation in accordance with Fig. 4.177, see Fig. 4.6.

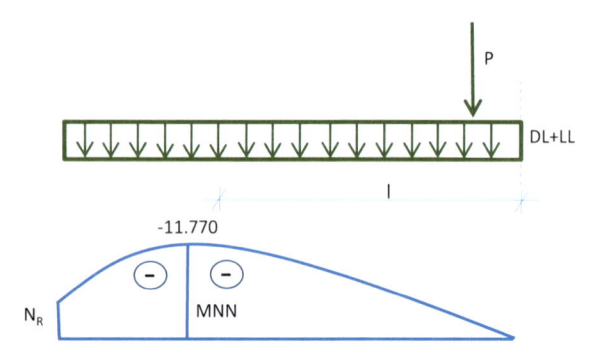

Figure 4.177 Normal forces from UDL

Full UDL
DL + LL = 128 + 11.5 = 139.5 kN/m
$N_{Tower} = \dfrac{(DL + LL)}{2h} \cdot l^2 = \dfrac{139.5 \cdot 67{,}5^2}{2 \cdot 27} = 11770$ kN/CP compression
Concentrated load in accordance with Fig. 4.178

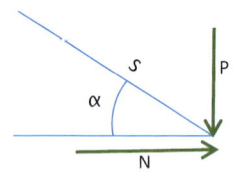

Figure 4.178 Normal force from CL

$N^{CL} = P \cdot \dfrac{1}{\tan \alpha} = 375 \cdot \dfrac{1}{\tan 24.2°} = 834$ kN/CP compression
min $N_{Tower} = 11770 + 834 = 12604$ kN/CP compression

C) Normal forces in tower
From DL plus LL (UDL + CL), see Fig. 4.179.

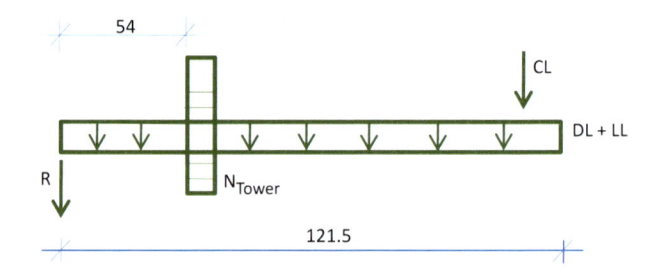

Figure 4.179 Normal forces, tower

$\Sigma M_R = 0$
$\rightarrow N_{Tower} = (128 + 11.5) \cdot 121.5 \cdot \dfrac{121.5/2}{54} + 375 \cdot \dfrac{114}{54}$
$= 19860$ kN/CP compression

4.5.2.3 Bending moments
A) For beam
For rigidly supported beam
For DL, see Fig. 4.180.

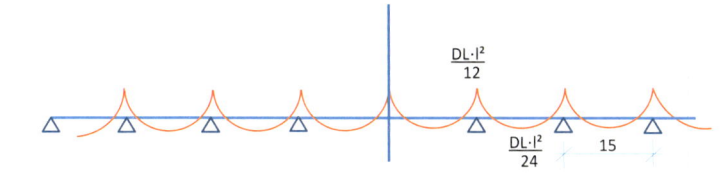

Figure 4.180 DL moments of rigidly supported beam

Positive moments at main span

$$\max M_{rigid}^{DL} = \frac{DL \cdot l^2}{24} = \frac{128 \cdot 15^2}{24} = 1200 \text{ kNm}$$

Beam bending moments for live load
For a beam on elastic foundation
Determination of elastic length L, see Fig. 4.181.

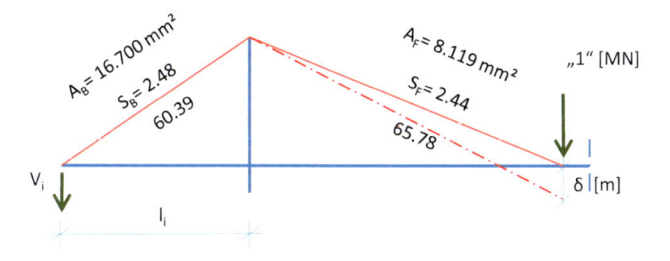

Figure 4.181 Parameters for calculation of beam deflection

Cable lengths $s_i = \dfrac{l_i}{\cos \alpha}$

Cable forces $S_i = \dfrac{V_i}{\sin \alpha}$

$$S_B = \frac{60}{54} \cdot \frac{1}{\sin 26.6°} = 2.48 \text{ MN}$$

$$S_F = \frac{1}{\sin 24.2°} = 2.44 \text{ MN}$$

$$\delta = \Sigma N_i^2 \cdot \frac{l_i}{E \cdot A}$$

$$\delta = 2.44^2 \cdot \frac{65.78 \cdot 10^6}{2 \cdot 10^5 \cdot 8119} + 2.48^2 \cdot \frac{60.39 \cdot 10^6}{2 \cdot 10^5 \cdot 16700}$$

$$= 0.24 + 0.12 = 0.352 \text{ m/MN}$$

Elastic support

$$c = \frac{1}{\lambda_n \cdot \delta} = \frac{1}{15 \cdot 0.352} = 0.189 \text{ MN/m}^2$$

Elastic length

$$L = \sqrt[4]{\frac{4 \cdot E_B \cdot I_B}{c}} = \sqrt[4]{\frac{4 \cdot 37 \cdot 10^3 \cdot 0.341}{0.189}} = 22.73 \text{ m}$$

$$M^{LL} = 0.161 \cdot p \cdot L^2 + 1/4 \cdot P \cdot L$$
$$= 0.161 \cdot 11.5 \cdot 22.73^2 + 1/4 \cdot 375 \cdot 22.73 \text{ m}$$
$$= 957 + 2131 = 3088 \text{ kNm/CP}$$

Proof of stresses
for min. N at tower

$$\min f_C = \frac{12\,604}{3.99} = 3.16 \text{ MN/m}^2 < 11.5 \text{ central compression,}$$
$$\text{reserve for mements available}$$

from max. M at bridge center

$$f_C = \frac{Mg + p}{S} \quad \text{with } S = \text{Section modulus}$$

$$f_C = \frac{1200 + 3087}{\begin{array}{c}-0.83\\+0.49\end{array}}$$

$$f_C = \begin{array}{c}-5.2 \text{ MN/m}^2\\+8.7 \text{ MN/m}^2\end{array}$$

Central prestress to cover tensile stresses

$$f_{C,prestr.} = -9 \text{ MN/m}^2$$

$$\rightarrow \Sigma f_C = \begin{array}{c}-14.2 \text{ MN/m}^2 < -17.0\\+0.3 \text{ MN/m}^2 < 0\end{array}$$

B) Bending moments in tower
The maximum tower head deflection is calculated in accordance with Fig. 4.182.

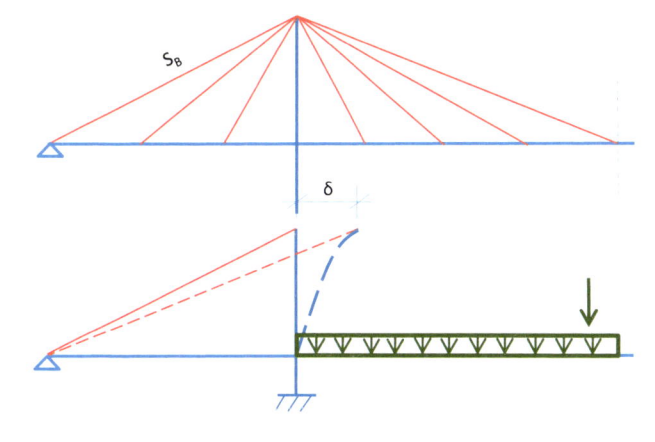

Figure 4.182 Deflection of tower head

Calculation of tower head deflection δ from backstay cable S_B elongation, Fig. 4.183.

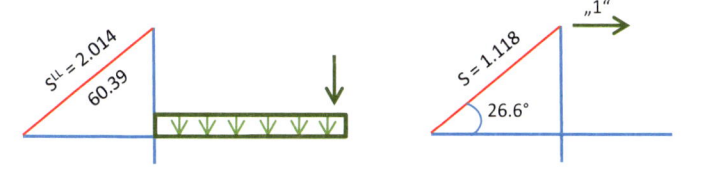

Figure 4.183 Calculation of tower head deflection

$$\delta = S_1 \cdot S_2 \cdot \frac{l_i}{E \cdot A} \qquad S_1 = \max S^{LL}$$
$$= 2014 \cdot 1118 \cdot \frac{60.39}{2 \cdot 10^2 \cdot 16\,700} = 0.04 \text{ m}$$

Calculation of tower moment from deflection δ, see Fig. 4.184.

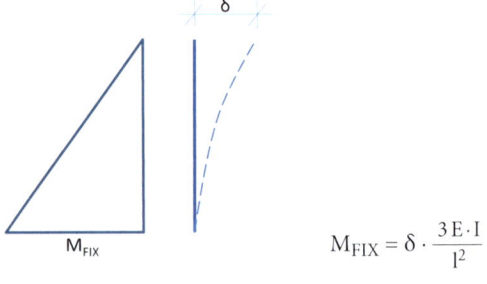

$$M_{FIX} = \delta \cdot \frac{3\,E \cdot I}{l^2}$$

Figure 4.184 Tower moment from tower head deflection δ

4.5.3 Typical cable-stayed steel bridge
4.5.3.1 General

The permanent load moments of steel beams for cable-stayed bridges cannot be preliminarily calculated in the same way as concrete beams because their permanent load moments may deviate distinctly from those of a rigidly supported beam. The aim is to minimize the required amount of material. For the determination of these moments the negative beam moments from the live load are as important as the positive live load moments. The negative beam loads, however, cannot be determined with sufficient precision for a beam on elastic foundation – see Fig. 4.9.

In the following, only the principal calculation steps are explained. As an example the pedestrian bridge Schillersteg in Stuttgart is used, Figs 4.185 and 4.186. In the comprehensive publication by Leonhardt and Andrä [5.40], the basic design assumptions are outlined.

4.5.3.2 System

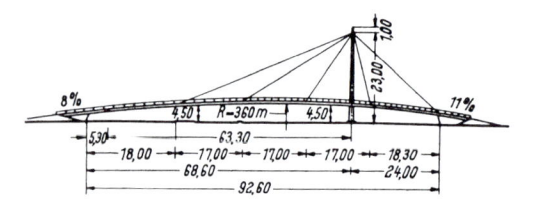

Figure 4.185 Elevation

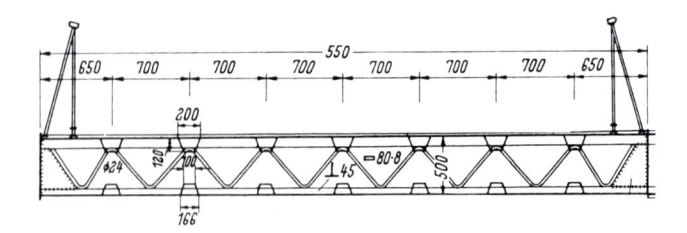

Figure 4.186 Cross-section

4.5.3.3 Section properties and loads
Beam
St 37, perm. $f_S = 140$ N/mm² (safety for buckling included)
Main span
$I_{eff} = 2100$ m² mm²/CP effective: participating width considered
$S_{top,eff} = -9020$ m mm²/CP
$S_{bot,eff} = +7950$ m mm²/CP

4.5.3.4 Beam moments from live load

The elastic system is loading spanwise in order to obtain the precise live load moment envelope, Fig. 4.187.

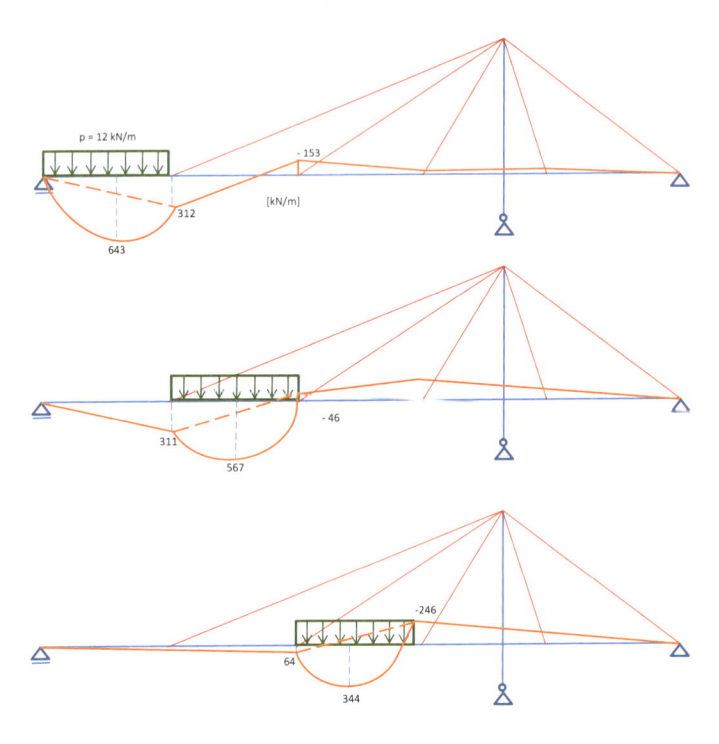

Figure 4.187 Moments from spanwise LL

The resulting live load moments from individual loading spans are combined in such a way that the moment envelopes for the largest positive span moments and the largest negative moments at the cable supports are determined, Fig. 4.188.

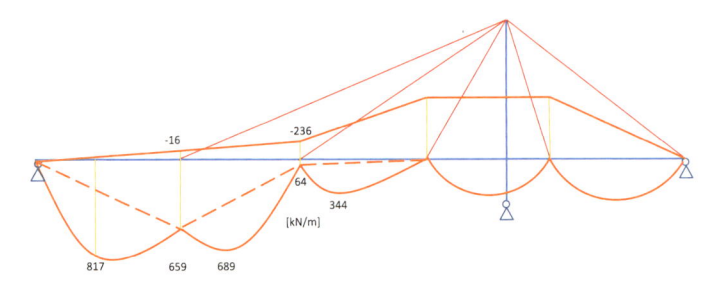

Figure 4.188 Superposition of max/min LL moments for LL envelopes

4.5.3.5 Permissible beam moments

The permissible beam moments result from the permissible stresses and corresponding section moduli, Fig. 4.189.

As a simplification the permissible stresses and section moduli are reduced by 25 % in order to cover local bending stresses and the compression stresses from normal forces.

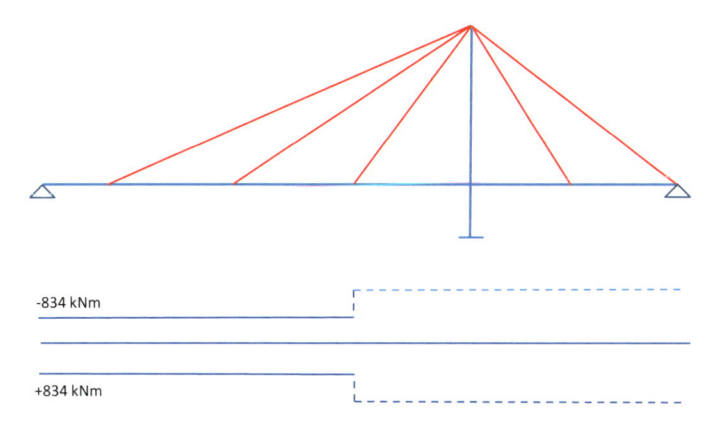

Figure 4.189 Moment capacity of beam

4.5.3.6 Moments from dead load for articulated system

In a first step the moments from dead load for an articulated system are used, Fig. 4.189.

By superpositioning the live load moment envelope from Fig. 4.188 with the dead load moments from Fig. 4.190, the restraint moments from cable shortenings can by determined in such a way that the sum of these moments remains smaller than the permissible moments from Fig. 4.189.

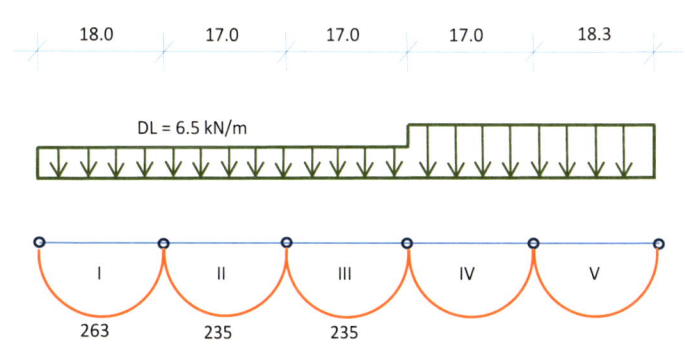

Figure 4.190 DL moments for articulated system

In Fig. 4.191 the dead load moments are added to the live load envelope. In span 1 the permissible positive moment is exceeded:

M_I^{DL} = 263 kNm/CP   Fig. 4.190
max M_I^{LL} = 817 kNm/CP   Fig. 4.188

ΣM_1 = 1080 kNm/CP
perm. M = 834 kNm/CP   Fig. 4.189

req'd ΔM_1 = – 246 kNm/CP

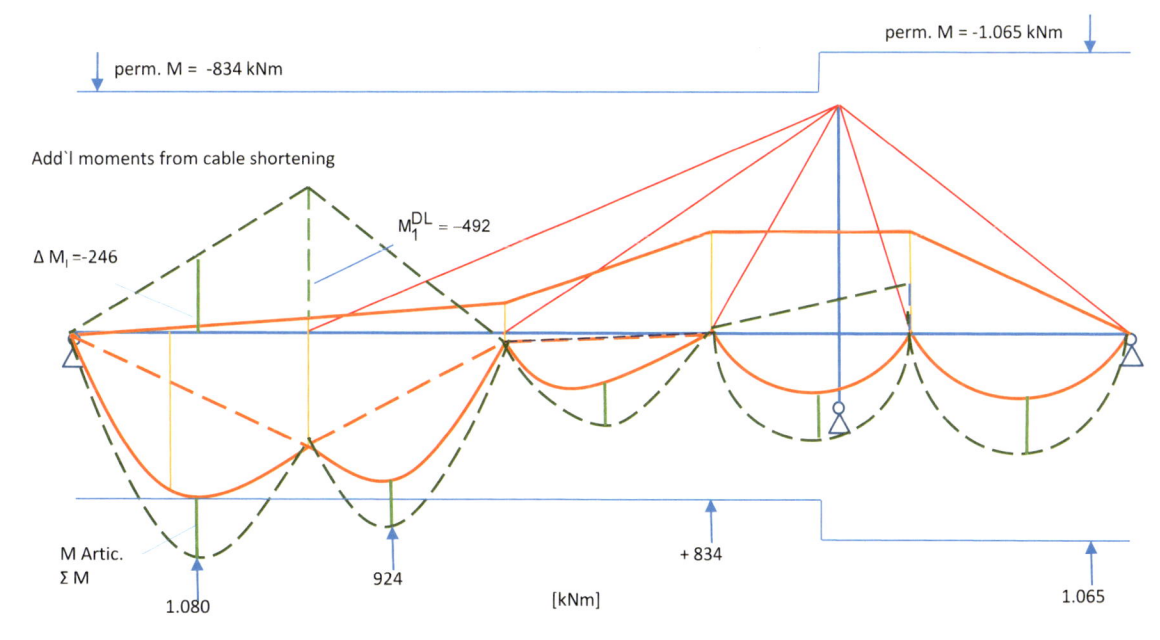

Figure 4.191 Superposition of LL moments with DL moments

The resultant support moment at the anchorage of cable 1 thus results in:

$M_1^{DL} = -2 \cdot 246 = -492$ kNm/CP Fig. 4.191
$\min M_1^{LL} = -16$ kNm/CP Fig. 4.188

$\Sigma M_1 \phantom{^{DL}} = -508$ kNm/CP < perm. $M_1 = -834$ kNm

The sum of the support moments of – 508 kNm at cable 1 is smaller than the permissible moment of – 434 kNm and may thus be created by cable shortenings.

Figure 4.191 indicates that the chosen cross-section is not fully utilized in bending. The support moments will be selected in such a way that the bending stresses in the top and bottom flange are about equal. All finally selected support moments for permanent loads (dead load + cable shortening) are shown in Fig. 4.192.

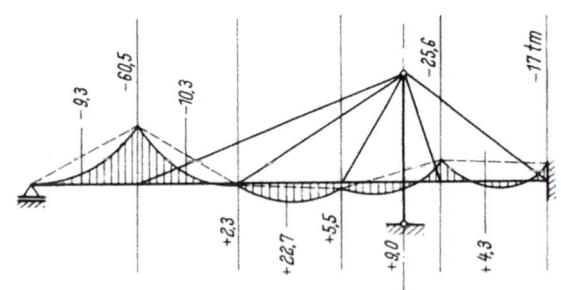

Moments in tm: tm · 10 = kNm

Figure 4.192 Moments from permanent loads PL

The moment envelope for the superimposed dead load and live load moments are shown in Fig. 4.193. It is apparent that the governing positive and negative beam moments are about equal over the length of the beam. Therefore, a sufficient reserve for the stresses from normal forces, which are constant over the depth of the beam, remains.

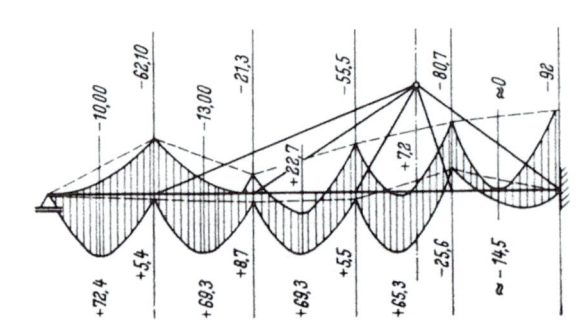

Moments in tm: tm · 10 = kNm

Figure 4.193 Moment envelope from PL + LL

The thus-determined support moments are then used as conditions for the cable shortenings under permanent loads in the detailed computer calculations. The corresponding deformations are superimposed with negative sign onto the desired gradient and thus result in the shop form.

Further examples for the selection of the run of moments under permanent loads for steel bridges are given in Section 4.2.1.3 for the Mannheim-Ludwigshafen Bridge, Fig. 4.33, the Rhine River Bridge Flehe, Fig. 4.34, and the steel alternate for the Brotonne Bridge, Fig. 4.48.

All these road bridges have box cross-sections with a large orthotropic deck top flange. The capacity of these top flanges is bigger than that of the bottom flanges, predominantly negative moments were thus selected for permanent loads.

4.5.4 Cable-stayed bridge with side spans on piers

4.5.4.1 System and loads

Consider a three-span cable-stayed bridge with side spans directly resting on piers, Fig. 4.194. The anchorage of the main span loads takes place via the forestays to a concentrated backstay anchored to the hold-down pier.

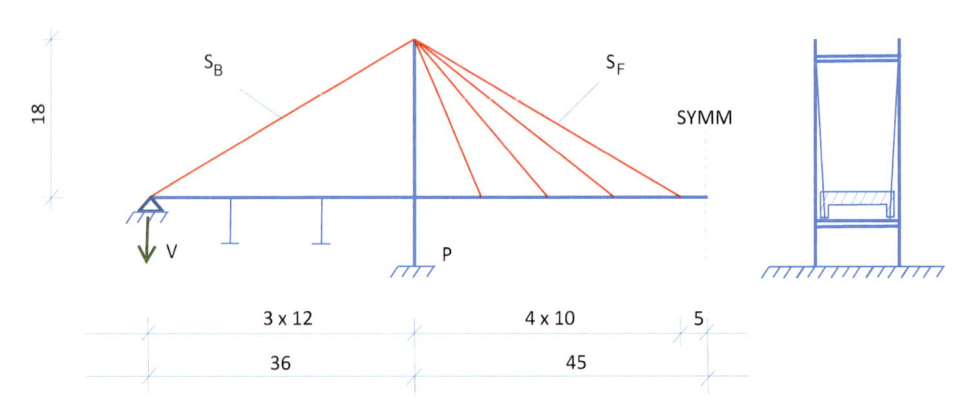

Figure 4.194 Elevation

4.5 Preliminary design calculations

The bridge beam comprises a solid twin t-beam cross-section with two outer cable planes, Fig. 4.195.

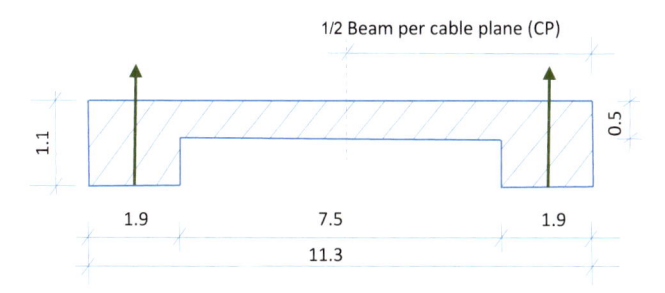

Figure 4.195 Cross-section

The cross-section properties per cable plane (CP) are those for the half bridge width for one cable plane.

Design parameters
Cables
Parallel wire cables: ø 7 mm, A = 38.48 mm^2
E_{eff} = 2 · 10^5 N/mm^2
Steel 1470/1670 (f_y/GUTS)
perm. f_S = 0.45 · 1650 = 752 N/mm^2, see Section 3.7.4, Table 3.3
perm. Δf_S = 200 N/mm^2, see Section 3.7.4, Table 3.3

Alternatively: Locked coil ropes
GUTS 1570 N/mm^2
E_{eff} = 1.7 · 10^5 N/mm^2
perm. f_S = 0.42 · 1570 = 659 N/mm^2, see Section 3.7.4, Table 3.3
perm. Δf_S = 150 N/mm^2, see Section 3.7.4, Table 3.3
Selection of rope diameters from manufacture catalog for example BRIDON [3.2].

Beam
Concrete B 45
E_C = 37 000 MN/m^2
Central compression perm. f_C = 11.5 MN/m^2
Edge compression perm. f_C = –17.0 MN/m^2
A_C = 3.99 m^2/CP
Vertical moment of inertia I_V = 0.341 m^4/CP
S_T = – 0.83 m^3/CP with S = Section modulus
S_B = + 0.49 m^3/CP

Dead load
Dead Load DL, including Super Imposed Dead Load SIDL:
DL = 128 KN/m, CP
Live load LL
Uniformly distributed load: UDL = 11.5 kN/m^3/CP
Concentrated Load CL: CL = 375 kN/CP

4.5.4.2 Cable forces of articulated system
Backstay cable SB

$\alpha_F = \arctan \dfrac{18}{40} = 24.22°$

$\alpha_B = \arctan \dfrac{18}{36} = 26.56°$

Dead Load, Fig. 4.196

$V^{DL} = 45 \cdot 128 \cdot \dfrac{22.5}{36} = 3600$ kN

$S^{DL} = \dfrac{3600}{\sin 26.56} = 8051$ kN

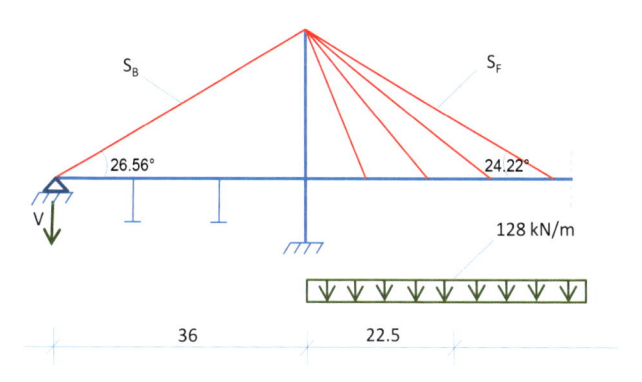

Figure 4.196 Permanent load arrangement for backstay

Live load, Fig. 4.197

$V_B = (45 \cdot 11.5 \cdot 22.5 + 375 \cdot 40) \cdot \dfrac{1}{36} = 740$ kN

$S_B = \dfrac{740}{\sin 26.56°} = 1654$ kN

$\max S_B = 8051 + 1654 = 9705$ kN

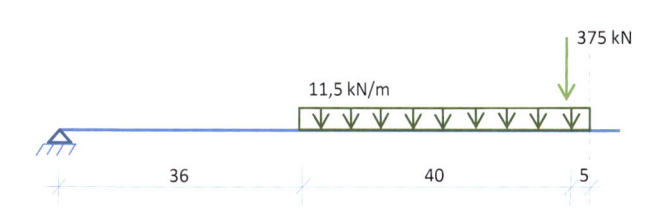

Figure 4.197 Permanent load arrangement for LL

Sizing for parallel wire cables

req'd $A_S = \dfrac{9705}{752} \cdot 10^3 = 12\,907$ mm^2

req'd n = $\dfrac{12\,907}{38.5} = 335$ wires ø 7 mm

Largest standard parallel wire cable in accordance with Fig. 4.222 has 325 wires. Therefore two cables with 169 wires each are selected.

act. $A_S = 13\,006$ mm^2/CP

Forestay cable s_F
Dead Load, Fig. 4.198

$$S_F = 10 \cdot 128 \cdot \frac{1}{\sin 24.22°} = 3120 \text{ kN}$$

Live load, Fig. 4.198

$$S_F = (10 \cdot 11.5 + 375) \cdot \frac{1}{\sin 24.22°} = 1194 \text{ kN}$$

(For the articulated system the full concentrated load is assumed, see note at the end of Section 4.5.2.2).

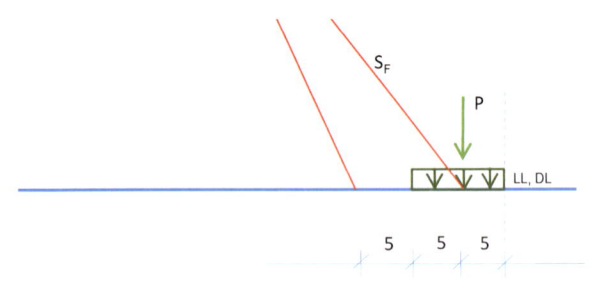

Figure 4.198 Dead load and live load

Sizing for parallel wire cables

max S_F = 4314 kN

req'd $n = \frac{4314 \cdot 10^3}{752 \cdot 38.5} = 149$ wires ø 7 mm

$n = 6x + 1$

act. $n = 151$

act. $A_S = 5813$ mm²/CP

Fatigue stresses
For 50 % of traffic
For forstay S_F

$$\Delta f_S = \frac{1/2 \cdot S^{LL}}{A_S} = \frac{1/2 \cdot 1194 \cdot 10^3}{5810} = 103 \text{ N/mm}^2 < 200 \text{ N/mm}^2$$

For backstay S_B

$$\Delta f_S = \frac{1/2 \cdot 1654 \cdot 10^3}{13\,006} = 64 \text{ N/mm}^2 < 200 \text{ N/mm}^2$$

Alternatively: locked coil rope
Backstay S_B
max S_B = 9705 kN
max ΔS_B = 1654 kN

req'd $A_S = \frac{9705}{659} \cdot 10^3 = 14\,727$ mm²

Selected: two ropes ø 105 mm, in accordance with BRIDON Catalogue [3.2]
act. $A_S = 2 \cdot 7710 = 15\,420$ mm²

Fatigue stresses

$$\Delta f_S = \frac{1/2 \cdot 1654 \cdot 10^3}{15\,420} = 54 \text{ N/mm}^2 < 150 \text{ N/mm}^2$$

Forestay SF
max S_F = 4314 kN
max ΔS_F = 1194 kN

req'd $A_S = \frac{4314 \cdot 10^3}{659} = 6546$ mm²

Selected: one rope ø 105 [3.2]
act. $A_S = 7710$ mm²

Fatigue stresses

$$\Delta f_S = \frac{1/2 \cdot 1194 \cdot 10^3}{7710} = 77 \text{ N/mm}^2 < 150 \text{ N/mm}^2$$

4.5.4.3 Bending moments for beam
For rigidly supported beam
The dead loads act on the rigidly supported system, Fig. 4.199.

$g = 128$ kN/CP

$$\frac{q \cdot l^2}{12} = \begin{cases} -1066 \text{ (Main span)} \\ -1536 \text{ (Side span)} \end{cases} \text{kNm/CP}$$

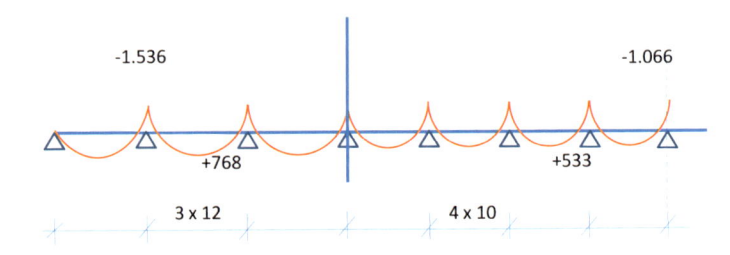

Figure 4.199 Dead load moment of rigidly supported beam

For live loads on beam on elastic foundation
The live loads act on the elastically supported system, Fig. 4.200.

$\delta = \Sigma S^2 \cdot \dfrac{1}{E \cdot A}$ with E for parallel wire stay cable

$\delta = 2.49^2 \cdot \dfrac{40.25}{2 \cdot 10^5 \cdot 13\,006 \cdot 10^{-6}} + 2{,}44^2 \cdot \dfrac{43.86}{2 \cdot 10^5 \cdot 5810 \cdot 10^{-6}}$

$ = 0.096 + 0.2246 = 0.32 \text{ m/MN}$

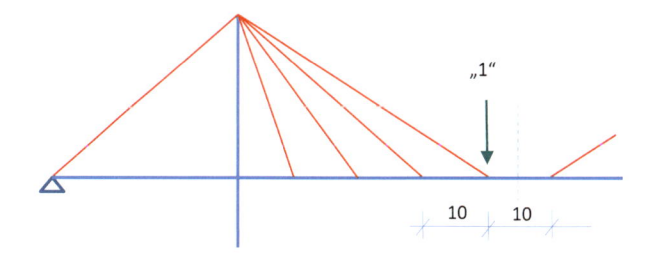

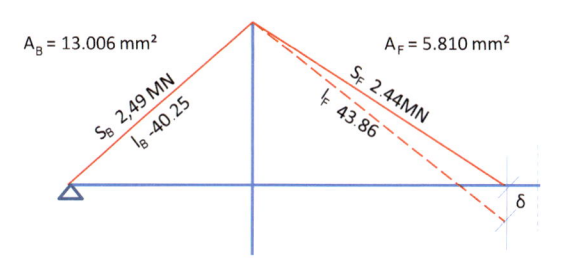

Figure 4.200 Parameters for the calculation of beam deflection

Elastic foundation

$c = \dfrac{1}{\delta \cdot \lambda_N} = \dfrac{1}{0.30 \cdot 10} = 0.31 \text{ MN/m}^2$

Elastic length

$L = \sqrt[4]{\dfrac{4 \cdot E_B \cdot I_B}{c}} = \sqrt[4]{\dfrac{4 \cdot 37 \cdot 10^3 \cdot 0.341}{0.31}} = 20.09 \text{ m}$ with E, I for beam

Live load moments

$M = 0.161 \cdot p \cdot L^2 + 1/4 \cdot P \cdot L$

$ = 0.161 \cdot 11.5 \cdot 20.09^2 + 1/4 \cdot 375 \cdot 20.09$

$ = 747 + 1883 = 2630 \text{ kNm/CP}$

Proof of stresses

$f = \dfrac{M}{S}$

$M = +533 + 2630 = 3163 \text{ kNm/CP}$

$f = 3163 / \begin{cases} -0.83 \\ +0.49 \end{cases} = \begin{cases} -3.8 \text{ MN/m}^2 \\ +6.5 \text{ MN/m}^2 \end{cases}$

Selected centrical prestress

$\Sigma f = \begin{cases} -10.8 < 17 \\ -0.5 < 0 \end{cases} \text{ MN/m}^2$

4.5.5 Cable-stayed bridge with harp arrangement
4.5.5.1 With regular side spans

Until now only cable-stayed bridges with fan arrangement have been used, for which all stay cables converge into a single point at the tower head.

For a harp cable arrangement there is an offset between the anchor points of the affected forestay cable and the anchor point for the backstay cable at the tower head, Fig. 4.201.

The cable forces are carried by shear and bending across this offset which means that cable-stayed bridges with harp arrangement require the bending stiffness of the tower for equilibrium. Therefore, the tower of a harp system has to carry higher moments than a tower of a fan system.

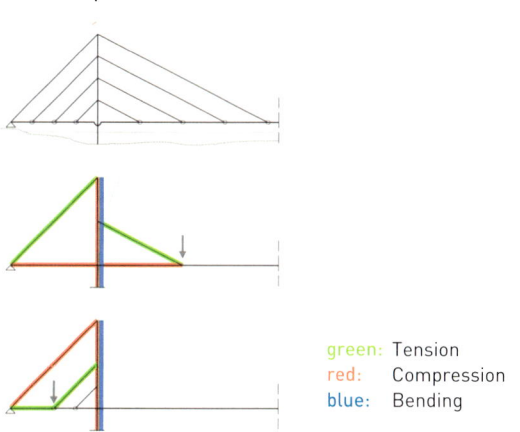

green: Tension
red: Compression
blue: Bending

Figure 4.201 Load distribution for a harp cable-stayed bridge

In the following, the inner flow of forces of a harp arrangement, Fig. 4.202, is compared with that of a fan system in accordance with Fig. 4.172. The overall dimensions are identical.

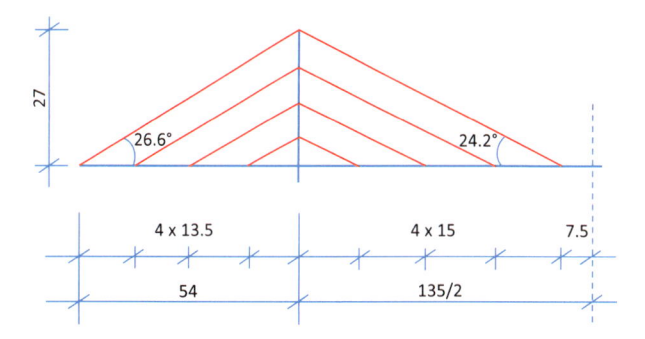

Figure 4.202 Elevation of harp system, cf. Fig. 4.172

For the investigation of the forces in the backstay cables the same loads are used as in Fig. 4.174. It is apparent that the support reactions of both systems are identical, including the anchor forces at the hold-down piers. Therefore, the forces in the backstay cables are identical. The inner flow of forces, however, is different because the inner stay cables in harp arrangement are flatter than those of a fan arrangement. The flow of forces is shown in Fig. 4.203.

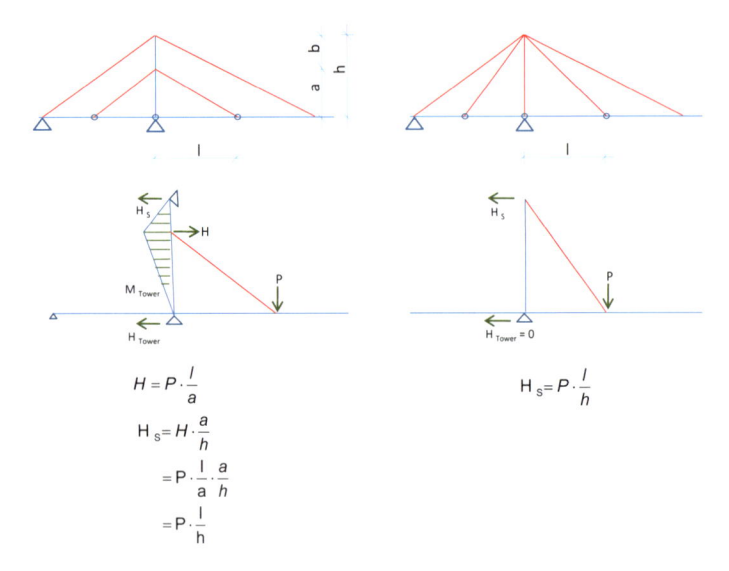

Figure 4.203 Comparison of load distributions for harp and fan systems

$$H = P \cdot \frac{l}{a}$$
$$H_s = H \cdot \frac{a}{h}$$
$$= P \cdot \frac{l}{a} \cdot \frac{a}{h}$$
$$= P \cdot \frac{l}{h}$$

$$H_s = P \cdot \frac{l}{h}$$

Exactly that part of the bigger horizontal force of the flatter harp cables runs into the tower foundation which is larger than the smaller horizontal component of the steeper fan cable. In this way the forces in the backstay cables, which have the same inclination for both systems, are identical.

A comparison of both systems shows:
– the forces in the backstay cables are identical, but bigger in the other harp cables,
– the beam and tower normal forces are bigger in the harp system,
– the harp tower receives bending from the distribution of the stay cable forces to the tower head and the foundation,
– the beam moments are about equal because the smaller elastic support of the flatter harp cables is somewhat compensated by their larger cross-section.

As a result it can be stated that the better appearance of a cable-stayed bridge with harp arrangement against that of a fan arrangement requires slightly more material for the stay cables and the towers.

4.5.5.2 With side spans on piers

If all side span cables of a harp system are directly anchored to piers in the side span the bending in the towers is strongly reduced, Fig. 4.204.

For loads in the main span, the tensile forces of the forestay cables try to find their way via the largest resistance, meaning with the smallest deformations. They thus transfer their loads to the nearest backstay cable at the tower, which is directly anchored at a side span. In this way the lever arm to the actual pier is smaller than the distance to the outside anchor pier and the backstay cable forces are larger than for a system without piers in the side span. The reduced tower moments thus result in increased backstay cable forces and uplift support reactions in the side span.

Loads in the side span are directly supported by the side span piers and thus do not produce compression in the backstay cables, which reduces their fatigue range. In summary, there are increasing and decreasing loads and stresses for harp and fan systems.

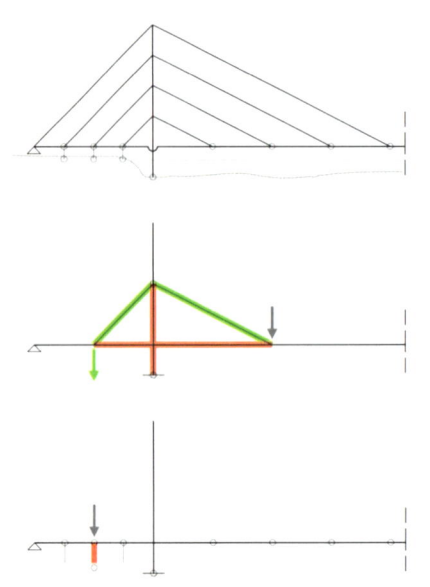

Figure 4.204 Load distributions of harp systems with piers in the side span

4.5.6 Cable-stayed bridge with longitudinal A-tower

The transfer of loads is fundamentally different for a cable-stayed bridge with a stiff A-frame in the longitudinal direction, Fig. 4.205. The tension in each cable is transferred to the tower head and from there in tension or compression of the tower legs directly into the foundations. This tower shape is used for cable-stayed bridges with equal lengths of main span and side spans, and for series of cable-stayed bridges.

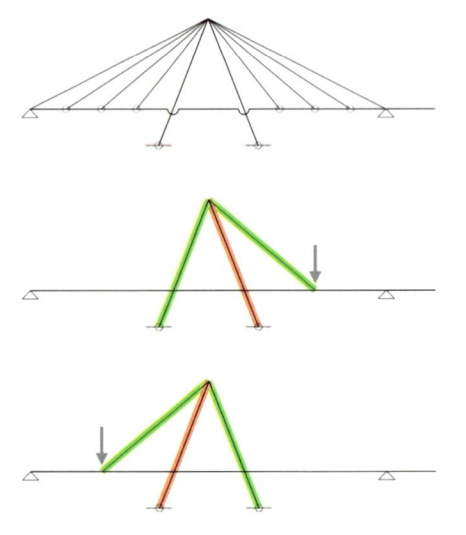

Figure 4.205 Load distribution for A-tower

4.5.6.1 System and loads

Consider an inner partial bridge with an A-tower of a series of cable-stayed bridges, Fig. 4.206. The load transfer from both cantilevers takes place via the A-tower into the foundations. The beam cross-section is shown in Fig. 4.207. The tower legs will be considered weightless and infinitely stiff.

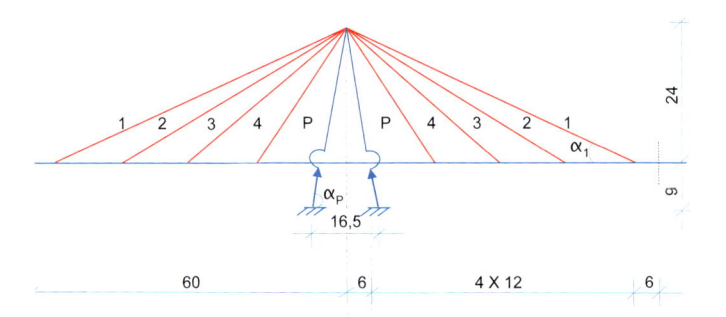

Figure 4.206 Elevation

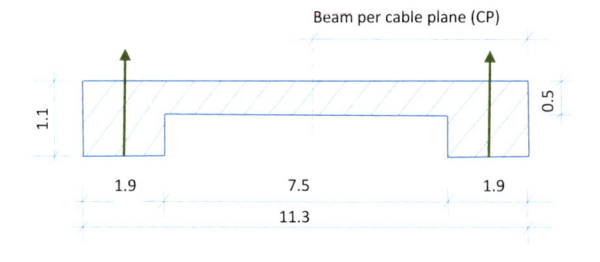

Figure 4.207 Cross-section

Design parameters
Cables
Parallel strand cables
ø 15 mm, A = 150 mm²
$E_{eff} = 1.9 \cdot 10^5$ N/mm²
Steel 1660/1860 (f_y/GUTS)
perm. $f_S = 0.45 \cdot 1860 = 837$ N/mm²
perm. $\Delta f_S = 200$ N/mm², see Section 3.7.4, Table 3.3

Beam
Concrete B 45
$E_C = 37\,000$ MN/m²
Central compression perm. $f_C = 11.5$ MN/m²
Edge compression perm. $f_C = -17.0$ MN/m²
$A_C = 3.99$ m²/CP
Vertical moment of inertia $I_V = 0.341$ m⁴/cable plane (CP)
$S_T = -0.83$ m³/CP with S = section modulus
$S_B = +0.49$ m³/CP

Dead load
Dead load DL, including superimposed dead load SIDL:
DL = 128 KN/m, CP
Live load LL
Uniformly distributed load: UDL = 11.5 kN/m³/CP
Concentrated load CL: CL = 375 kN/CP

4.5.6.2 Normal forces for articulated system
Dead load, Fig. 4.208

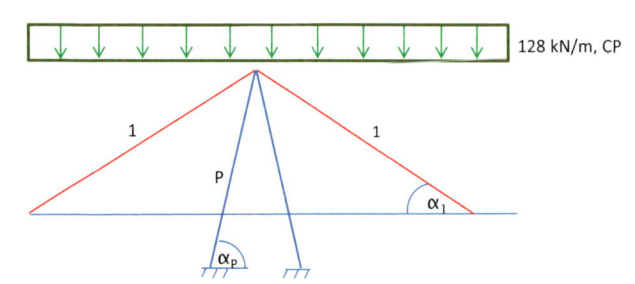

Figure 4.208 Arrangement of dead loads

$\alpha_1 = \arctan \dfrac{24}{54} = 23.96°$

$S_1 = 12 \cdot 128 \cdot \dfrac{1}{\sin 23.96°} = 3782$ kN

$\alpha_P = \arctan \dfrac{6}{33} = 10.3°$

$P = -60 \cdot 128 \cdot \dfrac{1}{\cos 10.3°} = -7805$ kN

Normal forces from full live load, Fig. 4.209

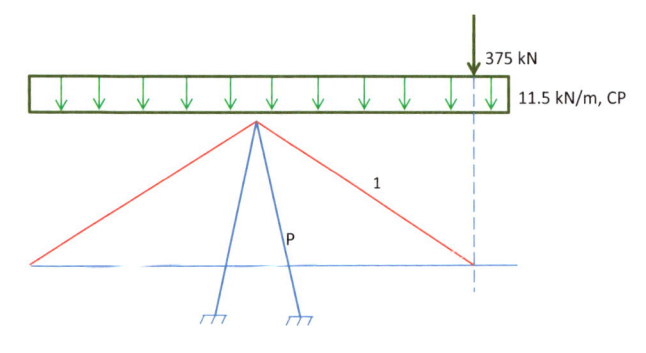

Figure 4.209 System with full live load

$\max S_1 = (12 \cdot 11.5 + 375) \cdot \dfrac{1}{\sin 23.96°} = 1263$ kN

$P = \left(-60 \cdot 11.5 - 375 \cdot 54 \cdot \dfrac{1}{16.5}\right) \cdot \dfrac{1}{\cos \alpha} = -1949$ kN

Tower with one-sided live load, Fig. 4.210

$P = \pm \left(\dfrac{1}{2} \cdot 60^2 \cdot 11.5 + 375 \cdot 54\right) \cdot \dfrac{1}{16.5 \cdot \cos 10.3°} = \pm 2522$ kN

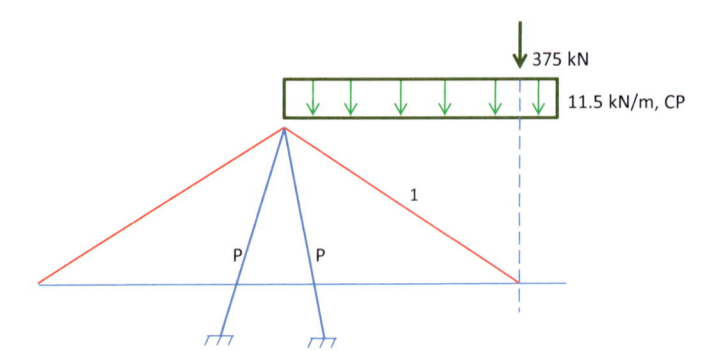

Figure 4.210 System with full live load

Max/min N in tower legs

min V = −7805 − 2522 = −10 327 kN/CP

max V = −7805 + 2522 = −5283 kN/CP < 0

4.5.6.3 Cable sizing

max S_1 = 3782 + 1263 = 5045 kN

Number of strands

req'd n = $\dfrac{5045}{837 \cdot 150} \cdot 10^3$ = 40.2

Chosen: n = 41 strands

$\Delta f = \dfrac{0.5 \cdot 1263}{41 \cdot 150} \cdot 10^3 = 102.7$ N/mm² < 200 N/mm²

4.5.6.4 Bending moments for beam

Dead load, Fig. 4.211

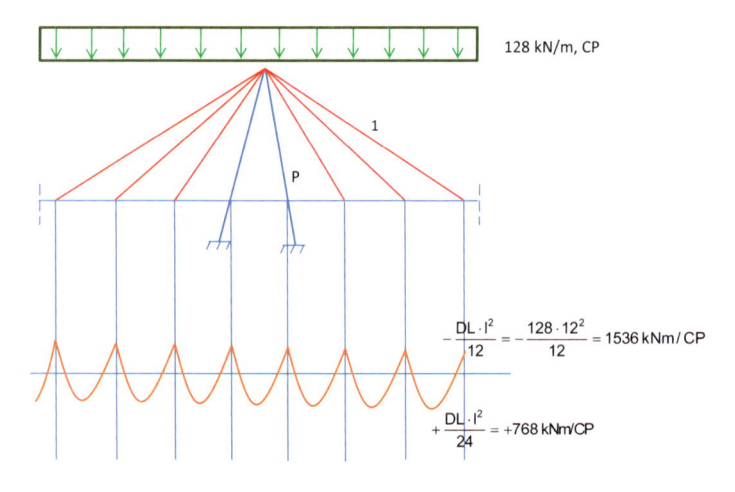

$-\dfrac{DL \cdot l^2}{12} = -\dfrac{128 \cdot 12^2}{12} = 1536$ kNm/CP

$+\dfrac{DL \cdot l^2}{24} = +768$ kNm/CP

Figure 4.211 Dead load moments for rigidly supported system

Live load for beam on elastic foundation, Fig. 4.212

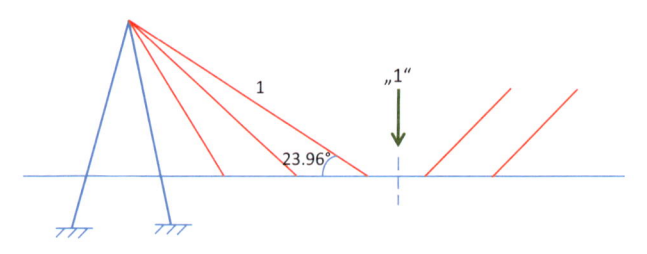

Figure 4.212 Calculation of the elastically supported moments

$S_1 = \dfrac{1}{\sin \alpha} = 2.46$

$A_1 = 41 \cdot 150 = 6150$ mm²

$I_1 = \dfrac{54}{\cos \alpha} = 59.09$

$\delta = S_1^2 \cdot \dfrac{I_1}{E \cdot A_1} = 2.46^2 \cdot \dfrac{59.09 \cdot 10^6}{1.9 \cdot 10^5 \cdot 6150} = 0.306$ m

$c = \dfrac{1}{12 \cdot 0.306} = 0.272$

$L = \sqrt[4]{\dfrac{4 \cdot 37 \cdot 10^3 \cdot 0.341}{0.272}} = 20.75$ m

$M_V^{elast} = 0.161 \cdot p \cdot L^2 + 1/4 \cdot P \cdot L$

$= 0.161 \cdot 11.5 \cdot 20.75^2 + 1/4 \cdot 375 \cdot 20.75$

$= 797 + 1945 = 2742$ kNm/CP

Proof of stresses

N ≅ 0

max M = $M_{rigid} + M_{elast.}$ = 768 + 2742 = 3510 kNm

$f = \dfrac{M}{S} = \dfrac{3510}{-0.83 \, / \, +0.49} = \begin{matrix} -4.22 \text{ MN/m}^2 \\ +7.16 \text{ MN/m}^2 \end{matrix}$

4.5.6.5 Post-tensioning

Chosen: $V_\infty = -8.0$ MN/m² centrically

$\Sigma f = \begin{cases} -4.22 - 8.0 \\ +7.16 - 8.0 \end{cases}$

$\Sigma f = \begin{cases} -12.22 > 17.0 \\ -0.84 \; < \; 0 \end{cases}$ MN/m²

4.5 Preliminary design calculations

4.5.7 Slender cable-stayed concrete bridge

Given is a cable-stayed bridge with a very slender concrete beam, beam depth to span $1.1:396 = 1:360$.

The non-linear increase of live load moments is estimated and the aerodynamic safety against flutter oscillations is proven.

4.5.7.1 System and loads
System

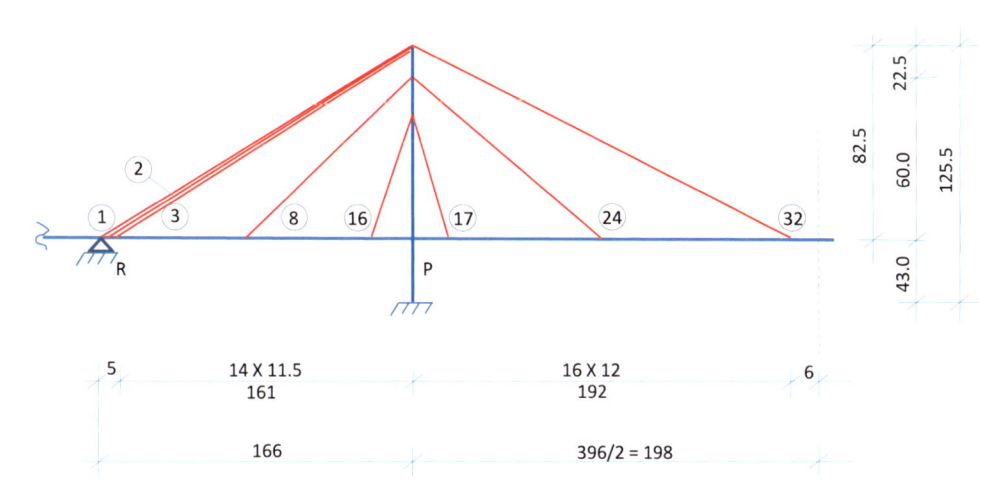

Figure 4.213 Elevation

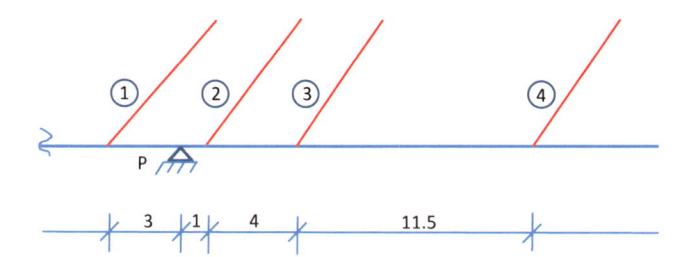

Figure 4.214 Cable anchorages at anchor pier

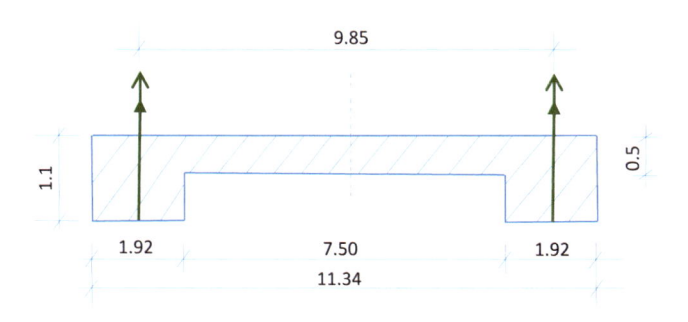

Figure 4.215 Simplified cross-section

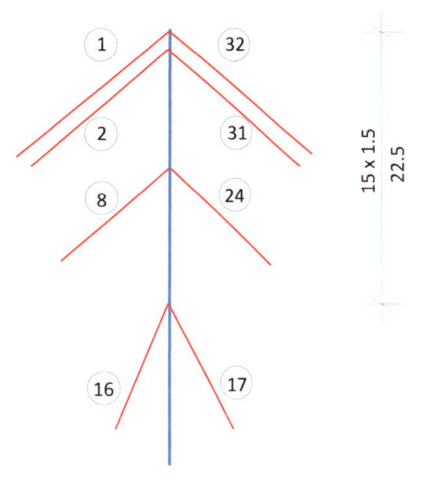

Figure 4.216 Cable anchorages at tower head

Check of geometry
Span lengths
actual $\dfrac{l_1}{l} = \dfrac{166}{396} = 0.42$ Ok, compare Fig. 4.2

Tower height
actual $\kappa = \dfrac{h}{l} = \dfrac{82.5}{396} = 0.21$ Ok, compare Fig. 4.3

Section properties
Parallel wire cables
St 1470/1670
$E_{eff} = 2 \cdot 10^5$ N/mm^2
permissible $f_S = 0.45 \cdot 1670 = 752$ N/mm^2
permissible $\Delta f_S = 200$ N/mm^2 fatigue strength
Wires ø 7 mm, $A_S = 38.48$ mm^2

Beam
Concrete B 45
$E_B = 37 \cdot 10^3$ MN/m guaranteed compression strength
central compression: permissible $f_C = 11.5$ MN/m^2
edge stresses: permissible $f_C = 17.0$ MN/m^2

Section properties per cable plane (CP)
Vertical per cable plane
$I_V = 0.341$ m^2/CP vertical beam moment of inertia
$S_T = -0.83$ m^3/CP section modulus at top
$S_B = +0.49$ m^3/CP section modulus at bottom
$A = 3.99$ m^2/CP area

Horizontal per bridge
$I_{Hor} = 112.6$ m^4/bridge
$S_{Hor} = \dfrac{I}{b/2} = \dfrac{112.6}{11.34/2} = 19.9$ m^3/bridge

Loads
Dead loads including SIDL: 128 kN/m, CP

Live loads per lane

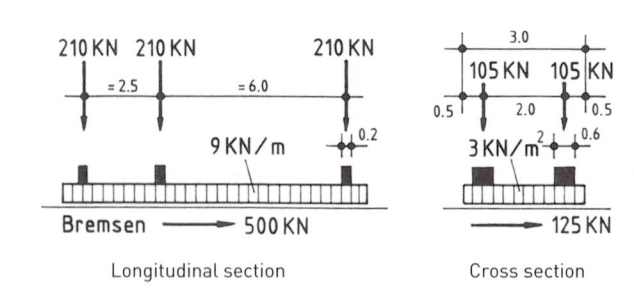

Figure 4.217 Live loads per lane

The traffic lanes are moved as far sideways as possible and distributed for a beam supported at the cable planes, the torsional beam stiffness neglected, impact factor 1.

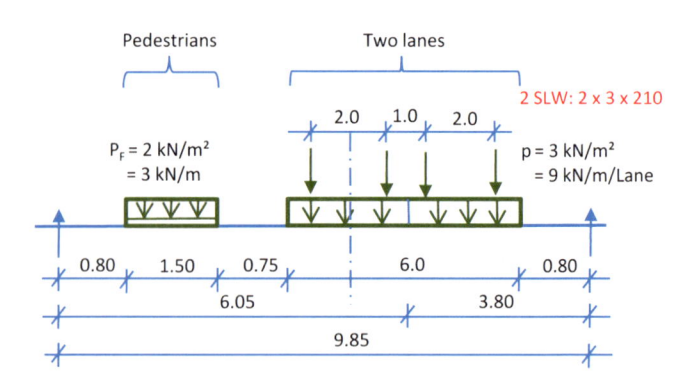

Figure 4.218 Transverse live load distribution

2 traffic lanes $\eta = 6.05/9.85 = 0.61$
pedestrians $\eta = 1.55/9.85 = 0.16$
UDL $p = 2 \cdot 9 \cdot 0.61 + 0.16 \cdot 3 = 11.5$ kN/m, CP
truck $P = 2 \cdot 630 \cdot 0.61 = 774$ kN/CP
$\cong$ 3 Axles with 258 kN/CP each

Summary of live loads
UDL = 11.5 kN/m, CP
truck = $3 \cdot 258 = 774$ kN/CP

Eccentricity factors
$UDL = \dfrac{11.5}{1/2\,(3+2\cdot 9)} = 1.095$

$truck = \dfrac{3\cdot 258}{1/2\cdot 2\cdot 3\cdot 210} = 1.23$

4.5.7.2 Stay cables
Backstay cables
Support reactions
Dead loads
$V^{DL} = 42.75 \cdot 128 \cdot \dfrac{176.63}{166} = 5822$ kN/CP

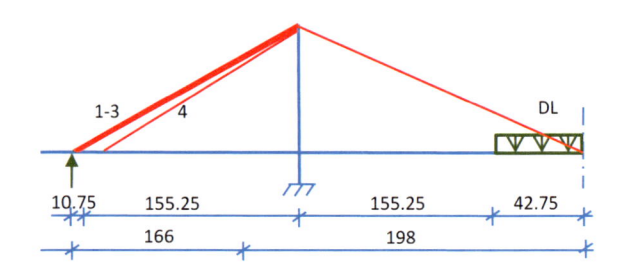

Figure 4.219 Governing dead load arrangement

4.5 Preliminary design calculations

Live load

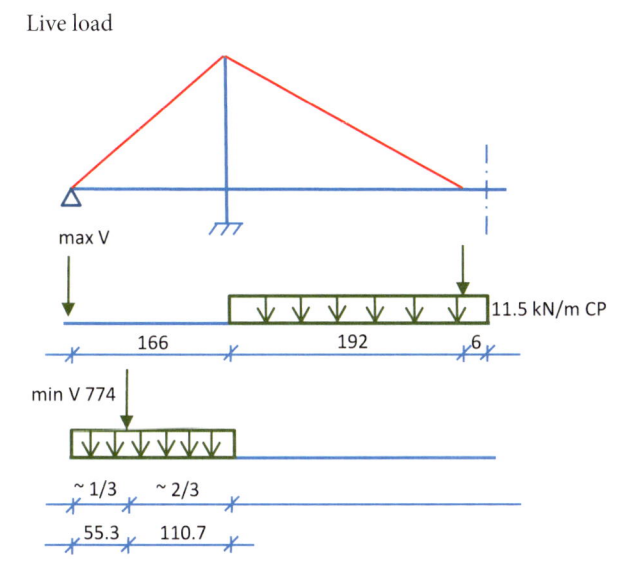

Figure 4.220 Governing live load arrangement

$\max V^P = \frac{1}{2} \cdot \frac{198^2}{166} \cdot 11.5 + 774 \cdot \frac{192}{166}$

$\qquad = 1358 + 895 = 2253$ kN/CP ↓

$\min V^{CL} = \frac{1}{2} \cdot 166 \cdot 11.5 + \frac{2}{3} \cdot 774$

$\qquad = 955 + 516 = 1470$ kN/CP ↑

with
CL   concentrated load is used for truck loads

Cable forces

$\alpha_2 = \arctan \frac{81}{165} = 26.15°$ at cable 2

For backstay cables 1–3

$S_R^g = 13\,210$ kN/CP

$\max S_R^P = 5112$ kN/CP

$\min S_R^P = -3335$ kN/CP

$\max S_R = 18\,322$ kN/CP

$\Delta S_R = 8447$ kN/CP   for 100% live load

Sizing
from $\max S_R$

req'd $n = \frac{18\,322\,000}{752 \cdot 38.48} = 633$

Chosen: $n = 3 \cdot 211$ for each cable S_1, S_2, S_3

$n = 633$ wires ø 7 mm

$A_S = 24\,358$ mm^2

Fatigue for 50% live load:

$\Delta f = \frac{1/2 \cdot 8447}{633 \cdot 38.48} = 173$ N/mm^2 < 200 N/mm^2

Forestays

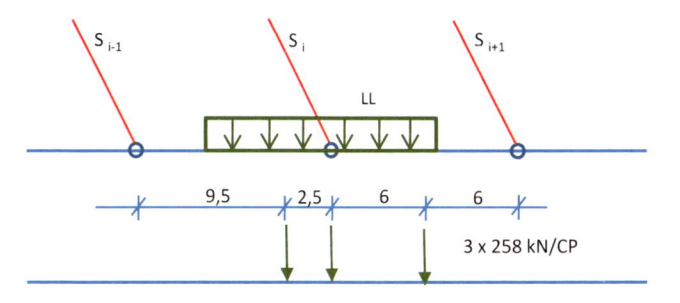

Figure 4.221 Live load distribution

$\max V_i^{II} = a_i \cdot \text{UDL} + \eta \cdot \text{CL}$

$\eta = \left(1 + \frac{9.5}{12} + \frac{6}{12}\right) \cdot \frac{258}{3 \cdot 258} = 0.76$

$\max V_i = 12 \cdot 139.5 + 0.76 \cdot 774 = 2262$ kN/CP

$\max S_i = V_i \cdot \frac{1}{\sin \alpha}$

req'd $n = \frac{\max S_i}{752 \cdot 38.48 \cdot 10^{-3}}$

Cable 32 in bridge center

$\alpha = \arctan \frac{82.5}{192} = 23.25°$

$\max S = 6743$ kN

$n = 235$ ø 7 mm

$A_S = 9043$ mm^2

Cable 24 in ¼ point

$\alpha = \arctan \frac{70.5}{96} = 36.3°$

$\max S = 4496$ kN

$n = 157$ ø 7 mm

$A_S = 6041$ mm^2

Cable 17 at tower

$\alpha = \arctan \frac{60}{12} = 78.7°$

$\max S = 2714$ kN

$n = 97$ ø 7 mm

$A_S = 3732$ mm^2

Cable dimensions, see Fig. 4.222.

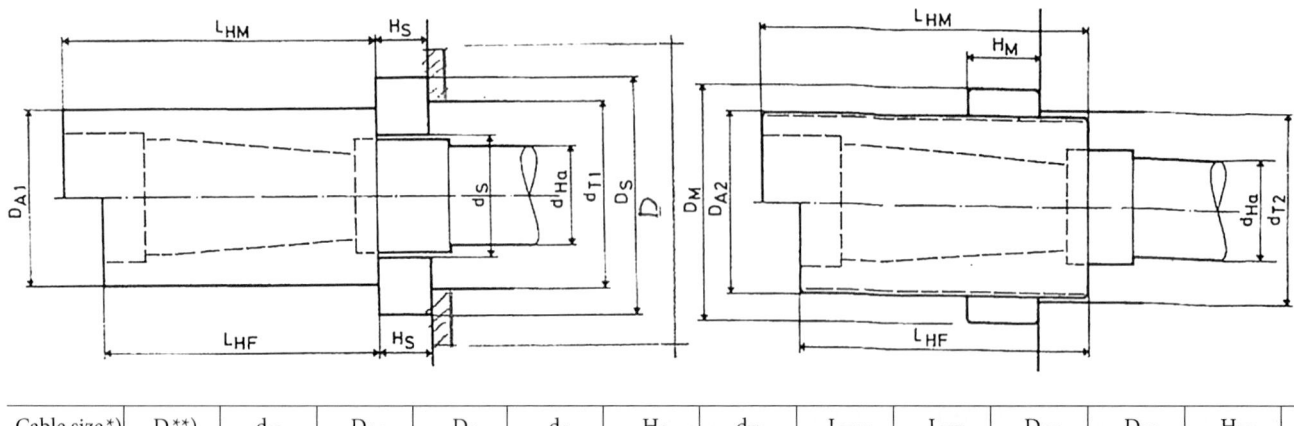

Cable size*)	D**)	d_{Ha}	D_{A1}	D_S	d_S	H_S	d_{T1}	L_{HM}	L_{HF}	D_{A2}	D_M	H_M	d_{T2}
19		75	125	165	100	35	140	230	230	135	175	45	150
31		90	145	190	115	40	160	270	255	155	200	55	170
61		110	190	250	135	55	205	340	310	195	255	75	210
91	400	125	220	295	150	65	235	390	350	225	295	90	240
121	450	140	245	330	165	75	260	445	390	250	325	100	265
151	500	160	275	370	185	85	290	490	425	275	360	110	290
187	550	160	295	400	185	90	315	535	460	305	395	125	325
211	600	180	320	430	205	95	340	570	485	320	415	135	340
241	650	180	330	450	205	105	350	600	510	340	445	140	360
253		180	335	455	205	105	355	615	525	350	460	145	370
277		200	355	485	225	110	375	645	545	365	480	150	385
295		200	365	500	225	115	385	660	555	375	495	155	395
313		200	370	510	225	120	390	675	565	385	505	160	405
325		200	375	515	225	120	395	705	595	395	515	165	415

dimensions in mm
*) based on use of dia. 7 mm wires with an u.t.s.: 1670 N/mm^2
**) type of polyethylene sheating is „Hostalen GM 5010 T2", Serie 4

Figure 4.222 Cable dimensions

4.5 Preliminary design calculations

Table 4.4 Cable diameters

Cable No.	ø PE Pipe [mm]
1–3	180
16, 17	140
8, 24	160
32	180

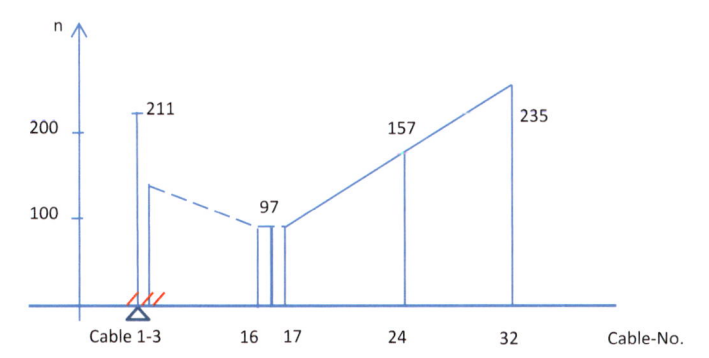

Figure 4.223 Cable sizes

Approximate determination of total steel quantity
In Fig. 4.3, formulas for the approximate calculation of the total steel quantity required are given. Using the nomenclature of Fig. 4.3, we find the total steel quantity.

q = g + p = 128 + 11.5 = 139.5 kN/m, CP
(uniformly distributed dead load + live load)

Only $1/3$ of the concentrated load is carried directly by each cable.
Δq = 1/3 · 774 · 1/12 = 21.5 kN/m, CP (15 % increase)

Thus the effective total uniformly distributed load becomes:

$\bar{q}$ = 139.5 + 21.5 = 161 kN/m, CP

γ = 0.785 kg/cm²m
σ = 752 N/mm²
l = 396 m
h = 82.5 m
κ = 0.21

$C_F = 2\kappa + \dfrac{1}{6\kappa} = 1.21$ see Fig. 4.3

Steel quantity for fan arrangemant

$$G_F = \dfrac{\bar{q} \cdot \gamma \cdot l^2}{633 \cdot 38{,}48} \cdot C_F \cdot 2$$

$$= \dfrac{161 \cdot 0.785 \cdot 396^2}{752} \cdot 10 \cdot 1.21 \cdot 2 = 638 \text{ t for total bridge}$$

A = (2 · 166 + 396) · 11.34 = 8256 m² total bridge area

g = 77.3 kg/m² unit weight of stay cables

4.5.7.3 Beam moments

For rigidly supported beam
From dead load
g = 128 kN/m, CP
Moments

$M_{Support} = -\dfrac{DL \cdot l^2}{12} = \begin{cases} -1536 \\ -1411 \end{cases}$ kNm/CP $\begin{vmatrix} \text{main span} \\ \text{side span} \end{vmatrix}$

$M_{span} = \dfrac{DL \cdot l^2}{24} = \begin{cases} +768 \\ +706 \end{cases}$ kNm/CP

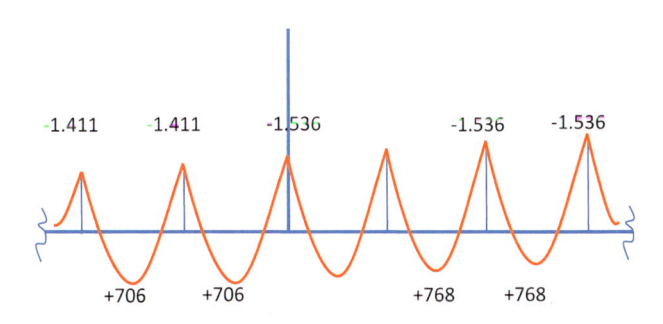

Figure 4.224 Bending moments of a beam rigidly supported at the cable anchor points

Normal forces at tower, see Fig. 4.6: Full load

h = 82.5 – 1/3 · 22.5 = 75.0

$N_{Tower} = -\dfrac{g+p}{2 \cdot h} \cdot l^2 = -\dfrac{139.5}{2 \cdot 75} \cdot 198^2 = -36460$ kN/CP

Normal forces at backstay – see Section 2.1.2 above.

$S_{1-3} = 13\,210 + 5112 - 3335 = 14\,987$ kN/CP

$N_{1-3} = S_{1-3} \cdot \cos\alpha_2 = -13\,452$ kN/CP

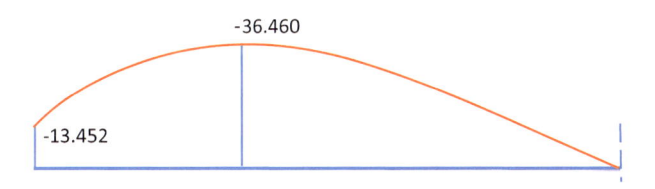

Figure 4.225 Beam normal forces from full load

For beam on elastic foundations
Deflections, see Fig. 4.226

$\delta = \Sigma S_i^2 \cdot \dfrac{l_i}{E_i \cdot A_i}$

at tower:

$\delta = 0.164^2 \cdot \dfrac{183.8 \cdot 10}{2.0 \cdot 24\,358} + 1.02^2 \cdot \dfrac{61.2 \cdot 10}{2.0 \cdot 3723}$

= 0.001 + 0.086 = 0.087 m/MN

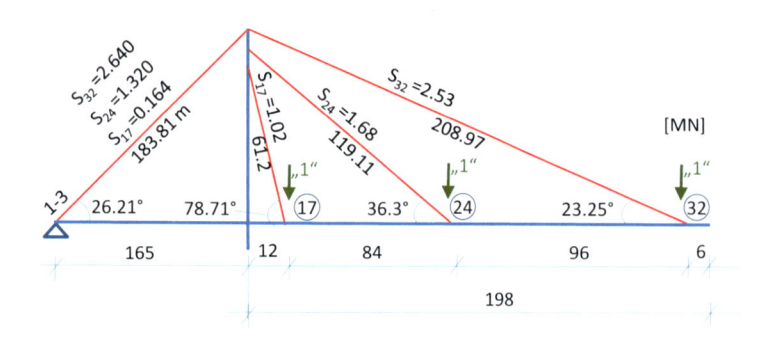

Figure 4.226 Design parameters for deflection calculations

at ¼ point:
$$\delta = 0.320^2 \cdot \frac{183.8 \cdot 10}{2.0 \cdot 24\,358} + 1.68^2 \cdot \frac{119.11 \cdot 10}{2.0 \cdot 6041}$$
$$= 0.066 + 0.278 = 0.344 \text{ m/MN}$$

at bridge center:
$$\delta = 0.640^2 \cdot \frac{183.8 \cdot 10}{2.0 \cdot 24\,358} + 2.53^2 \cdot \frac{208.97 \cdot 10}{2.0 \cdot 9043}$$
$$= 0.263 + 0.740 = 1.003 \text{ m/MN}$$

Elastic support
$$c = \frac{1}{\delta \cdot \lambda_i} \quad \lambda_i = 12 \text{ m} \text{ in main span}$$

at tower:
$$c = \frac{1}{0.087 \cdot 12} = 0.958$$

at ¼ point:
$$c = \frac{1}{0.344 \cdot 12} = 0.242$$

at bridge center:
$$c = \frac{1}{1.003 \cdot 12} = 0.083$$

Elastic length
$$L = \sqrt[4]{\frac{4\,E_B \cdot I_B}{c}} = \left(\frac{4 \cdot 37 \cdot 10^3 \cdot 0.341}{c}\right)^{1/4} = \frac{14.99}{\sqrt[4]{c}}$$

at tower: $L = 15.15$ m
at ¼ point: $L = 21.37$ m
at bridge center: $L = 27.92$ m

Moments from live load
UDL
$$M^P = 0.161 \cdot p \cdot L^2$$
$$= 0.161 \cdot 11.5 \cdot L^2 = 1.852 \cdot L^2$$

at ¼ point:
$L = 21.34$ m
$\frac{\pi}{2} \cdot L = 33.6$ m
$L/4 = 5.3$ m

Moments from truck load
Reduction factor for truck, see Figs 4.227 and 4.228.

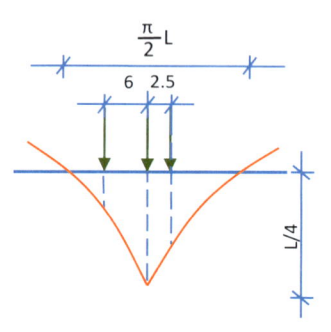

Figure 4.227 Truck loads

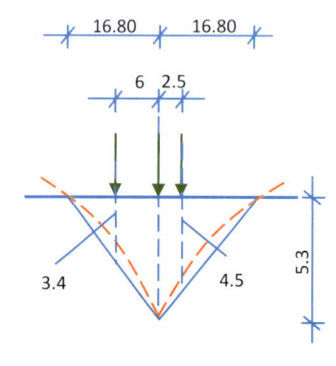

Figure 4.228 Approximation for truck load influence line

i. M. $\eta = (3.4 + 5.3 + 4.5) \cdot \frac{1}{3 \cdot 5.3} = 0.83$

Thus the moment from trick becomes
$$M_{EL}^{SLW} \cong 0.83 \cdot P \cdot \frac{L}{4} \cong 0.83 \cdot 774 \cdot \frac{L}{4} = 161 \cdot L$$

$$M_{EL} = 1.852 \cdot L^2 + 161 \cdot L$$

at ¼ point:
$M_{EL} = 845.77 + 3441 = 4286.77$ kNm/CP

at bridge center:
$M_{EL} = 1443.68 + 4495 = 5938.68$ kNm/CP

4.5 Preliminary design calculations

Non-linear moment increase (second order theory)
The non-linear moment increase is approximately determined by using the safety against buckling of the beam as outlined in Fig. 4.13.

Safety against buckling

$$\gamma = \frac{P_{Ki}}{N}$$

$P_{Ki} = 2 \cdot \sqrt{(E \cdot I \cdot c)}$ with c elastic beam support at tower

$P_{Ki} = 2 \cdot \sqrt{37 \cdot 10^3 \cdot 2 \cdot 0.341 \cdot 0.958} = 311$ MN

$$\gamma = \frac{311}{36.46} = 8.53$$

non-linear moment increase:

$$\frac{M^{II}}{M} = \eta = \frac{1}{1 - 1/\gamma}$$

$$\eta = \frac{1}{1 - 1/8.53} = 1.13$$

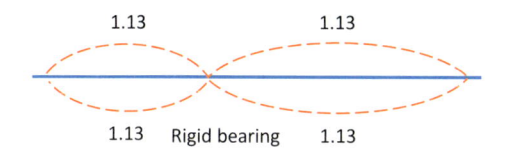

Figure 4.229 Non-linear moment increase

For the somewhat similar Helgeland bridge the exactly determined non-linear moment increase came to 1.29 in service state, see Fig. 4.47.

Superposition of moments

$\Sigma M = (M_{rigid} + M_{elastic}) \cdot \eta$

at bridge center:
$\eta \cong 1$
$\Sigma \max M = +768 + 5939 = 6707$ kNm/CP

at ¼ point:
$\eta \cong 1.13$
$\Sigma \max M = (768 + 4286) \cdot 1.13 = 5711$ kNm/CP

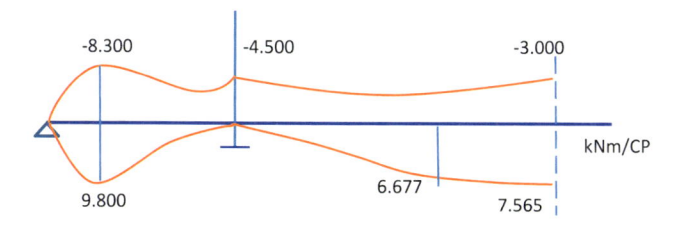

Figure 4.230 Approximate live load moment envelope, see Fig. 4.39

Proof of stresses

$$f = \frac{M}{S} + \frac{N}{A}$$

at bridge center:

$\min_{\max} f = \frac{6707}{\begin{array}{c}-0.83\\+0.49\end{array}}$ $N \cong 0$

$\min_{\max} f = \frac{-8.1}{+13.7}$ MN/m^2

Centrical post-tensioning with
$V_\infty = -5$ MN/m^2

$\Sigma f = \begin{array}{l}-8.1 - 5.0 = -13.1 \quad > -17 \text{ MN/m}^2\\+13.7 - 5.0 = +8.7\end{array}$

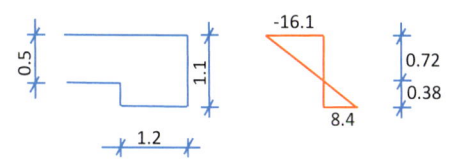

Figure 4.231 Reinforcement for concrete tensile stresses

Reinforcement for tensile stresses
$T = 1/2 \cdot 8.7 \cdot 0.44 \cdot 1.92 = 3.7$ MN

req'd $A_S = \frac{3.7}{210} \cdot 10^4 = 176$ cm$^2 \triangleq 34$ ø 25

with
T total Tension in concrete, compare Fig. 4.231

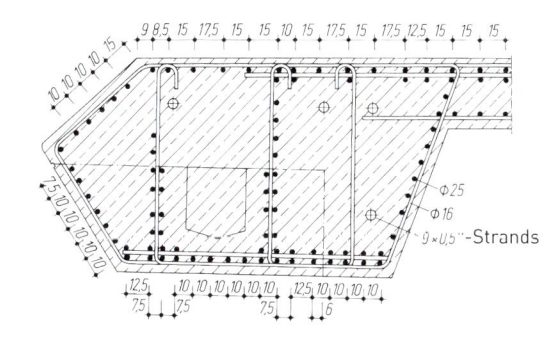

Figure 4.232 Typical reinforcement of edge beam

Transverse wind
Wind loads
Wind speed
v = 70 m/sec without traffic on bridge

Wind pressure
$$q = \frac{v^2}{1600} \cdot 3.1 \text{ kN/m}^2$$

Beam
Wind on beam, drag factor referred to beam depth h
$C_H \cong 1.0$ railing taken into account
$h \cong 1.16 + 0.26 = 1.42$ m
$w = 3.1 \cdot 1.0 \cdot 1.42 = 4.4$ kN/m

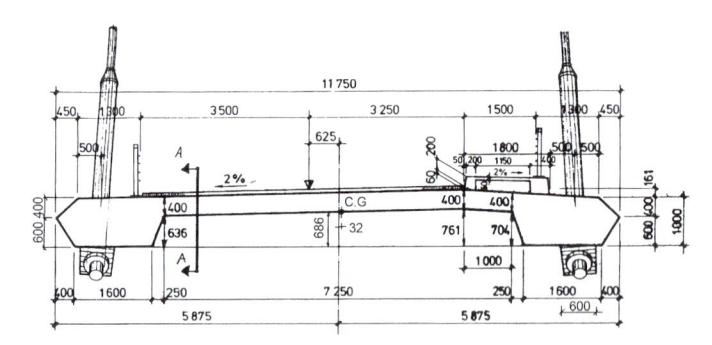

Figure 4.233 Beam cross-section

Wind on cables
Half each on beam and towers
Cable ø $0.14 \div 0.18$ m (see Table 4.4)
$C_w = 0.7$, see Fig. 3.90 and 3.91

Cable 32
$w_c = C_w \cdot q \cdot ø \cdot 1/2 \cdot l_{Cable}$
$= 0.7 \cdot 3.1 \cdot 0.18 \cdot 1/2 \cdot 209 \cdot 2$ (wind on both cable planes)
$= 81.6$ kN/12 m $= 6.8$ kN/m

Cable 17
$w_c = 0.7 \cdot 3.1 \cdot 0.14 \cdot 1/2 \cdot 61.2 \cdot 2$
$= 18.5$ kN/12 m $= 1.5$ kN/m

On average
$w_c \cong 1.5 + (6.8 - 1.5) \cdot 0.6 = 4.7$ kN/m constant

$\Sigma w = 4.4 + 4.7 = 9.1$ kN/m

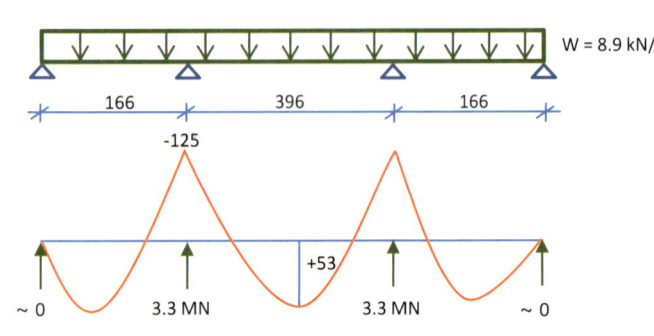

Figure 4.234 Loading and moments in beam from transverse wind

Action forces
For full bridge width
$M_{Tower} \cong -1/2 \cdot 9.1 \cdot 166^2 = -125$ MNm/Bridge

$M_{Span} = -125 + 1/8 \cdot 396^2 \cdot 9.1 \cdot 10^{-3} = +53$ MNm/Bridge

Stresses in beam from transverse wind on unloaded bridge
in bridge center: $N^{DL} \cong 0$
$f_{c, transv} = \pm \dfrac{53}{19.9} = \pm 2.7$ MN/m^2

at tower: $N^{DL} \cong -36460$ MN/CP
$f_{c, transv} = \pm \dfrac{125}{19.9} - \dfrac{36.5 \cdot 2}{3.99 \cdot 2} = \pm 6.3 - 9.1 = \dfrac{-15.4}{-2.8}$ MN/m^2 > -17 MN/m^2

4.5.7.4 Aerodynamic stability
As an approximation the critical wind speed for flutter is determined, Section 4.3.5.

Eigenfrequencies
In bending
$f_b = \dfrac{1}{2\pi} \cdot \sqrt{\dfrac{K_B}{m}} = \dfrac{0.58}{\sqrt{\delta}}$, see Section 4.2.7

with
K_B spring constant in bending
δ deflection in bridge center unter DL

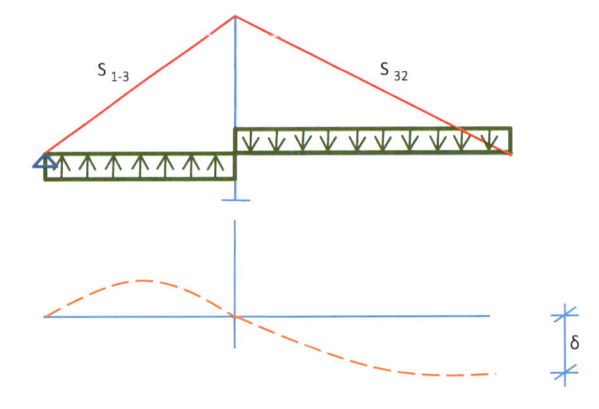

Figure 4.235 Determination of deflection δ

The cable forces S_{1-3} result from Section 4.5.7.2 above as follows
$V_{1-3} = (1358 + 955) \cdot \dfrac{128}{11.5} = 25744$ kN/CP

$S_{1-3} = 58413$ kN/CP

$S_{32} = 12 \cdot 128 \cdot \dfrac{1}{\sin 23.25°} = 3891$ kN/CP

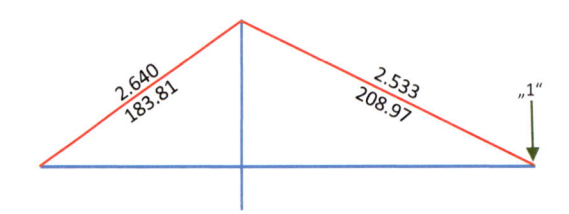

Figure 4.236 Calculation parameters

4.5 Preliminary design calculations

$$\delta = 2.64 \cdot 58.413 \cdot \frac{183.81 \cdot 10}{2.0 \cdot 24\,358} + 2533 \cdot 3891 \cdot \frac{208.97 \cdot 10}{2.0 \cdot 9043}$$

$$= 5.81 + 1.14 = 6.95 \text{ kN/CP}$$

$$f_B = \frac{0.58}{\sqrt{6.95}} = 0.22 \text{ 1/sec}$$

1. Torsional frequency

$$f_T = \frac{b_s}{2\,i_y} \cdot f_B$$

with

b_s distance between cable planes
i_y horizontal mass moment of inertia

$$i_y = \sqrt{\frac{I_y}{A}} = \sqrt{\frac{112.6}{2 \cdot 3.99}} = 3.76 \text{ m}$$

$$f_T = \frac{9.85}{2 \cdot 3.82} \cdot 0.22 = 0.29/\text{sec}$$

$$\varepsilon = \frac{f_T}{f_B} 1.32 \ll 2 \quad \text{with H-tower not sufficient}$$

This torsional frequency is valid for vertical cable planes. The A-tower in accordance with Fig. 4.238 will increase this frequency considerably. Since the independent movements at the tips of both legs is prevented for an A-tower, the tower head acts in torsion as if rigidly kept in place.

If the influence of the backstays on the deflection of the main span is thus omitted, the torsional frequency becomes:

$$\overline{\delta} = 1.14 \text{ m}$$

$$\overline{f_B} = \frac{0.58}{\sqrt{1.14}} = 0.54 \text{ 1/sec}$$

$$\overline{f_T} = \frac{9.85}{2 \cdot 3.76} = 0.71 \text{ 1/sec}$$

$$\varepsilon = \frac{0.71}{0.22} = 3.2 \ll 2 \quad \text{with A-tower probably sufficient}$$

Critical wind speed

In accordance with the method of Klöppel/Thiele the critical wind speed for the onset of flutter becomes approximately, see Section 5.4:
For whole bridge

$$v_{CRT} = \eta \cdot 2\pi \cdot b \cdot f_B \left[1 + \left(\frac{f_T}{f_B} - 0.5 \right) \cdot \sqrt{\frac{0.72 \text{ m} \cdot r}{\pi \cdot \rho \cdot b^3}} \right]$$

with

η reduction factor against airfoil
$b = 11.75$ m whole beam width

$$m = \frac{G}{g} = \frac{2 \cdot 128}{9.81} = 26\,100 \text{ kg/m} \quad \text{beam mass}$$

$$r = \sqrt{\frac{I_x + I_y}{A}} = \sqrt{\frac{2 \cdot 0.341 + 112.6}{2 \cdot 3.99}} = 3.76 \text{ m} \quad \text{radius of inertia}$$

$\rho = 1.225$ kg/m^3 specific weight of air
$f_B = 0.22$ 1/sec 1st bending eigenfrequency
$\overline{f_T} = 0.71$ 1/sec 1st bending eigenfrequency
$\varepsilon = \overline{f_T}/f_B = 3.2$ ratio of frequencies

we thus arrive at

$$v_{CRT} = \eta \cdot 2\pi \cdot 11.75 \cdot 0.22 \left[1 + \left(\frac{0.71}{0.22} - 0.5 \right) \cdot \sqrt{\frac{0.72 \cdot 26.1 \cdot 3.76}{\pi \cdot 1225 \cdot 10^{-3} \cdot 11.75^3}} \right]$$

$$v_{CRT} = \eta \cdot 16.2 \cdot (1 + 2.73 \cdot 3.36) = \eta \cdot 165 \text{ m/s}$$

reduction factor against airfoil, see diagram Fig. 4.237,
for $\varepsilon = 3.2$ and $d/2b = 1:12 \approx 0.08$: $\eta \approx 0.55$

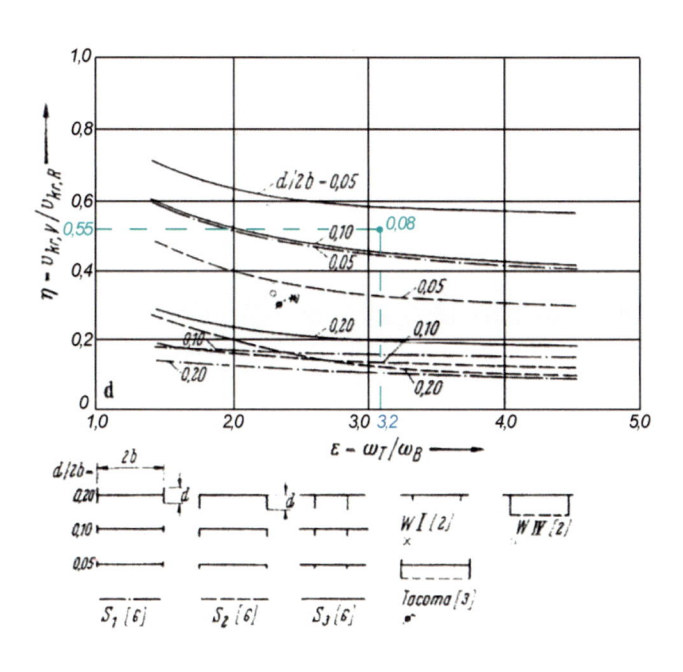

Figure 4.237 Reduction factor for actual cross-sections against an airfoil from tests

We thus arrive at the actual critical wind speed:

$$v_{CRT} = 0.55 \cdot 165 = 91 \text{ m/sec}$$

Safety against flutter

$$v = \frac{91}{52} = 1.75 \quad \text{appears acceptable}$$

4.5.7.5 Towers

System and section properties

The towers have diamond shape and concrete box sections, see Fig. 4.238.

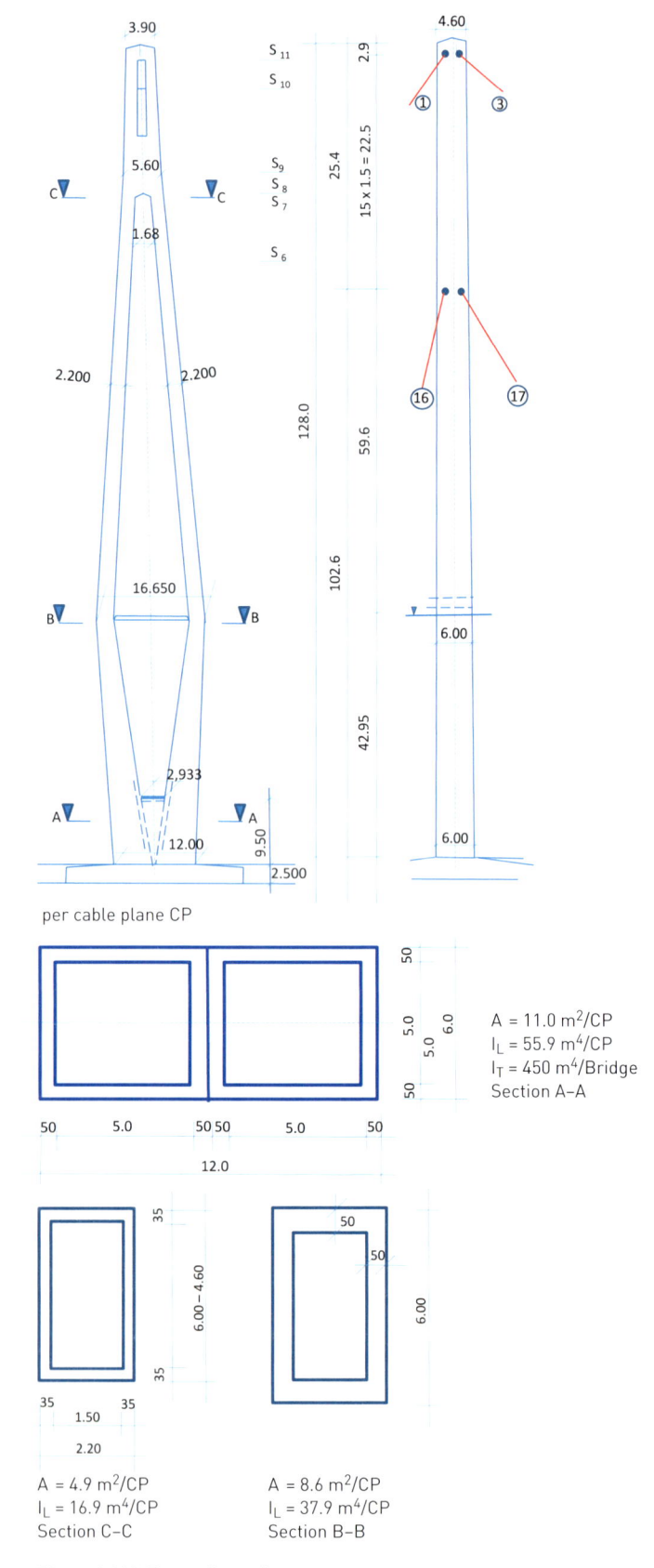

Figure 4.238 Tower dimension

Section A–A
A = 11.0 m²/CP
I_L = 55.9 m⁴/CP
I_T = 450 m⁴/Bridge

Section C–C
A = 4.9 m²/CP
I_L = 16.9 m⁴/CP

Section B–B
A = 8.6 m²/CP
I_L = 37.9 m⁴/CP

per cable plane CP

Permanent loads
Beam dead load

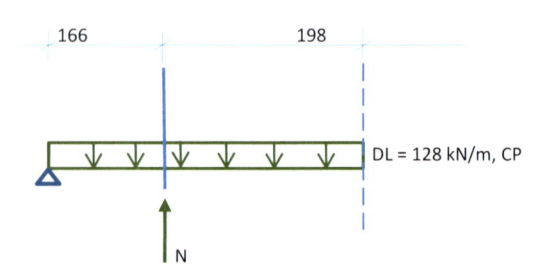

Figure 4.239 Beam dead load

from beam
$$\max P^{DL} = \frac{1}{166} \cdot 128 \cdot \frac{1}{2} \cdot 364^2 = 51.1 \text{ MN/CP}$$
$$M^{DL} = 0$$

from DW tower
a solid concrete column with the tower height is assumed
$$f_c^N \cong \frac{128 \cdot 25.0}{1} = -3.2 \text{ MN/m}^2$$

Live loads

Fixation moment of tower due to tip deflection from live load: The earlier determined live load for half the bridge is used for both halves because the reduction for the second half is rather small, see the eccentricity factors in Section 4.5.5.1 above.

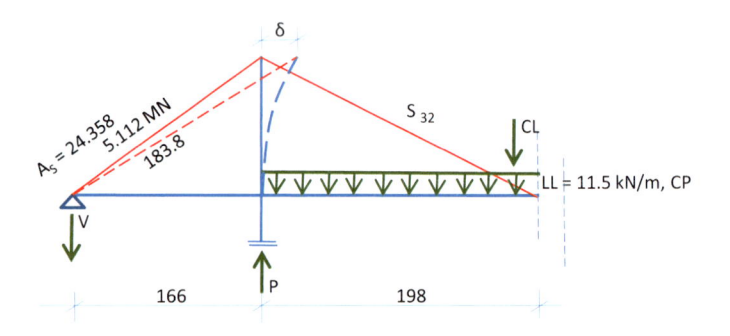

Figure 4.240 Live load arrangement for max M tower

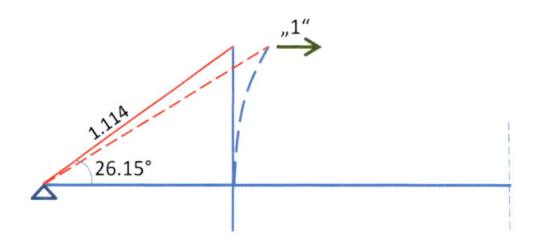

Figure 4.241 Unit load for tip deflection

$$\delta = S_B^F \cdot S_B^1 \cdot \frac{l_R}{E \cdot A_R} = 5.112 \cdot 1.114 \cdot \frac{183.8}{2.0 \cdot 10^2 \cdot 24.358} = 0.215 \text{ m}$$

4.5 Preliminary design calculations

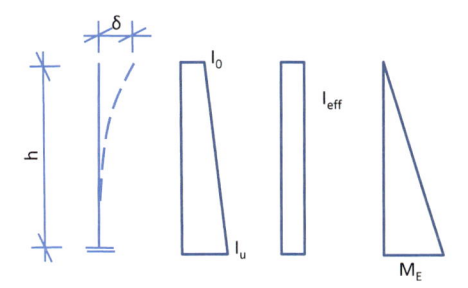

Figure 4.242 Tower moments

$$M_{FIX} = \frac{3 \, E \cdot I_{eff}}{h^2} \cdot \delta$$

$I_{eff} \cong c \cdot \max I$

$c = \sqrt{\dfrac{I_o}{I_u}} = \sqrt{\dfrac{16.9}{55.9}} = 0.55$

$I_{eff} = 0.55 \cdot 55.9 = 30.7 \text{ m}^4$

$M_{FIX} = \dfrac{3 \cdot 37 \cdot 10^3 \cdot 30.7}{125.5^2} \cdot 0.215 = 46.5 \text{ MN/CP}$

Corresponding normal force from LL

$\Sigma M_V = 0$

$N^{LL} = \dfrac{1}{166} \cdot (198 \cdot 11.5 \cdot 265 + 774 \cdot 358) = 5.3 \text{ MN/CP}$

Horizontal stiffness comparison tower versus backstays
Tower
$H = \dfrac{46.5}{125.5} = 0.37 \text{ MN/CP}$

Backstays
$H = S_H = 5.112 \cdot \cos 26.15° = 4.589 \text{ MN/CP}$

$\dfrac{H_{Tower}}{S_H} = \dfrac{0.37}{4.589} = 8.1\,\%$

The horizontal tower stiffness comes to only 8.1 % of the cable system: the articulated system is an acceptable approximation.

Transverse wind
On beam
W = 3.3 MN see Fig. 4.243

At tower head same wind load as on beam
$W \cong 4.7 \cdot (166 + 198) = 1.7 \text{ MN}$

Wind directly on tower
$c_W \cong 2.0$
upper: $w_o \cong 3.1 \cdot 4.6 \cdot 2.0 = 28.5 \text{ kN/m}$
down: $w_u \cong 3.1 \cdot 6.0 \cdot 2.0 = 37.2 \text{ kN/m}$

From transverse wind
$M = 1.7 \cdot 120 + 3.3 \cdot 42.95 + (1/2 \cdot 28.5 + 1/3 \cdot 8.7) \cdot 128^2 \cdot 10^{-3}$
  $= 204.0 + 141.7 + 281.0 = 626.7 \text{ MNm/Tower}$

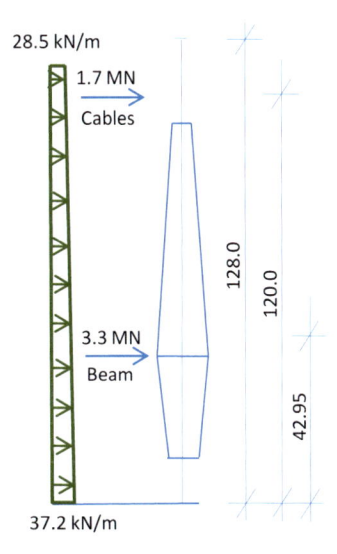

Figure 4.243 Transverse wind loads on tower

No wind to be assumed on leeward leg in accordance with DIN 1055. To cater for slightly skew wind, the tower moment is increased by 30 %.

M = 815 MNm/Tower

Proof of stresses
At base
$A = 11.0 \text{ m}^2/\text{CP}$
$I_L = 55.9 \text{ m}^4/\text{CP}$
$S_L = \dfrac{55.9}{3} = 18.63 \text{ m}^4$
$I_T = 450 \text{ m}^4/\text{Bridge}$
$S_T = 75 \text{ m}^3/\text{Bridge}$

From DL
$f_C = \dfrac{51.1}{11.0} - 3.2 = -7.8 \text{ MN/m}^2$

From LL
$f_C = \pm \dfrac{38.5}{18.63} - \dfrac{5.3}{11.0} = \pm 2.1 - 0.5 = \dfrac{+1.6}{-2.6} \text{ MN/m}^2$

From transverse wind
$f_C = \pm \dfrac{815}{75} = \pm 10.9 \text{ MN/m}^2$

Most unfavorable superposition of DL + LL + wind on unloaded bridge

$\min f_C = -7.8 - 2.6 - 10.9 = -21.3 \text{ MN/m}^2$ corner stress to edge stress
$\max f_C = -7.8 + 1.6 - 10.9 = +4.7 \text{ MN/m}^2$

For the unfavorable combination of wind on unloaded bridge with traffic on bridge the stresses appear to be acceptable.

5 Construction of cable-stayed bridges

Cable-Stayed Bridges. 40 Years of Experience Worldwide. First Edition. Holger Svensson.
© 2012 Ernst & Sohn GmbH & Co. KG. Published 2012 by Ernst & Sohn GmbH & Co. KG.

5.1 Examples
5.1.1 General

Cable-stayed bridges are especially construction-friendly because their intermediate systems during free cantilevering are stable by themselves, Fig. 5.1.

Major arches with their roadway above or underneath had generally to be supported by auxiliary piers during construction. A bridge construction on auxiliary piers, for example in the navigation channel of a river, means an obstacle to shipping which is hardly acceptable for the increased commercial shipping of today. So if an arch can not be built on auxiliary piers, expensive tie-backs with auxiliary towers and stays may be necessary. If, however, such auxiliary tie-backs are required for construction it appears more efficient to include these temporary measures into the final structure. In this way one arrives quite naturally at cable-stayed bridges which have proven to be more economic than arches for long spans. Self-anchored suspension bridges have to be erected on scaffolding.

Chapter 5 gives a systematical breakdown of erection systems for towers and beams in all materials. The installation of stay cables is treated separately in Section 3.9.

5.1.2 Tower construction
5.1.2.1 Steel towers

The Strelasund Bridge includes two steel towers which rest on concrete piers. Its construction is outlined in Section 6.3.

5.1.2.2 Concrete towers

Concrete towers are generally built cast-in-place. One example is the bridge across the Valley of the Obere Argen, Fig. 2.214 [2.127].

The A-shaped towers have box sections over their free lengths. They were built with jumping forms, Fig. 5.2. These are preferable to sliding forms in order to locate any built-in items securely. In order to reduce the moments due to their inclination auxiliary struts are used, Fig. 5.3.

5.1.2.3 Composite towers

Composite towers generally comprise concrete tower legs with a steel box inside the tower head which connects the horizontal cable forces and introduces the vertical cable components into the tower legs by composite action with the surrounding concrete. One example is the Elbe River Bridge at Niederwartha, [2.111].

The lower part of the A-shaped tower is built from concrete. The tower head is provided with an inner steel box in order to simplify the cable anchorages and to limit its size, Figs 5.4 and 5.5.

The transition of the tower legs to the tower head takes place 50 m above ground. The width of the tower head decreases from 3.5 m to 3 m at the top. The composite structure comprises a 3.3 m by 1.9 m steel box, cast into 0.7 m thick concrete. Composite action is created by dowels and shear studs. Inside the steel box the cable anchorages rest on stiffeners, Fig. 5.6.

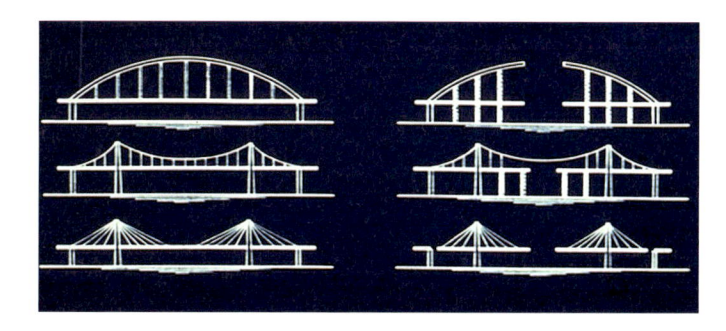

Figure 5.1 Comparison of erection systems

Figure 5.2 Free cantilevering

Figure 5.3 Temporary struts

After completion of the tower legs the steel anchor head is placed by a tower crane, Fig. 5.7. Afterwards the steel box is cast-in for composite action, Fig. 5.8.

Fig. 5.9 shows the shear studs and the reinforcement connection of the steel box with the surrounding concrete 'ears'.

5.1.3 Beam construction
5.1.3.1 General [1.19]

Beam construction generally starts with the building of the approach spans on auxiliary piers. The main span beam construction takes place by free cantilevering, whereby the cantilevering beam is tied back against the already completed approach span. The free cantilevering can continue as long as the weight of the engaged approach is approximately balanced. In this way a single-tower cable-stayed bridge is reached for which the tower represents the symmetry axis. If this symmetry is not provided by the span ratios, the approach span piers have to be tied down or to be additionally weighted in order to provide load balancing. Assuming for a single tower system a mirror-like image, a continuous symmetrical cable-stayed bridge with two towers and a main span with the double length of each cantilever is created. Thereby wide bodies of water or valleys can be bridged without auxiliary piers. This is the decisive reason for an economical construction because auxiliary piers in rivers would have to withstand high water, ship impact and ice pressure. Deep valleys would require expensive high auxiliary piers. These heavy auxiliary structures would make construction much more expensive. Free cantilevering is thus the most economic construction method for long spans.

If a tower is located in deep water so that construction of the side spans on scaffolding would be too expensive, free cantilevering of the beam to both sides of the tower is appropriate.

The success of an economic construction planning requires keeping the number of construction elements and the required auxiliary structures to a minimum.

The bridge should be designed in such way that no additional strengthening of the final structure is required for construction only, and that auxiliary piers and auxiliary cables are as much as possible avoided.

Early on, it has to be decided whether to use an erection traveler or a floating crane. Floating cranes do not load a structure with heavy construction equipment and permit the installation of heavy preassembled units. On the other hand, the use of a floating crane means important restrictions for the ship traffic and dependence of the construction progress on water level, which may cause delays. Therefore, an erection traveler running on the already completed beam should be preferred. This is generally also the more economic solution.

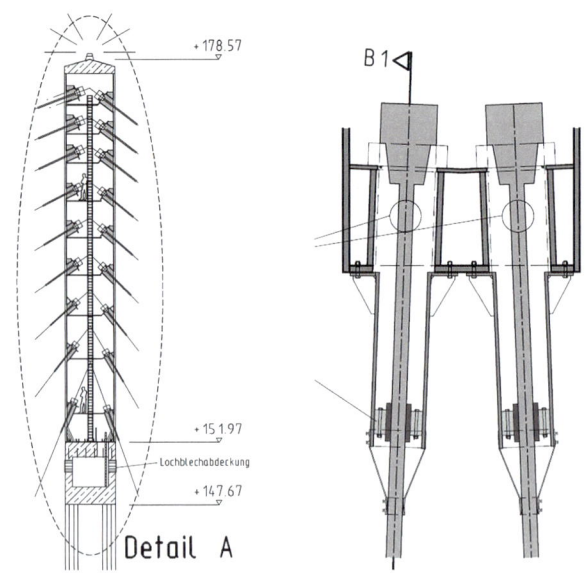

Figure 5.4 Composite tower head

Figure 5.5 Cable anchorages

Figure 5.6 Steel box for cable anchorages

Figure 5.7 Lifting of steel box

Figure 5.8 Tower head with formwork

Figure 5.9 Reinforcement from steel box into the concrete 'ears'

5.1.3.2 Concrete beam

Concrete beams are cast-in-place or assembled from precast elements. They are mostly built by free cantilevering, but launching or rotating of the beam into position has also been used.

Free cantilevering

During free cantilevering the beam is either strong enough to be supported by the final stay cables only, or auxiliary tie-backs have to be used.

Precast elements

The advantage of precast elements is their fabrication under controlled conditions, independent of adverse weather. Their disadvantage is their high weight for installation and their joints.

The fabrication of precast beam elements for the Pasco-Kennewick Bridge is described in Section 6.1.2, and for the East Huntington Bridge in Section 6.1.3.

The erection of precast elements with CIP concrete joints is outlined for the Posadas-Encarnación Bridge across the Paraná River in Argentina, Fig. 5.10 [1.24].

The precast elements were floated to site on barges and lifted by derricks located on each tip of the already erected beam, cantilevered to both sides, Figs 5.11 and 5.12. Each derrick was tied back to a temporary strut at the tower.

After lifting, the new elements are aligned to the already installed beam at a distance of 0.3 m. The CIP joint is bridged by overlapping reinforcement and is closed with a rapid-hardening concrete. Casting took place at night in order to minimize the temperature gradient from solar radiation during the day.

The elements were placed in such a way that they were either balanced about the tower, or that one element was ahead in the main span, Fig. 5.13. Thereby the tower required only one tie-back to the anchor pier, and a second tie-back across the main span to the other tower could be avoided. The construction progress was 6–8 elements per month.

Cast-in-place (CIP) concrete
– With auxiliary cables

The Neckarcenter Bridge for pedestrians crosses the Neckar River in Mannheim with a CIP beam, Fig. 5.14 [2.134]. The beam was built by free cantilevering in sections, Fig. 5.15.

In order not to overload the already constructed beam with the weight of the new section before the permanent stay-cables are installed, the formwork traveler was tied back to the tower head with auxiliary cables, Fig. 5.16. The adjustment of the auxiliary cables for each new section was time-consuming, and the auxiliary cables all had to be sized for the biggest force at bridge center.

– With final stay-cables

In order to avoid the disadvantages of auxiliary cables, the permanent stay-cables were used to support the 12 m long formwork traveler of the Helgeland Bridge in Norway [2.80], see Section 6.2.2.3.

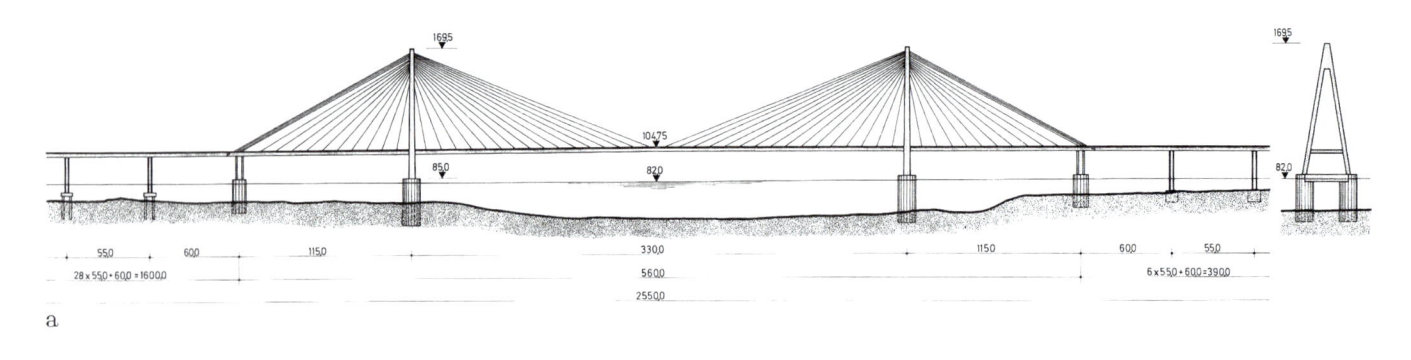

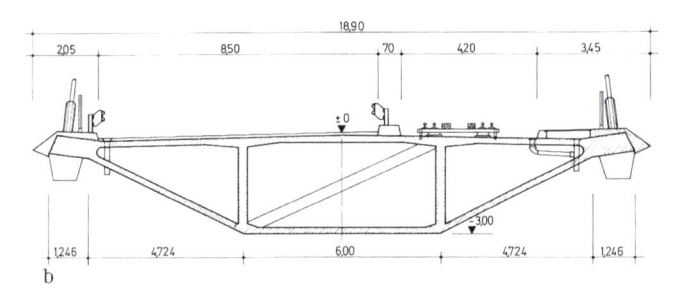

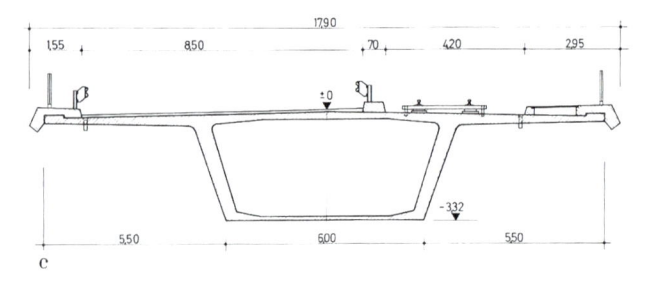

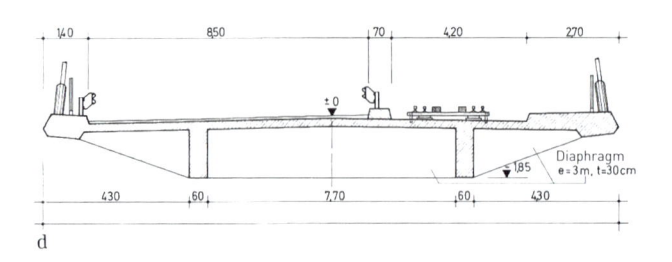

Figure 5.10 Posadas-Encarnación Bridge, general layout

Figure 5.11 Lifting of first element

Figure 5.12 Free cantilevering

5.1 Examples

Figure 5.13 Free cantilevering unbalanced to main span

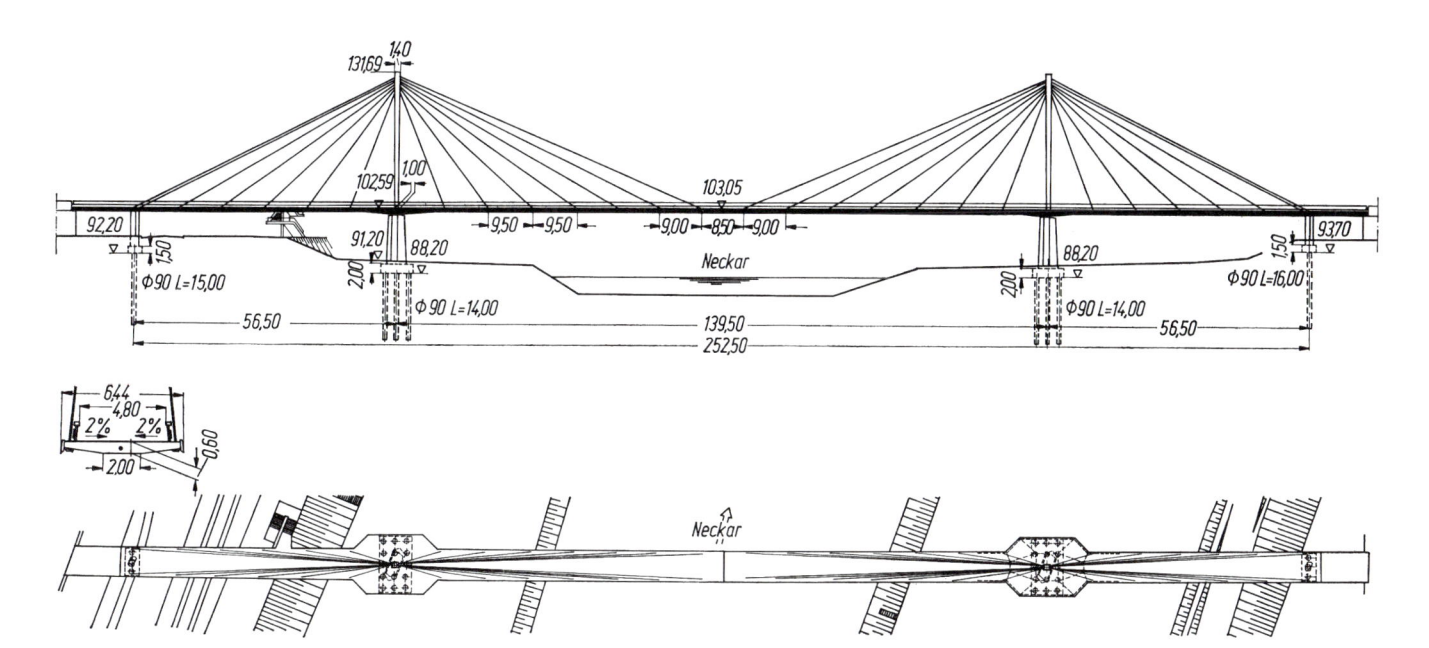

Figure 5.14 Neckarcenter Bridge; general layout

Figure 5.15 Free cantilevering with auxiliary cable

Figure 5.16 Auxiliary cables at form traveler

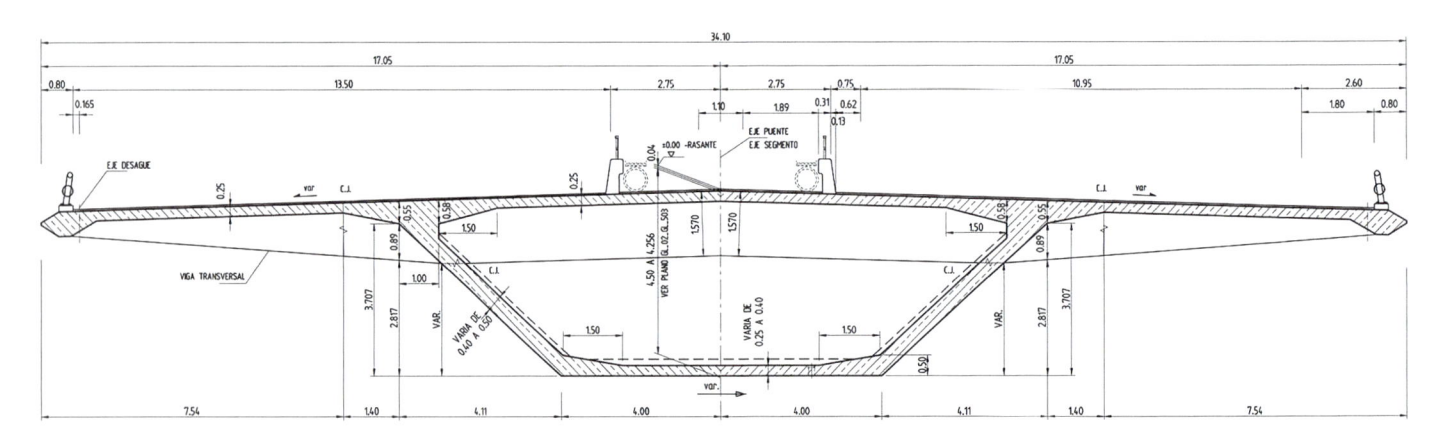

Figure 5.17 Second Bridge across the Panama Canal

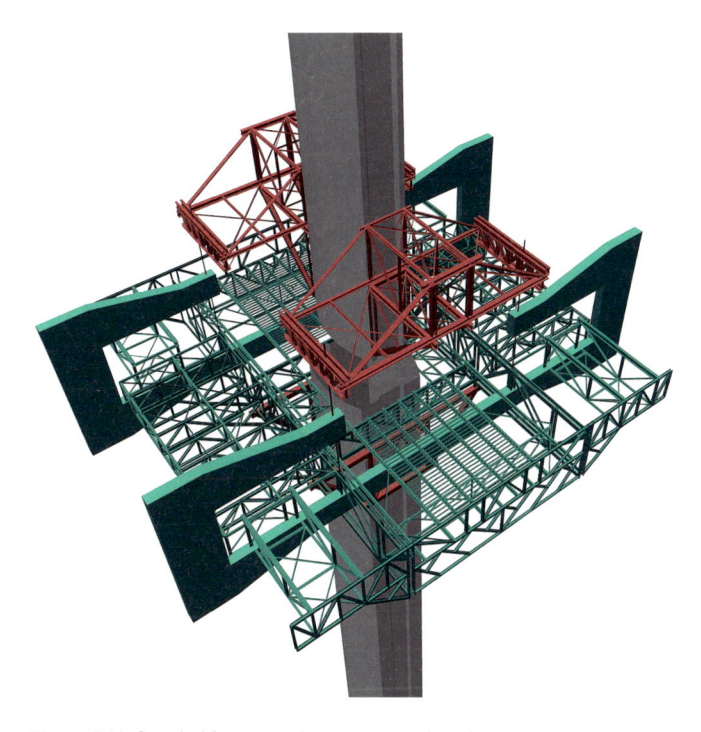

5.1 Examples

Figure 5.18 Coupled form travelers at tower, visualisation

Figure 5.20 Free cantilevering without auxiliary support

Figure 5.19 Installed form travelers

Figure 5.21 Balanced free cantilevering

Figure 5.22 Completed second bridge across the Panama Canal with LAP-Team

298　　5 Construction of cable-stayed bridges

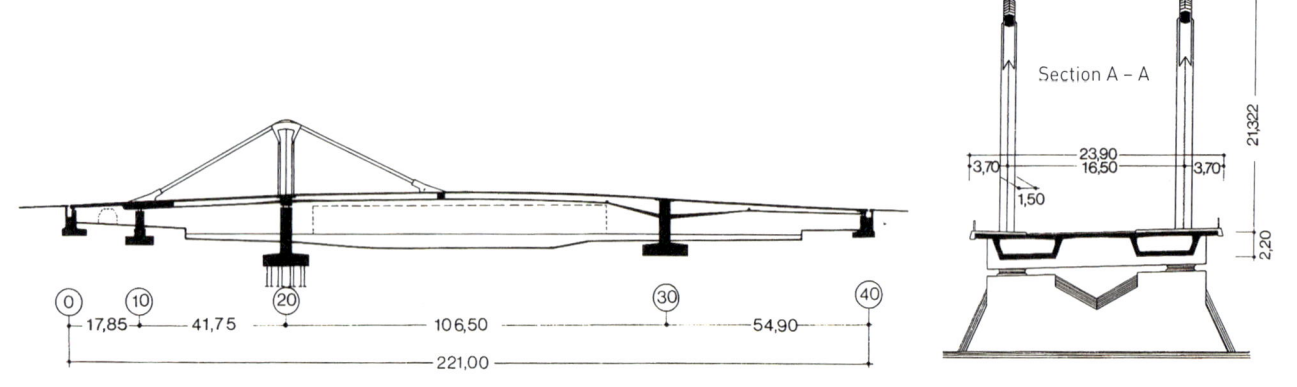

Figure 5.23 Flößer Bridge; general layout

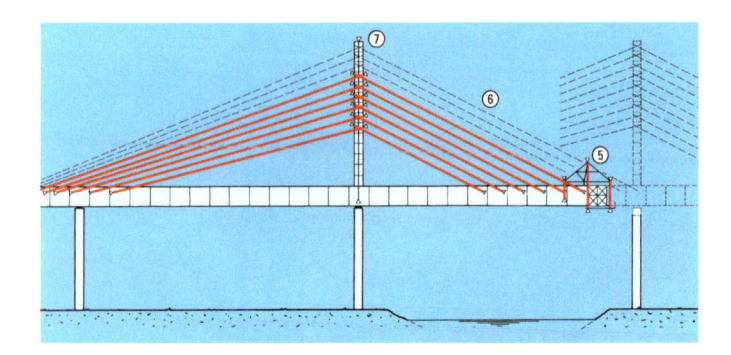

Figure 5.24 System of free cantilevering

Figure 5.25 End of free cantelevering

– Without auxiliary support

The second bridge across the Panama Canal is a central-plane cable-stayed bridge with a total length of 1052 m and a main span of 420 m. The 34.1 m wide beam with a trapezoidal box cross-section has a navigational clearance of 80 m above the canal, Fig. 5.17 [2.83].

The 6 m long beam sections were built in a weekly cycle. This could only be achieved by using a formwork traveler running underneath the beam and leaving the work platform completely unobstructed, Figs 5.18 and 5.19. For the starter sections at each tower the formwork travelers of the main and side span were coupled.

The large depth of the 4.5 m box enabled the cantilevering of the beam in 6 m sections, equal to the cable distance, without auxiliary support, Figs 5.20 and 5.21. Fig. 5.22 shows my colleagues from LAP under the guidance of Dr. Reiner Saul (second from left).

– With auxiliary stays

The beam of the Flößer Bridge across the Main River in Frankfurt was CIP, supported by auxiliary stays during free cantilevering. In the final stage the beam is supported by single pairs of forestays and backstays, Fig. 5.23 [2.92].

Free-cantilevering with auxiliary stays was required because of
– the long spans with about 50 m between piers and final stays,
– the small depth of the beam of only 2.2 m,
– the auxiliary piers which could not be placed in the Main River.

The general layout of the free cantilevering system is shown in Fig. 5.24 [5.1]:

⑤ Anchorage of the traveler to the beam by hold-downs, anchorage of forms to the traveler by ties.
⑥ Temporary stay-cables (adjustable) between an auxiliary tower and the cantilevering beam through which the formwork traveler can pass.
⑦ Vertical post-tensioning of the precast elements of the auxiliary tower.

Fig. 5.25 shows the construction stage during center joint closure. The formwork for the backstay is already installed.

Launching

The side spans of the Kap Shui Mun Bridge to Lantau Island in Hong Kong were built by incremental launching of the beam on final and auxiliary piers [2.107]. The segments were 18.3 m long, Figs 5.26 and 5.27. In order to permit this type of construction, the cross-girders underneath the beam had to be cast in a second step after the beam had reached its final position.

Each side span was built in two halves so that their weights and sizes did not exceed the range of current experience.

These steel transition sections had, therefore, to be sized for the loads during launching also, Fig. 5.28, especially for the high support reactions and corresponding bending moments.

Rotating

The cable-stayed bridge Ben Ahin across the Meuse River in Belgium has a main span of 168 m with only one tower, Fig. 5.29 [5.2].

The tower is 93.5 m high, and the legs are drawn together underneath the beam to rotate on a single foundation. The beam itself rests on a cross-girder between the kinks of the tower legs, Fig. 5.30.

The beam box girder is supported by a central cable plane, the cable forces being transversely distributed into the webs by inclined steel ties, Fig. 5.31.

The beam was built on scaffolding on the shore, parallel to the river, Fig. 5.32.

Afterwards the whole bridge was rotated by 70° and connected to the approach bridge by CIP concrete, Fig. 5.33.

The rotational device comprised two parts, the upper movable one and the lower stationary one, Figs 5.34 and 5.35. The lower part has a diameter of 4.5 m and is rigidly founded on limestone. On its top rest the sliding parts in the shape of a cross. Teflon-coated neoprene is sliding on polished stainless steel. The friction coefficient decreases from 5 % at the start to 2 % during rotation, equivalent to a friction force of 800 – 200 t. The rotational forces are exerted by radially located hydraulic jacks on both sides. Each jack stroke of 20 cm causes a rotation of the beam tip of 6 m.

At the end of construction the sliding device is removed with the use of vertical jacks, and the joint is finally grouted.

For the construction of the large spans – often over water or inaccessible terrain – of concrete beams free cantilevering with CIP concrete is the mostly used method worldwide today. Since the beam depth is often kept to a minimum in order to reduce the transverse windloads and for aesthetic reasons, the traveler has to be supported by a cable. In order to avoid its tediousre-installment for every new sections, the permanent cables should be used for temporary support of the traveler as done for the Helgeland bridge, see Section 6.2.2.3. All the other construction methods outlined above are rather rare exceptions.

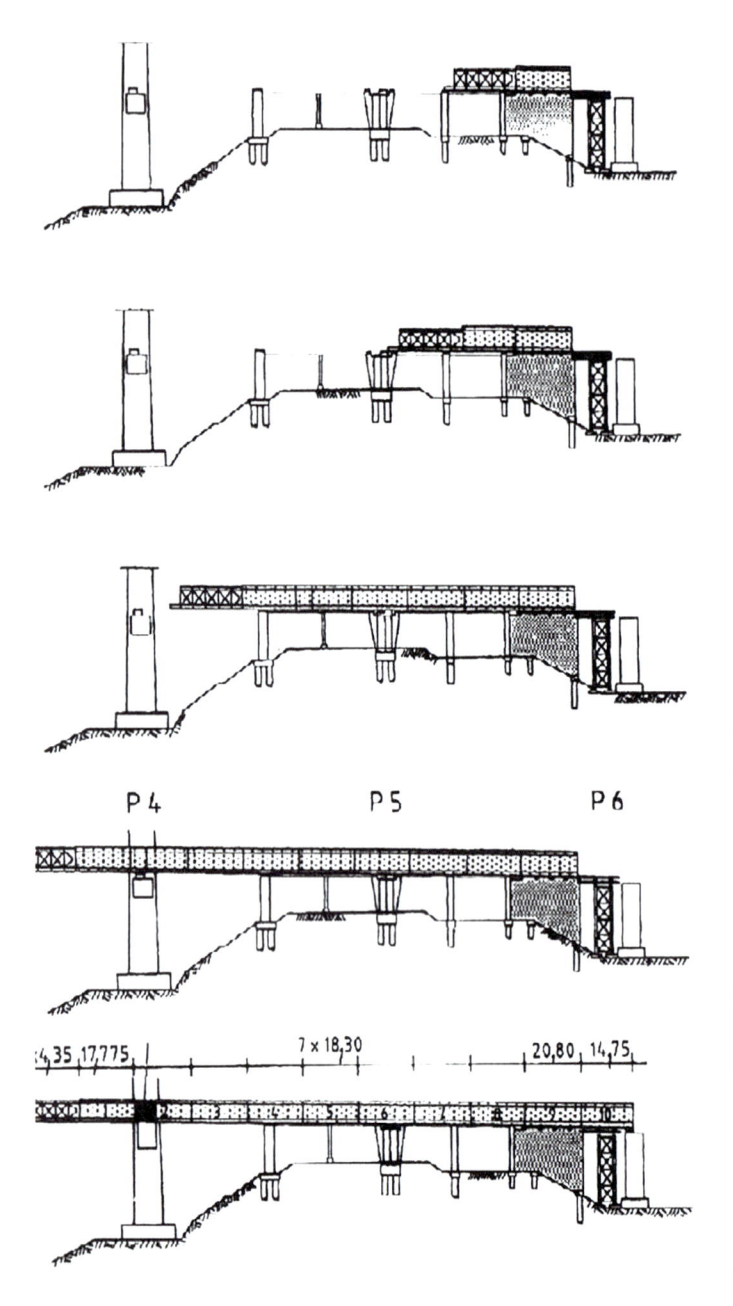

Figure 5.26 General layout of launching, Kap Shui Mun Bridge

Figure 5.27 Launching of the side spans

Figure 5.28 Launching nose

5.1 Examples

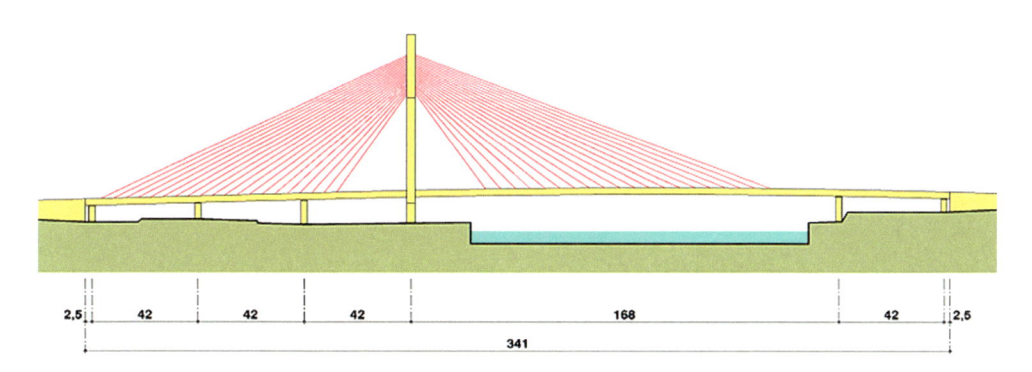

Figure 5.29 Elevation Ben-Ahin Bridge, Belgium

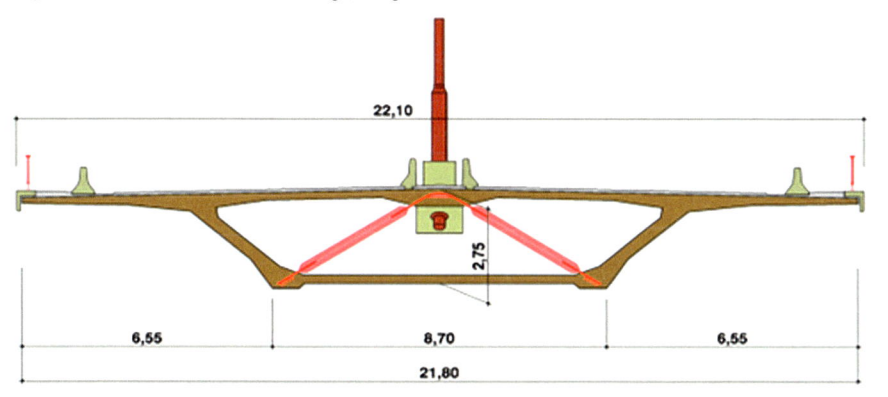

Figure 5.31 Beam cross-section

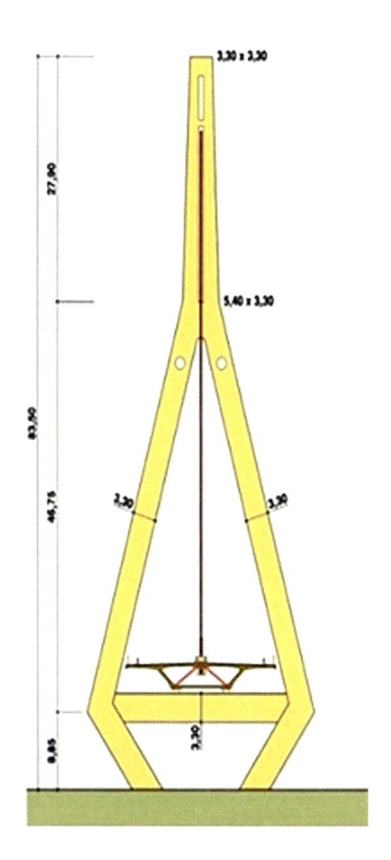

Figure 5.30 Tower

Figure 5.32 Beam construction on scaffolding

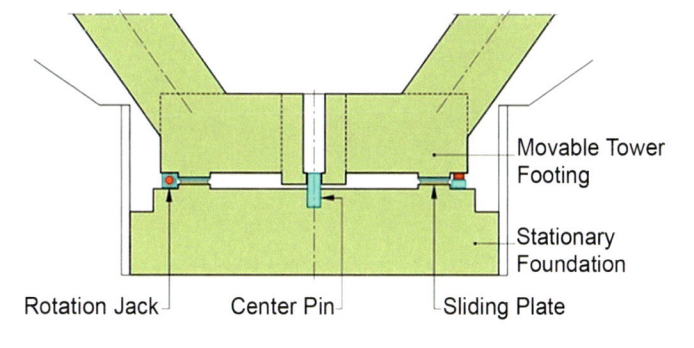

Figure 5.34 Rotational device at tower footing, sect. elev.

Figure 5.33 Rotating of complete bridge

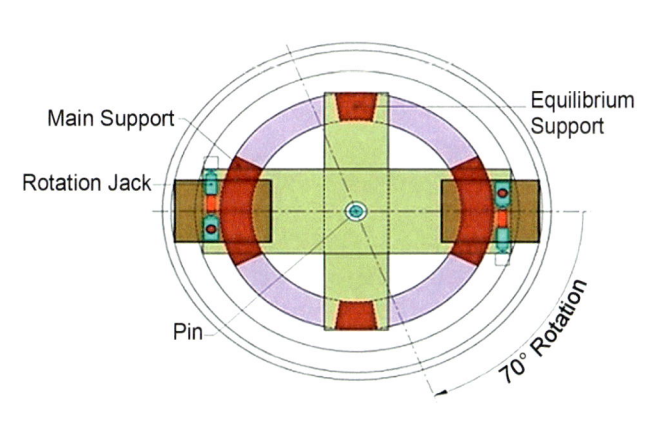

Figure 5.35 Rotational device at tower footing, plan

Rotating on scaffolding

The approach for the Rhine River Bridge Flehe, Fig. 5.36 [2.51], was built on spanwise launching formwork. In the regular region two launching girders were used, and an additional one was required in the region of the backstay anchorages, Fig. 5.37.

In the region of the backstay cable anchorages near the tower the beam was cast on auxiliary piers. Their load was initially equal to those of the other piers.

During installation of the backstays parallel to the free cantilevering progress in the main span the loads on these auxiliary piers were successively relieved. The support reactions were controlled by hydraulic jacks, and if necessary adjusted to the predetermined values. After installation of all stay cables and application of all superimposed dead loads the auxiliary piers were removed, Fig. 5.38.

For the Leven River Bridge in Scotland, Fig. 5.39, an unusual construction method for cable-stayed bridges was selected. The beam was completely cast on stationary scaffolding.

The tower was built simultaneously with the beam [2.82]. The scaffolding with about 40,000 m³ proved to be more economic than, for example, construction by incremental launching, Fig. 5.40.

With a main span of 1018 m the Stonecutters Bridge is the second largest cable-stayed bridge since 2009 and will be a new landmark of Hong Kong [1.22]. The bridge crosses the Rambler Channel between Stonecutters Island, west of Kowloon, and Tsing Yi Island and is thus centrally located and well visible from downtown.

The towers of the Stonecutters Bridge are 298 m high and have a circular, upwards tapering cross section. The bridge beam comprises two individual boxes which straddle the towers, Fig. 5.41. Both main girders are connected by cross beams, forming a girder grid. The side spans have continuous concrete main girders with 70 m spans. The steel beams of the main span protrude 50 m into the side spans where they are monolithically connected.

Characteristic of the Stonecutters Bridge are the comparatively short (289 m, 28 % of the main span) side spans from concrete which act as counterweight to the steel main span. The side spans are post-tensioned longitudinally and transversely. The geometry is rather complex due to the monolithic penetration with the pier heads and the anchorage of the backstays, all post-tensioned together, Fig. 5.42. The total volume of concrete on top of the pile caps comes to 22 000 m³ on each side.

The up-to-4 m deep multi-cell box girders are connected by three box cross beams in each span with typical wall thicknesses of 0.5 m, Fig. 5.43. The webs of the cross beams become diaphragms in the main girders. With their inclined bottom slabs they are aerodynamically shaped. In the outer solid edge beams the backstays and the transverse tendons are anchored.

The complex beam was appropriately cast on stationary scaffolding, Fig. 5.44. In a first step the cross beams were built which form the supports for the formwork of the main girders. The post-tensioning took place in steps in order to control the bending moments con-

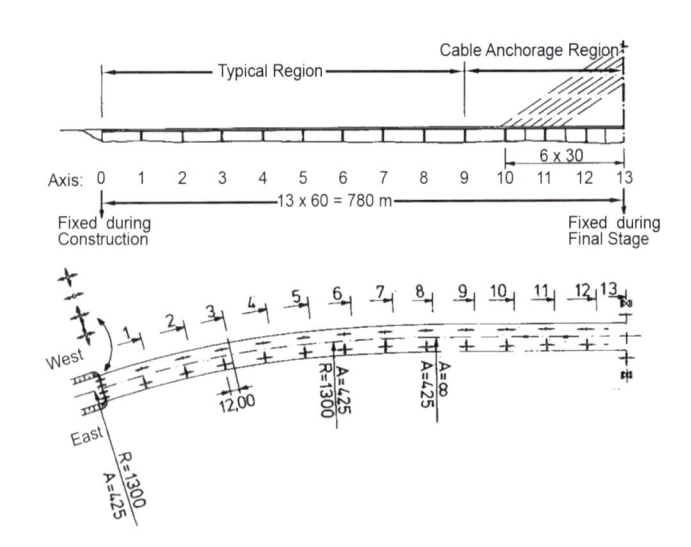

Figure 5.36 Approach of Flehe Bridge, Germany

Figure 5.37 Construction on launching formwork

Figure 5.38 View into the forms with cable anchorages

5.1 Examples

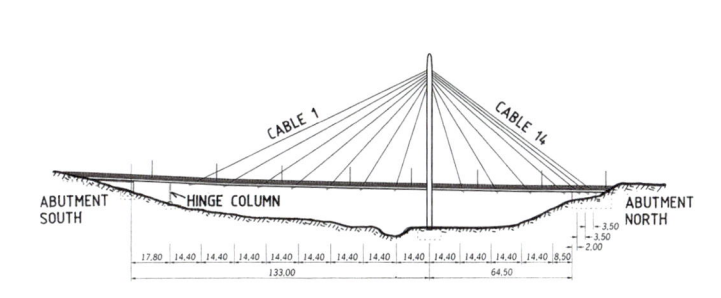

Figure 5.39 Leven River Bridge, GB; elevation

Figure 5.40 Scaffolding

Figure 5.41 Stonecutters Bridge; concrete side span on scaffolding

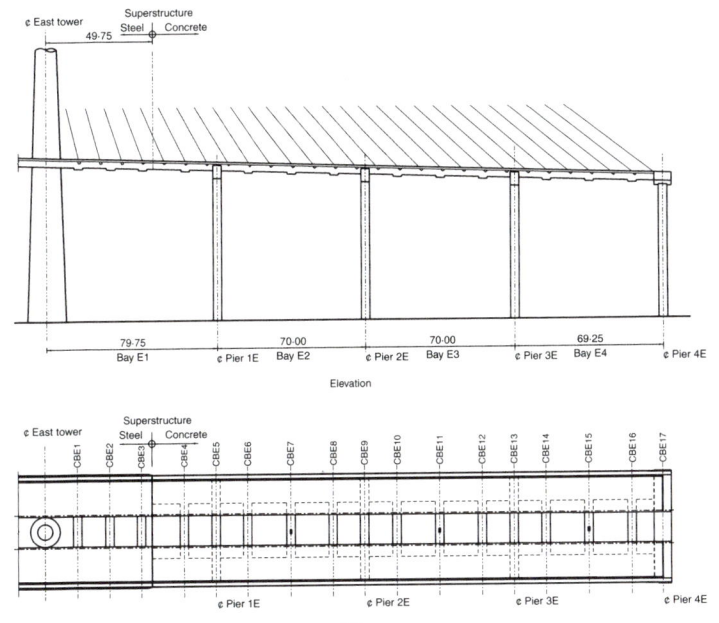

Figure 5.42 Approach bridge east; elevation and plan

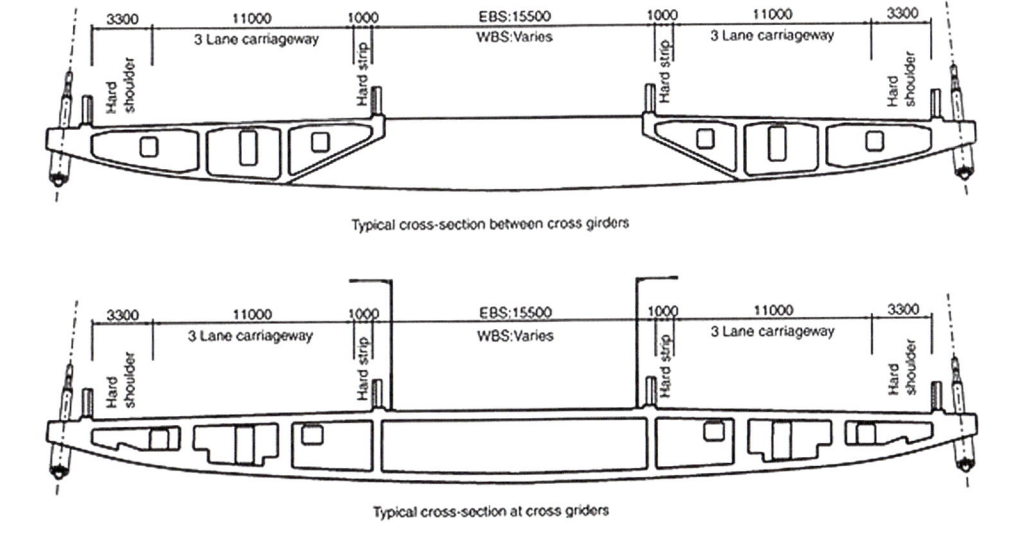

Figure 5.43 Typical cross sections at and between cross girders

tinuously. The first tendons were stressed after connecting the cross beams. After the subsequent casting of the main girders, which also create bending in the cross beams, the remaining tendons were stressed. The permanent tendons in the bottom chord had to be strengthened by temporary tendons in the outside region of the upper chords in order to avoid tensile stresses during construction.

Beam construction started in the middle of each span which was then closed by casting the sections between the last cross beams and the pier heads, Figs 5.45 and 5.46.

5.1.3.3 Steel beams

The usual method of building the main span steel beam is free cantilevering which is one of the main advantages of cable-stayed bridges. The beam sections with a length of 10 to 20 m, equal to the cable distance, are completely preassembled and transported to site on water, if possible, or on the already installed beam. The lifting takes place either by derrick from the already erected beam or by floating crane. The sections are then aligned and welded.

Free cantilevering

The steel beam of Stonecutters Bridge also comprises two separate box girders connected by cross beams at 18 m distance. The first steel sections straddle the tower and are lifted with winches fixed to the tower, Fig. 5.47. Unusually the steel beam protrudes into the concrete approach, which was done here to ease the construction of the complicated cross sections. More commonly the concrete beam extends into the main span, as done for example for the Normandy Bridge, Section 6.5.

The subsequent elements of the main span were lifted from the water with derricks at the tips, Fig. 5.48. Each 18 m long twin section had a weight of about 500 t. After welding, the corresponding cables were installed; 11 days were required for the erection of each twin section.

The exact location of the pontoon underneath the final beam position was determined by a GPS-supported dynamic positioning system and computer-controlled anchorages, Fig. 5.49. The final section for the center closure was adjusted to the actual geometry, Fig. 5.50.

The construction of the Sutong Bridge across the Yangtze River, about 100 km upstream from Shanghai, China, with a world record main span of 1088 m follows the methods used for the main spans of other large steel bridges. The sections are floated in, Fig. 5.51, and are lifted by derricks from the already erected beam, Fig. 5.52, and are aligned and welded [1.21]. Fig. 5.53 shows the fitting of the final beam section by aligning and spreading the center joint gap with auxiliary trusses on the outsides.

Launching

The beam of the Millau Bridge, high above the Tarn Valley in the south of France, was preassembled in increments behind each abutment and launched. The first tower for the final stage was used for

Figure 5.44 Construction of scaffolding

Figure 5.45 Formwork

Figure 5.46 Completed approach beam

supporting the beam tip during launching, Fig. 5.54 [2.123]. The construction of the Millau Bridge is outlined in detail in Section 6.6.2.

Transverse shifting

The Rhine River Bridge at Oberkassel forms part of the highly traveled inner city road system of Düsseldorf. In order to avoid interrupting the traffic at this central artery, the new bridge was built alongside the old one and later transversely shifted by 47.5 m, Figs 5.55 and 5.56 [2.45].

The sliding track rested on the old pier connected to the new one, Fig. 5.57.

The bridge is symmetric in elevation with a main and a side span of 257 m each, Fig. 5.58. For transverse shifting the cables were stressed in such way that the approach piers could be disconnected from the beam and almost the total bridge weight of 10 300 t was centered at the tower. The tensile force for transverse shifting was 37.5 MN.

The bridge was shifted on four launching tracks: two on the Düsseldorf abutment and one on the Oberkassel abutment. They consisted of lubricated polished stainless steel, Fig. 5.59.

The fourth main launching track was located on the central pier underneath the tower. The bearing there consisted of three Neopot bearings, the central one with a diameter of 3 m, the largest at that time, Fig. 5.60.

Underneath this bearing a continuous, 70 m long, 3.2 m wide and 18 mm thick high-strength steel plate was inserted by lifting the bridge by 38 mm. The sliding partner was a PTFE plate.

Four hydraulic center-hole jacks, two on each pulling stations, were stressing pull rods during launching, Fig. 5.61. The main pulling station was at the tower where the biggest friction forces occurred, secondary pulling stations were on the abutments.

5.1.3.4 Composite beam

Composite beams are built from a steel girder grid on which rests a concrete roadway slab, either CIP or precast.

Free cantilevering

During free cantilevering the beam is usually lifted and connected in segments spanning between the cable anchorages. Then the concrete roadway slab is installed. The steel beam should not run ahead of the concrete slab because in that case the action forces are very difficult to determine due to shrinkage and creep. In some cases complete cross-sections comprising the steel grid with the concrete roadway slab have been installed with lifting weights of up to 500 t, the same range as for precast beam elements.

Precast roadway slab

Precast slabs are advantageous for rapid erection and reduced shrinkage and creep values due to their advanced age at installation. One disadvantage is their joints, which are mostly in compression in the both directions under permanent loads.

Figure 5.47 Stonecutters Bridge; lifting of first steel sections

Figure 5.48 Free cantilevering of steel beam

Figure 5.49 Lifting of twin beam section

Figure 5.50 Final section

Figure 5.51 Sutong Bridge, China; floating-in of the beam section

Figure 5.52 Lifting of the final section

Figure 5.53 Fitting of the final beam element

Figure 5.54 Millau Bridge, France; launching of the steel beam

5.1 Examples

Figure 5.55 Oberkassel Bridge, Düsseldorf, Germany; in lateral position

Figure 5.56 Oberkassel Bridge, Düsseldorf, in final position

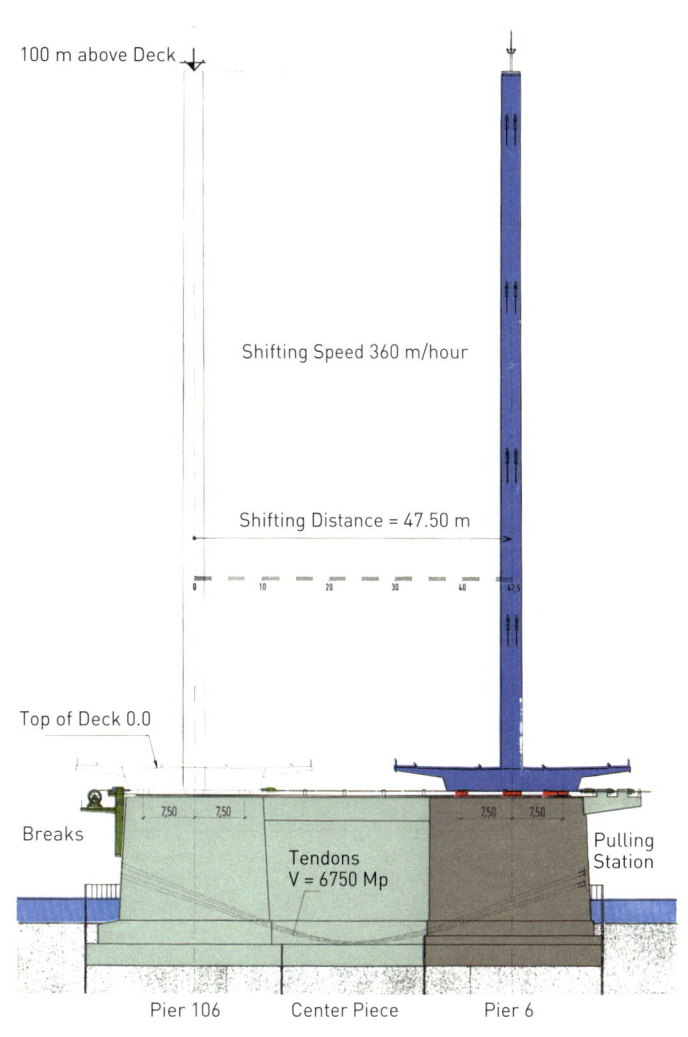

Figure 5.57 Transverse shifting

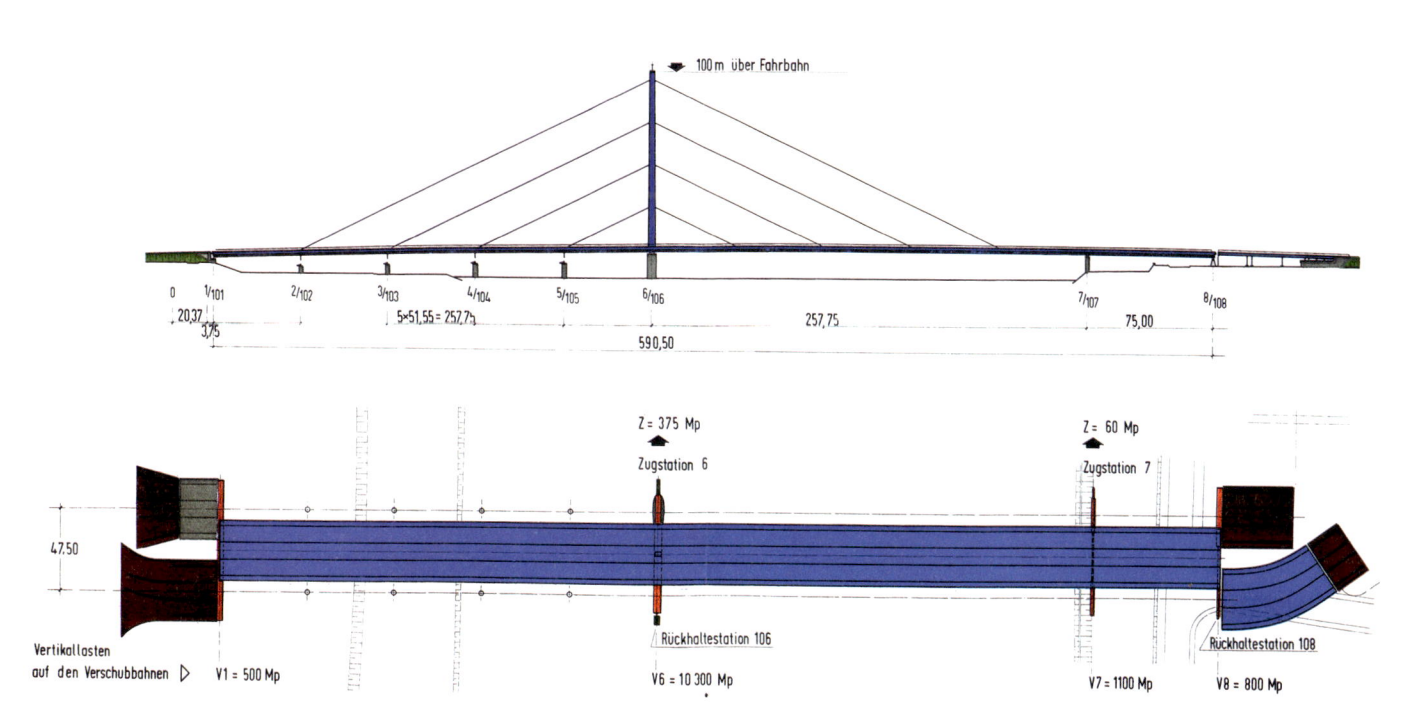

Figure 5.58 Elevation and plan

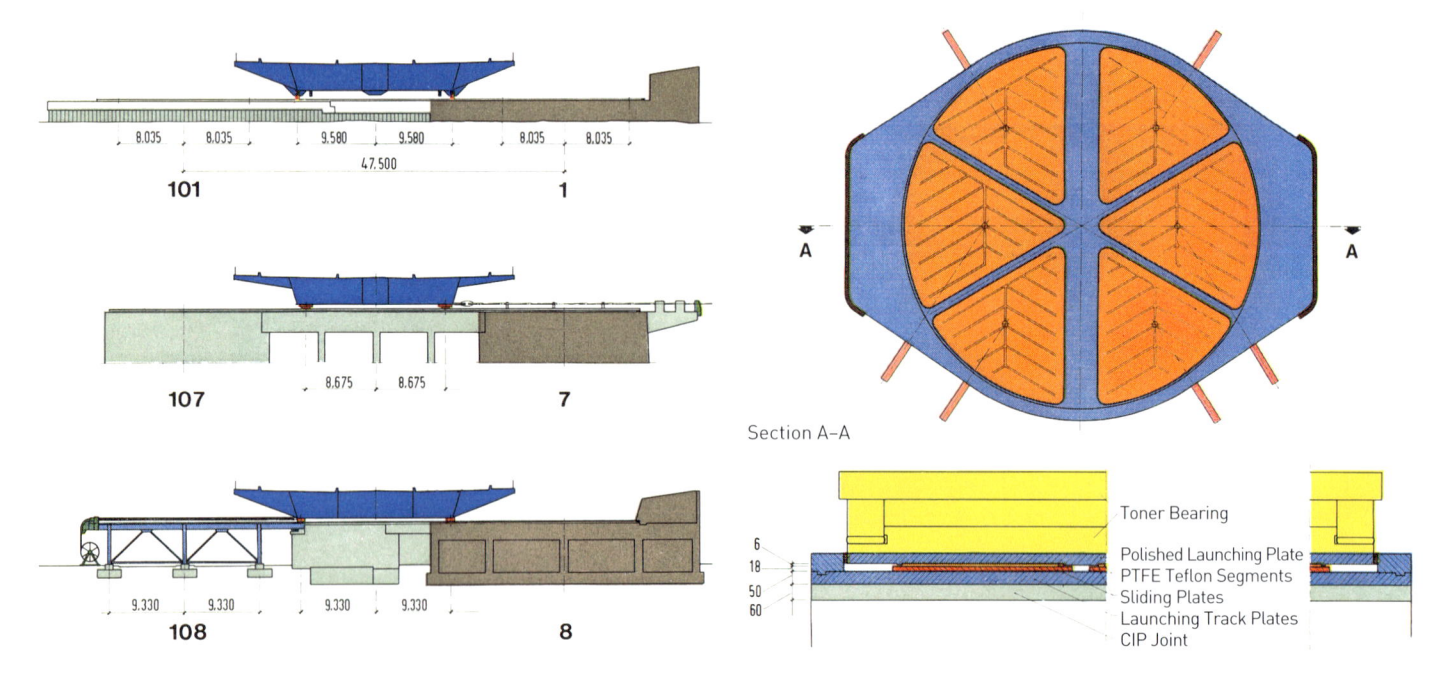

Figure 5.59 Launching track

Figure 5.60 Launching bearing

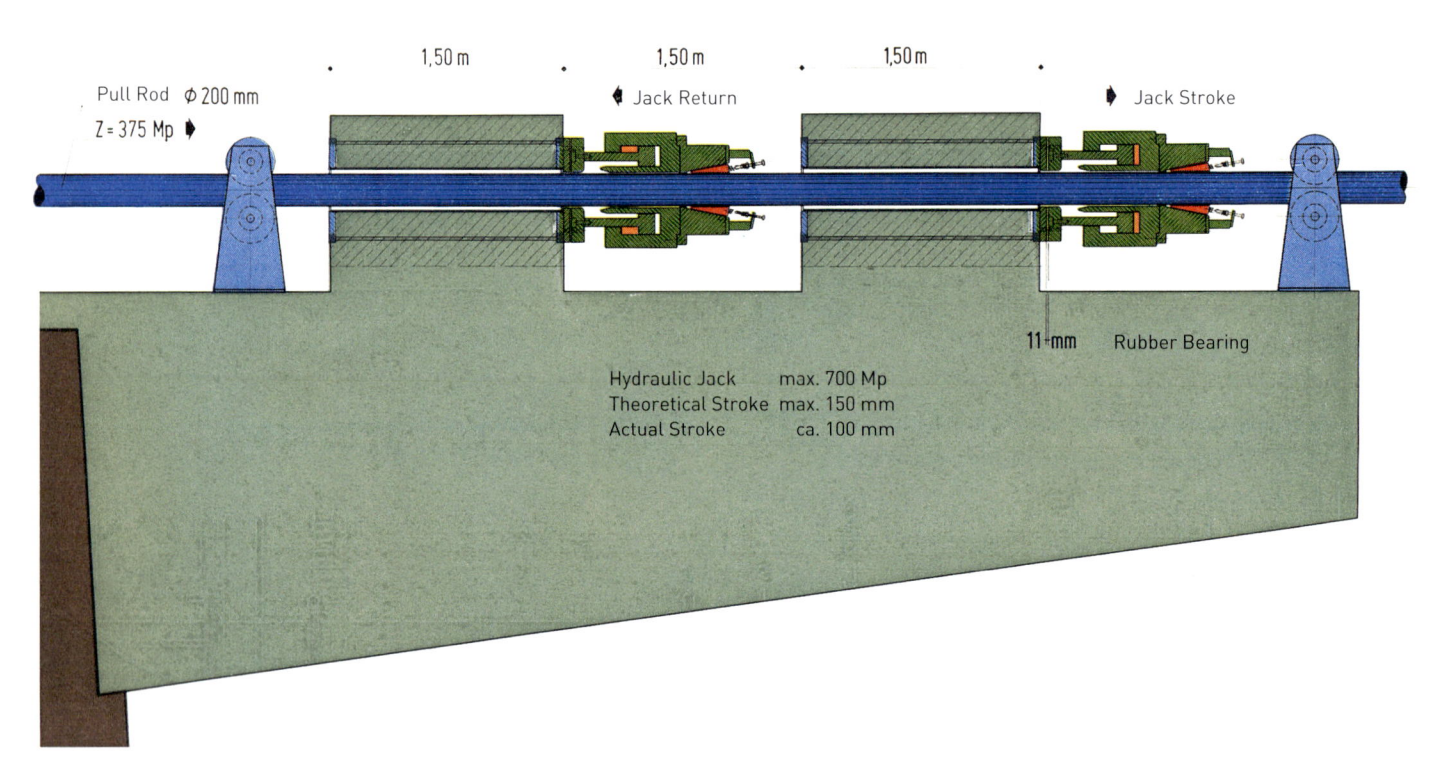

Figure 5.61 Pull station for transverse shifting

5.1 Examples

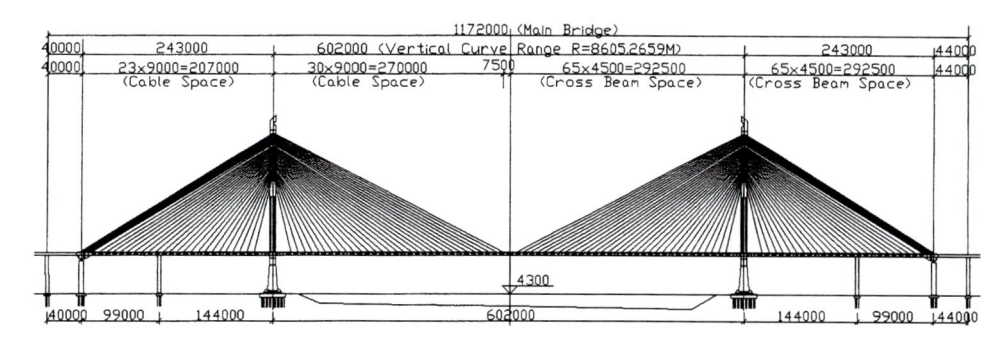

Figure 5.62 Yang Pu Bridge, China; elevation

The Baytown Bridge in Houston, Texas, has two separate, unconnected composite beams, each of which comprises a steel grid made from plate girders with a precast roadway slab on top. The beam construction is shown in detail in Section 6.4.2.3 [1.17].

For the Yang Pu Bridge in Hong Kong, Fig. 5.62, a special construction method was chosen to provide additional transverse compression for the longitudinal joint in the center of the beam [2.105].

Before casting of the central joint the cross girders were pushed upwards by means of a kingpost, Figs 5.63, 5.64 and 5.65. After curing, the kingpost was removed and in this way additional compression was introduced into the joint as top flange of the simply supported cross beam.

Cast-in-place roadway slab
The Elbe River Bridge Niederwartha near Dresden has a composite beam with a CIP roadway slab. The steel beam grid sections were delivered by barge, Fig. 5.66 [2.111].

The grid sections were lifted by derrick, Fig. 5.67, and welded to the already erected beam, Fig. 5.68.

The falsework for the roadway slab is installed, Fig. 5.69, the reinforcement placed, Fig. 5.70, and the roadway cast, Fig. 5.71.

Complete cross-section
As an exception the 12 m long composite beam sections for the Rion-Antirion Bridge in Greece were installed complete with the roadway slab [2.122]. The sections with a weight of about 270 t were lifted by floating crane, Fig. 5.72. The webs were connected with high-strength bolts and the joints of the roadway slab with overlapping reinforcement and low-shrink concrete, similarly to the Baytown Bridge. The Rion-Antirion is outlined in detail in Section 6.6.2.

The Kap Shui Mun Bridge in Hong Kong has a double-deck composite cross-section with webs and cross-frames in steel and top and bottom flange of concrete, Fig. 5.73 [2.107].

The free cantilevering took place by lifting each beam section from a barge with hydraulic jacks located on a traveler. Due to the large beam depth of 7.5 m the complete new elements with a weight of about 500 t each could be lifted without any auxiliary tie-backs, Figs 5.74 and 5.75.

Launching
The Heinola Bridge in Finland has an open composite cross-section and a single concrete tower with parallel strand cables in harp arrangement, Fig. 5.76 [2.106].

The steel grid was launched from both abutments towards the center, across the longer spans on auxiliary piers, Figs 5.77 and 5.78.

After casting the roadway slab intermittently, Fig. 5.79, the stay cables were installed and the auxiliary piers removed.

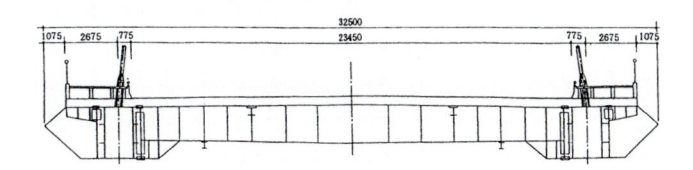

Figure 5.63 Cross girder

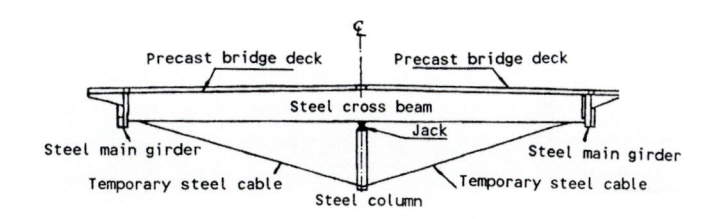

Figure 5.64 Cross girder with temporary king post

Figure 5.65 Front girder grid stayed from underneath

Figure 5.66 Transportation by barge

Figure 5.70 Reinforcement

Figure 5.67 Lifting from barge

Figure 5.71 Planing of roadway slab

Figure 5.68 Aligning new section

Figure 5.72 Rion-Antirion Bridge, Greece; lifting of a complete beam section

Figure 5.69 Falsework for roadway

Figure 5.73 Kap Shui Mun Bridge, Hong Kong, China; double-deck beam

5.1 Examples

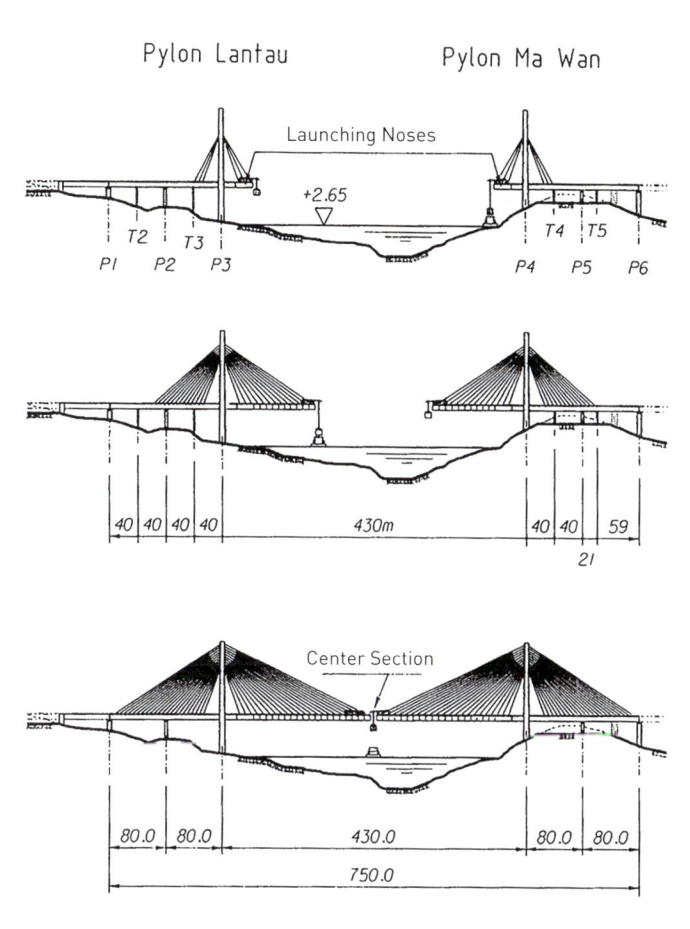

Figure 5.74 General layout of free cantilevering

Figure 5.75 Lifting of a complete double-deck beam section

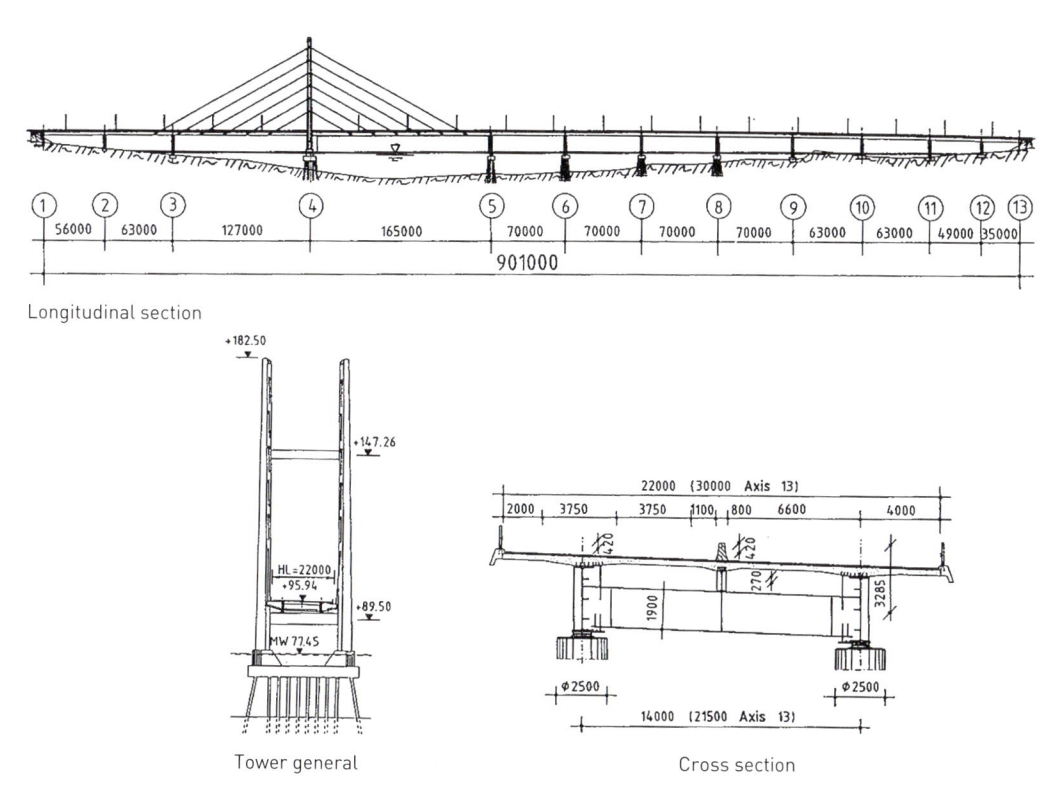

Figure 5.76 Heinola Bridge, Finland; elevation

Figure 5.77 Launching of beam steel grid

Figure 5.78 Beam in final position

Figure 5.79 Casting of roadway slab

5.2 Construction engineering
5.2.1 General
For the construction of bridges the design, fabrication, erection and construction engineering belong together. All these parts have to be completed successfully and in close cooperation to achieve a successful bridge construction, within time and within budget [2.37].

5.2.2 Construction engineering by dismantling
5.2.2.1 General [1.19]
Although a bridge has to be designed carefully for its final stage (i.e. the complete bridge), it must not be overlooked that a thorough construction engineering is of the same importance. For the bridge in its final stage, the assumed life load will normally not even be reached during test loadings. During construction, however, the assumed loads will actually occur and may even be exceeded. Accidents, rare as they are, normally only take place during construction, and hardly ever after construction is completed. This is highlighted by the fact that bridges are generally only insured during construction, but not in the final stage.

The sizing of a bridge is determined by the following conditions [2.71] and [2.36]:
– The initial sizing is done for the final stage and thus given for the construction stages.
– The construction stages have to be selected to minimize costs. The load-bearing capacity of the bridge is thus fully utilized during construction.
– The costs for local strengthening during construction are compared with those created by a different, more involved construction method which avoids additional strengthening.

The advantage of cable-stayed bridges for overcoming large spans mainly by normal forces (truss action) is especially advantageous for the construction of the main span. If the topography permits, the approach spans can be built ahead on scaffolding, and the towers also. (Building the tower in parallel to the free cantilevering may save time but causes additional interdependencies in addition to dangerous working overhead, and is therefore mostly avoided.) The main span can then be erected by free cantilevering. The stay cables are built-in with their predetermined forces or lengths and thus reduce the bending moments in the cantilevered beam. The towers are appropriately tied back, also to reduce their moments.

The distance of the cable anchorage points determines the beam section length which has to cantilever freely. The already installed beam is elastically supported by its stay cables, so that a region of negative moments is created behind the new element before the corresponding cables are installed and relieve the bending.

The stresses of the bridge beam and the cables are in the same range during construction as in the final stage. During each construction stage the capacities of the beam and the stay cables have to be checked carefully. This is also true for the stay cables which tie back the towers.

Only the comprehensive calculation of all stresses in each system guarantees the required safety during all construction stages.

At the end of construction and after all superimposed dead loads have been applied, the bridge gradient should be as required by the road design. Differences can be adjusted by restressing the cables. The corresponding deviating moments must not exceed those tight limits given for the final stage. Restressing is expensive. (It is a small point of honor for the construction engineers that restressing is not required.)

The requirement to construct correctly can only be fulfilled if the precise run of action forces and deformations is known for every construction stage. Starting from a given construction stage the free cantilevering of the next beam section and the installation of the corresponding cables leads to a new erection stage with different action forces and deformations.

If these values have been correctly determined,
– the new beam section can be aligned to the already installed beam so that its tip elevation corresponds with the theoretical value
– it is possible to install the new cables with their predetermined forces (or lengths) so that the new tip elevation also corresponds with the theoretical value.

If each step on site is done carefully as determined in the construction engineering, the desired run of forces, as well as the geometry, is achieved.

For the calculation of the individual construction stages it is therefore prudent to start from the well-known 'final stage' (the completed bridge) and to dismantle the bridge backwards. All loads and deformations are superimposed onto the final stage with negative sign. At the end of this dismantling process all action forces must be zero!

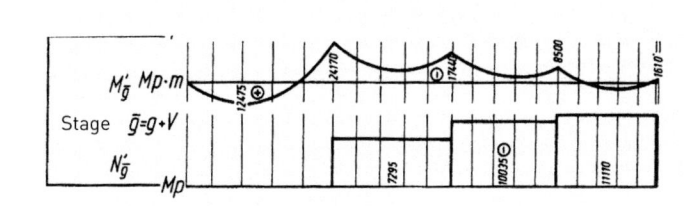

Figure 5.80 Start system at t = ∞

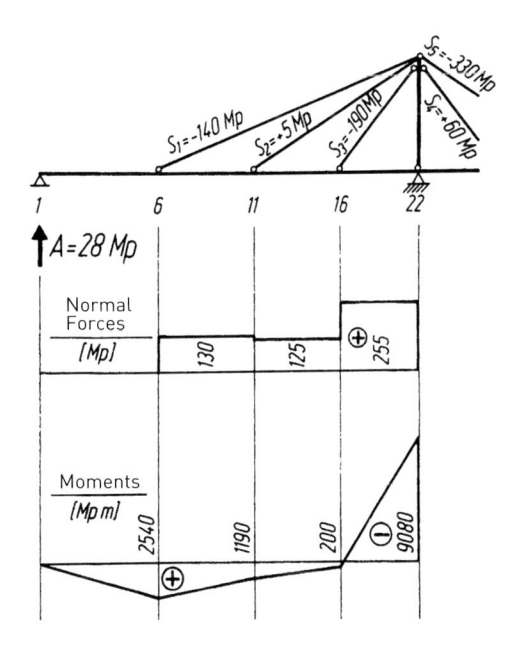

Figure 5.81 Remove shrinkage and creep

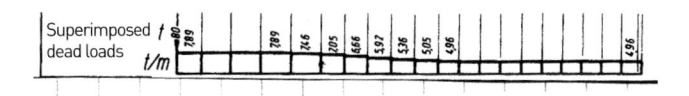

Figure 5.82 Remove superimposed dead loads

5.2.2.2 Dismantling from t = ∞ to t = 1

The final stage under permanent loads and after shrinkage and creep have taken place (t = ∞ after S+C) has been outlined in Chapter 4. In order to complete the first step of dismantling, the superimposed loads must be removed and S+C between t = ∞ and t = 1 must be applied with negative sign, Figs 5.80 to 5.82 [1.19].

In this way, the situation at the end of the structural beam construction on site is reached, when all checks on cable forces and beam geometry are executed and restressing of cables – hopefully not required – would take place. This stage has thus to be documented in some detail.

5.2.2.3 Dismantling of bridge
With floating crane
Single tower system

The construction with a floating crane places the least loads onto the beam and is the simplest method for calculation. Figs 5.83 and 5.84 show the individual steps and loads required to dismantle the bridge calculation-wise.

Start System

t=1 without S+C and SIDL

1. Pier Settlement A≠0
2. Remove Element 1

Elongate Cable S_R so that $M_{Tower} < M_{perm.}$

Intermediate Stage

3. Elongate Cable 1 so that $S_1 = 0$

Remove Cable 1

4. Remove Element 2

Remove Cable S_R

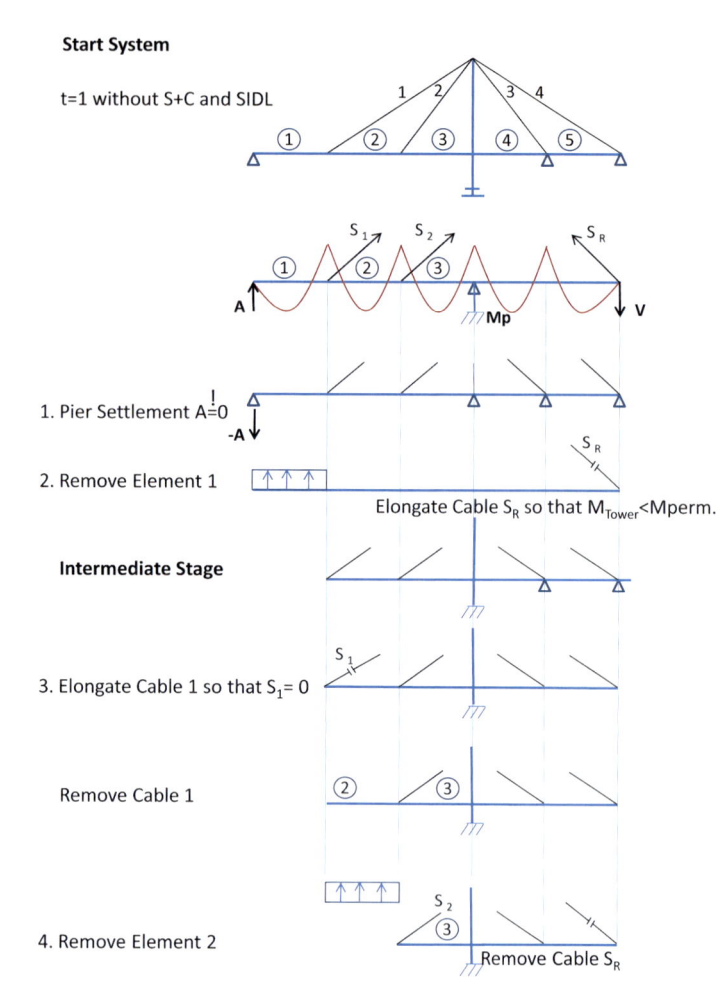

Figure 5.83 Beginning the dismantling with floating crane

5. Elongate Cables S_2 and S_3 so that S_2 and $S_3 = 0$

Remove Cables S_2 and S_3

6. Remove Element 3

7. Remove Approach Beam

8. Remove Tower

9. Remove Piers

All Action Forces $\stackrel{!}{=} 0$

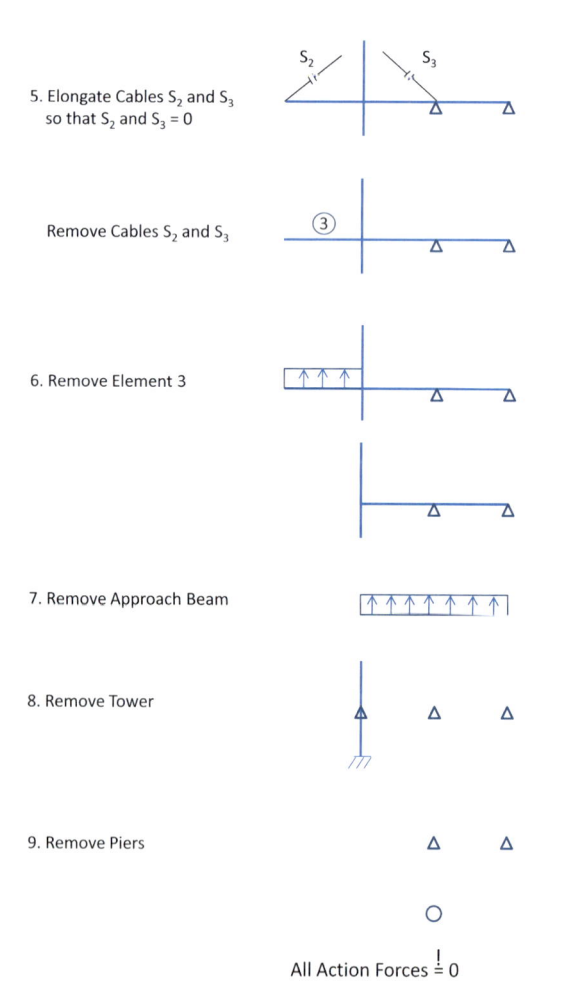

Figure 5.84 Completion of dismantling with floating crane

① Unload abutment A by apply a loading 'pier settlement' in such way that the support reaction A becomes zero. Remove support A from the statical system.
② Remove the dead weight (DW) from the now cantilevering element 1 results in M, N, V becoming zero. Remove element 1 from the statical system. Check whether the tower moments $M_{Tower} < M_{perm.}$, otherwise adjust backstay cable S_B (true for all construction stages).
③ Adjust cable 1, so that $S_1 = 0$. Remove cable 1.
④ Remove DW element 2. Remove element 2 and cable S2.
⑤ Adjust cables S_2 and S_3 and removal.
⑥ Remove element 3.
⑦ Remove approach bridge.
⑧ Remove tower.
⑨ Remove piers.

As an overriding check, all action forces in the removed structural elements must be zero and the sum of all deformations must correspond with the shop form.

Symmetrical system

For a symmetrical system dismantling starts by making the action forces M, N and V at both sides of the center element zero.

The required calculation steps for system dismantling are shown in Fig. 5.85:
① Remove weight of center element.
② Elongate 3 cables left and right of the center element, so that M, N and V in the joints become zero. (Two times three cable elongations are sufficient conditions for two times three action forces.)
③ Open joints.
④ Remove center element.

The further dismantling takes place similar to the single tower system.

Dismantling with derrick

If a derrick is used for dismantling, its weight has initially to be placed onto the beam, moved and finally taken off, always simulating the backward calculation.

The individual steps of calculation for dismantling the bridge are shown in Fig. 5.86 for the single tower system:
① After removal of the support the DW of the derrick is placed onto the bridge.
② Support element 1 by the derrick, and remove the DW of element 1 from the beam.
  The force couple from the moment of the supported element acts on the bridge beam via the two derrick supports. Remove simultaneously the force couple from the derrick and the weight of element 1. (Only the DW of the derrick remains on the beam.)
③ Remove element 1 from statical system.
④ Shift derrick to next element behind. (Remove DW derrick from old position and replace it at the new position.)
⑤ Elongate and remove cable S_1.

Finally, at the end of backward construction calculation:
– Remove derrick from starter element.
– Remove last cable and beam element.
– Again the condition $\Sigma M, N, V = 0$ has to be fulfilled for all removed structural elements of the bridge.

5.2 Construction engineering

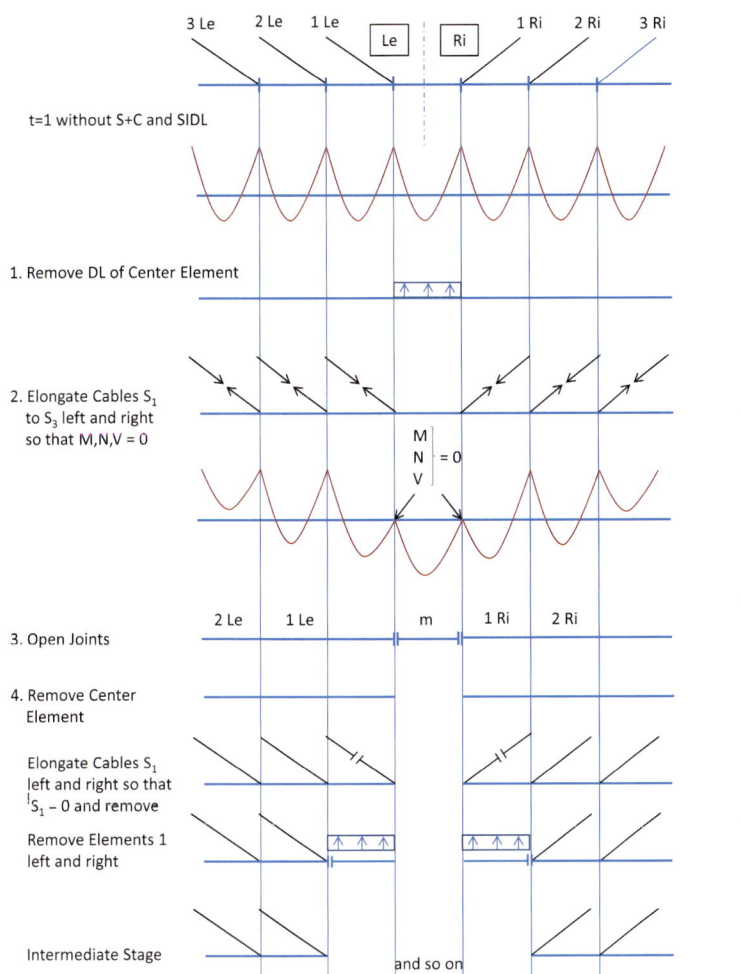

Figure 5.85 Dismantling symmetrical system with floating crane

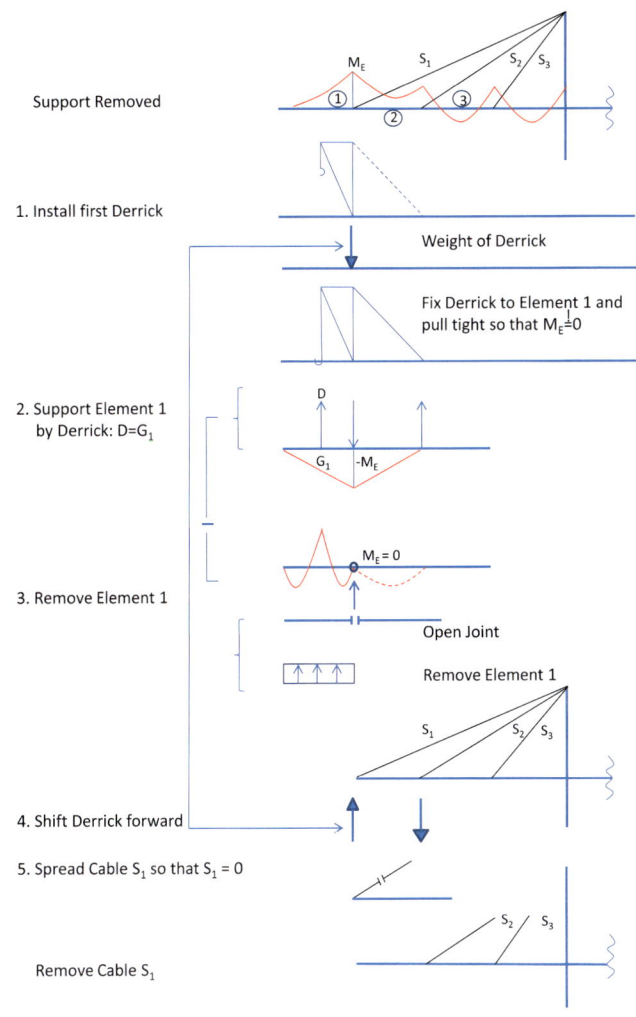

Figure 5.86 Dismantling with derrick

5.2.2.4 Aerodynamic stability

The aerodynamic stability must also be proven for all construction stages. Fig. 5.86a gives as an example the Zárate-Brazo Largo Bridge [2.60]. The flutter wind speeds were determined in accordance with Fig. 4.92.

Construction stage	14	12	12 Tied down	11	5
Flutter wind speed [m/s]	50	41	47	33	59

Figure 5.86a Flutter wind speeds during construction

5.2.3 Example for construction engineering

The construction and the construction engineering for the Rhine River Bridge Mannheim North has been outlined in detail by Erwin Volke and Carl-Heinz Rademacher in their comprehensive publication [1.19], which we follow.

5.2.3.1 Forward construction

The construction sequence as performed forward on site is shown in Fig. 5.87 and described in the following.

The basic concept for construction was free cantilevering across the river without interfering with ship traffic. The tender documents permitted the use of auxiliary piers in the river, but the preliminary construction investigation showed that free cantilevering was not only possible but even more economic. Any auxiliary piers in the river would have to be designed for ship collision, which would make them very expensive. Since the cable distances at the beam are moderate (for 1970, when the bridge construction was planned), the use of free cantilevering with a derrick running on the beam appeared to be most suitable.

After completion of the foundations of the approach bridge, the starter piece of the steel beam was installed by a crawler crane with the help of an auxiliary pier near the tower. The reinforcement and post-tensioning was threaded into the end diaphragm of the steel starter piece and the concrete approach beam cast against the steel starter piece.

Then the construction of the steel main span across the river started, during which the auxiliary pier at the quay was loaded with up to 1600 t.

The transportation of the first span took place by water transport and lifting the steel elements by a special loading derrick near the tower onto the already installed starter element. After very thorough alignment, the kink angle to the starter element was built in. The kink had to take into account the remaining S + C in the concrete approach bridge.

The same loading derrick installed the beam derrick in span 1, which was used for all following construction. The free cantilevering always took place in the same manner. The transportation of the new beam sections on barges directly to the tip of the cantilever would have been in conflict with the ship traffic, therefore, the loading derrick lifted the new beam section from the barge and placed it onto a carriage for transport to the beam tip. There it was hung into the beam derrick and installed.

In order to bridge the distance to stay 1 an auxiliary tower had to be installed, Fig. 5.87. Each auxiliary rope was stressed with up to 200 t in order to minimize the cantilever moment. Span 7 was installed and finally the auxiliary tower could be removed.

The final spans, 9 and 10, had to be built with partial cross sections to prevent overloading of the beam. Initially only the two steel boxes were cantilevered, and then the cross-section completed after they had reached the first pier.

5.2.3.2 Construction engineering

The construction planning and engineering was started early in order to include construction requirements into the design and sizing of the bridge. All construction stages had to be calculated very thoroughly in order to prove their safety and to provide the site with all cable forces and tower and beam deformations for construction control.

Starting from the final stage of the completed bridge under permanent loads at $t = \infty$, Fig. 5.80, the construction engineering was performed backwards by dismantling the bridge. The intermediate construction stages are shown in Fig. 5.88. The loads during construction were assumed on the safe side, because experience shows that the actual construction loads always tend to be bigger than the originally assumed ones.

Fig. 5.89 shows the deformations and cable forces during each construction stage together with special measures required. Deformations and cable forces were selected because they can be measured on site. The deformations of the beam are considerable, reaching 2.35 m downwards and 1.85 m upwards at the tip of the beam. The theoretical values coincided very well with those measured on site.

The erection systems A – G shown in Fig. 5.89 tally with those of Fig. 5.88. Given are
– the construction steps,
– the deformation line of the beam,
– the forces and jacking distances for each stay cable.

The run of moments is given for information only, but cannot be measured on site. The explanations to each construction system are given in detail in Fig. 5.89.

5.2.3.3 Construction manual

The actions to be taken on site are summarized in a so-called 'construction manual'.

For each construction stage the theoretical geometry and forces are given. The cable forces are of special importance because they can be measured directly. Only if the actual geometry of beam and towers and the cable forces coincide sufficiently, can the construction proceed. Fig. 5.89 is typical for a construction manual.

The actual construction systems on site are shown in Figs 5.90 to 5.93.

Finally, the completed bridge is shown in Fig. 5.94, which can be considered the last step of the construction engineering and construction on site.

5.2.3.4 Control measurements

The actual forces and deformations on site are controlled during each construction stage for agreement with the theoretical values from the construction manual.

The action forces are controlled by the cable forces only, see Section 3.9.5.2. (Moments and shear forces can not be practically measured on site.) The complete run of all action forces can be

5.2 Construction engineering

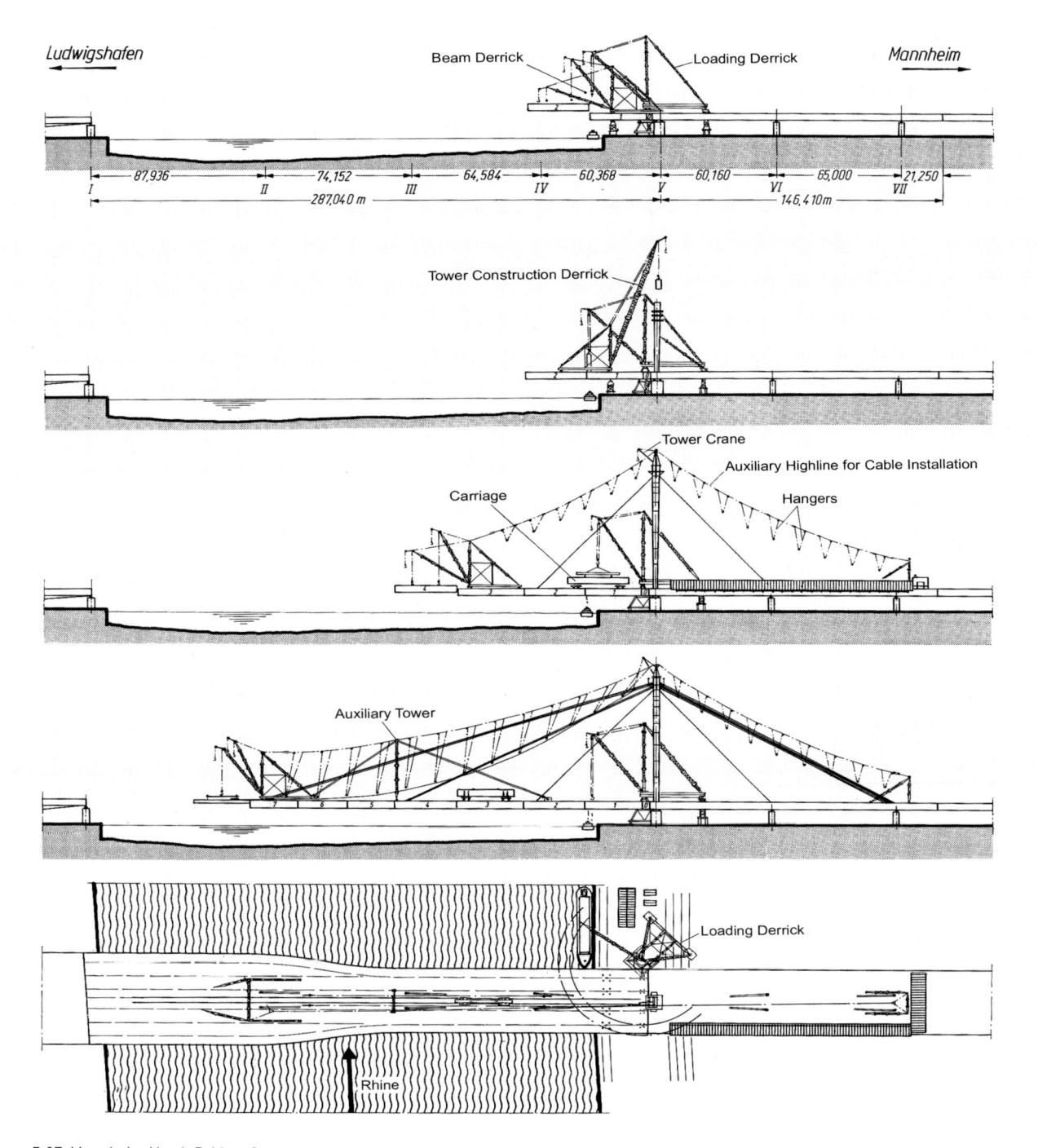

Figure 5.87 Mannheim North Bridge, Germany; forward construction sequence

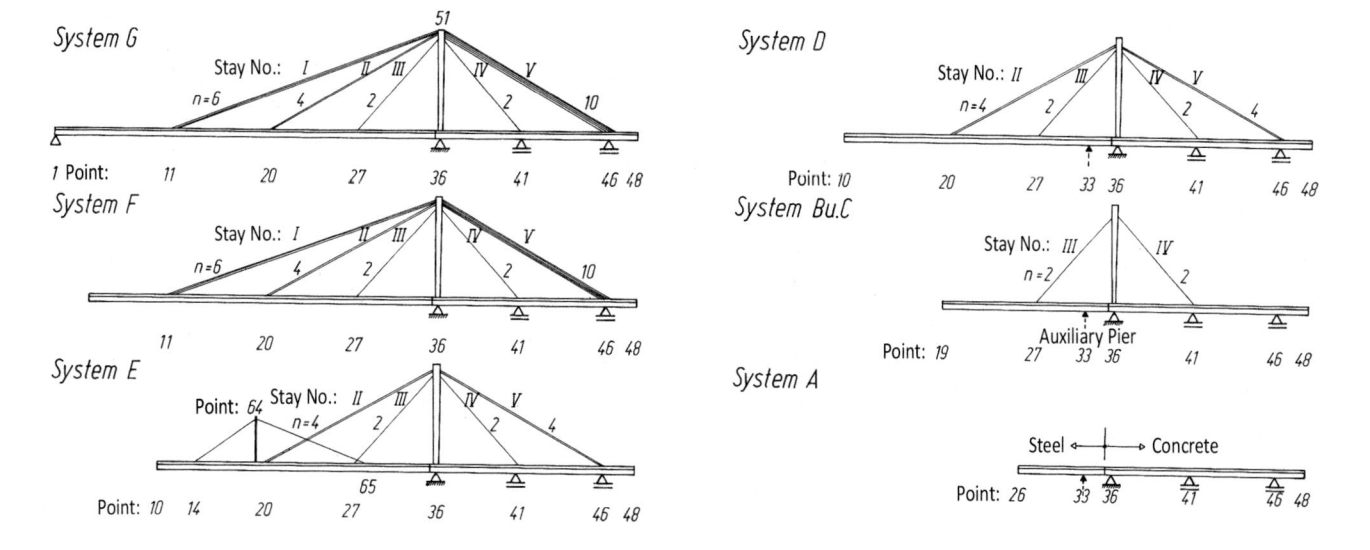

Figure 5.88 Dismantling of bridge, statical system of intermediate stages

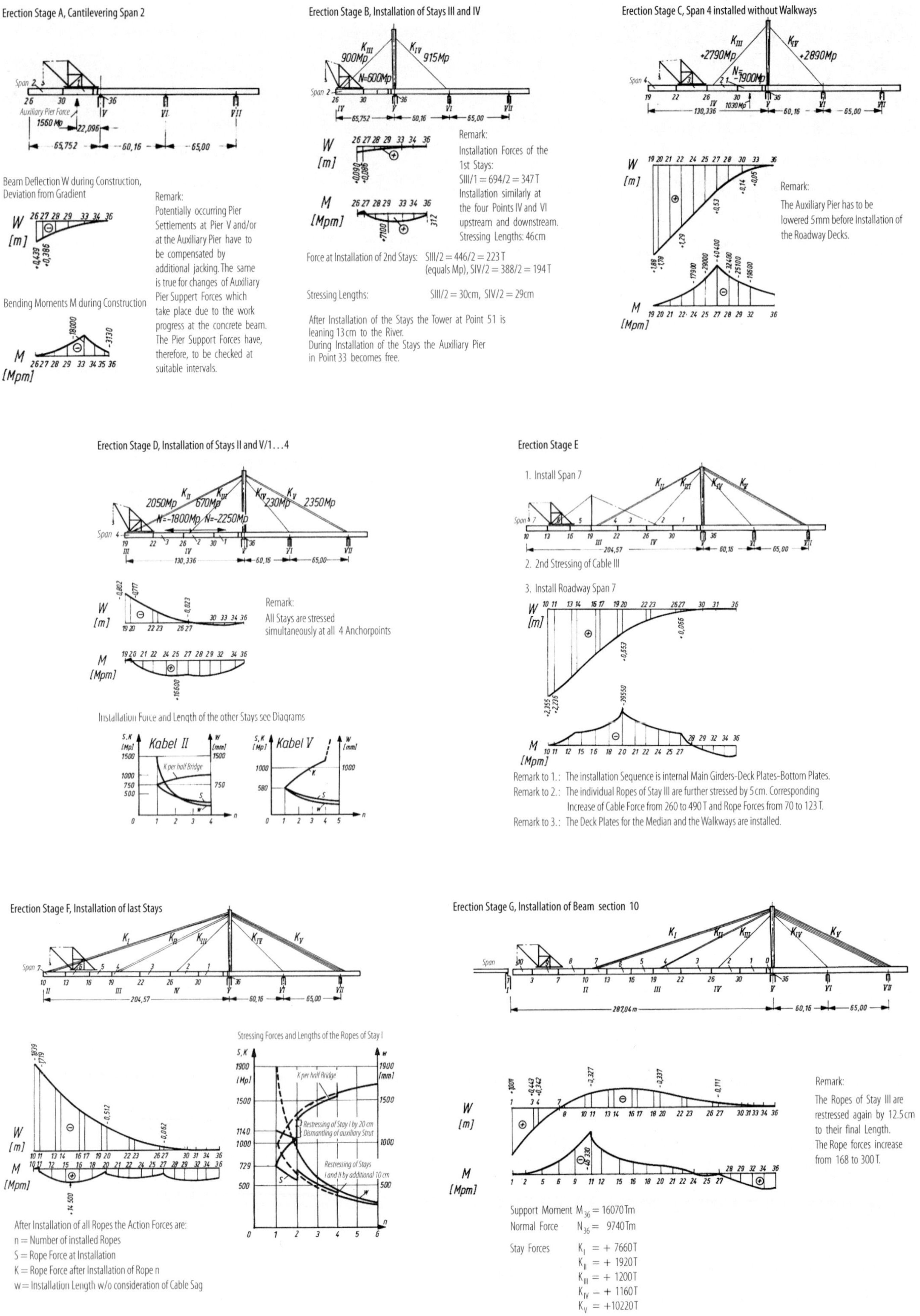

318 5 Construction of cable-stayed bridges

Figure 5.89 Construction manual

5.2 Construction engineering

Figure 5.90 Construction stage B

Figure 5.91 Construction stage C

Figure 5.92 Construction stage D

Figure 5.93 Construction stage G

Figure 5.94 Completed bridge Mannheim North

determined by the actual loadings and cable forces. Possible deviations can be corrected by stressing or releasing the stay cables.

5.2.4 Design of auxiliary stays

For the single-tower cable-stayed bridges which have been considered so far, the tower is normally tied back to the back span cables, anchored to the approach bridge constructed previously. If free cantilevering has to take place to both sides of the tower, for example to avoid expensive auxiliary piers in the side spans, it may be necessary to stabilize the towers temporarily with auxiliary stays, especially if precast concrete beam elements cause big eccentric moments about the tower.

5.2.4.1 Symmetrical auxiliary stays for towers

We use as an example the Pasco-Kennewick Bridge, Figs 5.95 and 5.96 [1.15]. The towers were not able to carry the eccentric weight of 270 t in bending and were therefore tied back to a hold-down pier and forward to the foundation of the opposite tower, Figs 5.95 and 5.96. The reddish cable in the foreground of Fig. 5.96 is a forward auxiliary stay to the opposite tower.

Such auxiliary stays are expensive, and, into those across the river to the opposite tower a ship may run. They should thus be avoided if possible.

Calculation

If auxiliary stays are used, their forces cannot be directly introduced into the backward calculation because they are not present anymore at the end of construction. A forward calculation of these auxiliary stay forces has to be introduced in order to size them and to find the cable forces at the time of dismantling them on site. These forces are then introduced into the backward calculation, at the time when they are actually dismantled, which means that they are 'installed' into the backward calculation. Fig. 5.97 shows the systems which have to be investigated.

Forward calculation

Stayed tower: the towers are tied backwards and forwards with auxiliary stays. They have the forces $S_{F,0}$ and $S_{B,0}$. The horizontal components equalize one another at the tower tips.

Determination of installation forces: the installation forces are determined in such a way that the auxiliary stays do not become slack if the eccentric beam element is located on their side, Fig. 5.97:
– min S_B in stage 2.1,
– min S_F in stage 2.2.

It is expedient to keep min S_F bigger than $\frac{1}{2} S_{F,0}$.

On the other hand, each cable has to be sized for its biggest force if the eccentric beam element is located on the other side:
– max S_F in stage 2.1,
– max S_B in stage 2.2.

During all construction stages the tower moments must not exceed the permissible ones. When calculating the geometry, the effective

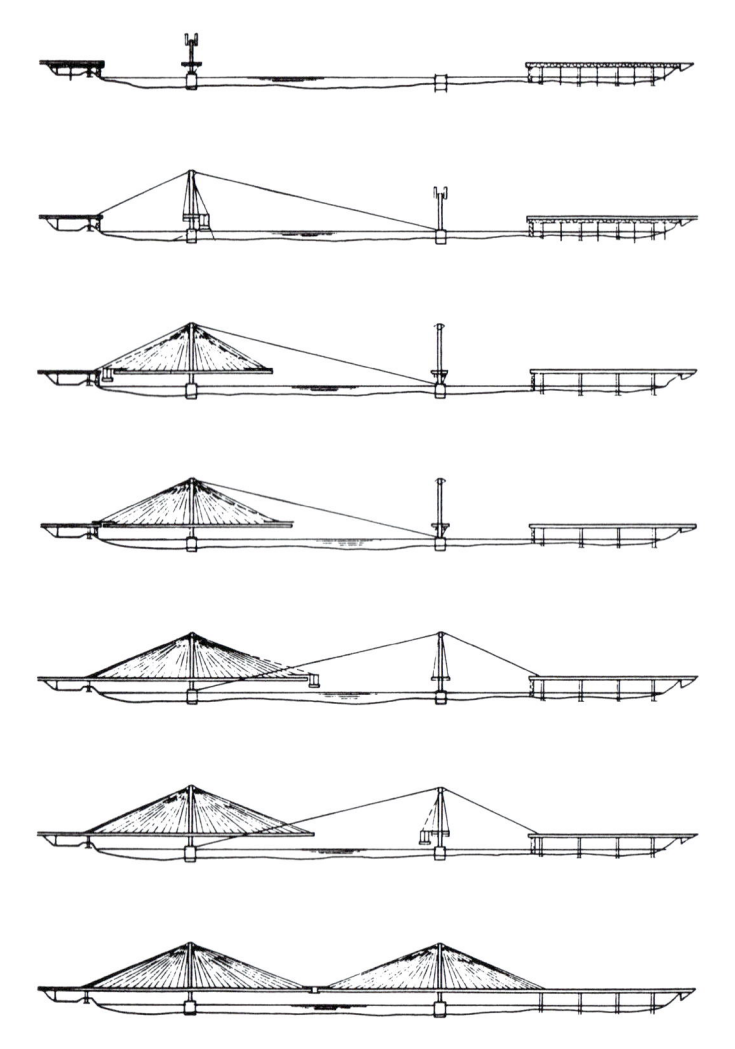

Figure 5.95 Auxiliary tower stays to both sides

Figure 5.96 Lifting of the last precast element of the Pasco-Kennewick Bridge, USA

moduli of elasticity have to be taken into account because of the strongly varying cable stresses – see Section 1.1.2.2.

Backwards calculation
Installing auxiliary stays: When calculating backwards, the auxiliary stays are 'installed' with that force which has been determined during the forward calculation for the time of their dismantling.

In the backward calculation the cable forces $S_{F,0}$ and $S_{B,0}$ at the end must be those chosen in the forward calculation – see above.

5.2.4.2 One-sided auxiliary stays for towers

If the use of forward auxiliary stays is to be avoided, the tower must be stabilized with auxiliary backstays as was done for the Posadas-Encarnación Bridge, Figs 5.98 and 5.99 [1.24].

The precast elements are lifted alternately between main span and side span, but the unsymmetrical stages are always towards the main span. The permanent backstay cables are used as tie-backs during construction, although with lesser forces than in the final stage.

The cable forces and stiffnesses are determined for the governing eccentric element as well as for the smallest cable force in balanced situations, where the unbalanced backstay force causes bending in the tower. The force in the backstays may have to be adjusted to the eccentric situation if the tower cannot carry the full constant backstay force.

The conditions for determining the forces in the backstays are given in Fig. 5.100.

Forward calculation
System 1: Tower moment together with smallest compression force has to be acceptable: $\min S_B$ should be at least $½ \max S_B$.
System 5: Biggest cable force during lifting of last precast element.

Backward calculation
The backstay cables are 'installed' with those forces which have been determined at the end of the forward calculation.

5.2.4.3 Auxiliary stays for beam

In order to secure a complete construction system against unusual loads, e.g. against hurricane, the beam may be tied down against the tower foundation, Fig. 5.101 [1.17]. The tie-downs are oriented longitudinally and transversely in order to reduce longitudinal and torsional beam oscillations. The bending moments due to eccentric construction loads could in this case be carried by the tower alone because the twin composite beam was built from light elements.

5.2.4.4 Without auxiliary stays for beam

If tie-downs are to be avoided, even for heavy construction loads, the beam elements can be installed starting with one half element. During further construction with full elements they are then eccentric only by one half of their length, either to the main span or to the side span, Fig. 5.102.

As a refinement, the element lengths can be varied in such way that during the governing situation before side span joint closure the

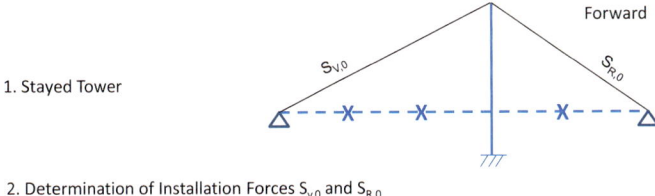

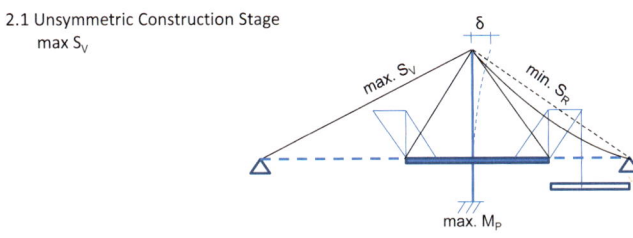

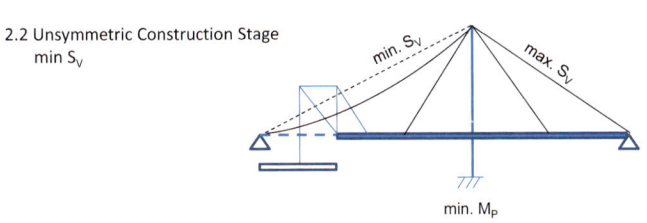

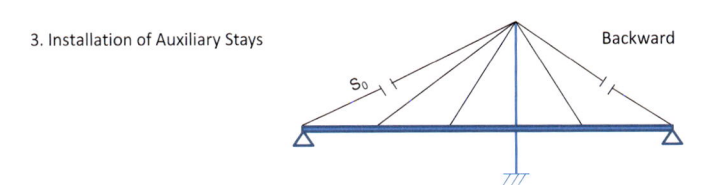

Figure 5.97 Symmetric auxiliary tower stays

Figure 5.98 Installation of first precast element for the Posadas-Encarnación Bridge, Argentina

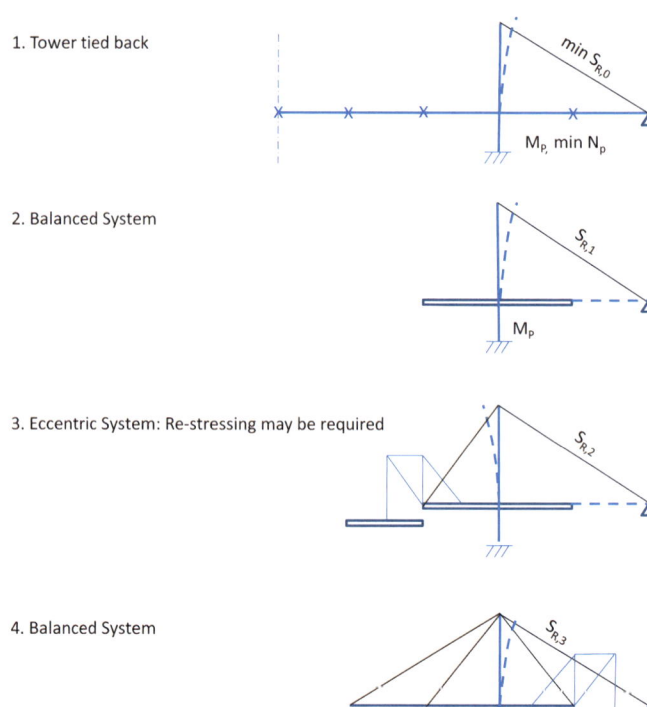

1. Tower tied back
2. Balanced System
3. Eccentric System: Re-stressing may be required
4. Balanced System
5. Completed Bridge

Figure 5.100 Construction with one-sided auxiliary tower stays

Figure 5.99 Unsymmetrical free cantilevering; eccentricity to main span

Figure 5.101 Tie-downs for the beam of the Baytown Bridge against hurricane

difference in lengths between the side span and the main span beam reaches one half, Fig. 5.103. This requires a difference in element lengths between side span and main span of

$$n \cdot a_s - n \cdot a_v = 1/2\, a_v$$
$$\rightarrow a_s = a_v \cdot \left(\frac{1}{2n} + 1\right)$$

5.2.5 Auxiliary tie-backs for travelers

During lifting, installation or concreting, the new beam element creates a negative moment at the tip of the already constructed beam. For steel beams with lighter sections this moment can often be taken by a force couple at the first two cable anchorages, and the corresponding moments can be carried by the beam itself.

If the beam capacity is not sufficient to carry the moment during lifting, the traveler or derrick has to be tied back to a tower, so that the beam moments are sufficiently reduced, Fig. 5.104.

In addition to the new element weight, the weight of the derrick also acts at the tip of the beam. The total weight can be distributed further backwards by spreading the distance of the derrick tie-back from a to n · a.

One example is the construction of the steel beam for the Zárate-Brazo-Largo Bridge, Fig. 5.105.

For the high weight of concrete beam elements the traveler for precast elements or the CIP forms may have to be tied back to the tower.

While the full weight of the new beam element is acting, the tip of the traveler can be tied back to the tower in such a way that its front support is relieved. A major part of the weight is then carried by the tie-back and the remainder acts at the second cable only where the beam is less sensitive.

The corresponding flow of forces is shown in Fig. 5.106. The beam moment is strongly reduced, but the important horizontal component of the tie-back has to be transferred into the beam, and the auxiliary ties have to be elongated for each new construction system. One example for this method is shown in Fig. 5.107 for the Pasco-Kennewick Bridge [1.15].

On Fig. 5.96 an orange auxiliary stay for the tower can be seen in the foreground, as well as the tie-backs of the traveler.

The formwork traveler for CIP beams can also be tied back to the tower. One example is the pedestrian bridge at Mannheim Neckarcenter, Figs 5.15 and 5.16 [2.134].

If the beam has a high bending capacity, even major weights can be supported without tie-backs. The flow of forces is shown in Fig. 5.108. In this case, the full DW acts between the first two cables.

An example is the Kap Shui Mun Bridge, which has a double-deck beam with a depth of about 7.5 m, Fig. 5.109 [2.107]. This permits lifting of complete composite beam sections including their top and bottom concrete slabs with a total weight of about 500 t.

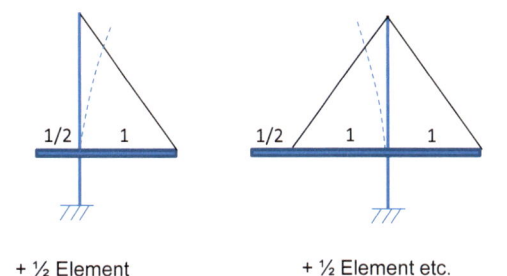

Figure 5.102 Construction with ± ½ elements

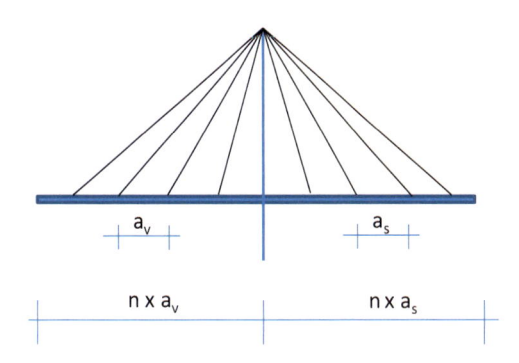

Figure 5.103 Construction with ± ½ element before side span closure

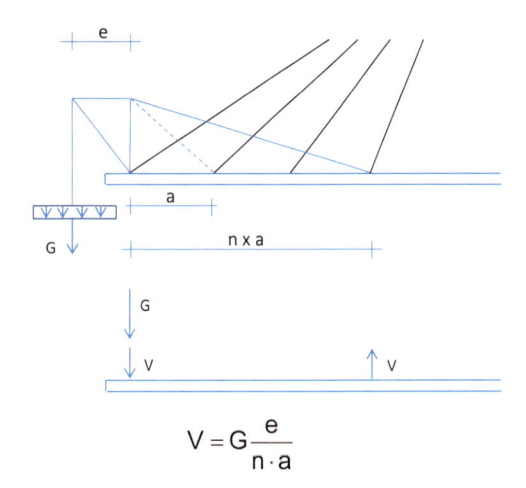

$$V = G \frac{e}{n \cdot a}$$

Figure 5.104 Derrick with tie-back to the beam

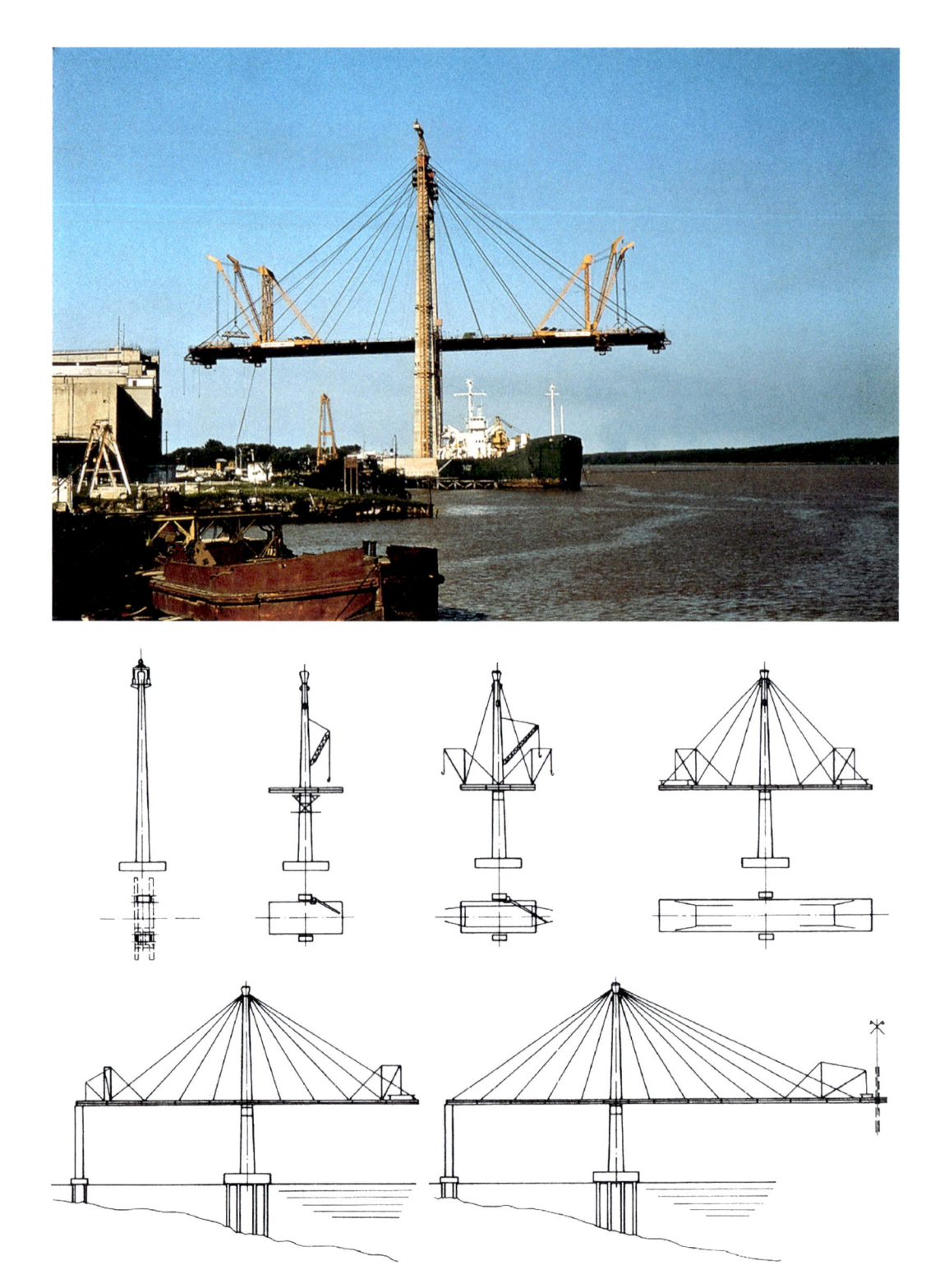

Figure 5.105 Derrick with tie-back to the beam, Zárate-Brazo Largo Bridge, Argentina

5.2 Construction engineering

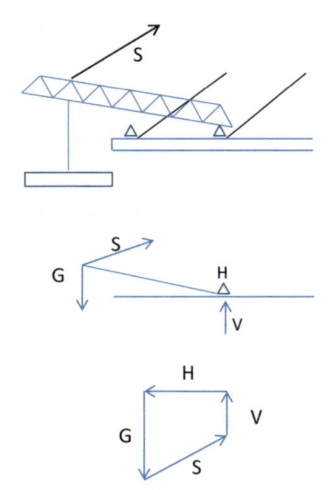

Figure 5.106 Traveler tie-back to the tower

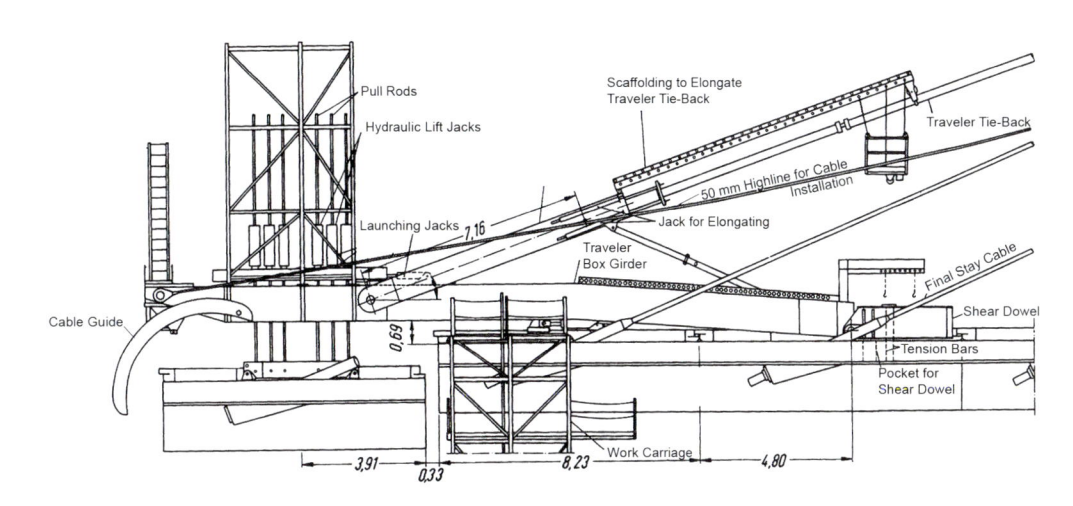

Figure 5.107 Traveler tie-back to the tower for the Pasco-Kennewick Bridge

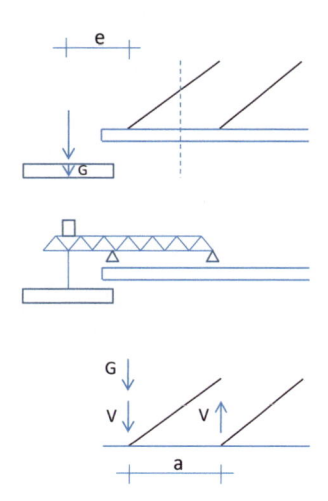

Figure 5.108 Beam loading during lifting without tie-backs

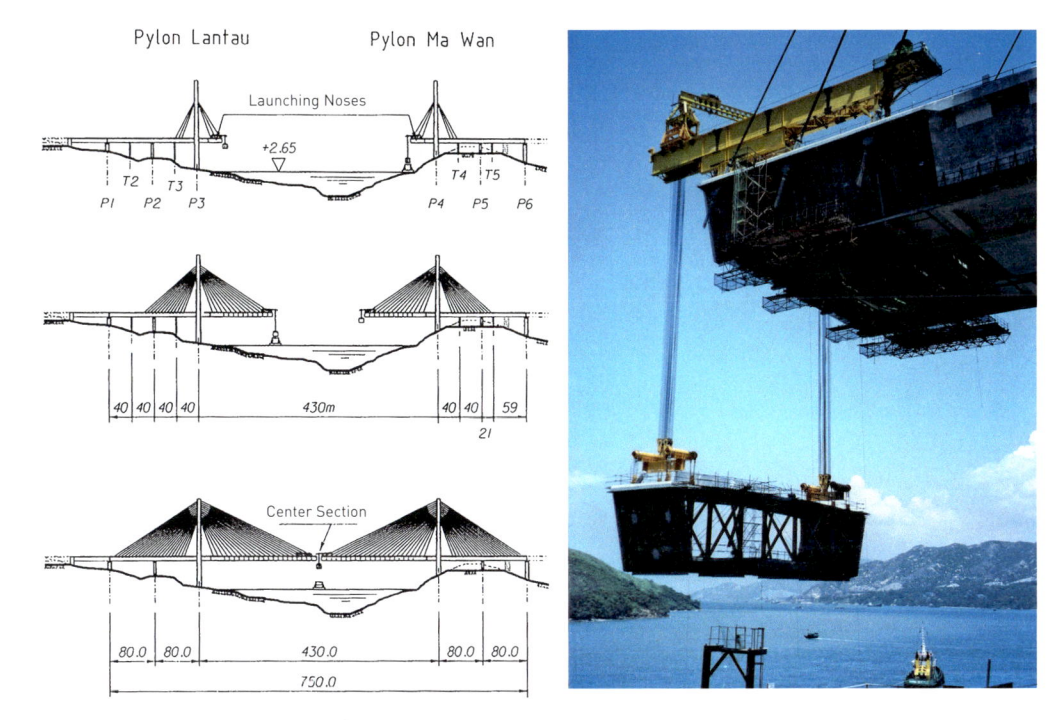

Figure 5.109 Lifting of a complete beam section with a weight of about 500 t without tie-backs

6 Examples for typical cable-stayed bridges

Cable-Stayed Bridges. 40 Years of Experience Worldwide. First Edition. Holger Svensson.
© 2012 Ernst & Sohn GmbH & Co. KG. Published 2012 by Ernst & Sohn GmbH & Co. KG.

6.1 Cable-stayed concrete bridges with precast beams
6.1.1 General
Cable-stayed concrete bridges with beams from precast elements have not been built very often. The first major examples are the Pasco-Kennewick Bridge and the East Huntington Bridge, both in the USA, which were completed in 1978 and 1985.

6.1.2 Pasco-Kennewick Bridge
6.1.2.1 General layout
The Pasco-Kennewick Bridge was the first cable-stayed bridge which the author had to design on his own during the years 1973 to 1978, see appendix. What he learned from this work he published in [1.15], which forms the basis for the following description.

The roadway bridge across the Columbia River between the cities of Pasco and Kennewick, WA, Fig. 6.1, replaces a steel truss built in 1921. The river is 732 m wide and up to 21 m deep. The flow velocity and the change in water level are small because the river is regulated by a system of dams. The required navigational clearance was 15 m.

The soil comprises very hard consolidated layers of clay with a thickness of 25–30 m, which are covered by sand and gravel. Below the clay, bedrock in the form of solid basalt is present.

The fan arrangement of the stay cables requires a minimum of cable steel, produces a high compression in the beam, which is favorable for concrete, and reduces the bending in the towers.

Parallel wire cables of high-strength steel permit high stresses and, in combination with their high modulus of elasticity, provide a high stiffness, which creates favorable live load moments in the beam. A small distance of cable anchorages at the beam reduces the cable sizes, simplifies their anchorages, reduces the beam moments from permanent loads, simplifies the construction and improves the aerodynamic stability.

Continuity of the bridge beam over the full length of the bridge, including the approaches, prevents kinks in the beam under live load and reduces the number of roadway joints, which improves the driving comfort. Even at the towers the beam is elastically supported by the cables in order to avoid the large negative moments which would be created by rigid supports at the towers.

By using two cable planes anchored at the outside of the bridge beam a torsionally weak open cross-section without bottom slab can be used, which simplifies beam fabrication and construction. With this cable arrangement the roadway slab acts as the top flange of a simply supported girder in the transverse direction and thus receives only compression from the dead load and live loads. The beam depth is primarily determined by the cross girders, and can be small. Consequently, the wind area of attack and the gradient of the approaches is reduced. By choosing suitable span lengths for the approach bridges the beam depth and shape can be kept constant over the total bridge length. Strong edge girders distribute the cable forces uniformly in the longitudinal direction and permit the same shape for the main and secondary cross girders.

Figure 6.1 Location of bridge

The roadway slab spans in the longitudinal direction between the closely spaced cross girders so that the overall high compression forces from the cables are superimposed onto the local tensile stresses from wheel loads.

The fabrication of the precast elements of the bridge beam permits good quality control and rapid erection. The high compression at the cable anchorages acts on completely cured concrete from mature precast elements. The remaining shrinkage and creep is small.

In addition to these technical considerations the desire to create an aesthetically pleasing bridge was equally important. For this purpose, it is especially important to have balanced proportions between all bridge members, a clear flowing outline over the complete bridge length and slender towers and piers.

The high slenderness of the bridge beam, 1:140, is visually increased by the fascia with a slenderness of 1:421, behind which the full beam depth is reduced by the inclined outer outside slabs, see Fig. 6.4. The large number of thin white cables has the tendency to blur against the sky and creates the impression of a veil, Fig. 6.2.

Overall system

The bridge comprises two approaches and the inner three-span symmetrical cable stayed bridge with a beam supported by 144 cables in two planes, Fig. 6.3. The cables converge closely in steel tower heads. The beam is continuous with a constant shape over the full length of the bridge. It is fixed in the longitudinal direction at abutment 1. In axes 1, 3, 4, 6 and 9 transversely fixed bearings are located. The uplift forces from the backstays are transmitted by pendulums into the foundations.

Cross-sections

The beam cross-section comprises two outer triangular boxes and the inner roadway slab supported by cross girders, Fig. 6.4. The shape of the boxes was confirmed in the wind tun-nel tests outlined in [6.1]. A longitudinal section through the cross girders and roadway slab is shown in Fig. 6.5.

The beam of the approach bridges has the same outer shape but a bottom slab and two additional inner longitudinal girders, underneath which the bearings are located in order to reduce the transverse widths of the piers. There are only cross girders over the piers and in mid span, so that the roadway slab carries in the transverse direction.

At the hold-down piers the beam is solid over a length of 9.45 m, in order to reduce the uplift forces and to carry the high bending moments in the longitudinal and transverse directions created by the three concentrated backstay cables.

Precast elements

The precast elements, which are 8.23 m long – equal to the cable anchorage distance – comprise the whole cross-section with a width of 24.3 m. In order to achieve the required perfect fit of the joints, the elements were match-cast against one another. The bulkheads of the precast elements were not provided with a profile for shear interlock, because the shear forces remained always below 5% of the overall compression forces.

Four conical steel dowels, with 51 mm diameters, protruding into steel plates were placed into the forms in order to facilitate the joining of the precast elements and the temporary shear during erection, Fig. 6.6. The upper roadway reinforcement was welded for sustainability. This was costly and has not been repeated. Some additional post-tensioning is more economic.

In order to reduce the local wheel load moments in the roadway slab, the joints were placed at the quarter point between cross girders.

Post-tensioning

The spans of the approach bridges were post-tensioned with 24 continuous draped tendons for 2.5 MN each. The precast elements were provided with straight bar tendons of at least 26 mm and 32 mm diameters which were coupled at each joint.

The epoxy resin in the joint required a minimum compression of 0.5 MN/m² during curing so that a minimum construction post-tensioning of 2.6 MN was selected. At the bridge center the number of longitudinal bars increases strongly because the normal force from the stay cables gradually tapers down to zero and the live load bending moments increase. The cast-in-place joints at the bridge center and at the tips of the approach bridges are post-tensioned with overlapping tendons.

Each cross girder of the main bridge is post-tensioned transversely with a 2.25 MN tendon. The stiff triangular edge boxes distribute the cable forces in the longitudinal direction so that at the cable anchorages only three short 1.0 MN tendons are additionally required in order to tie back the vertical cable components to the inner edge of the inclined slab from where they are distributed in strut action to the tendon anchorages of the adjacent cross girders, Fig. 6.7.

Stay cables and anchor heads

The tensile forces in the stay cables are carried by parallel wires with 6.35 mm (¼") diameter steel St 1450/1650 (f_y/GUTS) in accordance with ASTM A421. The wire bundles are surrounded by a ⅜ inch strand helix which keeps the wire in order and guarantees a minimum distance to the surrounding PE pipe, Fig. 6.8. The black PE pipes are wrapped with white UV-resistant PVF tapes for coloring.

The wires terminate in steel anchor heads with strengths of 380/580 N/mm² where they are anchored in a retainer plate with button heads, Fig. 6.9. The main anchorage force between the wires and the anchor head is created by the clamping effect of the so-called HiAm anchorage which uses small steel balls to fill the interstices between the wires and the inner cone. The steel balls are secured in place by epoxy resin filled with zinc dust.

6.1 Cable-stayed concrete bridges with precast beams – Pasco-Kennewick Bridge

Figure 6.2 Completed bridge

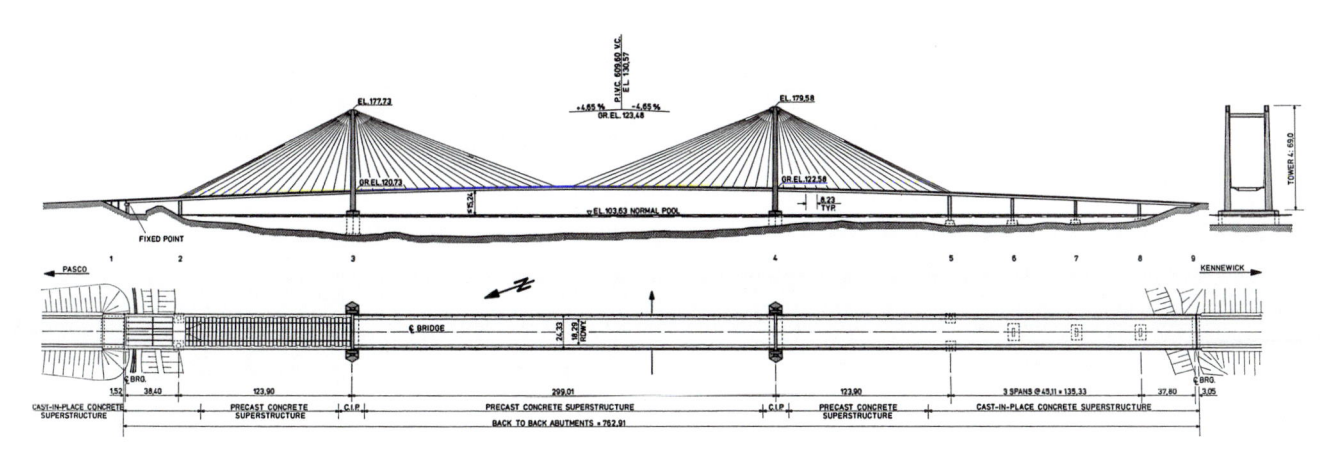

Figure 6.3 General layout

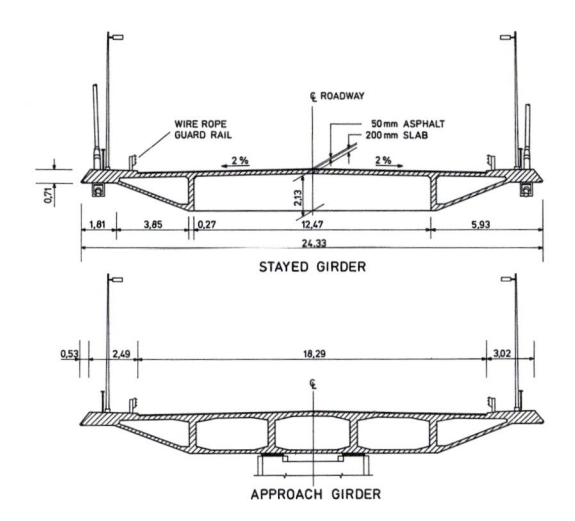

Figure 6.4 Cross-sections

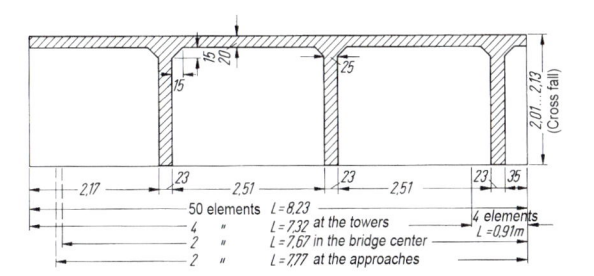

Figure 6.5 Longitudinal section along bridge center

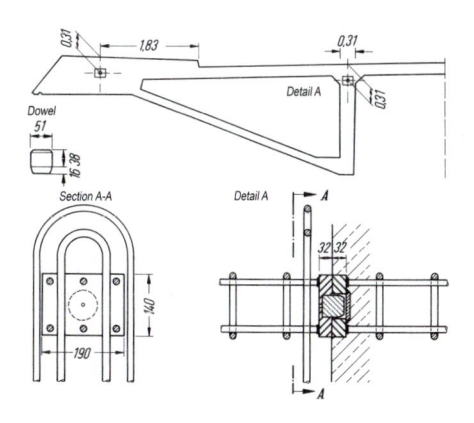

Figure 6.6 Steel dowels with sleeves in precast joints

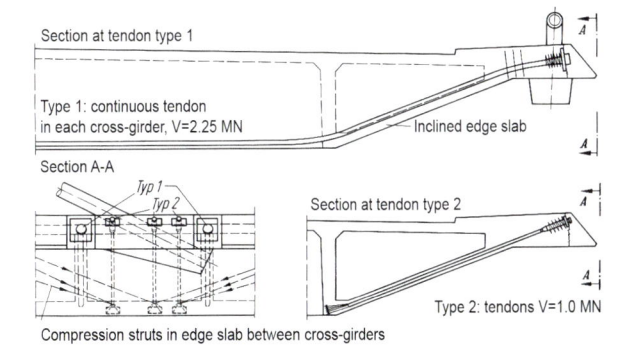

Figure 6.7 Transverse post-tensioning of main bridge

The transition between the inner HiAm casting and the cement grout of the free cable length is filled with epoxy resin plus zinc dust. A detailed description of this HiAm anchorage is provided in Section 3.4.

The short steel pipe at the tip of the anchor head serves for the airtight, tension and com-pression resistant anchorage of the PE pipe to the anchor head.

Stay cable anchorages

At the superstructure the stay cables are anchored into the outer edge beams, Fig. 6.10. Part of the cable force flows directly via the contact area into the concrete, the remainder running into the steel pipe and from there via the welded shear rings into the concrete. The distribution depends on the support area and the stiffness ratio between concrete and steel which is subject to change.

In the concrete, the horizontal component of the inclined cable force spreads as normal force over the complete beam cross-section, whereas the vertical component is carried in the inclined transverse tendons, Fig. 6.7. At the upper end of the steel pipe a neoprene ring centers the stay cable against the steel pipe.

Outside the tip of the steel pipe a neoprene boot seals the steel pipe against the intrusion of water. The boot is connected to the steel pipe and the stay cable with stainless steel straps. A hole in the lower steel plate serves as drainage in case the upper seal does not work or condensation water appears.

At the tower head the stay cables are individually anchored in the steel tower heads, Fig. 6.11. The large cable forces required thick steel plates, each steel tower head weighing 63 t.

In order to approach the ideal fan arrangement of the cables with a common point of intersection, the stay cables are anchored in three parallel vertical planes.

Cable tests

In order to prove the required characteristics of the cable anchorages two tests with 2.54 m long specimens with 83 wires each were executed [6.2]. The results of the fatigue tests and the tensile tests as well as the slip at room temperature and at 80 °C were satisfactory and in accordance with former tests outlined in [3.19 – 3.21].

Towers

The towers are designed as frames with vertical legs and struts, fixed to the foundations, Fig. 6.12. The legs consist of reinforced concrete, the struts are post-tensioned. The box cross-section of the legs has constant wall thickness and tapers upwards in both directions with vertical inclines. The steel tower heads rest on the tower legs. In addition, at their out-sides concrete 'ears' carry shear from different cable forces in the main span and the side span plus moments from transverse wind into the tower legs. In order to avoid deviating forces from the stay cables, each tower head axis has the same transverse inclination as the corresponding cable plane.

Bearings

The US neopot bearings which carry the horizontal and vertical loads are roughly similar to those fabricated worldwide.

For the safety of the bridge against possible moderate earthquakes it was not strengthened, but the beam was permitted to remain at rest against the horizontal oscillations of the soil and in this way to avoid inertia forces from earthquake accelerations [6.3]. For this purpose the longitudinal bearing at the abutment and the transverse bearings at the towers were provided with the desired failure joints, Fig. 6.13, which fail when earthquake forces occur which are larger than those assumed for service conditions. The relative movements between beam and piers are limited to 25 cm in all directions.

Between the beam and the abutments a movement of 25 cm is only possible in the longitudinal direction. This limitation is necessary to prevent the shearing-off of the pendulums and to protect the roadway joints as far as possible.

Tension pendulums

At the hold-down piers uplift forces occur, together with longitudinal movements of the superstructure, for which tension pendulums, Fig. 6.14, from parallel wire cables with 157 wires each are arranged. They stress the beam down in such a way that even under increased service loads no uplift from the bearings takes place.

In order to prevent a kink in the wires at the entrance into the anchor heads, these anchor heads can freely rotate on spherical bearings, Fig. 6.15. Since even the moment from the friction in the spherical surface would create too high additional bending stresses in the wires due to non-linear effects from tension – see Fig. 4.26 – a strong steel pipe with a longitudinal hinge at its center ensures the rotation of the anchor heads, Fig. 6.14.

In order to avoid the strong increase of compression forces in the bearings, which would be created by the elongation of the steel pipe for beam movements of ± 21 cm at pier 5 under service loads (during earthquake ± 25 cm), the steel pipes are provided with a longitudinal joint in the central point of counter-flexion.

The very limited depth in the anchorage region of the superstructure requires the cable anchor heads to be anchored with support nuts, Fig. 6.15.

Design calculations

The design calculations followed the principles outlined in Chapter 4. For the various static and dynamic calculations a modified STRUDL-program was used. The action forces for the final stage were determined at a plane frame with 111 nodes and 180 members. All stay cables received a slightly reduced effective modulus of elasticity of $2 \cdot 10^5$ N/mm^2 which was kept constant because the change of sag for live loads was negligible.

The concrete stiffness of the beam and the towers was calculated for uncracked sections, taking into account the reinforcement. The local beam moments were calculated with a girder grid by using the

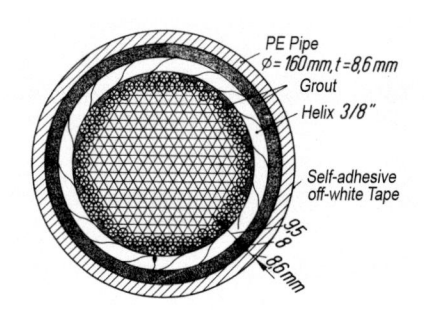

6.1 Cable-stayed concrete bridges with precast beams – Pasco-Kennewick Bridge

Figure 6.8 Stay cable cross-section with 283 wires

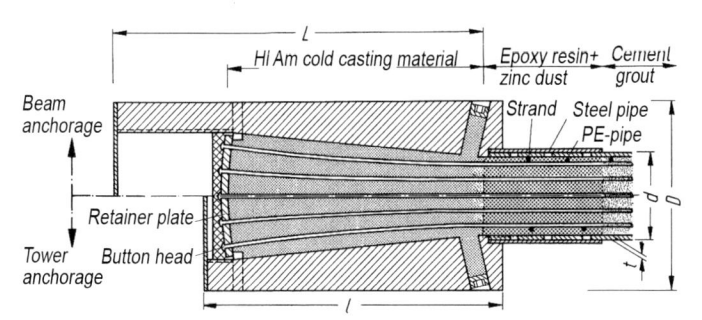

Figure 6.9 Longitudinal section of anchor head

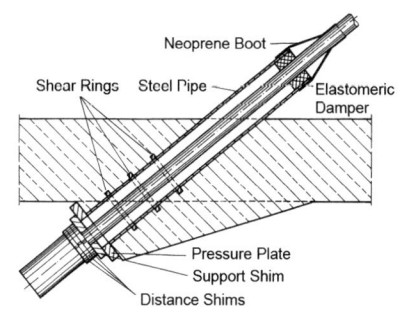

Figure 6.10 Cable anchorage at beam

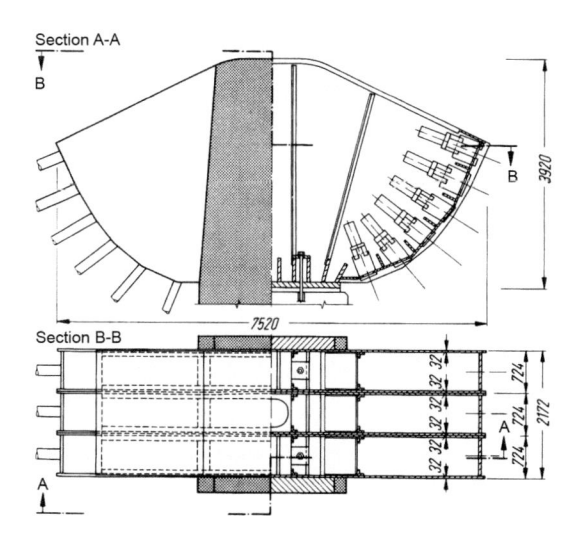

Figure 6.11 Steel tower head

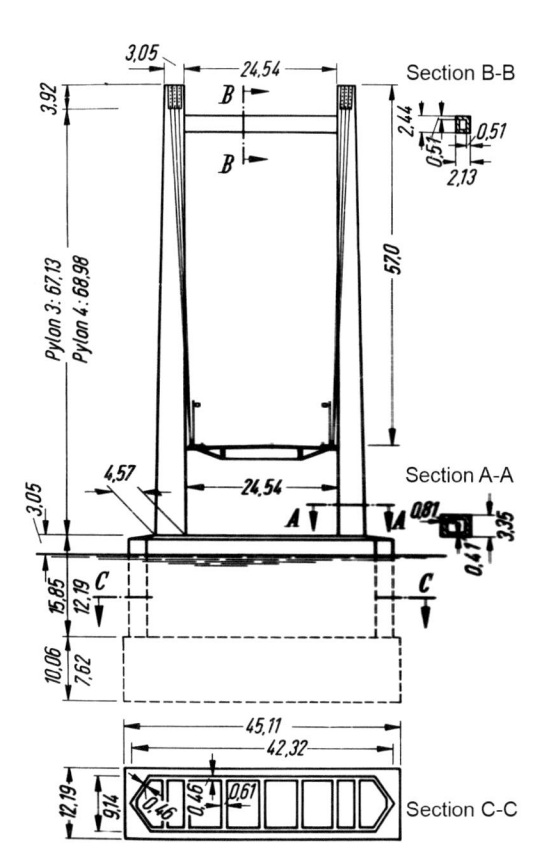

Figure 6.12 Tower layout

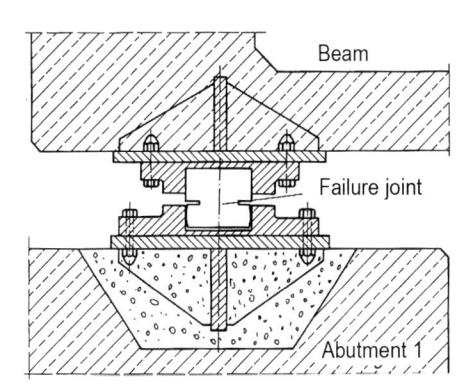

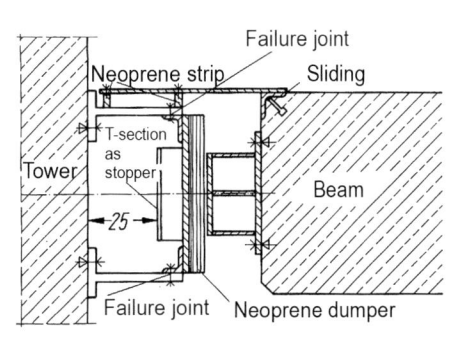

Figure 6.13 Longitudinal bearings at the abutment and transverse bearings at the towers with the desired failure joints

forces from the overall systems. The edge box girders were replaced by stiff members located in the shear centers.

The towers were investigated in a 3D-system, for which the cable forces and longitudinal deflections of the overall system were introduced with the exception of those loadings which cause torsion in the tower legs. Special local problems such as the introduction of the cable forces into the longitudinal steel plates of the tower heads were treated by means of finite elements.

Some of the difficulties due to the limited computer capacity in 1976 are mentioned in the Appendix, and the action forces of the overall system are given in Fig. 4.10.

Earthquake

The longitudinal oscillation period of the completed bridge comes to about 0.5 sec. As soon as earthquake forces shear off the desired failure joints, Fig. 6.13, the period increases to about 12 sec, which renders the system nearly insensitive to the rapid movements of an earthquake.

Static wind loads

The design wind speed for the unloaded bridge in accordance with AASHO was assumed as 160 km/h. For the determination of the static drag factors, wind tunnel tests were performed on a section model at a scale of 1:38.4 and length of 1.8 m [6.1]. Five different edge configu-rations were investigated but they did not give significantly different results. The aerodynamic shape factors are shown in Fig. 6.16.

Fig. 6.17 gives the relation between wind speed and wind angle of attack as measured for the Severn Bridge [6.4], and confirmed on other occasions. This results in the design wind speed with angles of attack up to ±2°. The corresponding drag factor in accordance with Fig. 6.16 comes to 1.17, referred to the beam depth. For larger wind angles of attack the wind speed decreases more strongly than the drag factors increase.

The drag factor for the stay cables was taken as 0.7, see Figs 3.90 and 3.91, and that for the bluff tower legs with 2.0, see Fig. 4.81.

Aerodynamic stability

Since the bridge is located in the vicinity of the infamous Tacoma Narrows Bridge, Fig. 6.1, the aerodynamic stability was investigated in depth. With the same section model used for the static wind tests the dynamic characteristics were investigated in the wind tunnel [6.5]. It was found that wind oscillations of any kind only occur outside the assumed wind spectrum as shown in Fig. 6.18.

When comparing the test results with flutter calculations in accordance with Klöppel/Thiele [4.17], the shape reduction factor against an air foil comes to about 0.6 for a wind angle of attack of about 4°, see Fig. 4.237. This tallies with earlier test results for similar cross-sections.

6.1.2.2 Construction engineering
General

The construction engineering was performed backwards by dismantling the final bridge as outlined in Section 5.2.

Desired shape in the final stage after shrinkage and creep

Beam: The shop form of the precast elements was determined from the following considerations:
- all precast elements are fabricated 3 mm longer than their final lengths in order to take into account one half of their later shortenings due to elastic and shrinkage and creep deformations
- all cast-in-place joints are cast in their final shape
- the gradient after shrinkage and creep must reach the theoretical value.

For the determination of the coordinates of the cable anchor points the following influences were taken into account:
- the change of the fixed points for the intermediate construction stages due to elas-ticity, shrinkage and creep determined the location of four characteristic points, Fig. 6.19
- the changes in the lengths of all precast elements due to elasticity, shrinkage and creep
- the thickness of all final joints between elements, taking into account sandblasting, comes to 3 mm (the actual thickness was finally measured at only 0.6 mm)
- the temperature during construction was assumed to be 13 °C, and the temperature during casting of the elements was estimated and considered in the bridge geometry.

The lengths of the precast elements were not influenced by the ambient temperature during casting because the steel forms expand similarly to the concrete. The temperature during closure of the side span and main span joints was taken into account by moving the cable suspended beam with jacks at the towers into that position which corresponds with the position in the final stage. In this way the joint closure temperature did not enter into the final geometry.

Towers: The towers were built in such a way that the locations of the cable anchor points at the tower heads are those in the final stage after shrinkage and creep. For this purpose, the tower heads were cast 44 mm higher for the first tower and 4 mm higher for the second tower. Their pier settlement was assumed to be 13 mm. The tower heads were built in and rotated by 0.066° (0.046°) in the direction of the side spans, in order to compensate for the different cable forces under permanent load in the main and side spans.

Cable lengths: The fabrication lengths of the stay cables were calculated between the coordinates of the cable anchor points at the beam and towers plus the following corrections:
- distance between the theoretical and actual distance (shims plus bearing plates), Fig. 6.10

6.1 Cable-stayed concrete bridges with precast beams – Pasco-Kennewick Bridge

Figure 6.14 Pendulum layout

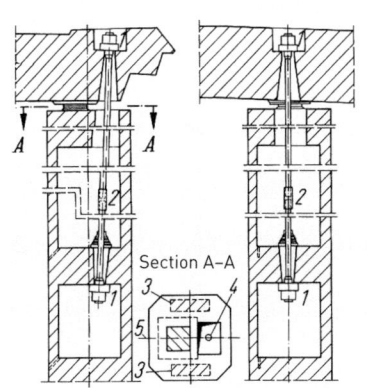

1 Spherical bearing
2 Longitudinal joint in point of counter flecture
3 Additional bearing during construction
4 Pendulum
5 Axis of end cross girder

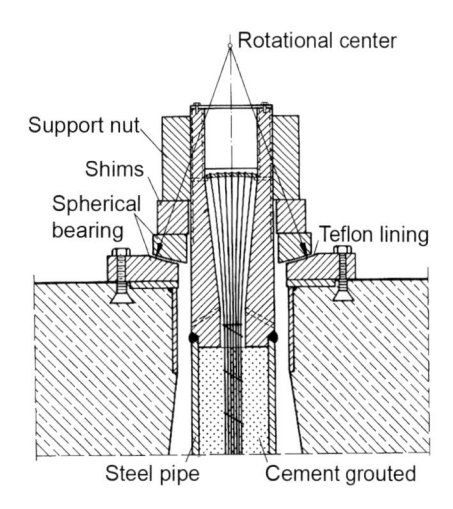

Figure 6.15 Rotating pendulum anchorage at top and bottom

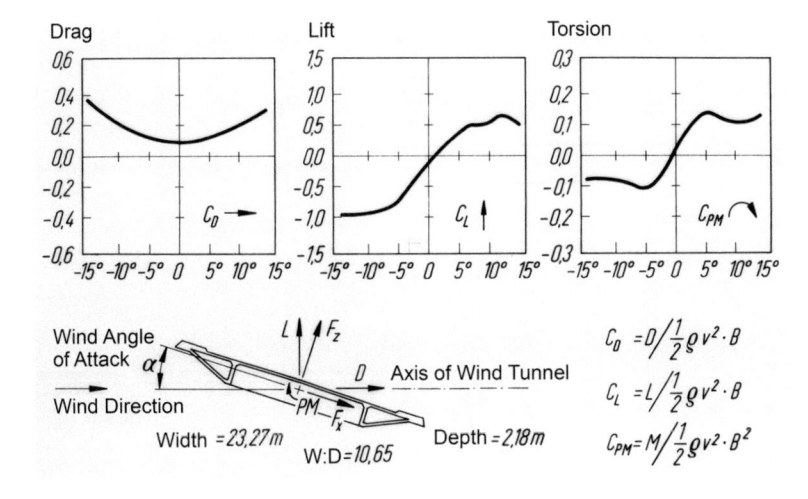

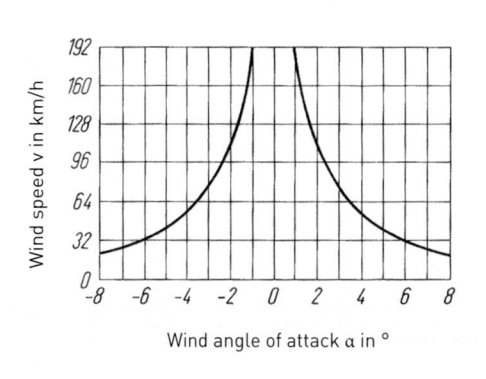

Figure 6.16 Aerodynamic shape factors

$$C_D = D / \tfrac{1}{2} \varrho v^2 \cdot B$$
$$C_L = L / \tfrac{1}{2} \varrho v^2 \cdot B$$
$$C_{PM} = M / \tfrac{1}{2} \varrho v^2 \cdot B^2$$

Width = 23,27 m   W:D = 10,65   Depth = 2,18 m

Figure 6.17 Correlation between wind speed and angle of attack

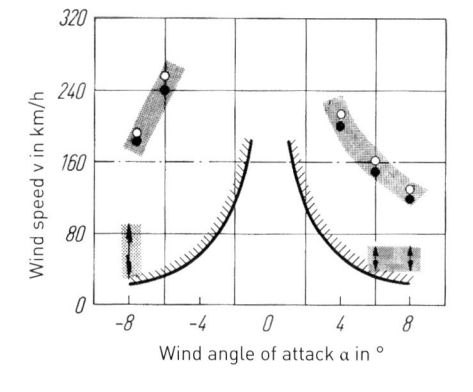

- Region of resonant vibrations
○ Flutter vibrations for rising wind speeds
● Flutter vibrations for decreaseing wind speeds
⋙ Statistical wind limitations
–·– Design wind speed
max v = 160 km/h

Figure 6.18 Results of the dynamic wind tunnel tests

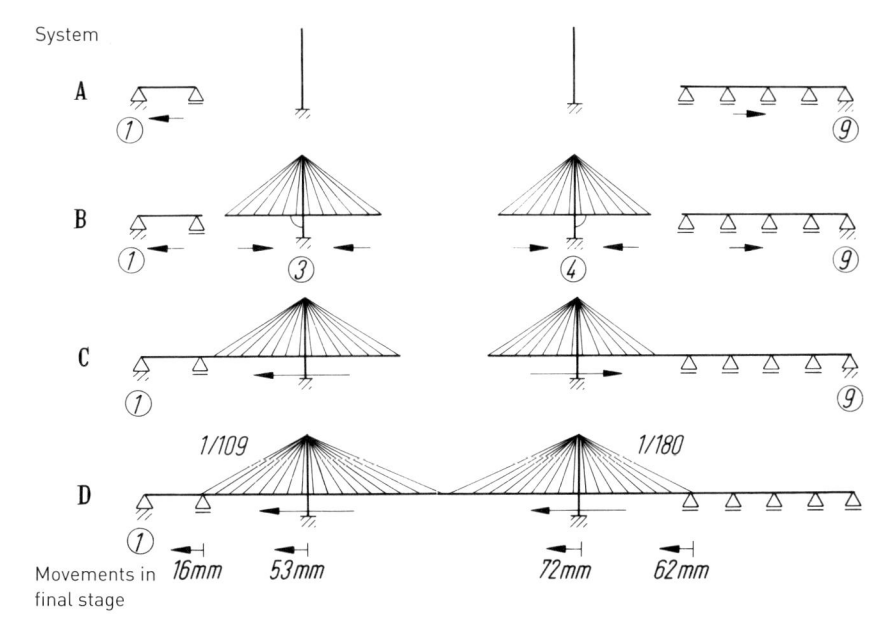

Figure 6.19 Change of fixed points during construction

Movements in final stage: 16 mm   53 mm   72 mm   62 mm

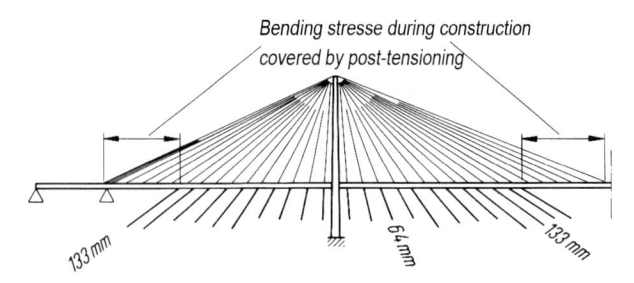

Figure 6.20 Initial overlength of stay cables at installation

– elastic elongations
– sag
– slip in both anchor heads, assumed 5 mm
– required overlength during construction
– difference between the construction temperature (13 °C) and the calibration temperature of the measuring tapes (20 °C).

The distance between the anchor heads determined in this way was adjusted for the wire cutting length for:
– distance between support plane and retainer plate, Fig. 6.9
– additional length for button heading the wires, 12.5 mm each
– additional 10 mm to avoid too short cables (the cable fabricator guaranteed the cable lengths to ±10 mm).

Geometry and action forces during construction: As mentioned earlier, the construction engineering was done backwards by dismantling the system, see Section 5.2.2.1. Onto the action forces in 'final stage' at t = ∞ shrinkage and creep were superimposed with negative sign in order to reach the stage 'opening for traffic' at t = 1. Then the superimposed dead loads were removed to reach the stage 'center joint closure' at t = 0.

To open the bridge by calculation one traveler was placed across the center joint, the post-tensioning was taken off and six cables on each side of the joint were shortened in such a way that all action forces in the nodes of both sides of the joints became zero. After that the beam was opened and each of the two bridge halves was dismantled, taking into account shrinkage and creep and the construction equipment, see Fig. 5.85.

Both side span joints were opened similarly to the center joint. At the end of the construction engineering the straight towers with their original heights remained. During dismantling, geometrical controls were applied and at the end the overriding condition was fulfilled that all action forces had become zero.

After this first global run for dismantling, complete erection cycles were calculated for several typical intermediate systems and the resulting stresses investigated. It became apparent that the tensile forces at the underside of the second last joint between precast elements required special measures. These tensile stresses were caused by the moment from the eccentric action of the horizontal support reaction on the beam during lifting of a precast element, see Fig. 6.35. In order to introduce additional compression into the critical joint during construction most stay cables were initially installed too long, Fig. 6.20, thus producing a temporary negative moment at the critical joints.

Tower construction: The tower foundations were built within sheet piles in 8 m and 15 m deep water respectively.

After installing the sheet piles and dredging down to the load-bearing soil the concrete base slabs were cast under water. After pumping out of the water the remainder of the foundations were built conventionally in the dry.

When the intended foundation level was reached for tower 4, it became apparent that the actual load-bearing soil layer was 0.6–3.0 m deeper. Since the sheet piles could not be elongated, 316 steel piles with double-T cross-section were driven, on which the base slab was supported.

The tower legs were cast with jumping forms in 4.27 m sections on a weekly cycle, Fig. 6.21.

The steel tower heads were fabricated in Japan. The up to 21 mm filled welds of the corbels for the cable anchorages were stressed-relieved. In order to keep the transportation weight small, each tower head was split into three compartments of 21 t weight each, which were later connected by high-strength bolts, Fig. 6.22.

Figure 6.23 shows an installed tower head with all cables after concreting the external concrete 'ears'.

Fabrication of precast elements: The cast-in-place beam of the approaches was built on scaffolding extending over the full length, cast spanwise and post-tensioned as complete units. At the tips of their cantilevers over the river auxiliary piers were left in place in order to adjust the moments (and geometry to a limited extent) in the beam before closing the joints to the main bridge.

The cast-in-place starter pieces at the towers were cast-in-place on scaffolding, Fig. 6.24. For their bulkheads, short precast elements were used, which had served as counter-planes for match-casting the first elements on both sides of the tower.

The precast elements were cast in a steel form on shore near the bridge on a weekly cycle, Fig. 6.25. Match-casting was used; a release agent was sprayed onto the joints to enhance the separation of the two elements and to improve the joint surfaces.

Fig. 6.26 shows the match-casting arrangement: after curing, each element was moved forward to serve as bulkhead for the next element. For the forming of each individual corbel against which the stay cables are later anchored, a special three-dimensional adjustable form was used. The completed element was very carefully aligned against the form because the correct run of geometry and action forces depended on the precise fit between the precast elements.

After steam curing and breaking the bond between concrete and the steel forms with com-pressed air, the precast elements were lifted out of the forms by a portal crane, Fig. 6.27, moved one length forward for the next casting operation, and finally transported to the storage area where they were kept wet for another two weeks. Shortly before installation the transverse tendons were post-tensioned. From then onwards the precast elements had to be supported at their edge girders in the axis of the stay cables, whereas before they rested underneath the inner longitudinal girders.

Beam installation: Large precast elements were selected because, amongst other reasons, the complete stayed beam is located above sufficiently deep water for floating-in the 270 t elements. Initially it was planned to lift the two elements symmetrical to a tower

6.1 Cable-stayed concrete bridges with precast beams – Pasco-Kennewick Bridge

Figure 6.21 Casting of tower legs

Figure 6.22 Tower heads before installation

Figure 6.23 Tower head

Figure 6.24 Starter piece

Figure 6.25 Steel form

Figure 6.26 Match-casting

Figure 6.27 Portal crane

simultaneously. In order to prevent progressive collapse in the case of one element crashing down it was necessary to stay the tower heads to both sides. The final backstays were used for temporarily staying the towers, and as forestay an auxiliary cable was anchored at a tower head between the location of the flattest final forward cable and at the foundation of the other tower, Fig. 6.28.

These temporary stays made it possible to install the elements staggered by a half weekly cycle with the advantage that the same crew could complete the joining works alternately on the river and shore side. (As a general rule, there is no 'simultaneous' during any structural erection.) Since the backstays were stronger than the temporary forestays, the river side elements went ahead by one element after the symmetric construction stages.

Before lifting the last precast element in the main span, the forward auxiliary tower stay cables were dismantled to make room for the last final stay cables. This was possible because the bridge beam on the land side was meanwhile connected to the approach bridge and thus each tower was safely retained.

After a certain learning curve a construction progress of 12 elements equal to about 100 m per month was achieved.

The large horizontal compression forces in the beam from the weight of each unbalanced element in the main span was supported by transversely post-tensioned corbels in the tower axis, Fig. 6.29.

Fig. 6.30 shows a temporary forward tower stay in the final top left tower cable anchorage, an erection cable for supporting the traveler at the left side of the tower strut and the first final stay cables anchored in the tower head.

The precast elements were transported to site on a pontoon, Fig. 6.31, and lifted with the help of a traveler to the already erected beam, Fig. 6.32.

Before closing, the joints were troweled with a two-component epoxy resin, Fig. 6.33. The uniformly distributed stress bars were threaded into their ducts and coupled with threaded sleeves. Some reinforcement bars cross the joint, welded to the short rebars protruding into the pockets on both sides of the joint. (This welding of rebars proved to be tedious and was not repeated for later bridges. Instead, the number of stressed bars was increased.)

After shifting the new precast element against the already installed ones the stress bars were post-tensioned, Fig. 6.34, so that a uniform pressure of 0.5 N/mm^2 ensured a tight fit of the joint and hardening of the epoxy under favorable conditions.

Lifting traveler: From early on it was planned to use some slab lifting equipment from building construction readily available in the USA to lift the element from the pontoon, Fig. 6.35 and 6.36. The hydraulic jacks with the pull rods rested longitudinally movable on the two traveler main girders which were tied back by two erection cables to the tower strut, cf. Fig. 6.30. Without these erection cables the previously installed stay cable and the already installed beam would have both been overloaded during lifting.

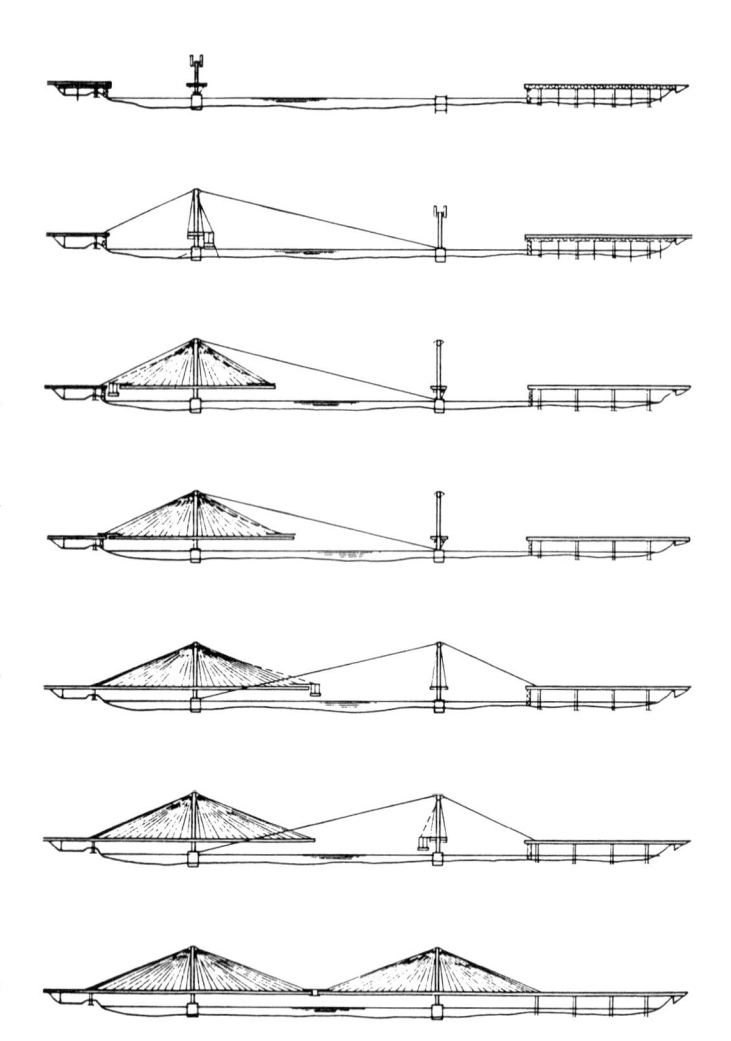

Figure 6.28 Temporary tower stays

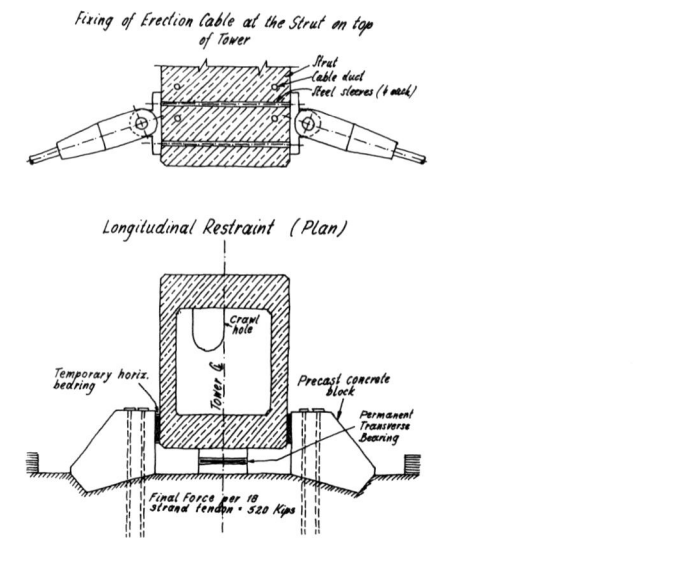

Figure 6.29 top: Connection of the traveler erection cables at the tower strut; bottom: Horizontal beam restraint at the tower

6.1 Cable-stayed concrete bridges with precast beams – Pasco-Kennewick Bridge

Figure 6.30 Temporary and final stay cables

Figure 6.32 Lifting of precast elements

Figure 6.33 Joint with epoxy and stressed bars

Figure 6.34 Post-tensioning of the stressed bars

Figure 6.31 Transportation on water

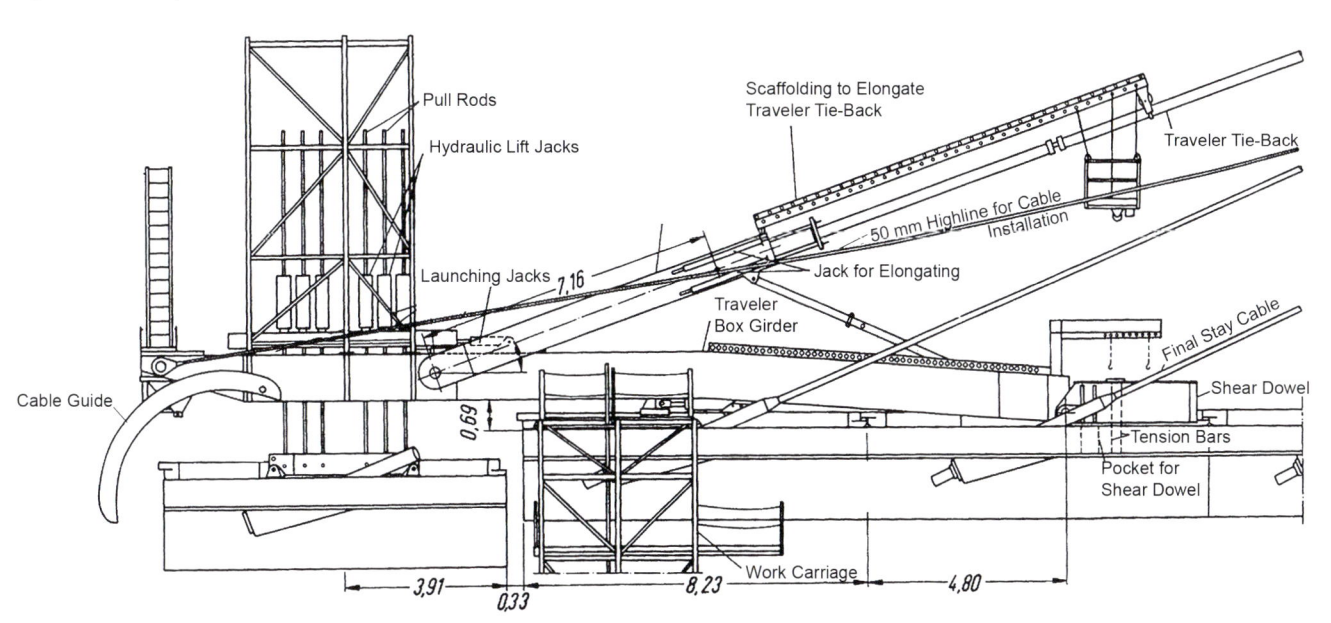

Figure 6.35 General layout of lifting traveler

The weight of the precast elements was transmitted in tension via the erection cables to the tower heads and in compression via the traveler main girders and shear corbels into the beam. The horizontal component of the erection cables was carried back to the approach beam, and the compression in the beam was carried via the temporary corbels at the towers, Fig. 6.29, into the foundation. These temporary corbels were active until the side span joint was closed.

During installation of the final stay cables the weight of the precast elements was stepwise transferred from the erection cables to the final cables by stressing the stay cable jacks and releasing the erection cable jacks. The criterion for the limit of each step was the change of inclination of the lifting traveler which was limited to ±1°.

For forwarding the lifting traveler the erection cables were disengaged from their jacks and elongated by inserting additional cable lengths. The lifting traveler itself was forwarded by jacks acting between the rails and spur racks. In the new position the shear corbels were anchored into their pockets in the beam. Finally the erection cables were restressed in such a way that the traveler was lifted off its front supports.

The precast elements were hung into the tie rods and these were 'simultaneously' (difference criterion ±1° change of inclination of the traveler) stressed. After the precast element had reached its final level it was shifted on heavy-duty rollers for joint closure together with the lift slab equipment.

Joint closure: The free-cantilevering took place from the towers outwards to both sides, Fig. 6.37, until the side span joint was closed by cast-in-place concrete, Fig. 6.38. The lifting of the last precast element is shown in Fig. 6.39.

The closure joints to the side spans and at the bridge center were closed with cast-in-place concrete. The forms were hung from the top, and the girders of one traveler were stressed across the joint as a splint so that the strains on the fresh concrete from relative movements between the two beam tips were reduced to a minimum, Figs 6.40 and 6.41. Casting took place at midnight, at dawn an initial post-tensioning was applied.

Cable installation: The completely shop-fabricated parallel wire cables in their PE pipes were delivered on reels to site, Fig. 6.42.

Stay cables were pulled up with a pull rope, supported by movable hangers from a highline as erection cable, Fig. 6.43. In front of the tower head the position of the anchor head was adjusted with the top hanger so that it could be pulled into its final anchorage.

Since PE pipes become brittle at low temperatures, the reels were placed in cubicles on the pontoons in which the air was heated during delivery to site. Then the PE pipes were pulled up via so-called 'bananas' to ensure a minimum bending radius, Fig. 6.44.

In order to limit the curvature of the PE pipes during lifting, they were supported by hangers from a highline, running on reels, connected by a pull rope, Figs 6.45 and 6.46.

Figure 6.36 Precast element at final level

Figure 6.38 Side span joint closure

6.1 Cable-stayed concrete bridges with precast beams – Pasco-Kennewick Bridge

Figure 6.37 Begin of free cantilevering

Figure 6.39 Lifting of last precast element

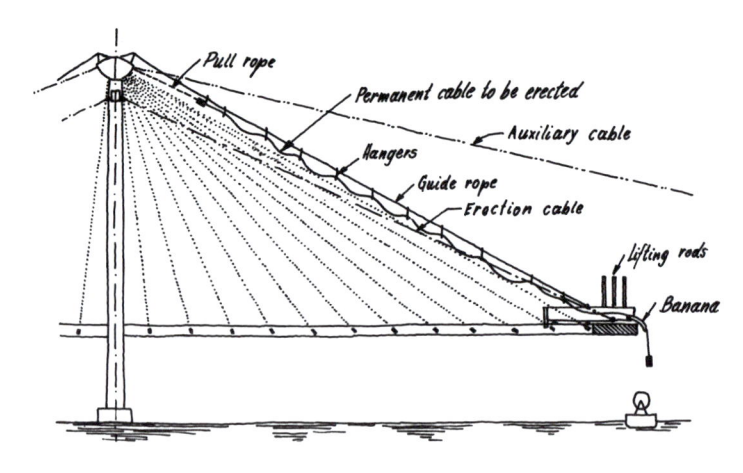

Figure 6.43 Stay cable installation

Figure 6.40 Center joint with splint

Figure 6.41 Traveler main girder across center joint

Figure 6.44 Begin of cable installation

Figure 6.42 Stay cables on reels

6.1 Cable-stayed concrete bridges with precast beams – Pasco-Kennewick Bridge

Figure 6.45 Supporting the cables during lifting

Figure 6.46 Hangers from highline

Figure 6.47 Cable anchor heads inside tower head

Figure 6.49 Steel pipe cast into beam

Figure 6.48 Cable anchorage underneath beam

Figure 6.50 Cement grouting of PE pipe

Fig. 6.47 shows the cable anchor heads inside the steel tower head. The lower anchor heads are supported by corbels underneath the beam, Fig. 6.48. They are guided to their anchorages through cast-in steel pipes, Fig. 6.49.

The lower anchor heads were led to the upper end of the cast-in steel pipe by adjusting the sag of the highline into such a position that the tension rod of a 10 MN center hole jack could be threaded-in. The jack was supported by a carriage underneath the beam, Fig. 6.39, which was also used for applying epoxy resin and to post-tension the stress bars.

The center hole jack pulled the cable anchor heads to the front of the lower anchorage where they were secured in place by shims. The hangers became slack and were pulled back for the installation of the next cable.

In order to create compression in the second joint back, most cables had to be installed longer than required in the final stage, Fig. 6.20. Only after two further beam elements were installed were the cables stressed to their final lengths.

After installation, the PE pipes were grouted for corrosion protection, Fig. 6.50.

The required grout pressure at the top was ensured by stand pipes, Fig. 6.51, through which the grout was wasted.

The grout had a water/cement ratio of 0.38 and contained 1 % by weight of expanding agent. The same equipment was used as for grouting the tendons.

The cable grouting took place in three steps:
1. The lower 2 m in front of the anchor head
2. Up below upper anchor heads. During this stage the PE pipe was reinforced with circular straps in front of the already hardened concrete.
3. After setting of the grout, the final part in front of the upper anchor heads with the help of the stand pipes.

The biggest hoop stresses in the PE pipes came to about 9 N/mm^2, so that safety against bursting of the PE pipes during grouting and afterwards came to at least 1.4 – compare Section 4.2.1.

For controlling the grout, a window was cut into the upper end of a PE pipe, Fig. 6.52: The PE pipe was tightly filled with grout also at this critical location.

For coloring, the black PE pipes were wrapped with self-adhesive off-white tape, Fig. 6.53. The off-white tape improved the appearance of the bridge considerably, compare Figs 6.54 to 6.58.

Construction by geometry

A large bridge from precast elements and shop-fabricated cables with well-known predetermined lengths is advantageously constructed by geometry (as opposed to construction by force). The measurements of forces and deformations during construction serve only as controls. If the weights and the lengths of all individual members and the assumptions for shrinkage and creep are correct, then the bridge with its desired shape and internal forces must result, independently of the selected construction method and intermediate construction steps. (The shrinkage and creep effects are partly dependent on the construction sequence). On the other hand, deviations from the theoretical geometry of the individual members result in a final bridge stage in which the desired geometry and action forces cannot be reached together. Therefore, the individual members have to be fabricated as accurately as possible or their deviations have to be measured precisely in order to permit appropriate corrective measures. During each erection stage the results of several control measurements were compared with the theoretical values. In the following some result are outlined.

Force measurements: The weight of each of the precast elements was measured by the precisely calibrated draft of the transport barges. On average, the elements were 3 % heavier than assumed. The measured weight deviations were placed as additional loads into the computer system for the final stage. By adjusting the thickness of the shims underneath the cable anchor heads the cable lengths were adjusted accordingly so that the final geometry with new cable forces was achieved at the end of construction.

The cable forces during installation were measured by the calibrated jack gauges and later controlled by measuring the eigenfrequency, see Section 3.9.5. The differences between the actual and theoretical values were within the measurement tolerances, see Section 5.2.3.4. Therefore, no re-stressing of stay cables was required at the end of construction.

Geometry controls: The thickness of the joints was measured by the distance between two bolts cast into the precast elements. It was determined that their distance immediately after concreting and at the end of construction increased on average by only 0.6 mm, which means that the additional thickness of the epoxy layer in the joints was nearly equal to the thickness of the cement skin and the separating agent, which were both removed by sandblasting. The shim thicknesses were adjusted accordingly.

During construction the theoretical and actual beam levels tallied well initially. Only towards the end did the beam become higher than assumed – at the center joint about 17 cm.

This geometry deviation could be reconciled with the following deviations against the theo-retical assumptions.
- The modulus of elasticity of the beam concrete was not 35 000 MN/m^2 as given by AASHTO, but for the average 28-day cylinder compression strength of 51.3 MN/m^2 it came to about 39 000 MN/m^2 due to the use of high-quality crushed granite aggregate.
- The shortenings due to shrinkage and creep were smaller than assumed, equal to code assumptions for a humidity of 90 %.
- The towers shortened by only $2/3$ of the assumed values due to the higher concrete strength and lower foundation settlement.

6.1 Cable-stayed concrete bridges with precast beams – Pasco-Kennewick Bridge

Figure 6.51 Stand pipes at tower heads

Figure 6.53 Wrapping of the PE pipes

Figure 6.54 Wrapped PE pipe

Figure 6.52 Control window in PE pipe

The resulting changes in gradient in the bridge center were

Higher concrete modulus of elasticity	1.4 cm
Smaller shrinkage	5.9 cm
Smaller creep	8.1 cm
Smaller tower shortening and pier settlement	2.0 cm
Total	17.4 cm

These geometry deviations were taken into account by the installation of the final cables only in so far that the angular deviations between the two beam tips before joint closure were taken out; otherwise the beam was left slightly too high.

The horizontal deviations of the beam axis remained within the small range of + 2 cm and – 0.4 cm.

6.1.2.3 Completed bridge

Figs 6.55 to 6.58 show the completed Pasco-Kennewick Bridge. Also because of its beauty, the bridge has won many important prizes in the USA, including the Presidential Award.

Participants

Client Federal Highway Administration Washington D.C. with Dr. Walter Podolny; Washington State Highway Department with Stewart Gloyd
Design and supervision on site Arvid Grant and Associates, Olympia, Washington, in collaboration with Leonhardt, Andrä and Partners, Stuttgart, Germany
General contractor Peter Kiewit Sons Co., Vancouver, Washington with Mr Martin Kelley
Cable fabricator The Prescon Corporation, San Antonio, Texas
Wind tunnel tests National Aeronautical Establishment, Ottawa, Canada with Mr R. L. Wardlaw
Cable testing University of Texas, Department of Engineering, Austin, Texas, with Prof. John Breen

Figure 6.55 Longitudinal view

Figure 6.56 Partial lateral view

6.1 Cable-stayed concrete bridges with precast beams – Pasco-Kennewick Bridge

Figure 6.57 Overall view

Figure 6.58 Bridge at night

6.1.3 East Huntington Bridge
6.1.3.1 General design considerations
The bridge crosses the Ohio river between the States of Ohio and West Virginia, Fig. 6.59, with a main span of 274 m and one tower only. The structurally equivalent system with two towers would have a main span of about 466 m which would have been a record at the time of construction. The design for this concrete alternate was only assigned after the design for the steel alter-nate was completed and its main foundations already built, Fig. 6.60.

In order to design a concrete bridge for the foundations of the lighter steel alternate, steel cross girders, high-strength concrete B 56 for the precast beam and 42 for the tower were used, Fig. 6.61.

The concrete alternate was bid in 1981 for 29 % less than the steel alternate with an orthotropic deck.

6.1.3.2 Construction
Approach bridge

The approach bridge on the West Virginia side was built as a haunched girder by free cantilevering, Fig. 6.62, in order to carry part of the loads in the main span directly via bending in the beam to the approach bridge and not via cables and towers to the hold-down pier which would have been more expensive.

Tower construction

The stay cables were connected in the tower head by overlapping, Fig. 6.63. In this way the tensile forces from main and side span are directly connected in compression. The geometry of the cable anchorages and the tower inspection ladders and platforms were controlled during the design phase with the help of a model, Fig. 6.64.

The reinforcement for the tower head was pre-assembled for each concreting step, Fig. 6.65. In order to ensure the precise location of the steel pipes for the cable anchorages, the pipes were fixed on the ground to the reinforcement cages. (For later bridges the steel pipes were fixed to cages from steel sections, for example for the Helgeland Bridge). Today composite anchorages with an inner steel box are used, for example for the Normandy Bridge.

The steel cages were connected to the already built tower in their precise location, Fig. 6.66.

Beam fabrication and erection

For the free cantilevering of the beam with 250 t precast elements the tower was again tem-porarily stayed forward and backwards, Figs 6.67 and 6.68. During construction, a barge ran into the forward stay, but no serious damage was done. (For later bridges, however, the temporary forward stays were avoided, for example for the Posadas-Encarnación Bridge).

The cross-section for the East Huntington Bridge was simplified compared to that of the Pasco-Kennewick Bridge, Fig. 6.69:
– Instead of the triangular outside boxes, solid main girders were used.

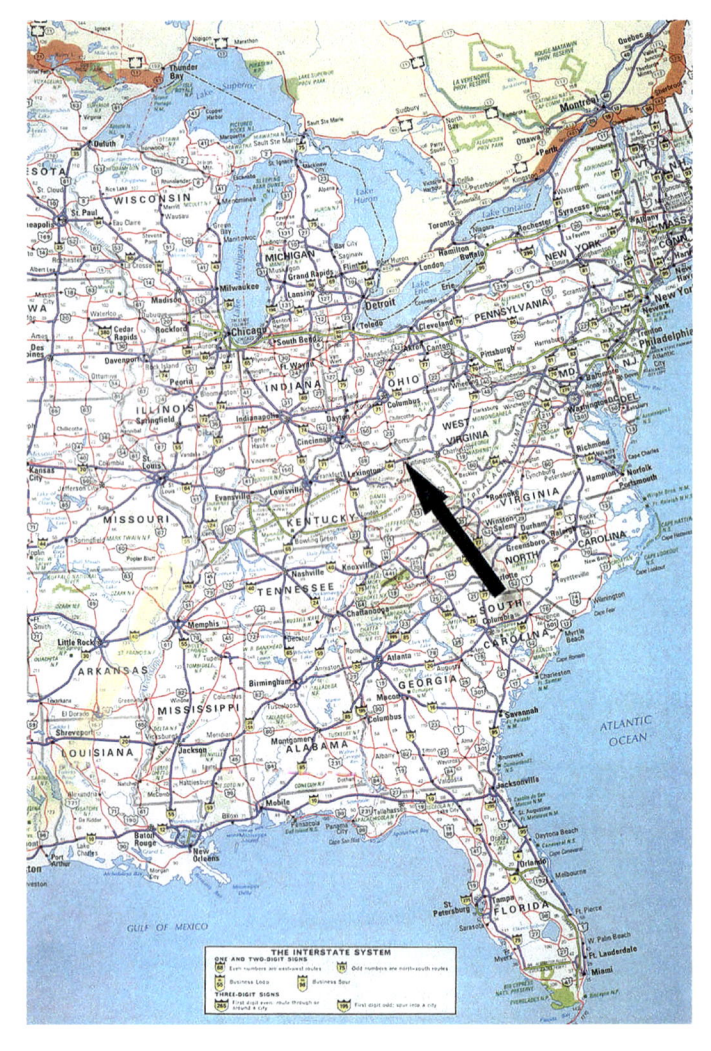

Figure 6.59 Bridge location

Figure 6.60 Already completed piers for the steel alternate

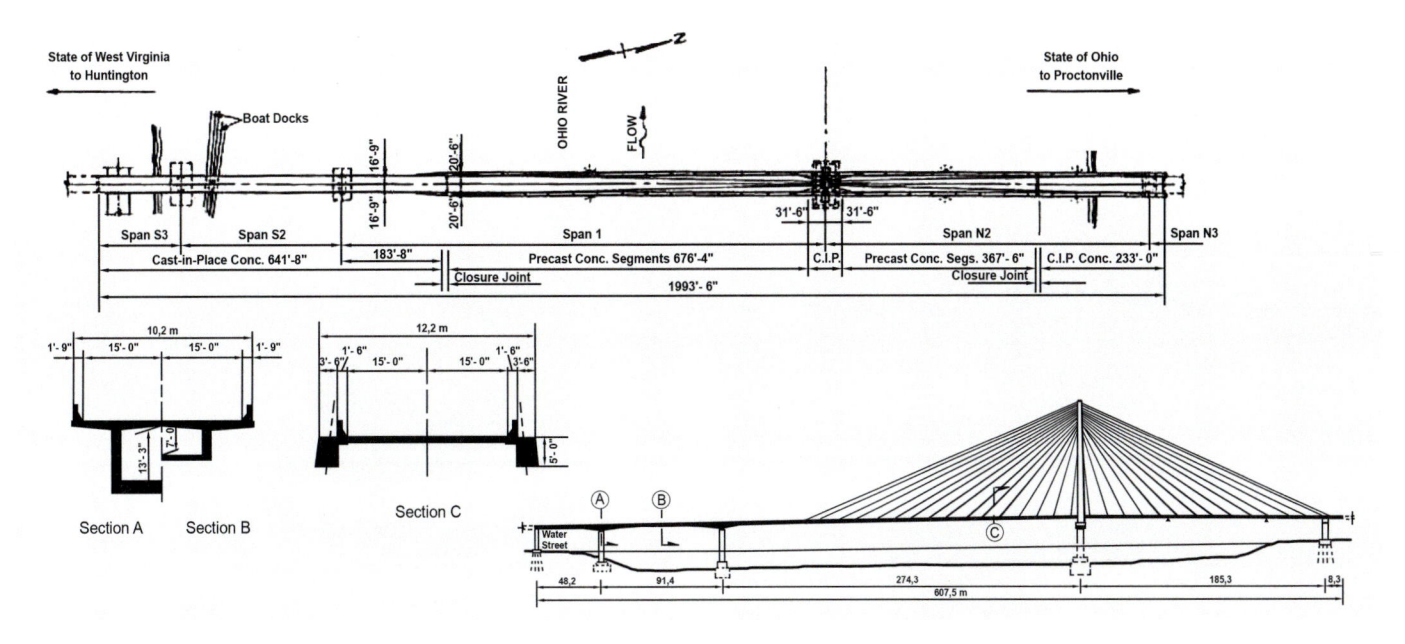

6.1 Cable-stayed concrete bridges with precast beams – East Huntington Bridge

Figure 6.61 General layout

Figure 6.62 Free cantilevering of approach span

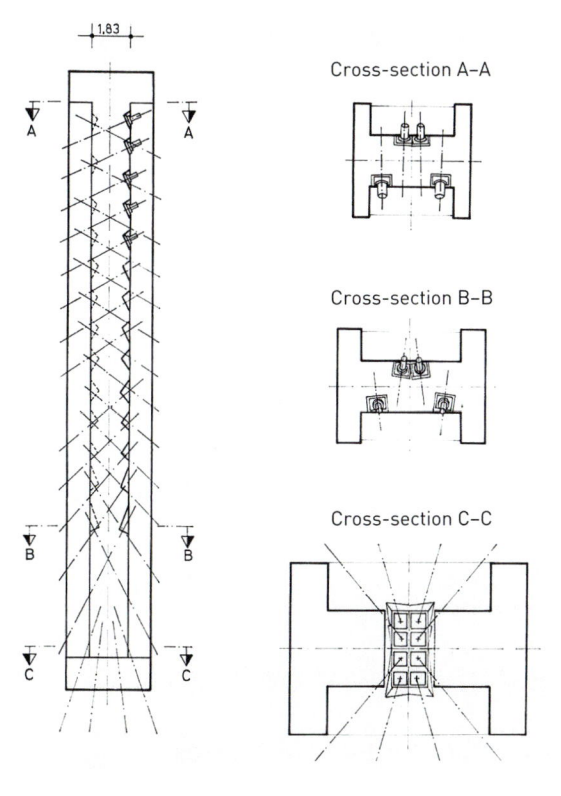

Figure 6.63 Cable anchorage at tower head

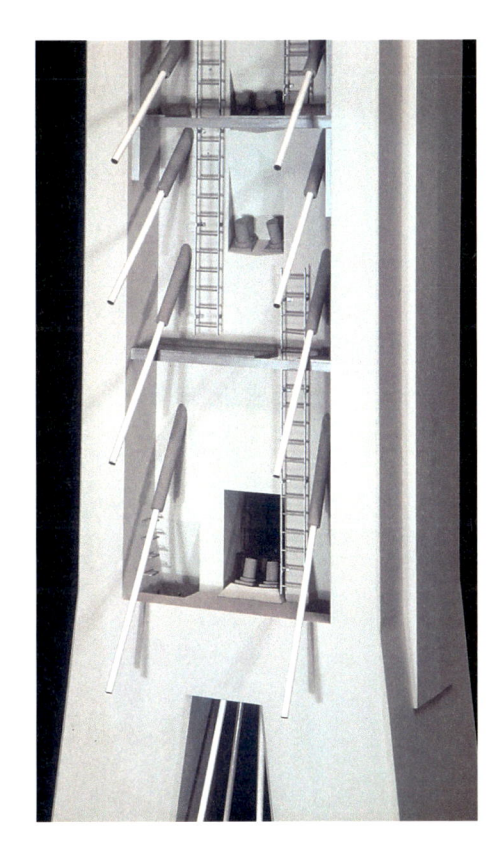

Figure 6.64 Model picture

Figure 6.65 Pre-assembled rebar cage with steel pipes

Figure 6.66 Installed rebar cage

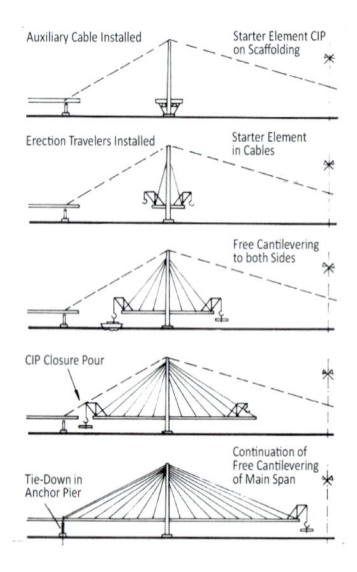

Figure 6.67 Erection sequence with temporary tower stays

Figure 6.68 Stayed tower at the begin of beam construction

6.1 Cable-stayed concrete bridges with precast beams – East Huntington Bridge

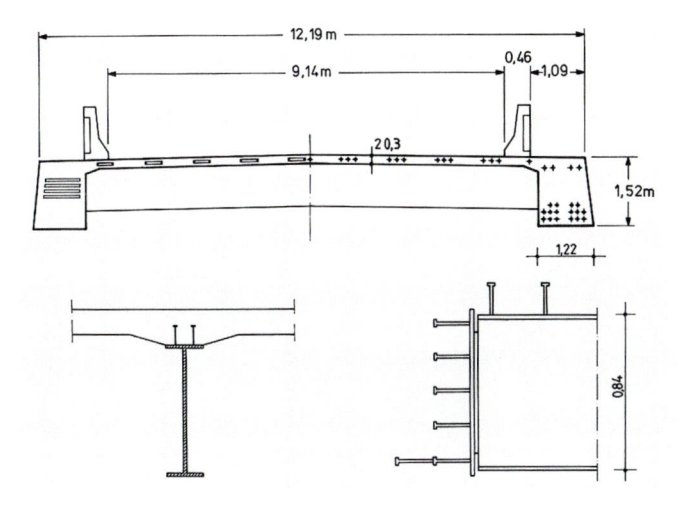

Figure 6.69 Beam cross-section

Figure 6.70 Precast element

Figure 6.71 Floating crane

Figure 6.72 Lifting of final element

– The concrete cross girders were replaced by steel cross girders, Fig. 6.69.
– Instead of the usual concrete B 35 'high-strength' concrete B 56 was used – quite unusual for the USA.
– Instead of the steel dowels the front ends of the main girders were profiled for shear transfer, Fig. 6.70.
– The precast elements were cast in a long line (all elements together) instead of a short line (one after the other).

The precast elements were not lifted by a traveler on the beam but with a floating crane. The lifting of the final element is shown in Fig. 6.71.

Cable Installation

At the beginning of the cable installation a tie rod was threaded into the inner thread of the lower anchor head, Fig. 6.73. The upper anchor head was lifted with the tower crane to the tower head and pulled into the anchorage pipes with grip hoists, Fig. 6.74.

The tie rod at the lower cable anchor head was initially pulled into the steel pipe with two strands, Fig. 6.75.

Once the tie rod had reached the jack located on a jack chair underneath the beam, Figs 6.76 and 6.77, the anchor head was pulled into its final position and the washers were installed. The actual cable length at this stage was determined by the thickness of the washers.

6.1.3.3 Completed bridge

The A-tower forms a tent-like space in the driver's view, which provides a certain feeling of security, Fig. 6.78.

In the skew view from the water the advantage of an A-tower is apparent, in that the stay cables in a fan arrangement hardly intersect visually, Fig. 6.79.

Participants

Client Federal Highway Administration Washington D.C. with Dr. Walter Podolny; Washington State Highway Department with Stewart Gloyd
Design and supervision on site Arvid Grant and Associates, Olympia, Washington, in collaboration with Leonhardt, Andrä and Partners, Stuttgart, Germany
General contractor Peter Kiewit Sons Co., Vancouver, Washington with Mr Martin Kelley
Cable fabricator The Prescon Corporation, San Antonio, Texas
Wind tunnel tests National Aeronautical Establishment, Ottawa, Canada with Mr R. L. Wardlaw
Cable testing University of Texas, Department of Engineering, Austin, Texas, with Prof. John Breen

Figure 6.73 Lifting of stay cable

Figure 6.74 Introduction of stay cable into the tower head

Figure 6.75 Pulling the tie rod near the beam anchorage

6.1 Cable-stayed concrete bridges with precast beams – East Huntington Bridge

Figure 6.76 Stressing jack underneath the beam

Figure 6.77 Pulling the tie rod into the steel pipe

Figure 6.78 Driver's view

Figure 6.79 View from the river

6.2 CIP concrete cable-stayed bridge – Helgeland Bridge
6.2.1 General layout
6.2.1.1 Introduction

The Helgeland Bridge is a slender cable-stayed concrete bridge with a main span of 425 m. The aerodynamically shaped beam has a depth of 1.2 m and is 12 m wide. The towers are founded on rock in 30 m depth. The bridge is exposed to severe storms with gusts of up to 77 m/s wind speed.

For the design of the governing loading 'wind', an analytical time-history investigation in the ultimate limit state was performed, which took into account the aerodynamic damping as well as geometrical and material non-linearities. The bridge was built by free cantilevering from the towers outwards to both sides and was opened after a two-year construction period in July 1991.

The beam was built with CIP concrete, which has the advantage that no heavy precast elements have to be transported and lifted. A disadvantage is that CIP construction requires about one to two weeks for each new beam section, whereas precast elements permit a construction progress of one to two elements per week.

In order to reduce the construction period, long beam sections of 12 m, equal to the cable distance, were used. During casting the unsupported weight of the new section would have caused too high bending moments at the tip of the already completed beam. Therefore, the formwork carriage had to be tied back. If auxiliary tie-backs are used, their elongation for the next beam section is tedious and time-consuming. For the Helgeland Bridge the final cables were used to support the form traveler during casting. They were anchored in precast elements as part of the final beam which were bolted to the form traveler. A detailed description of the bridge is given in [2.80].

The Helgeland Bridge is located on the west coast of Norway on the Arctic Circle near the city of Sandnessjöen, crossing the Leirfjord between the island of Alster and the mainland, Fig. 6.80.

The geology in the surroundings of the bridge consists of granite which is partly eroded by ice age glaciers. The fjord is up to 130 m deep, nearly 400 m wide and has steep slopes at both sides, Fig. 6.81.

The initially proposed main span of 400 m had to be increased to 425 m in order to remove the tower foundations sufficiently from the fjord edges to safely avoid sliding of the granite slopes.

The Gulf Stream prevents very low temperatures on site, but a real problem is the frequently occurring severe storms. Wind measurements taken on site over several years yield the wind characteristics outlined in Table 6.1.

In addition to the high gust speed of 77 m/s at beam level (+ 50 m) the turbulence intensity of 21 % is remarkable, caused by a mountain range close by (the 'Seven Sisters') in the main direction of wind.

Ship traffic to the industrial harbor of Mosjöen in the hinterland requires a vertical navigational clearance of 45 m. The towers were designed for ship impact with an equivalent static load of 5 000 t. For live load the Norwegian codes require a 600 kN truck in addition to a uniformly distributed load of 3 kN/m².

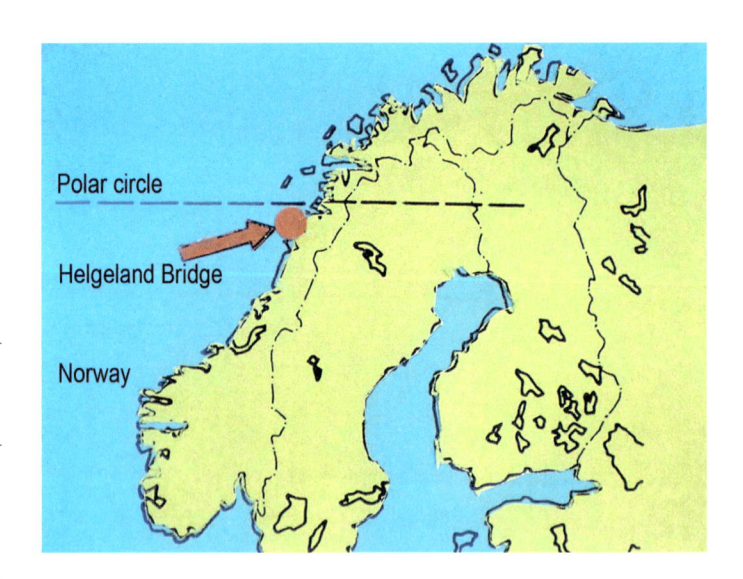

Figure 6.80 Bridge location

Figure 6.81 Geology

Table 6.1 Wind characteristics

Wind speed	Height + 10 m	Height + 50 m
50-year-probability, final stage		
10 min. mean	40 m/s	50 m/s
Gust (3–5 s)	60 m/s	70 m/s
10-year-probability, construction stage		
10 min. mean	36 m/s	45 m/s
Gust	54 m/s	63 m/s

Turbulence intensity: $I = \sigma/V$ for 10 min. mean
  Horizontal: $I_V = 25\%$
  Vertical: $I_H = 10\%$

6.2 CIP concrete cable-stayed bridge – Helgeland Bridge

Figure 6.82 Layout

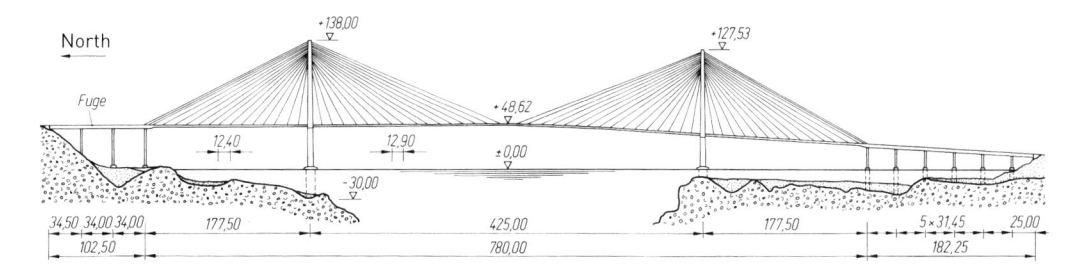

Regular high-strength concrete B 65 was used throughout. As alternates the use of light-weight concrete LB 65 was investigated, which is often used in Norway for free cantilevering construction, as well as a composite beam cross-section. It turned out that the additional beam costs for the alternates outweighed the savings in cable steel. Also, more loads would have been required for the foundations.

As a result, the layout shown in Fig. 6.82 was selected as the most economic solution. After an international tender the construction was awarded in April 1989 for about € 25 million to a Norwegian contractor.

6.2.1.2 Bridge system

The topography required a main span of 425 m with appropriate side spans of 177.5 m. The low traffic level at this location near the Arctic Circle needed only two traffic lanes and one walkway. This led to a 12 m wide beam with a transverse slenderness of 1 : 36.

The requirements for low wind resistance, aerodynamic stability and suitability for CIP construction led to an open cross-section with two solid edge girders and a beam depth of 1.2 m, giving a vertical slenderness of 1 : 354.

The towers are A-shaped above the roadway in order to increase the torsional stiffness by coupling the two tower legs. Below the beam the tower legs merge on top of the single foundations.

The beam is continuous between joints in the second approach span, Fig. 6.82, and is monolithic with all slender side span piers. At both towers the beam rests on 22 cm deep neoprene bearings which can deform horizontally. In this way the breaking forces and temperature changes of the beam are taken up by both towers simultaneously. Only small differences in tower action forces are caused by the different lengths of the approaches and tower heights. Both towers are therefore sized identically.

Bridge beam

For the very slender bridge beam, Fig. 6.83, partial post-tensioning in both directions was selected in order to ensure sufficient ductility. In the edge girders, four Type 1 continuity tendons were arranged over the full bridge length, and these were coupled at each construction joint, Fig. 6.84. At the end of beam construction additional continuity tendons, Types 2 and 3, were threaded into empty ducts and grouted in the center of the bridge and in the end regions.

Cross girders are located at the cable anchorage points at a distance of 12.9 m which contain the only transverse post-tensioning. The 40 cm thick roadway slab spans 12.4 m longitudinally and 7.5 m transversely between the main and cross girders. The sizing of the beam was governed by the following loadings:
a) permanent loads plus live loads
b) turbulentwind
c) loads during construction.

For loading a) the non-linear characteristics of the bridge (P-Δ effect) and the material (elastic-plastic) were taken into account. The

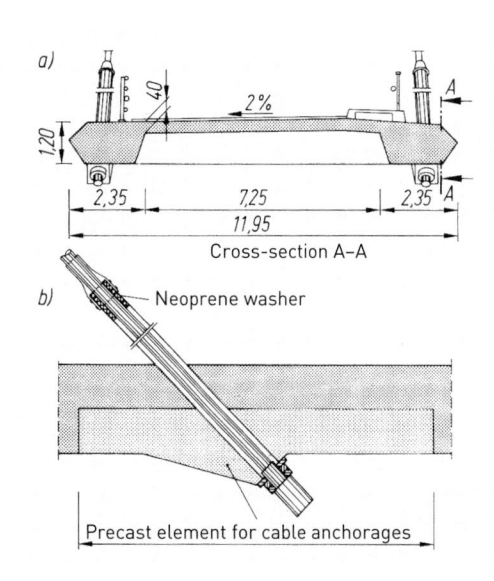

Figure 6.83 Beam; a) cross-section, b) sectional elevation

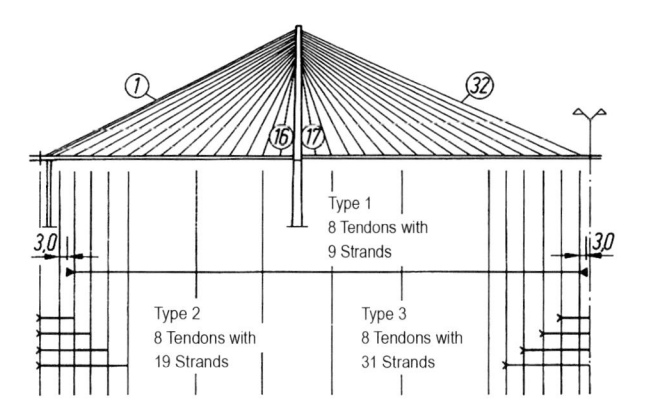

Figure 6.84 Post-tensioning of beam

increase of live load moments reached up to 50% against the linearly calculated moments because of the extreme slenderness of the beam (1:354), Fig. 6.85.

A typical arrangement of the reinforcement in the edge girders is shown in Fig. 6.86.

The governing loadings during construction for positive beam moments were stressing of the stay cables to final lengths and, for negative moments, the situation after forwarding the formwork carriage, before installation of the corresponding stays. The envelopes of the governing beam moments in the final stage and during construction are given in Fig. 6.87.

It was taken into account that the compression forces in the beam during construction in the most forward beam section are caused only by the first pair of stay cables. The sizing was done in the ultimate limit state. Under service conditions the crack width was limited to 0.2 mm generally and to 0.1 mm at the tendons.

This is true under 60% of live load as well as for wind alone, or during all construction stages. At the same time, the steel stresses in the cracked tensile zone are limited to 190 N/mm². The chosen post-tensioning is sufficient to avoid longitudinal tensile stresses under permanent loads plus frequently occurring live loads.

The four tendons with nine 12 cm strands in each edge girder (Type 1) were mainly required during construction. This partial pre-stress permitted the optimization of the amount of reinforcement for the final stage and during construction. The arrangement of reinforcement shown in Fig. 6.84 allowed good accessability for concreting and compaction as well as simple and economic lap-splices and easy arrangement of the tendons.

The stay cables rest on corbels underneath the edge beams and run through cast-in steel pipes to the tower heads, Fig. 6.83. This type of stay-cable anchorage has proved to be very expedient since the opening of the Pasco-Kennewick Bridge, USA, in 1978.

Towers

For the towers, H-, A- and diamond-shapes were investigated. The diamond shape, as shown in Fig. 6.88, was chosen because of economy, aesthetics and improved torsional resistance of the beam. The towers are stiff transversely, and in the longitudinal direction the tower head is restrained by the backstay cables. Because independent movements of both A-tower legs are prevented, the influence of the backstay cables during rotational oscillations of the beam is omitted and the torsional frequency is thus importantly increased.

During construction auxiliary piers were required in both side spans in order to reduce the weather-vaning horizontal moments on the tower by the beam, created by differential wind on the two cantilevers. These auxiliary piers were stressed with tendons onto the rock. The connection with the beam was vertically sliding in order to avoid moment peaks in the beam.

The tower legs are solid underneath the beam to provide sufficient downloads for the solid foundations in up to 30 m water depth.

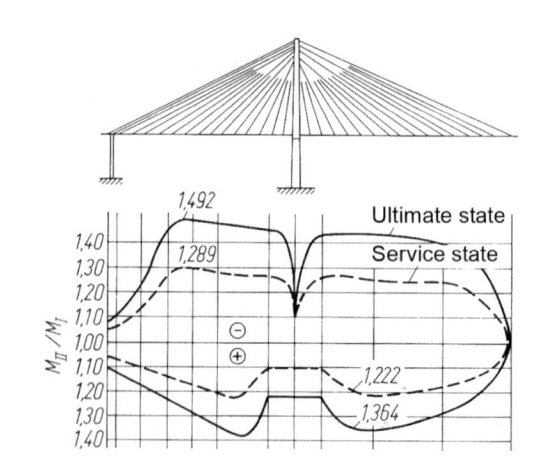

Figure 6.85 Increase of live load moments due to non-linearity

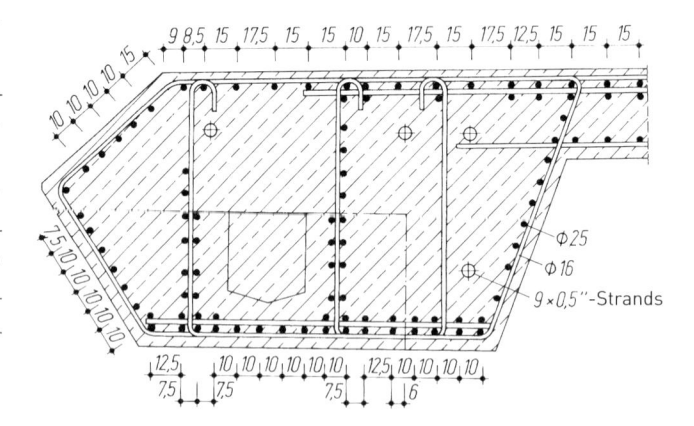

Figure 6.86 Typical edge girder reinforcement

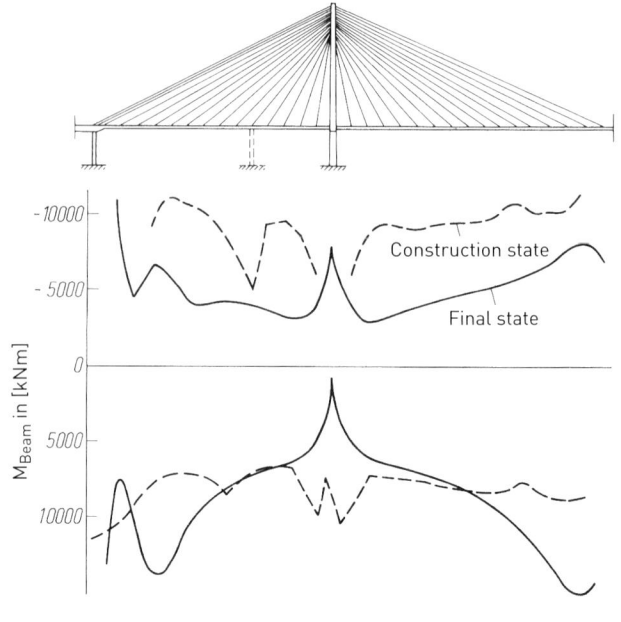

Figure 6.87 Governing moment envelopes for beam

Above the beam they have box girder cross-sections with 40 cm wall thickness. The stay cables are anchored inside the concrete tower heads, Fig. 6.89.

The horizontal cable components are carried by loop tendons, which also permit exchange of any single stay cable. This system was first used for the Baytown Bridge across the Houston Ship Channel in Texas, [1.17]. A test with a full-size model showed that by using thick, smooth and lubricated pipes for tendons the usual friction coefficients were not exceeded even in the regions of high curvature with a radius of 0.8 m. In this way, expensive additional transverse post-tensioning is avoided. Vertical cracks did not occur.

The loop tendons and the transverse reinforcement were sized in the ultimate limit state by a strut-and-tie model, and, in addition, for bending and normal force during service. The cable forces are spread transversely into the longitudinal box walls, Fig. 6.90:
- one half of the cable force is transmitted by two struts into the loop anchorages, Fig. 6.90 a;
- one half of the cable force is taken up by the radial forces of the loops, Fig. 6.90 b.

Stay cables

The stay cables were sized in accordance with the PTI Guidelines [3.47] for galvanized wires, ø 7 mm, steel 1450/1650.

The 4 × 32 stay cables with lengths between 64 m and 225 m required between 67 and 231 wires. The governing loading was permanent loads plus live load for a permissible stress of 0.45 ultimate under service conditions.

The wires are covered by PE pipes, Fig. 6.91, and end in anchor heads with a cold casting from steel balls and epoxy resin, called HiAm, Fig. 6.92.

Cement grouting of the PE pipes on site was not possible because the temperatures would have been too low. Therefore, a petroleum wax was injected into the PE pipes during shop fabrication. This material is flexible enough to permit coiling and uncoiling on reels, yet it is stiff enough to prevent the build-up of hydrostatic pressure after installation under direct solar radiation, which could cause bursting of the pipes during the life of the bridge [3.9].

Aerodynamic stability

The exposed location of the bridge with regard to wind, and its great slenderness, required an especially careful investigation of the wind response, both during construction and in the final stage, which was done to a large extent analytically.

An in-house developed time-history method was used, which simulated the wind in such way that not only the spectrum but also all the other characteristics were fulfilled such as
- the non-linear moment-curvature relation of the beam together with all interactions between horizontal, vertical and torsional moments

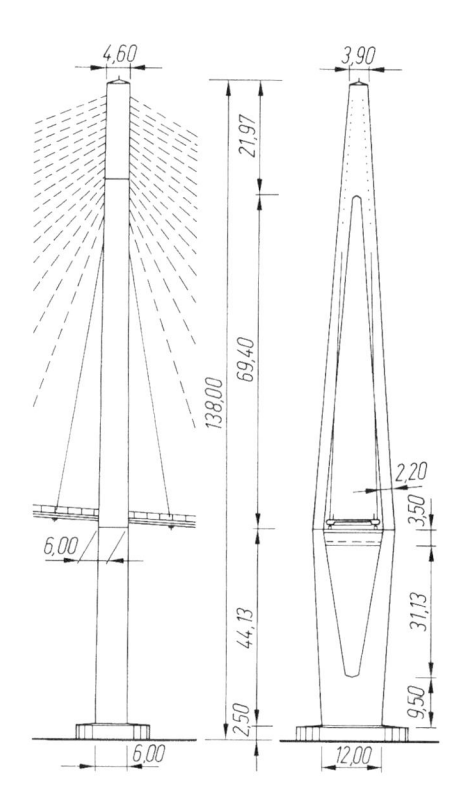

Figure 6.88 North tower

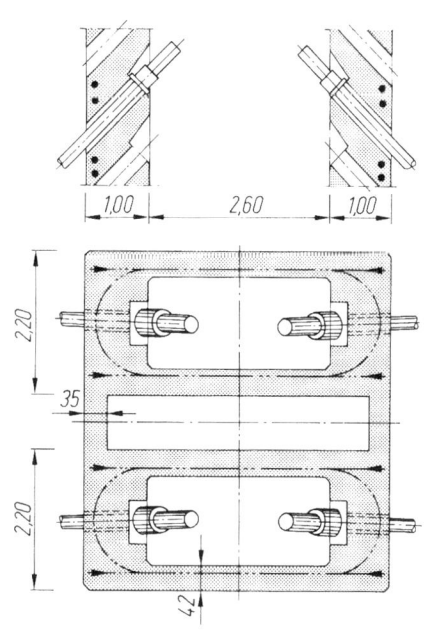

Figure 6.89 Cable anchorages in tower heads

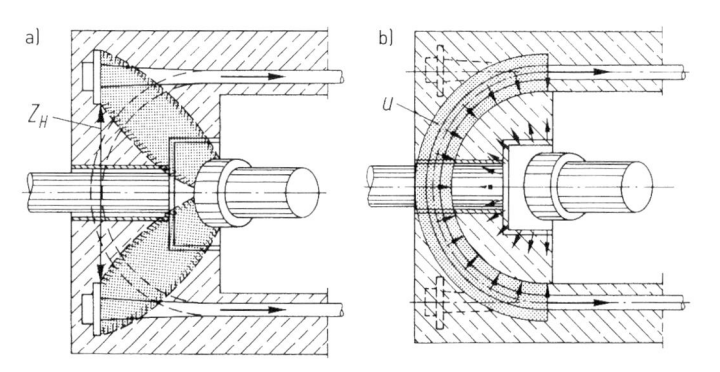

Figure 6.90 Strut-and-tie model
a) compression struts, b) loop deviation forces

- the non-linear stability effects (P-Δ effects). In this connection the tilting of the beam under transverse bending in the center of the bridge is important, where the restoring due to the cables is small because of their flat angle.

A detailed description of this calculation is given in [4.35].

Wind simulation

The wind speed at ultimate limit state is generated from those in service limit state by a $\sqrt{\gamma_\omega}$ increase. The 10-min mean at beam level for a 50-year probability in the ultimate limit state thus resulted in a wind speed $v_{(ultimate)} = \sqrt{1.6} \cdot 50 = 63.25$. The tender documents prescribed the turbulence spectrum in accordance with ESDU 1974, [4.26, 4.27]

Fig. 6.93 gives the turbulence components from the ESDU spectrum simulated by 1/12-octave-components. The exactness with which the turbulences were generated by 128 harmonic components was considered sufficient for the wind simulation.

Fig. 6.94 gives as an example of the development of the longitudinal wind components during a 3-second time period simulation. The individual lines show the momentary wind speed distribution in 0.2 second steps. Fig. 6.95 shows a temporary situation of the oscillating calculation model under the analytically created wind spectrum.

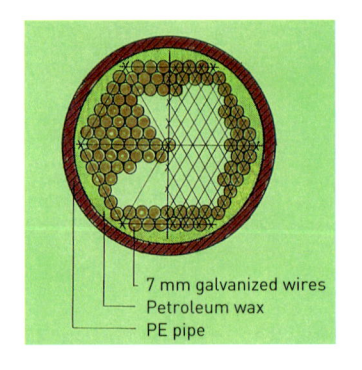

Figure 6.91 Cross-section of parallel wire cable

Figure 6.92 Cable anchorage

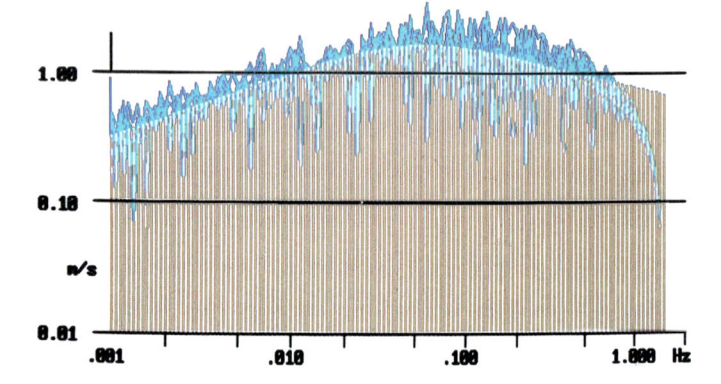

Figure 6.93 Wind speed spectrum simulation

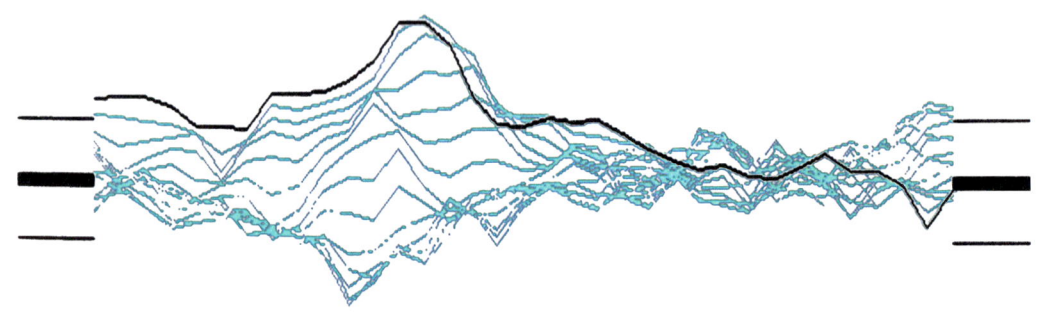

Figure 6.94 Longitudinal momentary with distribution

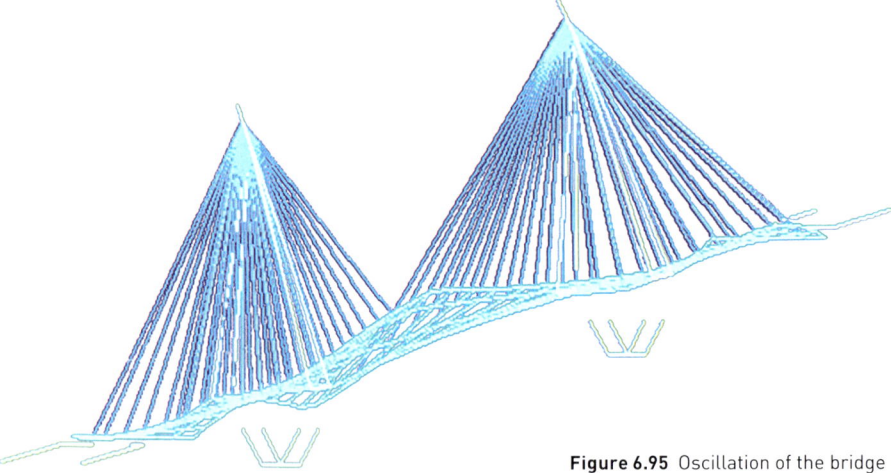

Figure 6.95 Oscillation of the bridge under dynamic wind load

Sizing

The non-linear effects were calculated for in a sizing program, taking into account the geometrical stiffness of the cable system with the chosen forces as well as buckling and tilting effects. The force–deformation behavior at the nodes was corrected continuously in an iteration process. In order to account for the revised reinforcement in each step a parallel program for skew bending was used for parametric studies. The resulting spring characteristics are closely enough reproduced in the non-linear moment–curvature diagram, Fig. 6.96.

All this resulted, for example, in the run of horizontal moments shown in Fig. 6.97, where the moments in the non-linear ultimate limit state are given in comparison to the γ_ω times moments of the linear system.

The high moments in the tower region during construction and the corresponding high amount of reinforcement there resulted in plastic rearrangement of moments which reduced the moments in the center of the span. This effect was further enhanced by the resulting reduction of the span reinforcement.

Windtunnel tests

The wind tunnel tests performed on an elastically supported, dynamic section models showed that aerodynamic instabilities are not probable. The same was true for vortex excitation. The tests with a full model in a boundary layer wind tunnel, Figs 6.98 and 6.99, showed as usual the largest amplitudes in turbulent wind.

The Helgeland Bridge is a very daring design with a vertical beam slenderness of 1:354 and a horizontal slenderness of 1:36 in an area with very strong turbulent storms.

The detailed aerodynamic investigation proved a sufficient safety not only for the final stage but also during construction. The bridge was indeed stable during construction, although the anticipated severe storms for a 50-year probability did indeed occur. The control measurements showed a good correlation between calculated and actual deformations – see Section 3.5.

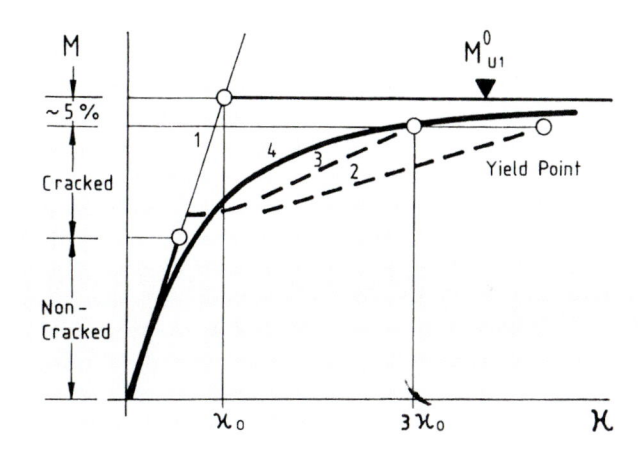

Figure 6.96 Non-linear moment – curvature diagram

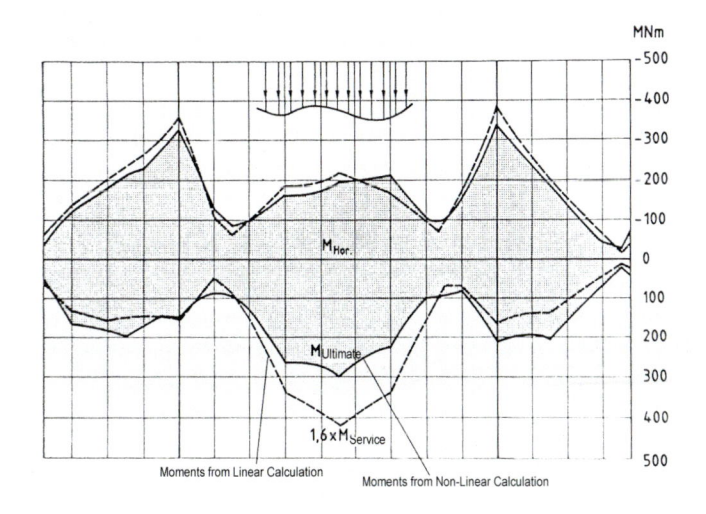

Figure 6.97 Envelope of the horizontal moments

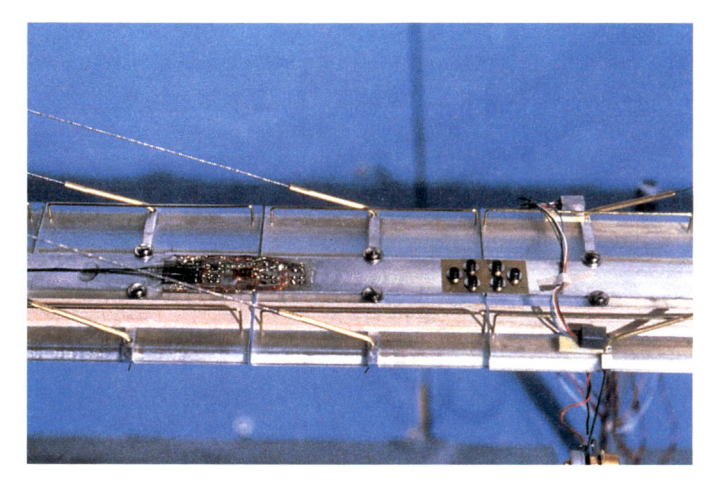

Figure 6.98 Detail of cable anchorage at beam

Figure 6.99 Full model for boundary layer wind tunnel

6.2.2 Construction
6.2.2.1 Climate
Due the location of the bridge on the west coast of Norway near the Arctic Circle, Fig. 6.80, difficult weather conditions had to be taken into account for the construction planning. The problem was not low temperatures, as these were prevented by the influence of the Gulf Stream, but severe storms which regularly occur during winter months. The free cantilevering for 210 m from each tower with a beam depth of only 1.2 m was a very daring undertaking. Fig. 6.100 shows a storm with wind speeds of up to 70 m/s (252 km/h) during free cantilevering which threw the spray up to beam level.

The contractor planned the job site installation in such a way that continuous work, partly around the clock, was possible even during strong winds. Large pontoons (40 m × 100 m) were anchored at each tower as working platforms. The batching plant with a capacity of 60 m^3/h, together with the aggregates and the cement silo, were located here. So not only the bridge but also the working pontoon with the equipment were endangered by the storms, Fig. 6.101.

6.2.2.2 Towers
The tower foundations were built 30 m under water with tremie concrete for which a special mix was developed to prevent segregation during batching. Foundation construction started with placing a base layer of concrete by divers on the prepared rock surface, Fig. 6.102. The precast elements for the underwater parts of the tower were cast on shore and floated to the site, Fig. 6.103.

The precast elements were lowered onto the prepared foundation with a floating crane, Fig. 6.104, stressed together under water and filled with CIP concrete. In this way, the required DW was created to prevent tension underneath the foundation and to create sufficient resistance against ship collision.

Above beam level the tower legs consist of concrete boxes with a wall thickness of 40 cm. They were concreted with sliding forms.

The steel forms were enclosed and provided with an insulating skirt including heating elements to protect the fresh concrete against low temperatures, Figs 6.105 and 6.106.

The progress of the sliding forms came to 1.5 m/day for the lower solid tower legs and reached up to 3 m/day above deck. In the cable anchorage region the tower were built in 3 m sections. Even in this difficult area with many built-in items, including the cable anchorages, sliding forms continued to be used, because the contractor considered it too risky to use a 3 m free cantilevering jumping form in strong winds.

The steel pipes with their support plates were welded in shop to steel frames which were lifted in 3 m sections to the tower head and bolted to the previously installed frame, Fig. 6.107. In this way the cable anchorages were safely kept in place during sliding.

The completed tower head shows the cast-in steel pipes, in which later the stay cables are anchored, and at the outsides the anchorages of the horizontal loop tendons, Fig. 6.108.

Figure 6.100 Storm during free cantilevering

Figure 6.101 Storm with working pontoon

6.2 CIP concrete cable-stayed bridge – Helgeland Bridge

Figure 6.102 Foundation under water

Figure 6.103 Floating of the precast tower leg elements

Figure 6.104 Sinking of foundation elements

6.2.2.3 Beam

The beam was designed from the beginning in such a way that it could be built CIP by free cantilevering from the towers outwards to both sides in sections of 12.90 m, equal to the cable distances. It was also planned during the design phase to support the travelers by permanent stay cables. Different types of construction and sequences were investigated, including the possibility of using water ballast in order to permit stressing of the cables to a minimum force before concreting. Finally a stiff traveler was used which was stressed onto the already completed beam. Its lateral trusses carried an excess cable force before casting, back to the completed beam, Fig. 6.109. Each traveler weighed about 115 t.

As the tower legs, the traveler had to be provided with an enclosure for protection of the fresh concrete against cold winds, Fig. 6.110. The additional wind load on this area would have created too high transverse bending moments for the beam. It was therefore decided to use light plastic sheets for protection, which would have been blown away by a severe storm. This illustrates that the beam was used to its limits during construction.

In order to fix the position and the direction of each cable anchorage precisely against the beam, the steel pipes with their base plates were cast into a short precast element, Fig. 6.111.

Fig. 6.112 shows the transport of a precast anchorage element with blockouts for the transverse tendons. (In later bridges the precast elements were avoided and the cable anchorage part of the beam was cast ahead in the form traveler. In this way the transportation was not required, but a longer construction period resulted.)

Each of these precast elements was bolted in its final position to the form traveler so that the final stay cables could be installed and preliminarily stressed to a low level to support the form traveler during casting. The horizontal compression component of the inclined stay cable was carried by a precast strut from the precast anchorage element to the bulkhead of the already completed beam, Fig. 6.113.

Fig. 6.114 gives a view into the form before casting. The steel pipe can be seen protruding from the precast element. The final stay cables have already been installed. Each precast element weighed 15 t and 24 t of reinforcement were placed before installation of the stay cables.

Fig. 6.115 gives a view from underneath the beam onto the lowered bottom forms during launching to the next section.

The approach bridges were built with a spanwise self-launching form, together with the towers, Fig. 6.116.

The bottom form of the traveler alone was supported directly from the towers for casting the beam starter section, Fig. 6.117.

After completion the travelers were assembled on the starter sections and free cantilever construction could proceed, Fig. 6.118.

It was necessary to build auxiliary piers at the quarter-points of the side span. These were to take the horizontal forces from unbalanced transverse wind on the beam, Fig. 6.119. The towers would not have been able to carry the weather-vaning moment by itself.

Figure 6.105 Sliding forms for tower legs

Figure 6.106 Insulating skirts for protection

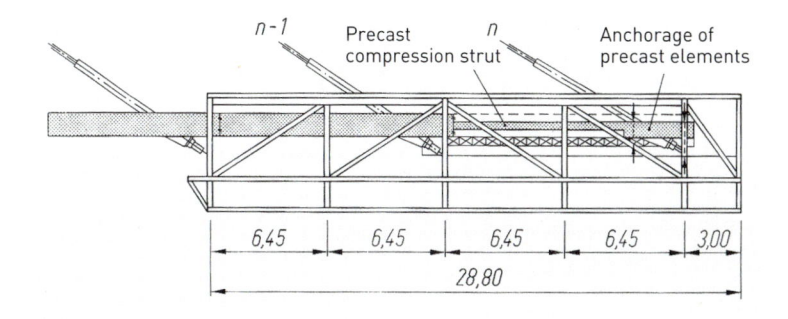

6.2 CIP concrete cable-stayed bridge – Helgeland Bridge

Figure 6.109 Traveler view

Figure 6.107 Pre-assembled cable anchorages in tower head

Figure 6.110 Traveler elevation

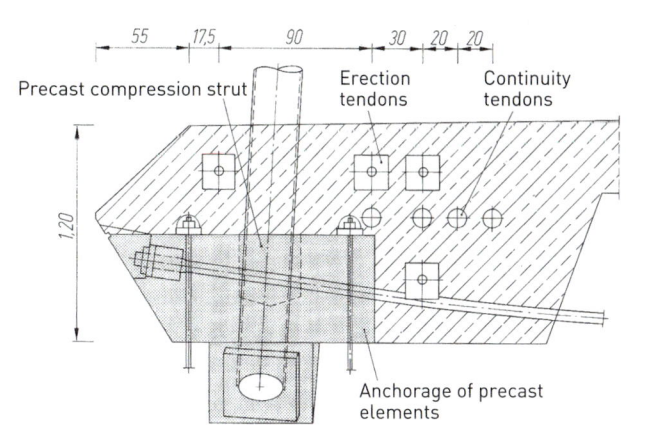

Figure 6.111 Cable anchorage in precast element at beam

Figure 6.108 Tower head with cable and tendon anchorages

Figure 6.112 Transportation of a precast anchorage element

The connection to the beam was vertically sliding in order to prevent the high moments created by a rigid support in between the elastic cable supports. At the end of construction the auxiliary piers were blasted and served as a new habitat for the fish in the fjord.

The free cantilevering took place only slightly staggered for both towers, Fig. 6.120.

Shortly before closure of the center joint with the longest cantilevers of about 210 m each, a severe storm took place again. The bridge stood firm and the measured beam deflections were as calculated, Fig. 6.121.

Finally one formwork traveler was moved across the center joint and the closure pour took place, Fig. 6.122.

The geometry and action forces were adjusted in such a way that, after shrinkage and creep, the desired gradient and run of moments appeared. The change in bridge geometry between opening for traffic ($t = t_1$) and after shrinkage and creep ($t = \infty$) is shown in Fig. 6.123.

6.2.2.4 Stay cables

Cable installation

The cables were transported on reels by ship from their place of fabrication in Zurich, Switzerland, to the site and lifted up to the beam, Fig. 6.124. The tower crane then lifted each cable anchor to the tower head where it was threaded into the steel pipe and anchored inside the tower head with shims, Fig. 6.125.

The lower cable head was pulled with grip hoists near the steel anchor pipe in the beam. Then a tensioning rod was screwed into the inside thread of the anchor head, Fig. 6.126. It can be seen clearly that the beam at this stage was not yet concreted, but that the reinforcement was already in place, see also Fig. 6.113.

The tensioning rod protrudes far enough beyond the underside of the beam that a center hole jack could be placed onto the jack chair in order to pull the cable into its final position, Fig. 6.127. At this stage the stay cable had a greater length than in the final stage because the concrete weight was still missing.

Cable oscillations

During free cantilevering some of the foremost cables oscillated during strong winds with amplitudes the size of several cable diameters. This is a phenomenon well-known from the construction of other bridges. The reasons are the large sag of the stay cables due top missing beam loads and the flexibility of the beam tips during free cantilevering which both encourage oscillations in the wind. These may lead to large cable amplitudes caused by so-called anchorage excitation, see Section 3.8.3.

As a counter-measure the stay cables can be tied down to the beam at a height of about 3 m with simple hemp ropes, which provide a degree of damping by internal friction. This measure proved effective, as it had for other bridges. During the first winter after completion, cable oscillations took place under very strong winds,

Figure 6.113 Precast element in form

Figure 6.114 Traveler with final stay cables

6.2 CIP concrete cable-stayed bridge – Helgeland Bridge

Figure 6.115 Launching of bottom form of traveler

Figure 6.116 Start of construction

Figure 6.117 Beam starter section at tower

Figure 6.118 Start of free cantilevering

364 6 Examples for typical cable-stayed bridges

Figure 6.119 Reaching an auxiliary pier in the side span

Figure 6.120 Free cantilevering beyond the auxiliary piers

Figure 6.121 Storm at the end of free cantilevering

Figure 6.122 Center joint closure

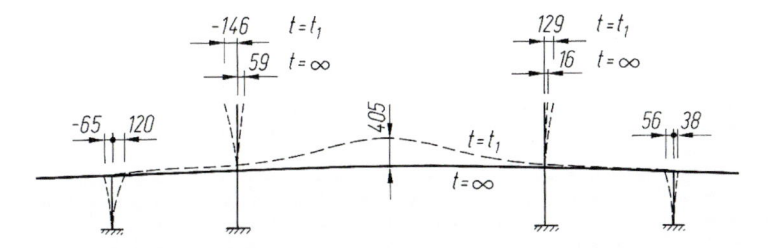

6.2 CIP concrete cable-stayed bridge – Helgeland Bridge

Figure 6.123 Change of geometry of the completed bridge due to shrinkage and creep

Figure 6.124 Parallel wire cable on reel

Figure 6.125 Pulling up an anchor head

Figure 6.126 Pulling a cable into its lower beam anchorage

Figure 6.127 Stressing a stay cable from underneath the beam

probably also caused by anchorage excitations at the beam or at the towers. These oscillations became so strong that even the small movements between PE pipes and the neoprene washers at the tips of the steel pipes, Fig. 6.83, partly destroyed the neoprene. This phenomenon could be repeated in tests at the EMPA in Zurich, [6.6]. These tests also showed that a stainless steel sheet between PE pipe and neoprene ring prevents damage of the neoprene. The PE pipes were therefore locally thickened and protected by steel sheets.

Analysis of videos of the oscillating stay cables showed that an amplitude of up to 0.67 m was reached at a wind speed of 30.2 m/s. A theoretical investigation assuming anchorage excitations led to the same results. In order to suppress these cable oscillations, hydraulic and friction dampers were investigated. For easier maintenance, however, 4 × 3 tie-down strands of stainless steel ø 15 mm per cable plane, Fig. 6.128, were installed with a force of 220 kN in order to prevent slackening.

Unfortunately, these tie-downs disturb the appearance of the bridge and make it difficult to inspect the cables with carriages running on them, Fig. 6.129.

6.2.2.5 Instrumentation

During the winters between 1992 and 1994 the bridge was instrumented as indicated in Fig. 6.130 [6.7].

A comparison between theoretical and actual deformations of the beam measured in the center and the quarter-points is shown in Table 6.2. The calculated values were a little bit too low for the horizontal deformations and a little bit too high for vertical deformations. The horizontal and vertical beam accelerations varied only slightly.

Figure 6.129 Completed bridge with stay cable tie-downs

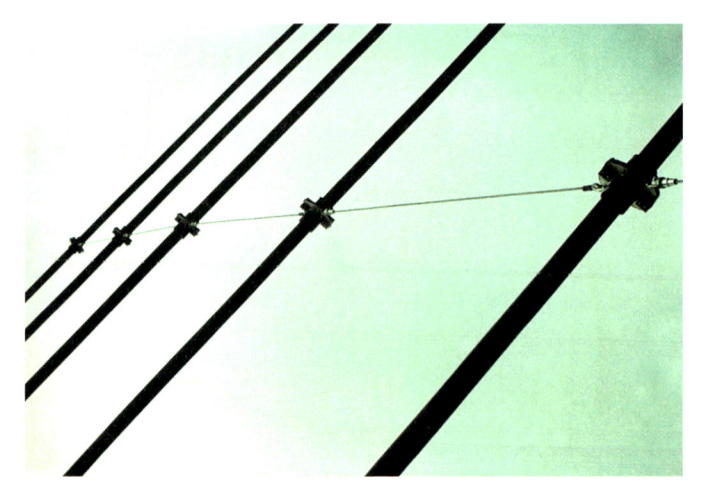

Figure 6.128 Tie-downs of stay cables against oscillations

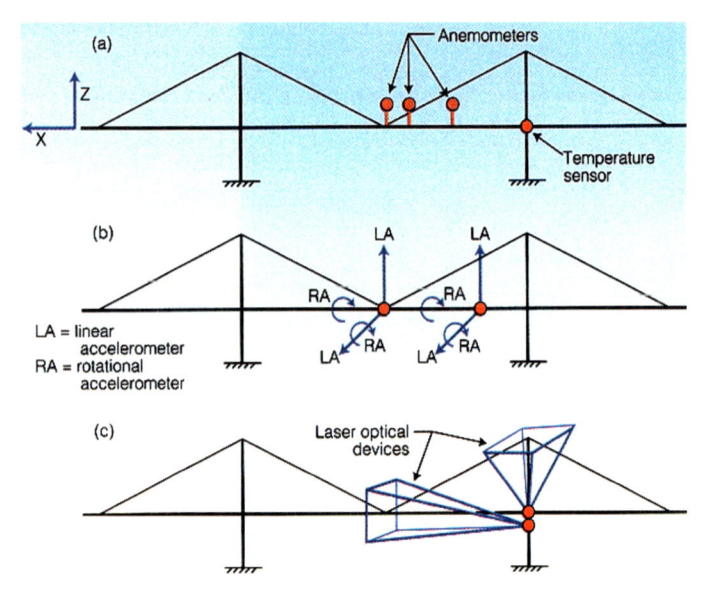

Figure 6.130 Instrumentation
a) wind speed and direction with air temperature, b) acceleration, c) deflections

6.2 CIP concrete cable-stayed bridge – Helgeland Bridge

Table 6.2 Ratio between theoretical and actual bridge deformations and accelerations

Storm date	Centre main span			
	Horizontal response		Vertical response	
	Displacement	Accelerations	Displacement	Accelerations
3 Feb 1993	0.99	0.96	1.18	1.02
18 Feb 1993	0.76	1.00	1.25	1.27
9 Mar 1993	0.73	0.79	1.07	1.01
21 Jan 1994	0.63	1.87	1.03	1.16
	Response – quarter-span			
21 Jan 1994	–	1.15	–	1.19
	Dynamic cable forces			
	Mid span		quarter-span	
21 Jan 1994	1.34	–	1.28	–

6.2.2.6 Completed bridge

The bridge was opened on midsummer night, 21 June 1991, with the Norwegian Crown Prince and his wife present. The accompanying folk festival went on through the night, during which the sun did not set. Normally the road traffic is small in the thinly settled areas near the Arctic Circle, but during the opening festivities the bridge experienced probably its only traffic queue during its lifetime, Fig. 6.131.

The two inclined cable planes form a sort of tent for the drivers which gives a feeling of safety, Fig. 6.132.

In spite of its large size the slender beam and towers give a delicate appearance to the bridge which does not disturb the view of the surrounding imposing Norwegian mountains, Fig. 6.133.

6.2.3 Summary

At 425 m the Helgeland Bridge has the third longest concrete mainspan after the Skarnsundet Bridge, Norway, at 530 m, [2.80], and the Barrios de Luna Bridge in Spain at 440 m, [2.97]. In spite of its record slenderness of 1:354 and the occurrence of powerful storms during construction the Helgeland Bridge was completed within just two years.

This success was made possible by the close collaboration between client, contractors and designers. It was especially advantageous that the design, detailing and construction engineering, including geometry control during construction, was done by the same engineers.

Figure 6.131 The only evertraffic queue on the bridge – during its opening

Participants

Client Statens Vegvesen Norway, Wilhelm B. Klaveness
Contractor Aker Entreprenor SA
Subcontractors
Towers Gleitbaugesellschaft mbH, Salzburg
Stay cables Stahlton AG, Zurich
Wind tunnel tests University of Western Ontario, Prof. A. Davenport
Design, detailing, construction engineering and control
A. Aas Jakobsen AS, Oslo, Elljarn Jordet, and Leonhardt, Andrä and Partners, Stuttgart

Figure 6.132 Driver's view

Figure 6.133 Completed Helgeland Bridge

6.3 Cable-stayed steel bridge – Strelasund Crossing
6.3.1 Design considerations

The second bridge across the Strelasund, in Germany, connects the island of Rügen with the mainland, Fig. 6.134.

The complete Strelasund Crossing with a total length of 2830 m runs across the east part of the city of Stralsund, the Ziegelgraben waterway, the island of Dänholm and the Strelasund itself. It consists of six individual bridges, four prestressed concrete bridges, one steel composite bridge and the cable-stayed bridge with a steel box girder supported by one single, 128 m high steel tower, Fig. 6.135.

Remarkable from an engineering point of view are – in addition to the variety of different structures within the total bridge – the unusual number of structural innovations, including the first use of parallel-strand cables for a major German bridge.

We follow the publications [2.65] by Dr. K. Kleinhanß, the project manager of the client, DEGES, the checking engineer, Dr. R. Saul, and his collaborator Dipl.-Ing. M. Romberg of Leonhardt, Andrä and Partners and the designer Dr. B. Schmidt-Hurtienne from EHS Consulting Engineers.

The second Strelasund Crossing runs parallel to the Rügen Causeway, some 100 m away from the old quarter of the city of Stralsund which is part of a UNESCO World Heritage Site. The required navigational clearance of 42 m results in a gradient 48 m above ground. The bridge dominates the view of the city and the designers, therefore, tried to develop an optically unobtrusive and elegant bridge.

Special care was provided for the 'Guide Structure', the roughly 200 m wide crossing of the Ziegelgraben waterway which was the starting point for the development of the design in several steps.

6.3.1.1 Bridge alternates

For a span of 200 m, alternative designs for arch bridges, girder bridges and cable-stayed bridges were compared for aesthetic, technical and economic criteria, Fig. 6.136.

Most important for judging the urban impact at the given location were visualizations. Initially the arch bridge was favored, but later it became apparent that the large dimensions in reality dominated the medieval silhouette of Stralsund too much. Furthermore, the arch required the biggest span for keeping the navigational clearance completely free, which led to very substantial cross-sections with the corresponding higher costs.

The steel-composite haunched girder bridge had economic advantages but had to be discarded from an aesthetic viewpoint because the unbalanced proportions and the optical barrier effect of the up to 9 m deep haunches were not compatible with the landscape and the city skyline. Finally the cable-stayed bridge with harp arrangement was selected in spite of concerns that the stay cables might obstruct the flight path of migrating birds. The cable-stayed bridge was considered to fit the maritime surroundings, its contours reminiscent of a sail ship.

Figure 6.134 Bridge location

Figure 6.135 General view of complete bridge

A~ 5.500 m²

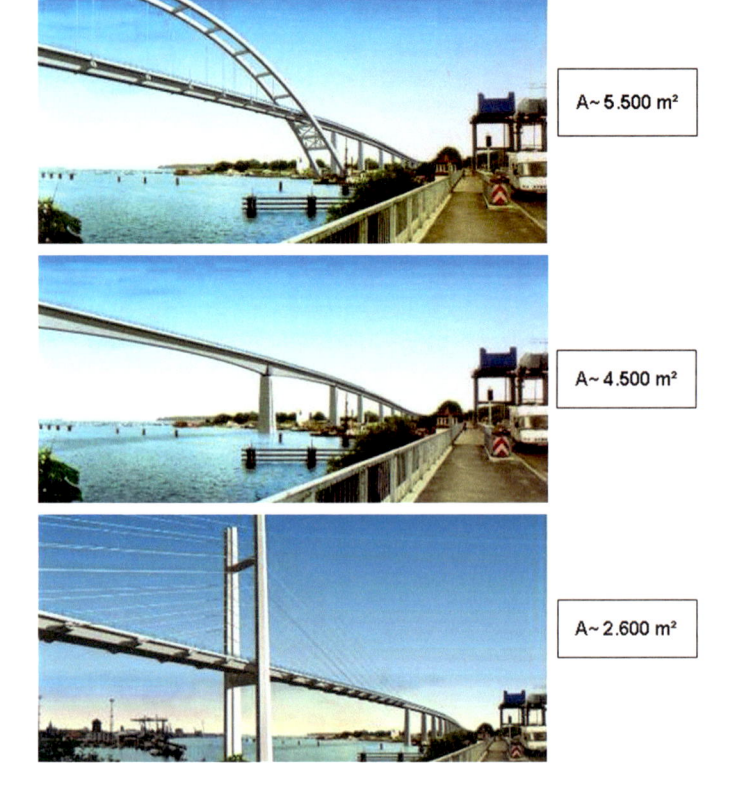

A~ 4.500 m²

A~ 2.600 m²

Figure 6.136 Comparison of arch, girder and cable-stayed bridge

Figure 6.137 Typical tear-shaped cross-section

Figure 6.138 Steel tower on concrete pier

6.3 Cable-stayed steel bridge – Strelasund Crossing

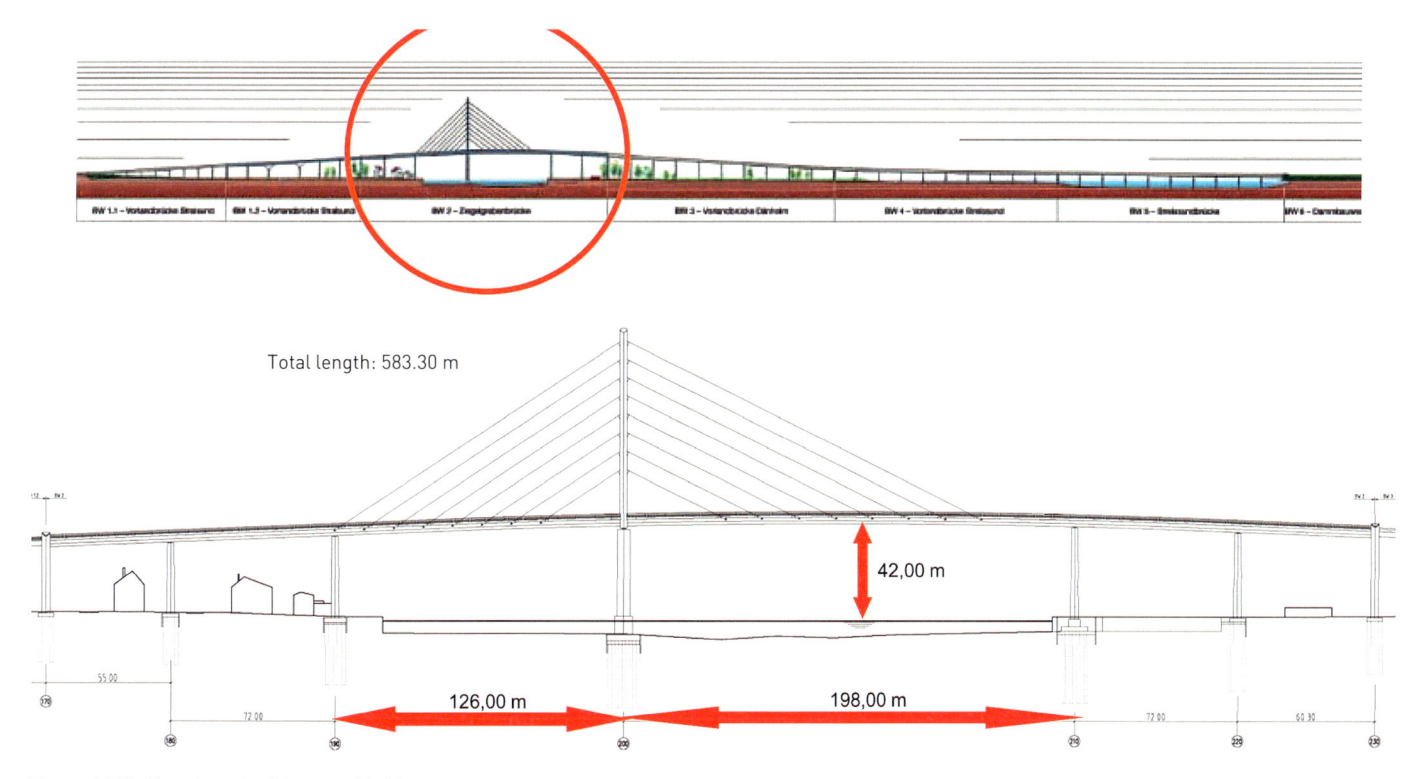

Figure 6.139 Elevation of cable-stayed bridge

In the end the comparison of the summed-up areas in elevation was decisive, which indicates the amount of visual barrier and thus the optical impact on the landscape and the city. In this respect the cable-stayed bridge was the most advantageous.

Although an inquiry made to all members of IABSE (International Association of Bridge and Structure Engineers, Zurich) indicated that no accident with birds had been documented, the environmental activists requested a cable distance of at least 8 m and a cable diameter in excess of 120 mm to minimize the obstruction to migrating birds. This also increased the overall transparency of the structure, while making the individual cables more clearly visible.

6.3.1.2 Optimizing the cable-stayed solution

In a second step the shape of the 128 m high tower was very carefully optimized, because the tower exceeds the heights of the churches in Stralsund and dominates the city skyline from all directions. A wide variety of tower shapes were investigated from engineering and aesthetic points of view with regard to different load-bearing systems, different types of materials and different structural details until a convincing solution was arrived at. The selected tower, with a concrete substructure and a light upper steel part steel, dissolves the complete massive tower structure into two parts with the steel, avoiding the heaviness of a tower completely built from concrete. The beam, as a light cable-supported structure, is separated from the solid concrete substructure, including the upper steel tower, and this gives it a floating appearance.

6.3.1.3 Structural details

A tear-shaped cross-section for all load-bearing members over the whole length of the bridge was introduced as common bridge characteristic. All piers were built as frames to improve their visual transparency, Fig. 6.137. The tear-shaped cross-section was initially used for shaping with regard to the maritime surroundings. During the detailed design phase wind tunnel tests confirmed that this shape is also aerodynamically and structurally advantageous for the 128 m high tower. Aerodynamically shaped airfoil cross-sections which are traditionally used for vehicles and airplanes were rediscovered for large engineered structural components under high wind loads.

The piers with their uniform layout of two separate tear-shaped legs determine the overall appearance of the complete bridge which looks light and transparent in spite of its major dimensions, Fig. 6.138.

6.3.2 The cable-stayed bridge

6.3.2.1 Span lengths

The spans for the cable-stayed bridge are 198 m for the main span and 126 m for the side span, Fig. 6.139. The main span length resulted from the navigational requirements, the location of the existing adjacent first Strelasund Crossing and the request not to expose the piers of the new bridge to possible ship collision.

6.3.2.2 Beam cross-section

The beam cross-section comprises an aerodynamically shaped, three-cell steel box girder with an orthotropic deck, Fig. 6.140. Its details are in accordance with modern requirements for fabrication, for example by using the same stiffeners for the top and bottom flange.

The wearing surface, with a total thickness of 8 cm, represents the state of the art, Fig. 6.141. The deflection of the orthotropic deck under wheel loads was limited in accordance with the German codes in such a way that separation of the asphaltic surface from the orthotropic deck is prevented.

The usual ratio of 0.4 between side span and main span of cable-stayed bridges results in uplift forces at the bridge ends. (For further details on the span ratio see Fig. 4.2.)

These can be carried by steel pendulums or by a combination of bearings with tensioned cable hold-downs. Alternatively, the side span may be built in concrete. The span ratio of the Strelasund Crossing with $126/(2 \times 198) = 0.32$ results in a distinctly more advantageous situation due to the continuity of the bridge beam at both cable-stayed bridge ends.

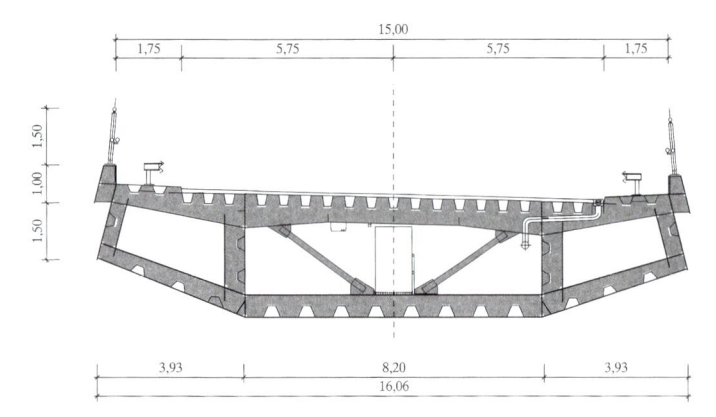

Figure 6.140 Cross-section of the cable-stayed beam

Figure 6.141 Composition of the roadway wearing surface

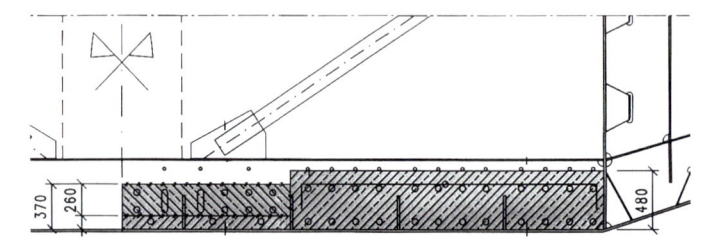

Figure 6.142 Composite concrete at bottom flange

Figure 6.143 Traffic railing with glass cladding

The uplift forces from steel dead load and superimposed dead load are small: 460 kN/cable plane. The only significant upload forces are due to live load, temperature etc., and these result in 3100 kN/cable plane from live load. These can be counteracted with 520 m³ of regular concrete (23.5 kN/m³) and 50 m³ of heavyweight concrete (30.6 kN/m³). The ballast concrete with a total weight of 13 750 kN is distributed in the box girder over a length of 116 m, with a concentrated load at the hold-down axis, Fig. 6.142.

The normal weight concrete not only acts as a counterweight but is connected with shear studs to the steel beam in composite action as a load-bearing part of the beam cross-section. During construction the concrete was placed in segments in order to provide the necessary counterweight for the free-cantilevering of the main span, and in order to equalize deviations of the steel deadweight.

6.3.2.3 Wind barriers

Special attention was paid to the traffic safety on the bridge. At the high-level beam the increased wind speeds were taken into account for determining the lateral drag forces. It was shown by numerical simulation, as well as by wind tunnel testing, that the 1.5 m high glass cladding integrated into the railing provides sufficient wind shielding. It is thus expected that cars and loaded trucks can use the bridge even during strong winds. This was proven during the test phase before opening of the bridge by comprehensive measurements with the extensive instrumentation, Fig. 6.184.

In order to rule out exposure of pedestrians beneath the bridge to crashing vehicles, a strengthened truck retaining system (German category H 4 b) is installed over the full cable-stayed length. The base plates for introducing the loads from the railing posts into the steel deck were mostly preinstalled in the shop.

In accordance with current German codes the strengthening was not strictly required because the road is only classified as 'federal', but it was permitted for this exposed situation by the German Federal Highway Authority at the instigation of the client DEGES, Fig. 6.143. Later experiences with bridges along the A 71 'Thuringia Forest' freeway as well as the A 20 'Baltic Sea Coast' freeway proved the expedience of this measure.

6.3.2.4 Tower

The steel tower, which comprises two parallel vertical legs with tear-shaped cross-section connected by three struts, rests on a concrete pier, Figs 6.144 and 6.145. The transition between steel towers and concrete pier was a demanding task.

The chosen solution – to fix the steel tower to the steel beam and support both on hinged bearings – was selected for aesthetic reasons. During detailed design, this solution with a visually separated beam also proved to be structurally and economically advantageous, for example because of the robustness against pier settlements and tiltings. The stiffening of the tower legs in the transverse direction required struts over their height.

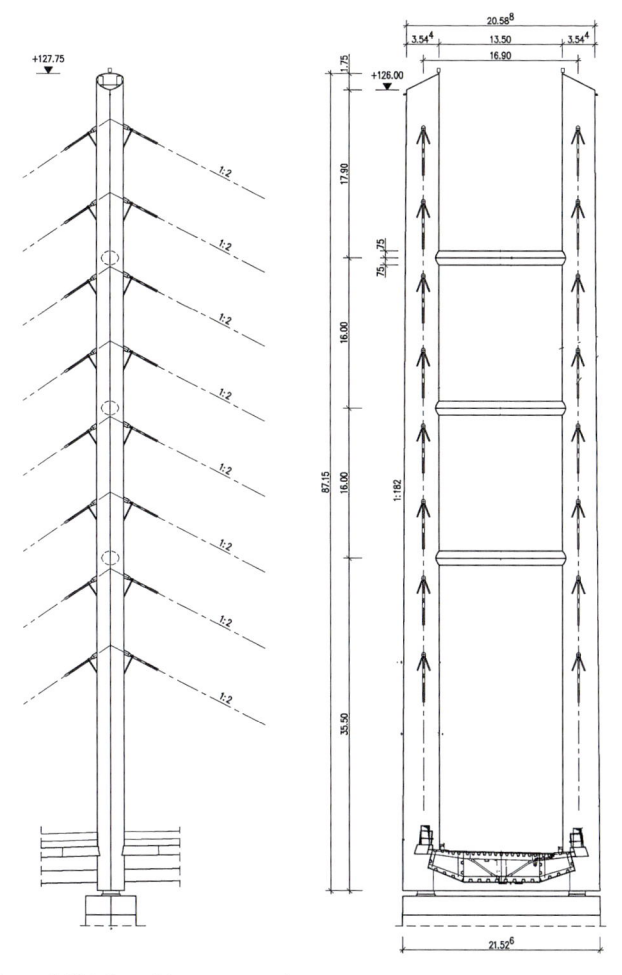

Figure 6.144 Transition tower to pier

Figure 6.145 Tower view

Both legs of the steel tower are supported on Neopot bearings, Fig. 6.146.

6.3.2.5 Stay cables

The stay cables have a harp arrangement, cf. Fig. 6.138. The client promised in the official planning permit to increase the stay cable diameters to at least 12 cm diameter to make the cables more clearly visible for the protection of the migrating birds. This would have required a substantial oversizing of locked coil ropes, which led to the first use of parallel-strand cables for a major German bridge. The tender documents detailed the generally licensed locked coil ropes, but included detailed additional specifications for contractors' alternates for parallel-strand cables.

The parallel-strand system DYNA-Grip, Fig. 6.147, was accepted on this basis with a special license, based on an evaluation by the German Institute for Structural Design (DIBt) [3.36]. The experience from the comprehensive quality control program for the parallel-strand cables of this pilot project from fabrication to control measurements at the completed structure indicates that parallel-strand cables are at least equal to locked coil ropes with regard to corrosion protection, exchangeability, installation and economy. The sizing of parallel-strand cables is outlined in Section 3.7.

6.3.2.6 Aerodynamic investigation

General

The Strelasund Bridge is the first German cable-stayed bridge designed in accordance with the new Eurocode. In 2001 the client DEGES provided an expert evaluation in agreement with the German Federal Highway Authority and the State of Mecklenburg-Vorpommern on the application of the Eurocode and its consequences for the required quantities [6.8].

Since wind loadings are not governing for most bridges they are covered rather globally in Appendix N. For example, the wind loads in accordance with Table N.1 are independent of the wind zone and the cross-section but only depend on the height above ground and the width-to-depth ratio of the beam. The longitudinal forces – for the first time in German codes – are also globally given with 25 % or 50 %, independent of the surface roughness of the beam, for example due to cross girders. For the wind-exposed Strelasund Bridge on the Baltic Sea coast 50 m above ground with aerodynamically shaped cross-sections of beam, towers and piers, detailed investigations were required.

Transverse wind

For the protection of trailers and unloaded trucks from strong winds the already-mentioned installation of 1.5 m high wind barriers at both beam edges was decided when the detailed design was nearly completed, Fig. 6.143. By numerical flow simulation and wind tunnel tests the influence of these wind barriers on the tilting moment of a 4 m high truck and on the loads of beam and piers was investigated.

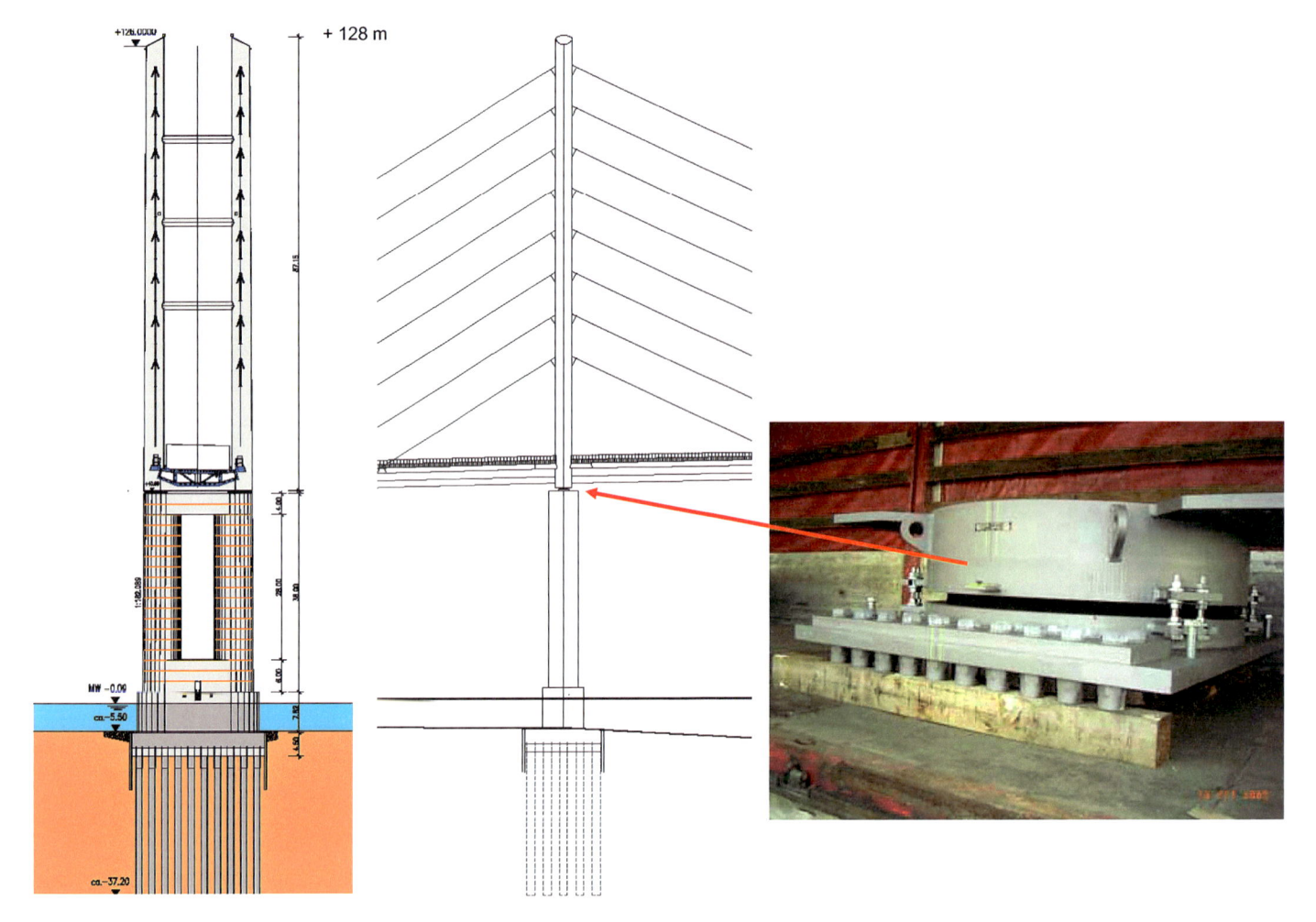

374 6 Examples for typical cable-stayed bridges

Figure 6.146 Tower bearings

The numerical calculations showed the shielding of the wind barriers, Fig. 6.148. The tilting truck moment was reduced by about 50%, which means the critical wind speed against tilting is increased by about 40%.

Wind on beam and tower

The wind loads of the Eurocode in comparison with the wind tunnel test are shown in Fig. 6.149. An exact determination of the drag factor resulted in 20% lower wind loads in spite of the wind barriers. This opens up a promising perspective for the retrofitting of other bridges. For the sizing of the wind barrier itself a drag factor of 1.7 was found, which resulted in a wind load of 3.67 kN/m² (50 m height, 59 m/second wind speed), significantly more than required by Eurocode.

A non-linear finite element calculation showed that 1.3×2.3 m composite glass safety panels made from 2×6 mm thick plates provide sufficient resistance. The non-linear calculation resulted in a decrease of stresses against a linear calculation of 87% and a reduction of deflections of 55%.

Superposition of longitudinal wind and temperature

DIN FB 101 provides different rules for superpositioning of wind and temperature:
– Appendix C.2.1.1 requires road bridges not to take into account wind and temperature simultaneously
– Appendix O.1(2), however, requests to fully superimpose wind and temperature.

This coincidence of extreme temperatures with heavy storms is not expected at this bridge location. A combination factor of 0.6 was used. In this way the total movement at one roadway joint was reduced from 933 mm to 787 mm, which simplified the joints. (Meanwhile this contradiction in DIN FB 101 is omitted.)

6.3.3 Construction

6.3.3.1 Construction engineering

The beam was assembled from a total of 18 sections, Fig. 6.150. The section lengths varied from 16.1 m to 95.3 m with weights between 145 t and 850 t. The sections were pre-assembled on the ground and lifted into position by a crawler crane or a floating crane. The erection started with four sections for the three approach spans on the Stralsund side, Fig. 6.150.

The 128 m long side span of the main bridge acted as a side span of the continuous three-span approach girder without the stay cables in place. This construction stage did not govern the sizing. The main span was built by free cantilevering with 16.1 m long sections. The tower construction took place parallel to the beam erection and the installation of the stay cables.

The tower was subdivided into 15 segments (2×6 tower legs and three cross girders). After each newly installed beam section the corresponding forestay and backstay cables were installed. The biggest

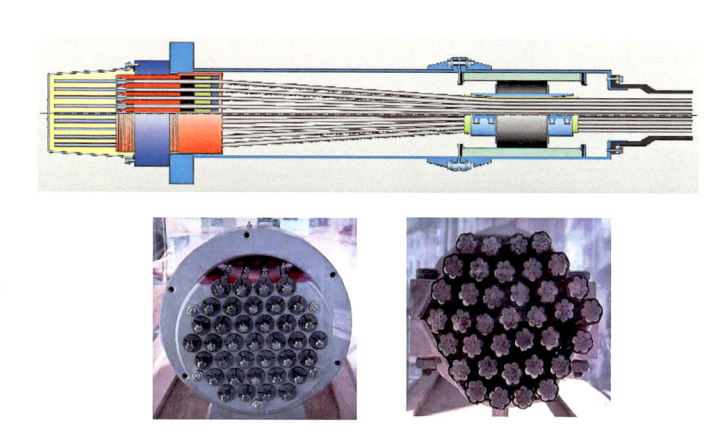

Figure 6.147 Parallel-strand cable DYNA-Grip

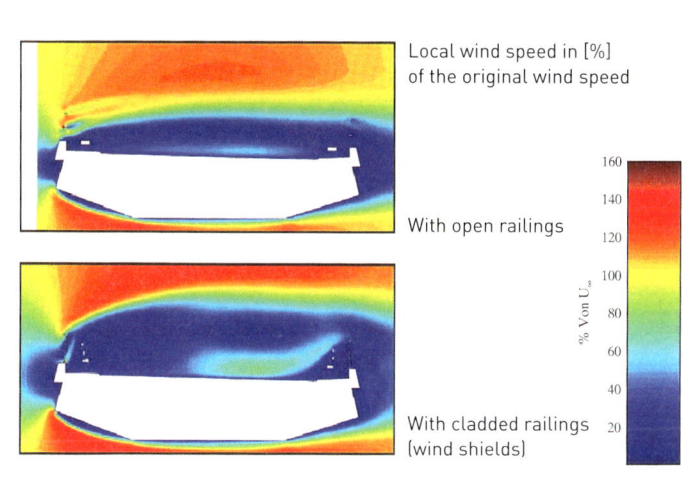

Figure 6.148 Wind field with and without wind barrier

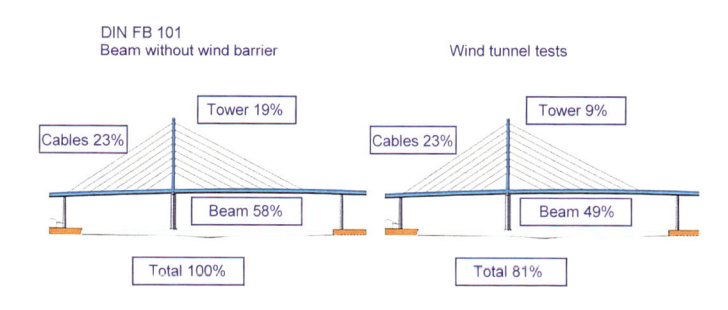

Figure 6.149 Comparison of wind loads in accordance with DIN FB 101 and from wind tunnel tests

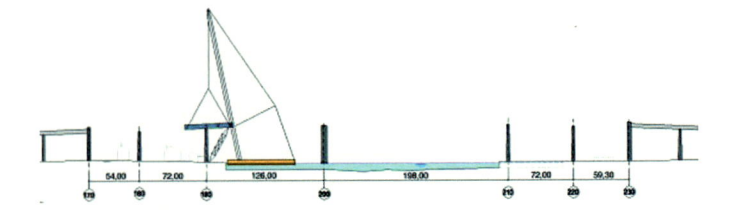

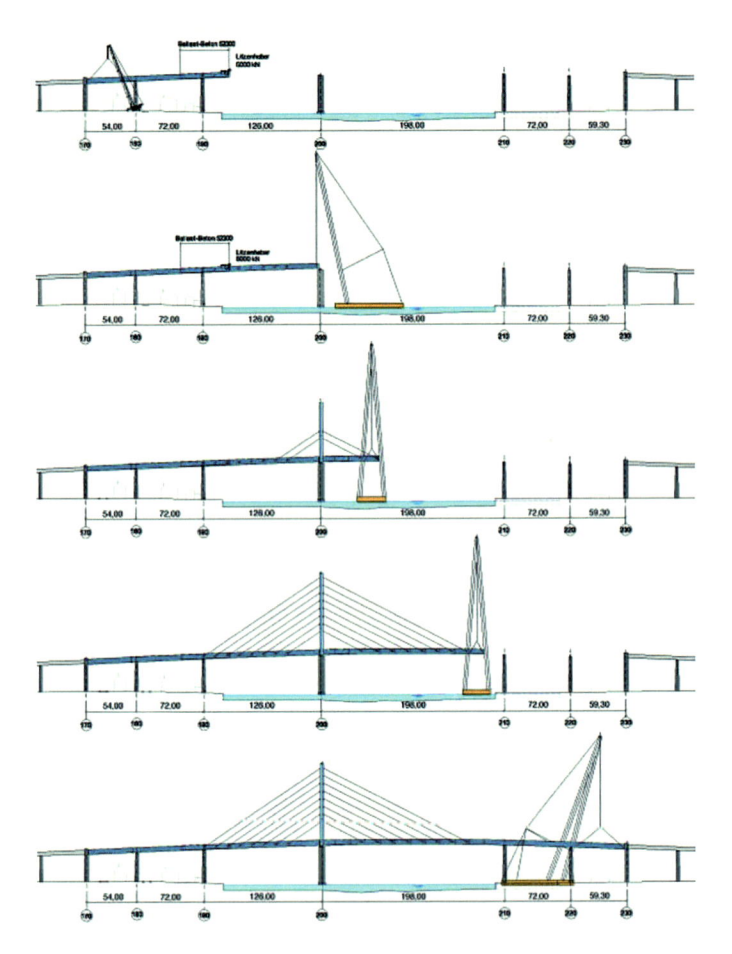

Figure 6.150 Construction sequence

main span cantilever was 159.3 m. The two side spans on the Rügen side were crossed with three 50 m long sections.

For each individual erection stage the construction engineering provided the theoretical geometry and cable forces, Fig. 6.151. Furthermore unit loadings for scaffolding, wind and temperature were precalculated in order to estimate the influence on site quickly. The deflections were systematically measured and listed numerically and graphically, Fig. 6.151.

6.3.3.2 Construction of the main bridge

Substructure

The piers were founded on bore piles with 1.5 m diameter. In water they were built within sheet piles, Figs 6.152 and 6.153. An overview on the construction phases and types of foundations is provided by Fig. 6.154.

For proof of the pile load-bearing capacity test loadings were executed on shore, Fig. 6.155.

The piers with their characteristic tear-shaped cross-section were built with jumping forms, Figs 6.156 to 6.158.

Side spans

The beam sections of the side spans were pre-assembled in various lengths and lifted in. Fig. 6.159 shows the installation of one section on shore, were the limited space required an additional transverse movement.

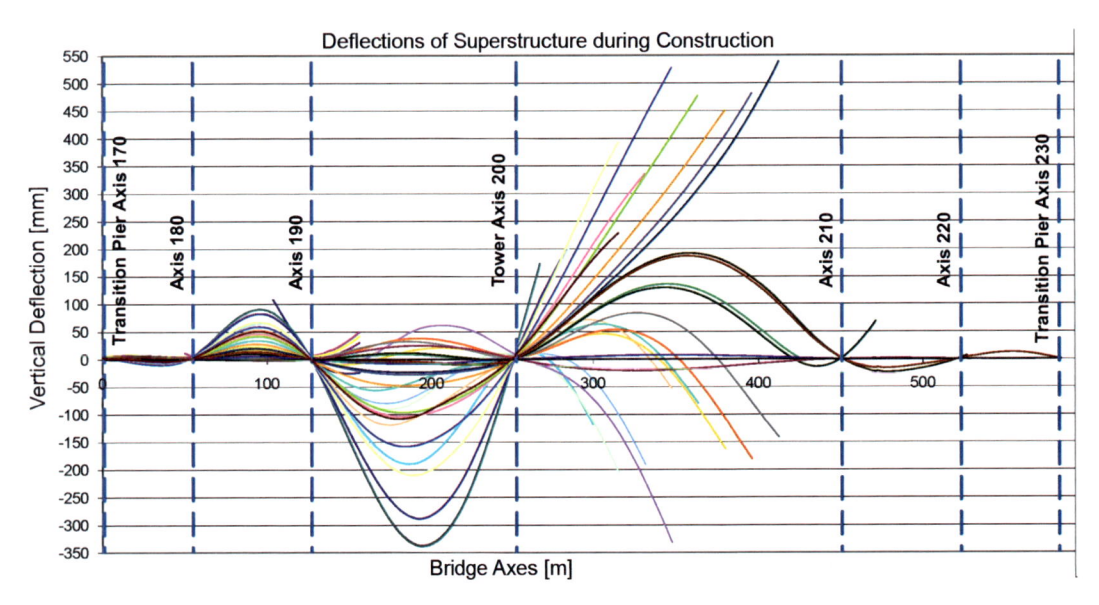

Figure 6.151 Beam deflections during construction

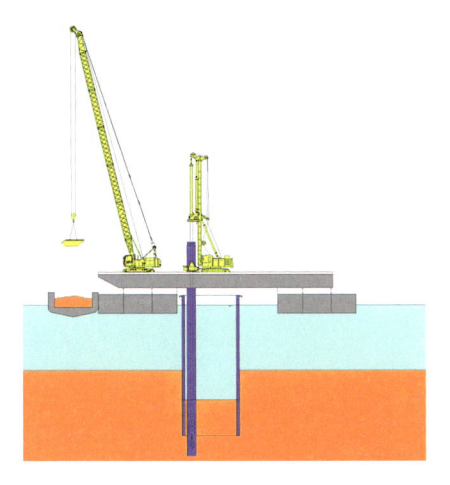

6.3 Cable-stayed steel bridge – Strelasund Crossing

Figure 6.152 Construction of bore piles in water

Figure 6.153 Use of a catamaran for bore pile installation

Figure 6.154 Overview on substructure construction

Figure 6.155 Test loading with 1480 t

Figure 6.156 Main pier construction

Figure 6.157 Anchor pier construction

Figure 6.158 Construction of approach piers

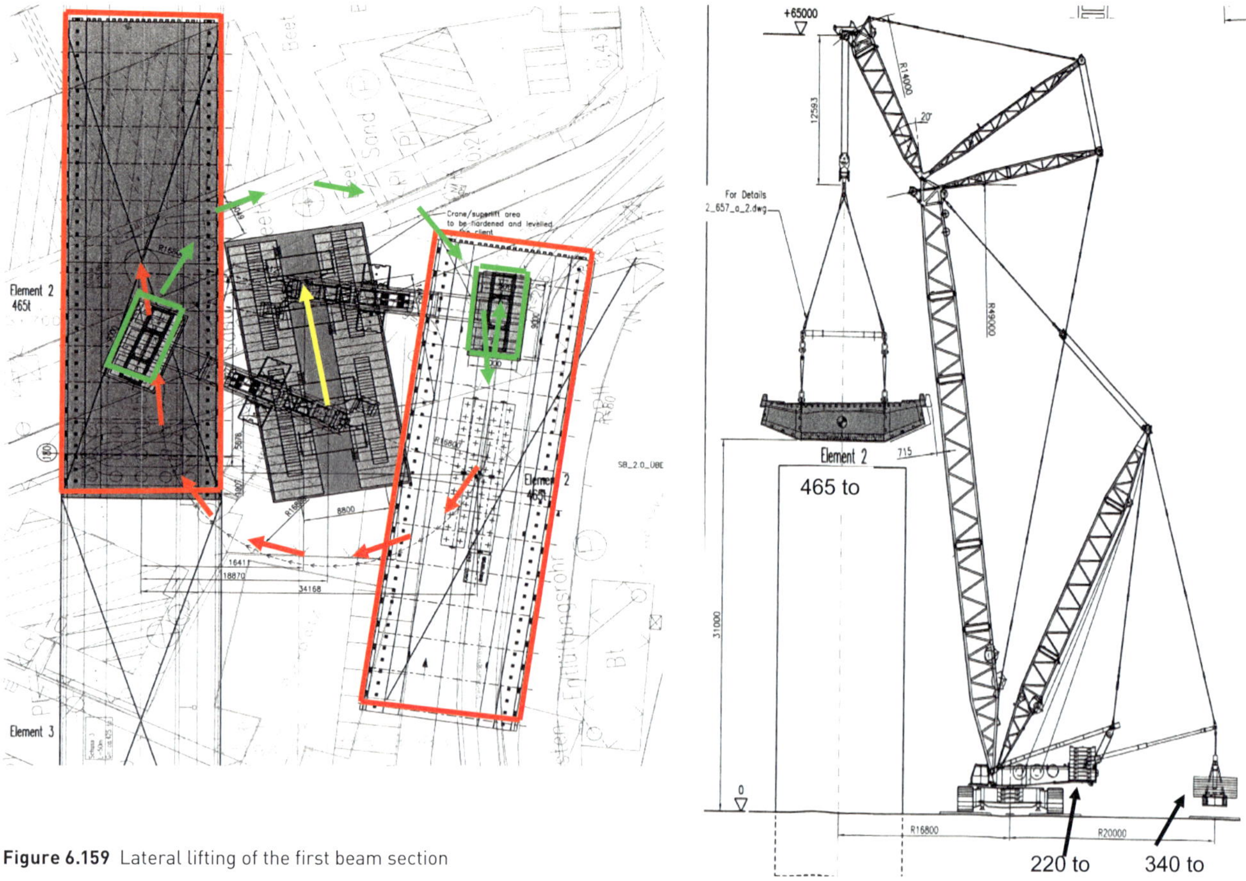

Figure 6.159 Lateral lifting of the first beam section

Figure 6.160 Lifting with crawler crane at night

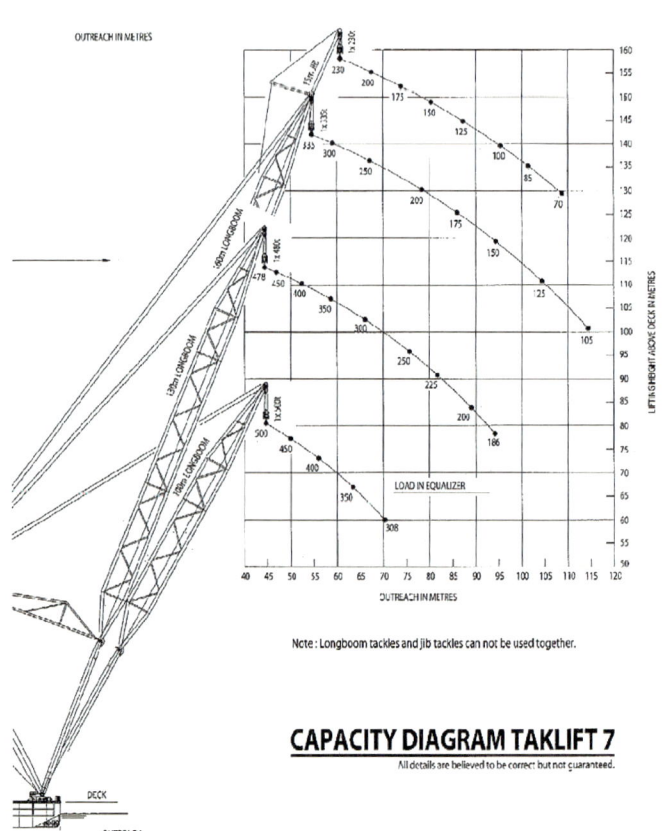

Figure 6.161 Lifting with Taklift 7

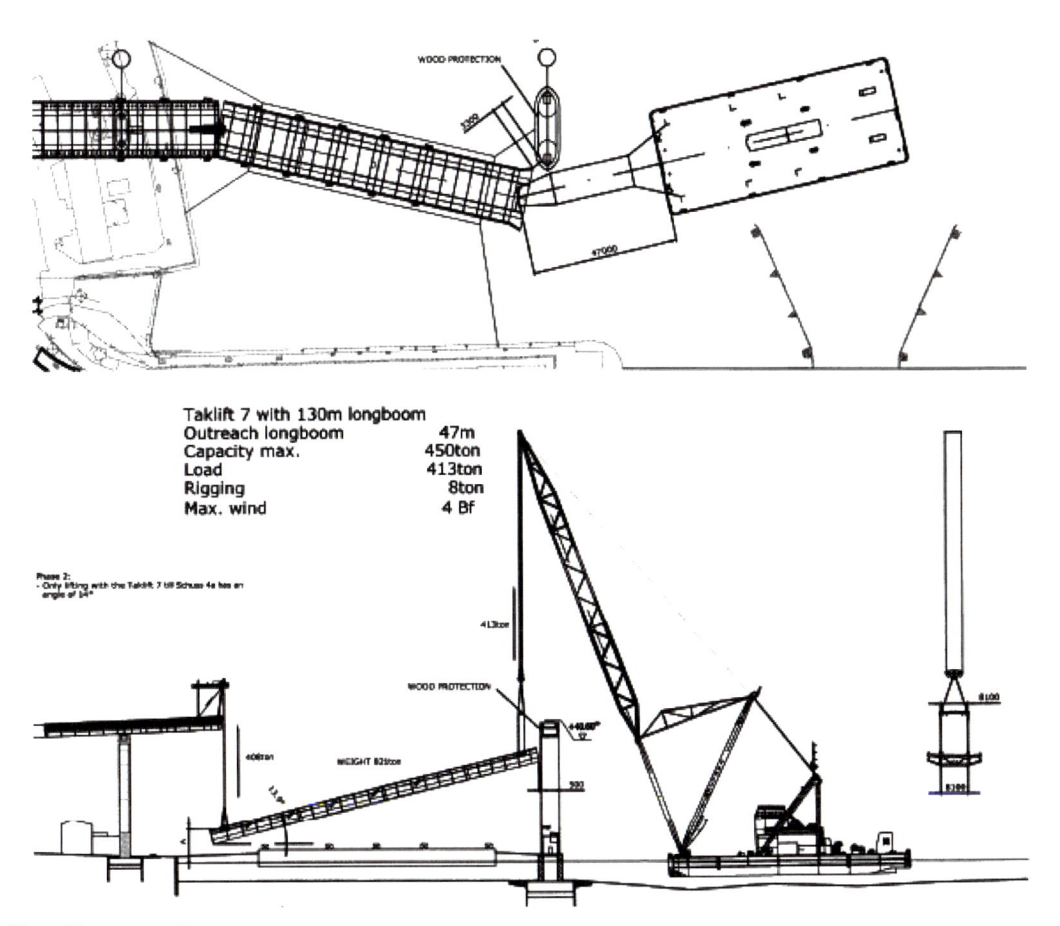

6.3 Cable-stayed steel bridge – Strelasund Crossing

Figure 6.162 Planning the lifting of the 90 m side-span section

In order to interrupt the road traffic underneath the bridge as little as possible, some sections were installed at night, Fig. 6.160. For the lifting of beam sections over water the floating crane, Taklift 7 was used, Fig. 6.161.

The main span was constructed from full sections with the same floating crane Taklift 7 (capacity 500 t) and 1750 t crawler crane. Most impressive was the installation of the 90 m long section with a total weight of 850 t. The beam tip at the tower was lifted with the floating crane, and at the land side a hydraulic strand jack resting on the already installed beam was used, Figs 1.162 and 6.163. The geometry required this section to be lifted with an angle of 13° until it could slide onto the tower pier.

Tower construction

The steel tower was also delivered in sections to the site, pre-assembled and lifted, Figs 6.164 to 6.166. The lower tower leg sections were lifted together with the first main span section, Fig. 6.167.

Then the remaining tower sections were lifted by the floating crane, and all sections were welded together in place, Figs 6.167 to 6.170.

Free cantilevering of the main span

The main span was erected by free cantilevering. The pre-assembled sections were transported by barge and lifted with the floating crane, Fig. 6.171.

The sections were then hung into a derrick located at the tip of the already erected beam, Fig. 6.172. There they could be aligned precisely and welded. The floating crane would not have been steady enough for this purpose. The already installed beam was strong enough to support the new section without tie-backs.

Figure 6.163 Actual lifting of the 90 m side-span

Figure 6.164 Delivery of the lower tower section

Figure 6.165 Transportation by road

Figure 6.166 Transportation on water

Figure 6.167 Lifting of the lower tower section

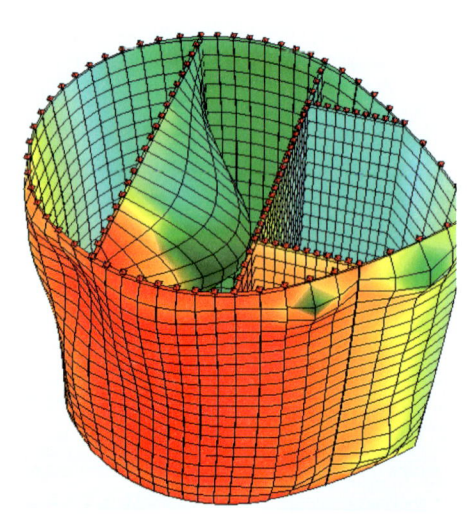

Figure 6.168 Tower leg cross-section

6.3 Cable-stayed steel bridge – Strelasund Crossing

Figure 6.169 Lifting of a tower tip

Figure 6.170 Installation of tower tip

Figure 6.171 Lifting with floating crane

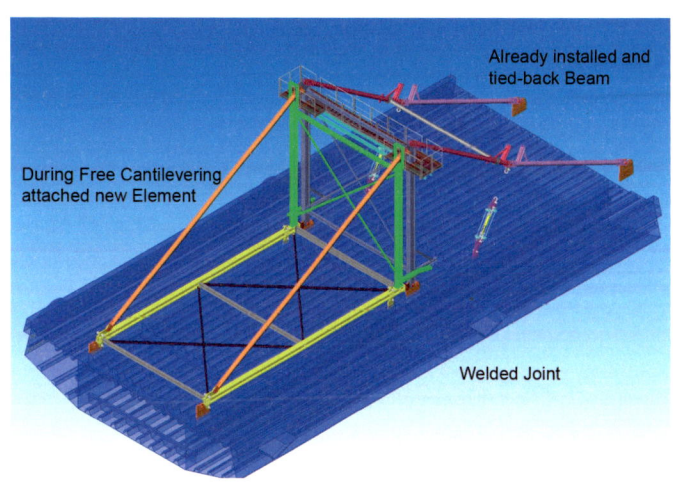

Figure 6.172 Derrick for alignment of new beam section

The welding of the new section took place under sheltered conditions from a traveler, Fig. 6.173.

After reaching the side span pier on the Stralsund side the steel beam installation was continued across the first side spans, Fig. 6.174.

Cable installation

The Strelasund Bridge was the first major German bridge to use parallel-strand cables. In accordance with international practice the cables were fabricated on site from their individual components. Fig. 6.175 shows the lifting of a welded PE pipe to the tower head with the lift cable attached to a temporary collar at its tip.

The individual monostrands were then pulled into the PE pipe, Fig. 6.176. Fig. 6.177 shows an intermediate fabrication stage in which only a few of the monostrands have yet been installed.

The wedges which clamp the monostrands into the anchor heads are pre-wedged at the passive tower head anchorage in order to prevent their slipping during the stressing from the active beam anchorage, Fig. 6.178.

The beam anchorages cantilever sideways out of the beam, Figs 6.179 and 6.180.

The monostrands were stressed individually with small monojacks to their theoretical force, Fig. 6.181. The development of the forces in individual strands during installation is shown in Fig. 6.182. The Isotensioning (Freyssinet) or ConTen (DSI) stressing system ensures that all strands reach the same final theoretical force [3.36].

The strands were stressed in stages with hydraulically coupled monostrands. In this way, all strands were stressed to the same force during the second stressing stage. This was not precisely possible during the first stressing stage because during the long period required for installing one strand after another, the temperature could vary considerably. Another advantage of stressing in two stages was that the second stage could be adjusted to the deflections of the beam during the first stage. This happened because the deflection behavior of the monolithic connection between beam and tower differed slightly from the design assumptions. Fig. 6.183 shows the installed stay cables of the completed bridge.

Corrective measures

Beam: The ballast concrete required against uplift in the beam at the anchor pier was structurally exploited in composite action to influence the deflections of the beam. In order to correct possible deviations from the theoretical geometry, a part of the concrete in the side span was only placed after the construction of the main span – including the superimposed dead loads – was completed.

Another possible corrective measure was the adjustment of the stay cable forces. When checking the deflections of the cantilever after installation of every new beam section an increase of 7 % of the deadweight was detected. This increase was confirmed by measuring the draft of the transportation barge. Consequently all cable forces

Figure 6.173 Welding under sheltered conditions

Figure 6.174 Erection of steel beam on the Stralsund approach

Figure 6.175 Lifting of a PE pipe to the tower head

Figure 6.176 Pulling-in of strands

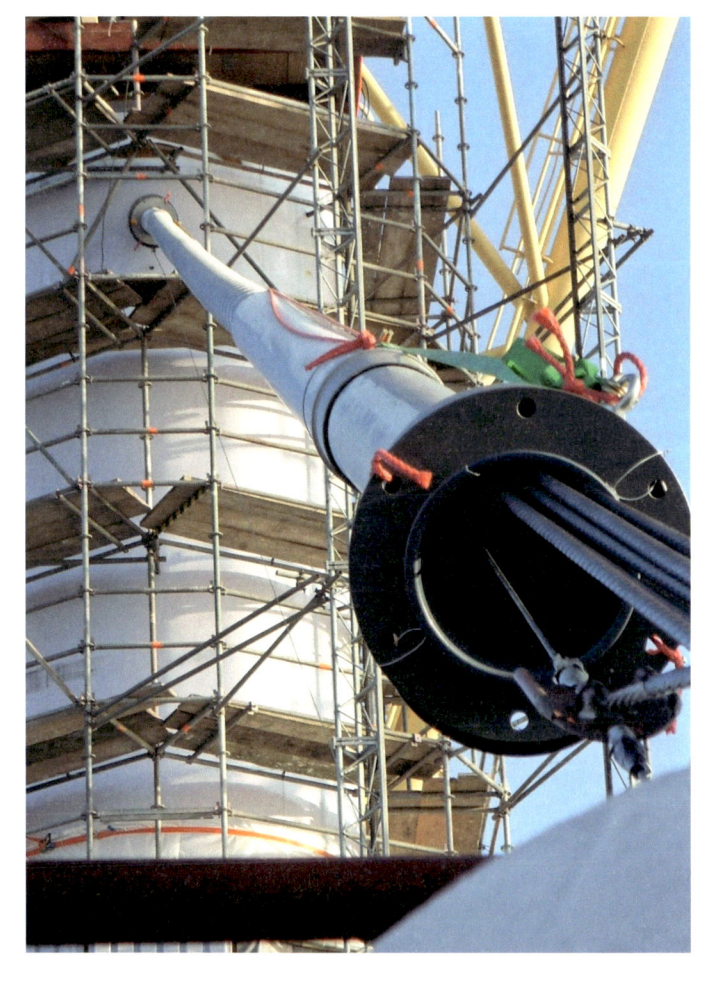

6.3 Cable-stayed steel bridge – Strelasund Crossing

Figure 6.177 Cable with the first monostrands installed

Figure 6.178 Pre-wedging of monostrands at the tower anchorage

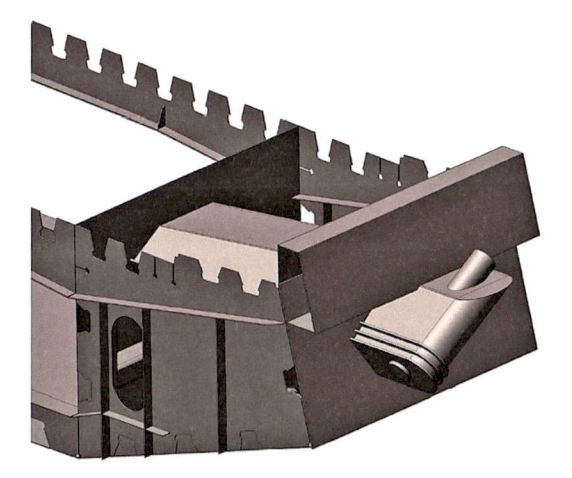

Figure 6.179 3-D visualization of a beam anchorage

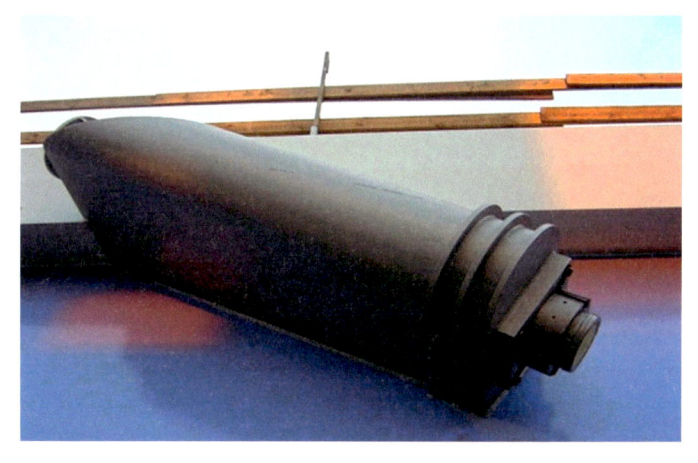

Figure 6.180 View of a beam anchorage

Figure 6.181 Stressing of individual strands with monojacks

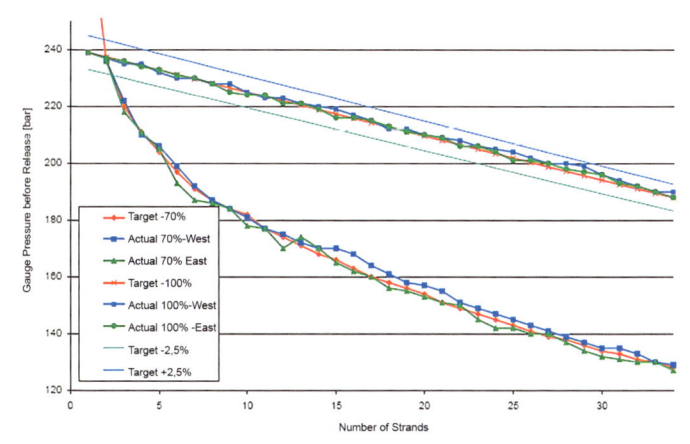

Figure 6.182 Control of hydraulic jack pressure during stressing

Figure 6.183 Finally installed stay cables

were adjusted accordingly. Further adjustments were made to compensate small geometrical deviations during free cantilevering. The corresponding change in structural stresses was always checked and found to remain within the allowable range.

After completion of the steel erection the geometry at each cross girder was checked and no further corrective measures were required. Therefore, the remaining ballast concrete was placed in the box girder before all superimposed dead loads were applied.

Instrumentation: In order to account for the special loads and the structural innovations a detailed control and measuring program was installed. This permitted the control of the assumptions made for the detailed design during the period between completing the bridge and opening it for traffic. Also a database for the life of the bridge was assembled.

The measuring program comprises the load-bearing and stiffness behavior of the structure as well as the compilation of the climatic conditions for traffic guidance. For bridges of this size verification of the design assumptions are required because, for example, cable oscillations cannot be predicted precisely so that they may or may not occur.

The assumptions of DIN FB 101 with regard to superpositions of wind and temperature as well as the reduction of the wind loads found in wind tunnel tests against those stipulated were also checked by measurements.

The measuring was executed over a period of ten month before and after placement of the wearing surfaces, so that the influence of the asphalt on the temperature in the box girder and the eigendamping of the beam and the stay cables could be investigated. Fig. 6.184 gives an overview of the executed control measurements.

All the data is centrally recorded in a computer. The results are downloaded by remote control and as well as the tuning of the individual installations. Fig. 6.185 shows the recording of the horizontal oscillations of cable 15 with an amplitude of 90 mm at the center of the cable which was recorded during a longitudinal wind with a speed of about 30 km/h.

Initial records show that vertical oscillations are also excited by moderate wind speeds, but not in the transverse direction. A coincidence of strong winds and cable oscillations has not yet been observed.

Tower: The control measurements of the tower geometry showed only very small deviations within the tolerances, Fig. 6.186.

Cable forces: At the end of the construction all cable forces were measured with hydraulic jacks by 'lift-off-tests'. Taking into account the adjustments during construction, the cable forces varied only by ±5% against the theoretical values, which is within the measuring tolerances. Based on the results of the actual cable forces and the adjusted dead load the beam moments were recalculated, with the result that the changes corresponded with those already taken into account during construction. All stresses remain within the allowable limits, and a reduction of the partial load factor for permanent loads was not required. This would have been considered acceptable because the actual deadweight and the geometry were measured precisely.

Forces and geometry after completion: By precise planning of all construction stages and permanent controlling, including corrective measures for example due to deviations in the deadweight, it was ensured that at the end of construction all structural forces and the geometry coincided with the theoretical values. This is in accordance with the requirements of DIN FB 103, Section II-A.2.4 'Prestressing of structural elements', which permits the use of the same partial load factors for the forces from 'permanent loads acting on the elastic system' and for the forces for 'cable shortenings'.

In order to determine the stiffness of the beam, the tower and the stay cables static load tests were executed. The oscillation behavior of the beam in the main span was determined by dynamic drive tests of trucks across sleepers. The excitation and especially the decay spectrum were of interest. This test was repeated after placement of the wearing surfaces.

A continuous measurement of the oscillations for the longest seaward stay cables formed the basis for deciding whether additional damping against cable oscillation was required. During the design phase, structural preparations were made for the later installation of dampers near the cable anchorages at the beam. For this purpose, the horizontal and vertical cable accelerations were recorded at a point one tenth of the cable length away from the anchorage.

Whether parametric excitation can be identified as the cause of the cable oscillations can be determined with additional acceleration measurements at both ends of the cables. Dampers were installed afterwards on the four longest cables.

The influence of the wind on the cable oscillation was observed by a continuous recording of wind speed and direction. Wind sensors are installed at the footing of the tower pier (2 m above sea level), at the beam (about 44 m above sea level) and at the tower tip (128 m above sea level). This allows account to be taken of the influence of the topographical roughness on the vertical development of wind speed.

In addition to the permanent recording of the wind, the wind profile on the bridge deck before and after the installation of the wind barriers was measured. These measurements are to verify the theoretical investigations on the wind barriers. They also provide data on the wind forces during the operation of the bridge and thus form the basis for the decision when to close the bridge during strong winds.

The deflections due to temperature and longitudinal wind are measured at both roadway joints in order to confirm the applied combination factor.

6.3 Cable-stayed steel bridge – Strelasund Crossing

Figure 6.184 Overview of control measurements

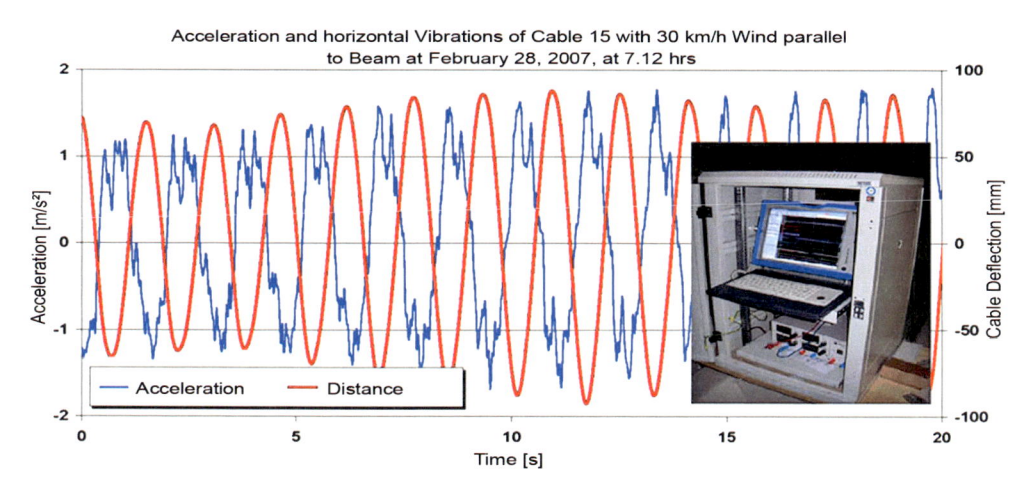

Figure 6.185 Recording of cable oscillations

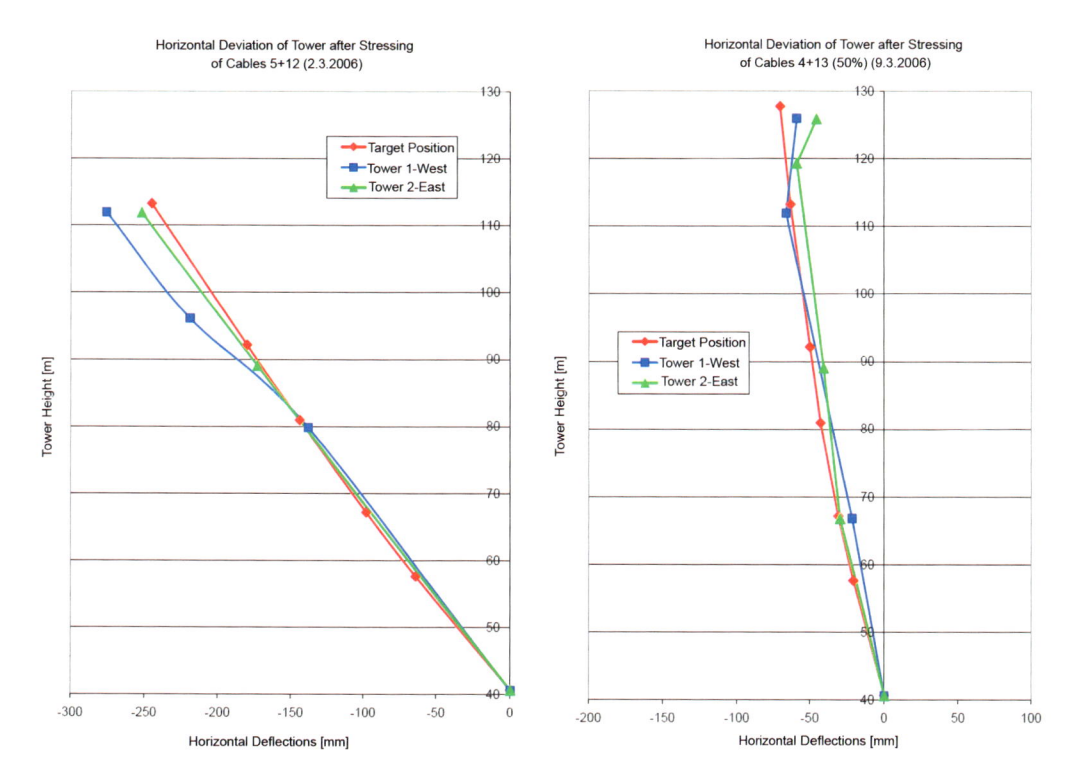

Figure 6.186 Control of tower geometry

Figure 6.187 Completed Strelasund Bridge

The outside temperature is measured with a temperature sensor in the tower axis. The steel temperature at the top and bottom flange is measured in the center of the main span. In addition, the air temperature inside the box girder is also recorded. The temperature records will provide a correlation between the movements of the roadway joints and of the top and bottom flange temperatures of the box girder after placement of the wearing surface.

6.3.3.3 Completed bridge

The completed bridge was generally accepted by the public, Fig. 6.187, who participated in large numbers at the bridge opening by walking over the full bridge length, Fig. 6.188.

Figure 6.188 Pedestrians at the bridge opening

Participants

Client Federal Republic of Germany represented by the State of Mecklenburg-Vorpommern, with DEGES in charge
Design SPI Schüßler-Plan
Detailed design SSF, Büchting + Streit
Checking engineers Dr.-Ing. E. h. R. Saul, Stuttgart,
Dr.-Ing. H.-P. Andrä, Berlin,
Prof. Dr.-Ing. Ulrike Kuhlmann, Ostfildern
Dipl.-Ing. Peter Otte, Neustrelitz,
Dipl.-Ing. Winfried Koldrack, Rostock
Construction supervision Joint venture EHS/VSE
General contractor Joint venture 2. Strelasundquerung Max Bögl Bauunternehmung GmbH & Co. KG
Substructure Joint venture 2. Strelasundquerung Schmitt Stumpf Frühauf and Partners (SSF), Büchting + Streit
Steel fabrication Max Bögl Stahl- und Anlagenbau
Stay cables DYWIDAG Systems International DSI

6.4 Composite cable-stayed bridge – Baytown Bridge
6.4.1 General layout

During the mid 1980s the existing tunnel underneath the Houston Ship Channel in Texas, USA, could not cope with the increased volume of road traffic and had to be replaced by a bridge. This large amount of traffic required four lanes with full shoulders in the each direction. A detailed investigation resulted in two independent beams as the most economic solution.

The bridge crosses the Houston Ship Channel 32 km east of Houston between the cities of Baytown and LaPorte, Fig. 6.189. The busy ship traffic between the Gulf of Mexico and Houston Harbor required a navigational clearance of 53 m. The towers were protected against ship collision by locating one of them on shore and the other one in shallow water and surrounding it with an artificial island.

As usual for major bridges in the USA complete designs for a concrete and a steelcomposite alternate were tendered. The project was awarded for $ 91.3 m to the steelcomposite alternate outlined here.

The bridge was designed in accordance with AASHTO, amended by other US and international codes. All structural parts were designed in service limit state and ultimate limit state. The concrete for the roadway slab was C 50 standard and for the tower C 42, the reinforcement had a yield strength of 420 N/mm^2 and the structural steel ultimate strength of 520 N/mm^2. The design and construction of this unique cable-stayed bridge was outlined in [1.17].

6.4.1.1 Bridge system
The cable-stayed bridge with a main span of 381 m has a beam continuous over 674.8 m, Fig. 6.190, and is symmetrically supported.

The vertical and horizontal support loads are taken at the two towers and the hold-down piers, Fig. 6.191

The beam and the anchor piers are connected by 305 mm diameter pins which also transmit uplift forces, Figs 6.192 and 6.193.

The fixed point of the bridge is thus located in the center of the main span. The beam deformations due to change of temperature and shrinkage and creep are taken up by the neoprene bearing with horizontal shear deformations, and the anchor piers are slender enough to follow by deformations in bending.

The beam rotations at the anchor piers are taken up by hinged roadway joints, the longitudinal movements of ± 380 mm are carried by additional movement joints at the ends of the short first approach spans.

By distributing the rotations and translations onto two roadway joints on each side of the bridge their detailing is simplified and they become more robust. The movements of the superstructures are limited at the towers, longitudinally between short concrete corbels and transversely between buffers, Fig. 6.194. The transverse wind forces are carried by central wind bearings in the middle of each end cross girder.

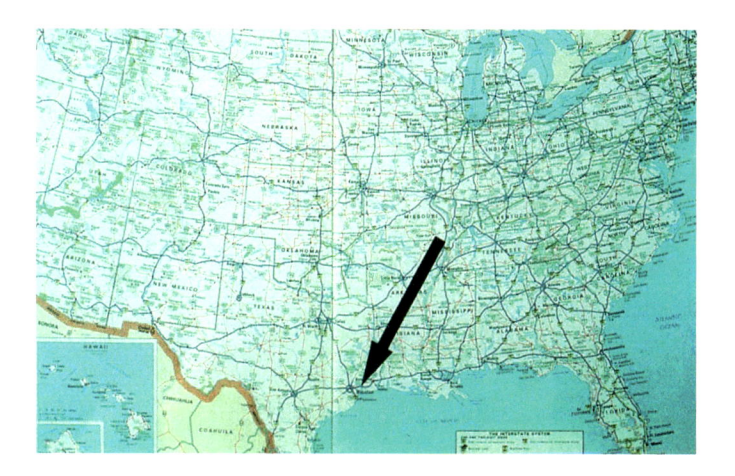

Figure 6.189 Location of bridge

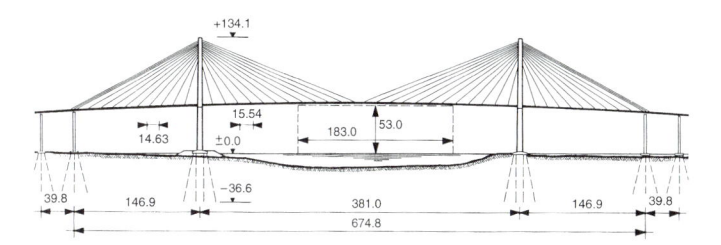

Figure 6.190 Elevation

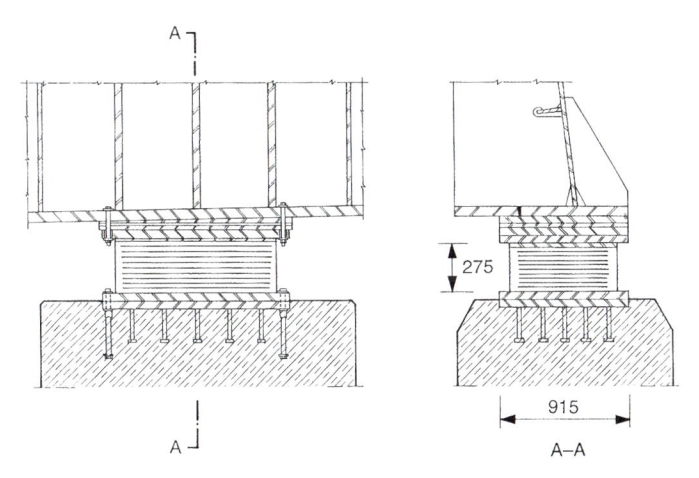

Figure 6.191 Neoprene bearings at the towers

6.4.1.2 Composite beam
Structural detailing

Because of the large beam width required for four lanes in each direction of travel, systems with two, three and four cable planes were investigated. For two outer cable planes the amount of steel required in the transverse direction was as big as required longitudinally. Two independent beams with two cable planes each required only half the steel in the transverse direction and were thus selected as the most economic solution, Fig. 6.195.

Each composite beam comprises a concrete deck resting on a steel grid from open outside main girders and cross girders. For economic reasons the webs were selected in such a way that only one longitudinal stiffener was required in the middle of the main girders, Fig. 6.196.

The vertical stiffeners of the main girders at a distance of 5.2 m are welded to their top and bottom flanges in order to transmit the end moments of the cross girders, which are created by their torsional stiffness. Only in the region of high moments with changing signs in the bridge center and at the bridge ends are the vertical stiffeners bolted to the bottom flanges of the main girders for improved fatigue resistance.

The concrete roadway slab consists of a 200 mm thick structural concrete layer covered by a 100 mm wearing layer, requested by the Texas State Highway Department for protection against a burning petrol truck in case of an accident.

The cables are anchored at a distance of 15.2 m in welded boxes bolted to the webs of the main girders. Composite action between the steel grid and the concrete roadway slab is provided by shear studs protruding into the CIP joints, Fig. 6.197. Together with the stay cables the anchorages are also exchangeable.

The eccentricity moment between each cable axis and the main girder webs is carried by a force couple at the ends of the cable anchor box. The compression component is transmitted into the roadway slab, the tension component is transmitted into the bottom flange of a cross girder. The bottom of the anchor boxes is thus always located at a bottom flange of the cross girder, Fig. 6.196. The cable distances thus vary slightly with the cable inclination.

For the concentrated anchorage of the three backstay cables the main girder is widened to a box at the bridge ends, Figs 6.192 and 6.193. These boxes connect the regular open cross girders after a distance of 15 m, Fig. 6.196. The box girder was necessary to achieve the required torsional stiffness in the 5.4 m long cantilevers protruding into the first approach span. They have a width of only 0.45 m and can thus not be inspected. They are therefore filled with low-shrink concrete.

Design calculations

The run of moments under permanent loads was chosen as that of a beam rigidly supported at the cable anchor points. In the center of the bridge and at the bridge ends positive moments were additionally introduced to create additional compression in the concrete roadway slab. The beam moments from life load increase up to 26 % from non-linear effects (P-Δ effect) because of the high beam slenderness of 1:200, Fig. 6.198.

In case any cable has to be exchanged two loadings were investigated. For an anticipated loading, e.g. in the case of corrosion, the adjacent lane would be closed to traffic and increased stresses are permissible.

In the case of a sudden loss of a cable, e.g. during an accident, no structural part is to suffer permanent deformations. The corresponding action forces were calculated by taking into account the torsional stiffness of the main girders supported transversely by the cross girders and the roadway slab.

The roadway slab is permanently under compression from dead loads: transversely as bottom chord of the cross girders, longitudinally from the compression component of the inclined stay cables. In the center of the bridge and at the bridge ends, where the compression forces from the stay cables are small, frozen-in positive moments are introduced into the beam by cambering the shop form.

As top flange of the composite beam these cambered moments cause additional permanent compression in the roadway slab. The crack widths from transient loads are limited by a substantial minimum reinforcement, consisting of bars ø 18 mm and at a distance of 115 mm crosswise, equal to a minimum content of 2.2 % reinforcement.

The shear studs were designed for the combined action of shear from overall and local loads, introduction of cable force and local transverse bending.

Assuming limited slip and plastic deformations in the ultimate limit state led to a uniform distribution of the shear studs over the length of the edge girder.

The introduction of the shear forces from the 300 mm thick concrete edge slab into the 200 mm thick inner roadway slab, Fig. 6.196, proved to be critical and was designed with a strut and tie model in the ultimate limit state.

Contractor's alternate

In the tender design it was proposed to cast the roadway slab onto the girder grid in the pre-assembly yard before installation, in order to carry the concrete weight in the transverse direction by composite action. The contractor, however, shied away from lifting the corresponding 250 t elements and proposed to install the roadway slab from precast segments after the installation of the girder grid.

The CIP joints were arranged on top of the cross girders; the shape of the joints, Fig. 6.199, was developed in collaboration between the engineers and the contractor.

This joint makes it possible to connect the upper layer of reinforcement with straight lap splices for good crack distribution. In

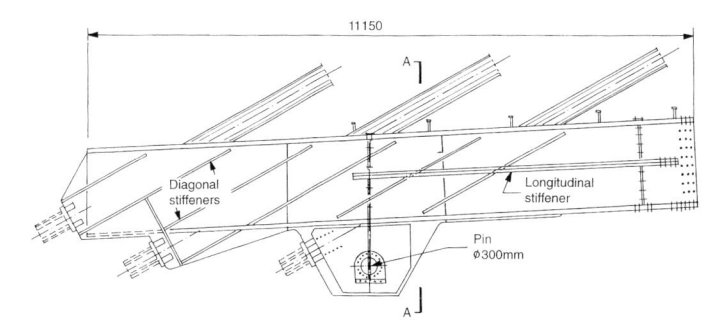

6.4 Composite cable-stayed bridge – Baytown Bridge

Figure 6.192 Box girder for anchorage of backstay cables – elevation

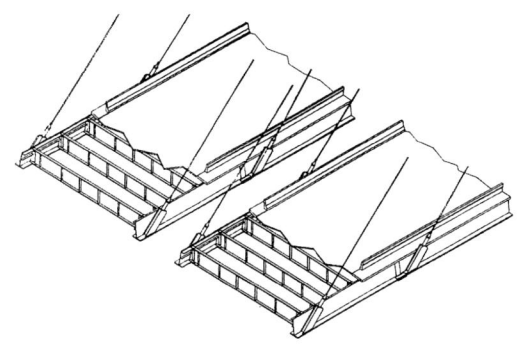

Figure 6.195 View of the twin beams

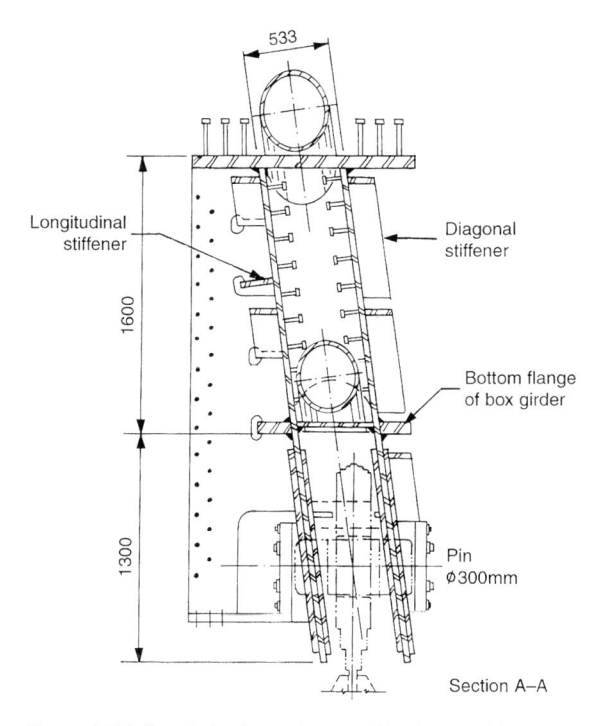

Figure 6.193 Box girder for anchorage of backstay cables – section

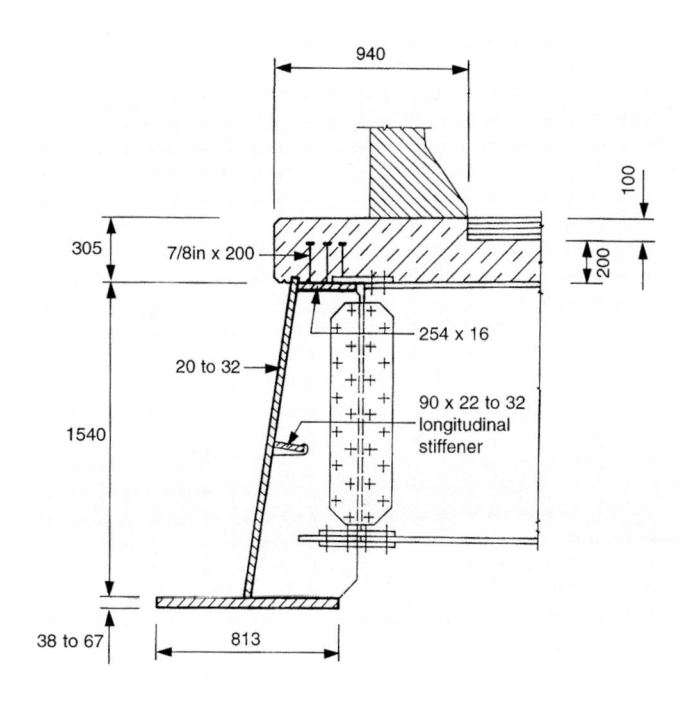

Figure 6.196 Main girder

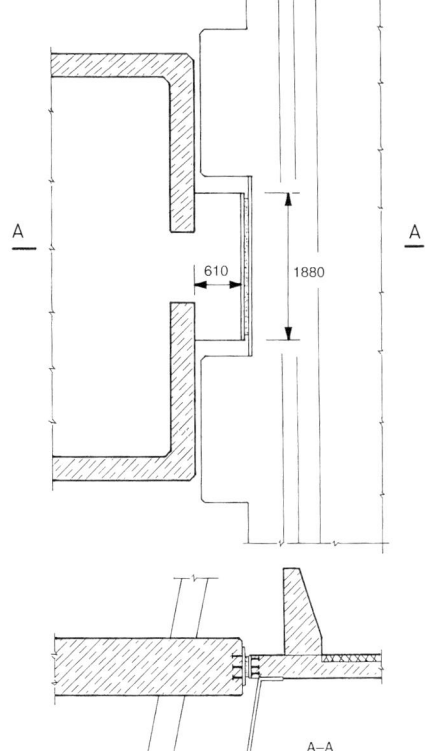

Figure 6.194 Movement limitations of beams at the towers

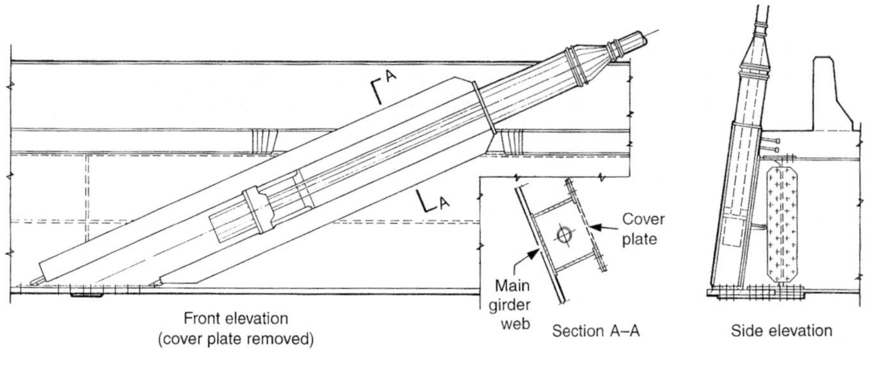

Figure 6.197 Cable anchorages at beam

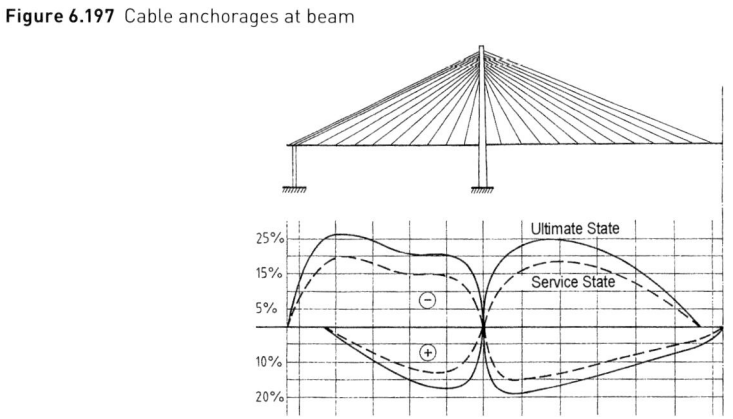

Figure 6.198 Non-linear increase of beam moments due to life load (P-Δ effect)

order to use the top flanges of the cross girders just as joint formwork, loop splices were used for the bottom layer of reinforcement, Fig. 6.199.

Due to the loss of composite action for the weight of the fresh concrete, auxiliary stiffeners were required for the installation of the girder grid. Their additional weight of 640 t increased the total quantity of steel for the girder by 17 % from 121 kg/m² to 141 kg/m².

6.4.1.3 Towers

The concrete towers in double diamond shape have box cross-sections with minimum wall thicknesses of 305 mm over most of their height, Fig. 6.200.

The shape of the towers resulted from the two beams. The upper A-frames increase the rotational stiffness of the beams by forming a stiff three-dimensional truss together with the stay cables and the beam. This increases the important first torsional frequency of the beam considerably.

By merging each pair of tower legs underneath the beam, the number of foundations is reduced to two.

In this way each twin tower forms a complete truss which carries the horizontal loads from wind by tension and compression only. The widths of the tower legs can thus be reduced to 2.13 m, constant over the full height of the towers.

In the longitudinal direction the towers act as a cantilever, especially during free cantilevering construction of the beam, and thus substantially greater dimensions are required. The ties underneath the beams carry the deviation forces from the kinks of the tower legs. They are fully post-tensioned.

In the concrete tower head the stay cables run through cast-in steel pipes. The anchor heads rest on steel base plates and are secured in place by shims, Fig. 6.201.

The horizontal components of the cable forces are carried by staggered loop tendons. This design had already been successfully used for the Burlington and Helgeland Bridges, [1.28, 2.80].

6.4.1.4 Stay cables

The stay cables comprise 20–61 strands ø 15 mm with wedge anchorages at both ends. They are protected by thick-walled PE pipes and were grouted after installation. Wrapping of the PE pipes with UV-resistant self-adhesive yellow PVF tape reduces the temperature changes of the stay cables from direct solar radiation [3.9] and improves the aesthetics of the bridge.

In order to avoid simultaneous failure of the three closely spaced backstay cables at each end of the bridge, for example by a burning petrol truck, these cables were additionally protected by outer steel pipes with a length of 15 m, the 25 mm clear space between the PE pipes and the steel pipes was filled with grout.

The stay cables were designed in accordance with the PTI guidelines, [3.47]. The cables were pre-assembled on the approach bridges.

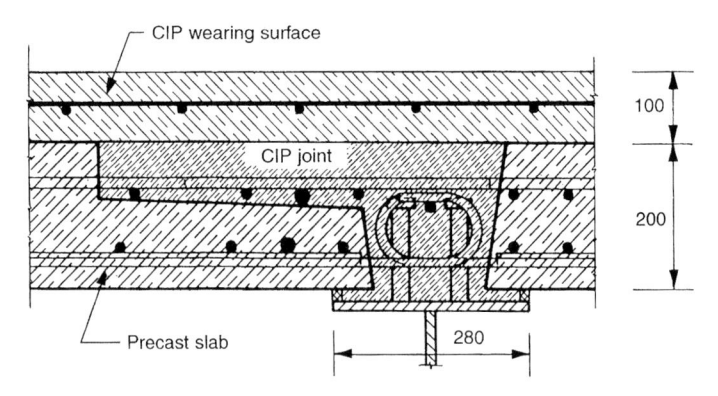

Figure 6.199 Transverse joint of the roadway slab

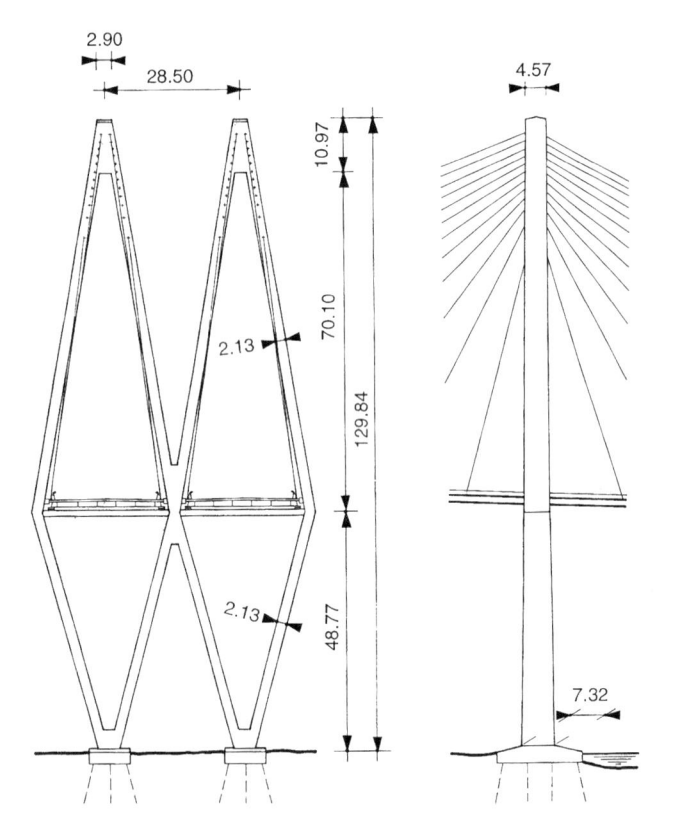

Figure 6.200 Towers

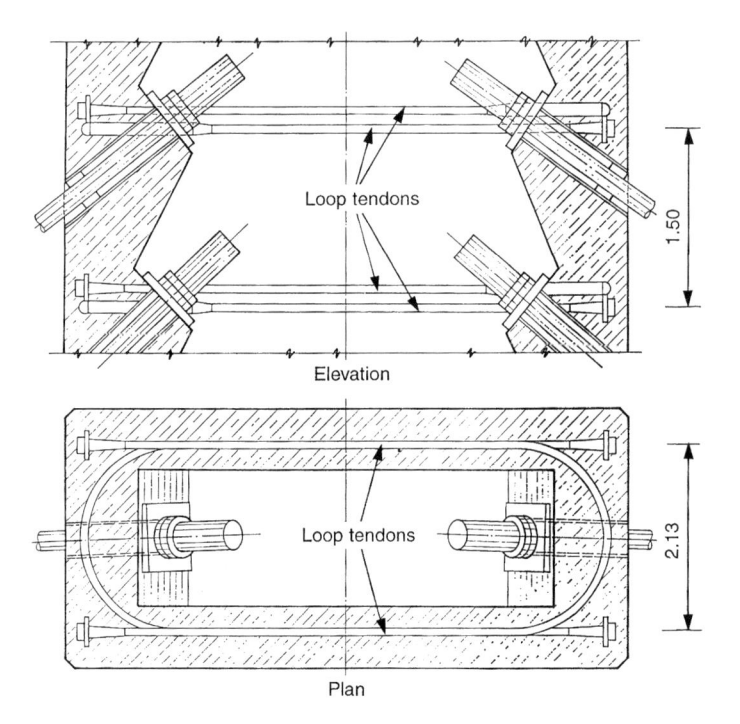

Figure 6.201 Cable anchorages in tower heads

6.4.1.5 Aerodynamic stability

The bridge is located in a hurricane-prone area. The wind design speed for a 100-year probability at 10 m height for a duration of 10 min. was assumed as 50 m/s. The ratio between the first eigenfrequencies in torsion and bending comes to $f_T/f_B = 0.670/0.273 = 2.45$. This favorably high value, in spite of the torsionally weak open beam cross-sections, was achieved by the use of A-towers, see Section 4.5.7.4.

Wind tunnel tests with a section model scale 1:96 [6.9] and a full model scale 1:250 [6.10], as well as an independent analytical investigation by Bob Scanlan [6.11] resulted in a critical flutter wind speed of more than 67 m/s. For a turbulence intensity of 12%, the bridge beam oscillates with a double amplitude of 1.65 m from buffeting.

The governing construction stage during cantilevering just before side span joint closures had critical wind speeds of only 47 m/s which was considered too low, taking into account the high hurricane probability. The beams were thus tied down during construction, Fig. 6.202. A hurricane actually crossed the bridge during this critical construction stage and the bridge behaved as anticipated.

The aerodynamic investigations showed that an energy transfer takes place between the oscillating windward beam and the leeward one, whereby energy is consumed, [6.10, 6.11]. For the actual geometric and dynamic characteristics of this bridge the unconnected twin beams are less sensitive against wind excitation than the individual beams would be by themselves.

6.4.2 Construction

6.4.2.1 Foundations

Floating foundations are used in the alluvial soils around the Houston Ship Channel, where rock is found only in great depth. They consist of 130 Nos, 50 cm square and 40 m long prestressed driven piles. Load tests indicated a bearing capacity of 200 t during service, Fig. 6.203.

The 3.6 m deep pile cap was reinforced with several layers of bars with ø 40 mm, Fig. 6.204.

The pouring of the more than 1000 m³ of concrete for each pile cap went through a day and a night.

6.4.2.2 Towers

The tower legs were built by free cantilevering with jumping forms in 4.5 m sections. For the first tower a formwork truss was used covering its full width, on which two cranes were running, Fig. 6.205.

The relocating of the large steel truss proved to be difficult. Therefore, smaller individual jumping forms for each tower leg were used for the second tower. The inclined tower legs were supported against one another by horizontal struts or ties, Fig. 6.206.

Finally, the steel truss was used again for casting the upper parts of the tower legs, Fig. 6.207.

All tower legs had wall thicknesses of 305 mm and two layers of reinforcement at 150 m, Figs 6.208 and 6.209.

Figure 6.202 Auxiliary tie-downs during construction

Figure 6.203 Pile foundations

Figure 6.204 Pile cap reinforcement

At the intersection of the inner tower legs at beam level 220, reinforcement bars ø 25 mm from each side had to cross. This reinforcement was pre-assembled in a cage with a weight of 18 t on the ground and then lifted. The bars were coupled with squeezed sleeves. In spite of the dense reinforcement a successful casting was achieved by the use of gravel with ø < 16 mm and a super-plasticizer, Fig. 6.210.

In the tower head the stay cables run through cast-in steel pipes and their anchor heads are retained on compression plates resting on concrete, Fig. 6.211.

The forms for the cable anchorage area were built in sections on the ground with the steel pipes precisely located, Fig. 6.212.

Around the inner forms dense reinforcement was placed, so that the precise location of the cable anchorage points in space was ascertained, Fig. 6.213.

After concreting, the short overlapping loop tendons were stressed against the hardened concrete. Fig. 6.214 gives a view of a completed tower head with anchorage steel pipes and pockets for the loop tendons.

An overall view of the site with the completed approach bridges and nearly completed towers is shown in Fig. 6.215.

6.4.2.3 Beam

The steel grids for the beams were, for economic reasons, fabricated in Cape Town, South Africa, Fig. 6.216. The supervision was done by members of the Texas Department of Transportation and the designers.

After shipment, the final corrosion protection was applied in the pre-assembly yard in Texas and the grid sections were then test bolted and surveyed, Fig. 6.217.

The first short starter elements of the beam in the tower axis were lifted with tower cranes, Fig. 6.218. On them the derricks were installed with which all other beam elements were lifted, Fig. 6.219.

Above water the grid elements were delivered from the pre-assembly yard by barges and lifted to the beam by the derricks, Fig. 6.220.

Since no welding on site is acceptable in the USA even today, the main girders were connected with cover plates and high-strength bolts, Fig. 6.221. (For this reason orthotropic decks are still not used in the USA today because they require field welding.)

The precast slabs were also delivered by barges, Fig. 6.222, and lifted up by the derricks, Fig. 6.223.

The precast slabs span between the cross girders. The shear studs protrude from their top flanges into the slab joints, Fig. 6.224.

The joints were cast with low-shrink rapid-hardening joint concrete, Fig. 6.225. They had cured before the next beam element was installed.

The final beam sections in the side span are cantilevered into the first approach span. In them the concentrated three backstay cables are anchored, Fig. 6.226.

Fig. 6.227 shows the bridge beam before side span joint closure. In Fig. 6.228 only the center joint is left.

Figure 6.205 Concreting of the first tower legs

Figure 6.206 Casting of the lower tower legs for first and second towers

Figure 6.207 Casting of upper tower legs

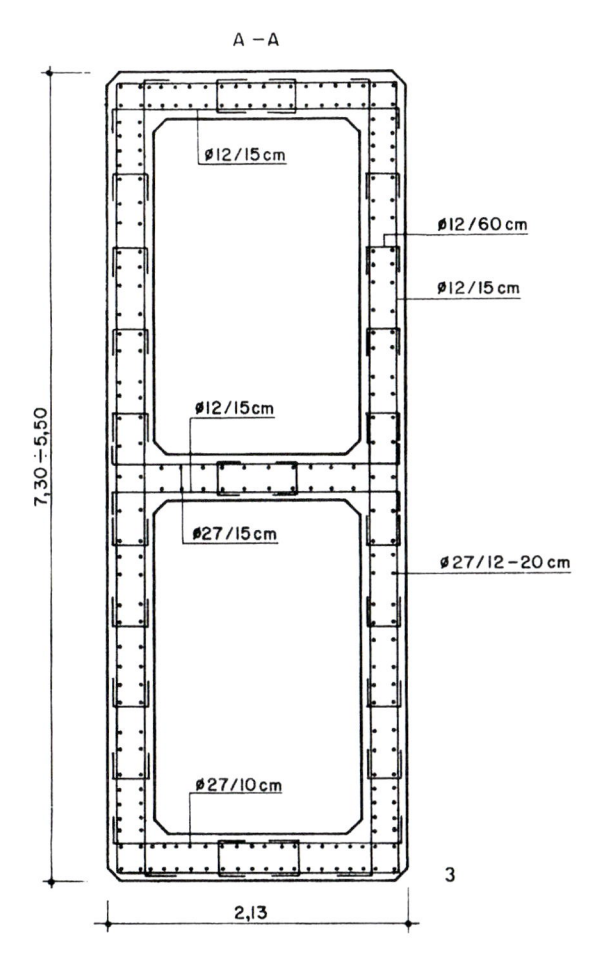

6.4 Composite cable-stayed bridge – Baytown Bridge

Figure 6.208 Typical reinforcement of tower legs

Figure 6.210 Pre-assembled reinforcement cage

Figure 6.209 Reinforcement of tower legs

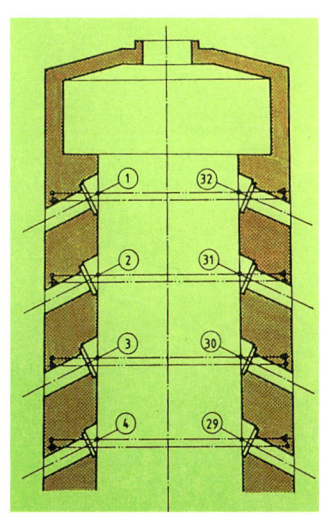

Figure 6.211 Cable anchorage in tower head

Figure 6.212 Formwork with cable anchor pipes

Figure 6.213 Anchorage pipes with reinforcement

Figure 6.214 Scaffolding for cable installation at tower head

Figure 6.215 Finalizing the towers

Figure 6.216 Shop fabrication in Cape Town, South Africa

Figure 6.217 Pre-assembly yard in Texas, USA

6.4 Composite cable-stayed bridge – Baytown Bridge

Before center joint closure the two cantilever tips were carefully surveyed to adjust the end piece to the actual conditions, Fig. 6.229. This survey takes best place shortly before sunrise when the temperature gradient in the beam is equalized.

Finally, the closure piece was lifted-in and bolted, Fig. 6.230. During free cantilevering the fixed points of the beams were located at each corresponding tower. Before side span joint closure the beam was shifted outwards with jacks between the corbels at the towers, so that the neoprene bearings, Fig. 6.191, were deformed to the outsides. Simultaneously the anchor piers were pushed 75 mm to the outside. After shrinkage and creep, the neoprene bearings and the anchor piers are supposed to end up vertically.

The construction of the first half of the bridge was delayed substantially due to difficulties during fabrication of the steel grids. Originally the 'Buy American Clause' applied for the steel, which was lifted after these costs were too high, and the steel was procured and fabricated in South Africa. The second beam half, however, was built in only five months.

Figure 6.218 Lifting of starter grids

Figure 6.219 Free cantilevering of the twin beams to both sides

Figure 6.220 Lifting of girder grid from barge

Figure 6.221 Joining the main girders with high-strength bolts

Figure 6.222 Delivery of precast slabs

Figure 6.223 Lifting of precast slabs onto girder grid

Figure 6.224 Joint between precast slabs

Figure 6.225 Casting of joints

6.4 Composite cable-stayed bridge – Baytown Bridge

Figure 6.226 Lifting of the main girder ends in which the backstay cables are anchored

Figure 6.227 Before first side span joint closure

Figure 6.228 Center joint

Figure 6.229 Surveying of the center joint before sunrise

Figure 6.230 Inserting the closure piece

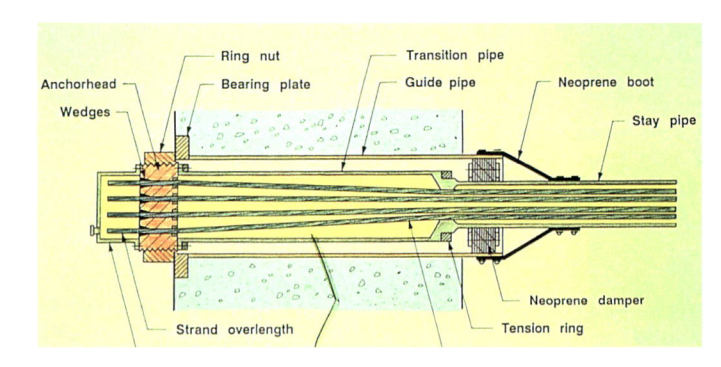

Figure 6.231 Parallel strand cable system VSL

6.4.2.4 Stay cables

For economic reasons, parallel strand cables were selected by the contractor. They were delivered in their individual components. These were assembled on site and the completed stay cables were installed and stressed like completely shop-fabricated cables.

The assembled stay cables are in accordance with the state-of-the-art of that time: 15 mm galvanized strands with anchorages in thick-walled black PE pipes, Fig. 6.231, which are filled after installation with cement grout, Fig. 6.232.

The PE pipes were butt welded to their final lengths on the approach bridges and the strands were pulled into them, Fig. 6.233.

The ends of the strands were fixed with wedges to the anchor heads, Fig. 6.234. At the passive anchorage at the beam, the wedges were finally pressure wedged and secured, whereas at the active stressing ends for the tower heads they were only pre-wedged.

The pre-assembled stay cables were pulled to the main bridge supported by a highline, Fig. 6.235.

The pull rope up to the tower head was connected to a guide at the stay cable in order not to damage the PE pipe by too sharp bends, Fig. 6.236. The cables were then lifted up to the tower heads, Fig. 6.237.

Since the anchor box and the bottom flange of the main girder did not allow the stay cables to be stressed at the beam anchorage, the anchor heads were inserted there and secured against slipping by shims, Fig. 6.238.

The inclination of the cable tip was precisely adjusted to the steel pipe at the end of lifting, Fig. 6.239.

Then the stay cables could be pulled into the tower head, Fig. 6.240.

Sufficient space was provided inside the tower head to install the jack chair and the hydraulic jack, Fig. 6.241. A grip hoist at the end of the jack pulled the cable close enough so that the protruding strands could be gripped by the jack.

Fig. 6.242 shows the stressing of a complete stay cable by gripping the protruding strands with a hydraulic jack.

After final stressing and grouting of the stay cables they were wrapped with a yellow, UV-resistant PVF tape for aesthetic reasons, Fig. 6.243.

The yellow color forms a strong contrast to the hot sun and the deep blue sky in Texas, Fig. 6.244.

6.4.2.5 Completed bridge

The completed bridge, Fig. 6.245, was generally acclaimed, and was honored with several awards, including the Federal 'Presidential Award', the highest bridge prize in the USA.

From a driver's view the four A-shaped cable planes create a tent like impression which gives confidence and a feeling of security to the users, Fig. 6.246. During sunset at the Gulf of Mexico the large bridge appears quite delicate, Fig. 6.247. At night the towers and cables are uniformly floodlit from the beam, Fig. 6.248.

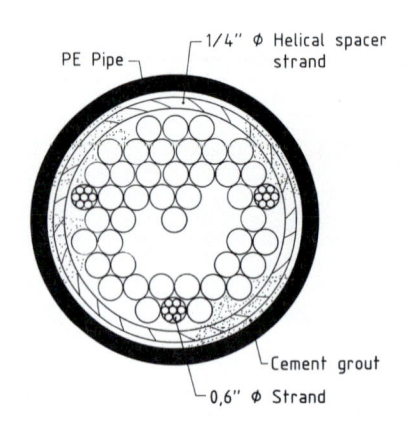

Figure 6.232 Stay cable cross-section

Figure 6.233 Cable fabrication on approach bridge

Figure 6.234 Inserting strands into anchor heads

6.4 Composite cable-stayed bridge – Baytown Bridge

Figure 6.235 Transportation of stay cables with highline

Figure 6.238 Cable anchorage at beam

Figure 6.236 Attaching the guide for lifting the stay cables

Figure 6.239 Before inserting the stay cable into the tower head

Figure 6.240 Inserting the stay cable into the tower head

Figure 6.237 Lifting of the stay cables

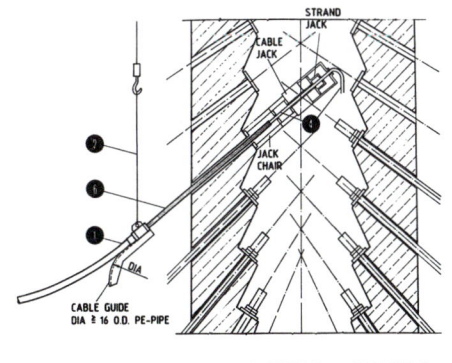

Figure 6.241 Stressing stay cables inside the tower heads

400 6 Examples for typical cable-stayed bridges

Figure 6.243 Wrapping the stay cables with yellow PVF tape

Figure 6.242 Stressing a complete stay cable

Figure 6.244 Wrapped stay cables

Figure 6.245 Aerial view

6.4 Composite cable-stayed bridge – Baytown Bridge

Figure 6.246 Longitudinal view

Figure 6.247 Sunset at the Gulf of Mexico

6.4.3 Summary

The bridge was opened for traffic on 27 September 27 1995, and named the Fred-Hartmann-Bridge. In the 'Outstanding Civil Engineering Award' for 1996 it says: 'The bridge with its succinct appearance seems to float above the water. It uses new characteristics in design and construction. The double diamond towers with a height of 44 stories have not been built before. The bridge is an important contribution to the progress in structural engineering.'

Participants

Client State of Texas
Designers Greiner, Inc., Tampa, Florida, and Leonhardt, Andrä and Partners GmbH, Stuttgart, Germany, who also supervised construction
Consultant for aerodynamics Dr. Robert Scanlan
Contractors Williams Brothers Construction Co. and Traylor Bros. Inc., for whom DRC Consultants, New York, did the construction engineering, which was checked by Greiner-LAP
Stay cables VSL Corp.

Figure 6.248 Floodlit towers and stay cables

Figure 6.249 Normandy Bridge, France

6.5 Hybrid cable-stayed bridge – Normandy Bridge
6.5.1 Design considerations

The Normandy Bridge was completed on 8 August 1994. Its main span length of 856 m made it by far the largest cable-stayed bridge of the world, Figs 6.249 and 6.250.

Three precursors in Germany with one only tower had previously indicated that such spans were possible: The Severins Bridge in Cologne with 302 m main span in 1959, Fig. 2.76, the Knie Bridge with 320 m main span in 1969, Fig. 2.71, and the Flehe Bridge with 368 m main span in 1979, Fig. 1.91. During the design period of the Normandy Bridge (originally called Honfleur Bridge) the span record for bridges with two towers was twice exceeded: In November 1991 by the Skarnsundet Bridge in Norway with a main span of 530 m, Fig. 2.145, and by the Yang Pu Bridge in Shanghai in 1993 with a main span of 602 m, Fig. 2.165a. (The current record holder is the Sutong Bridge in China with 1088 m from 2008, Fig. 2.105.)

The governing design assumptions for the Normandy Bridge are [1.29, 6.12–6.14]:
- streamlined cross-sections for main and side spans in order to reduce the wind loads and to improve the aerodynamic stability
- closed box girder cross-sections in order to achieve a high torsional stiffness, which, together with A-shaped towers, increases the torsional frequency and thus the aerodynamic stability
- A-towers also for achieving a high transverse stiffness
- a light steel girder in the main span and heavy concrete girders in the side spans as counterweights (for this combination of steel main span and concrete side spans, the author uses the designation 'hybrid bridge')
- the concrete side span beams protruding 116 m into the main span for improved stiffness and economy
- use of high-strength concrete B 60
- composite anchorage of the parallel strand cables in the tower heads
- special provisions for improving the aerodynamic stability of the stay cables which are sensitive because of the large span and the light steel cross-section.

The engineer in charge of the design and construction supervision was Michel Virlogeux, Fig. 2.101.

6.5.1.1 Structural design

The complete bridge beam – approaches, side spans and main span – has a similar box girder cross-section with a depth of 3 m (slenderness = 3 : 624 = 1 : 208) and inclined outer bottom slabs, Figs 6.251 and 6.252.

The 203 m high concrete A-towers also have box girder cross-sections, Fig. 6.253. The stay cables are anchored at the upper part of the tower in composite boxes, Fig. 6.254. The horizontal tensile forces between forestays and backstays are directly connected by the longitudinal steel plates. The vertical cable components are introduced into the tower legs via the vertical steel plates.

The 8 × 23 = 184 parallel strand stay cables use monostrands, Fig. 6.255. They are anchored with wedges in retainer plates at the ends of the anchor heads, Fig. 6.256.

6.5 Hybrid cable-stayed bridge – Normandy Bridge

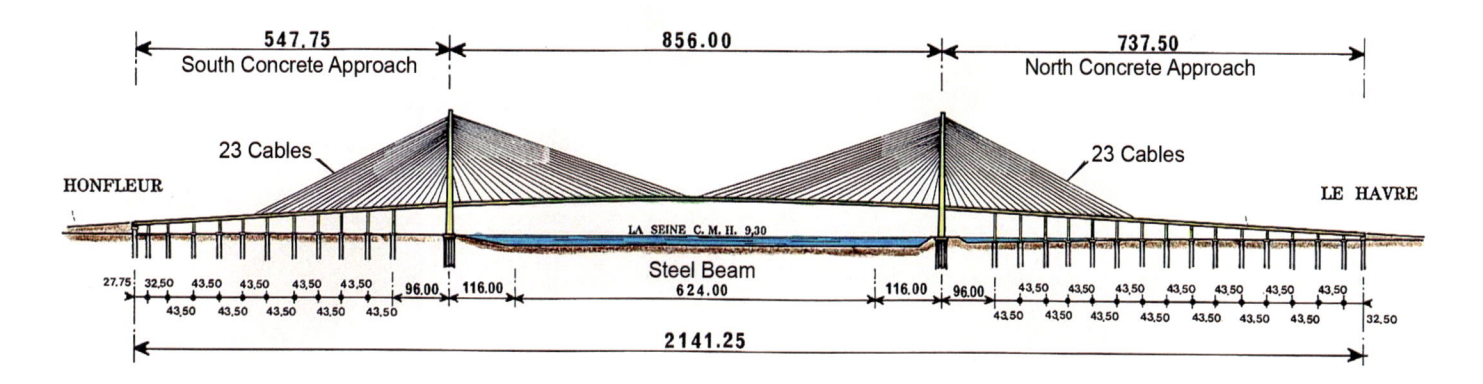

Figure 6.250 Elevation

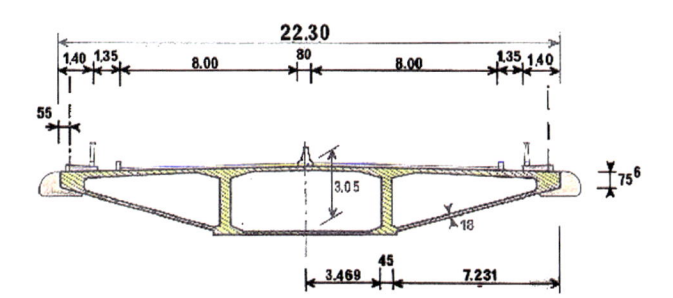

Figure 6.251 Concrete cross-section of approach bridges

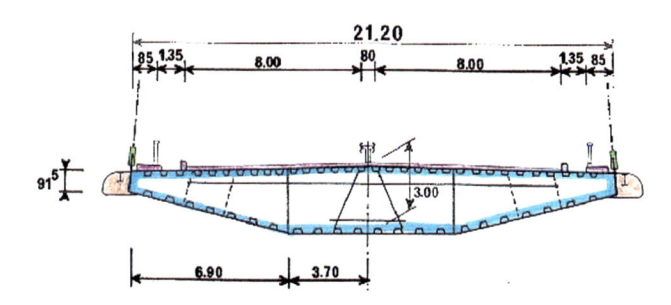

Figure 6.252 Steel cross-section of main span

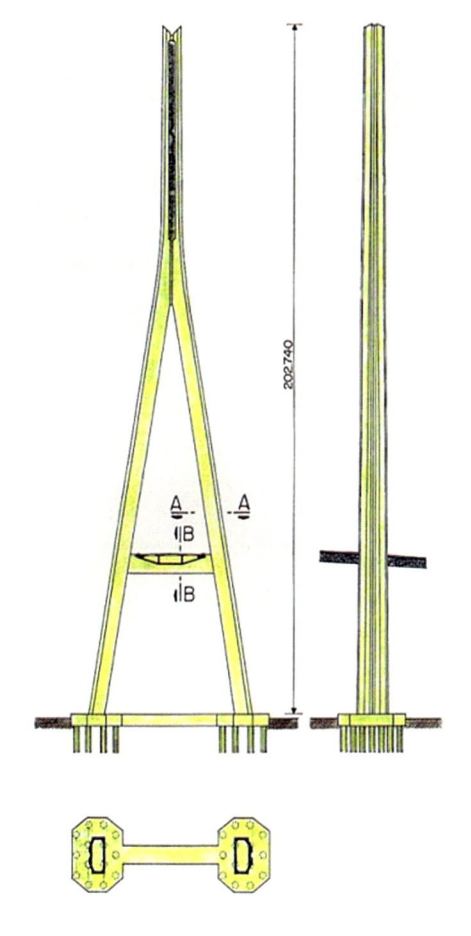

Figure 6.253 Towers

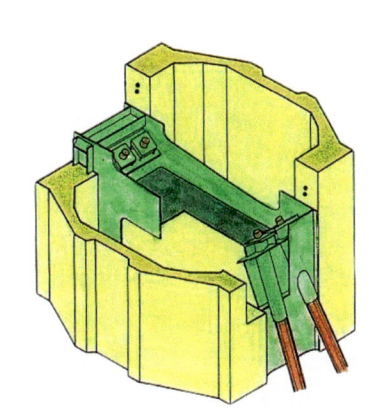

Figure 6.254 Cable anchorages in the tower

6.5.1.2 Cable dynamics

Since the longest cables have a record length and because the steel beam in the main span is rather light, several provisions were selected to stabilize the cables against oscillations.

First provision: In order to prevent rain–wind-induced cable oscillations the PE pipes were provided with outer helical fillets, Fig. 6.257, which prevent the formation of upper and lower rivulets that can cause substantial cable oscillations due to galloping. Fig. 6.258 indicates how rain–wind-induced cable oscillations can be suppressed with such helical fillets.

Second provision: Between the stay cables ties were installed, Fig. 6.259, which increase the eigenfrequency of the stay cables by providing additional elastic supports. In this way their sensitivity against parametric and anchorage excitation is strongly reduced.

Third provision: Near the cable anchorages at the beam hydraulic dampers are installed which dampen all types of cable oscillations, Fig. 6.260.

The driver's view along the bridge shows clearly the tie-downs and the hydraulic dampers, Fig. 6.261. The appearance of the bridge is in this way slightly impaired. (For this reason sometimes bridge photographs are used which were taken before installation of the tie-downs!)

6.5.2 Construction

6.5.2.1 Tower

Load-bearing limestone is found in 40 m depth only. All foundations thus rest on concrete piles with 1.5 to 2.1 m diameter and lengths of up to 55 m, Fig. 6.262.

Underneath the north approach a 4 m thick layer of silt is located which required the construction of a temporary construction bridge from which all foundation works were executed.

The pile caps are 3.5 m deep, transversely post-tensioned and highly reinforced, Fig. 6.263.

The tower legs are monolithic with the pile caps. They were built by free cantilevering with sliding forms, Fig. 6.264. Up to the cross girders underneath the deck the bending from transverse inclination did not require lateral supports.

Between deck level and the point where the two tower legs meet one temporary steel truss strut was required, Fig. 6.265.

The steel elements for the cable anchorage in the tower top were shop-fabricated, Fig. 6.266, and transported to the site.

They were lifted into their final position with the tower crane, Fig. 6.267.

Finally the steel anchor boxes were surrounded with concrete in order to achieve the desired composite action for bending and transfer of the vertical cable components into the tower legs, Fig. 6.268.

Fig. 6.269 shows the completed tip of the tower which combines excellent flow of forces with good appearance.

Figure 6.255 Monostrand

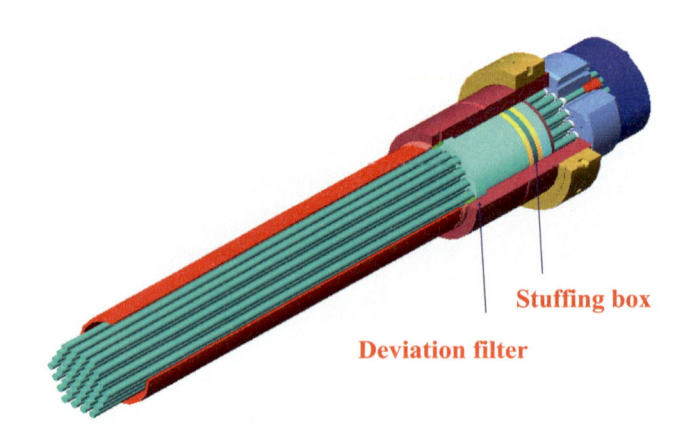

Figure 6.256 Parallel strand cable (Freyssinet)

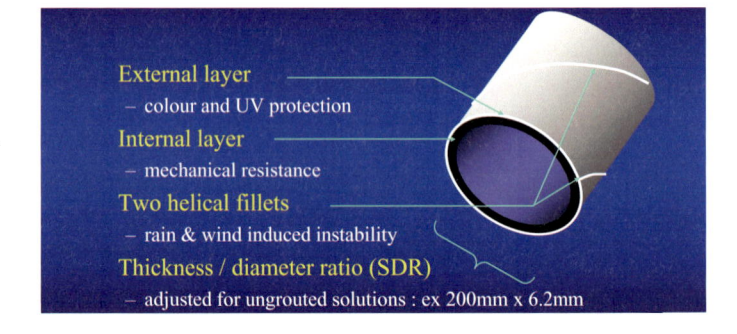

Figure 6.257 PE pipe with outer helices

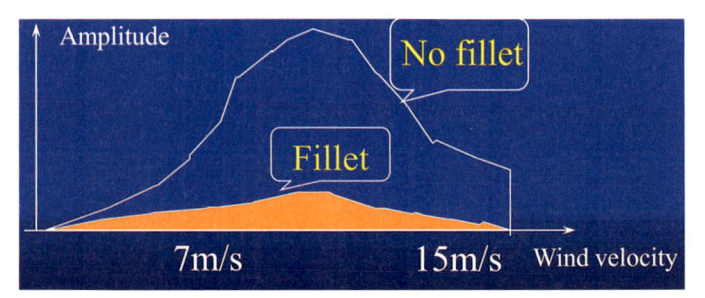

Figure 6.258 Suppression of cable oscillations with helical fillets

6.5 Hybrid cable-stayed bridge – Normandy Bridge

Figure 6.259 Stay cable tie-downs

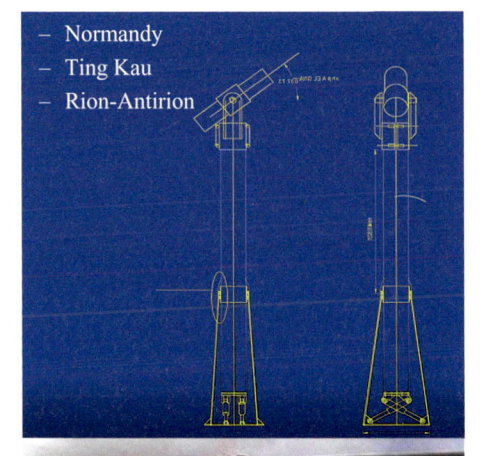

Figure 6.260 Hydraulic cable dampers at beam

Figure 6.261 Stay cables with tie-downs and hydraulic dampers

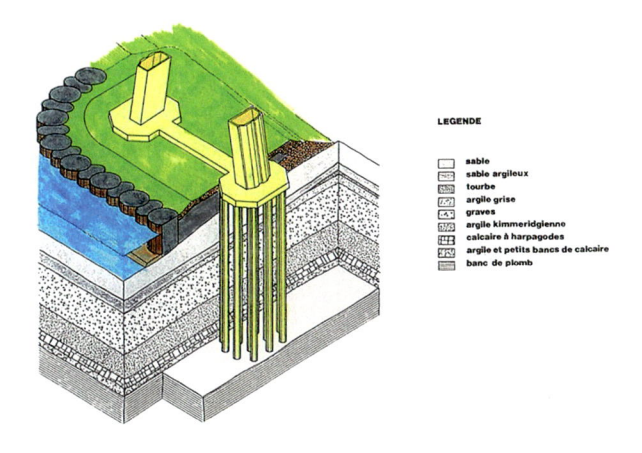

Figure 6.262 Tower foundation

Figure 6.263 Pile cap during construction

Figure 6.264 Start of free cantilevering of the tower legs

Figure 6.265 Free cantilevering with transverse temporary strut

6.5.2.2 Concrete approach bridges

The construction sequence of the beam is shown in Fig. 6.270. After casting of the towers, the outer approaches were built by incremental launching. The last part of each side span which protrudes into the main span was built by free-cantilevering to both sides of the towers outwards. Then followed the free-cantilevering of the remaining steel main span.

The seagoing vessels on the Seine River required a navigational clearance of 56 m. In order to keep the lengths of the approach bridges to a minimum, they are given the maximum possible slope of 6 %. This steep grade required a modification of the standard incremental launching method which would have resulted in too large horizontal forces of 30 MN at the abutments and high friction forces at the pier heads, Fig. 6.271. The so-called 'lift-launch method' was used which separates every inclined launching increment of the deck into its 150 mm horizontal and 9 mm vertical movements.

Thereby each concrete box girder is launched on wedges which slide across the pier heads. The undersides of the wedges are horizontal, their inclined surfaces have the bridge slope of 6 %. The box girder is shifted alternately 150 mm horizontally with jacks located at the abutments and lifted by 9 mm with jacks located on each pier head outside of the wedges.

Special sensors for the control of the horizontal and vertical movements were located at each pier head. Their results were combined in a central computer which controlled every single movement. It was especially important that the vertical movements were synchronized.

Fig. 6.272 shows the usual steel launching nose at the tip of the concrete beam for the reduction of the cantilever moments.

The final part of the concrete beam at both sides of the towers was built by free-cantilevering, Fig. 6.273. The 116 m long part protruding into the main span was built in 42 sections, the backwards parts in 31 sections. The concrete was placed alternately in the river and shore side in order to keep the cantilevers in balance. Each new section was supported by temporary stays.

After every five sections supported by temporary stays a final stay cable was installed into the sixth section and the temporary stays were removed for reuse, Fig. 6.274.

In a general investigation the author has shown that for major steel main spans it is advantageous to build the outside parts in concrete with lengths in the range of 120 m, independent of the main span length. In this region it is economical to carry the high compression forces by concrete, and the stiffer concrete beam increases the eigenfrequencies of the main span, thereby improving its aerodynamic stability. The cost advantage has to be compared with the additional costs required for another construction method, in this case free cantilevering in addition to incremental launching. As a consequence, the protruding concrete cantilever in many cases is only extended as far as the original construction method for the approach beams permit. For the Normandy bridge that would be about 45 m,

Figure 6.266 Fabrication of the steel cable anchorages for the tower

Figure 6.267 Installation of steel cable anchorages

Figure 6.268 Casting-in of the steel tower anchorages

6.5 Hybrid cable-stayed bridge – Normandy Bridge

Figure 6.269 Completed tower top

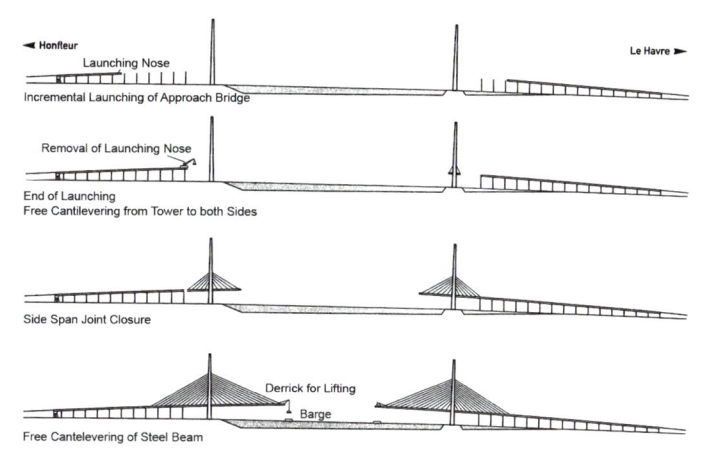

Figure 6.270 Construction sequence

Figure 6.271 Fabrication of an approach bridge by the lift-launch method

Figure 6.272 Launching nose

Figure 6.273 Free-cantilevering of the concrete beam in main and side span

Figure 6.274 Completed concrete beam

but at least one auxiliary pier to cross the last side span with 96 m would have been required additionally on each side. The most economic solution has thus to be carefully investigated for every bridge.

6.5.2.3 Steel main span
The 624 m long steel beam of the main span comprises 32 sections with lengths of 19.65 m. In order to save time the fabrication took place in two different shops, Fig. 6.275.

After successful pre-assembly, the sections were transported to the bridge on barges, Fig. 6.276.

The elements were lifted with erection derricks located at the beam tips, Fig. 6.277. The beam was strong enough and the elements light enough so that no tie-backs were required for the derricks.

During installation of the final sections a dynamic tuned mass damper (TMD) with a total mass of 50 t was installed at both cantilever tips to reduce the transverse oscillations due to strong wind. A similar system is shown for the Millau Bridge in Fig. 4.100.

Fig. 6.278 shows the lifting of the final element before installation of the final short fitting piece.

6.5.2.4 Cable installation
Modern parallel strand cables are fabricated on site from their individual components, thereby avoiding the transportation of heavy shop-fabricated cables whose weights and reel diameters increase with the cable length. Fig. 6.279 shows the fabrication process in principle.

The monostrands are delivered to the site on specially built reels with brakes, Fig. 6.280.

The strands are pulled up inside the PE pipe to the tower head with the help of a to and fro running shuttle element, Fig. 6.281. Fig. 6.282 shows a PE pipe nearly filled with monostrands.

For stressing, the strands protrude well beyond the anchor plates, Fig. 6.283. Since each strand is stressed individually only small monojacks are required, Fig. 6.284. This is an additional advantage against shop-fabricated cables which have to be stressed as a unit with several-ton heavy hydraulic jacks.

By means of the so-called isotensioning method (Freyssinet) all strands receive the same force. The first pilot strand of each cable is stressed to a precalculated value which is continuously controlled by a load cell. Each newly installed strand is stressed to that level which the pilot strand has at that moment. In the end all individual strands and thus the complete cable have their theoretical force, independent of temperature conditions and random erection loads. The development of the forces in each strand during the installation process is shown in Fig. 6.285.

The PE pipes of the Normandy bridge consist of two half-shells which are connected by a snap-in tongue-and-groove joints, Fig. 6.286.

Finally the tie-down cables were installed using abseiling techniques, Fig. 6.287.

Figure 6.275 Shop-fabrication of a steel section

Figure 6.276 Transportation on a barge

Figure 6.277 Erection derrick

6.5 Hybrid cable-stayed bridge – Normandy Bridge

Figure 6.278 Lifting of last beam section

Figure 6.282 PE pipe nearly filled with monostrands

Figure 6.283 Protruding strands

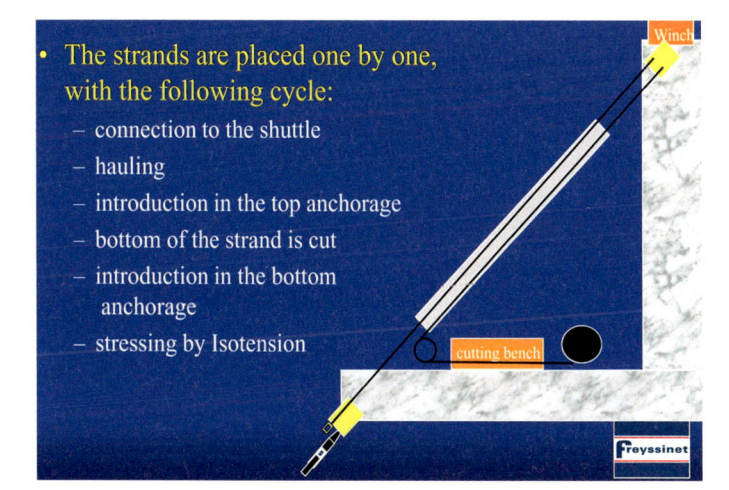

Figure 6.279 Installation of the individual strands

- The strands are placed one by one, with the following cycle:
  - connection to the shuttle
  - hauling
  - introduction in the top anchorage
  - bottom of the strand is cut
  - introduction in the bottom anchorage
  - stressing by Isotension

Figure 6.284 Monojacks for stressing individual strands

Figure 6.280 Monostrands on installation reels

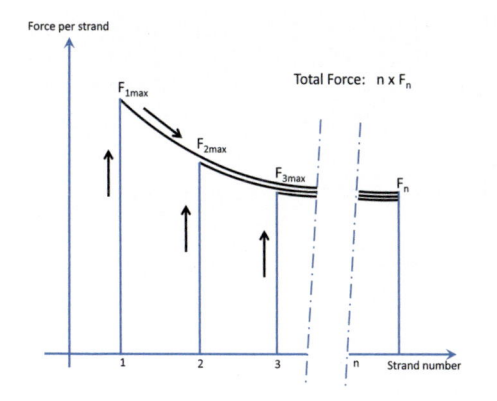

Figure 6.285 Isotensioning of strands (Freyssinet)

Figure 6.281 Feeding of a strand into the PE pipe

Figure 6.286 PE pipe with tongue and groove joint

Figure 6.287 Installation of tie-downs using abseiling techniques

6.5.2.5 Completed bridge

The completed bridge looks slender and elegant, Fig. 6.288. Seen against the setting sun the bridge appears very fragile, the cables giving the impression of a veil, Fig. 6.289.

Participants

Client Chamber of Commerce and Industry Le Havre, France
Designer Michel Virlogeux, Bonnelles, France
Contractor JV Monberg & Thorsen, Bouygues, Campenon Bernard
Aerodynamic Consultants Robert H. Scanlan and Allan Davenport
Construction Engineering COWI
Stay Cables Freyssinet

Figure 6.288 Overall view

Figure 6.289 Normandy Bridge, Sunset

6.6 Series of cable-stayed bridges
6.6.1 Millau Bridge
6.6.1.1 General

The Millau Bridge and the Rion-Antirion Bridge are the most modern and most important examples for series of cable-stayed bridges.

They answer the main questions for series individually:
– How can the inner towers be stabilized to avoid too large beam deflections due to spanwise live loads?
– How can the changes in beam lengths of these long bridges, e.g. due to temperature,be taken into account, together with longitudinal forces, e.g due to breaking, together with the forces from earthquake?

In 2004 the large viaduct of Millau was opened for traffic, thereby closing the last gap in the second north–south highway A 75, Fig. 6.290, [2.123, 6.15 – 6.18].

The bridge has earned its title 'Record Bridge' not only because it is the highest in the world, crossing the Tarn River at a height of 270 m, but also the construction period of only 38 months and the costs of about € 400 m are outstanding, Fig. 6.291.

For the second north–south highway the main obstacle was the wide valley of the Tarn River near the city of Millau, well-known for its seemingly endless queues during the summer. It was decided to cross the valley by a 2 460 m long viaduct.

In 1996 a series of cable-stayed bridges with two side spans of 204 m, and sixmain spans of 342 m was selected. For the financial success of the concession company it was important to keep the construction period to a minimum in order to receive tolls as early as possible. For this reason the French contractor Eiffel Construction Métallique proposed an alternate with steel beam and steel towers above deck against the tender design of a post-tensioned concrete bridge. This alternate was selected in March 2001 and construction started in October 2001, Fig. 6.291.

The advantages of the steel design against the original concrete design are:
– light-weight and slenderness of the beam (36 t against 120 t)
– reduction of the depth of the beam to 4.2 m, meaning smaller wind loads
– improved safety: less work at great height due to pre-assembly on ground and incremental launching construction
– minimizing the number of stay cables and the size of foundations
– reduction in total costs – the over-riding advantage.

Within two and a half years nearly 43 t of steel were fabricated for the beam, the towers and the auxiliary piers.

For the complete design phase Michel Virlogeux was the engineer-in-charge for the French Highway Authority SETRA, and during the construction phase he was an important advisor, Fig. 2.101.

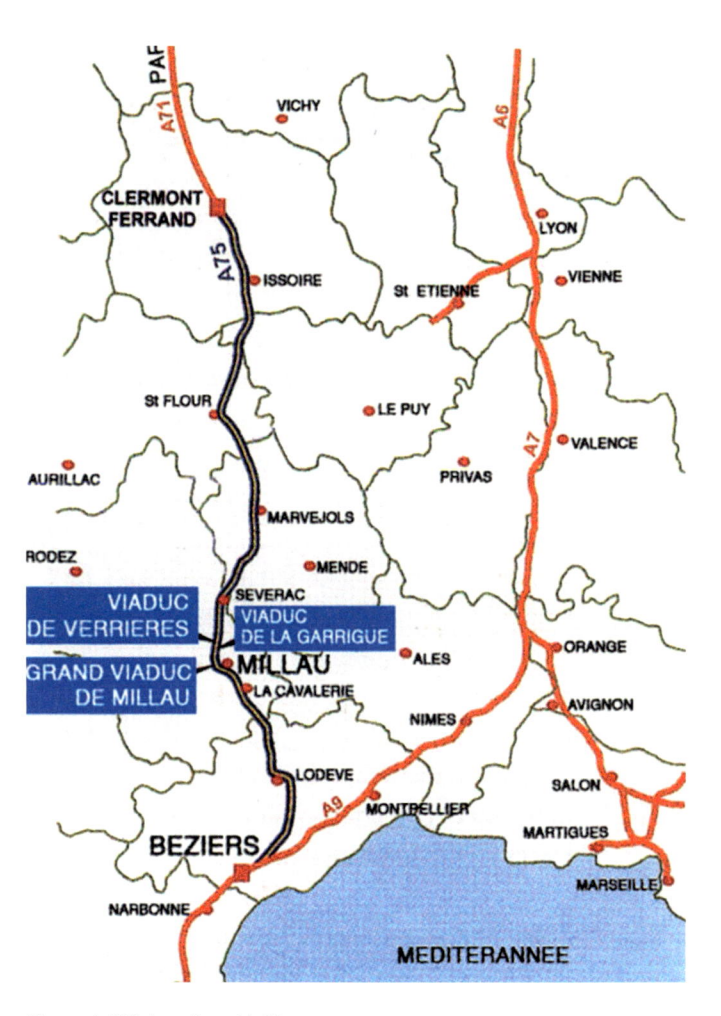

Figure 6.290 Location of bridge

Figure 6.291 Overall view

Figure 6.292 Elevation

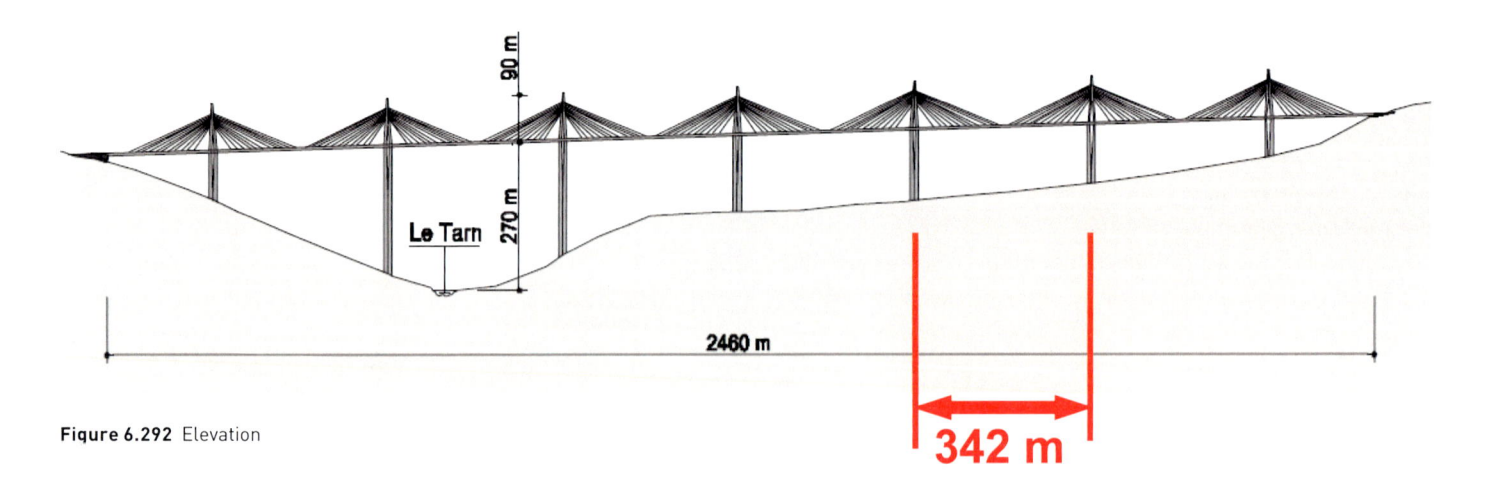

6.6.1.2 Design

The Millau Bridge has a total length of 2 460 m and comprises eight spans: two side spans with 204 m and six inner spans with 342 m, Fig. 6.292.

The cross-section consists of a steel box with orthotropic deck, two vertical inner webs and two inclined outer plates. The vertical webs are required for construction by incremental launching, and the triangular outside boxes create a streamlined cross-section which reduces the wind load on the bridge, Fig. 6.293. On the outside of the beam, wind shields are installed which prevent the overturning of high-sided vehicles. The rounded edges on both outsides improve the aerodynamic behavior and the appearance.

Due to the great height above the valley a central girder, with one cable plane only, was the most effective design for avoiding twin piers. The required torsional stiffness is provided by the box girder.

Governing the design of the bridge was the condition that the up to 230 m high piers are stiff enough to carry unsymmetrical loads in the longitudinal direction. Also they have to be flexible enough to follow the beam's longitudinal changes due to temperature. The solution is a strong concrete box girder for the piers with a vertical slot underneath the beam, Fig. 6.294. This bisectioning of the box provides for the required flexibility.

The 87 m high towers above deck form stiff A-frames in the longitudinal direction. The spread legs of the towers meet at the beam the spread upper pier halves. In this way an overall system from pier and tower is created which converts the moment from loads acting at the tower tips into a couple of tension and compression and thus restrains the tower head. The 90 m long pier shafts are vertically post-tensioned to counteract the tensile stresses due to wind in the inner piers and due to temperature changes in the outer piers.

Bearings are located between the steel girder and the concrete piers, which are stressed down against uplift. Unsymmetrical loads and extreme winds cause support reactions on each pier of up to 100 MN. A new type of spherical bearing was used. The towers above deck are made from steel to make them as light and slender as possible, Fig. 6.295. The central cable plane is anchored at the tower heads between longitudinal steel plates.

Due to architectural reasons the towers extend beyond the upper cables.

6.6.1.3 Construction

Piers

The piers have varying cross-sections, Fig. 6.294, but dimensions were selected in such a way that they could be formed easily. Four sides have constant dimensions, and the others vary uniformly in each construction section. This permits construction with outer jumping forms and inner forms which are stepwise lifted by the tower crane, Fig. 6.296. Sliding forms could have made the correct positioning of the built-in items difficult.

Beam

About 2100 stiffened panels, four per work day, were fabricated from plates and stiffeners in the shops of Eiffel Construction Métallique in Lauterbourg (Alsace). After transportation to site they were welded in the two 170 m long pre-assembly yards, Fig. 6.297. The central boxes were prewelded ahead in Fos-sur-Mer. Up to 75 welders were working in each pre-assembly yard. Fig. 6.298 shows a complete cross-section during installation.

Launching the beam

The detailed design and construction engineering of the steel parts – beam, towers and auxiliary piers – was provided by the consultant Greisch from Liège, Belgium. The engineer-in-charge was Jean-Marie Crémer, Fig. 6.299.

During launching from both abutments several special measures were introduced to control the large cantilever moments. The spans were reduced by half with temporary telescopic piers, a launching nose was used and the front tower with some of the final stay cables was launched together with the beam. These cables were not adjusted during launching! After joint closure on 28 May 2004, the remaining towers and stay cables were installed and the auxiliary piers removed.

The steel beams were launched from both ends at the same time with a closure joint above the Tarn River. In the center of the large span trussed auxiliary truss piers were installed with the exception of the center span which was bridged by cantilevering from both sides. The launching bearings on top of the piers with a distance of 20 m, Fig. 6.300, reduced the beam moments quite significantly in the ratio of span lengths $(151/171)^2 = 0.78$, Fig. 6.301.

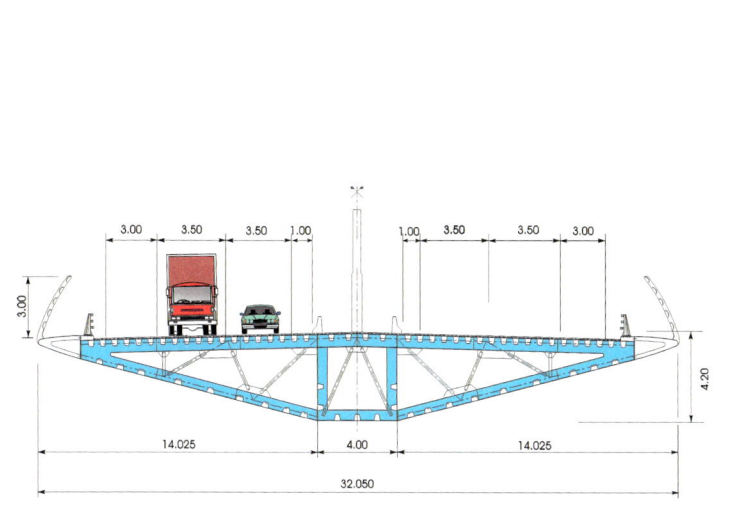

6.6 Series of cable-stayed bridges – Millau Bridge

Figure 6.293 Beam cross-section

Figure 6.294 Piers and towers

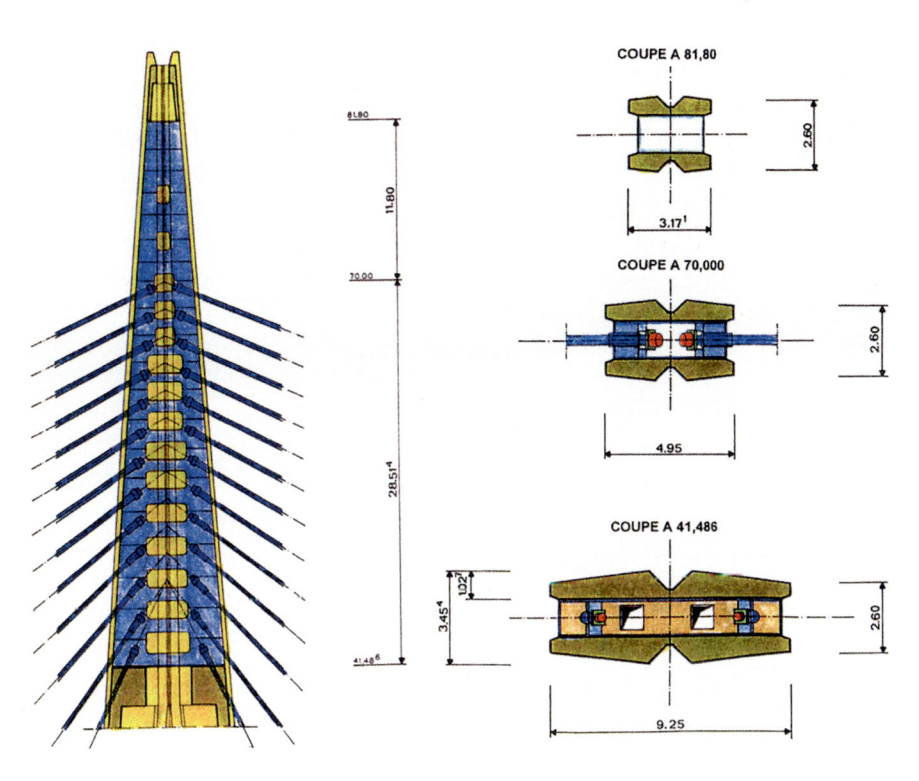

Figure 6.295 Towers

Figure 6.296 Pier construction

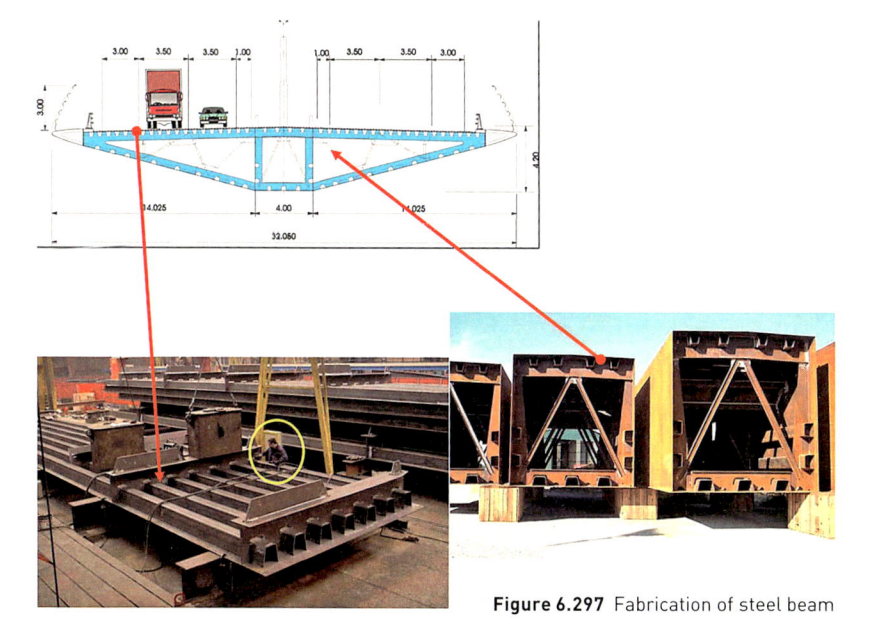

Figure 6.297 Fabrication of steel beam

Figure 6.298 Steel beam during installation

born 1945 in the Ardennes, Belgium
1968 Diplôme d'Ingénieur Civil des Constructions, Université de Liège
since 1973 Bureau d'Études Greisch, Chief Executive Officer
Ben Ahin Bridge
Viaduc de l'Eau Rouge
Millau Bridge

Figure 6.299
Jean-Marie Crémer

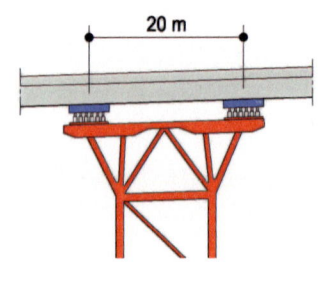

Figure 6.300 Launching bearings on top of auxiliary piers

The auxiliary piers are built from prefabricated sections with 12 m heights. They are designed like a crane.

The elements prefabricated on the ground from steel sections are lifted with an inside lifting device in such way that the next sections could be installed from underneath and also lifted until these telescoping auxiliary piers reached a height of up to 175 m. Their highest load came to 7000 t roughly equal to the total weight of the Eifel Tower, Fig. 6.302.

The final piers were only provided with an auxiliary scaffolding for launching, Fig. 6.303.

A unique characteristic of this launching was that, for cost reasons, the pre-assembly yard was located at the level of the future road gradient, 4.8 m above the final bridge deck elevation, Fig. 6.304.

A shortened view during launching shows the elasticity of the steel beam by overcoming the gradient offset, Fig. 6.305.

Each of the two beam halves was launched over an auxiliary pier and a launching nose across the outer side spans, Fig. 6.306.

After that the final tower was installed at the front end as auxiliary support, but with a reduced height of 70 m instead of 87 m to minimize the transverse wind loads during launching, Fig. 6.307. The highest wind speed permitted during launching was 3 km/h.

Due to the extreme heights of the piers the friction forces during launching had to be equalized, so on top of each pier two active launching bearings were installed in each support axis. Horizontal hydraulic jacks acted between the beam and the piers in such a way that the pier tips remained in place during launching, centrally controlled by sensors, Fig. 6.308.

Finally the center span was bridged from each side with cantilevers, supported by towers, Figs 6.309 and 6.310.

The design and construction on site was an extraordinary engineering achievement.

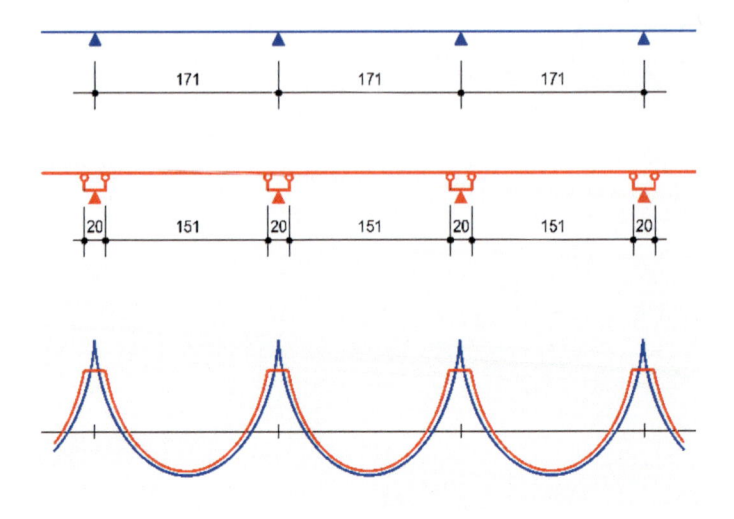

Figure 6.301 Reduced beam moments during launching

6.6 Series of cable-stayed bridges – Millau Bridge

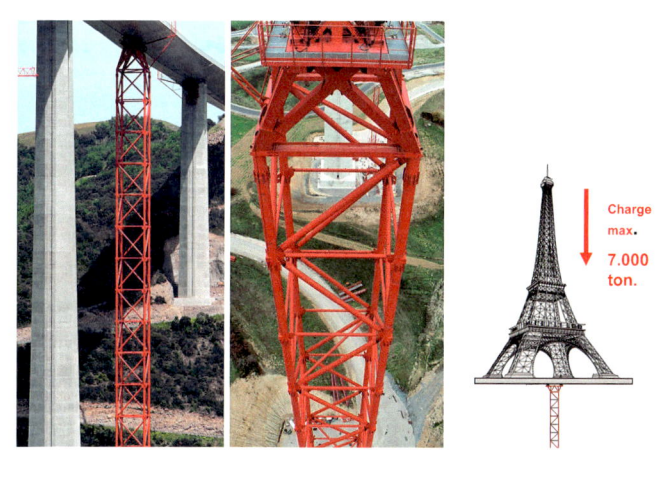

Figure 6.302 Telescoping auxiliary piers

Figure 6.303 Final pier with launching equipment

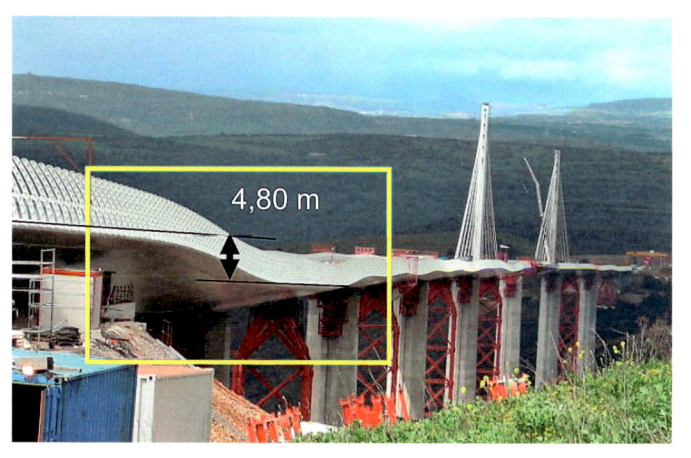

Figure 6.305 Gradient offset during launching

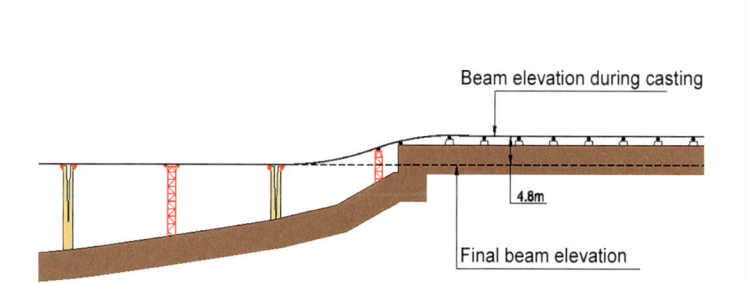

Figure 6.304 Gradient offset between pre-assembly yard and bridge

Figure 6.306 Start of launching without tower

Figure 6.307 Proceeding launching

Figure 6.308 Further launching

Figure 6.309 Crossing the Tarn

Figure 6.310 Center joint

Figure 6.311 Substructure in place

Figure 6.312 Transportation of a steel tower on the beam

Towers

After the steel beam was in place, Fig. 6.311, the remaining steel towers were pre-assembled behind the abutments.

Each tower was then moved with crawler cranes over the bridge beam to its final position. The total weight of such a convoy came to 8 MN, thus acting as a test loading, Fig. 6.312.

The towers were lifted from their horizontal position with the help of a temporary guyed tower, Fig. 6.313. Finally, they were connected with the beam, and the stay cables were installed.

6.6.1.4 Completed bridge

The Millau Bridge is an important example for a series of cable-stayed bridges. A structure with impressive elegance has been built in an impressive landscape, where it gracefully crosses the valley at great height, Fig. 6.314.

Participants

Authorities Direction des Routes, France, Arrondissement interdépartemental des ouvrages d'art de l'autoroute A75, Millau, France
Design Michel Virlogeux, Bonnelles, France
Detailed design and construction engineering Greisch, Liège, Belgium; EEG Simescol; Arcadis, Sèvres, France; Thales E et C, Rungis, France; Serf; STOA Eiffage
Consulting architect Foster and Partners, London, UK
Preliminary design SETRA
Client – concessionist Compagnie Eiffel du Viaduct de Millau, Millau, France
Contractors in JV Eiffage TP, Neuilly sur Marne, France
Concrete construction Eiffage Construction
Steel construction Eiffel Construction Métallique, Colombes, France
Plate fabricator Dillinger Hütte GTS

6.6 Series of cable-stayed bridges – Millau Bridge

Figure 6.313 Tower installation

Figure 6.314 Completed Millau Bridge

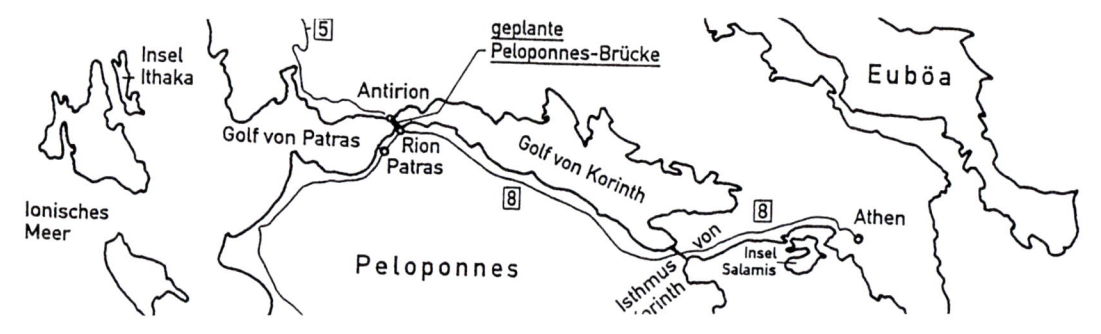

Figure 6.315 Location of bridge

Figure 6.316 Overall view

6.6.2 Rion-Antirion Bridge
6.6.2.1 General
The Rion-Antirion Bridge crosses the Gulf of Korinth at its narrowest westerly location, Fig. 6.315, [6.19].

The design had to overcome some unusual problems. The difficult geology requires a minimum of piers. In the bridge region the sea is an average of 60 m deep, at places more than 65 m. The ground consists of a 20–30 m thick layer of clay, covered by a layer of sand and gravel of varying thickness. Rock is estimated only at about 800 m depth.

The whole region is subject to earthquakes with an intensity of 6.5 on the Richter scale. The maximum ground acceleration can reach 0.48 g, the corresponding maximum response acceleration in the bridge is 1.2 g in the eigenfrequency range of 1–5 Hz. In addition, horizontal and vertical tectonic dislocations of up to 2 m have to be considered.

Although commercial ship traffic has a low density, the piers have to resist an impact of a 180 000 t oil tanker with a speed of 16 knots, [2.122].

The main bridge has a length of 2 252 m with 2 880 m approaches on each side, Fig. 6.316.

The Rion-Antirion Bridge was designed by Jacques Combault, Fig. 2.209. Michel Virlogeux, Fig. 2.101, was an important advisor.

6.6.2.2 Design
Based on the above conditions several alternate designs were considered in order to find the most economic solution. This led to a five-span cable-stayed bridge with three inner spans of 560 m and two side spans of 286 m, Figs 6.317 and 6.318.

A series of cable-stayed bridges of this size means a record for this type of bridge. The tower heads are stabilized by the four stiff tower legs with A-shapes in both directions. The beam is supported by 8 × 23 pairs of stay cables uniformly distributed over the length of the bridge. Vertically the beam is only supported on the two outer common piers with the approaches.

A detailed analysis of the interaction between foundations and superstructure showed that the beam, continuous over its full length and fully supported by stay cables, is able to adjust without permanent damage to the large possible horizontal and vertical tectonic dislocations of the towers.

Beam
The 27.2 m wide composite bridge beam comprises two open steel main girders with a depths of 2.2 m (h:l = 2.2:560 =1:252), steel cross girders at a distance of 4 m, and a 24 cm thick concrete slab, Figs 6.319 and 6.320. All changes in lengths, due to temperature and tectonic dislocations, are only taken up at the ends by roadway joints with ±2 m dilatation. In the transverse direction the beam is connected to each tower with four dampers.

Foundations
The hollow reinforced concrete foundations with 89.5 m diameter are 9 m high at the outside and 13.5 m at the connection to the cone-shaped tower piers, Fig. 6.321. At the inside they are stiffened by a torsion ring and radial beams. The foundations of the first three towers on the Rion side, which have a depths of 35 m, are directly resting on a gravel base layer which is strengthened by steel piles. These piles are 25–30 m long and placed on a 7 m by 7 m grid on a circular area of 130 m diameter.

The base layer on top of the steel piles consists of a precisely leveled layer of gravel. The steel pipes are not directly load-bearing. They assist in distributing the tower loads into the ground and in limiting differential settlements. The gravel layer has to transfer the horizontal loads from the bridge onto the steel piles and surround-

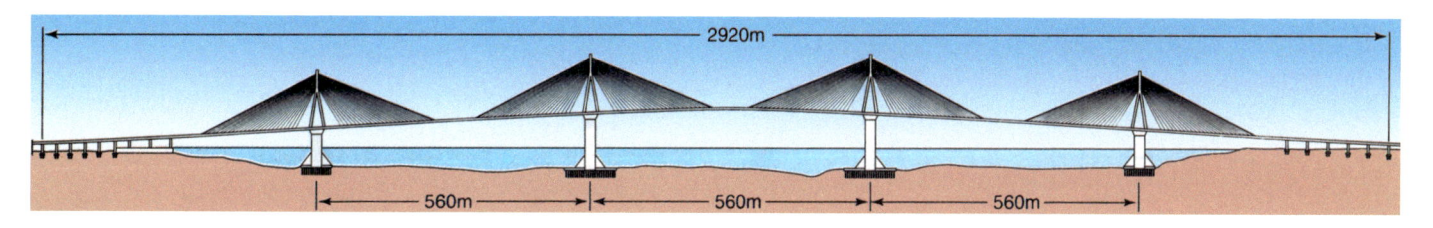

6.6 Series of cable-stayed bridges – Rion-Antirion Bridge

Figure 6.317 Elevation of complete bridge

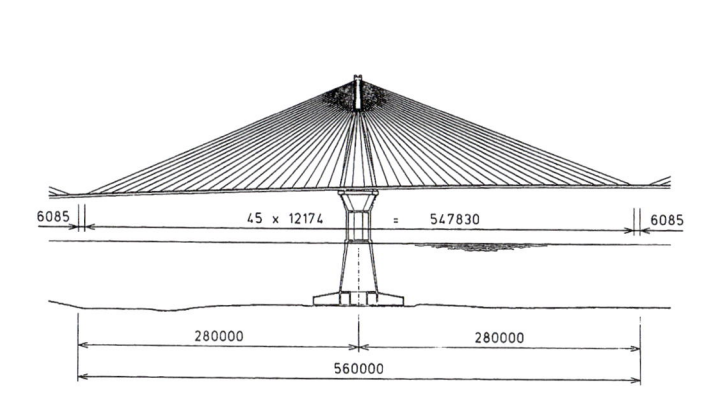

Figure 6.318 Elevation of single main span

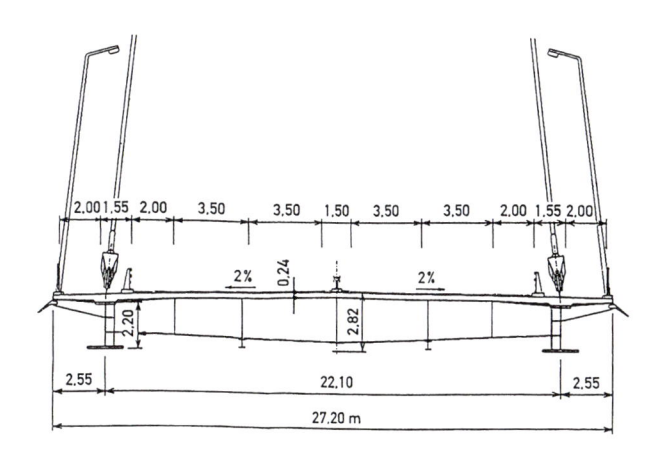

Figure 6.319 Cross-section

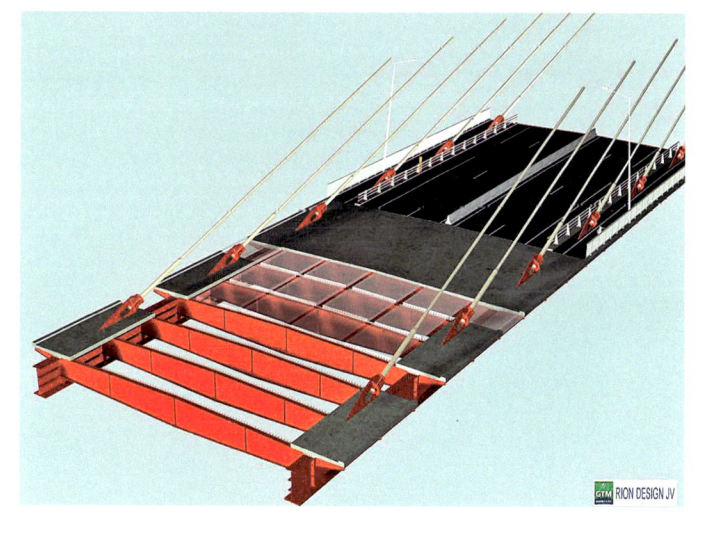

Figure 6.320 Composite beam girder grid

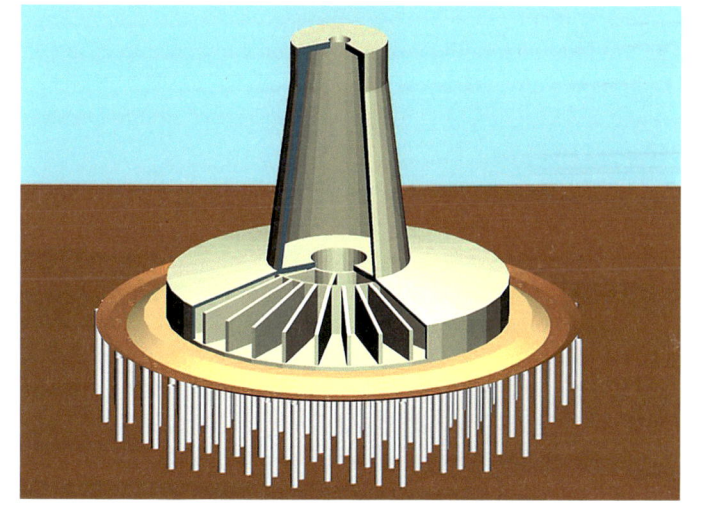

Figure 6.321 Tower foundation

ing ground plastically, thereby avoiding a failure plane in the clay layer.

Towers
The concrete towers comprise a three-part pier with the tower legs above, Figs 6.322 and 6.323. A cone-shaped lower part with a diameter of 37.99 m at the base and 26.93 m at the top is fixed to the foundation. On it rests an octagonal shaft, 28.4 m high, with a pyramid-shaped 19.3 m high part on top. The tower legs have cross-sections of 4 m × 4 m. The tower head for the cable anchorage is composite with a steel box cast into concrete.

Stay cables
For the Rion-Antirion Bridge parallel strand cables were, as usual today, selected. At the beam they are anchored above the roadway at a web extension, Fig. 6.320. At the tower heads they converge into composite anchorages, Fig. 6.324.

Earthquake
The conditions for earthquake are based on a response spectrum on the ground with a 2000-year return period, Fig. 6.325. The highest ground acceleration comes to 1.2 g over a range of eigenfrequencies between 1 and 5 Hz.

There is no connection between the steel piles on ground and the foundation of the towers. The tower foundations can thus move against the layer of gravel strengthened by the steel piles. This new foundation concept was presented in [6.20] and extended here to earthquake-prone regions. The upper border of the load-bearing capacity of the strengthened foundations was developed by applying yield theory with appropriate kinematic mechanisms, Figs 6.326 and 6.327.

Non-linear finite element calculations were executed based on the above. They resulted in the force-displacement relations of Fig. 6.328 and the moment-rotation relations of Fig. 6.329 which were introduced into the overall bridge design as foundation characteristics.

The dynamic investigation of the bridge showed that the largest oscillations from earthquakes, Fig. 6.330, resulted in a multitude of cracks along the tower legs under the combined action of bending and tension. This cracked stage is, on one hand, helpful because it increases the flexibility of the legs below yield. On the other hand it is difficult to define precisely the cracked and uncracked regions during tower oscillations. As a consequence, 13 cross-sections were calculated in time steps of 0.02 sec which resulted in 130 000 different stages which all had to be investigated.

An amplification factor of Fig. 6.331 at the tower tip made it possible to differentiate the individual steps by taking into account the behavior of the whole group of towers. At the same time the safety against progressive collapse of the whole group of towers was proved, [6.21], in order to prevent a collapse of the whole bridge should one pier fail.

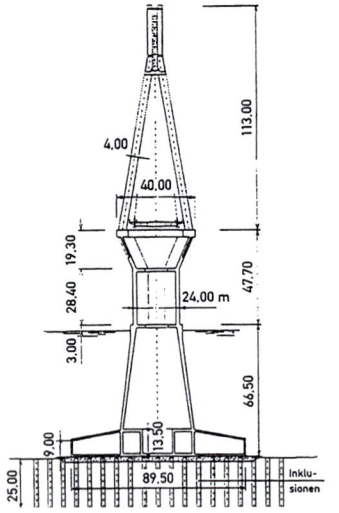

Figure 6.322 Tower section

Figure 6.323 Tower – beam system

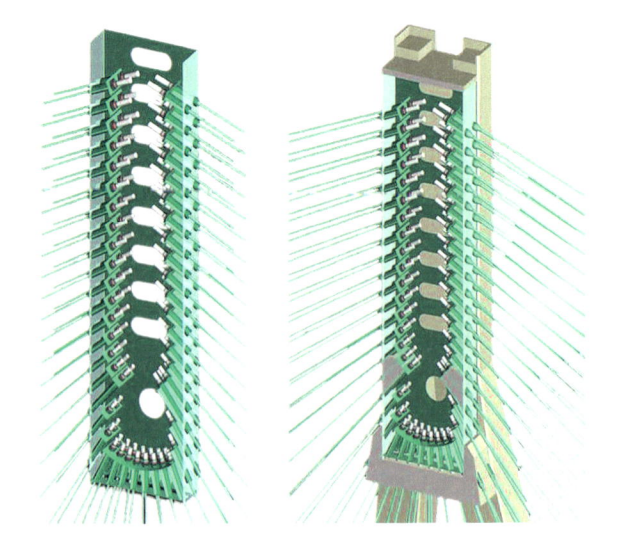

Figure 6.324 Cable anchorage in tower head

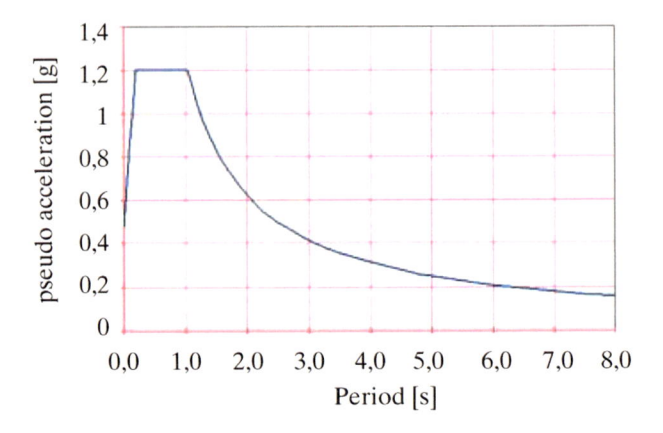

Figure 6.325 Response spectrum due to earthquake

6.6 Series of cable-stayed bridges – Rion-Antirion Bridge

Figure 6.326 Foundation kinematics during earthquake

The beam is continuous over the complete bridge length. At the towers an earthquake protection system is located transversely between the beam and the tower legs. It comprises dynamic dampers and hold-back systems with desired failure joints, Fig. 6.332 [6.22]. These desired failure joints have to fail for earthquake forces above the highest wind loads in order to activate the hydraulic dampers which dissipate the energy and limit the transverse oscillations of the beam.

The capacity of each of the four dampers at each tower comes to 3 500 kN in tension and compression. The relative movements between beam and towers during a designed earthquake reach ± 1.3 m with corresponding accelerations above 1 m/sec.

For the approach bridges a combination of elastic isolators and hydraulic dampers was used, Fig. 6.333.

6.6.2.3 Construction

Placing gravel layer

The particular difficulties for the construction of the piers are the large water depth of up to 65 m for the main piers, and the poor soils conditions. A combination of the latest technologies for off-shore oil platforms and submerged tunnels was used.

The foundation works started with dredging the upper ground layer, placing a 19 cm thick layer of sand and ramming of the steel piles. The piles protrude 1.5 m above the sand layer and were covered with another 2 m thick layer of rounded river aggregates with a 50 cm layer of broken aggregates on top. This grading of gravel with decreasing inner friction from above to below provides the desired plastic behavior during earthquake.

All these works were done from a 60 m long and 40 m wide floating platform which was anchored with chains to removable concrete blocks on the sea bottom, Fig. 6.334. The equipment for piling was located on a submersible pontoon connected to one side of the platform with steel levers.

Foundations

The foundations for the towers were built in two steps near the city of Antirion. In a dry dock, 230 m long and 100 m wide, two circular foundations were concreted at the same time, Figs 6.335 and 6.336 [6.23]. The rear part of the dry dock had a depth of 8 m, the front part of 12 m. After completing the front foundation with a first pier section of 3.2 m the dry dock was opened by removing the front coffer dam and by towing foundation out.

The second foundation was then towed forward and initially used to close the dry dock again, Figs 6.337 and 6.338. With this trick a lot of time was saved against the standard method of closing the coffer dam again.

The construction of the pier shaft on top of the towed foundation was built in sections with jumping forms, Fig. 6.339.

After reaching the required height each foundation was towed to its final location and lowered onto the prepared gravel foundation,

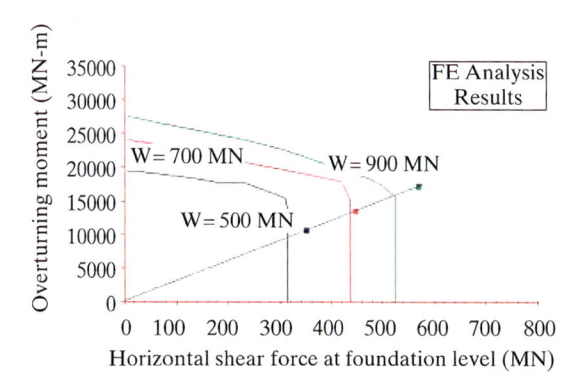

Figure 6.327 Horizontal shear forces in the foundation joint

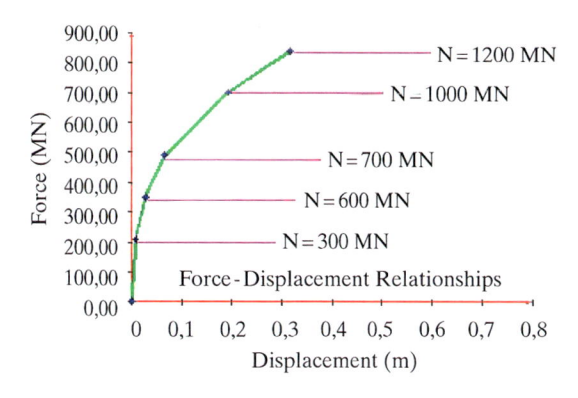

Figure 6.328 Force-displacement relationships

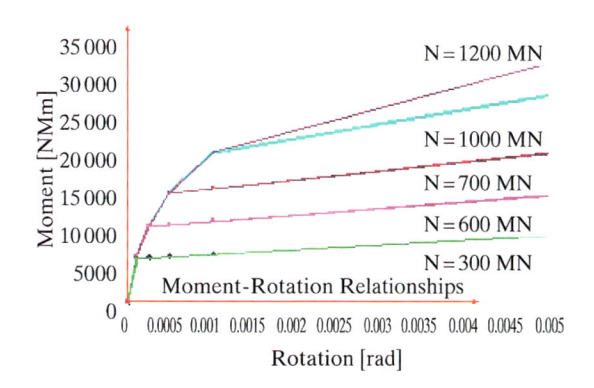

Figure 6.329 Moment-rotation relationships

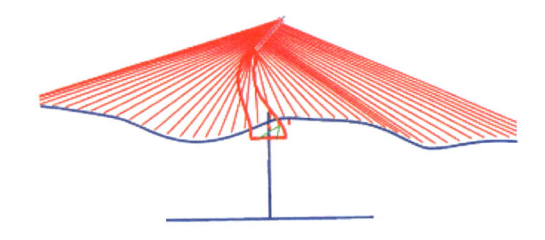

Figure 6.330 Typical deflection of a tower during earthquake

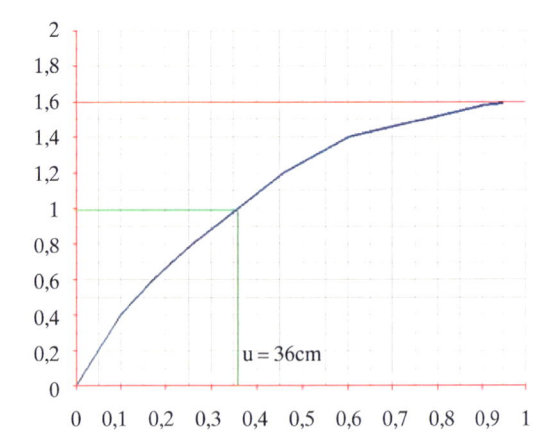

Figure 6.331 Deformations of tower tip with amplification factor

Fig. 6.340. The whole foundation was then flooded in order to accelerate the expected settlements of 20–30 cm.

Towers

The four octagonal shafts were built in 4.8 m high sections with jumping forms. Heavy truss girders between the freestanding legs provided the required safety against earthquake during construction. The steel core for the cable anchorages in the tower heads was lifted in prefabricated units by a floating crane, Fig. 6.341.

Beam

The composite beam was pre-assembled in 12 m long units, and the concrete roadway slab was cast on top, which is quite unusual. These 270 t elements were lifted by floating crane to the already built-in beam, Fig. 6.342. The main girders were then connected with high-strength bolts, and the joints in the concrete roadway slab were closed with overlapping reinforcement and CIP concrete.

Fig. 6.343 shows the free cantilevering of the beam to both sides, simultaneously for all towers.

Figure 6.332 Arrangement of hydraulic dampers at towers

Figure 6.333 Elastomeric isolaters and hydraulic dampers at the approach bridges

6.6 Series of cable-stayed bridges – Rion-Antirion Bridge

Figure 6.334 Ramming of steel piles

Figure 6.335 Foundations in dry dock

Figure 6.337 Floating of foundations

Figure 6.336 Preparation of two foundations in the dry dock

Figure 6.338 Floating of the front foundation out of the dry dock

Figure 6.339 Concreting of a pier shaft

Figure 6.340 Lowering a foundation onto the gravel layer

Figure 6.341 Building the four tower legs by free cantilevering

Figure 6.342 Lifting of a complete beam element

6.6 Series of cable-stayed bridges – Rion-Antirion Bridge

6.6.2.4 Completed bridge

The Rion-Antirion Bridge represents a milestone in the development of cable-stayed bridges. Very difficult foundation conditions together with the high danger of earthquakes had to be overcome. Similar wide waterways may in the future be tackled with confidence following the example of the Rion-Antirion Bridge, Fig. 6.344.

Participants

Authorities Ministry of Environment and Public Works, Athens, Greece
Engineers VINCI Construction Grands Projets, Paris, France; INGEROP, Paris, France; DOMI, Athens, Greece; Buckland and Taylor, Vancouver, Canada; DENCO, Athens, Greece
Consulting architect B. Mikaelian, Paris, France
Contractors VINCI, Paris, France; Elliniki Technodomike – TEV, Athens, Greece; J & P – AVAX, Athens, Greece; Proodeftiki, Athens, Greece; Pantechniki, Athens, Greece

Figure 6.343 Free cantilevering

Figure 6.344 Completed bridge

7 Future development

The development of cable-stayed bridges with spans up to about 1100 m is more or less completed. How will their spans develop in future? The span record is currently held by the Sutong Bridge in China, Fig. 2.105, with a main span of 1088 m. In the summer of 2012 the Russki Bridge in Russia, Fig. 2.105a, with a new record span of 1104 m will open.

Longer spans have been designed earlier, but not yet realized. A cable-stayed alternate for the Great Belt Bridge was proposed with a main span of 1204 m [7.1]. Leonhardt proposed a cable-stayed bridge with a main span of 1 800 m for the Messina Crossing in Italy [7.2], for road and rail traffic, Fig. 7.1.

Such spans can be built with currently used materials, steel for the box girder of the main span with a streamlined cross-section, and possibly concrete for the side span girder, which might protrude about 120 m into the main span. The towers will have an A-shape above the deck and will consist of concrete if the foundation conditions are favorable.

The parallel strand stay cables will be fabricated on site from their components. Crossties may reduce the sag economically, so that the effective modulus of elasticity is increased. Theoretically carbon cables, with their low specific weight, could help in this respect, but they are still rather expensive.

The fabrication and installation of steel box girders is well known; even large units would not go beyond proven experience.

The aerodynamic stability of future record spans will require special attention, for the stay cables as well as for the beam. Until now a torsionally stiff steel beam with two cable planes and A-towers has proved to be sufficiently stable, and dampers were only required as an exception. For the aerodynamic stability of the stay cables, especially against wind–rain-induced galloping, the PE pipes have been profiled, and dampers have been provided as well as crossties.

For very long future spans adaptive installations may have to be provided, which adapt their action to the actual stage of aerodynamic excitation. For the Russki Bridge, Maurer has developed magnetic-rheological (MR) fluid dampers, Fig. 7.2 [7.3]. The damper fluid displays the characteristic that its shear strength can be varied under the influence of a magnetic field. Within the damper, coils are arranged that can create a variable magnetic field, adjusted to actual changing cable frequencies and amplitudes.

All dampers are connected by cables with the current drivers and the control units located in switch boxes which may be remotely controlled via the internet, Fig. 7.3 [7.4].

Damping of the main girder with tuned mass dampers with constant characteristics has been described for the Kehl-Straßburg Bridge against torsional oscillations, Fig. 4.87, and for the Millau Bridge against lateral oscillations during construction, Fig. 4.100.

Ostenfeld and Larsen proposed an active control system for streamlined box girders as used for aircraft wings [7.4]. Their system

Figure 7.1 Proposal for the Messina Crossing by Leonhardt with a main span of 1800 m, in 1982

is based on the idea of constantly monitoring the movements of the deck and of using control surfaces (winglets, flaps) to generate stabilizing aerodynamic forces which counteract, for example, the flutter mechanism shown in Fig. 4.89. Fig. 7.4 displays the potential increase in critical wind speed.

The actively controlled flaps are installed at the leading and trailing edges of streamlined box girders, outside the turbulent boundary layer, Fig. 7.5.

Control rods located inside the box girder are moved by hydraulic cylinders activated by computer-controlled servo pumps.

For all actively controlled damping systems it has to be ensured that the bridge would remain stable if they failed, possibly with a reduced factor of safety. For the actively controlled flaps sectioning and duplication of the driving and steering system as well as the power supply is proposed. For the two tuned mass dampers (TMD) used for the Kehl-Straßburg Bridge it has been shown that if one TMD is completely stuck and the other one is out of tune by 10%, a safety factor of 1.1 remains, Fig. 4.88.

The technical problems connected with longer main spans are thus solvable. Their actual construction will depend, however, whether they are needed so much that their construction costs can be recovered by toll.

The future development of cable-stayed bridges will remain fascinating.

Figure 7.2 Arrangement of coils around the piston in a magneto-rheological damper

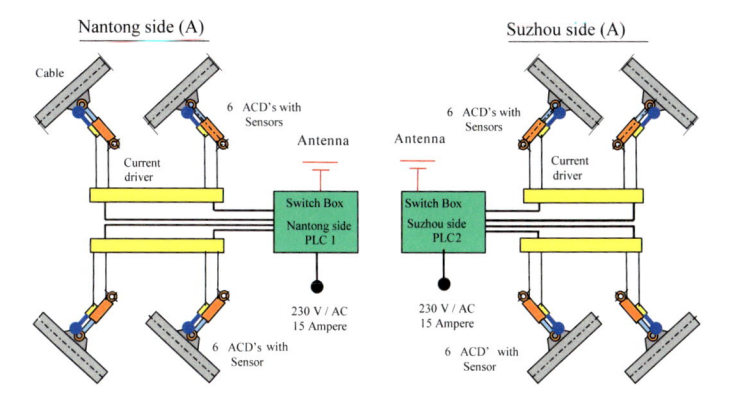

Figure 7.3 Networking of adaptive magneto-rheological cable dampers (ACDs)

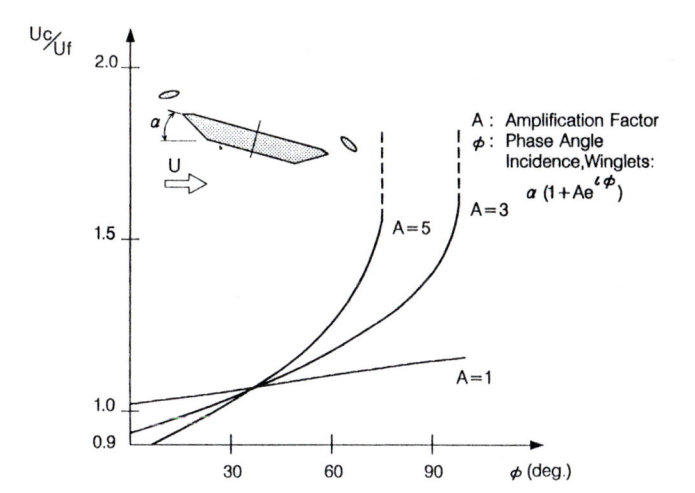

Figure 7.4 Potential increase of critical wind speed through actively controlled flaps

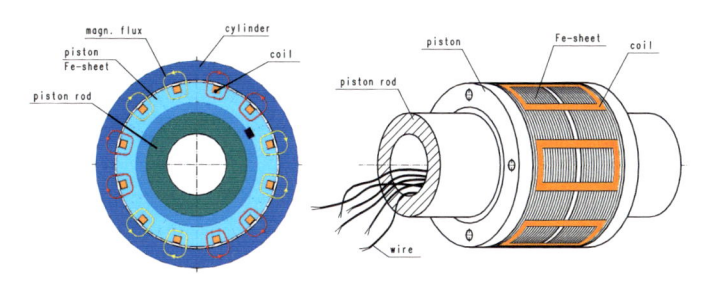

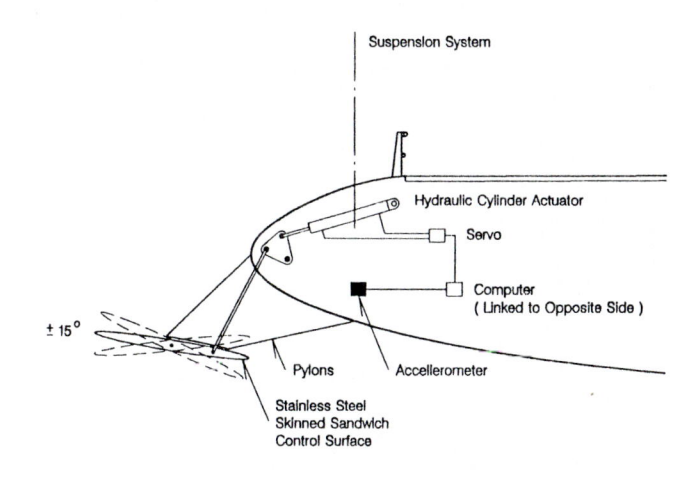

Figure 7.5 Use of active control flaps at streamlined box girder

Index

Bridge Index

Akashi Bridge 154, 212, 243
Akkar Bridge 108
Ala Habat Bridge 119 f.
Alamillo Bridge 41, 43, 73 f.
Albert Bridge 52
Allegheni River Bridge 51
Annacis Bridge 102, 106, 109
Annapolis Bridge 32, 34 f., 36, 449
Anzac Bridge 164
Arenas Viaduct 121, 123
Audubon Bridge 244

Badajoz Bridge 164, 168, 172
Barrios de Luna Bridge 98, 100, 367
Baytown Bridge 40 ff., 103, 111 ff., 195, 206, 208, 212, 309, 322 f., 355, 387 ff., 448
Ben-Ahin Bridge 299, 301, 414
Berlin Bridge Halle 115 f.
Beska Bridge 452
Bickensteg Villingen 133 f.
Blaubeurer Tor Bridge 94 f.
Bluff Dale Bridge 55
Bonhomme Bridge 52, 54
Bonn North Bridge *see* Rhine River Bridge Bonn North
Bratislava Bridge 70 f.
Britz Canal Bridge 445
Brooklyn Bridge 51
Brotonne Bridge 25, 98 f., 208 f.,
Büchenau Bridge 102, 105
Burlington Bridge 36, 39, 102, 111, 164, 172, 447

Cassagne Bridge 55
Caroni River Bridge 30 f.
Carpineto Bridge 92
Chao-Phraya River Bridge 67
Chenab Bridge 452
Cincinatti Bridge 51
Clark Bridge 103
Cologne-Deutz Bridge *see* Rhine River Bridge Cologne-Deutz
Confederation Bridge 109
Cooper River Bridge 109

Dame Point Bridge 80, 147
Danube Bridge Metten 92, 95
Danube Canal Bridge Vienna 81, 84
Dartford Bridge 67, 104

Dee River Bridge 36 f.
Diepoldsau Bridge 94, 97
Dittenbrunn Railway Bridge 36
Dominion Bridge 109
Donziére-Montragon Bridge 81 f.
(First) Dryburgh Abbey Bridge 48, 50
(Second) Dryburgh Abbey Bridge 48
Düsseldorf Bridge Family 59 f.
Düsseldorf North Bridge 59 f., 76

East Huntington Bridge 35, 80, 87, 103 f., 346 ff., 446
El Cañon 103
El Zapote 103
Elbe River Bridge, System Dischinger 57
Elbe River Bridge Niederwartha 115 f., 175 f., 291 f., 309 f.
Elbe River Bridge Pirna 37
Elbe River Bridge Tangermünde 31 f., 36, 39, 44
Elorn Bridge 171
Emscher Bridge 66
Enz River pedestrian bridge Mühlacker 42, 44
Erskine Bridge 79
Evripos Bridge 97 f., 108

Faroe Bridge 164
Felsenau Bridge 96
Femer Bridge 129 f., 453
Flehe Bridge
 see Rhine River Bridge Flehe
Flößer Bridge 94, 96, 300 f.
Forth Bridge 120 f., 154, 453
Franz Josef Bridge 52
Freeway triangle Neukölln 444 f.
Friedrich-Ebert-Bridge
 see Rhine River Bridge Bonn North

Galata Bridge 261, 264
Ganter Bridge 94, 96
Gdansk Bridge 164
Geo Geum Bridge 72, 212, 244 f., 246
Glebe Island Bridge 35, 37
Golden Ears Bridge 132 f., 243, 450 f.

Cable-Stayed Bridges. 40 Years of Experience Worldwide. First Edition. Holger Svensson.
© 2012 Ernst & Sohn GmbH & Co. KG. Published 2012 by Ernst & Sohn GmbH & Co. KG.

Golden Horn Bridge Vladivostok 222, 453
Great Belt Bridge 122 f., 164, 214, 426

Hammerbrook Railway Bridge 33
Harbor Bridge Riesa 42 f.
Heer-Agimont Bridge 102
Heinola Bridge 34, 41 ff., 102, 111 ff., 311 ff.
Helgeland Bridge 35 ff., 39, 84, 89, 151, 164, 193, 206, 212, 220 f., 241, 262, 265, 346, 352 ff., 449
Hoechst Bridge 82, see (Second) Main River Bridge Hoechst
Höga Kusten Bridge 30 f., 450 f.
(Second) Hooghly River Bridge 33 f., 102, 106 ff.
Honfleur Bridge 402, see Normandy Bridge
Houston Ship Channel Crossing 34 f., 36 f., 41, 212, 253, 258
Humboldt Harbor Bridge 34, 36

Ikuchi Bridge 66
Ilm River Bridge Oberroßla 35, 37
Ilverich Bridge see Rhine River Bridge Ilverich
Incheon Bridge 234 f.

Kap Shui Mun Bridge 41, 43, 66, 103, 111, 113, 253, 258, 299, 302, 311 f., 325
Karnali River Bridge 102
Kehl-Straßburg pedestrian bridge 136, 138, 212, 215, 230 ff., 262 f.
Kemjoki Bridge 102
Kiehl-Hörn pedestrian bridge 236 ff.
Kings Meadow Bridge 48
Knie Bridge 60 ff., 63 f. 76, 402, see Düsseldorf Bridge Family
Kocher Valley Bridge 31 f., 38, 40, 443
Köhlbrand Bridge 141 f., 170
Kolbäcks Bridge 103, 115
Kuang Fu Bridge 121, 123

Kurt-Schumacher-Bridge see Rhine River Bridge Mannheim-Ludwigshafen

Leven River Bridge 38, 41, 85, 90, 304 f., 449 f.
Lions Gate Bridge 109
Lockmeadow Bridge 79
Luling Bridge 151 f.
Lumberjack's Candle Bridge 112
Lézardrieux Bridge see Pont de Lézardrieux

Ma-Chang Bridge 164
Macau Bridge 125
Magliana Bridge 92
(Second) Main River Bridge Hoechst 25, 81 f., 146
(Second) Manama Bridge 94, 97
Mannheim North Bridge see Rhine River Bridge Mannheim-Ludwigshafen
Maracaibo Bridge 80 f., 92, 98, 122 f., 145 f.
Masséna Bridge 104
Metten Bridge see Danube Bridge Metten
Mezcala Bridge 103, 124 f.
Millau Bridge 78, 118, 128, 238 f., 306 f., 411 ff.
M0 Bridge Budapest 164, 169 f., 216 ff.
Monterrey Bridge 164, 172
Mosel Valley Bridge Winningen 31 f.
Mulden Bridge Siebenlehn 444
Munksjö Bridge 119, 121
My Thuan Bridge 262, 265

Nan Pu Bridge 41 f., 102, 110
Neckar Bridge Kirchheim 253, 258
Neckar River Bridge Untertürkheim 70 f.
Neckarcenter Bridge Mannheim 135 f., 198 f., 204 f., 293, 295, 323
Neckar Valley Bridge see Weitingen Bridge
Neuwied Bridge see Rhine River Bridge Neuwied
Nhattan Bridge 236 f., 246 f.

Niagara Falls Bridge 51
Niederwartha Bridge see Elbe River Bridge Niederwartha
Nienburg Bridge 47, 48 f.
Nordhordland Bridge 85, 89
Normandy Bridge 26 f., 38, 66, 76, 78, 158 f., 171, 174, 232 f., 346, 402 ff.
North Bridge see Düsseldorf North Bridge
Nuevo Leon Bridge 173

Obere Argen Valley Bridge 130 f., 291
Oberkassel Bridge see Rhine River Bridge Oberkassel, Düsseldorf Bridge Family
Oder River Bridge Frankfurt 36
Öresund Bridge 26 f., 103, 114 f.
(Second) Orinoco River Bridge 41, 43, 118 f., 128 f., 263, 265
(Third) Orinoco River Bridge 26, 28, 116 f.

Papineau Bridge 109
Paraná River Bridge Chaco-Corrientes 81, 83
Paraná River Bridge Posadas-Encarnación 25, 38, 40, 88, 293 f., 321 f.
(Second) Panama Canal Bridge 87, 90, 226, 246 f., 255 f., 298 f., 301
Pasco-Kennewick Bridge 25, 33, 42, 85 f., 163, 191, 195, 210, 212, 322 f., 323, 325, 327 ff., 346, 445
Patna Bridge 120 f.
Penang Bridge 85, 88, 147, 164, 166, 450 f.
Pforzheim pedestrian bridge 135 f.
Phu My Bridge 242
Polcevera Viaduct 92, 122 f.
Pont de Lézardrieux 55 f.
Pont de Saint Ilpize 50, 53
Pont des Iles 102
Poole Harbor Bridge 79, 119, 121
Posadas-Encarnación Bridge see Paraná River Bridge Posadas-Encarnación

Quincy Bridge 102

Railway Bridge Gemünden 32
Raippaluoto Bridge 103
Rama 8 Bridge 40, 42
Ravensburg pedestrian bridge 36, 38, 136 f.
Rhine River Bridges, Overview 64
Rhine River Bridge A 42 64
Rhine River Bridge Bonn North 19 f., 62, 64, 66 f.
Rhine River Bridge Cologne-Deutz 31 f.
Rhine River Bridge Duisburg-Homberg 58 f.
Rhine River Bridge Duisburg-Neuenkamp 64, 76 f.
Rhine River Bridge Flehe 26 f., 37, 39, 60 f., 66, 69 f., 80, 146, 200 f., 304, 402, 446
Rhine River Bridge Ilverich 20, 40, 42, 64, 69 f., 142
Rhine River Bridge Kehl see Kehl-Straßburg pedestrian bridge
Rhine River Bridge Kleve-Emmerich 64
Rhine River Bridge Leverkusen 62, 64 f.
Rhine River Bridge Mannheim-Ludwigshafen 64, 66 f., 184, 200, 210, 316 ff.
Rhine River Bridge Maxau 64
Rhine River Bridge Neuwied 64, 66, 68, 70
Rhine River Bridge Oberkassel 34, 60, 63, 66, 206, 305, 309
Rhine River Bridge Rees-Kalkar 64
Rhine River Bridge Speyer 61, 64, 66, 68
Rhine River Bridge Wesel 70 f.
Rhône River Bridge Avignon 52
Rio Magdalena Bridge 92 f.
Rion-Antirion Bridge 125, 127 f., 158 f., 311 f., 418 ff.
Roosevelt Dam Bridge 32, 450
Rosario-Victoria Bridge 91, 258, 261
Russki Bridge 76, 80, 426, 453

Saale River Bridge Nienburg
  see Nienburg Bridge
Saame Bridge 112
Saint Maurice Bridge 102
Sancho-El-Mayor Bridge 85 f.
Saone River Bridge 53
Sava Bridge Ada Ciganlija 25, 66, 73, 75, 452
Save Bridge Belgrad 70, 72
Schatten Bridge Stuttgart 36, 38
Schillersteg pedestrian bridge Stuttgart 35, 37, 133, 147, 270 ff.
Severins Bridge 62, 64 f., 76
(Second) Severn Bridge 103, 332 f.
Sitka Harbor Bridge 102
Skarnsundet Bridge 85, 98, 100, 367, 402
South Elbe River Bridge 19 f.
St. Nazaire Bridge 36, 38, 76
Steyregger Danube Bridge 102
Stöbnitztal Valley Bridge 42, 44
Stonecutters Bridge 18, 24, 30 f., 34, 41, 44, 76, 79, 166, 212, 240 f., 252, 256, 304 ff.
Strelasund Bridge (Crossing) 75, 160, 291, 369 ff.
Strömsund Bridge 58, 102
Sunniberg Bridge 125 f.
Sunningesund Bridge 103
Sunshine Skyway Bridge 40 f., 102, 106, 108, 194, 212, 241, 249 f., 258, 260, 447
Sutong Bridge 24, 58, 76, 79, 80, 306, 308, 402, 426
Svinesund Bridge 30 f., 451 f.
Swansea Sail Bridge 79

Tacoma Narrows Bridge 211, 213, 332
Tamar Bridge 52, 54
Tatara Bridge 66, 76, 78
Tähtiniemi Bridge *see* Heinola Bridge

Tempul Aqueduct 55
Teresina Bridge 224 f.
Thames River Bridge Dartford
  see Dartford Bridge
Theodor-Heuss-Bridge
  see Düsseldorf Bridge Family
Ting Kau Bridge 41, 43, 61, 103, 120 f., 125 f., 228
Tisza River Bridge 130, 132
(New) Tjörn Bridge 66, 252, 254
Transporter Bridge Marseille 54
Transporter Bridge Nantes 54
Transporter Bridge Portugalete/Bilbao 54
Twerton Bridge 50

Uddevalla Bridge 114 f., 164, 262, 265, 450 f.
Ume Älv Bridge 450 f.
Upper Havel River South Bridge 34 f.
Utsjoki Bridge 102

Van Troi-Tran Thi Ly Bridge 112
Vasco da Gama Bridge 104
Villingen-Schwenningen pedestrian bridge 42, 44

Wadi el Kuf Bridge 81, 92, 98
Warnow Bridge 444
Weil der Stadt pedestrian bridge 136 f., 139
Weirton-Steubenville Ohio River Bridge 102
Weitingen Bridge 30 f., 42, 44, 130 f., 444
Wye Bridge 79

Yang Pu Bridge 35, 103, 106, 110, 252, 255, 311, 402, 450 f.

Zárate-Brazo Largo Bridges 36, 39, 70, 72, 104, 178, 248, 258 f., 325 f., 447

References

[1.1] Ewert, S.: Brücken, Die Entwicklung der Spannweiten und Systeme, Ernst & Sohn, Berlin 2003, pp. 147–188

[1.2] Zellner, W. and Leonhardt, F.: Cable-Stayed Bridges: Report on latest developments. Canadian Structural Engineering Conference, 1970

[1.3] Leonhardt, F. and Zellner, W.: Vergleiche zwischen Hängebrücken und Schrägkabelbrücken für Spannweiten über 600 m, IVBH Abhandlung 32-I, Zürich 1972, pp. 127–165

[1.4] Saul, R.: On Frontiers of Cable-Stayed Bridges, Proceedings of Bridges into the 21st Century, Hong Kong Institution of Engineers 1995, pp. 203–210

[1.5] Leonhardt, F.: Schrägkabelbrücken. Deutscher Betontag 1979 in Berlin

[1.6] Leonhardt, F. and Zellner, W.: Cable-Stayed Bridges, IABSE Surveys S-13/80, pp. 21–48

[1.7] Zellner, W., Saul, R. and Svensson, H.: Recent Trends in the Design and Construction of Cable-Stayed-Bridges, IABSE 12th Congress, Vancouver, BC, 3–7 Sept 1984, pp. 279–284

[1.8] Leonhardt, F.: Cable-Stayed Bridges, FIP Congress, New Delhi 1986

[1.9] Svensson, H.: The Development of Cable-Stayed Bridges in Europe, International Symposium on Cable-Stayed-Bridges, Shanghai, 1994

[1.10] Havemann, H. K.: Spannungs- und Schwingungsmessungen an der Brücke über die Norderelbe, Stahlbau 1964, pp. 289–297

[1.11] Thul, H.: Die Friedrich-Ebert-Brücke über den Rhein in Bonn, Bauingenieur 1971, pp. 327–333

[1.12] Autorenkollektiv: Rheinquerung Ilverich, Stahlbau 2002, Vol. 6, pp. 385–439

[1.13] Ernst, H. J.: Der E-Modul von Seilen unter Berücksichtigung des Durchhanges, Bauingenieur 1965, pp. 52–55

[1.14] Leonhardt, F., Zellner, W. and Saul, R.: Zwei Schrägkabelbrücken für Eisenbahn-und Straßenverkehr über den Rio Paraná (Argentinien), Stahlbau 1979, pp. 225–236, 272–277

[1.15] Leonhardt, F., Zellner, W. and Svensson, H.: Die Spannbeton-Schrägkabelbrücke über den Columbia zwischen Pasco und Kennewick im Staat Washington, USA, Beton- und Stahlbetonbau 1980, pp. 29–36, 64–70, 90–94

[1.15 a] Leonhardt, F., Zellner, W. and Svensson, H: The Columbia River Bridge at Pasco-Kennewick, WA, USA, Eighth Congress Poceedings FIP, 1978

[1.16] Dittmann, G. and Bondre, K. G.: Rheinbrücke Düsseldorf-Flehe, Planung, Entwurf, Ausschreibung, Vergabe und Überblick über den Ausführungsentwurf. Bauingenieur 1979, pp. 59–66

[1.17] Svensson, H., Hopf, S. and Humpf, K.: Die Zwillingsverbundschrägkabelbrücke über den Houston Ship Channel bei Baytown, Texas, Stahlbau 1997, pp. 57–63

[1.17 a] Svensson, H. and Lovett, T. G.: The twin cable-stayed composite bridge at Baytown, Texas. IABSE Symposium Mixed Structures, Brussels, 1990, pp. 317–322.

[1.17 b] Svensson, H.: The twin cable-stayed Houston Ship Channel Bridge. The Structural Engineer (IStructE), March 1992, pp. 13–20.

[1.18] Schlaich, J. and Bergerman, R.: Die Zweite Hooghly Brücke in Kalkutta. Bauingenieur 1996, pp. 7–14

[1.19] Volke, E. and Rademacher, C.-H.: Nordbrücke Mannheim-Ludwigshafen, Stahlbau 1973, pp. 97–105, 138–152, 161–172

[1.20] Leonhardt, F., Andrä, W. and Wintergerst, L.: Entwurfsbearbeitung und Versuche. In Tamms, F., Beyer, E.: Kniebrücke Düsseldorf, Beton-Verlag, Düsseldorf 1969, pp. 53–72

[1.21] You, Q., et al.: Sutong Bridge – The longest cable-stayed bridge in the World. Structural Engineering International 4/2008, pp. 390–395

[1.22] Morgenthal, G., Sham, R. and Jamane, K.: Montageplanung und Herstellung der Seitenfelder der Stonecutters Bridge. Beton- und Stahlbeton 2008, pp. 766–773

[1.23] Hopf, S.: The Main Design of the Sava Bridge in Belgrade. Vortrag Belgrad, 01.10.2010

[1.24] Heckhausen, C. F. and Cabjolsky, F. H.: The Posadas-Encarnación Bridge over the Paraná River between Argentina and Paraguay, L'industria italiana del Cemento, Feb 1995, pp. 78–97

[1.25] Köppel. J., Buchs and Bacchetta: Rheinbrücke Diepoldsau. Schweizer Ingenieur und Architekt 1984, pp. 1–6, 818–821

[1.26] Brault, J.-L. and Mathivat, J.: Le Pont de Brotonne. Travaux Feb 1976, pp. 22–43

[1.27] Schambeck, H.: Bau der zweiten Mainbrücke der Farbwerke Hoechst AG – Konstruktion und Ausführung. Deutscher Betontag 1973, p. 359–372

[1.28] Svensson, H. and Humpf, K.: Die Schrägkabelbrücke über den Mississippi bei Burlington, USA. Stahlbau 1994, pp. 193–199

[1.29] Virlogeux, M.: Design and Construction of the Normandy Bridge. FIP notes, 1994, pp. 4–16

[1.30] Öresundskonsortiet: The Öresund Fixed Link, Design and Construction, December 1998, pp. 1–40

[1.31] Saul, R., Humpf, K. and Schiele, I.: Die dritte Brücke über den Orinoco/Venezuela – Eine zweistöckige Schrägkabelbrücke für Straße und Eisenbahn mit Verbundfachwerk. Stahlbau 2010, p. 63

[1.32] Company brochure DSI

[1.33] Taylor, P.: Hybrid Design for the World's Longest Span Cable-Stayed-Bridge. IABSE 1984, pp. 319–324

[1.34] Leonhardt, F.: Brücken. Bridges. DVA, Stuttgart 1982

[1.35] Schiller, F.: Über Anmut und Würde. Gesammelte Werke in 3 Bänden, Band 2, Hanser Verlag, München 1966, pp. 382–424

[1.36] Bundesministerium für Verkehr, Bau und Stadtentwicklung: Brücken und Tunnel der Bundesfernstraßen 2009, Schrägseilbrücke Bremen Neustadt pp. 7–21

[1.37] Moro, J. L.: Symbiose Ingenieur Jörg Schlaich, Architekt Volkwin Mark: Ingenieurbaukunst in Deutschland. Jahrbuch 2001, Junius Verlag, Hamburg, pp. 142–157

[1.38] Tamms, F.: Ingenieurbauten-Gestaltung und städtebauliche Bedeutung. In Beyer/Lange: Verkehrsbauten. Beton-Verlag, Düsseldorf 1974, pp. 23–32

[2.1] Troitsky, M. S.: Cable-Stayed Bridges, Second Edition BSP Professional Books 1988

[2.2] Pelke, E.: Leistungsfähiger Konkurrent? – Die Entwicklung der Schrägkabelbrücke. In: Fachhochschule Potsdam (Publ.): Johann

August Röbling 1806 – 1869. Vom preußischen Baukondukteur zum Konstrukteur der Brooklyn Bridge. Ingenieurbau zwischen Kunst und Wissenschaft, Potsdam 2006: Proceedings

[2.3] Pelke, E., Ramm, W. and Stiglat, K.: Geschichte der Brücken, Zeit der Ingenieure, Deutsches Straßenmuseum Germersheim 2005

[2.4] Faustus Verantius: Machinae novae Fausti, Venedig 1617

[2.5] Löscher, C. T.: Angabe einer ganz besonderen Hängewerksbrücke, die mit wenigen und schwachen Hölzern, sehr weit über einen Fluss gespannt werden kann, die größten Lasten trägt und vor den stärksten Eisfahrten sicher ist. Leipzig 1784

[2.6] Navier, M.: Mémoires sur les Pont Suspendus, Imprimerie Royale, Paris 1823

[2.7] Stephenson, R.: Description of Bridges of Suspension. The Edinburgh Philosophical Journal 1821, pp. 237 – 256

[2.8] Birnstiel, C.: The Nienburg Cable-Stayed Bridge Collapse: An Analysis 18 Decades later. Proceedings of the 5th International Conference on Bridge Management, University of Surrey, 2005, pp. 179 – 186

[2.9] Gewerbefleiß Verein: Über die Nienburger Brücke. Verhandlungen des Vereins zur Beförderung des Gewerbefleißes in Preußen, Duncker and Humblot, Berlin 1826

[2.10] Motley, T.: Twerton Bridge over the River Avon near Bath, The Civil Engineer and Architect Journal Volume I 1838, pp. 253 – 254

[2.11] Rüntheroth, N. and Kahlow, A.: Johann August Roebling, Ingenieurbaukunst Jahrbuch 2005/2006, Junius Verlag, Hamburg 2005, pp. 124 – 137

[2.12] Sayenga, D.: Roebling als Drahtseilpionier – die entscheidenden Jahre. Mühlhauser Beiträge, Sonderheft 15, Mühlhausen/Thür. 2006

[2.13] Roebling, J. A.: American Railroad-Journal and Mechanics Magazine. 1841, pp. 161 – 166.

[2.14] Weigold, M. E.: Silent Builder, Emily Warren Roebling and the Brooklyn Bridge, Associated Faculty Press, Port Washington, New York 1984

[2.15] Tyrrell, H. G.: History of Bridge Engineering, Chicago 1911, pp. 230

[2.16] Häseler, E.: Die eisernen Brücken – ein Handbuch zum Gebrauche beim Entwerfen eiserner Brücken, 4. Lieferung, Vieweg Verlag, Braunschweig 1908

[2.17] de Nansouty, M.: Road-Bridge over the Rhône at Lyons, Proc. Inst. Civil Engineers, 108, London 1892, pp. 430 – 432

[2.18] Leinekugel le Cocq, G.: Bonhomme Suspension Bridge over the Blavet, Proc. Instn. Civ. Engineers, 1905, 162 (4), pp. 436 – 437

[2.19] Stiglat, K.: Schwebefähren: Triumphbögen zwischen Festland und Meer – Versuch einer Chronologie, Stahlbau 2008, pp. 575 – 587

[2.20] Leinekugel le Cocq, G.: Ponts à Transbordeur, Le Génie Civil 1903, pp. 33 – 36 and Nr. 4, pp. 49 – 55

[2.21] N. N.: Le Pont Tranbordeur et la Vision moderniste, Ausstellungskatalog. Musées de Marseille – Musée Cantini 1991

[2.22] Buonopane, S. G. and Brown, M. M.: History and engineering analysis of the cable-stayed Bluff Dale Bridge, Madrid: Proceedings 1. Congress on Construction History 20. – 24. January 2003, Volume I, pp. 433 – 442

[2.23] Gisclard, A. V.: Note sur un nouveau type de pont suspend rigide, Annls Ponts Chauss., 1899 – 1900

[2.24] Leinekugel le Cocq, G.: Pont Suspendus, tomes 1 and 2, Octave Doin et Fils, Paris 1911

[2.25] Grattesat, G.: Pont de France, Paris: Presses Pont et Chaussées, 1982

[2.26] Stiglat, K.: Brücken am Weg, Ernst & Sohn, Berlin 1977

[2.27] Torroja; E.: L'emploi des cables d'acier dans les constructions en béton armé, Internationale Vereinigung für Brücken- und Hochbau: Kongress- und Schlussbericht Paris 1932, pp. 683 – 688

[2.28] Torroja, E.: Logik der Form, Georg D. W. Callwey, München 1961

[2.29] Dischinger, F.: Hängebrücken für schwerste Verkehrslasten, Bauingenieur 1949, pp. 65 – 75, 107 – 113

[2.30] Cornelius, W.: Die Berechnung der Ebenflächentragwerke mit Hilfe der Theorie der orthogonal-anisotropen Platte, Stahlbau 1952, pp. 21, 43, 60

[2.31] Fiedler, E.: Die Entwicklung der orthotropen Fahrbahnplatte in Deutschland. Stahlbau 2009, pp. 562 – 576

[2.32] Weitz, F. R.: Schrägseilbrückensysteme als Beispiel für Entwicklungstendenzen im modernen Großbrückenbau, Thyssen Technische Berichte, Vol. 1/83, pp. 40 – 59

[2.33] Sievers, H. and Görtz, W.: Der Wiederaufbau der Straßenbrücke über den Rhein zwischen Duisburg-Ruhrort und Homberg (Friedrich-Ebert-Brücke), Stahlbau 1956, pp. 77 – 88

[2.34] Görtz, W.: Zügelgurtbrücken, VDI-Z. Bd. 98 Nr. 35, 1956, pp. 1909 – 1948

[2.35] Wenk, H.: Die Strömsundbrücke, Stahlbau 1954, pp. 73 – 76

[2.36] Ernst, H. J.: Montage eines seilverspannten Balkens im Groß-Brückenbau, Stahlbau 1956, pp. 101 – 108

[2.37] Beyer, E. and Lange, K.: Verkehrsbauten, Beton-Verlag, Düsseldorf 1974

[2.38] Leonhardt, F.: Baumeister in einer umwälzenden Zeit. DVA, Stuttgart 1984

[2.39] Stiglat, K.: Bauingenieure und ihr Werk, Andrä, H.-P., Wolfhart Andrä. Ernst & Sohn, Berlin 2004, pp. 38 – 50

[2.40] Stiglat, K.: Bauingenieure und ihr Werk, Hans Grassl. Ernst & Sohn, Berlin 2004, pp. 164

[2.41] Beyer, E.: Nordbrücke Düsseldorf Gesamtanlage und Montage. Stahlbau 1958, pp. 1 – 6

[2.42] Lange, K.: Moderner Stahlbrückenbau. Vortrag gehalten am 15.11.1972 an der Uni Karlsruhe anlässlich der Verleihung der Ehrendoktorwürde, Stumm Journal Nr. 3, 1972

[2.43] Rademacher, C.-H. und Ramberger, G.: Erwin Volke gestorben. Stahlbau 2006, pp. 247 – 248

[2.44] Schreier, G.: Die Strombrücke über das Rheinknie in Düsseldorf: Konstruktion, Berechnung, Fertigung und Montage. Acier-StahlSteel 1972, Vol. 5

[2.45] Beyer, E., Volge, E., Grassl, H., v. Gottstein, F., Andrä, W., Wintergerst, L.: Die Oberkasseler Rheinbrücke und der geplante Querverschub. Betonverlag 1974

[2.45a] Hein, Lehmann AG: Düsseldorf Oberkassel Rhinebridge, Construction and displacement, Düsseldorf, 1976

[2.46] Heß, H.: Die Severinsbrücke Köln, Entwurf und Fertigstellung der Strombrücke. Der Stahlbau 1960, Vol. 8, pp. 225 – 261

[2.47] Daniel, H. and Schumann, H.: Die Bundesautobahnbrücke über den Rhein bei Leverkusen. Stahlbau 1967, pp. 225

[2.48] Kurrer, K.-E., Pelke, E. and Stiglat, K.: Einheit von Wissenschaft und Kunst im Brückenbau: Hellmut Homberg (1919–1990). Bautechnik 2009, pp. 647–655, 794–809, Bautechnik 2010, pp. 86–115

[2.49] Epple, G., Rössing, E., Schaber, E. und Wintergerst, L.: Die neue Rheinbrücke für die Bundesautobahn bei Speyer. Stahlbau 1977, Vol. 10, 11, 12

[2.50] Idelberger, K.: Die Schrägseilbrücke mit A-Pylon über den Rhein bei Neuwied. Stahlbau 1978, pp. 302–307

[2.51] Modemann, J. and Thönnissen, K.: Rheinbrücke Düsseldorf-Flehe – Planung, Entwurf Ausschreibung, Vergabe und Überblick über den Ausführungsentwurf. Bauingenieur 1979, pp. 1–12

[2.52] Zellner, W. and Schmidts, P.: Rheinbrücke Düsseldorf-Flehe – Planung, Entwurf Ausschreibung, Vergabe und Überblick über den Ausführungsentwurf. Bauingenieur 1979, pp. 85–93

[2.53] Schambeck, H., Foerst, H., Honnefelder, N.: Rheinbrücke Düsseldorf-Flehe – Planung, Entwurf Ausschreibung, Vergabe und Überblick über den Ausführungsentwurf. Bauingenieur 1979, pp. 111–117

[2.54] Kahmann, R. and Koger, E.: Koordination der Gesamtbaumaßnahme und Beschreibung des Stahlüberbaus. Bauingenieur 1979, pp. 177–187

[2.55] Gebert, G., et al.: Die neue Rheinbrücke Wesel – Entwurfsplanung und Ausschreibung. Stahlbau 2007, pp. 657–670

[2.55a] Löckmann, H. and Marzahn, G. A.: Spanning the Rhine River with a new cable-stayed bridge, Structural Engineering International 3/209, pp. 271–275.

[2.56] Anistoroaiei, Ch., et al.: Rheinbrücke Wesel – Konstruktion und statische Berechnung. Stahlbau 2008, pp. 473–488

[2.57] Heeb, A., Gerold, W. and Dreher, W.: Die Stahlkonstruktion der Neckarbrücke Untertürkheim. Stahlbau 1967, pp. 33–38

[2.58] Freudenberg, G.: Die Stahlhochstraße über den neuen Hauptbahnhof in Ludwigshafen/Rhein. Stahlbau 1970, pp. 257–267, 306–314

[2.59] Tesár, A.: Das Projekt der neuen Straßenbrücke über die Donau in Bratislava/ČSSR. Bauingenieur 1968, pp. 189–198

[2.60] Leonhardt, F., Zellner, W. and Saul, R.: Zwei Schrägkabelbrücken für Eisenbahn- und Straßenverkehr über den Rio Paraná (Argentinien). Stahlbau 1979, pp. 225–236, 272–277

[2.61] Leonhardt, F., Zellner, W. and Saul, R.: Die Betonpylonen und Unterbauten der Schrägkabelbrücken Zárate-Brazo Largo über den Rio Paraná (Argentinien). Bauingenieur 1980, pp. 1–10

[2.62] Leonhardt, F., Zellner, W. and Saul, R.: Modellversuche für die Schrägkabelbrücken Zárate-Brazo Largo über den Rio Paraná (Argentinien). Bauingenieur 1979, pp. 321–327

[2.63] Hajdin, N. and Jevtovic, L.: Eisenbahnschrägseilbrücke über die Save in Belgrad. Stahlbau 1978, pp. 97–106

[2.64] Calatrava Valls, S., et al.: Paso del Alamillo, Junta de Andalucia, Dirección General de Carreteras

[2.65] Kleinhanß, K., Romberg, M. and Saul, R.: Die zweite Strelasundquerung mit der Schrägseilbrücke über den Ziegelgraben. Bauingenieur 2007, pp. 159–169

[2.65a] Kleinhanß, K. and Saul, R.: The Second Strelasund Crossing, Structural Engineering International 1/2007, pp. 30–34

[2.66] Daniel, H.: Die Rheinbrücke Duisburg-Neuenkamp. Stahlbau 1971, pp. 193–200, Stahlbau 1972, pp. 7–14, 73–78

[2.67] Tschemmernegg, F.: Zur Berechnung der Pylonen der Rheinbrücke Duisburg-Neuenkamp. Stahlbau 1971, pp. 337–343

[2.68] Gräfe, R.: Brücke über die Mündung der Loire. Stahlbau 1977, pp. 120–122

[2.69] Manabe, Y., Mukasa, N., Hirahara, N., Yabuno, M.: Accuracy Control on the Costruction of the Tatara Bridge, Proceedings IABSE Conference „Cable-Stayed Bridges", Malmö, 2–4 June 1999

[2.70] Zellner, W. and Svensson, H.: Zur Entwicklung der Schrägkabelbrücken aus Beton. Proceedings 9. Internationaler Spannbetonkongress, Spannbetonbau in der Bundesrepublik, Stockholm 1982

[2.71] Roik, K., Albrecht, G. and Weyer, U.: Schrägseilbrücken. Ernst & Sohn, Berlin 1986

[2.72] Caquot, A.: Les ouvrages d'art du canal de Donzière-Montragon (fin.). La Technique Modernes-Construction VI No. 7 (1954), pp. 239–243

[2.73] Kerisel, J.: Albert Caquot 1881–1976 Savant, soldat et batisseur, Paris: Presses Ponts et Chausées, 2001

[2.74] Rothmann, H. B. und Chang F.-K.: Longest precast-concrete box-girder bridge in Western-Hemisphere, Civil Engineering-ASCE, March 1974, pp. 56–60

[2.75] Pauser, A. and Beschorner, K.: Betrachtungen über seilverspannte Massivbrücken, ausgehend vom Bau der Schrägscilbrücke über den Donaukanal in Wien. Beton- und Stahlbetonbau 1976, pp. 261–265, 298–304

[2.76] Troyano, L. F.: Tierra Sobre El Agua, Visión Histórica Universal de Los Puentes, Colegio de Ingenieros de Caminos, Canales y Puertos, Madrid, 1999

[2.77] Zellner, W. and Svensson, H.: Mitarbeit an der Planung von Schrägkabelbrücken in den USA. In: Verband Beratender Ingenieure (VBI): „Konstruktiver Ingenieurbau", Ernst & Sohn, Berlin 1985, pp. 87–95

[2.78] Globig, H.: The Penang Bridge. Tenth International Congress of the FIP, New Delhi 1986, pp. 43–50

[2.79] Cabjolsky, H., Saul, R. and Schwarz G.: Zur Größe des Windwiderstandes bei sehr hohen Windgeschwindigkeiten (Tornados). Bauingenieur 1984, pp. 253–260

[2.80] Svensson, H. and Hopf, S.: Die Spannbeton-Schrägkabelbrücke Helgeland. Beton- und Stahlbeton 1993, pp. 247–250, 279–281

[2.80a] Svensson, H. and Hopf, S.: The concrete cable-stayed Helgeland Bridge, Norway. Proceedings of the ACI Spring Convention, Washington DC, 1992

[2.80b] Svensson, H. and Jordet, E.: The concrete cable stayed Helgeland bridge in Norway. Civil Engineering (ICE), Vol. 114, May 1996, pp. 54–63.

[2.81] Statens Vegvesen: The Nordhordland Bridge, Europe's longest floating Bridge crosses the Salhus Fjord. Bergen, 22 Sept 1994

[2.82] Svensson, H., Humpf, K. and Straub, W.: Die River-Leven-Stahlbeton-Schrägkabelbrücke. Beton- und Stahlbetonbau 1996, pp. 127–131

[2.83] Saul, R., Humpf, K., Hopf, S. and Patsch, A.: Die zweite Brücke über den Panamakanal – eine Schrägkabelbrücke mit 420 m Mittelöffnung und Rekordbauzeit. Beton- und Stahlbetonbau 2005, pp. 225–235

[2.84] Moormann, Chr., Svensson, H. and Humpf, K.: Gründungsoptimierung im internationalen Großbrückenbau – Neue Entwicklungen und aktuelle Projekterfahrungen. Vorträge der 31. Baugrundtagung, München, 03.–06.11.2010, Deutsche Gesellschaft für Geotechnik, Essen, pp. 211–218

[2.85] Saul, R., Hopf, S., Humpf, K., Patsch, A. and Bacher, A.: Innovativer Schutz gegen Schiffsanprall für die Brücke Rosario-Victoria über den Paraná (Argentinien). Stahlbau 2003, pp. 469–484

[2.86] Longest concrete cable-stayed span cantilevered over tough terrain – Wadi Kuf. Engineering News Record, 15 July 1971, pp. 28, 29

[2.87] Morandi, R.: Il Viadotto dell Ansa della Magliana per la Autostrada Roma-Aeroporta di Fiumicino. L'Industria Italiana del Cemento 1968, pp. 147–162

[2.88] Morandi, R.: Il ponte sul fiume Magdalena a Barranquilla (Colombia). L'Industria Italiana del Cemento 1974, pp. 383–406

[2.89] Morandi, R.: Il viadotto Carpineto 1 per la strada di grande comunicazione Basentana. L'Industria Italiana del Cemento 1977, pp. 817–830

[2.90] Schambeck, H. and Kroppen, H.: Die Zügelgurtbrücke aus Spannbeton über die Donau in Metten. Beton- und Stahlbetonbau 1982, pp. 131–136, 156–161

[2.91] Blaubeurer Torbrücke, Ulm – interne Unterlagen LAP 1988

[2.92] Schambeck, H.: Die Flößerbrücke in Frankfurt. Bauingenieur 1987, pp. 151–157

[2.93] Menn, Ch. and Rigendinger, H.: Ganterbrücke. Schweizer Ingenieur und Architekt 1979, pp. 733–738

[2.94] Plain sailing in Bahrain. Civil Engineer International, Nov 1996, pp. 31, 32

[2.95] Walther, R.: Schrägseilbrücken mit dünner Fahrbahnplatte. 23. Forschungskolloquium des DAfStb, ETH Zürich, 1990, pp. 37–42

[2.96] Bögle, A.: leicht weit Light Structures, Ernst & Sohn Berlin 2004, pp. 197

[2.97] Manterola Armisen, J. and Troyano, L. F.: The Ing. Carlos Fernandez Casado Bridge across Barrios de Luna Lake. L'Industria Italiana del Cemento 3/1985, pp. 150–175

[2.98] Statens Vegvesen Nord-Trondelag: Skarnsundet Bridge in Norway, 1991

[2.99] Svensson, H.: The Development of Composite Cable-Stayed Bridges – IABSE Conference-Malmö, Sweden 2–4 June 1999

[2.100] Prime Piling, New Civil Engineer 17 Nov 1994, pp. 38–42

[2.101] Dussart, R.: Le Pont Massená à Paris. TRAVAUX, May/Sept 1970

[2.102] Knox, H. S. G. and Walther, F.: The Open Steel Deck Cable-Stayed Bridge. Strait Crossings 94, Krokeborg (ed.), 1994 Balkema, Rotterdam, pp. 123–131

[2.103] Kunz, R., Trappmann, H. und Tröndle, E.: Die Büchenauer Brücke, eine neue Schrägseilbrücke der Bundesstraße 35 in Bruchsal. Stahlbau 1957, pp. 98–102

[2.104] Saul, H., Svensson, H., Andrä, H.-P. and Selchow, H.-J.: Die Sunshine-Skyway Brücke in Florida USA – Entwurf einer Schrägkabelbrücke mit Verbundüberbau. Bautechnik 1984, pp. 1–16

[2.104a] Svensson, H., Christopher, B. G. and Saul, R.: Design of a cable-stayed composite bridge, Journal of Structural Engineering, ASCE (1986), pp. 489–504

[2.105] Lin Yuanpei: Yangpu Bridge, Shanghai, China. Structural Engineering International 3/95, pp. 143, 144

[2.105a] Yang Xiao-lin: Yangpu Bridge, Shanghai Scientific and Technological Education Publishing House

[2.106] Saul, R., Braun, M., Järvenpää, E. and Pulkkinen, P.: Die Tähtiniemi-Brücke in Finnland – eine Schrägkabelbrücke mit gekrümmtem Verbundüberbau. Stahlbau 1995, pp. 161–167

[2.107] Saul, R. and Hopf, S.: Die Kap Shui Mun Brücke in Hongkong – eine zweistöckige Schrägkabelbrücke für Straßen- und Eisenbahnverkehr. Beton- und Stahlbetonbau 1997, Vol. 10, pp. 261–265, Vol. 11, pp. 308–312

[2.108] Vägverket: Sunningeleden – the loveliest short cut! 2000

[2.109] Jordet, E. A. and Karlsson, M.: Kolbaecksbron, a Composite Cable-Stayed Bridge in Umeaa, Sweden. 16th Congress of IABSE, Lucerne, 2000, pp. 106 ff.

[2.110] Ingenieurbau-Preis von Ernst & Sohn 2006: Neubau Berliner Brücke, Halle (Saale)

[2.111] Eilzer, W., Richter, F., Wille, T., Heymel, U. and Anistoroaiei, Ch.: Die Elbebrücke Niederwartha – die erste Schrägseilbrücke in Sachsen. Stahlbau 2006, pp. 93–104

[2.112] Virlogeux, M.: Bridges with Multiple Cable-Stayed Spans, Structural Engineering International 1/2001, pp. 61–82

[2.113] Carter, M., et al.: Forth Replacement Crossing. IABSE Bangkok 2009

[2.114] Die Brücke über den Maracaibo-See in Venezuela. Bauverlag Berlin 1963

[2.115] Morandi R.: Il Viadotto sul Polcevera per l'autostrada Genova-Savona. Industria Ital. del Cemento 1967, pp. 849–872

[2.116] Freyssinet International: Kwang Fu Bridge, Taiwan, FI. 1092 A 12/77

[2.117] King, C., et al.: The Four Cable-Stayed Bridges of the Mexico-Acapulco Highway

[2.118] The new Macau-Taija Bridge: The Friendship Bridge. Port and Bridge Office, Macau, 1994

[2.119] Bergermann, R. and Schlaich, M.: Die Ting-Kau-Schrägkabelbrücke in Hongkong – Entwurf und Konstruktion. Bauingenieur 1999, pp. 413

[2.120] Bergermann, R., Schlaich, M. and Näher, F.: Die Ting-Kau-Schrägkabelbrücke in Hongkong – Bau. Bauingenieur 1999, pp. 480–484

[2.121] Menn, Ch., et al.: Sunnibergbrücke. Schweizer Ingenieur und Architekt No. 44, 1998

[2.122] Combault, J. et al.: Rion-Antirrion Bridge, Greece – Concept, Design and Construction. SEI 2005, pp. 22–27

[2.123] Virlogeux, M.: Der Viadukt über das Tarntal bei Millau. Bautechnik 2006, pp. 96

[2.124] Saul, R., Humpf. K. and Lustgarten, M.: Die Orinoco-Brücke in Ciudad Guayana/Venezuela – Doppel-Schrägkabelbrücke mit Verbundüberbau für Straßen- und Eisenbahngüterverkehr. Stahlbau 2006, pp. 82 – 92

[2.125] Femern Sund-Baelt: Feste Fehmarnbeltquerung, Aktueller Planungsstand, Bericht, 25. 11. 2010

[2.126] Wößner, K., Andrä, W., Kahmann, R., Schumann, H. and Hommel, D.: Die Neckartalbrücke Weitingen. Stahlbau 1983, pp. 65 – 77, pp. 113 – 124

[2.127] Schlaich, J., Seidel, J. and Sandner, D.: Teilweise unterspannte Schrägkabelbrücke über die Obere Argen. IABSE Congress Report, Vol. XIII, 1988, pp. 863 – 868

[2.128] Otsuka, H., et al.: Comparison of structural characteristics for different types of cable supported prestressed concrete bridges. Structural Concrete 2002, pp. 3 – 21

[2.129] Bergman, D. W., Radojevic, D. and Ibrahim, H.: Design of the Golden Ears Bridge, IABSE Symposium Weimar 2007, pp. 58 – 61

[2.130] Schlaich, J. and Bergermann, R.: Fußgängerbrücken. Katalog zur Ausstellung an der ETH Zürich, 1992

[2.131] Baus, U. and Schlaich, M.: Fußgängerbrücken. Birkhäuser, Basel Boston Berlin 2007

[2.131a] Idelberger, K.: The World of Footbridges. Ernst & Sohn, Berlin, 2011

[2.132] Leonhardt, F. and Andrä, W.: Fußgängersteg über die Schillerstraße in Stuttgart. Bautechnik 1962, pp. 110 – 116

[2.133] BBRV: Cable-Stayed Structures, Rapport No. 7802

[2.134] Völkcl, E., Zellner, W. and Dornecker, A.: Die Schrägkabelbrücke für Fußgänger über den Neckar in Mannheim. Beton- und Stahlbeton 1977, pp. 29 – 35, 59 – 64

[2.135] Morgenthal, G. and Saul, R.: Die Geh- und Radwegbrücke Kehl-Strasbourg. Stahlbau 2005, pp. 121 – 125

[2.136] Andrä, H.-P., Burghagen, K., Häberle and Svensson, H.: Geh- und Radwegbrücke Weil der Stadt. Bauingenieur 2007, pp. 341 – 345

[2.137] Kurrer, K.-E.: The history of the theory of structures, Ernst & Sohn, Berlin 2008

[2.138] Kolyushev, I.: Two cable-stayed bridges in the East of Russia, Verbal presentation at the Bridge Symposium Leipzig, 2010

[3.1] Bechtold, M., Mordue, B. and Rentmeister, F. E.: Locked Coil Cables and their End Connections. EUROSTEEL 2008, Graz, Austria

[3.2] Bridon: Structural Systems 11/2007

[3.3] Roik, K.: Vorlesungen über Stahlbau, Ernst & Sohn, Berlin 1983

[3.4] Krüger, U.: Berechnung von Seilköpfen zur Verankerung patentverschlossener Drahtseile. Bauingenieur 52, 1977

[3.5] Gabriel, K. and Helmes, F.: The mechanics of socketing: the zinc alloy cast cone as a special compound structure. Proceedings of the First International Offshore and Polar Engineering Conference, 1991

[3.6] Prehn, W. and Mertens, M.: Die Rheinquerung der A 44 – Darstellung der Gesamtmaßnahme. Stahlbau 2002, pp. 386 – 392

[3.7] EN 10264:2002 Steel wire and wire products – Steel wire for ropes

[3.8] EN 12385-10:2003 Steel wire ropes – Safety – Part 10: Spiral ropes for general structural use

[3.9] Saul, R. and Svensson, H.: On the Corrosion Protection of Stay Cables. Stahlbau 1990, pp. 165 – 176

[3.10] Martin, M., Stromberg, H. and Tins, J.: Verhalten vollverschlossener Spiralseile bei Zuschwellbeanspruchung. Die Bautechnik 60, 1983, pp. 369 – 372

[3.11] Funke, W.: Neuere Erkenntnisse über die Wirkungsweise von Aktivpigmenten und Barriereprinzip. Lechler-Chemie, Symposium 84

[3.12] British Patent Specification No. 835883, 1064973

[3.13] Zellner, W. and Saul, R.: Über Erfahrungen beim Umbau und Sanieren von Brücken. Die Bautechnik 62, 1985, pp. 51 – 65

[3.14] Saul, R.: Auswechseln der Tragseile von Brücken im In- und Ausland. Lindauer Bauseminar 1982

[3.15] Gurtmann, S., Hamme, M., Marzahn, G. und Sieberth, S.: Seiltausch unter Verkehr an der Rheinbrücke Flehe. Stahlbau 2010, pp. 682 – 688

[3.16] Finsterwalder, K.: Schrägseile. Bauwirtschaft, Heft 9, 24 Feb 1973

[3.17] Globig, H.: The Penang Bridge. FIP, New Delhi 1986, pp. 43 – 47

[3.18] Jungwirth, D.: Requirements for Stay Cables in International Competition. DSI Stay Cable Report 4/1995

[3.19] Andrä, W. and Zellner, W.: Zugglieder aus Paralleldrahtbündeln und ihre Verankerung bei hoher Dauerschwellbelastung. Die Bautechnik, 1969

[3.20] Firmenschrift BBR: DINA System

[3.21] Andrä, W. and Saul, R.: Versuche mit Bündeln aus parallelen Drähten und Litzen. Die Bautechnik, 1974, pp. 289 – 298, 332 – 340, 371 – 373

[3.22] „Recommendations for Stay Cable Design and Testing" by Post-Tensioning Institute Committee on Cable-Stayed Bridges, USA, Fifth Edition, 2007

[3.23] Walder, C. and De Coste, H.: Weathering Studies on Polyethylene, Industrial and Engineering Chemistry, 1950, pp. 2320 – 2325

[3.24] Howard, G.: Natural and Artifical Weathering of Polyethylene Plastics, Polymer Engineering and Science, 1969, pp. 286 – 294

[3.25] Shinko Wire Company: Test Results for Weather Resistance of Polyethylene. Amagasaki, 25 Feb 1983

[3.26] Hoechst AG: PE-Rohre (PE-Pipes), Frankfurt/Main, 1983

[3.27] Svensson, H.: Investigation on the Coiling of Stay Cables with PE-Pipes. Leonhardt, Andrä and Partners GmbH, Stuttgart, June 1989 (unpublished)

[3.28] Shinko Wire Company: Test Report on Weather-Resistance of Wrapping Tapes for Cable-Stayed Bridges. Report No. SWENH-8603. Amagasaki, 30 Apr 1984

[3.29] Institut für Kunstoffprüfung an der Universität Stuttgart, IKP (Testing Laboratory for Plastics at Stuttgart University): Untersuchungen zur Verstärkung von PE-Rohren mit Kunststoffbinden. Report M 980, 17. 11. 1980 – Report 980/2, 1. 4. 1981

[3.30] Eidgenössische Materialprüfungs- und Versuchsanstalt Dübendorf, EMPA (Swiss Federal Testing Institute): Vergleichende

Korrosionsschutz-Untersuchungen mit Schrägseilen. Report 20-707/1, 3.4.1973 – Report 29-707/2, 24.11.1973

[3.31] Shinko Wire Company: Fatigue Tests for Grouted Cables with Hi-Am Anchor Sockets. Research Report No. DRG 8203, Part 3, Feb 1982

[3.32] Amtliche Forschungs- und Materialprüfungs- und Versuchsanstalt für das Bauwesen an der Universität Stuttgart : Untersuchungen des Korrosionsschutzes der Schrägkabel des Schillersteges. Report S 11828, 21.10.1970, Report S 12410, 6.2.1974, Report S 13594, 10.12.1979

[3.33] Ohlemutz, A.: Neue Schrägseilbrücke über den Mississippi (New Cable-Stayed Bridge across the Mississippi). Der Stahlbau 48, 1979, pp. 151

[3.34] Svensson, H.: Investigation of Cracks in the PE-Pipes of the Luling Bridge. Leonhardt, Andrä and Partner GmbH, Stuttgart, June 1986 (unpublished)

[3.35] Company information, BBR Schrägkabel

[3.36] Company information, DSI Schrägkabel

[3.37] Company information, Freyssinet Schrägkabel

[3.38] Company information, VSL Schrägkabel

[3.39] Tall in the Saddle, Bridge design and engineering 2008, pp. 38, 39

[3.40] DIN 1073: Stählerne Straßenbrücken; Berechnungsgrundlagen. Ausgabe Sept 1987, Abschnitt 6.5

[3.41] DIN 18809: Stählerne Straßen- und Wegebrücken; Bemessung, Konstruktion, Herstellung. Ausgabe Sept 1987, Abschnitt 6.1

[3.42] DIN 18800, Teil 1: Stahlbauten; Bemessung und Konstruktion. Ausgabe Mar 1981, Abschnitt 6.2

[3.43] DIN Fachbericht 103: Stahlbrücken. Ausgabe Mar 2009, Anhang II-A

[3.44] DIN 18800, Teil 1: Stahlbauten; Bemessung und Konstruktion, Ausgabe Nov 1990, Abschnitt 9

[3.45] ZTV-ING, Teil 4 Stahlbau, Stahlverbundbau – Abschnitt 4 Brückenseile, Anhang B

[3.46] Eurocode 3: Bemessung und Konstruktion von Stahlbauten – Teil 1-11: Bemessung und Konstruktion von Tragwerken mit Zuggliedern aus Stahl. Dec 2006

[3.47] PTI: Recommendations for Stay Cable Design, Testing and Installation. Fifth Edition. Phoenix, AZ, USA. Oct 2007

[3.48] FIB, bulletin 30: „Acceptance of stay cable systems using prestressing steels". Lausanne, Jan 2005. Munich, Dec 2006

[3.49] Andrä, W. and Saul, R.: Versuche mit Bündeln aus parallelen Drähten und Litzen für die Nordbrücke Mannheim-Ludwigshafen und das Zeltdach in München. Die Bautechnik, Vol. 9, 10, 11, 1974, pp. 289 – 298, 332 – 340, 371 – 373

[3.50] Andrä, W. and Saul, R.: Die Festigkeit, insbesondere Dauerfestigkeit langer Paralleldrahtbündel. Die Bautechnik 56, 1979, Vol. 4, pp. 128 – 130

[3.51] Saul, R. and Andrä, W.: Zur Berücksichtigung dynamischer Beanspruchungen bei der Bemessung von verschlossenen Seilen stählerner Straßenbrücken. Sonderdruck, Die Bautechnik 4, 1981

[3.52] Saul, R. and Romberg, M.: Bericht zur Ermüdungsbeanspruchung der Litzenbündel der Ziegelgrabenbrücke. LAP Stuttgart, Dec 2006

[3.53] DIN 1055-4: Einwirkungen auf Tragwerke: Windlastannahmen

[3.54] Matsumoto, M., Shiraishi, N. and Shirato, H.: Rain-wind induced vibration of cables of cable-stayed bridges. J. of Wind Engineering and Industrial Aerodynamics, Vol. 44, 1992, Elsevier

[3.55] Yamaguchi, H. and Fujino, Y.: Damping of Cables in Cable-Stayed Bridges with and without Vibration-XControl Measures. Symposium Deauville, 1994

[3.56] Miyata, T., Yamada, H. and Hojo, T.: Aerodynamic Response of PE Stay Cables with Pattern-Indented Surface. Symposium Deauville, 1994

[3.57] Ruscheweyh, H.: Dynamische Windwirkung an Bauwerken. Bauverlag, Berlin 1982

[3.58] Kovacs, I.: Zur Frage der Seilschwingungen und der Seildämpfung. Die Bautechnik 10, 1982, pp. 325 – 331

[3.59] Kovacs, I. and Oba, N.: Einfluss der Längsresonanz auf die Schwingungsanfälligkeit von Tragkabeln. Die Bautechnik 74, 1997, pp. 369 – 375

[3.60] Davenport, A. G.: A simple representation of the dynamic of a massive stay cable. Symposium Deauville, 1994

[3.61] Kovacs, I., Strömmen, E. and Hjorth-Hansen, E.: Damping devices against cable oscillations on Sunningesund Bridge. Cable Dynamic Symposium, Trondheim 1998

[3.62] Flamand, O.: Rain-wind induced vibration of cables. Symposium Deauville, 1994

[3.63] Tanaka, Y.: Installation of Semi Parallel Wire Cables, Shinko Wire Presentation, Japan 2009

[4.1] Orlov, G. und Saxenhofer, H.: Balken auf elastischer Unterlage, Verlag Leemann, Zürich 1963

[4.2] Pflüger, A.: Stabilitätsprobleme der Elastostatik, Springer Verlag Berlin Heidelberg New York 1975

[4.3] Tang, M. C.: Buckling of Cable-Stayed Girder Bridges. Journal of the Structural Division, ASCE, Sept 1976, pp. 1675 – 1684

[4.4] Klöppel, K., Eßlinger, M. and Kollmeier, H.: Die Berechnung eingespannter und fest mit dem Kabel verbundener Hängebrückenpylonen bei Beanspruchung in Brückenlängsrichtung. Stahlbau 1965, pp. 358 – 361

[4.5] Hartwig, H.-J.: Die Kaiserleibrücke. Stahlbau 1965, pp. 97 – 110

[4.6] Saul, R. and Svensson, H.: Zur Behandlung des Lastfalls „ständige Last" beim Tragsicherheitsnachweis von Schrägkabelbrücken. Bauingenieur 1983, pp. 329 – 335

[4.7] Völkel, E., Zellner, W. and Dornecker, A.: Die Schrägkabelbrücke für Fußgänger über den Neckar in Mannheim. Beton- und Stahlbetonbau 1977, pp. 29 – 35, 59 – 64

[4.8] Homberg, H.: Schrägseilbrücken, Vielseilsysteme, Le Pont der Brotonne. Stahlbau 1975, pp. 235 – 243

[4.9] Homberg, H.: Einflusslinien von Schrägseilbrücken. Stahlbau, 1955, pp. 40 – 44

[4.10] Szabo, I.: Einführung in die Technische Mechanik. Springer Verlag Berlin Heidelberg New York 1963, pp. 337

[4.11] Herzog, M.: Näherungsberechnung von Schrägkabelbrücken. Bautechnik 1987, pp. 348 – 356.

[4.12] Farquharson, F. B., Vincent, G. S. et al.: Aerodynamic Stability of Suspension Bridges with special reference to the Tacoma Narrow Bridge. Bulletin No. 16, University of Washington Engineering Experimental Station, Parts I to V, pp. 1949–54
[4.13] Theodorsen, T.: General Theory of Aerodynamic Instability and the Mechanism of Flutter. NACA-Report 496, 1935
[4.14] Bleich, F.: Dynamic Instability of Truss-Stiffened Suspension Bridges under Wind Action. Trans. Am. Soc. Civ. Engrs., 114, pp. 1177–1232, 1949
[4.15] Steinman, D. B.: Aerodynamic Theory of Bridges Oscillations. Trans. Am. Soc. Civ. Engrs., 115, pp. 1180–1260, 1950
[4.16] Leonhardt, F.: Die Autobahnbrücke über den Rhein bei Köln-Rodenkirchen. Stahlbau 1951, Hefte 7, 9, 11
[4.17] Klöppel, K. and Thiele, F.: Modellversuche im Windkanal zur Bemessung von Brücken gegen die Gefahr winderregter Schwingungen. Der Stahlbau, Vol. 12, pp. 353–365, Dec 1967
[4.18] Davenport, A. G. and King, J. P. C.: Dynamic Wind Forces of Long-Span Bridges. Final Report, 12th Congress, International Association for Bridges and Structural Engineering, Vancouver, BC, Sept 1984, pp. 705–712
[4.19] Scanlan, R. and Simiu, E.: Wind Effects on Structures. New York: John Whiley & Sons 1986
[4.20] Svensson, H. and Kovacs, I.: Examples of analytical aerodynamic investigation of long-span bridges. Proceedings of the first International Symposium on Aerodynamics of large bridges. Copenhagen 1992
[4.21] Wardlaw: National Research Council Canada, National Aeronautical Establishment: A Wind Tunnel Study of Aerodynamic Stability of the Proposed Pasco-Kennewick Intercity Bridge. LTR-LA-163, 1974
[4.22] Davenport, A. G. and King, J. P. C.: A Study of Wind Effects for the Sunshine Skyway Bridge, Steel Alternate, BLWT-SS25-1982. The University of Western Ontario. London, Canada, 1982
[4.23] Scanlan, R. H. and Jones, N. R.: Aeroelastic analysis of cable-stayed bridges. Journal of Structural Engineering, ASCE, Vol. 116, No. 2, Feb 1990, pp. 279–297
[4.24] Petersen, Ch.: Dynamik der Baukonstruktionen. Vieweg, Braunschweig/Wiesbaden 1996
[4.25] DIN 1055: Einwirkungen auf Tragwerke – Teil 4: Windlasten. March 2005
[4.26] Engineering Science Data Unit (ESDU) No 85020. Characteristic of Atmosphaeric Turbulence Near the Ground. Single Point Data. London 1985
[4.27] Engineering Science Data Unit (ESDU) No 86010. Characteristic of Atmosphaeric Turbulence Near the Ground. Variations in Space and Time for Strong Winds. London 1986
[4.28] Eurocode 1: Actions on structures – Part 1–4: General actions – Wind actions. 2005
[4.29] Ruscheweyh, H.: Dynamische Windwirkung an Bauwerken, Bd. 2 (Praktische Anwendungen). Bauverlag, Wiesbaden 1982
[4.30] Rosemeier, G.: Winddruckprobleme bei Bauwerken. Springer, Berlin 1976

[4.31] Kovacs, I.: Zur Frage der Seilschwingungen und der Seildämpfung (Aspects of cable vibrations and cable damping). Die Bautechnik 10, 1982
[4.32] Kovacs, I.: Computer simulation of the dynamic response of wind-loaded bridges under limit load conditions. Symposium on Computational Wind Engineering CWE'96, Colorado 1996
[4.33] Hjorth-Hansen, E.: Fluctuationg drag, lift and overturning moment from static, mean loads. Lectures, Trondheim 1988
[4.34] Kovacs, I.: Synthetic wind for investigations in time domain. Paper, ASCE Symposium Atlanta 1994
[4.35] Kovacs, I. and Svensson, H.: Analytische aerodynamische Untersuchung der Schrägkabelbrücke Helgeland. Beton und Stahlbeton 1994, pp. 149–153, pp. 201–203
[4.35a] Kovacs, I., Svensson, H. and Jordet, E.: Analytical aerodynamic investigation of the cable-stayed Helgeland bridge, Journal of Structural Engineering ASCE, Jan 1992, pp. 147–168.
[4.36] Kovacs, I., Hjorth-Hansen, E. and Stroemmen: Damping Devices against Cable Oscillations on Sunningesund Bridge. Paper, Trondheim 2000
[4.37] Kovacs, I.: C-Bridge, S-Bridge, two programs for the numerical simulation of buffeting effects on cable-stayed and suspension bridges. Information brochure, Büro für Baudynamik, 1993
[4.38] Katz, C., Kovacs, I. and Morgenthal, G.: Dreidimensionale aerodynamische und aeroelastische Analyse der Fußgängerbrücke Kehl-Strasbourg (Coauthor). Vortrag Zürich, D-A-CH-Tagung 2003
[4.39] Morgenthal, G., Kovacs, I. and Saul, R.: Analysis of Aeroelastic Bridge Deck Response to Natural Wind. SEI 2005
[4.40] Försching, H. W.: Grundlagen der Aeroelastik. Springer, Berlin 1974
[4.41] Minorsky, V. U.: An Analysis of Ship Collision with Reference to Protection of Nuclear Power Plants. Journal of Ship Research 3, 1959
[4.42] Woisin, G.: Die Kollisionsversuche der GKSS (The Collision Tests of the GKSS). Jahrbuch der Schiffbautechnischen Gesellschaft, Volume 17, 1976, Berlin, Heidelberg, New York, pp. 465–487
[4.43] Saul, R. and Svensson, H.: Zum Schutz von Brückenpfeilern gegen Schiffsanprall (On the Protection of Bridge Piers against Ship Collision). Die Bautechnik 58, 1981, pp. 326–335, 374–388
[4.44] Saul, R. and Svensson, H.: On the Theory of Ship Collision against Bridge Piers. IABSE Proceedings P-51/82, pp. 29–40
[4.45] Saul, R. and Svensson, H.: Ship Collision with Bridges and Offshore Structures. Colloquium Copenhagen 1983, Introductory Report, pp. 165–179
[4.45a] Saul, R. and Svensson, H.: Means of reducing the consequences of ship collision with bridges and offshore structures. IABSE Introductory Report, Copenhagen 1983, pp. 165–179.
[4.46] AASHTO 1991. Guide Specification and Commentary for Vessel Collision Design of Highway Bridges. American Association of State Highway and Transportation Officials, Washington DC
[4.47] Pederson, P. T.: Ship Impacts: Bow Collisions. Proceedings from the Third International Symposium on Structural Crashworthiness and Failure, University of Liverpool 1993

[4.48] Larsen, O. D.: Ship Collision with Bridges. IABSE Structural Engineering Documents 4, 1993
[4.49] Wang, J., et al.: Comparison of Design Formula of Ship Collision for Bridges based on FEM Simulations. (in Chinese) Journal of Highway and Transportation Research and Development, Vol. 23, No. 2, pp. 68–73
[4.50] Brodin, S.: Tjörn Bridge. Swedish National Road Administration, 1984, ISBN 91-7810-124-7
[4.51] Yang Xiao-lin: Yang Pu Bridge. Shanghai Scientific and Technological Publishing House, 1994. ISBN 7-5428-0941-5/J1
[4.52] Stonecutters Bridge Design Competition – Stage 2. Design Documents by Leonhardt, Andrä and Partners, Stuttgart 2000
[4.53] Andrä, H.-P., Dietsch, B. and Sandner, D.: Formfindung und Entwurfsplanung des Spinaker Tower in Portsmouth (Shaping and Designing the Spinaker Tower in Portsmouth). Bautechnik 2004, pp. 507–515
[4.54] Saul, R., Hanke, R. and Kusch, G.: Die neue Neckarbrücke zwischen Kirchheim und Gemrigheim. Stahlbau 1998, pp. 36–45
[4.55] Knott, M.: Pier Protection System for the Sunshine Skyway Bridge Replacement. Proceedings at 3rd Annual International Bridge Conference, Pittsburgh, PA, 1986
[4.56] Schreier, G.: Beiträge zur Anwendung von baustatischen Methoden auf Probleme der Verformungstheorie. Dissertation Karlsruhe 1961
[4.57] Svensson, H.: Investigation on the relations between beam depth and live load stresses in cable-stayed bridges. Stuttgart, 1974 (unpublished)

[5.1] Company Information ALLSPANN, München: DYWIDAG – Stähle für Bauhilfsmaßnahmen
[5.2] Crémer, J.-M.: Advanced Methods for Cable-Stayed Bridge Erection. IABSE Structures 51/90, pp. 4 ff.

[6.1] National Research Canada, National Aeronautical: Wind Tunnel Studies on the Aerodynamic Stabilisy of Bridge Sections for the Proposed New Burrard Inlet Crossing. LTR-LA-31, 53, 54; 1969, 1970
[6.2] Test Report Stay Cable Anchors, Prepares by the University of Texas, Civil Engineering Structures Research Laboratory, Austin, Texas, and The Prescon Corporation, San Antonio, Texas, 23 July 1976
[6.3] Leonhardt, F.: Improving the Seismic Safety of Prestressed Concrete Bridges. Journal of the PCI 17, No. 6, 1972
[6.4] Roberts, G.: Severn Bridge – Design and Contract Arrangements. Proc. Inst. Civ. Eng. pp. 41, Sept 1958, pp. 1–58
[6.5] National Research Council Canada, National Aeronautical Establishment: A Wind Tunnel Study of the Aerodynamic Stability of the Proposed Pasco-Kennewick Intercity Bridge. LRT-LA-163, 1974
[6.6] EMPA: Gleit- und Druckschwellversuche mit Neporenblöcken. Bericht Nr. 141-173, Zürich 1992
[6.7] Svensson, H. and Jordet, E.: The concrete cable-stayed Helgeland Bridge in Norway. Proceedings of the Institution of Civil Engineers, Civil Engineering 1996, pp. 54–63

[6.8] Leonhardt, Andrä and Partners, Beratende Ingenieure VBI, GmbH: B 96 n, Zubringer Stralsund/Rügen: Anwendung der DIN-Fachberichte und Auswirkung auf die Massen. Stuttgart, 18. 03. 2002
[6.9] Raggett, J. D.: Report on Section Flutter Derivatives, Baytown Bridge. West Wind Laboratory, Camel, CA, 3/14/1998
[6.10] Reinhold, T. A.: Wind Tunnel Aeroelastic Model Study of the Baytown Bridge Steel Alternative. Applied Research Engineering Services, Inc., Report 5101-1, 29 May 1986
[6.11] Scanlan R. H. and Jones, N. R.: Aeroelastic Analysis of Cable-Stayed Bridges. Journal of Structural Engineering, ASCE, Vol. 116, No. 2, Feb 1990, pp. 279–297
[6.12] Virlogeux, M.: The Normandie Bridge, France: A New Record for Cable-Stayed Bridges. Structural Engineering International, 4/1994, pp. 208–213
[6.13] Scheuch, G.: Die Normandie Brücke. Bautechnik 1995, pp. 546–548
[6.14] Virlogeux, M.: Wind Design and Analysis for the Normandie Bridge. In: „Aerodynamics of Large Bridges". Balkema, Rotterdam, 1992, pp. 183–216
[6.15] Virlogeux, M., et al.: Millau Viaduct, France. Structural Engineering International 2005, pp. 4–7
[6.16] Dillinger Hütte GTS: Viadukt von Millau. Company information
[6.17] Virlogeux, M.: The Millau cable-stayed bridge. Recent development in bridge engineering. Edited by K. M. Mahmoud – Balkema, 2003, pp. 3–18
[6.18] Martin, J.-P., Servant, Cl., Crémer, J. M. and Virlogeux, M.: The design of the Millau viaduct. Fib Avignon Symposium Proceedings, 2004, pp. 83–107
[6.19] Scheuch, G.: Projekt der Peloponnes-Brücke Rion-Antirion. Bautechnik 2000, pp. 73–75.
[6.20] Pecker, A.: A Seismic Foundation Design Process. Lessons learned from two major projects: The Vasco da Gama and the Rion-Antirion Bridges. ACI International Conference on Seismic Bridge Design and Retrofit. La Jolla, California 2003
[6.21] Infanti, S.: Protective Measures. Bridge design & engineering, 2nd quarter 2004, pp. 54–55
[6.22] Teyssandier, J.-P. and Combault, J.: Le Pont de Rion-Antirion. Un ouvrage exceptionnel à vocation européenne. Travaux (1998) Nr. 748 (Dec), pp. 24–31
[6.23] Russell, H.: Greek triumph. Bridge design & engineering, 4th quarter 2000, pp. 29–36

[7.1] Petersen, A., Larsen, A. and Eilzer, W.: Outline design and special studies for a 1200 m cable-stayed bridge (unpublished)
[7.2] Gruppo Lambertini: Attraversamento Stabile Viario e Ferroviario dello Stretto di Messina, C. Lotti & Associati, Roma, 1982
[7.3] Maurer stay cable dampers, 2005
[7.4] Information about Maurer cable damper systems, 2010
[7.5] Ostenfeld, K. and Larsen, A.: Bridge engineering and aerodynamics. In: Aerodynamics of Large Bridges, Balkema, 1992, pp. 3–22

Figure Origins

The figures not listed here originate from the archive of the author or of Leonhardt, Andrä and Partners.

Figure 1.2 [1.5]
Figure 1.3 [1.5]
Figure 1.5 [1.3]
Figure 1.6 [1.4]
Figure 1.7 [1.1]
Figure 1.8 [1.5]
Figure 1.9 [1.1]
Figure 1.10 [1.5]
Figure 1.11 [1.5]
Figure 1.12 [1.5]
Figure 1.13 [1.10]
Figure 1.14 [1.11]
Figure 1.15 [1.12]
Figure 1.16 [1.5]
Figure 1.17 [1.5]
Figure 1.18 [1.6]
Figure 1.19 [1.3]
Figure 1.21 [1.15]
Figure 1.22 [1.16]
Figure 1.23 [1.17]
Figure 1.24 [1.18]
Figure 1.25 [1.6]
Figure 1.26 [1.6]
Figure 1.28 [1.20]
Figure 1.29 [1.21]
Figure 1.30 Guido Morgenthal
Figure 1.31 [1.12]
Figure 1.32 [1.19]
Figure 1.33 [1.23]
Figure 1.34 [1.15]
Figure 1.35 [1.24]
Figure 1.36 [1.25]
Figure 1.37 [1.26]
Figure 1.38 [1.27]
Figure 1.39 [1.28]
Figure 1.40 [1.28]
Figure 1.41 Michel Virlogeux
Figure 1.42 Michel Virlogeux
Figure 1.43 [1.16]
Figure 1.44 [1.16]
Figure 1.45 [1.30]
Figure 1.46 [1.30]
Figure 1.47 [1.31]
Figure 1.48 [1.32]
Figure 1.49 [1.15]
Figure 1.50 [1.33]
Figure 1.51 [1.6]
Figure 1.52 [1.6]
Figure 1.53 [1.15]

Figure 1.54 [1.3]
Figure 1.92 Michel Virlogeux
Figure 1.119 Michael Fehlauer, Conceptimage

Figure 2.1 [2.1]
Figure 2.3 [2.3]
Figure 2.4 [2.3]
Figure 2.5 [2.1]
Figure 2.6 [2.1]
Figure 2.7 [2.1]
Figure 2.8 [2.2]
Figure 2.9 [2.2]
Figure 2.10 [2.2]
Figure 2.11 [2.8]
Figure 2.12 [2.8]
Figure 2.13 [2.8]
Figure 2.14 [2.8]
Figure 2.15 [2.7]
Figure 2.16 [2.6]
Figure 2.17 [2.3]
Figure 2.18 [2.1]
Figure 2.19 [2.10]
Figure 2.20 [2.12]
Figure 2.21 [2.12]
Figure 2.22 [2.12]
Figure 2.25 [2.12]
Figure 2.26 [2.16]
Figure 2.27 [2.2]
Figure 2.28 Structurae
Figure 2.29 [2.16]
Figure 2.30 Klaus Stiglat
Figure 2.31 [2.17]
Figure 2.32 [2.1]
Figure 2.35 [2.19]
Figure 2.36 [2.19]
Figure 2.37 [2.19]
Figure 2.38 [2.19]
Figure 2.39 [2.15]
Figure 2.40 [2.3]
Figure 2.41 [2.23]
Figure 2.43 [2.1]
Figure 2.44 Klaus Stiglat
Figure 2.45 [2.3]
Figure 2.46 Structurae
Figure 2.47 Structurae
Figure 2.48 Structurae
Figure 2.49 Structurae
Figure 2.50 [2.3]

Figure 2.51 [2.29]
Figure 2.53 [2.31]
Figure 2.54 – 2.57 [2.34]
Figure 2.86 Eva Gassmann
Figure 2.92 – 2.94 [2.64]
Figure 2.100 Michel Virlogeux
Figure 2.104 Guido Morgenthal
Figure 2.105 XIANG, Haifan
Figure 2.105a [2.138]
Figure 2.106 Klaus Stiglat, [2.72]
Figure 2.111 – 2.114 [2.75]
Figure 2.116 [2.76]
Figure 2.127 [2.87]
Figure 2.128 [2.88]
Figure 2.129 [2.89]
Figure 2.130 – 2.132 Herbert Schambeck
Figure 2.143 [2.97]
Figure 2.145 [2.98]
Figure 2.147, 2.148 [2.104]
Figure 2.154 Schlaich Bergermann und Partner
Figure 2.155 [1.18]
Figure 2.160 – 2.164 Buckland and Taylor [1.33]
Figure 2.165 LIN, Yuanpei
Figure 2.167 Esko Järvenpää
Figure 2.174 Elljarn Jordet
Figure 2.178 – 2.180, 2.182, 2.184, 2.187, 2.188, 2.191 – 2.197, 2.199 [2.112]
Figure 2.185 Ian Firth
Figure 2.186 Dissing & Weitling
Figure 2.189, 2.190 [2.113]
Figure 2.201 T. Y. Lin
Figure 2.206 Schlaich Bergermann und Partner
Figure 2.207 Christian Menn
Figure 2.208 Jacques Combault
Figure 2.210 Michel Virlogeux
Figure 2.215 Bilfinger Berger
Figure 2.221 [2.133]

Figure 3.2, 3.3, 3.5, 3.9 [3.2]
Figure 3.20 – 3.22 Werner Brand, DSI
Figure 3.23, 3.25, 3.28 Marcel Poser, BBR
Figure 3.42 – 3.45 Yoshito Tanaka, Shinko
Figure 3.46 – 3.48 [3.36]

Figure 3.49 – 3.50 [3.37]
Figure 3.51 – 3.61 [3.38]
Figure 3.62 [1.6]
Figure 3.63, 3.64 [1.15]
Figure 3.65 [1.33]
Figure 3.66, 3.68, 3.69 [3.2]
Figure 3.70, 3.72, 3.73 [3.38]
Figure 3.75, 3.76 Michel Virlogeux
Figure 3.77 Jacques Combault
Figure 3.83, 3.84 [2.65]

Figure 4.1 [2.112]
Figure 4.2 [1.6]
Figure 4.3 [1.3]
Figure 4.9 [1.3]
Figure 4.18 [1.15]
Figure 4.21, 4.22 [4.4]
Figure 4.26 [4.5]
Figure 4.27, 4.32 [4.6]
Figure 4.29 – 4.31, 4.38 [4.7]
Figure 4.33, 4.42 – 4.45 [1.19]
Figure 4.34 – 4.36 [1.16]
Figure 4.37, 4.48, 4.49 [4.9]
Figure 4.50, 4.51 [1.19]
Figure 4.52 [1.15]
Figure 4.55 – 4.116 Imre Kovacs
Figure 4.90, 4.91 Michel Virlogeux
Figure 4.118 [4.41]
Figure 4.121 [4.42]
Figure 4.122, 4.123 [4.44]
Figure 4.124, 4.125 [4.46]
Figure 4.126 – 4.130, 4.138, 4.139 [4.49]
Figure 4.135, 4.136 [4.51]
Figure 4.140, 4.142 [4.53]
Figure 4.149, 4.151 [4.43]
Figure 4.154 [4.55]

Figure 5.2, 5.3 Dywidag
Figure 5.4; 5.5 [2.111]
Figure 5.6 – 5.9 Rolf Jung, LAP
Figure 5.23 – 5.25 Herbert Schambeck
Figure 5.26 [2.107]
Figure 5.29 – 5.35, 5.54 Jean-Marie Crémer
Figure 5.41, 5.44 – 5.53 Guido Morgenthal
Figure 5.42, 5.43 [1.22]
Figure 5.55 – 5.61 Hein, Lehmann

Cable-Stayed Bridges. 40 Years of Experience Worldwide. First Edition. Holger Svensson.
© 2012 Ernst & Sohn GmbH & Co. KG. Published 2012 by Ernst & Sohn GmbH & Co. KG.

Figure 5.63, 5.64 [2.105]
Figure 5.65 LIN, Yuanpei
Figure 5.66 – 5.71 Rolf Jung,
 LAP
Figure 5.72 Jacques Combault
Figure 5.74 [2.107]
Figure 5.77 – 5.79 Esko Järvenpää
Figure 5.80 – 5.82 [1.19]
Figure 5.86a [2.60]
Figure 5.87 – 5.94 [1.19]

Figure 6.1 – 6.20 [1.15]
Figure 6.61, 6.69 [2.77]
Figure 6.82 – 6.92, 6.111,
 6.123 [2.80]
Figure 6.93 – 6.97 [4.35]
Figure 6.130 [6.7]
Figure 6.134 – 6.151 [2.65]
Figure 6.152 – 6.188
 Martin Romberg, LAP
Figure 6.249 – 6.289
 Michel Virlogeux
Figure 6.290 – 6.314
 Jean-Marie Crémer
Figure 6.315 – 6.344
 Jacques Combault

Figure 7.1 [7.2]
Figure 7.2 [7.3]
Figure 7.3 [7.4]
Figure 7.4, 7.5 [7.5]

A35 Martin Steinkühler

List of Advertisers	Page
Alpin Technik und Ingenieurservice GmbH, 04179 Leipzig, Germany	A3
BBR VT International Ltd, 8603 Schwerzenbach, Switzerland	A2
Bridon International GmbH, 45881 Gelsenkirchen, Germany	A5
DMT GmbH & Co. KG, 45307 Essen, Germany	142a
DSD Brückenbau GmbH, 66740 Saarlouis, Germany	70a
DSI Holding GmbH, 80796 Munich, Germany	26a
Leonhardt, Andrä und Partner, 70469 Stuttgart, Germany	6
mageba sa, 8180 Bülach, Switzerland	A5
Maurer Söhne GmbH & Co. KG, 80807 Munich, Germany	A7
Mavis Cable Service GmbH, 52076 Aachen, Germany	144a
VSL International Ltd., 3098 Könitz, Switzerland	10

Appendix: 40 years of experience with major bridges all over the world

Cable-Stayed Bridges. 40 Years of Experience Worldwide. First Edition. Holger Svensson.
© 2012 Ernst & Sohn GmbH & Co. KG. Published 2012 by Ernst & Sohn GmbH & Co. KG.

Beginnings

At the end of school, I decided that I wanted to become a structural engineer in order to design major bridges all over the world. Consequently, I studied at the University of Stuttgart under Prof. Dr.-Ing. Fritz Leonhardt, Fig. A1 [1], at that time the most famous German bridge engineer.

At the end of my studies, in 1969, Prof. Leonhardt provided me with my first employment with the contractor Grinaker in South Africa. I started in the design office in Johannesburg, initially at the drawing board and later as a design engineer. I had to design in detail all types of structures, including industrial plants, posttensioned tanks and silos, retaining walls, foundations and small prestressed concrete bridges from precast elements with cast-in-place roadway slabs. British units were used.

After one year in Johannesburg, I went with Grinaker to a site near Pikwe/Selebi in the adjacent state of Botswana, north of South Africa, where a new copper-nickel mine was to be built. Our joint venture, Cosgrin, comprising the contractors Richard Costain, London, and Grinaker Construction, had the task of building all the plant above ground for processing the ore. Since our site was in the bush, right in the middle of nowhere, we started by building our camp with trailers and tents, Fig. A2. We drilled our own wells, installed our diesel power station and found suitable accessible rock to brake for the aggregates for concrete. The sand came from a (mostly) dry river bed. Later we built ourselves a tennis court from the red clay of termite mounds.

In 1972 I came back to Germany, as planned, and started to work as a design engineer with Leonhardt, Andrä and Partners (LAP) in Stuttgart.

Bridges in Germany

Initially I worked on some important German bridges. For new German freeways several large bridges crossing complete valleys were required, whose design was given to LAP.

The slopes of the Kocher Valley Bridge, consisting of badly decomposed shell limestone, were declared to be prone to sliding by the state geologist, and thus not suitable for bridge foundations. To bridge them without piers, Prof. Leonhardt proposed a cable-stayed bridge with a main span of 600 m, which would have been a world record at that time, Fig. A3. For this design, I evaluated Kernpoint moment influence lines 'by hand'. But then the state geologist retired and his successor considered foundations in the slopes possible. Contractor's alternates using these foundations were thus permitted and were, of course, more economic than the cable-stayed bridge. The design of the contractor Wayss & Freytag provided foundations in the slopes which consisted of large diameter wells in which the bridge piers were located with ample clear space to the well walls, thus permitting sliding of the slopes without affecting the piers. This permitted a constant depth continuous girder, 180 m above ground, resting on elegant, parabola-shaped piers, Fig. A4. Prof. Leonhardt

A1 Fritz Leonhardt

A2 Camp in Botswana

A3 Design for a cable-stayed Kocher Valley Bridge, Germany

A4 Built Kocher Valley Bridge, Germany

was appointed as checking engineer, and I was involved in the checking.

For the *Neckar Valley Bridge*, near Weitingen, similar slopes had indeed to be kept free of foundations. After many preliminary design alternates the solution was found for which the 260 m end spans were stayed from underneath with kingposts, Fig. A5. Unfortunately, the locked coil ropes supporting the kingposts did not creep as anticipated so that until today the end spans are hogging so much that a speed limitation is required. In order to obstruct as little as possible of the view through the beautiful Neckar valley Prof. Leonhardt decided at the end of the preliminary design phase to use single piers in the center of the bridge. The torsional beam moments are carried by the outside twin piers at the slopes. We design engineers were not very happy with his decision, because it came so late, and we had to change our design considerably.

The construction took place by free-cantilevering from both sides. The required auxiliary piers with heights of up to 120 m used steel parts of unknown heritage which caused a considerable amount of additional checking, and finally test loadings were executed.

In 1975, LAP was awarded the detailed design and construction engineering for the approach bridge of the *cable-stayed Rhine River Bridge at Flehe* on behalf of the contractor, Fig. A6. I was originally in charge of this project until I had to go to the USA on behalf of the Pasco-Kennewick Bridge, see later. The 600 m long concrete beam was built spanwise on self-launching formwork. The backstay cables in the vicinity of the tower were anchored in steel pipes cast into the beam. During installation of the formwork for the last span it became apparent that the tower was located wrongly by 25 cm in the bridge axis, although its position had been surveyed several times.

Only after the German reunification in 1990 did I again design bridges in Germany. One was the *Mulden Bridge near Siebenlehn* in Saxony as part of the A 4 freeway, Fig. A7. Prof. Leonhardt succeeded in preserving the massive piers which were built in the 1930s under his supervision, although the lower strength of their manually tamped concrete had a strength below current codes, but the corresponding stresses are also very low. The new twin composite beams were built laterally on auxiliary piers and, after demolition of the old beam, sidewise shifted [2]. In this way, four lanes of traffic with reduced widths were kept open during the complete construction period which is generally required in Germany.

The Baltic Sea coastal freeway, A 20, required a bridge across the *Warnow Valley* near Rostock, Fig. A8. I tried hard to win the design contract for LAP, if only to speak my north-German dialect again after 30 years in Stuttgart, south Germany. We did indeed win the contract and I participated in all meetings on site. The main span of the haunched concrete girder with a span in the range of 70 m had to be widened on both sides in order to avoid the local beavers colliding with the piers.

The *freeway triangle Neukölln*, which connects the Berlin inner-city freeways A 100 and A 113 was planned without intersections.

A5 Neckar Valley Bridge, Germany

A6 Cable-stayed Rhine River Bridge at Flehe, Germany

A7 Mulden Bridge near Siebenlehn, Germany

A8 Warnow Bridge, Germany

This resulted in a structure with six continuous composite girders and a tunnel with five levels, Fig. A9. Adjacent, a twin-arch bridge across the Britz Canal was built whose central arches we combined into a single one for aesthetical reasons, Fig. A10. The deck cross girders are continuous over the width of the bridge, so that a single arch bridge with three planes was created [3].

Cable-stayed bridges abroad
Pivotal task: Construction engineering

After my return from South Africa in 1972, Prof. Leonhardt repeatedly sought my collaboration in projects abroad in which he was very much personally interested, not least because of my knowledge of English. Initially he drew up the preliminary design and I tried to calculate it.

An early example is the *Pasco-Kennewick Bridge* across the Columbia river, WA, USA, near Seattle in the Pacific Northwest, Fig. A11. The small local consultant Arvid Grant had asked for advice, and Leonhardt designed a cable-stayed concrete bridge with 300 m main span, a world record at that time. I was entrusted with the design, initially in Stuttgart and then, during 1976, in Olympia, the state capital of Washington. During that time we completed the detailed design and the final construction engineering, including the construction equipment.

A concrete beam was selected because of the fluctuating high steel prices on the west coast. Since American contractors at that time were not prepared to build a beam by free-cantilevering with spans up to 150 m, large precast elements with glue joints and weights up to 270 t were tendered. At that time, this was quite an unusual new design for which no experience was available in the USA.

The towers were cast-in-place within jumping forms. The beam starter piece between the tower legs was also built cast-in-place on scaffolding, Fig. A15 [4]. The precast elements comprising the full beam width of 24.3 m with a length of 8.2 m, equal to the cable distance, were built by match-casting adjacent elements near the site and then floated in. During lifting of the heavy precast elements the towers had to be temporarily stayed with auxiliary forestay and backstay cables. The hydraulic lift slab equipment at the tip of the cantilevers also had to be tied back to the tower in order not to overload the beam.

Although my visit in early 1976 was originally planned only for a short period, it soon became clear that I had to supervise the detailed design and the construction engineering in Olympia and thus had to stay for nearly a year. I explained the situation to my wife Meg and she arrived two weeks later with my six-week-old son. Meanwhile I had rented a fully furnished house and borrowed baby equipment from my colleagues in the office. We felt at home quickly and found many friends. There were no language problems for my wife, as a professional translator.

It was not easy for me to lead the American team completely on my own, especially since I had not yet performed a similar construc-

A9 Freeway Triangle Intersection Neukölln, Berlin, Germany

A10 Britz Canal Bridge, Berlin, Germany

A11 Pasco-Kennewick Intercity Bridge, WA, USA

tion engineering task. Valuable advice was given to me by my colleague Reiner Saul during a short visit on his way to Buenos Aires, Argentina, where he had to perform a similar task for the Zárate-Brazo Largo bridges.

In the tender documents, we had proposed an optional construction method including the construction equipment. The construction engineering itself, however, was left to the contractors as usual in Germany. Since American contractors do not have their own design office and no US consultant had experience in this respect, it was later decided that we would perform the complete construction engineering, the design of the equipment and the construction controls on site.

The construction engineering calculations provided some difficulties because of the state of the computer techniques at that time for a system with 2×72 stay cables and a corresponding number of statical indeterminates. The load stage 'permanent loads' as a starting system had to be calculated by hand, because the only available program, STRUDL, did not provide a sufficient number of defined decimal places. (The matrix of the cable shortenings does not possess a dominant main diagonal, and the results were numerically unstable.)

The forces of the different erection systems could not be summed electronically so that their summation for all intermediate construction systems had to be done by hand. The computer calculations themselves were performed at night in the data processing center of the McDonnell-Douglas aircraft factory in St Louis, MO. Normally one construction system per day was calculated, the results printed out in Seattle and delivered by courier service to Olympia.

The bridge construction took place by geometry with shop-fabricated cables and precast elements, both with fixed lengths. The beam was erected by free-cantilevering to both sides of the tower. The lifting of the final precast element is shown in Fig. A12. The results of the construction engineering were confirmed by the fact that after the beam center joint closure no cable had to be restressed.

It was quite satisfying to realize this unusual bridge from preliminary design through detailed design to construction engineering and controls on site. From Prof. Leonhardt's side it was courageous to send me as a beginner alone into the USA to supervise the design of a world-record cable-stayed bridge.

New developments by competition

In 1978 the design for a *steel cable-stayed bridge across the Ohio river near Huntington*, WVA, USA, was completed and its river foundations already built when the Federal Highway Administration decided that two designs, one in steel and one in concrete, should be tendered for all major bridges [5]. In this way, bridge design competitions should be improved because otherwise the contractors were not prepared to accept the liability for own alternate designs.

Together with Arvid Grant, we were given the task of designing an inherently heavier concrete cable-stayed bridge with a record

A12 Lifting of last precast element

A13 East Huntington Bridge, West Virginia, USA

A14 Transportation of a precast element

main span of 270 m with one tower only for the existing foundations of a steel cable-stayed bridge, Fig. A13. This was achieved by designing a slender A-tower with high-strength concrete B60, which was quite unusual at that time in the USA. Furthermore, the cross girders for the beam were designed as steel composite members to save weight, Fig. A14. Our design bid was distinctly lower than the steel alternate. For this bridge the precast elements were lifted by a floating crane.

In 1980, a freighter collided with the *Sunshine Skyway Bridge* near Tampa, Florida, USA [6]. The design of the steel alternate was awarded to Greiner Engineering together with Leonhardt, Andrä and Partners, Fig. A15. My wife Meg packed our suitcases again and we moved for one year to Tampa, FL. Our two children, three and five years old, attended a small American kindergarten in which they soon felt at home. This time I had two colleagues with me and we worked in the offices of Greiner Engineering. In order to be competitive against the concrete alternate from Jean Muller we developed the following design principles, taking into account that no site welding is permissible in the USA and that all joints have to be closed on site with cover plates and high-strength bolts:

– An orthotropic deck would have been too expensive, whereas a reinforced but not posttensioned roadway slab is more economic.
– A steel box would also have been too expensive – a girder grid from open steel sections is advantageous.
– The stay cables are connected directly to the outside main girders.
– Inner load distributing longitudinal girders are avoided because their expensive joints are more costly than the savings in steel.

We developed diamond-shaped towers with footings which fitted the central towers of the concrete alternate as well.

Although we lost the bidding against the concrete alternate by a small margin due to additional costs in the foundations of the long approach bridges, the design principles proved to be trendsetting. They are used today worldwide, including in Germany after the limitation of tensile stresses under service conditions were omitted.

Later we designed the cable-stayed bridge across the Mississippi at Burlington, Missouri, USA, Fig. A16 [7], together with Sverdrup. For this project, our team was awarded both steel and concrete alternates and the composite alternate won.

After the *bridges across the Paraná river*, between the cities of Zárate and Brazo Largo in Argentina, were already completed in 1980, their piers needed to be protected against ship collision. We found that no comprehensive design rules existed anywhere for this task and we thus had to investigate it ourselves [8]. The impact force of a ship against a rigid wall forms the basis for the design of bridge piers against ship collision. For their determination we evaluated the results of collision tests which were performed in Germany in preparation for the design of the nuclear vessel Otto Hahn under the direction of Gerhard Woisin. I determined that the impact force increases with the square root of the ship size and varies by ±50% depending on the particular bow structure, Fig. A17 [8]. This

A15 Sunshine Skyway Bridge, Tampa, Florida, USA; composite alternate

A16 Burlington Bridge across the Mississippi, USA

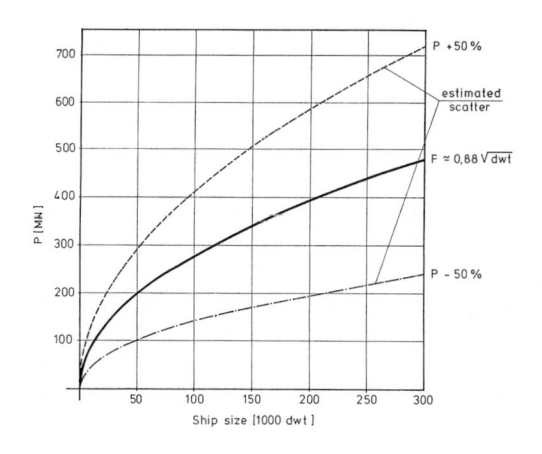

A17 Equivalent impact loads as a function of the ship size according to Svensson

correlation is meanwhile used worldwide, where for simplification the 75 % fractile is used.

After the old *Sunshine Skyway Bridge* was destroyed by ship collision, as mentioned above, the collision protection of the new bridge was done particularly thoroughly, Fig. A18. Both towers were made unreachable for ships by artificial islands around their foundations and, in addition, were protected by dolphins. The sheet piles of the circular cells are welded together, filled with lean concrete and covered by a 1.5 m thick concrete cap. For smaller collisions timber fenders are installed, whereas heavy collisions are expected to destroy a dolphin, thereby stopping the ship.

More bridges in the USA

Between 1984 and 1986 further designs of cable-stayed bridges in the USA had to be done. Meg again routinely packed, each time less because meanwhile we felt at home in the USA. This time our children briefly refreshed their knowledge of English and then attended the first and third grade of a local American school where they participated without problems and with good success. After a time you could not pick them out of a crowd by their accent.

After some unsuccessful competitive designs, our team with Greiner Engineering was awarded the steel alternate for the *Houston Ship Channel Crossing at Baytown* (Fred Hartmann Bridge), Texas, USA, Fig. A19, with a main span of 381 m and a navigational clearance of 53 m [9]. The bridge is unusually wide with eight lanes plus full shoulders. We found that two independent beams with four cable planes was the most economic solution. For the towers we selected a double-diamond shape which carried transverse load due to wind by tension and compression into the foundations. The beam design followed the principles we developed for our design of the Sunshine Skyway Bridge, see above. Since the concrete alternate was again given to Jean Muller we were set to win this time and thus tried to save on materials as far as possible. The 134 m high towers, for example, have a constant transverse width of 2.1 m and a wall thickness of 30 cm only. We did indeed win the competition.

The two independent beams were built by free cantilevering in parallel to both sides of the tower. Against hurricane winds they were tied down temporarily to the tower foundations, Fig. A20. In fact, a hurricane went directly across the bridge during construction without causing any damage. Since then almost every year a hurricane crosses the bridge.

The steel grid sections with a weight of 60 t were transported by barge to the site and lifted with derricks located on the deck and connected with cover plates and high-strength bolts. The precast concrete roadway slabs with weights of 20 t were also lifted and connected with cast-in-place joints. The PE pipes were finally colored yellow using UV-resistant tapes.

In the center of Columbus, Ohio, a true arch bridge had to be replaced. Leonhardt proposed using three plate arches which give the impression of an arch but were less expensive. His idea was realized

A18 Sunshine Skyway Bridge with protections against ship collision

A19 Baytown Brücke Bridge across the Houston Ship Channel, Texas, USA

A20 Free cantilevering with temporary tie-downs

A21 Broad-Street-Bridge, Columbus, Ohio, USA

Checking of bridges

in our team with Burgess & Niple Engineers and H2L2 Architects, Fig. A21 [10].

For a new bridge across the *Severn River at Annapolis*, Maryland, a US-wide design competition was tendered in 1992. In order to respect the view onto the famous naval academy on one shore our team with Greiner proposed a rather simple haunched continuous girder on slender piers, Fig. A22. We won the first prize against more complicated designs of arches and cable-stayed bridges with structural members above deck [11].

Back in Europe

In Norway, all inhabited major islands were to be connected by bridges or tunnels to the mainland in the early 1990s. The Lejrfjord in Helgeland near the Arctic Circle required a main span of 425 m and a beam width of only 12 m for the low expected traffic. Together with the Norwegian consultant Aas-Jakobsen we designed a cable-stayed concrete bridge with a depth of only 1.2 m, Fig. A23, which resulted in the remarkable slenderness of 1:354 vertically and 1:35 horizontally.

The major design problem was the regularly occurring heavy storms with speeds of up to 250 km/h and turbulence intensities up to 25%. For this bridge all stages of the design from preliminary to detailed design to construction engineering and supervision on site were in our hands, as it was for the Pasco-Kennewick Bridge above, which is especially satisfying for an engineer [12]. Dr. Kovacs developed computer design methods to ensure the aerodynamic stability of the beam during free cantilevering with up to 210 m spans, Fig. A24 [13]. The assumed heavy storms really occurred during construction and the bridge was stable as expected. The Helgeland Bridge was our boldest design so far.

For the crossing of the *River Leven valley*, north of Edinburgh, we designed a cable-stayed concrete bridge together with Babtie from Glasgow, Fig. A25. The single reinforced but not posttensioned beam was completely cast on scaffolding [14].

Checking of bridges

In order to be permitted to work as an independent checking engineer for checking the design and construction method for bridges abroad, I took professional examinations in several countries.

In 1985, I passed the two eight-hour written examinations back to back for a Professional Engineer (PE) in Florida, USA. The registration was later extended to cover 12 states. Later I took the examinations in Canada (PEng) and Great Britain (CEng, FICE, FIStructE) and further registrations in Hong Kong (MHKIE), Australia (RPEQ) and Malaysia (TPEng), based on the former examinations in the USA and UK. The English and American registrations are also accepted in other countries with Anglo-Saxon engineering tradition, for example in the Middle and Far East. The checking is normally awarded personally to a checking engineer who has to sign and seal all execution drawings.

A22 Annapolis Bridge across the Severn River, Maryland, USA

A23 Helgeland Bridge, Norway

A24 Free cantilevering

In the late 1980s we were awarded, together with Greiner, the checking and construction supervision of a steel arch with a main span of 325 m and a composite beam across the *Roosevelt Lake*, near Phoenix, Arizona, Fig. A26 [15]. The contractor initially had problems with the free cantilevering and we had to help a lot. Our engineers spent nearly two years on site.

The E6 between Göteborg and Oslo crosses the Swedish–Norwegian border across the *Svinesund* with a new four-lane bridge. The navigational clearance of 170 × 70 m required an arch fixed high to the slopes. The central arch comprises two orthotropic steel boxes. The concrete arch was built by free cantilevering, the steel beams were launched from both sides and its center piece lifted in, Fig. A27.

With a main span of 1210 m, the *Höga Kusten Bridge* on the east coast of Sweden is one of the largest suspension bridges in the world, Fig. A28. The main cables consist of prefabricated strand cables individually anchored in rock.

The *Yang Pu Bridge in Shanghai* connects Old Shanghai with the new suburb of Pudong, Fig. A29. The bridge was financed by the World Bank in 1994 which asked for a checking team comprising an American engineer, a Japanese engineer and myself. With a main span of 601 m, the bridge held the overall world record at the time of construction and is still has the longest composite span in the world. The design of the Chinese engineers followed the state of the art, using parallel wires with concrete A-towers and a composite beam. The checking did not reveal any design deficits. The development of Pudong is breathtaking; in the early 1990s, cows were grazing there, while today a cluster of impressive skyscrapers surrounds the famous TV towers.

In Sweden I had to check several major bridges. The *Uddevalla Bridge*, Fig. A30, crosses the Sunningesund near Göteborg with a main span of 414 m. Its beam comprises an open steel grid with a concrete deck of precast slabs.

The Bridge across the *Ume Älv* has a main span of 130 m with just one tower, Fig. A31. The steel grid was launched across auxiliary piers, the roadway slab cast-in-place and the parallel strand cables installed. Only then were the auxiliary piers removed.

The *Penang Bridge*, opened in 1985, Fig. A32, connects the mainland of Malaysia with the island of Penang. The main bridge is a concrete cable-stayed bridge with a main span of 225 m. We had to investigate the stability of the bridge for new live loads in accordance with four different international codes. As a result we proposed to introduce bearings for the beam at the towers, among others.

The *Golden Ears Bridge* across the Annacis River in Vancouver, BC, Canada, is a five-span, 968 m long extradosed cable-stayed bridge with a composite beam, Fig. A33. We checked the bridge during construction and in the final stage on behalf of the Canadian Bridge Authority employed by the contractor Bilfinger Berger. The major problems are earthquakes and the bad soil conditions which could cause significant pier settlement over a long period of time. To cover these, adjustable bearings were installed [16].

A25 Leven River Bridge at Glenrothes near Edinburgh, Scotland

A26 Roosevelt Lake Bridge, USA; free cantilevering with auxiliary tie-backs

A27 Svinesund Bridge between Norway and Sweden, lifting of center span

Checking of bridges

A28 Höga Kusten Bridge, Sweden

A29 Yang Pu Bridge, Shanghai, China

A30 Uddevalla Bridge across the Sunningesund, Sweden

A31 Ume Älv Bridge, Sweden

A32 Penang Bridge, Malaysia

A33 Golden Ears Bridge, Vancouver, B.C., Canada

Participation in Code Commissions

In the mid 1980s the Post-Tensioning Institute (PTI) in the USA was awarded the assignment of preparing a code for stay cables. I contributed the experience of LAP to the 'Recommendations for Stay Cable Design, Testing and Installation' which has since gone through several editions [17].

In connection with the design of protections against ship collision for our two Zárate-Brazo Largo bridges in Argentina, I had developed a formula for the calculation of the impact forces as a function of ship size, Fig. A17, which today is applied worldwide [8], see above. Based on this work, I was appointed a member of the IABSE Working Commission on Ship Collision [18], the US Marine Board [19], as well as the AASHTO Committee on Pier Protection [20].

I contributed European knowledge on the design of concrete box girders and segmental construction to the Transportation Research Board (TRB) Committee which developed the first specifications in the USA for concrete box girders for AASHTO [21].

Current projects

At the end of my work as executive director of LAP at the beginning of 2010, several major projects of which I was in charge were not complete.

The *new Beska Bridge across the Danube river near Belgrade, Serbia*, forms part of the road from Novi Sad to Belgrade, Fig. A34, with a total length of 2.2 km. The main bridge consists of a three-span haunched box girder built by free cantilevering. The approach bridges were incrementally launched. We advised the joint venture headed by Alpine Construction on their tender and prepared the detailed design including construction engineering.

The *Ada Ciganlia Bridge also crosses the Sava river in Belgrade*. The main bridge has a span of 375 m with a single tower and a beam width of 45 m, which makes it the biggest of its kind in Europe, Fig. A35 [22]. The 200 m high, slender A-tower permits two railway tracks to run in between the tower legs and three traffic lanes on each side, outside the tower legs. The main span uses a steel box, while the side span has a concrete beam as counterweight to the main span, for the first time not supported on piers but by stay cables only. For this bridge we advised the joint venture headed by PORR Construction for their tender, and performed the detailed design and construction engineering.

The *Chenab Bridge* forms part of the new railway line between Katra and Laole at the base of the Himalayas in India, Fig. A36. The steel beam is supported by steel piers on two steel arches with a record span of 465 m. Both arches are steel boxes, connected vertically by trusses and horizontally by plates. The steel boxes will be filled with concrete in order to improve their durability and aerodynamic behavior. The detailed design is completed for Afcons Infrastructure Ltd. on behalf of the Konkan Railway Company. Work on site has started, but the completion is delayed – not unusual for India.

A34 New Beska Bridge, Serbia

A35 Sava Bridge Ada Ciganlija, Serbia, free cantilevering

A36 Railway bridge across the Chenab River, India

A37 Bridge across the Golden Horn in Vladivostok, Russia

Current projects

The *Golden Horn Bridge* across an inner-city bay in Vladivostok in the far east of Russia connects two parts of the city, with a main span of 736 m in steel and side spans in concrete, Fig. A37 [25]. The V-shaped concrete towers without cross girders make the bridge architecturally unique but require special considerations on tower and cable oscillations. We advised the Russian consultants Giprostroymost, from St. Petersburg, and checked the detailed design of the towers.

A 17.6 km long permanent connection across the *Femer Belt in the Baltic Sea* will replace the current ferry connection [23]. A tunnel and bridge alternate are being investigated, with a cable-stayed main bridge comprising three spans of 724 m each plus side spans with a total length of 2414 m, Fig. A38 [12]. The ships are to sail through the outer spans, the inner span providing a safety distance. The double-stock truss will be continuous with the fixed point located in the bridge center.

The tender design for the new *Forth Road Bridge* comprises a five-span cable-stayed bridge. In the middle of the central span the stay cables overlap in order to stabilize the inner towers, Fig. A39 [24]. Loads in one of the inner spans thus create compression in the stay cables of the adjacent spans. This system is appropriate if the beam is sufficiently stiff. We are currently preparing the detailed design and construction engineering with other consultants for the joint venture headed by HOCHTIEF.

In the fall of 2008, I was called at short notice to the city of Omsk in Siberia, Russia, to advise the Russian consultant Mostovik on their preliminary design for the *Russki Bridge*, near Vladivostok, Fig. A40 [25]. We discussed the aerodynamics, especially the prevention of stay cable oscillations. This bridge has a new record span of 1104 m and will open in the summer of 2012.

Summary

I am very thankful that I have been able to realize my dream of designing major bridges all over the world during my 40 years of professional service. Prof. Leonhardt supported me in this regard by engaging me early on in the design of important bridges. I am also thankful to my partners and colleagues from whom I learnt and who have always supported me. A large bridge is never the work of one single person but is always the result of a team effort. The prerequisite for the design was the confidence of our clients to engage us for major projects which is also highly appreciated.

A38 Femer Belt Bridge across the Baltic Sea, Bridge Alternate 1999

A39 Design for the new Forth Road Bridge, Scotland

A40 Russki Bridge near Vladivostok, Russia, 2012, 1104 m

References

[1] Svensson, H.: Holger Svensson. In: Klaus Stiglat (Publ.): Bauingenieure und ihr Werk. Ernst & Sohn, Berlin 2004, pp. 404–415

[2] Svensson, H., Wange, G., Korte, R.-P., Eilzer, W. and Humpf, K.: Entwurf und Ausschreibung der Autobahnbrücke Siebenlehn. Beton- und Stahlbeton 92 (1997), pp. 29–36

[3] Svensson, H., Foth, E., Burkhardt, H.-G. and Fischer, M.: Neubau der BAB A 113 (neu), Entwurf, Ausschreibung und Vergabe des Neuköllner Autobahndreiecks und der Bogenbrücke über den Britzer Verbindungskanal. Stahlbau 69 (2000), pp. 823–832

[4] Leonhardt, F., Zellner, W. and Svensson, H.: Die Spannbeton-Schrägkabelbrücke über den Columbia zwischen Pasco und Kennewick im Staat Washington, USA. Beton- und Stahlbetonbau 75 (1980), Vol. 2, 3, 4

[4a] Leonhardt, F., Zellner, W. and Svensson, H.: The Columbia River Bridge at Pasco-Kennewick, WA, USA. FIP Eighth Congress Proceedings, 1978

[5] Zellner, W. and Svensson, H.: Mitarbeit an der Planung von Schrägkabelbrücken in den USA. In: VBI (Editor) Konstruktiver Ingenieurbau. Ernst & Sohn, Berlin 1985, pp. 87–95

[6] Saul, R., Svensson, H., Andrä, H.-P. and Selchow, H.-J.: Die Sunshine Skyway Brücke in Florida – Entwurf einer Schrägkabelbrücke mit Verbundbau. Die Bautechnik 61 (1984), pp. 230–238, 305–309

[6a] Svensson, H., Christopher, B. G. and Saul, R.: Design of a cable-stayed composite bridge. Journal of Structural Engineering, (ASCE), 1986, pp. 489–504

[7] Svensson, H. and Humpf, K.: Die Schrägkabelbrücke über den Mississippi bei Burlington, USA. Stahlbau 63 (1994), pp. 193–199

[7a] Svensson, H. and Petzold, E.: The cable-stayed bridge over the Mississippi at Burlington, USA, Strait Crossings 94, Balkema, Rotterdam, 1994, pp. 239–246

[8] Saul, R. and Svensson, H.: Zum Schutz von Brückenpfeilern gegen Schiffsanprall, dargestellt am Beispiel der Brücken Zárate-Brazo Largo über den Paraná, Argentinien. Die Bautechnik (1981), pp. 326–335, 374–388

[8a] Saul, R. and Svensson, H.: On the theory of ship collision against bridge piers. IABSE Proceedings P-52/82, 1982

[8b] Saul, R. and Svensson, H.: Means of reducing the consequences of ship collision with bridges and offshore structures. IABSE Introductory Report, Copenhagen 1983, pp. 165–179

[9] Svensson, H., Hopf, S. and Humpf, K.: Die Zwillings-Verbundschrägkabelbrücke über den Houston Ship Channel bei Baytown, Texas. Stahlbau 66 (1997), pp. 57–63

[9a] Svensson, H. S. and Lovett, T. G.: The twin cable-stayed composite bridge at Baytown, Texas. IABSE Symposium Mixed Structures, Brussels, 1990, pp. 317–322

[9b] Svensson, H.: The twin cable-stayed Houston Ship Channel Bridge. The Structural Engineer (IStructE), March 1992, pp. 13–20

[10] Svensson, H., Humpf, K. and Hopf, S.: Die neue Broad-Street-Brücke in Columbus, Ohio, USA. Beton- und Stahlbetonbau 89 (1994), pp. 192–196

[11] Svensson, H., Eilzer, W. and Patsch, A.: Die neue Straßenbrücke über den Severn River bei Annapolis, USA. Stahlbau 66 (1997), pp. 64–69

[12] Svensson, H. and Hopf, S.: Die Spannbeton-Schrägkabelbrücke Helgeland. Beton- und Stahlbeton 88 (1993), pp. 247–250, 279–281

[12a] Svensson, H. and Hopf, S.: The concrete cable-stayed Helgeland Bridge, Norway. Proceedings of the ACI Spring Convention, Washington, DC, 1992

[12b] Svensson, H. and Jordet, E.: The concrete cable-stayed Helgeland Bridge in Norway. Civil Engineering (ICE), Vol. 114, May 1996, pp. 54–63

[13] Svensson, H. and Kovács, I.: Analytische aerodynamische Untersuchung der Schrägkabelbrücke Helgeland. Beton- und Stahlbetonbau 98 (1994), pp. 149–153, 201–203

[13a] Kovács, I., Svensson, H. S. and Jordet, E.: Analytical aerodynamic investigation of the cable-stayed Helgeland bridge, Journal of Structural Engineering ASCE, Jan 1992, pp. 147–168.

[14] Svensson, H., Humpf, K. and Straub, W.: Die River Leven Stahlbeton-Schrägkabelbrücke. Beton- und Stahlbetonbau 100 (1996), pp. 127–131

[15] Svensson, H., Humpf, K. and Roesler, H.: Die Montage der Bogenbrücke über den Roosevelt Lake. Der Stahlbau 60 (1991), pp. 131–138

[16] Heerdt, M.: Golden Ears, Vancouver-Extradosed-Brücke in effizienter Verbundbauweise. Deutscher Beton- und Bautechnik-Tag, May 2011, pp. 39–40

[17] Recommendations for Stay Cable Design, Testing and Installation. PTI Guide, Specification, 4th Edition 2001

[18] Larsen, O. D.: Ship Collision with Bridges. IABSE: Structural Engineering Documents, 1993

[19] Ship Collision with Bridges. Marine Board, National Research Council, Washington DC, 1983

[20] Guide Specification and Commentary for Vessel Collision Design of Highway Bridges. US Department of Transportation, FHWA, 1990

[21] Design and Construction Specification for Segmental Concrete Bridges. PTI 1988. AASHTO Interim Specifications

[22] Steinkühler, M., Minas, F. and Hopf, S.: Ein neues Wahrzeichen für Belgrad: Schrägseilbrücke mit 200 m hohem Pylon in Belgrad, Serbien. Deutscher Beton- und Bautechnik-Tag, May 2011, pp. 93–94

[23] Lykke, S., Jönsson, U. and Christensen, H.: Fehmarnbelt Fixed Link-Status Report Dezember 2010. Deutscher Beton- und Bautechnik-Tag, May 2011, pp. 43–44

[24] Carter, M., et al.: Forth Replacement Crossing. IABSE Bangkok, 2009

[25] Kolyushev, I.: Two cable-stayed bridges in the east of Russia. Bridge Symposium Leipzig, 2010

Prof. Dipl.-Ing. Holger Svensson
Consulting Engineer
Niederlausitzstraße 22
15738 Zeuthen
Germany
holger.svensson@gmx.de

Translated by: Prof. Dipl.-Ing. Holger Svensson, Zeuthen, Germany
Prof. Dr. Guido Morgenthal, Weimar, Germany
(sections on dynamics)
Review and Improvement: Paul Beverley, U.K.

Cover: Helgeland Bridge over Leirfjord, Sandnessjøen, Norway
© Helga Rutzen, Düsseldorf, Germany

© 2012 Wilhelm Ernst & Sohn, Verlag für Architektur
und technische Wissenschaften GmbH & Co. KG,
Rotherstr. 21, 10245 Berlin, Germany

Production: HillerMedien, Berlin, Germany
Coverdesign and Layout: Sophie Bleifuß, Berlin, Germany
Typesetting: Uta-Beate Mutz, Leipzig, Germany
Printing: Medialis, Berlin, Germany
Binding: Stein + Lehmann, Berlin, Germany

Printed in the Federal Republic of Germany.
Printed on acid-free paper.

Print ISBN: 978-3-433-02992-3
ePDF ISBN: 978-3-433-60229-4
oBook ISBN: 978-3-433-60104-4

Library of Congress Card No.: applied for

British Library Cataloguing-in-Publication Data
A catalogue record for this book is available from the British Library.

**Bibliographic information published by
the Deutsche Nationalbibliothek**
The Deutsche Nationalbibliothek lists this publication in the Deutsche Nationalbibliografie; detailed bibliographic data is available in the Internet at <http://dnb.ddb.de>.

All rights reserved (including those of translation into other languages). No part of this book may be reproduced in any form – by photoprinting, microfilm, or any other means – nor transmitted or translated into a machine language without written permission from the publishers. Registered names, trademarks, etc. used in this book, even when not specifically marked as such, are not to be considered unprotected by law.

Lectures

Lectures on cable-stayed bridges

The video recordings of Lectures 1 to 30, given by the author in the summer semester 2011 and the winter semester 2011/2012 at the University of Dresden, Faculty of Structural Engineering, for students of the Diploma degree program in Structural Engineering, are enclosed as two DVDs at the end of this book.

7th Semester: Cable-stayed bridges in the final stage

Lecture 1
  Introduction to cable-stayed bridges
Lecture 2
  The precursors of cable-stayed bridges
Lecture 3
  Cable-stayed steel bridges
Lecture 4
  Cable-stayed concrete bridges
Lecture 5
  Composite cable-stayed bridges
Lecture 6
  Special systems of cable-stayed bridges
Lecture 7
  Stay cables 1. Ropes, bars and wires
Lecture 8
  Stay cables 2: Strands, anchorages and design criteria
Lecture 9
  Action forces for equivalent systems
Lecture 10
  Action forces for actual systems
Lecture 11
  Preliminary design calculations for basic systems
Lecture 12
  Preliminary design calculations for a slender cable-stayed bridge
Lectures 13, 14
  Cable dynamics
Lectures 15, 16
  Bridge dynamics

8th Semester: Construction of cable-stayed bridges

Lectures 17, 18
  Examples for the construction of cable-stayed bridges
Lectures 19, 20
  Construction engineering
Lecture 21
  Cable installation
Lecture 22
  Examples for precast cable-stayed concrete bridges
Lecture 23
  Example for cast-in-place cable-stayed concrete bridges
Lecture 24
  Example for cable-stayed steel bridges
Lecture 25
  Example for composite cable-stayed bridges
Lecture 26
  Example for hybrid cable-stayed bridges
Lecture 27
  Examples for series of cable-stayed bridges
Lecture 28
  Protection against ship collision
Lecture 29
  Aesthetic guidelines for bridges
Lecture 30
  Personal experiences with cable-stayed bridges worldwide